GROUP THEORY FOR PHYSICISTS

GROUP THEORY FOR PHYSICISTS

Zhong-Qi Ma

Institute of High Energy Physics, Beijing, China

World Scientific

NEW JERSEY • LONDON • SINGAPORE • BEIJING • SHANGHAI • HONG KONG • TAIPEI • CHENNAI

Published by

World Scientific Publishing Co. Pte. Ltd.

5 Toh Tuck Link, Singapore 596224

USA office: 27 Warren Street, Suite 401-402, Hackensack, NJ 07601

UK office: 57 Shelton Street, Covent Garden, London WC2H 9HE

Library of Congress Cataloging-in-Publication Data

Ma, Zhongqi, 1940–

Group theory for physicists / by Zhong-Qi Ma. -- 1st ed.

p. cm.

Includes bibliographical references and index.

ISBN-13: 978-981-277-141-4 (hardcover : alk. paper)

ISBN-10: 981-277-141-7 (hardcover : alk. paper)

ISBN-13: 978-981-277-142-1 (pbk. : alk. paper)

ISBN-10: 981-277-142-5 (pbk. : alk. paper)

1. Group theory. 2. Mathematical physics. I. Title.

QC20.7.G76M29 2007

512'.2--dc22

2007027914

British Library Cataloguing-in-Publication Data

A catalogue record for this book is available from the British Library.

Printed in Singapore.

Preface

Group theory is a powerful tool for studying the symmetry of a physical system, especially the symmetry of a quantum system. Since the exact solution of the dynamic equation in the quantum theory is generally difficult to obtain, one has to find other methods to analyze the property of the system. Group theory provides an effective method by analyzing the symmetry of the system to obtain some precise information of the system verifiable with observations. Now, Group Theory is a required course for graduate students majored in physics and theoretical chemistry.

The course of Group Theory for the students majored in physics is very different from the same course for those majored in mathematics. A graduate student in physics needs to know the theoretical framework of group theory and more importantly to master the techniques in the application of group theory to various fields of physics, which is actually his main objective for taking the course. However, no course or textbook on group theory can be expected to include explicitly the solution to every problem of group theory in his research field of physics. A student of physics has to know the fundamental theory of group theory, otherwise he may not be able to apply the techniques creatively. On the other hand, the student of physics is not expected to completely grasp all the mathematics behind group theory due to the breadth of the knowledge required.

I first taught the group theory course in 1962. Since 1986, I have been teaching for 20 years the course of Group Theory to graduate students mainly majored in physics at the Graduate School of Chinese Academy of Sciences. In addition, most of my research work has been related to applications of group theory to physics. In 1996, the Chinese Academy of Sciences decided to publish a series of textbooks for graduate students. I was invited to write a textbook on Group Theory for the series. In the textbook, based on my experience in teaching and research work, I explained the fundamental concepts and techniques of group theory progressively and systematically using the language familiar to physicists, and also emphasized the ways with which group theory is applied to physics. The textbook (in Chinese) has been widely used for the Group Theory course in China since it was published by Science Press (Beijing) in 1998. The second edition of this textbook was published in 2006 after systematic revision. Up to the seventh printing in December 2006, 16800 copies of this textbook were printed altogether. This textbook is the foundation for the present textbook, although some new materials are included and some are moved to another exercise book [Ma and Gu (2004)].

By the request of the readers, an exercise book on group theory, by the same author, was published in 2002 by Science Press (Beijing) to form a complete set of textbooks on group theory. In order to make the exercise book self-contained, a brief review of the main concepts and techniques is given before the problems in each section. The reviews can be used as a concise textbook on group theory. Cooperated with my student, the English version of the exercise book was published in 2004 by World Scientific named "Problems and Solutions in Group Theory for Physicists" ([Ma and Gu (2004)]). A great deal of new materials drawn from teaching and research experiences is included. The reviews of each chapter has been extensively revised. Last four chapters are essentially new. Some useful results are listed in the exercise book for reference. Some materials are moved from the textbook to the exercise book such that the textbook can concentrate the main subjects on group theory and include some new developments in group theory. This exercise book serves as a supplement for the present textbook.

The present textbook consists of 10 chapters. Chapter 1 is a short review on linear algebra. The reader is required not only to be familiar with its basic concepts but also to master its applications, especially the similarity transformation method. In Chap. 2, the concepts of a group and its subsets are introduced from the physical problems and explained through examples of some finite groups. The importance of the group table of a finite group is emphasized. The group table of the symmetric groups of regular polyhedrons are introduced in a new way (see §2.5.2). The theory of representations of a group is studied in Chap. 3. The transformation operator P_R for the scalar functions bridges the gap between the representation theory and physical application. The subduced and induced representations of groups are used to construct the character tables of finite groups in

Chap. 3, and to calculate the outer product of representations of the permutation groups in Chap. 6. The method of idempotents is systematically studied in §3.7. Based on this method the standard irreducible basis vectors in the group algebras of some finite groups in common use in physics are calculated. The method of Young operators studied in Chap. 6 is its development to the permutation groups.

The classification and representations of semisimple Lie algebras are introduced in Chap. 7 and partly in Chap. 4 by the language familiar to physicists. The methods of block weight diagrams and dominant weight diagrams are recommended for calculating the representation matrices of the generators and the Clebsch–Gordan series in a simple Lie algebra. The readers who are interested in the strict mathematical definitions and proofs in the theory of semisimple Lie algebras are recommended to read the more mathematically oriented books (e.g. [Bourbaki (1989)]).

The remaining part of the book is devoted to the important symmetric groups of physical systems. In Chap. 4 the symmetric group SO(3) of a spherically symmetric system in three dimensions is studied. The study on the symmetry of crystals is a typical example of the physical application of group theory. In Chap. 5, through a systematic analysis by the method of group theory, only based on the translation symmetry of crystals, the crystals are classified completely that there are 11 proper crystallographic point groups, 32 crystallographic point groups, 7 crystal systems, 14 Bravais lattices, 73 symmorphic space groups, and 230 space groups. The international symbols of the space groups are recommended in Chap. 5. The analysis method for the symmetry of a crystal from its International symbols are emphasized (see §5.4.6). The permutation groups S_n are the symmetric groups of the identical particles and are widely used in the decomposition of the tensor spaces. In Chap. 6 the permutation groups S_n are studied by the method of Young operators. The irreducible basis vectors in the group space of S_n are explicitly given based on the Young operators, and the calculating method for the similarity transformations from them to the orthonormal basis vectors is proved and demonstrated by examples. The matrix groups in common use in physics, such as the SU(N) groups, the SO(N) groups, the USp(2ℓ) groups, and the Lorentz group L_p are studied in some detail in the last three chapters.

The systematic examination of Young operators is an important characteristic of this book. We calculate the characters, the representation matrices, and the outer product of the irreducible representations of the permutation groups using the method of Young operators. For the matrix groups SU(N), SO(N), and USp(2ℓ), which are related to four classical Lie algebras, the basis states of the irreducible representations can be explicitly calculated using the method of Young operators. The dimensions of the irreducible representations of the permutation groups, the SU(N) groups, the SO(N) groups, and the Sp(2ℓ) groups are all calculated by the hook rule, a method based on the Young patterns.

An isolated quantum n-body system is invariant in the translation of space–time and the spatial rotation so that the energy, the momentum, and the total angular momentum of the system are conserved. The motion of the center-of-mass and the global rotation of the system should be separated from its internal motion, and its Schrödinger equation is reduced to the radial equation, depending only on the internal degrees of freedom. The method of the Jacobi coordinate vectors is summarized in §4.9.1 to separate the motion of the center-of-mass. The generalized harmonic polynomials are presented to separate the global rotation of an isolated quantum n-body system from the internal motion in §4.9 and generalized to arbitrary N-dimensional space in §9.4, where the Dirac equation in $(N+1)$-dimensional space–time is also studied.

In conclusion, I must express my cordial gratitude to my supervisor Prof. Ning Hu for his guidance, under which I entered the research field of the symmetric theory of particle physics. I am indebted to Prof. Yi-Shi Duan who made me an abecedarian to study and to teach the group theory. I am very grateful to my wife, Ms Xian Li, for her continuous support. This book was supported by the National Natural Science Foundation of China under Grant Nos. 10475082 and 10675050.

Institute of High Energy Physics
Beijing, China
June 2007

Zhong-Qi Ma

List of Symbols

$\boldsymbol{a}$	vector,
$\hat{\boldsymbol{n}}$, $\hat{\boldsymbol{x}}$, $\boldsymbol{e}_\mu$	unit vector,
$\underline{a}$	column matrix,
Γ, Λ	diagonal matrix,
$D(R)^{-1}$, $D(R)^T$	inverse and transpose of a matrix $D(R)$,
$D(R)^*$, $D(R)^\dagger$	complex matrix and conjugate matrix,
$\mathcal{L}$, $\mathcal{L}_1$	linear space or subspace,
$\mathcal{L}_1 + \mathcal{L}_2$	sum of two subspaces,
$\mathcal{L}_1 \oplus \mathcal{L}_2$	direct sum of two subspaces,
AB	product of two digits, two matrices etc.,
$\times$	direct product of matrices, representations,
$\otimes$	direct product of groups, outer product,
E, e	identical element of a group,
R, S	transformations, elements of a group,
$\mathcal{R}$	set, complex of elements,
$\mathcal{C}_\alpha$	class in a group,
$n(\alpha)$	number of elements in $\mathcal{C}_\alpha$,
C_α	class operator,
g	order of a group,
g_c	number of classes in a group,
$\rho \bmod n$	ρ is an integer and $\rho + n = \rho$,
$G' \approx G$	G is isomorphic onto G',
$G' \sim G$	G is homomorphic onto G',
C_N, D_N	cyclic group, dihedral group,
$\mathbf{T}$, $\mathbf{O}$, $\mathbf{I}$	symmetric groups of polyhedrons,
$D(G) \simeq \overline{D}(G)$	equivalent representations,
e_a, e^j_μ	idempotent,

$b^j_{\mu\nu}$	standard irreducible basis vectors in the group space of a finite group,
$\mathcal{L}^j_\mu$, $\mathcal{R}^j_\mu$, $\mathcal{I}^j$	left ideal, right ideal, two-side ideal,
T_a, $T^{(r)}_{ab}$, T_{ab}	generators in self representation,
I_j, I_A, $D(I_A)$	generators of a Lie group,
$Y^\ell_m(\hat{\boldsymbol{x}})$	spherical harmonic function,
$\mathbf{Y}^\ell_m(\boldsymbol{x})$	harmonic polynomial,
$Q^{\ell\lambda}_q(\boldsymbol{R}_1, \boldsymbol{R}_2)$	generalized harmonic polynomial,
$C^{jk}_{\mu\nu,JMr}$	Clebsch–Gordan coefficients,
$\langle j, \mu, k, \nu \vert J, (r), M\rangle$	
$\vert j, \mu\rangle \vert k, \nu\rangle$, $\vert\mu\rangle\vert\nu\rangle$,	basis vectors before similarity transformation,
$\vert\boldsymbol{M}_1, \boldsymbol{m}_1\rangle \vert\boldsymbol{M}_2, \boldsymbol{m}_2\rangle$	
$\vert\boldsymbol{m}_1\rangle\vert\boldsymbol{m}_2\rangle$,	
$\vert\vert J, (r), M\rangle$	basis vectors after similarity transformation,
$\vert\vert\boldsymbol{M}, (r), \boldsymbol{m}\rangle$	
$C_{jk}{}^\ell$, $C_{AB}{}^D$	structure constant of a Lie group,
$\boldsymbol{V}(x)_a$, $\boldsymbol{T}(x)_{a_1 \ldots a_n}$	components of vector and tensor,
$\boldsymbol{\theta}_d$, $\boldsymbol{\theta}_{d_1 \ldots d_n}$	basis vectors, basis tensors,
$\mathcal{T}$, $\mathcal{T}^*$, $\mathcal{T}^{[\lambda]}_\mu$	tensor space and subspace,
P_R, Q_R,	transformations operators for scalars, tensors
$O_R = P_R Q_R$	and spinors,
P and Q	horizontal and vertical permutations,
P_0 and Q_0	horizontal and vertical transpositions,
$\mathcal{Y}$, $\mathcal{Y}^{[\lambda]}_\mu$	Young operator,
$\mathcal{P}$, $\mathcal{Q}$	horizontal and vertical operators,
$[\lambda]$	Young pattern,
$\mathcal{L}$, $\mathcal{L}_R$	Lie algebra, real Lie algebra,
$\mathcal{H}$,	Cartan subalgebra of a Lie algebra,
g_{AB}, g_{jk}	Killing form,
H_j and $E_{\boldsymbol{\alpha}}$	Cartan–Weyl bases,
H_μ, E_μ and F_μ	Chevalley bases,
$\boldsymbol{r}_\mu$	simple roots,
$\boldsymbol{M}$, $\boldsymbol{m}$	the highest weight, weights,
$\boldsymbol{w}_\mu$	fundamental dominant weight
$A_{\mu\nu}$	Cartan matrix,
$C_2(\boldsymbol{M})$	Casimir invariant of order two,
$[(\pm)\lambda]$	self-dual and anti-self-dual representations of SO(N),

$[s, \lambda]$	spinor representation of $\mathrm{SO}(2\ell+1)$,
$[\pm s, \lambda]$	spinor representation of $\mathrm{SO}(2\ell)$,
Δ	set of all roots in a simple Lie algebra,
Δ_+	set of all positive roots in a simple Lie algebra.

Contents

Chapter 1

REVIEW ON LINEAR ALGEBRAS

The main mathematical tool in group theory is linear algebras. In this chapter, we will review some fundamental concepts and calculation methods in linear algebras, which are often used in group theory.

1.1 Linear Space and Basis Vector

Let $H(x)$ be the Hamiltonian of a system. Suppose that the eigenvalue E of $H(x)$ is m-degenerate,

$$H(x)\psi_\mu(x) = E\psi_\mu(x), \qquad \mu = 1,\ 2,\ \cdots,\ m, \tag{1.1}$$

where x briefly denotes the set of coordinates for all degrees of freedom. $\psi_\mu(x)$ are linearly independent to one another. Any linear combination of $\psi_\mu(x)$ is an eigenfunction of $H(x)$ with the same eigenvalue

$$\phi(x) = \sum_{\mu=1}^{m} \psi_\mu(x) a_\mu, \qquad H(x)\phi(x) = E\phi(x). \tag{1.2}$$

Conversely, any eigenfunction of $H(x)$ with the eigenvalue E can be expressed as a linear combination of $\psi_\mu(x)$ like (1.2). Two eigenfunctions satisfy the following calculation rule,

$$c \left(\sum_{\mu=1}^{m} \psi_\mu(x) a_\mu + \sum_{\mu=1}^{m} \psi_\mu(x) b_\mu \right) = \sum_{\mu=1}^{m} \psi_\mu(x) \left(c\, a_\mu + c\, b_\mu \right). \tag{1.3}$$

The set of $\phi(x)$ is called a linear space $\mathcal{L}$ of dimension m, generated by m basis vectors of $\psi_\mu(x)$. $\phi(x)$ is an arbitrary vector in $\mathcal{L}$ and a_μ is the μth-component of the vector $\phi(x)$ with respect to the basis vectors $\psi_\mu(x)$.

Generally, m objects $\boldsymbol{e}_\mu$ are said to be linearly independent if there do not exist m coefficients c_μ which are not vanishing simultaneously such that

$$\sum_{\mu=1}^{m} \boldsymbol{e}_\mu c_\mu = 0. \tag{1.4}$$

$\boldsymbol{e}_\mu$ satisfy the following linear formulas:

$$\begin{aligned} &\boldsymbol{e}_\mu a_\mu + \boldsymbol{e}_\nu a_\nu = \boldsymbol{e}_\nu a_\nu + \boldsymbol{e}_\mu a_\mu, \\ &c\left(\sum_\mu \boldsymbol{e}_\mu a_\mu + \sum_\mu \boldsymbol{e}_\mu b_\mu\right) = \sum_\mu \boldsymbol{e}_\mu \left(ca_\mu + cb_\mu\right), \end{aligned} \tag{1.5}$$

where c, a_μ, a_ν, and b_μ are arbitrary complex numbers. The m objects $\boldsymbol{e}_\mu$ generate a linear space $\mathcal{L}$ of dimension m, which is the set of all possible complex combinations $\boldsymbol{a}$ of $\boldsymbol{e}_\mu$

$$\boldsymbol{a} = \sum_{\mu=1}^{m} \boldsymbol{e}_\mu a_\mu. \tag{1.6}$$

$\boldsymbol{a}$ is called a vector in $\mathcal{L}$, $\boldsymbol{e}_\mu$ is a basis vector, and a_μ is the μth component of $\boldsymbol{a}$ with respect to the basis vectors $\boldsymbol{e}_\mu$. The space $\mathcal{L}$ is called a real space if the components a_μ of all vectors $\boldsymbol{a}$ in $\mathcal{L}$ are real. A vector is called a null vector if its components are all vanishing. Two vectors $\boldsymbol{a}$ and $\boldsymbol{b}$ are said to be equal to each other if and only if their components are respectively equal, $a_\mu = b_\mu$. In linear algebras, the concepts of vectors and linear space are independent of the physical content of the objects.

For a given space $\mathcal{L}$ and a given set of basis vectors, vector $\boldsymbol{a}$ is completely described by the m components a_μ. Usually, the m ordered numbers are arranged as a column-matrix $\underline{a}$,

$$\underline{a} = \begin{pmatrix} a_1 \\ a_2 \\ \vdots \\ a_m \end{pmatrix}. \tag{1.7}$$

The column-matrix $\underline{a}$ is another form to denote vector $\boldsymbol{a}$. Sometimes, we do not distinguish two symbols $\underline{a}$ and $\boldsymbol{a}$.

A basis vector is a special vector where only one component is nonvanishing and to be one,

$$\left(\boldsymbol{e}_\mu\right)_\nu = \delta_{\mu\nu} = \begin{cases} 1 & \text{when } \mu = \nu, \\ 0 & \text{when } \mu \neq \nu, \end{cases} \tag{1.8}$$

where $\delta_{\mu\nu}$ is the Kronecker δ function.

n vectors $\boldsymbol{a}^{(1)}$, $\boldsymbol{a}^{(2)}$, $\cdots$, $\boldsymbol{a}^{(n)}$ are linearly dependent if there exists a linear relation

$$\sum_{i=1}^{n} \boldsymbol{a}^{(i)} c_i = 0, \tag{1.9}$$

where n coefficients c_i are not vanishing simultaneously. Otherwise, they are linearly independent. In an m-dimensional space $\mathcal{L}$ the number n of linearly independent vectors is not larger than m.

In $\mathcal{L}$, n linearly independent vectors generate a subspace $\mathcal{L}_1$ of dimension n. A subspace is called a null space $\emptyset$ if it contains only the null vector. The whole space $\mathcal{L}$ and the null space $\emptyset$ are two trivial subspaces. Usually, we only consider nontrivial subspaces.

The sum of two subspaces $\mathcal{L}_1$ and $\mathcal{L}_2$ is a subspace, denoted by $\mathcal{L}_1+\mathcal{L}_2$, which contains all linear combinations of the vectors belonging to $\mathcal{L}_1$ and $\mathcal{L}_2$. The intersection of two subspaces is a subspace, denoted by $\mathcal{L}_1 \cap \mathcal{L}_2$, which contains all vectors belonging to both subspaces.

$\mathcal{L}$ is said to be the direct sum of two subspaces, $\mathcal{L} = \mathcal{L}_1 \oplus \mathcal{L}_2$, if $\mathcal{L} = \mathcal{L}_1 + \mathcal{L}_2$, and one of the following three equivalent conditions is satisfied.

(1) The intersection of $\mathcal{L}_1$ and $\mathcal{L}_2$ is a null space.

(2) The dimension of $\mathcal{L}$ is equal to the sum of the dimensions of $\mathcal{L}_1$ and $\mathcal{L}_2$.

(3) Each vector in $\mathcal{L}$ can be expressed uniquely as the sum of two vectors, respectively belonging to two subspaces $\mathcal{L}_1$ and $\mathcal{L}_2$.

$\mathcal{L}_2$ is called the complement of $\mathcal{L}_1$ in $\mathcal{L}$ if $\mathcal{L} = \mathcal{L}_1 \oplus \mathcal{L}_2$. $\mathcal{L}_1$ is also the complement of $\mathcal{L}_2$. The complement of $\mathcal{L}_1$ in $\mathcal{L}$ is not unique. A space $\mathcal{L}$ can be decomposed as a direct sum of some subspaces more than two.

1.2 Linear Transformations and Linear Operators

A transformation gives a rule, with which a function changes to another function. An operator is the mathematical symbol for a transformation. An operator $R(x)$ is linear if it satisfies

$$R(x)\left\{c_1\phi_1(x) + c_2\phi_2(x)\right\} = c_1 R(x)\phi_1(x) + c_2 R(x)\phi_2(x), \tag{1.10}$$

where the coefficients c_1 and c_2 are constant. A linear operator describes a linear transformation. The operators used in this textbook are linear if without special indication. The multiplication $R(x)S(x)$ of two operators

$R(x)$ and $S(x)$ is an operator, defined as a successive application to the function first with $S(x)$ and then with $R(x)$. Namely, if $S(x)\psi(x) = \phi(x)$, then $R(x)S(x)\psi(x) = R(x)\phi(x)$. Generally, the order of two operators in multiplication cannot be changed, namely $R(x)S(x) \neq S(x)R(x)$.

If an operator $R(x)$ is commutable with the Hamiltonian $H(x)$,

$$[H(x), R(x)] \equiv H(x)R(x) - R(x)H(x) = 0, \tag{1.11}$$

the application of $R(x)$ to the eigenfunction $\psi_\mu(x)$ of $H(x)$ is still an eigenfunction of $H(x)$ with the same eigenvalue,

$$H(x)\left\{R(x)\psi_\mu(x)\right\} = R(x)\left\{H(x)\psi_\mu(x)\right\} = E\left\{R(x)\psi_\mu(x)\right\}. \tag{1.12}$$

Thus, $R(x)\psi_\mu(x)$ belongs to the space $\mathcal{L}$ generated by m eigenfunctions ψ_μ of $H(x)$ with the same eigenvalue E, and can be expressed as Eq. (1.2),

$$R(x)\psi_\mu(x) = \sum_\nu \psi_\nu(x)D_{\nu\mu}(R). \tag{1.13}$$

$\mathcal{L}$ is called an invariant space to $R(x)$. The coefficients $D_{\nu\mu}(R)$ are arranged as a matrix $D(R)$ of dimension m, called the matrix of an operator $R(x)$ in the basis functions $\psi_\mu(x)$ of $\mathcal{L}$, or simply called the matrix of $R(x)$. Note that $D(R)$ depends on the operator $R(x)$, but not on x. The action of $R(x)$ to any function $\phi(x)$ in $\mathcal{L}$ can be calculated by the matrix $D(R)$. Namely, if $\phi(x) = \sum_\mu \psi_\mu(x)a_\mu$, and $R(x)\phi(x) = \phi_1(x) = \sum_\nu \psi_\nu(x)b_\nu$, one has

$$R(x)\phi(x) = \sum_{\mu=1}^{m} \left[R(x)\psi_\mu(x)\right] a_\mu = \sum_{\nu\mu} \psi_\nu D_{\nu\mu}(R)a_\mu,$$

$$b_\nu = \sum_{\mu=1}^{m} D_{\nu\mu}(R)a_\mu. \tag{1.14}$$

Generally, a linear operator R describes a transformation of vectors in a linear space $\mathcal{L}$ satisfying

$$R\left\{c_1\boldsymbol{a} + c_2\boldsymbol{b}\right\} = c_1R\ \boldsymbol{a} + c_2R\ \boldsymbol{b}. \tag{1.15}$$

$\mathcal{L}$ is invariant to R if the application of R to any vector $\boldsymbol{a}$ in $\mathcal{L}$ is still a vector in $\mathcal{L}$,

$$R\boldsymbol{a} = \boldsymbol{b} \in \mathcal{L}, \qquad \forall\ \boldsymbol{a} \in \mathcal{L}. \tag{1.16}$$

The matrix $D(R)$ of R in its invariant space $\mathcal{L}$ is calculated from the application of R to the basis vectors $\boldsymbol{e}_\mu$

$$R\,\boldsymbol{e}_\mu = \sum_\nu \boldsymbol{e}_\nu D_{\nu\mu}(R). \tag{1.17}$$

The action of R to any vector $\boldsymbol{a}$ in $\mathcal{L}$ can be calculated by $D(R)$. If

$$\boldsymbol{a} = \sum_\mu \boldsymbol{e}_\mu a_\mu, \qquad R\,\boldsymbol{a} = \boldsymbol{b} = \sum_\nu \boldsymbol{e}_\nu b_\nu, \tag{1.18}$$

then

$$R\,\boldsymbol{a} = \sum_\mu \left(R\boldsymbol{e}_\mu\right) a_\mu = \sum_{\nu\mu} \boldsymbol{e}_\nu D_{\nu\mu}(R) a_\mu,$$

$$b_\nu = \sum_\mu D_{\nu\mu}(R) a_\mu, \qquad \underline{b} = D(R)\ \underline{a}. \tag{1.19}$$

It is worthy to emphasize the difference between Eqs. (1.17) and (1.19). In Eq. (1.17) a basis vector $\boldsymbol{e}_\mu$ transforms in the operator R to a combination of basis vectors, where the combination index of the basis vectors is the row index ν of $D_{\nu\mu}(R)$. Equation (1.19) is a component equation for vector $\boldsymbol{a}$ transformed by the operator R to another vector $\boldsymbol{b}$, where the combination index of the vector components is the column index μ of $D_{\nu\mu}(R)$. Two equations are consistent because the basis vector $\boldsymbol{e}_\mu$ is a special vector, where only one component is nonvanishing but equal to one,

$$\left(R\boldsymbol{e}_\mu\right)_\rho = \sum_\lambda D_{\rho\lambda}(R) \left(\boldsymbol{e}_\mu\right)_\lambda = D_{\rho\mu}(R) = \sum_\nu \left(\boldsymbol{e}_\nu\right)_\rho D_{\nu\mu}(R). \tag{1.20}$$

1.3 Similarity Transformation

For a given set of basis vectors $\boldsymbol{e}_\mu$ in a linear space $\mathcal{L}$ of dimension m, there is a one-to-one correspondence between vector $\boldsymbol{a}$ and its column-matrix $\underline{a}$, and there is a one-to-one correspondence between an operator R and its matrix $D(R)$. However, the basis vectors in $\mathcal{L}$ are not unique. Any set of m linearly independent vectors can be chosen to be basis vectors. In this section we will discuss how the column-matrix of a vector and the matrix of an operator change when the basis vectors are changed.

Let $\boldsymbol{e}'_\nu$ be m linearly independent vectors with the components $S_{\mu\nu}$ in the original basis vectors $\boldsymbol{e}_\mu$,

$$\boldsymbol{e}'_\nu = \sum_\mu \boldsymbol{e}_\mu S_{\mu\nu}, \qquad \underline{e'_\nu} = \underline{S_{\cdot\nu}}. \tag{1.21}$$

Since $\boldsymbol{e}'_\nu$ are linearly independent, S is a nonsingular matrix (det $S \neq 0$) and has its inverse matrix S^{-1}.

$$\boldsymbol{e}_\mu = \sum_\nu \boldsymbol{e}'_\nu \left(S^{-1}\right)_{\nu\mu}. \tag{1.22}$$

Choosing $\boldsymbol{e}'_\nu$ to be new basis vectors, the components a'_ν of the vector $\boldsymbol{a}$ and the matrix $\overline{D}(R)$ of the operator R can be calculated as follows:

$$\boldsymbol{a} = \sum_\mu \boldsymbol{e}_\mu a_\mu = \sum_{\nu\mu} \boldsymbol{e}'_\nu \left(S^{-1}\right)_{\nu\mu} a_\mu = \sum_\nu \boldsymbol{e}'_\nu a'_\nu,$$

$$a'_\nu = \sum_\mu \left(S^{-1}\right)_{\nu\mu} a_\mu, \qquad \underline{a}' = S^{-1}\, \underline{a}. \tag{1.23}$$

$$R\boldsymbol{e}'_\nu = \sum_\rho \left(R\boldsymbol{e}_\rho\right) S_{\rho\nu} = \sum_{\mu\rho} \boldsymbol{e}_\mu D_{\mu\rho}(R) S_{\rho\nu},$$
$$R\boldsymbol{e}'_\nu = \sum_\rho \boldsymbol{e}'_\rho \overline{D}_{\rho\nu}(R) = \sum_{\mu\rho} \boldsymbol{e}_\mu S_{\mu\rho} \overline{D}_{\rho\nu}(R),$$

$$\sum_\rho D_{\mu\rho}(R) S_{\rho\nu} = \sum_\rho S_{\mu\rho} \overline{D}_{\rho\nu}(R), \qquad \overline{D}(R) = S^{-1} D(R) S. \tag{1.24}$$

The relation (1.24) between $\overline{D}(R)$ and $D(R)$ is called a similarity transformation and S is the matrix of the similarity transformation. In literature, $\overline{D}(R)$ and $D(R)$ are said to be equivalent to each other if Eq. (1.24) holds. Obviously, the matrix S has the same matrix form with respect to both the original set and the new set of basis vectors. If the new set of basis vectors is the same as the original one except for the order of basis vectors, the similarity transformation is called the simple one. Note that, the similarity transformation for two equivalent matrices is not unique. If X is commutable with $D(R)$ and Y is commutable with $\overline{D}(R)$, both XS and SY satisfy the similarity transformation relation (1.24).

Since $\underline{S_{\cdot\nu}}$ is nothing but the column matrix of new basis vector $\boldsymbol{e}'_\nu$ in the original basis vectors, Eq. (1.24) can be written as

$$D(R)\, \underline{S_{\cdot\nu}} = \sum_\rho \underline{S_{\cdot\rho}}\, \overline{D}_{\rho\nu}(R), \qquad R\boldsymbol{e}'_\nu = \sum_\rho \boldsymbol{e}'_\rho \overline{D}_{\rho\nu}(R). \tag{1.25}$$

It is nothing but the definition of the matrix of R in the new basis vectors $\boldsymbol{e}'_\nu$. The different choices of the basis vectors do not change the action of an operator on a vector. If $\boldsymbol{b} = R\boldsymbol{a}$, one has $\underline{b} = D(R)\underline{a}$ in the original basis vectors $\boldsymbol{e}_\mu$. In the new set of basis vectors $\boldsymbol{e}'_\nu$ one has

$$\underline{b}' = S^{-1}\underline{b} = S^{-1}D(R)\underline{a} = S^{-1}D(R)S\underline{a}' = \overline{D}(R)\underline{a}'. \tag{1.26}$$

Let $\mathcal{L}$ be an m-dimensional space, $\mathcal{L}_1$ be its n-dimensional subspace, invariant to the operator R, and $\mathcal{L}_2$ be the complement of $\mathcal{L}_1$. Choose a new set of basis vectors in $\mathcal{L}$ such that the first n basis vectors belong to $\mathcal{L}_1$, and the next $(m-n)$ ones belong to $\mathcal{L}_2$. Arrange the new basis vectors $\boldsymbol{e}'_\nu$ to be the column matrices of S, $\underline{S_{.\nu}} = \underline{e'_\nu}$. Through the similarity transformation S the matrix $D(R)$ of R is changed to $\overline{D}(R)$. Since $\mathcal{L}_1$ is invariant to R, one has

$$R\boldsymbol{e}'_\mu = \sum_{\nu=1}^{n} \boldsymbol{e}'_\nu \overline{D}_{\nu\mu}(R), \qquad 1 \leq \mu \leq n. \tag{1.27}$$

Namely, the down-left corner of $\overline{D}(R)$ is vanishing,

$$\overline{D}_{\rho\mu}(R) = 0 \qquad \text{when} \ \ \mu \leq n < \rho,$$

$$S^{-1}D(R)S = \overline{D}(R) = \begin{pmatrix} D^{(1)}(R) & M \\ 0 & D^{(2)}(R) \end{pmatrix}. \tag{1.28}$$

This matrix $\overline{D}(R)$ is called a ladder one. Furthermore, if $\mathcal{L}_2$ is also invariant to R, one has $M = 0$,

$$\overline{D}(R) = \begin{pmatrix} D^{(1)}(R) & 0 \\ 0 & D^{(2)}(R) \end{pmatrix} = D^{(1)}(R) \oplus D^{(2)}(R). \tag{1.29}$$

This matrix $\overline{D}(R)$ is a direct sum of two submatrices, and is called a block one in the type $[n, (m-n)]$. Generally, a matrix is also called a block one if it can be changed into the direct sum of submatrices by a simple similarity transformation. In order to determine whether or not a matrix is a block one, one may separate the indices of the matrix into two parts and check whether the matrix entries with indices respectively belonging to different parts all are vanishing. If Λ is diagonal and $X\Lambda = \Lambda X$, X is a block matrix.

1.4 Eigenvectors and Diagonalization of a Matrix

In quantum mechanics, the eigenequation of a physical operator $R(x)$ is

$$R(x)\psi(x) = \lambda\psi(x).$$

The eigenvalue λ is the possible observed value for the physical quantity. The eigenvalues describe the characteristic of the physical quantity and are independent of the choice of basis functions. In linear algebras, the eigenequation of an operator R is

$$R\,\boldsymbol{a} = \lambda\,\boldsymbol{a}. \tag{1.30}$$

If $\mathcal{L}$ is an m-dimensional space invariant to R and $\boldsymbol{a} \in \mathcal{L}$, one has

$$D(R)\ \underline{a} = \lambda\ \underline{a}, \qquad \sum_{\nu} D_{\mu\nu}(R)a_{\nu} = \lambda a_{\mu}. \tag{1.31}$$

The eigenequation (1.31) is a set of coupled linear homogeneous equations for m variables a_{ν}. The condition for the existence of nonvanishing solution is its coefficient determinant to be vanishing,

$$\det\left[D(R) - \lambda\mathbf{1}\right] = 0. \tag{1.32}$$

Equation (1.32) is called the secular equation for $D(R)$, and its roots are the eigenvalues of $D(R)$. The secular equation is invariant in similarity transformation, so that the eigenvalues are independent of the choice of the basis vectors. It is easy to see from Eq. (1.32) that the sum of the eigenvalues of $D(R)$ is its trace, Tr $D(R)$, and the product of the eigenvalues of $D(R)$ is its determinant, $\det D(R)$ (see Prob. 1).

Equation (1.31) means that the eigenvector generates a one-dimensional subspace which is invariant to R. If in $\mathcal{L}$ there exist m linearly independent eigenvectors $\boldsymbol{a}^{(\nu)}$ of R with the eigenvalues λ_{ν}, respectively, one may choose a new set of basis vectors $\boldsymbol{e}'_{\nu} = \boldsymbol{a}^{(\nu)}$. Then, the matrix of R in the new basis vectors $\boldsymbol{e}'_{\nu}$ is a diagonal one.

$$S^{-1}D(R)S = \Lambda, \qquad \underline{S_{\cdot\nu}} = \underline{a^{(\nu)}}, \qquad \Lambda_{\nu\mu} = \delta_{\nu\mu}\lambda_{\nu}. \tag{1.33}$$

Therefore, the key for diagonalizing an m-dimensional matrix is to find its m linearly independent eigenvectors. Since the eigenequation (1.30) is linearly homogeneous with respect to the eigenvector $\boldsymbol{a}^{(\nu)}$, $\boldsymbol{a}^{(\nu)}$ can be multiplied with a constant c_{ν}. When the eigenvalue is degenerate, its eigenvectors can be made a nonsingular linear combination. This is the reason why the similarity transformation matrix S is not unique. The number of the arbitrary parameters in S is equal to $\sum_\nu n_\nu^2$, where n_ν is the multiplicity of the eigenvalue λ_ν. Those parameters play an important role in calculating a common similarity transformation for a few pairs of equivalent matrices, which is often used in group theory.

Substituting an eigenvalue into Eq. (1.31), one is always able to solve at least one eigenvector. The eigenvectors for different eigenvalues are linearly independent. However, when an eigenvalue is a root of Eq. (1.32) with multiplicity n, it is not certain to obtain n linearly independent eigenvectors with the given eigenvalue by solving Eq. (1.31). The following matrix is the simplest example which has the eigenvalue 1 with multiplicity 2, but only one linearly independent eigenvector,

$$\begin{pmatrix} 1 & b \\ 0 & 1 \end{pmatrix} \begin{pmatrix} 1 \\ 0 \end{pmatrix} = \begin{pmatrix} 1 \\ 0 \end{pmatrix}, \qquad b \neq 0. \tag{1.34}$$

It cannot be diagonalized by a similarity transformation.

The eigenvalues are invariant in a similarity transformation. Therefore, two equivalent matrices can be diagonalized into the same matrix,

$$X^{-1}D(R)X = \Lambda = Y^{-1}\overline{D}(R)Y,$$

such that the similarity transformation related to them is easy to be calculated,

$$\overline{D}(R) = \left(XY^{-1}\right)^{-1} D(R) \left(XY^{-1}\right). \tag{1.35}$$

It can be proved that the sufficient and necessary condition for a matrix $D(R)$ which can be diagonalized by a unitary similarity transformation is that $D(R)^\dagger$ is commutable with $D(R)$ (see Prob. 11). A matrix H is called Hermitian if $H^\dagger = H$. A matrix u is called unitary if $u^\dagger = u^{-1}$. Both a Hermitian matrix and a unitary matrix can be diagonalized by a unitary similarity transformation. A real unitary matrix is called real orthogonal. A real Hermitian matrix is called real symmetric. A real symmetric matrix can be diagonalized by a real orthogonal similarity transformation. Generally, a real orthogonal matrix can be diagonalized by a unitary similarity transformation, but not a real orthogonal one.

1.5 Inner Product of Vectors

There are three types of products of vectors, depending on the products to be a scalar, a vector, or a tensor. The product of two vectors is called the inner product if the product is a scalar. In quantum mechanics the inner product of two wave functions is defined as

$$\langle \phi(x)|\psi(x)\rangle = \int (dx)\phi(x)^*\psi(x), \tag{1.36}$$

where the sign $\int$ denotes an integral for the continuous coordinate and a sum for the discrete coordinate. The form on the left-hand side of Eq. (1.36) is called the Dirac symbol. Generally, the inner product of two vectors $\langle \boldsymbol{a}|\boldsymbol{b}\rangle$ in the linear algebras satisfies

$$\begin{aligned}
&\langle c_1\boldsymbol{a}^{(1)} + c_2\boldsymbol{a}^{(2)}|\boldsymbol{b}\rangle = c_1^*\langle \boldsymbol{a}^{(1)}|\boldsymbol{b}\rangle + c_2^*\langle \boldsymbol{a}^{(2)}|\boldsymbol{b}\rangle, \\
&\langle \boldsymbol{a}|c_1\boldsymbol{b}^{(1)} + c_2\boldsymbol{b}^{(2)}\rangle = c_1\langle \boldsymbol{a}|\boldsymbol{b}^{(1)}\rangle + c_2\langle \boldsymbol{a}|\boldsymbol{b}^{(2)}\rangle, \\
&\langle \boldsymbol{b}|\boldsymbol{a}\rangle = \langle \boldsymbol{a}|\boldsymbol{b}\rangle^*, \qquad \langle \boldsymbol{a}|\boldsymbol{a}\rangle = |\boldsymbol{a}|^2 > 0, \quad \text{if } \boldsymbol{a} \neq 0.
\end{aligned} \tag{1.37}$$

The inner product is linear for the second vector and antilinear for the first vector. The inner product becomes its complex conjugate if changing the order of two factors. The self-inner product of a nonvanishing vector is real positive, called the square module of the vector. Denote by a Hermitian matrix Ω the inner product of two basis vectors

$$\langle \boldsymbol{e}_\mu|\boldsymbol{e}_\nu\rangle = \Omega_{\mu\nu}, \qquad \Omega_{\nu\mu} = \Omega_{\mu\nu}^* = \left(\Omega^\dagger\right)_{\nu\mu}. \tag{1.38}$$

Let $\boldsymbol{a}$ be an nonvanishing eigenvector of Ω with the eigenvalue λ,

$$\boldsymbol{a} = \sum_\nu \boldsymbol{e}_\nu a_\nu \neq 0, \qquad \sum_\nu \Omega_{\mu\nu} a_\nu = \lambda a_\mu,$$

$$\lambda \sum_\mu |a_\mu|^2 = \sum_{\mu\nu} a_\mu^* \Omega_{\mu\nu} a_\nu = \sum_{\mu\nu} a_\mu^* \langle \boldsymbol{e}_\mu|\boldsymbol{e}_\nu\rangle a_\nu = \langle \boldsymbol{a}|\boldsymbol{a}\rangle > 0. \tag{1.39}$$

Hence, the eigenvalue λ of Ω is positive, namely, Ω is positive definite. The inner product of two arbitrary vectors and the matrix entry of an operator can be calculated with Ω,

$$\boldsymbol{a} = \sum_\mu \boldsymbol{e}_\mu a_\mu, \qquad \boldsymbol{b} = \sum_\nu \boldsymbol{e}_\nu b_\nu, \qquad \langle \boldsymbol{a}|\boldsymbol{b}\rangle = \sum_{\mu\nu} a_\mu \Omega_{\mu\nu} b_\nu, \tag{1.40}$$

$$\sum_\rho \left(\Omega^{-1}\right)_{\mu\rho} \langle \boldsymbol{e}_\rho|R\boldsymbol{e}_\nu\rangle = \sum_{\rho\tau} \left(\Omega^{-1}\right)_{\mu\rho} \langle \boldsymbol{e}_\rho|\boldsymbol{e}_\tau\rangle D_{\tau\nu}(R) = D_{\mu\nu}(R). \tag{1.41}$$

A vector is called normalized if its module is one. Two vectors are orthogonal if their inner product is zero. Two nonvanishing orthogonal vectors must be linearly independent.

The basis vectors are called orthonormal if

$$\langle \boldsymbol{e}_\mu | \boldsymbol{e}_\nu \rangle = \Omega_{\mu\nu} = \delta_{\mu\nu}. \tag{1.42}$$

In the orthonormal basis vectors the formulas for the inner product become simpler,

$$\langle \boldsymbol{a} | \boldsymbol{b} \rangle = \sum_\mu a_\mu b_\mu, \qquad \langle \boldsymbol{e}_\mu | R\boldsymbol{e}_\nu \rangle = D_{\mu\nu}(R). \tag{1.43}$$

There are a few different definitions for the inner product of vectors. Another inner product is defined to be linear for both factors,

$$\begin{aligned}
&\langle c_1\boldsymbol{a}^{(1)} + c_2\boldsymbol{a}^{(2)} | \boldsymbol{b} \rangle = c_1 \langle \boldsymbol{a}^{(1)} | \boldsymbol{b} \rangle + c_2 \langle \boldsymbol{a}^{(2)} | \boldsymbol{b} \rangle, \\
&\langle \boldsymbol{a} | c_1\boldsymbol{b}^{(1)} + c_2\boldsymbol{b}^{(2)} \rangle = c_1 \langle \boldsymbol{a} | \boldsymbol{b}^{(1)} \rangle + c_2 \langle \boldsymbol{a} | \boldsymbol{b}^{(2)} \rangle, \\
&\langle \boldsymbol{e}_\mu | \boldsymbol{e}_\nu \rangle = \Omega_{\mu\nu} = \Omega_{\nu\mu}, \qquad \det \Omega \neq 0.
\end{aligned} \tag{1.44}$$

In this definition, the self-inner product of a vector may not be real.

The inner product of the column matrices has been defined, namely

$$\underline{a}^\dagger \ \underline{b} = \sum_\mu a_\mu^* b_\mu, \qquad \underline{a}^T \ \underline{b} = \sum_\mu a_\mu b_\mu. \tag{1.45}$$

In comparison with Eq. (1.40) the inner product (1.45) means that the basis vectors for the column matrices are orthonormal.

At last, we discuss the concept of the adjoint operator $R^\dagger$ of an operator R. The conjugate matrix $D(R)^\dagger$ of a matrix $D(R)$ was well defined,

$$\left[D(R)^\dagger\right]_{\mu\nu} = D_{\nu\mu}(R)^*, \tag{1.46}$$

$$\left[\underline{a}^\dagger \ D(R) \ \underline{b}\right]^* = \left[D(R)\underline{b}\right]^\dagger \ \underline{a} = \underline{b}^\dagger \ D^\dagger(R) \ \underline{a}. \tag{1.47}$$

The definition for the adjoint operator in quantum mechanics is the generalization of Eq. (1.47),

$$\langle \boldsymbol{a} | R\boldsymbol{b} \rangle^* = \langle R\boldsymbol{b} | \boldsymbol{a} \rangle = \langle \boldsymbol{b} | R^\dagger \boldsymbol{a} \rangle. \tag{1.48}$$

The adjoint relation between two operators is mutual. Note that the matrices of two adjoint operators are not necessary to be conjugate if the basis vectors are not orthonormal. Denote by $D(R)$ and X the matrices of two operators R and $R^\dagger$, respectively

$$R\boldsymbol{e}_\mu = \sum_\rho \boldsymbol{e}_\rho D_{\rho\mu}(R), \qquad R^\dagger \boldsymbol{e}_\nu = \sum_\rho \boldsymbol{e}_\rho X_{\rho\nu},$$

$$\begin{gathered}\sum_{\rho} D^*_{\rho\mu}(R)\langle \boldsymbol{e}_\rho|\boldsymbol{e}_\nu\rangle = \langle R\boldsymbol{e}_\mu|\boldsymbol{e}_\nu\rangle = \langle \boldsymbol{e}_\mu|R^\dagger \boldsymbol{e}_\nu\rangle = \sum_{\rho} \langle \boldsymbol{e}_\mu|\boldsymbol{e}_\rho\rangle X_{\rho\nu}, \\ D^\dagger(R)\Omega = \Omega X, \qquad X = \Omega^{-1}D^\dagger(R)\Omega.\end{gathered} \tag{1.49}$$

Conversely, if the basis vectors are orthonormal, two mutual conjugate matrices correspond to two operators adjoint to each other. Namely, in this case, a Hermitian (or an unitary) matrix corresponds to a Hermitian (or a unitary) operator.

1.6 The Direct Product of Matrices

If a quantum system consists of two subsystems, the wave function of the system is expressed as the product of two wave functions of the subsystems, or the combination of the products. Suppose that the two functional spaces $\mathcal{L}_1$ and $\mathcal{L}_2$ for two subsystems are respectively invariant to the operator R,

$$R\psi_\mu = \sum_{\nu=1}^{m} \psi_\nu D^{(1)}_{\nu\mu}(R), \qquad R\phi_i = \sum_{j=1}^{n} \phi_j D^{(2)}_{ji}(R). \tag{1.50}$$

For the composed system, the functional space $\mathcal{L}$ generated by

$$\psi_\mu\phi_i, \qquad 1 \le \mu \le m, \;\; 1 \le i \le n \tag{1.51}$$

is called the product of two subspaces, $\mathcal{L} = \mathcal{L}_1\mathcal{L}_2$, which is (mn)-dimensional. The space $\mathcal{L}$ is also invariant to R, i.e.,

$$\begin{gathered}R\left(\psi_\mu\phi_i\right) = \sum_{\nu\ j} \left(\psi_\nu\phi_j\right)\left[D^{(1)}(R) \times D^{(2)}(R)\right]_{\nu j,\mu i}, \\ \left[D^{(1)}(R) \times D^{(2)}(R)\right]_{\nu j,\mu i} = D^{(1)}_{\nu\mu}(R)D^{(2)}_{ji}(R).\end{gathered} \tag{1.52}$$

The matrix $D^{(1)}(R) \times D^{(2)}(R)$ of R in $\mathcal{L}$, which is (nm)-dimensional, is called the direct product of two submatrices $D^{(1)}(R)$ and $D^{(2)}(R)$. The row (column) of the direct product matrix is denoted by two indices μ and i. The order of indices is usually arranged such that the second index i increases for a given μ, and then the first index μ increases. For example, the direct product of two 2-dimensional matrices X and Y is

$$X \times Y = \begin{pmatrix} X_{11}Y & X_{12}Y \\ X_{21}Y & X_{22}Y \end{pmatrix} = \begin{pmatrix} X_{11}Y_{11} & X_{11}Y_{12} & X_{12}Y_{11} & X_{12}Y_{12} \\ X_{11}Y_{21} & X_{11}Y_{22} & X_{12}Y_{21} & X_{12}Y_{22} \\ X_{21}Y_{11} & X_{21}Y_{12} & X_{22}Y_{11} & X_{22}Y_{12} \\ X_{21}Y_{21} & X_{21}Y_{22} & X_{22}Y_{21} & X_{22}Y_{22} \end{pmatrix}.$$

The product of two matrices $D^{(1)}(R) \times D^{(2)}(R)$ and $D^{(1)}(S) \times D^{(2)}(S)$ is

$$\begin{aligned} &\left[D^{(1)}(R) \times D^{(2)}(R)\right] \left[D^{(1)}(S) \times D^{(2)}(S)\right] \\ &\quad = \left[D^{(1)}(R)D^{(1)}(S)\right] \times \left[D^{(2)}(R)D^{(2)}(S)\right] . \end{aligned} \tag{1.53}$$

Thus,

$$\begin{aligned} \left[D^{(1)}(R) \times D^{(2)}(R)\right]^{-1} &= D^{(1)}(R)^{-1} \times D^{(2)}(R)^{-1}, \\ \left[D^{(1)}(R) \times D^{(2)}(R)\right]^{T} &= D^{(1)}(R)^{T} \times D^{(2)}(R)^{T}, \\ \left[D^{(1)}(R) \times D^{(2)}(R)\right]^{\dagger} &= D^{(1)}(R)^{\dagger} \times D^{(2)}(R)^{\dagger}. \end{aligned} \tag{1.54}$$

The trace and the determinant of direct product $D^{(1)}(R) \times D^{(2)}(R)$ are

$$\begin{aligned} \text{Tr}\ \left[D^{(1)}(R) \times D^{(2)}(R)\right] &= \left[\text{Tr}\ D^{(1)}(R)\right] \left[\text{Tr}\ D^{(2)}(R)\right], \\ \det\left[D^{(1)}(R) \times D^{(2)}(R)\right] &= \left[\det\ D^{(1)}(R)\right]^{n} \left[\det\ D^{(2)}(R)\right]^{m}. \end{aligned} \tag{1.55}$$

If two matrices $D^{(1)}(R)$ and $D^{(2)}(R)$ depend on a continuous parameter α, one has

$$\begin{aligned} &\frac{d}{d\alpha}\left[D^{(1)}(R) \times D^{(2)}(R)\right] \\ &\quad = \left[\frac{dD^{(1)}(R)}{d\alpha}\right] \times D^{(2)}(R) + D^{(1)}(R) \times \left[\frac{dD^{(2)}(R)}{d\alpha}\right]. \end{aligned} \tag{1.56}$$

The direct product reduces to the product of a number and a matrix if one of the two factor matrices is one-dimensional. Generally, $D^{(1)}(R) \times D^{(2)}(R)$ is not equal to $D^{(2)}(R) \times D^{(1)}(R)$, but their difference is only a simple similarity transformation. For example, when $n = m = 2$, the similarity transformation matrix for the two direct products is

$$\begin{pmatrix} 1 & 0 & 0 & 0 \\ 0 & 0 & 1 & 0 \\ 0 & 1 & 0 & 0 \\ 0 & 0 & 0 & 1 \end{pmatrix}. \tag{1.57}$$

1.7 Exercises

1. Prove that the sum of the eigenvalues of a matrix is equal to the trace of the matrix, and the product of eigenvalues is equal to the determinant of the matrix.

2. Calculate the eigenvalues and eigenvectors of the Pauli matrices,

$$\sigma_1 = \begin{pmatrix} 0 & 1 \\ 1 & 0 \end{pmatrix}, \qquad \sigma_2 = \begin{pmatrix} 0 & -i \\ i & 0 \end{pmatrix}.$$

3. Calculate the eigenvalues and eigenvectors of the matrix

$$R = \begin{pmatrix} 0 & 0 & 0 & 1 \\ 0 & 0 & 1 & 0 \\ 0 & 1 & 0 & 0 \\ 1 & 0 & 0 & 0 \end{pmatrix}.$$

4. Calculate the eigenvalues and eigenvectors of the matrix

$$R = \begin{pmatrix} 0 & 0 & 1 \\ 1 & 0 & 0 \\ 0 & 1 & 0 \end{pmatrix}.$$

5. If $\det R \neq 0$, prove that both $R^\dagger R$ and $RR^\dagger$ are positive definite Hermitian matrices.

6. Prove: (1) if $R^\dagger R = \mathbf{1}$, then $RR^\dagger = \mathbf{1}$;
   (2) if $R^{-1}R = \mathbf{1}$, then $RR^{-1} = \mathbf{1}$;
   (3) if $R^T R = \mathbf{1}$, then $RR^T = \mathbf{1}$.

7. Find the independent real parameters in a 2×2 unitary matrix, a real orthogonal matrix, and a Hermitian matrix, respectively, and give their general expressions.

8. Find the similarity transformation to diagonalize the following matrices:

$$(1) \quad \begin{pmatrix} 1 & -\sqrt{2} & 1 \\ \sqrt{2} & 0 & -\sqrt{2} \\ 1 & \sqrt{2} & 1 \end{pmatrix}, \qquad (2) \quad \begin{pmatrix} \cos\theta & -\sin\theta \\ \sin\theta & \cos\theta \end{pmatrix}.$$

9. Find a similarity transformation matrix M which satisfies

$$M^{-1} \begin{pmatrix} 0 & -\cos\theta & \sin\theta\sin\varphi \\ \cos\theta & 0 & -\sin\theta\cos\varphi \\ -\sin\theta\sin\varphi & \sin\theta\cos\varphi & 0 \end{pmatrix} M = \begin{pmatrix} 0 & -1 & 0 \\ 1 & 0 & 0 \\ 0 & 0 & 0 \end{pmatrix}.$$

10. Find a similarity transformation matrix M which satisfies the following three equations simultaneously

$$M^{-1}\begin{pmatrix} 0 & -i & 0 \\ i & 0 & 0 \\ 0 & 0 & 0 \end{pmatrix} M = \begin{pmatrix} 1 & 0 & 0 \\ 0 & 0 & 0 \\ 0 & 0 & -1 \end{pmatrix},$$

$$M^{-1}\begin{pmatrix} 0 & 0 & 0 \\ 0 & 0 & -i \\ 0 & i & 0 \end{pmatrix} M = \frac{1}{\sqrt{2}}\begin{pmatrix} 0 & 1 & 0 \\ 1 & 0 & 1 \\ 0 & 1 & 0 \end{pmatrix},$$

$$M^{-1}\begin{pmatrix} 0 & 0 & i \\ 0 & 0 & 0 \\ -i & 0 & 0 \end{pmatrix} M = \frac{i}{\sqrt{2}}\begin{pmatrix} 0 & -1 & 0 \\ 1 & 0 & -1 \\ 0 & 1 & 0 \end{pmatrix}.$$

11. Let

$$R = \begin{pmatrix} 1 & 0 \\ 0 & -1 \end{pmatrix}, \qquad S = \frac{1}{2}\begin{pmatrix} -1 & -\sqrt{3} \\ \sqrt{3} & -1 \end{pmatrix}.$$

Find the common similarity transformation matrix X satisfying

$$X^{-1}(R \times R)X = \begin{pmatrix} 1 & 0 & 0 & 0 \\ 0 & -1 & 0 & 0 \\ 0 & 0 & 1 & 0 \\ 0 & 0 & 0 & -1 \end{pmatrix},$$

$$X^{-1}(S \times S)X = \frac{1}{2}\begin{pmatrix} 2 & 0 & 0 & 0 \\ 0 & 2 & 0 & 0 \\ 0 & 0 & -1 & -\sqrt{3} \\ 0 & 0 & \sqrt{3} & -1 \end{pmatrix}.$$

12. Find the similarity transformation matrix X to diagonalize the following three matrices simultaneously,

$$\begin{pmatrix} 0&0&0&1&0&0 \\ 0&0&0&0&1&0 \\ 0&0&0&0&0&1 \\ 1&0&0&0&0&0 \\ 0&1&0&0&0&0 \\ 0&0&1&0&0&0 \end{pmatrix}, \quad \begin{pmatrix} 0&0&0&1&0&0 \\ 0&0&0&0&0&1 \\ 0&0&0&0&1&0 \\ 1&0&0&0&0&0 \\ 0&0&1&0&0&0 \\ 0&1&0&0&0&0 \end{pmatrix}, \quad \begin{pmatrix} 0&0&0&1&1&1 \\ 0&0&0&1&1&1 \\ 0&0&0&1&1&1 \\ 1&1&1&0&0&0 \\ 1&1&1&0&0&0 \\ 1&1&1&0&0&0 \end{pmatrix}.$$

13. Show the general form of an $m \times m$ matrix, both unitary and Hermitian.

14. Prove that any unitary matrix R can be diagonalized by a unitary similarity transformation, and any Hermitian matrix R can be diagonalized by a unitary similarity transformation.

15. Prove that R and $R^{\dagger}$ can be diagonalized by a common unitary similarity transformation if $R^{\dagger}$ is commutable with R. Further prove that the necessary and sufficient condition for a matrix R to be diagonalized by a unitary similarity transformation is that $R^{\dagger}$ is commutable with R.

16. Prove that any matrix can be transformed into a direct sum of the standard Jordan forms, each of which is in the form

$$R_{ab} = \begin{cases} \lambda & \text{when } a = b, \\ 0 \text{ or } 1 & \text{when } a + 1 = b, \\ 0 & \text{the remaining cases.} \end{cases}$$

Chapter 2

GROUP AND ITS SUBSETS

Group theory is a powerful tool for studying the symmetry of a physical system. In this chapter, the mathematical definition of a group will be abstracted from the common property of the sets of the symmetric transformations of physical systems. Some simple examples are explained to give the readers a concrete understanding on groups. The group table, which is very important for a finite group, is emphasized in this chapter. The concepts of the subsets in a group, of isomorphism and homomorphism of two groups, and of the direct product of groups will be discussed.

2.1 Symmetry

Symmetry becomes more and more important in the modern sciences. What is symmetry? One often says that a scalene triangle is asymmetric, an isosceles triangle and a regular triangle are more symmetric, and a circle is the most symmetric among them. How does one estimate the symmetry of a system?

The concept of symmetry is tied up with transformations. A transformation which preserves the system invariant is called a symmetric transformation of the system. The symmetry of a system is described by the set of all its symmetric transformations. The identical transformation which preserves anything invariant is a trivial symmetric transformation of any system. A scalene triangle is not invariant in any transformation except for the identical transformation. An isosceles triangle is invariant in the reflection with respect to the perpendicular bisectrix plane to its bottom edge. A regular triangle is invariant in three reflections with respect to the perpendicular bisectrix plane to its edges and in the rotations through $2\pi/3$ around the axis perpendicular to the triangle at its center. A circle is invariant in any perpendicular plane containing its diameter and in any rotation around the axis perpendicular to the circle at its center.

A quantum system is described by its Hamiltonian. A symmetric transformation of a quantum system preserves its Hamiltonian invariant. For example, the Hamiltonian of an isolated n-body system is

$$H = -\frac{\hbar^2}{2} \sum_{j=1}^{n} m_j^{-1} \nabla_j^2 + \sum_{i<j} U(|\boldsymbol{r}_i - \boldsymbol{r}_j|),$$

where $\boldsymbol{r}_j$ and m_j respectively are the position vector and the mass of the jth particle, ∇_j^2 is the Laplace operator with respect to $\boldsymbol{r}_j$, and U is the interaction potential depending on the distance between two particles. Obviously, the Hamiltonian is invariant in the spatial translation, rotations, and inversion. If the particles are identical, the Hamiltonian is also invariant in the permutations among particles. Those symmetric transformations characterize the symmetry of the quantum system.

In terms of the method of group theory, one is able to analyze symmetry of the system and to obtain some precise information of the system verifiable with observations. First of all, we would like to give the readers an intuitive estimation on the power of group theory through a very simple example, a quantum system with the symmetry of spatial inversion. Denote by P the transformation operator of the wave function in the spatial inversion,

$$P\psi(\boldsymbol{r}_1,\ \boldsymbol{r}_2,\ \ldots) = \psi(-\boldsymbol{r}_1,\ -\boldsymbol{r}_2,\ \ldots).$$

Let ψ be an eigenfunction of the Hamiltonian of the system with the eigenvalue E. Since the Hamiltonian is invariant in the inversion, $P\psi$ is also an eigenfunction of the Hamiltonian with the same eigenvalue, so is any combination of ψ and $P\psi$. Take the following combinations:

$$\begin{aligned} &\phi_S \sim \psi + P\psi, \qquad &&\phi_A \sim \psi - P\psi, \\ &P\phi_S = \phi_S, &&P\phi_A = -\phi_A. \end{aligned} \tag{2.1}$$

In the spatial inversion, ϕ_S and ϕ_A are the wave functions with the even and odd parities, respectively. No matter how the detail property of a system is, only if the system is invariant in the spatial inversion, the wave function of any stationary state of the system (the eigenfunction of the Hamiltonian) can be chosen with a definite parity, namely, the parity is conserved and can be used to classify the wave functions of stationary states of the system. Furthermore, the electric dipole transition between two states with the same parity is suppressed. This is the selection rule for the electric dipole transition in quantum mechanics.

The simple example shows that some precise information of a system can be obtained from its symmetry although the Schrödinger equation is hard to be solved.

2.2 Group and its Multiplication Table

A transformation is called the symmetric transformation of a system if it preserves the system invariant. The set of the symmetric transformations of a system characterizes the symmetry of the system. There are some common properties in the sets of symmetric transformations. In physics, the multiplication RS of two transformations R and S is usually defined as successive applications first by S and then by R. Thus, the multiplication of two symmetric transformations is still a symmetric one of the system. The multiplication rule satisfies the associative law. The identical transformation E, which preserves everything invariant, is a symmetric one of any system. For any symmetric transformation R of the system, the multiplication ER is equal to R. The inverse of a symmetric transformation is also a symmetric one of the system. Those properties are essential and common for the set of symmetric transformations of every system, and can be used to define a group.

Definition 2.1 A group G is a set of elements R satisfying the following four axioms with respect to the given multiplication rule of elements.

(a) The set is closed to this multiplication,

$$RS \in G, \qquad \forall\ R \text{ and } S \in G. \tag{2.2}$$

(b) The multiplication rule of elements satisfies the associative law,

$$R(ST) = (RS)T, \qquad \forall\ R,\ S \text{ and } T \in G. \tag{2.3}$$

(c) The set contains an identical element $E \in G$ satisfying

$$ER = R, \qquad \forall\ R \in G. \tag{2.4}$$

(d) For any element $R \in G$, the set contains its inverse R^{-1} satisfying

$$R^{-1}R = E, \qquad \forall\ R \in G. \tag{2.5}$$

In Definition 2.1, the elements can be any objects and the multiplication rule of elements can be defined arbitrarily. The main thing for a group is that the set satisfies four axioms with respect to the given multiplication rule. The set of symmetric transformations of any system with respect to the multiplication rule of transformations satisfies four axioms, so that the set is a group, called the symmetric group of the system. In most cases in physics, the elements of a group are transformations, operators, or matrices. If without specification, the multiplication rule is defined as successive applications when the elements are transformations or their operators, and as the matrix multiplication when the elements are matrices.

It can be shown from Definition 2.1 that the inverse relation is mutual (see Prob. 1 of Chap. 2 in [Ma and Gu (2004)])

$$RR^{-1} = E. \tag{2.6}$$

Then, the identical element in a group is unique and the inverse element of any element $R \in G$ is unique,

$$\begin{aligned}
&T = E, && \text{if} \ \ TR = R, \\
&S = R^{-1}, && \text{if} \ \ SR = E, \\
&RE = R, && \\
&(RS)^{-1} = S^{-1}R^{-1}. &&
\end{aligned} \tag{2.7}$$

In the proof one cannot use any calculation rule except for Definition 2.1. From now on, the conclusions (2.6) and (2.7) can be used because they have been proved in Prob. 1.

Generally speaking, the multiplication of elements is not commutable, $RS \neq SR$. The group is called Abelian if multiplication of every two elements in the group is commutable. A finite group contains finite number g of elements. g is called the order of a finite group. An infinite group contains infinite number of elements. If the elements in an infinite group can be described by a set of continuous parameters, the group is called a continuous group.

In Definition 2.1 for a group, the important thing is the multiplication rule of elements, namely, how the elements construct the group. The property of objects, which are chosen to be elements in the group, are not important. If there is a one-to-one correspondence between elements of two groups, and the correspondence is invariant to the multiplication of elements, the two groups are the same in the viewpoint of group theory. Those two groups are called isomorphic.

Definition 2.2 A group G' is called isomorphic onto another group G if there is a one-to-one correspondence between elements of two groups, and the multiplication of two elements in G' is in the same way to map onto the multiplication of two corresponding elements in G. In mathematical symbols, $G' \approx G$ if

$$\begin{aligned} &\forall \ R \in G, \quad S \in G, \quad R' \in G', \quad \text{and} \quad S' \in G', \\ &\exists \ R' \longleftrightarrow R, \quad S' \longleftrightarrow S, \quad \text{and} \quad R'S' \longleftrightarrow RS, \end{aligned} \tag{2.8}$$

where "$\longleftrightarrow$" denotes one-to-one correspondence.

There are different one-to-one correspondences between elements in two groups. Two groups are isomorphic if and only if there exists a one-to-one correspondence which is invariant to the multiplication of elements. One cannot make conclusion that two groups are NOT isomorphic only based on that there is a one-to-one correspondence between two groups which is not invariant to the multiplication of elements.

A subset in the group G is called a complex of elements. Two complexes are equal to each other, $\mathcal{R} = \mathcal{S}$, if $\mathcal{R} \subset \mathcal{S}$ and $\mathcal{S} \subset \mathcal{R}$. Let $\mathcal{R} = \{R_1, R_2, \cdots, R_m\}$ and $\mathcal{S} = \{S_1, S_2, \cdots, S_n\}$ be two complexes of elements in a group G. The product of $\mathcal{R}$ and $\mathcal{S}$ is a complex $\mathcal{RS}$ which is the set of all elements R_jS_k. Similarly, the product of an element T and a complex $\mathcal{R}$ is a complex $T\mathcal{R}$ (or $\mathcal{R}T$) which consists of the elements TR_j (or R_jT). The set of complexes forms a group if their products satisfy the four axioms for a group.

Theorem 2.1 (Rearrangement theorem) For any element T in a group $G = \{E, R, S, \cdots\}$, the following three sets all are the same as the original group G:

$$\begin{aligned} TG &= \{T, TR, TS, \cdots\}, \\ GT &= \{T, RT, ST, \cdots\}, \\ G^{-1} &= \{E, R^{-1}, S^{-1}, \cdots\}. \end{aligned}$$

In the nomenclature of complex of elements,

$$G = TG = GT = G^{-1}. \tag{2.9}$$

Proof The proofs for three equilities are similar. We prove $TG = G$ as example. Due to Eq. (2.2) any element in the set TG is an element in the group G, so $TG \subset G$. Conversely, any element R in G can be expressed as $R = T(T^{-1}R)$. Since $T^{-1}R \in G$, $R \in TG$ and $G \subset TG$. □

For a finite group, the multiplication RS of each pair of elements R and S can be listed in a table, called the multiplication table of a group, or briefly the group table. In a group table, the right-multiplication element S is listed on the top, the left-multiplication element R on the left, and the multiplication is filled in the content of the table. Rearrangement theorem means that there are no repetitive elements in each row and in each column of the group table. Two groups are isomorphic if their multiplication tables are the same. Conversely, two isomorphic groups may have different group tables if the enumerations are unsuitable. Rearrangement theorem restricts strongly the number of non-isomorphic groups with a given order g. For example, rearrangement theorem completely determines the group tables of a group with the order $g = 2$ or 3, as given in Tables 2.1 and 2.2.

Table 2.1 The group table of V_2

V_2	e	σ
e	e	σ
σ	σ	e

Table 2.2 The group table of C_3

C_3	e	ω	ω'
e	e	ω	ω'
ω	ω	ω'	e
ω'	ω'	e	ω

V_2, called the inversion group of order two, is the symmetric group of a system which is invariant in the spatial inversion σ. V_2 is isomorphic onto C_2 constructed by two numbers 1 and -1 with respect to the multiplication rule of numbers. Group C_3 is constructed by three complex numbers $e = 1$, $\omega = \exp(-i2\pi/3)$, and $\omega' = \exp(i2\pi/3)$ with respect to the multiplication rule of complex numbers.

In V_2 one has $e = \sigma^2$. Each element in C_3 is equal to a power of ω, because $\omega' = \omega^2$ and $e = \omega^3$. Generally, a finite group constructed by the powers of one element R is called the cyclic group C_N

$$C_N = \{E,\ R,\ R^2,\ \cdots,\ R^{N-1}\}, \qquad R^N = E, \tag{2.10}$$

where N is the order of C_N. C_N is an Abelian group. The group table of C_N can be filled as follows. Arrange the elements in the order: E, R, R^2, $\ldots, R^{N-1}$. The elements in the first row of the table are the same as those in its top line (right-multiplication elements), respectively. The elements in each row are obtained by moving the elements in the preceding row leftward by one box, and moving the leftmost element to the rightmost position.

A typical physical example for the cyclic group C_N is the N-fold proper axis in the theory of crystals. A pure rotation in three-dimensional space is called a proper rotation, and a proper rotation multiplied by a spatial inversion σ is called the improper rotation. If the rotation R around a given axis through the angle $2\pi/N$ is a symmetric transformation of a crystal, the cyclic group C_N generated by R belong to the symmetric group of the crystal, the axis is its N-fold proper axis, and R is its N-fold proper rotation. The 1-fold proper rotation is the identical transformation E, and the group C_1 is trivial. The direction of the 2-fold proper axis does not matter because $R = R^{-1}$. The direction of the N-fold proper axis with $N > 2$ is essence for determining the generator R.

Since the number of elements in a finite group is limited, the powers of each element R will appear repeatedly when the power is high enough, say $R^{a+n} = R^a$, namely, $R^n = E$. The lowest n for $R^n = E$ is called the order of the element R. The subset of the powers of R in G is called the period of R. The order of the identical element E is one, and the order of other element $R \neq E$ must be larger than one. Do not confuse the nomenclature "order" for a group and for an element. Only for the cyclic group C_N generated by R, the order of R is equal to the order of the group C_N.

Taking another element $S \in G$ which does not belong to the period of R, one constructs a larger subset by multiplying R and S arbitrarily. For a finite group, one can enlarge the subset by adding more elements until the subset is equal to the group G. The elements are called the generators of a finite group G if each element in G can be expressed as a multiplication of the generators, and each generator is not equal to the multiplication of other generators. The choice of generators of a finite group is not unique. Choose the generators whose number is as small as possible. The least number of the generators is called the rank of a finite group. The rank of the cyclic group C_N is one, and at least, both R and R^{-1} can be chosen to be its generator.

Now, we turn to group G with order four. If there is an element with order four in G, G is the cyclic group C_4. The group table of C_4 is given in Table 2.3. It is impossible for G of order four to have an element with order three. Otherwise, repetitive elements would appear in one row or in one column of its group table. In the remaining case the order of each element in G is two except for the identical element. The typical example for that group is the inversion group V_4 of order four composed of the identical element e, the spatial inverse σ, the time inverse τ, and the space–time inverse ρ. The group table of V_4 is given in Table 2.4. A group G with order four is isomorphic onto C_4 if it contains an element with the order larger than two, and it is isomorphic onto V_4 if the number of elements with order two in it is more than one.

Table 2.3 The group table of C_4

C_4	E	R	S	T
E	E	R	S	T
R	R	S	T	E
S	S	T	E	R
T	T	E	R	S

Table 2.4 The group table of V_4

V_4	e	σ	τ	ρ
e	e	σ	τ	ρ
σ	σ	e	ρ	τ
τ	τ	ρ	e	σ
ρ	ρ	τ	σ	e

The simplest non-Abelian group is the symmetric group D_3 of a regular triangle, which contains six elements. The axis perpendicular to a regular triangle at its center O is a 3-fold axis of the triangle. Denote by E, D, and F the rotations around this axis through 0, $2\pi/3$, and $4\pi/3$, respectively. The three vertices of the triangle are denoted by A', B', and C' (see Fig. 2.1). Three axes OA', OB', and OC' all are the 2-fold axes of the triangle, where the 2-fold rotations are denoted by A, B, and C, respectively. The orders of D and F are three, and those of A, B, and C are two.

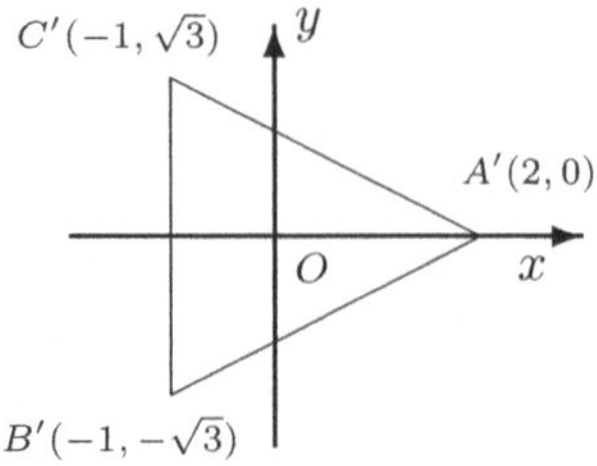

Fig. 2.1 The regular triangle.

We are going to introduce two typical methods for establishing the group table of D_3. One is the graphic method. Another is the method of coordinate transformation.

Draw a rectangular coordinate system OXY (see Fig. 2.1). The center of $\Delta A'B'C'$ coincides with the origin O, and A' is located at the x-axis. The coordinates of three vertices are $A'(2,0)$, $B'(-1,-\sqrt{3})$, and $C'(-1,\sqrt{3})$. Put the same triangle ΔABC on the plane such that A, B, and C coincide with A', B', and C', respectively. Now, fix $\Delta A'B'C'$ and move ΔABC with the six symmetric transformations. After a symmetric transformation, each vertex A, B, or C must coincide with one of the three vertices A', B', and C', respectively. List the coincidence for each symmetric transformation in Table 2.5. One is able to calculate the multiplication of two transformations, say DA, with the table. According to Table 2.5, the vertex A is invariant in the transformation A, and changes to C' in the transformation D. The vertex B changes to C' in the transformation A, then being a new vertex, C changes to B' in the transformation D. Namely, in the transformation DA, the vertex A changes to C' and the vertex B changes to B'. Obviously, the vertex C has to change to A' in the transformation DA. In fact, the vertex C changes to B' in the transformation A, and then changes to A' in the transformation D. In comparison with Table 2.5, the transformation DA is the same as the transformation B.

Table 2.5 Symmetric transformations of a regular triangle

	E	D	F	A	B	C
A	A'	C'	B'	A'	C'	B'
B	B'	A'	C'	C'	B'	A'
C	C'	B'	A'	B'	A'	C'

Table 2.6 The group table of D_3

D_3	E	D	F	A	B	C
E	E	D	F	A	B	C
D	D	F	E	B	C	A
F	F	E	D	C	A	B
A	A	C	B	E	F	D
B	B	A	C	D	E	F
C	C	B	A	F	D	E

In order to establish the group table of D_3, one does not need to calculate

$g^2 = 36$ multiplications of elements. Most multiplications can be obtained easily. The first row and the first column in Table 2.6 are obvious. Since three elements E, D, and F construct a cyclic group of order three, whose group table was given in Table 2.2, the left-up part of Table 2.6 is obtained. Due to 2-fold rotations one has $A^2 = B^2 = C^2 = E$. Then, it follows from $DA = B$ that $D = BA$ and $BD = A$. The remaining part of Table 2.6 can be determined from the rearrangement theorem. Therefore, in addition to the orders of elements and the rearrangement theorem, the group table of D_3 is established only from one formula $DA = B$.

It will be proved in Prob. 6 that any group of order six must be isomorphic onto D_3 or C_6. The group is isomorphic onto D_3 if it contains the elements of order two more than one, and it is isomorphic onto C_6 if it contains an element of order six.

In the method of coordinate transformation, each symmetric transformation of the triangle is denoted by a matrix transforming the planar coordinates of vertices from $(x,\ y)$ to $(x',\ y')$:

$$\begin{pmatrix} x' \\ y' \end{pmatrix} = \begin{pmatrix} a & b \\ c & d \end{pmatrix} \begin{pmatrix} x \\ y \end{pmatrix}. \tag{2.11}$$

Substituting the coordinates of each vertex into Eq. (2.11), one is able to calculate the matrix entries. For example, since the transformation D changes the vertex A to C' and the vertex B to A', one has

$$\begin{pmatrix} -1 \\ \sqrt{3} \end{pmatrix} = \begin{pmatrix} a & b \\ c & d \end{pmatrix} \begin{pmatrix} 2 \\ 0 \end{pmatrix}, \qquad \begin{pmatrix} 2 \\ 0 \end{pmatrix} = \begin{pmatrix} a & b \\ c & d \end{pmatrix} \begin{pmatrix} -1 \\ -\sqrt{3} \end{pmatrix}.$$

From the first equation, one obtains $a = -1/2$ and $c = \sqrt{3}/2$. Then, solving the second equation, one has $b = -\sqrt{3}/2$ and $d = -1/2$. The matrices for six elements are calculated as follows:

$$\begin{aligned} E &= \begin{pmatrix} 1 & 0 \\ 0 & 1 \end{pmatrix}, & D &= \frac{1}{2}\begin{pmatrix} -1 & -\sqrt{3} \\ \sqrt{3} & -1 \end{pmatrix}, & F &= \frac{1}{2}\begin{pmatrix} -1 & \sqrt{3} \\ -\sqrt{3} & -1 \end{pmatrix}, \\ A &= \begin{pmatrix} 1 & 0 \\ 0 & -1 \end{pmatrix}, & B &= \frac{1}{2}\begin{pmatrix} -1 & \sqrt{3} \\ \sqrt{3} & 1 \end{pmatrix}, & C &= \frac{1}{2}\begin{pmatrix} -1 & -\sqrt{3} \\ -\sqrt{3} & 1 \end{pmatrix}. \end{aligned} \tag{2.12}$$

The group table 2.6 can be obtained again from the multiplications of those six matrices, namely, the matrix group constructed by six matrices given in Eq. (2.12) is isomorphic onto D_3.

Generalize the method to a regular polygon of N sides. Put a regular polygon on a planar coordinates flame OXY where its center coincides with the origin and one vertex A is located on the x-axis. The z-axis is an N-fold axis of the polygon, where the N-fold rotation is denoted by T. In addition, the axis on the $x\ y$ plane with the angle $j\pi/N$ to the x-axis is a 2-fold axis of the polygon, where the 2-fold rotation is denoted by S_j. There are N 2-fold axes. $2N$ rotations, T^j and S_j, $0 \leq j \leq N-1$, consist of the symmetric group D_N of a regular polygon of N sides. D_N is called the Nth-dihedral group. In comparison with D_3, T is denoted by D in D_3, and S_0 by A.

Consider the multiplication rule in D_N. It is obvious that $T^N = S_j^2 = E$, and $T^aT^b = T^{a+b}$. The key is to calculate TS_j. Let a point A be located on the 2-fold axis OA related to S_j. A is invariant in the rotation S_j, but changes to B in the rotation T. The angle between the two axes OA and OB is $2\pi/N$. On the other hand, B changes to a symmetric position to the axis OA in S_j, and then, moves to A in T. Since points A and B transform to each other, TS_j is the 2-fold rotation around the angular bisector of $\angle AOB$, namely, $TS_j = S_{j+1}$. It is a generalization of $DA = B$ in D_3. Thus, one obtains the multiplication rule in D_N

$$\begin{aligned} &T^N = S_j^2 = E, \qquad T^aT^b = T^bT^a = T^{a+b}, \qquad T^aS_j = S_{a+j}, \\ &T^a = S_{a+j}S_j = S_jS_{j-a}, \quad S_jT^a = S_{j-a}, \quad j \text{ and } a \bmod N, \end{aligned} \tag{2.13}$$

where $j \bmod N$ is a mathematical symbol, denoting $S_{j+N} = S_j$. Equation (2.13) is another way to show the multiplication rule for elements where the generators of D_N are T and S_0.

2.3 Subsets in a Group

2.3.1 *Subgroup*

A subset H in a group G is a subgroup of G if the subset is a group under the multiplication rule of elements in G. Note that the multiplication rule for the subgroup H must be the same as that for G. Otherwise, there is no relation between two groups from the viewpoint of group theory.

A subset is not a subgroup if it does not contain the identical element E. The period of an element R is a subgroup of G, called the cyclic subgroup. A subgroup contains the whole period of its every element. For a finite group, a subset H in G is a subgroup if H is closed with respect to the multiplication rule in G. In fact, if it is closed, H contains the period of its every element R, including the inverse $R^{-1} = R^{n-1}$ and the identical element $E = R^n$, where n is the order of R. The associative law for the multiplication rule is obviously satisfied.

There are two trivial subgroups for any group G. One is the identical element E, and the other is the whole group G. We will only discuss nontrivial subgroups. The first step for finding subgroups of a finite group G is to list the period of each element in G, then to check whether the union of some periods is closed with respect to the multiplication rule of elements.

The group V_4 contains three cyclic subgroups: $\{e,\ \sigma\}$, $\{e,\ \tau\}$, and $\{e,\ \rho\}$. The group C_6 contains two cyclic subgroups: $\{E,\ R^3\}$ and $\{E,\ R^2,\ R^4\}$. The group D_3 contains four cyclic subgroups: $\{E,\ A\}$, $\{E,\ B\}$, $\{E,\ C\}$, and $\{E,\ D,\ F\}$. In addition to the nine cyclic subgroups, the group D_6 contains three D_2 subgroups and two D_3 subgroups:

$$\mathrm{D}_2 = \{E,\ R^3,\ S_0,\ S_3\}, \qquad \mathrm{D}_2' = \{E,\ R^3,\ S_1,\ S_4\},$$
$$\mathrm{D}_2'' = \{E,\ R^3,\ S_2,\ S_5\}, \qquad \mathrm{D}_3 = \{E,\ R^2,\ R^4,\ S_0,\ S_2,\ S_4\},$$
$$\mathrm{D}_3' = \{E,\ R^2,\ R^4,\ S_1,\ S_3,\ S_5\}.$$

2.3.2 *Cosets*

Let H of order h be a subgroup of a group G of order g:

$$H = \{S_1,\ S_2,\ S_3,\ \cdots,\ S_h\}, \qquad S_1 = E.$$

Taking an arbitrary element $R_j \in G$, which does not belong to H, one obtains two subsets by left- and right-multiplying R_j to H:

$$\begin{aligned} R_jH &= \{R_j,\ R_jS_2,\ R_jS_3,\ \cdots,\ R_jS_h\}, \\ HR_j &= \{R_j,\ S_2R_j,\ S_3R_j,\ \cdots,\ S_hR_j\}, \end{aligned} \qquad R_j \in G, \qquad R_j \overline{\in} H. \tag{2.14}$$

R_jH is called the left coset of H, and HR_j the right coset.

The following conclusions for cosets are very easy to prove, and are left to the readers as exercise. A coset does not contain any element belonging to the subgroup. A coset is not a subgroup. Two left cosets must be the same if there is at least one common element in them, namely, there is no common element between two different left cosets. Similar conclusions hold for the right cosets. The necessary and sufficient condition for the two elements R and S belonging to the same left coset is $R^{-1}S \in H$, and that for them belonging to the same right coset is $RS^{-1} \in H$. For a finite group, the number of elements in any coset is equal to the order h of the subgroup H, and the group G is equal to the union of the subgroup H and its left cosets (or right cosets),

$$G = \bigcup_{j=1}^{d} R_j H = \bigcup_{j=1}^{d} H R_j, \qquad R_1 = E. \tag{2.15}$$

Thus, the order h of a subgroup H is a whole number divisor of the order g of the group G (Lagrange Theorem), $g = hd$. d is called the index of a subgroup, and the number of the left cosets (or right cosets) of H in G is $d - 1$. A group G does not contain any nontrivial subgroups if its order g is a prime number. It leads that two groups are isomorphic if their orders are of the same prime number.

It is easy to find the cosets of a subgroup of a finite group G in terms of its group table. In the group table, the set of elements in each row of the columns related to the elements in the subgroup is the subgroup itself or its left coset $(R_j H)$, and the set of elements in each column of the rows related to the elements of the subgroup is the subgroup itself or its right coset (HR_j). For example, it can be seen from Table 2.6 that the subgroup $\{E,\ A\}$ in the group D_3 contains two left cosets, $\{D,\ B\}$ and $\{F,\ C\}$, and two right cosets, $\{D,\ C\}$ and $\{F,\ B\}$. The left cosets of the subgroup in D_3 are not equal to its right cosets. Another subgroup $\{E,\ D,\ F\}$ with index 2 has only one coset $\{A,\ B,\ C\}$.

2.3.3 *Conjugate Elements and the Class*

The element $S' = TST^{-1}$ is said to be conjugate to S in a group G where S, S', and T all belong to G. The conjugate relation of two elements is mutual. If two elements are both conjugate to a third element, they are also conjugate to each other:

$$S' = TST^{-1}, \qquad S'' = RSR^{-1} = \left(RT^{-1}\right) S' \left(RT^{-1}\right)^{-1}.$$

The subset of all mutually conjugate elements in a group G is called a class $\mathcal{C}_\alpha$ in G

$$\mathcal{C}_\alpha = \left\{S_1,\ S_2,\ \cdots,\ S_{n(\alpha)}\right\} = \left\{S_k | S_k = TS_jT^{-1}, T \in G\right\}, \tag{2.16}$$

where $n(\alpha)$ is the number of elements contained in the class $\mathcal{C}_\alpha$. There is no common element in two different classes. The identical element E itself forms a class. Therefore, the other class does not contain the identical element and is not a subgroup. For an Abelian group, each element itself forms a class, and the number g_c of classes in an Abelian group is equal to the order g of the group. For a given element $T \in G$, $TS_jT^{-1} \neq TS_kT^{-1}$ if $S_j \neq S_k$, so that

$$TC_\alpha T^{-1} = C_\alpha, \qquad \forall\, T \in G. \tag{2.17}$$

Theorem 2.2 Denote by $n(\alpha)$ the number of elements in a class $\mathcal{C}_\alpha$ of a finite group G with the order g, and denote by $m(\alpha)$ the number of $T \in G$ satisfying $TS_jT^{-1} = S_k$, where S_j and S_k are two given elements in $\mathcal{C}_\alpha$, then $g = n(\alpha)m(\alpha)$, and $m(\alpha)$ is independent of S_j and S_k.

The key for the proof is to define a subgroup H whose elements are commutable with an element S_j in $\mathcal{C}_\alpha$. Any element in the left coset TH satisfies $TS_jT^{-1} = S_k$. The detailed proof is left as exercise (see Prob. 14).

The inverse S_j^{-1} of an element S_j in a class $\mathcal{C}_\alpha$ is conjugate to the inverse S_k^{-1} of another element $S_k \in \mathcal{C}_\alpha$:

$$S_k = TS_jT^{-1}, \qquad S_k^{-1} = TS_j^{-1}T^{-1}.$$

Thus, the set of S_j^{-1} also constructs a class, called the reciprocal class $\mathcal{C}_\alpha^{(-1)}$ of $\mathcal{C}_\alpha$. $\mathcal{C}_\alpha^{(-1)}$ contains the same number of elements as $\mathcal{C}_\alpha$. The reciprocal relation of two classes is mutual. If a class $\mathcal{C}_\alpha$ contains the inverse S_j^{-1} of its element S_j, then $\mathcal{C}_\alpha = \mathcal{C}_\alpha^{(-1)}$, called the self-reciprocal class.

There are some common properties among elements in one class. For example, two conjugate elements in a finite group have the same order. In fact, if $S^n = E$, then $(TST^{-1})^n = TS^nT^{-1} = E$. However, two elements with the same order are not necessary to be conjugate. The necessary and sufficient condition for two elements conjugate to each other is that they can be expressed as two different products of other two elements, TR and RT. In fact, $TR = T(RT)T^{-1}$, and two conjugate elements satisfy $S' = T(ST^{-1})$ and $S = (ST^{-1})T$. This property can be used to judge whether two elements in a finite group are conjugate or not in terms of the group table. For a group table of a finite group where the arrangement order for the rows (left-multiplication elements) is the same as that for the columns (right-multiplication elements), two elements are conjugate to each other if and only if they appear at least once at the two symmetric positions with respect to the diagonal line of the group table.

Consider a system (e.g., a crystal) whose symmetric group G consists of proper rotations around the axes with a common point, taken as the origin. Let an axis along the direction $\hat{\boldsymbol{n}}$ be an N-fold proper axis of the system, whose generator is denoted by S. Hereafter, a vector $\hat{\boldsymbol{n}}$ with a hat represents a unit vector. If $T \in G$ changes the direction $\hat{\boldsymbol{n}}$ to the direction $\hat{\boldsymbol{m}}$, then TST^{-1} is a rotation around the direction $\hat{\boldsymbol{m}}$ through the same angle as S. In fact, T^{-1} changes $\hat{\boldsymbol{m}}$ to $\hat{\boldsymbol{n}}$ first, then S makes a rotation around $\hat{\boldsymbol{n}}$

through $2\pi/N$, and at last T changes $\hat{\boldsymbol{n}}$ back to $\hat{\boldsymbol{m}}$. Thus, the axis along $\hat{\boldsymbol{m}}$ does not change in the transformation TST^{-1}, and it is also an N-fold proper axis of the system just like that along $\hat{\boldsymbol{n}}$. Generally speaking, if a symmetric transformation T of a system relates two directions, and there is an N-fold proper axis along one direction, there must be a proper axis of the same fold along the other direction. Those two axes are called the equivalent axes. Two symmetric transformations around two equivalent axes through the same angle are conjugate to each other. An N-fold proper axis is called a nonpolar axis if there is a symmetric transformation related to two directions of the axis. Two rotations around a nonpolar N-fold proper axis through angles $\pm 2m\pi/N$ are conjugate. It does not matter whether a 2-fold proper axis is polar or not. No symmetric transformation of the system changes one proper axis to another with different folds. Two symmetric rotations, even through the same rotational angle, are not conjugate if they are around two axes of different folds.

2.3.4 *Invariant Subgroup*

Let H be a subgroup of a group G. The left coset R_jH of H in G is not necessary to be equal to its right coset HR_j. A subgroup is called invariant or normal if its left coset R_jH is always equal to its right coset HR_j:

$$R_jH = HR_j, \qquad R_jHR_j^{-1} = H. \tag{2.18}$$

Note that the element S_μ in an invariant subgroup H is not necessary to be commutable with any element R_j in G. Equation (2.18) means that an invariant subgroup H contains all elements conjugate to every element S_μ in it, $R_jS_\mu R_j^{-1} = S_\nu \in H$, namely, an invariant subgroup consists of some whole classes. Conversely, any subgroup in an Abelian group is invariant. The subgroup with index 2 is invariant, because it has only one coset such that its left coset has to be equal to its right coset. The main method for finding the invariant subgroups in a finite group G is first to find the classes in G by its group table, and then to check whether the union of some classes forms a subgroup. Before checking whether the union is closed to the multiplication rule of elements, one may first check whether it contains the identical element E, whether the number of elements in it is a whole number divisor of the order g of G, and whether it contains the whole period of every element in it.

An invariant subgroup H of G and its cosets construct a group with respect to the multiplication rule of the complexes of elements. In fact, from Eq. (2.18) one has

$$\begin{aligned}
&R_j H R_k H = R_j R_k H H = (R_j R_k) H, \\
&H R_j H = R_j H H = R_j H, \\
&R_j^{-1} H R_j H = R_j^{-1} R_j H H = H.
\end{aligned}$$

This group is called the quotient group G/H of an invariant subgroup H in G. The identical element in G/H is H and the order of G/H is the index $n = g/h$ of H in G. One cannot define a quotient group G/H if H is not an invariant subgroup of G.

The cyclic group C_N generated by R is an Abelian group, where each element forms a class, namely, the number g_c of classes in C_N is equal to the order N of C_N. If $N = nm$, C_N contains two cyclic invariant subgroups C_m and C_n, respectively generated by R^n and R^m.

A regular triangle contains one nonpolar 3-fold axis and three equivalent 2-fold axes. The symmetric group of a regular triangle is D_3 where the elements with the same order are conjugate to each other. D_3 contains three classes: the identical element $\{E\}$, the 3-fold rotations $\{D, F\}$, and the 2-fold rotations $\{A, B, C\}$. Each class is self-reciprocal.

A regular polygon of N sides contains a nonpolar N-fold axis, called the principal axis, and N proper 2-fold axes, distributing equably in the plane perpendicular to the principal axis. Denote by T the N-fold rotation, and by S_j the 2-fold rotations, respectively. The symmetric group of the polygon is D_N. When $N = 2n+1$, all 2-fold axes are equivalent to each other so that the D_{2n+1} group contains $n+2$ self-reciprocal classes,

$$\{E\}, \quad \{S_0, S_1, \ldots, S_{2n+1}\}, \quad \{T^m, T^{-m}\}, \quad 1 \leq m \leq n. \tag{2.19}$$

The cyclic subgroup C_{2n+1} generated by T is an invariant subgroup of D_{2n+1}, so are the possible subgroups in C_{2n+1}. When $N = 2n$, the 2-fold axes are divided into two parts. The 2-fold axes in the same part are equivalent to each other, but those in different parts are inequivalent. The D_{2n} group contains $n+3$ self-reciprocal classes,

$$\begin{aligned}
&\{E\}, \quad \{T^n\}, \quad \{T^m, T^{-m}\}, \quad 1 \leq m \leq n-1, \\
&\{S_0, S_2, \ldots, S_{2n-2}\}, \quad \{S_1, S_3, \ldots, S_{2n-1}\}.
\end{aligned} \tag{2.20}$$

The geometric meanings of the 2-fold rotations, S_{2m} and S_{2m+1}, in the last two classes are different. The axis for S_{2m} connects two opposite vertices and the axis for S_{2m+1} connects two midpoints of two opposite edges. In addition to the cyclic subgroup C_{2n} and its possible subgroups, D_{2n} contains another two invariant subgroups,

$$\begin{aligned} \mathrm{D}_n &= \{E, T^2, T^4, \ldots, T^{2n-2}, S_0, S_2, \ldots, S_{2n-2}\}, \\ \mathrm{D}'_n &= \{E, T^2, T^4, \ldots, T^{2n-2}, S_1, S_3, \ldots, S_{2n-1}\}. \end{aligned}$$

For example, the group D_5 contains one nontrivial invariant subgroup

$$\mathrm{C}_5 = \{E, T, T^2, T^3, T^4\}.$$

The group D_6 contains five nontrivial invariant subgroups

$$\begin{aligned} &\mathrm{C}_2 = \{E, T^3\}, \quad \mathrm{C}_3 = \{E, T^2, T^4\}, \quad \mathrm{C}_6 = \{E, T, T^2, T^3, T^4, T^5\}, \\ &\mathrm{D}_3 = \{E, T^2, T^4, S_0, S_2, S_4\}, \quad \mathrm{D}'_3 = \{E, T^2, T^4, S_1, S_3, S_5\}. \end{aligned}$$

2.4 Homomorphism of Two Groups

We have introduced the concept of isomorphism of two groups. If there is a one-to-one correspondence between elements of two groups, and the correspondence is invariant to the multiplication rule of elements, two groups are isomorphic. If the correspondence between elements of two groups is not one-to-one, but many-to-one, the relation of the two groups becomes homomorphism.

Definition 2.3 A group G is said to be homomorphic onto another group G', $G' \sim G$ if there is a many-to-one correspondence between elements of two groups, and the correspondence is invariant to the multiplication rule of elements. Namely, if every element R in G maps to one and only one element R' in G', every element R' in G' maps to at least one element R in G, and the product of two elements in G is in the same way to map onto the product of two corresponding elements in G' and vice versa. In mathematical symbols, $G' \sim G$ if

$$\begin{aligned} &\forall \ R \in G, \quad S \in G, \quad R' \in G', \ \text{ and } \ S' \in G', \\ &\exists \ R \longrightarrow R', \ \ S \longrightarrow S', \ \text{ and } \ RS \longrightarrow R'S'. \end{aligned}$$

The symbol "$R \longrightarrow R'$" means the many-to-one correspondence between R in G and R' in G'.

There are different many-to-one correspondences between elements of two groups. A group G is homomorphic onto another group G' if and only if there exists a many-to-one correspondence which is invariant to the multiplication rule of elements. One cannot make conclusion that two groups are NOT homomorphic only based on that there is a many-to-one correspondence between two groups which is not invariant to the multiplication rule of elements.

If two groups are isomorphic, each group reflects the full property of another group. If a group G is homomorphic onto a group G', the group G' describes only part property of the group G. Theorem 2.3 will show that G' describes the part property of G if $G' \sim G$.

Theorem 2.3 If $G' \sim G$, the subset H of elements in G which maps to the identical element E' in G' forms an invariant subgroup of G, and the subset of elements in G which maps to an element R' in G' is a coset of H in G. The quotient group G/H of H in G is isomorphic onto the group G', $G/H \approx G'$. H is called the kernel of homomorphism.

Proof The key of the proof is that the element R' in G' to which the element R in G maps is unique. We will first prove that the subset H of elements in G which maps to the identical element E' in G' forms a subgroup of G. Then, we will prove that H is an invariant subgroup of G. At last, the subset of elements in G which maps to an element R' in G' forms a coset RH of H. Thus, it is obvious from Definition 2.3 that G' is isomorphic onto the quotient group G/H.

Let $H = \{S_1, S_2, \cdots, S_h\}$, where $S_\mu \longrightarrow E'$. Because $G' \sim G$, $S_\mu S_\nu \longrightarrow E'E' = E'$. Thus, H is closed to the multiplication rule of elements. If the identical element E in G maps to an element T' in G', $ES_\mu \longrightarrow T'E' = T'$. On the other hand, $ES_\mu = S_\mu \longrightarrow E'$. Now, from the uniqueness, $T' = E'$, namely, E belongs to H. For an arbitrary element R in G and its inverse R^{-1}, letting $R \longrightarrow R'$ and $R^{-1} \longrightarrow P'$, one has $R^{-1}R = E$ which maps to both $P'R'$ and E'. From the uniqueness, $P' = R'^{-1}$. The inverse S_μ^{-1} of S_μ maps to $E'^{-1} = E'$ in G', and then also belongs to H. Therefore, H satisfies four axioms and is a subgroup of G. Since $RS_\mu R^{-1}$ maps to $R'E'R'^{-1} = E'$, H is an invariant subgroup of G.

Each element in a coset RH of H maps to $R'E' = R'$ in G'. Conversely, if $R_\mu \longrightarrow R'$, then, $R^{-1}R_\mu \longrightarrow R'^{-1}R' = E'$. Namely, $R^{-1}R_\mu \in H$, and both R and R_μ belong to the same coset RH. Therefore, we proved that there is a one-to-one correspondence between the complexes of elements in G and elements in G' that $H \longleftrightarrow E'$ and $RH \longleftrightarrow R'$. Their product satisfies the same one-to-one correspondence. Thus, $G/H \approx G'$. □

Theorem 2.3 shows that if $G' \sim G$, G' describes the property of the quotient group G/H, but the difference among elements in the kernel H of homomorphism is not described.

Theorem 2.4 Let G be a group and G' be a set which is closed with respect to the given multiplication rule of elements in G'. If there is a correspondence that every element R in G maps to one element R' in G' uniquely and every element R' in G' maps to at least one element R in G, and if this correspondence is invariant with respect to the multiplication rule of elements, then G' is a group and G is isomorphic or homomorphic onto G', depending on the correspondence to be one-to-one or many-to-one.

Proof Since the set G' is closed, one only needs to show that G' satisfies the next three axioms for a group. The key of the proof is the uniqueness of the element R' in G' to which the element R in G maps. If $R \longrightarrow R'$, $S \longrightarrow S'$, and $T \longrightarrow T'$, then $RS \longrightarrow R'S'$, $ST \longrightarrow S'T'$, $(RS)T \longrightarrow (R'S')T'$, and $R(ST) \longrightarrow R'(S'T')$. From $(RS)T = R(ST)$, one has $(R'S')T' = R'(S'T')$ and the multiplication rule of elements in G' satisfies the associative law. In the same reason, from $E \longrightarrow E'$, $R \longrightarrow R'$, and $ER = R \longrightarrow E'R'$ one has $E'R' = R'$, and the set G' contains the identical element E'. From $R^{-1} \longrightarrow P'$, and $R^{-1}R = E \longrightarrow P'R'$, one has $P'R' = E'$, and the inverse P' of R' belongs to G'. The proof is also effective if the many-to-one correspondence is replaced with a one-to-one correspondence. □

2.5 Proper Symmetric Group of a Regular Polyhedron

A regular polyhedron is a geometric figure with high symmetry. Let the origin of the coordinate system coincide with the center of the regular polyhedron, so that its symmetric transformations remain with the origin unchanged. A group is called a point group if it consists of rotations remaining with a common point unchanged. A point group is proper if it contains only the proper rotations. The proper symmetric group of a regular polyhedron is a proper point group.

Consider a regular polyhedron with N side-faces, L edges, and V vertices. Each side-face in the regular polyhedron is a regular polygon with n edges. m edges as well as m side-faces meet at one vertex. Since each edge connects two vertices and is shared by two side-faces, one has

$$nN = 2L = mV. \tag{2.21}$$

The internal angle of a regular polygon with n edges is $(n-2)\pi/n$. For a solid figure, m has to be larger than 2 and less than $2n/(n-2)$ because $m(n-2)\pi/n < 2\pi$. Thus, $m = 3$, 4, or 5 when $n = 3$, and $m = 3$ when $n = 4$

or 5. The regular polyhedron is a dodecahedron when $m = 3$ and $n = 5$. It is a cube when $m = 3$ and $n = 4$. When $n = 3$, the regular polyhedron is a tetrahedron if $m = 3$, an octahedron if $m = 4$, and an icosahedron if $m = 5$. If a regular polyhedron is obtained by connecting the centers of all side-faces to another regular polyhedron, the polyhedron is said to be conjugate to another polyhedron. The conjugate relation between two regular polyhedrons are mutual. They have the same symmetric group. A dodecahedron and an icosahedron are mutual conjugate. An octahedron and a cube are mutual conjugate. The tetrahedron is a self-conjugate figure.

2.5.1 *Tetrahedron, Octahedron, and Cube*

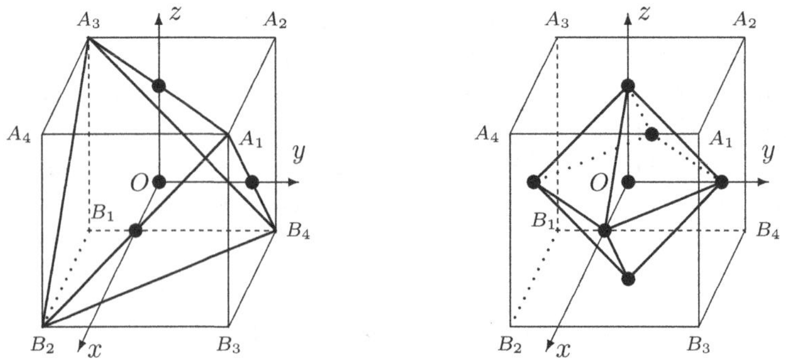

Fig. 2.2 A cube, a tetrahedron, and an octahedron.

A cube is shown in Fig. 2.2 with a rectangular coordinate system. The origin of the coordinate system coincides with the center of the cube, and the coordinate axes point from the origin to the centers of three squares of the cube. Four vertices above the x y plane are denoted in order by A_j, $1 \leq j \leq 4$, and their opposite vertices are denoted by B_j, respectively. An octahedron is obtained by connecting the centers of the six squares in the cube. The cube and the octahedron have the same proper symmetric group, denoted by **O**. Connecting four un-neighbored vertices A_1, B_2, A_3, and B_4 in the cube, one obtains a tetrahedron, whose proper symmetric group is denoted by **T**. **T** is a subgroup of **O**.

From Fig. 2.2, three coordinate axes are the 4-fold proper axes of the cube, but the 2-fold axes of the tetrahedron. Denote by T_μ, $\mu = x,\ y,\ z$, the rotations around the coordinate axes through the angle $\pi/2$, respectively. Only T_μ^2, not T_μ, belongs to the group **T**. Three diagonals, respectively connecting two opposite vertices, are the 3-fold axes of both cube and tetrahedron. The directions of the four axes is defined as follows for determining the four 3-fold rotations through the angle $2\pi/3$.

$$\begin{aligned} R_1 &: \quad \text{from } B_1 \text{ to } A_1, \quad (\boldsymbol{e}_x + \boldsymbol{e}_y + \boldsymbol{e}_z)/\sqrt{3}, \\ R_2 &: \quad \text{from } A_2 \text{ to } B_2, \quad (\boldsymbol{e}_x - \boldsymbol{e}_y - \boldsymbol{e}_z)/\sqrt{3}, \\ R_3 &: \quad \text{from } B_3 \text{ to } A_3, \quad (-\boldsymbol{e}_x - \boldsymbol{e}_y + \boldsymbol{e}_z)/\sqrt{3}, \\ R_4 &: \quad \text{from } A_4 \text{ to } B_4, \quad (-\boldsymbol{e}_x + \boldsymbol{e}_y - \boldsymbol{e}_z)/\sqrt{3}. \end{aligned} \tag{2.22}$$

R_j and R_j^2, $1 \le j \le 4$, belong to both group $\mathbf{O}$ and group $\mathbf{T}$. Six lines, respectively connecting the midpoints of two opposite edges, are the 2-fold axes of the cube, but not of the tetrahedron. Denote by S_k, $1 \le k \le 6$, six rotations around those 2-fold axes through π angle.

$$\begin{array}{llll} S_1: & (\boldsymbol{e}_x + \boldsymbol{e}_y)/\sqrt{2}, & S_2: & (\boldsymbol{e}_x - \boldsymbol{e}_y)/\sqrt{2}, \\ S_3: & (\boldsymbol{e}_y + \boldsymbol{e}_z)/\sqrt{2}, & S_4: & (\boldsymbol{e}_y - \boldsymbol{e}_z)/\sqrt{2}, \\ S_5: & (\boldsymbol{e}_x + \boldsymbol{e}_z)/\sqrt{2}, & S_6: & (\boldsymbol{e}_x - \boldsymbol{e}_z)/\sqrt{2}. \end{array} \tag{2.23}$$

A tetrahedron has four equivalent proper 3-fold axes and three equivalent proper 2-fold axes. The 3-fold axes are polar. The proper symmetric group $\mathbf{T}$ of the tetrahedron has 12 elements and four classes: $\mathcal{C}_1 = \{E\}$, $\mathcal{C}_2 = \{T_\mu^2 | \mu = x, y, z\}$, $\mathcal{C}_3 = \{R_j | 1 \le j \le 4\}$, and $\mathcal{C}_3^{(-1)} = \{R_j^2 | 1 \le j \le 4\}$. Two self-reciprocal classes $\mathcal{C}_1$ and $\mathcal{C}_2$ compose an invariant subgroup D_2 of $\mathbf{T}$. The quotient group $\mathbf{T}/\mathrm{D}_2$ is isomorphic onto the cyclic group C_3. Denote by (2θ) the angle between two neighbored 3-fold axes, whose angular bisector is a 2-fold axis. If the length of the edge of a tetrahedron is taken to be the unit, the radius R of the circumcircle and the radius r of the inscribed circle of the tetrahedron are as follows.

$$\begin{aligned} & R = \sqrt{3/8}, \quad r = \sqrt{1/24}, \\ & \cos\theta = \sqrt{1/3}, \quad \cos(2\theta) = -1/3, \quad \theta = 54.73^\circ. \end{aligned} \tag{2.24}$$

A cube has four proper 3-fold axes, three proper 4-fold axes, and six proper 2-fold axes. All proper axes are nonpolar, and any two axes with the same fold are equivalent to each other. The proper symmetric group $\mathbf{O}$ of the cube contains 24 elements and five classes: $\mathcal{C}_1 = \{E\}$, $\mathcal{C}_2 = \{T_\mu^2 | \mu = x, y, z\}$, $\mathcal{C}_3 = \{R_j, R_j^2 | 1 \le j \le 4\}$, $\mathcal{C}_4 = \{T_\mu, T_\mu^3 | \mu = x, y, z\}$, and $\mathcal{C}_2' = \{S_j | 1 \le j \le 6\}$. The subgroup $\mathbf{T}$ of $\mathbf{O}$ is invariant due to its index 2, and the quotient group $\mathbf{O}/\mathbf{T}$ is isomorphic onto C_2. The invariant subgroup D_2 of $\mathbf{T}$ is also an invariant subgroup of $\mathbf{O}$. Two cosets $R_1\mathrm{D}_2$ and $R_2\mathrm{D}_2$ belong to the subgroup $\mathbf{T}$ and the other three cosets $T_x\mathrm{D}_2$, $T_y\mathrm{D}_2$, and

$T_z D_2$ do not belong to **T**. The squares of the last three cosets, under the multiplication rule of complexes of elements, are all equal to the subgroup D_2 so that the quotient group $\mathbf{O}/D_2$ is isomorphic onto D_3. If the length of the edge of the cube is taken to be the unit, the radius R of the circumcircle and the radius r of the inscribed circle of the cube are

$$R = \sqrt{3}/2, \qquad r = 1/2. \tag{2.25}$$

2.5.2 *The Group Table of* T

The method for filling the group table of D_3 given in §2.2 can be generalized to fill the group tables of **T** and **O**. However, here we will introduce a new way for filling the group table of a finite group G.

Let H be a subgroup of G with index $d = g/h$. Denote by S_μ the elements in H. The group table of H is known. If there exists an element R with the order d, $R \in G$ and $R \overline{\in} H$, it is convenient to express the cosets of H as $R^j H$ and HR^j. Calculate the elements $X_{j\mu} = R^j S_\mu$ in the left cosets. Then, taking their inverse, one obtains the elements $Y_{\nu k} = S_\nu R^k$ in the right cosets. $Y_{\nu k}$ with $k \neq d$ has to belong to one left coset, say $Y_{\nu k} = X_{i\rho}$. Thus, $R^j S_\nu R^k = R^{j+i} S_\rho$. List the products $R^j S_\mu$, $S_\nu R^k$, and $R^j S_\nu R^k$ in a table, called the coset table of H in G. Any product of elements can be looked up from the group table of H and its coset table. For example, for a product XY, one first looks up, say $X = R^j S_\mu$ and $Y = S_\nu R^k$, then, $XY = R_j \left(S_\mu S_\nu\right) R^k$.

If g is not so large, one may divide the group table of G to be $d \times d$ blocks, where the block at the first row and the first column is the group table of H. Replacing the element S_μ in the group table of H with $R^j S_\mu R^k$, one obtains the block at the jth row and the kth column of the group table of G. If the order g of G is very large, the group table of the subgroup H and its coset table can be used to replace the group table of G.

Now, we apply this method to the group table of **T**. Based on the geometric meaning of any element R in **T**, one lists the new positions of four vertices of the tetrahedron after the transformation R in Table 2.7.

Table 2.7 Symmetric transformations of a tetrahedron

	E	T_x^2	T_y^2	T_z^2	R_1	R_1^2	R_2	R_2^2	R_3	R_3^2	R_4	R_4^2
A_1	A_1	B_2	B_4	A_3	A_1	A_1	A_3	B_4	B_4	B_2	B_2	A_3
A_3	A_3	B_4	B_2	A_1	B_2	B_4	B_4	A_1	A_3	A_3	A_1	B_2
B_2	B_2	A_1	A_3	B_4	B_4	A_3	B_2	B_2	A_1	B_4	A_3	A_1
B_4	B_4	A_3	A_1	B_2	A_3	B_2	A_1	A_3	B_2	A_1	B_4	B_4

Choose the subgroup D_2 of $\mathbf{T}$, composed of E, T_x^2, T_y^2, and T_z^2. Since D_2 is isomorphic onto V_4, its group table is given in Table 2.4. The left cosets of D_2 in $\mathbf{T}$ is expressed as $R_1\mathbf{T}$ and $R_1^2\mathbf{T}$ and can be calculated by Table 2.7, such as $R_1 T_x^2 = R_3$, $R_1^2 T_x^2 = R_4^2$, and so on. List them in the first two rows of Table 2.8. Then, taking the inverse of those equalities, one obtains the expressions for the right cosets, such as $T_x^2 R_1^2 = R_3^2$, $T_x^2 R_1 = R_4$, and so on. List them in the next two rows of Table 2.8. At last, calculate the remaining part of Table 2.8, such as

$$R_1\left(T_x^2 R_1\right) = R_1\left(R_4\right) = R_1\left(R_1 T_z^2\right) = R_2^2,$$
$$R_1\left(T_x^2 R_1^2\right) = R_1\left(R_3^2\right) = R_1\left(R_1^2 T_y^2\right) = T_y^2.$$

Table 2.8 The coset table of the subgroup D_2 in T

Left-multiplication	E	T_x^2	T_y^2	T_z^2	Right-multiplication
R_1	R_1	R_3	R_2	R_4	
R_1^2	R_1^2	R_4^2	R_3^2	R_2^2	
	R_1	R_4	R_3	R_2	R_1
	R_1^2	R_3^2	R_2^2	R_4^2	R_1^2
R_1	R_1^2	R_2^2	R_4^2	R_3^2	R_1
R_1	E	T_y^2	T_z^2	T_x^2	R_1^2
R_1^2	E	T_z^2	T_x^2	T_y^2	R_1
R_1^2	R_1	R_2	R_4	R_3	R_1^2

Table 2.9 The group table of the group T

	E	T_x^2	T_y^2	T_z^2	R_1	R_4	R_3	R_2	R_1^2	R_3^2	R_2^2	R_4^2
E	E	T_x^2	T_y^2	T_z^2	R_1	R_4	R_3	R_2	R_1^2	R_3^2	R_2^2	R_4^2
T_x^2	T_x^2	E	T_z^2	T_y^2	R_4	R_1	R_2	R_3	R_3^2	R_1^2	R_4^2	R_2^2
T_y^2	T_y^2	T_z^2	E	T_x^2	R_3	R_2	R_1	R_4	R_2^2	R_4^2	R_1^2	R_3^2
T_z^2	T_z^2	T_y^2	T_x^2	E	R_2	R_3	R_4	R_1	R_4^2	R_2^2	R_3^2	R_1^2
R_1	R_1	R_3	R_2	R_4	R_1^2	R_2^2	R_4^2	R_3^2	E	T_y^2	T_z^2	T_x^2
R_3	R_3	R_1	R_4	R_2	R_2^2	R_1^2	R_3^2	R_4^2	T_y^2	E	T_x^2	T_z^2
R_2	R_2	R_4	R_1	R_3	R_4^2	R_3^2	R_1^2	R_2^2	T_z^2	T_x^2	E	T_y^2
R_4	R_4	R_2	R_3	R_1	R_3^2	R_4^2	R_2^2	R_1^2	T_x^2	T_z^2	T_y^2	E
R_1^2	R_1^2	R_4^2	R_3^2	R_2^2	E	T_z^2	T_x^2	T_y^2	R_1	R_2	R_4	R_3
R_4^2	R_4^2	R_1^2	R_2^2	R_3^2	T_z^2	E	T_y^2	T_x^2	R_2	R_1	R_3	R_4
R_3^2	R_3^2	R_2^2	R_1^2	R_4^2	T_x^2	T_y^2	E	T_z^2	R_4	R_3	R_1	R_2
R_2^2	R_2^2	R_3^2	R_4^2	R_1^2	T_y^2	T_x^2	T_z^2	E	R_3	R_4	R_2	R_1

Divide the group table of $\mathbf{T}$ into 3×3 blocks. The block at the first row and the first column is the group table of D_2. Replacing the element $S_\mu \in D_2$ in the group table of D_2 with $R_1^j S_\mu R_1^k$, one obtains the block at the jth row and the kth column in the group table of $\mathbf{T}$. Recall that only the formulas for the left cosets need to be calculated by Table 2.7 and the remaining part in Table 2.8 is derived from the formulas. This is the main merit of the new way for filling the group table.

2.5.3 *The Group Table of* **O**

Some symmetric transformations of **O** are listed in Table 2.10. Choose the subgroup **T** in **O**. Calculate the formulas in the left coset $S_1\mathbf{T}$ by Table 2.10 and list them in Table 2.11. The remaining part in Table 2.11 is derived. The product of any two elements in **O** can be calculated easily from Tables 2.9 and 2.11, such as $T_xS_3 = S_1\left(R_3R_4^2\right)S_1 = S_1\left(T_z^2\right)S_1 = T_z^2$. The group table of **O** can be constructed by dividing it into 2×2 blocks, where the block at the first row and the first column is the group table of **T** and the remaining blocks are obtained from the group table of **T** by replacing the element $R_\mu \in \mathbf{T}$ with S_1R_μ, $R_\mu S_1$, and $S_1R_\mu S_1$, respectively. We omit the group table of **O** here.

Table 2.10 Some symmetric transformations of a cube

	T_x	T_x^3	T_y	T_y^3	T_z	T_z^3	S_1	S_2	S_3	S_4	S_5	S_6
A_1	A_4	B_3	B_3	A_2	A_2	A_4	B_3	B_1	A_2	B_1	A_4	B_1
A_3	B_1	A_2	A_4	B_1	A_4	A_2	B_1	B_3	B_3	A_4	B_3	A_2

	A_1	A_2	A_3	A_4	B_1	B_2	B_3	B_4
S_1	B_3	B_2	B_1	B_4	A_3	A_2	A_1	A_4

Table 2.11 The coset table of T in O

	E	T_x^2	T_y^2	T_z^2	R_1	R_4	R_3	R_2	R_1^2	R_3^2	R_2^2	R_4^2	
S_1	S_1	T_z	T_z^3	S_2	T_x^3	S_3	T_x	S_4	T_y	T_y^3	S_5	S_6	
	S_1	T_z^3	T_z	S_2	T_y^3	S_6	T_y	S_5	T_x	T_x^3	S_4	S_3	S_1
S_1	E	T_y^2	T_x^2	T_z^2	R_3^2	R_4^2	R_1^2	R_2^2	R_3	R_1	R_2	R_4	S_1

2.5.4 *Regular Icosahedron and the Group Table of* **I**

The regular icosahedron is shown in Fig. 2.3. The origin O of the coordinate frame coincides with the center of the regular icosahedron. The opposite vertices are denoted by A_j and B_j, $0 \leq j \leq 5$. A_j are located above the $x\,y$ plane. The z-axis is in the direction from the origin O to the vertex A_0. The y-axis is in the direction from O to the midpoint of the edge A_2B_5.

The regular icosahedron contains six 5-fold axes, ten 3-fold axis, and fifteen 2-fold axes. The 5-fold axes are along the directions from B_j to A_j with the generators T_j, $0 \leq j \leq 5$. One 5-fold axis ($j = 0$) is along the positive z-axis. The polar angles of the remaining 5-fold axes all are θ_1, and their azimuthal angles are $2(j-1)\pi/5$, respectively. The 3-fold axes are along the lines connecting the centers of two opposite triangles with the generators R_j, $1 \le j \le 10$. The polar angles of the 3-fold axes are θ_2 when $1 \le j \le 5$, and θ_3 when $6 \le j \le 10$. Their azimuthal angles respectively are $(2j-1)\pi/5$. The 2-fold axes are along the lines connecting the midpoints of two opposite edges with the generators S_j, $1 \le j \le 15$. The polar angles of the 2-fold axes are θ_4 when $1 \le j \le 5$, θ_5 when $6 \le j \le 10$, and $\pi/2$ when $11 \le j \le 15$. Their azimuthal angles are $2(j-1)\pi/5$ when $1 \le j \le 5$, $(2j-1)\pi/5$ when $6 \le j \le 10$, and $(4j-3)\pi/10$ when $11 \le j \le 15$, respectively. All proper axes are nonpolar, and any two axes with the same fold are equivalent to each other. Those polar angles are calculated in Prob. 12 of Chap. 4 of [Ma and Gu (2004)].

$$\begin{array}{lll}
\tan\theta_1 = 2, & \tan\theta_2 = 3-\sqrt{5}, & \tan\theta_3 = 3+\sqrt{5}, \\
\tan\theta_4 = (\sqrt{5}-1)/2, & \tan\theta_5 = (\sqrt{5}+1)/2, & \theta_1 = 2\theta_4 \approx 63.43^\circ, \\
\theta_2 \approx 37.38^\circ, & \theta_3 \approx 79.19^\circ, & \theta_5 \approx 58.28^\circ.
\end{array} \tag{2.26}$$

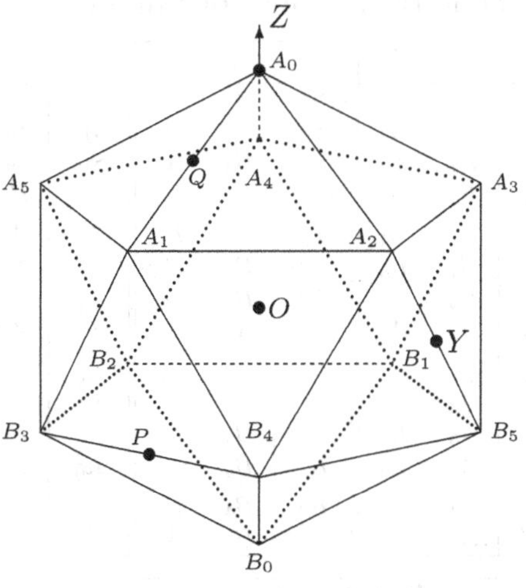

Fig. 2.3 The regular icosahedron.

The proper symmetric group of an icosahedron, denoted by **I**, contains 60 elements and five self-reciprocal classes: $\mathcal{C}_1 = \{E\}$, $\mathcal{C}_5 = \{T_j, T_j^4 | 0 \le j \le 5\}$, $\mathcal{C}_5' = \{T_j^2, T_j^3 | 0 \le j \le 5\}$, $\mathcal{C}_3 = \{R_j, R_j^2 | 1 \le j \le 10\}$, and $\mathcal{C}_2 = \{S_j | 1 \le j \le 15\}$. The numbers of elements in those classes are 1, 12, 12, 20, and 15. There is no nontrivial invariant subgroup in **I**. If the length of the edge of an icosahedron is taken to be the unit, the radius R of its circumcircle and the radius r of its inscribed circle are (also see Prob. 12 of Chap. 4)

$$R = \left(\frac{5+\sqrt{5}}{8}\right)^{1/2} = 0.9511, \qquad r = \frac{3+\sqrt{5}}{4\sqrt{3}} = 0.7558. \tag{2.27}$$

Table 2.12 The group table of the subgroup T in I

	E	S_8	S_{12}	S_1	R_6	R_{10}^2	R_4	R_2^2	R_6^2	R_4^2	R_2	R_{10}
E	E	S_8	S_{12}	S_1	R_6	R_{10}^2	R_4	R_2^2	R_6^2	R_4^2	R_2	R_{10}
S_8	S_8	E	S_1	S_{12}	R_{10}^2	R_6	R_2^2	R_4	R_4^2	R_6^2	R_{10}	R_2
S_{12}	S_{12}	S_1	E	S_8	R_4	R_2^2	R_6	R_{10}^2	R_2	R_{10}	R_6^2	R_4^2
S_1	S_1	S_{12}	S_8	E	R_2^2	R_4	R_{10}^2	R_6	R_{10}	R_2	R_4^2	R_6^2
R_6	R_6	R_4	R_2^2	R_{10}^2	R_6^2	R_2	R_{10}	R_4^2	E	S_{12}	S_1	S_8
R_4	R_4	R_6	R_{10}^2	R_2^2	R_2	R_6^2	R_4^2	R_{10}	S_{12}	E	S_8	S_1
R_2^2	R_2^2	R_{10}^2	R_6	R_4	R_{10}	R_4^2	R_6^2	R_2	S_1	S_8	E	S_{12}
R_{10}^2	R_{10}^2	R_2^2	R_4	R_6	R_4^2	R_{10}	R_2	R_6^2	S_8	S_1	S_{12}	E
R_6^2	R_6^2	R_{10}	R_4^2	R_2	E	S_1	S_8	S_{12}	R_6	R_2^2	R_{10}^2	R_4
R_{10}	R_{10}	R_6^2	R_2	R_4^2	S_1	E	S_{12}	S_8	R_2^2	R_6	R_4	R_{10}^2
R_4^2	R_4^2	R_2	R_6^2	R_{10}	S_8	S_{12}	E	S_1	R_{10}^2	R_4	R_6	R_2^2
R_2	R_2	R_4^2	R_{10}	R_6^2	S_{12}	S_8	S_1	E	R_4	R_{10}^2	R_2^2	R_6

Table 2.13 Symmetric transformations in an icosahedron

	A_0	A_1		A_0	A_1		A_0	A_1		A_0	A_1
E	A_0	A_1	S_4	A_4	B_1	S_8	B_1	B_0	S_{12}	B_0	B_1
S_1	A_1	A_0	S_5	A_5	A_4	S_9	B_2	B_1	S_{13}	B_0	B_3
S_2	A_2	A_3	S_6	B_4	A_2	S_{10}	B_3	A_5	S_{14}	B_0	B_5
S_3	A_3	B_1	S_7	B_5	B_1	S_{11}	B_0	B_4	S_{15}	B_0	B_2
T_1	A_5	A_1	T_2	A_1	B_4	T_3	A_2	B_4	T_4	A_3	A_2
T_1^2	B_3	A_1	T_2^2	B_4	B_5	T_3^2	B_5	B_0	T_4^2	B_1	B_5
T_1^3	B_4	A_1	T_2^3	B_5	A_3	T_3^3	B_1	B_2	T_4^3	B_2	B_0
T_1^4	A_2	A_1	T_2^4	A_3	A_0	T_3^4	A_4	A_5	T_4^4	A_5	B_3
T_5	A_4	A_0	R_2	A_2	B_5	R_5	A_5	A_0	R_8	B_5	B_4
T_5^2	B_2	A_4	R_2^2	A_3	A_4	R_5^2	A_1	A_5	R_8^2	B_2	B_3
T_5^3	B_3	B_2	R_3	A_3	B_5	R_6	B_3	B_4	R_9	B_1	A_3
T_5^4	A_1	B_3	R_3^2	A_4	B_2	R_6^2	B_5	A_2	R_9^2	B_3	B_0
R_1	A_1	A_2	R_4	A_4	A_3	R_7	B_4	B_0	R_{10}	B_2	A_5
R_1^2	A_2	A_0	R_4^2	A_5	B_2	R_7^2	B_1	A_4	R_{10}^2	B_4	B_3

	A_0	A_1	A_2	A_3	A_4	A_5	B_0	B_1	B_2	B_3	B_4	B_5
T_0	A_0	A_2	A_3	A_4	A_5	A_1	B_0	B_2	B_3	B_4	B_5	B_1
T_0^2	A_0	A_3	A_4	A_5	A_1	A_2	B_0	B_3	B_4	B_5	B_1	B_2
T_0^3	A_0	A_4	A_5	A_1	A_2	A_3	B_0	B_4	B_5	B_1	B_2	B_3
T_0^4	A_0	A_5	A_1	A_2	A_3	A_4	B_0	B_5	B_1	B_2	B_3	B_4

Obviously, the group **I** has a subgroup D_5, composed of T_0^m $(0 \le m \le 4)$ and S_j $(11 \le j \le 15)$. The index of D_5 in **I** is 6. Since there is no element with order 6 in **I**, it is inconvenient to construct a coset table based on the subgroup D_5. From Fig. 2.3 one finds three orthogonal 2-fold axes, $\overline{OP}$, $\overline{OY}$, and $\overline{OQ}$. Taking these axes as the axes of a rectangular coordinate system, one constructs a subgroup **T** in **I** with the index 5. The elements of the subgroup **T** in **I** correspond to the elements of **T** as follows,

$$R_6^{\pm 1} \longleftrightarrow R_1^{\pm 1}, \quad R_2^{\mp 1} \longleftrightarrow R_2^{\pm 1}, \quad R_4^{\pm 1} \longleftrightarrow R_3^{\pm 1}, \quad R_{10}^{\mp 1} \longleftrightarrow R_4^{\pm 1},$$
$$S_8 \longleftrightarrow T_x^2, \qquad S_{12} \longleftrightarrow T_y^2, \qquad S_1 \longleftrightarrow T_z^2.$$

Replacing the elements in Table 2.9 with the correspondence, one obtains the group table of the subgroup **T** in **I**, listed in Table 2.12. The symmetric transformations of the icosahedron are listed in Table 2.13, from which the formulas for the left cosets $T_0^j\mathbf{T}$ are calculated. The remaining part in the coset table 2.14 is derived. Then, the multiplications of elements in **I** can be calculated in terms of Table 2.12 and Table 2.14. For example,

$$S_{10}T_3^3 = T_0^3 \left(R_6^2 S_8\right) T_0 = T_0^3 \left(R_{10}\right) T_0 = R_6^2.$$

Table 2.14 The coset table of the subgroup T in I

Left	E S_8	S_{12} S_1	R_6 R_{10}^2	R_4 R_2^2	R_6^2 R_4^2	R_2 R_{10}	Right
T_0	T_0 T_4^3	S_{15} R_1^2	T_2^2 R_8	S_5 T_3^4	R_9 T_5^4	S_3 T_1^2	
T_0^2	T_0^2 R_9^2	S_{13} T_2^4	S_7 T_4^2	R_5^2 T_1	T_5^2 T_3	R_3^2 S_6	
T_0^3	T_0^3 R_7	S_{11} T_5	T_3^3 S_9	T_1^4 R_1	S_{10} R_3	T_4^4 T_2^3	
T_0^4	T_0^4 T_3^2	S_{14} R_5	R_8^2 T_5^3	T_4 S_2	T_1^3 S_4	T_2 R_7^2	
	T_0 T_3^3	S_{14} R_5^2	T_1^2 R_7	S_4 T_2^4	R_8 T_4^4	S_2 T_5^2	T_0
	T_0^2 R_7^2	S_{11} T_5^4	S_{10} T_2^2	R_3^2 T_4	T_3^2 T_1	R_1^2 S_9	T_0^2
	T_0^3 R_9	S_{13} T_2	T_5^3 S_6	T_3^4 R_3	S_7 R_5	T_1^4 T_4^3	T_0^3
	T_0^4 T_4^2	S_{15} R_1	R_9^2 T_1^3	T_5 S_3	T_2^3 S_5	T_3 R_8^2	T_0^4
T_0	T_0^2 R_8^2	S_{12} T_1^4	S_6 T_3^2	R_4^2 T_5	T_4^2 T_2	R_2^2 S_{10}	T_0
T_0	T_0^3 R_{10}	S_{14} T_3	T_1^3 S_7	T_4^4 R_4	S_8 R_1	T_2^4 T_5^3	T_0^2
T_0	T_0^4 T_5^2	S_{11} R_2	R_{10}^2 T_2^3	T_1 S_4	T_3^3 S_1	T_4 R_9^2	T_0^3
T_0	E S_9	S_{13} S_2	R_7 R_6^2	R_5 R_3^2	R_7^2 R_5^2	R_3 R_6	T_0^4
T_0^2	T_0^3 R_6	S_{15} T_4	T_2^3 S_8	T_5^4 R_5	S_9 R_2	T_3^4 T_1^3	T_0
T_0^2	T_0^4 T_1^2	S_{12} R_3	R_6^2 T_3^3	T_2 S_5	T_4^3 S_2	T_5 R_{10}^2	T_0^2
T_0^2	E S_{10}	S_{14} S_3	R_8 R_7^2	R_1 R_4^2	R_8^2 R_1^2	R_4 R_7	T_0^3
T_0^2	T_0 T_5^3	S_{11} R_2^2	T_3^2 R_9	S_1 T_4^4	R_{10} T_1^4	S_4 T_2^2	T_0^4
T_0^3	T_0^4 T_2^2	S_{13} R_4	R_7^2 T_4^3	T_3 S_1	T_5^3 S_3	T_1 R_6^2	T_0
T_0^3	E S_6	S_{15} S_4	R_9 R_8^2	R_2 R_5^2	R_9^2 R_2^2	R_5 R_8	T_0^2
T_0^3	T_0 T_1^3	S_{12} R_3^2	T_4^2 R_{10}	S_2 T_5^4	R_6 T_2^4	S_5 T_3^2	T_0^3
T_0^3	T_0^2 R_{10}^2	S_{14} T_3^4	S_8 T_5^2	R_1^2 T_2	T_1^2 T_4	R_4^2 S_7	T_0^4
T_0^4	E S_7	S_{11} S_5	R_{10} R_9^2	R_3 R_1^2	R_{10}^2 R_3^2	R_1 R_9	T_0
T_0^4	T_0 T_2^3	S_{13} R_4^2	T_5^2 R_6	S_3 T_1^4	R_7 T_3^4	S_1 T_4^2	T_0^2
T_0^4	T_0^2 R_6^2	S_{15} T_4^4	S_9 T_1^2	R_2^2 T_3	T_2^2 T_5	R_5^2 S_8	T_0^3
T_0^4	T_0^3 R_8	S_{12} T_1	T_4^3 S_{10}	T_2^4 R_2	S_6 R_4	T_5^4 T_3^3	T_0^4

2.6 Direct Product of Groups and Improper Point Groups

2.6.1 *The Direct Product of Two Groups*

Definition 2.4 A group G is called the direct product $G = H_1 \otimes H_2$ of its two subgroups H_1 and H_2

$$H_1 = \{R_1,\ R_2,\ \cdots,\ R_{h_1}\}, \qquad H_2 = \{S_1,\ S_2,\ \cdots,\ S_{h_2}\}, \tag{2.28}$$

if the following conditions are satisfied.

a) Except for the identical element $E = R_1 = S_1$, there is no other common element in H_1 and H_2.

b) Any two elements respectively belonging to two subgroups H_1 and H_2 are commutable, namely $R_j S_\mu = S_\mu R_j$ if $R_j \in H_1$ and $S_\mu \in H_2$.

c) The group G consists of all products $R_j S_\mu$.

Obviously, if $G = H_1 \otimes H_2$, H_1 and H_2 are both the invariant subgroups of G. Since there is only one common element E in two subgroups H_1 and H_2, there is no repetitive element in the set $\{R_j S_\mu\}$. In fact, if $R_j S_\mu = R_k S_\nu$, then $R_k^{-1} R_j = S_\nu S_\mu^{-1}$ is the identical element E, namely, $R_j = R_k$ and $S_\mu = S_\nu$. Thus, the order of G is $g = h_1 h_2$.

The situation of the direct product of two groups one often meets in the real problems is that the groups H_1 and H_2 are two sets of operators respectively affecting two different subsystems so that the elements in the two groups are commutable, $R_j S_\mu = S_\mu R_j$. Define two groups $H_1 S_1 \approx H_1$ and $R_1 H_2 \approx H_2$, where R_1 and S_1 are the identical elements of H_1 and H_2, respectively. The element $R_1 S_1$ is the only common element in two groups $H_1 S_1$ and $R_1 H_2$. The set

$$G = \{R_j S_\mu | R_j \in H_1, S_\mu \in H_2\} \tag{2.29}$$

satisfies four axioms under the multiplication rules of elements in H_1 and H_2, where the identical element is $R_1 S_1$. This group G is the direct product of two groups, H_1 and H_2.

2.6.2 *Improper Point Groups*

An improper point group G contains both improper and proper rotations. An improper rotation S' is a product of a proper rotation S and a spatial inversion σ. σ is commutable with any rotation and its square is equal to the identical element E,

$$S' = \sigma S = S\sigma, \qquad \sigma^2 = E. \tag{2.30}$$

The subset H, composed of all proper rotations in G, is a subgroup of G because the product of two proper rotations is a proper rotation so that the identical element E and the inverse R^{-1} of a proper rotation R are also the proper rotations. Any improper rotation in G belongs to the same coset of H because the product of two improper rotations is a proper rotation. Thus, the index of H in G is 2, and H is an invariant subgroup of G. H is called the proper subgroup of an improper point group. Based on this property, one concludes that there are two types of improper point groups. The improper point group of I-type contains the spatial inversion σ, and the improper point group of P-type does not contain σ.

An improper point group G of I-type consists of its proper subgroup H and the coset σH. G is the direct product of H and the inversion group V_2 of order 2

$$G = H \otimes \mathrm{V}_2, \qquad \mathrm{V}_2 = \{E, \sigma\}. \tag{2.31}$$

According to the Schoenflies notations, some improper point groups of I-type are listed as follows.

$$\begin{array}{ll}
\mathrm{C}_i \approx \mathrm{C}_1 \otimes \mathrm{V}_2, & \mathrm{C}_{(2n)h} \approx \mathrm{C}_{2n} \otimes \mathrm{V}_2, \\
\mathrm{C}_{(2n+1)i} \approx \mathrm{C}_{(2n+1)} \otimes \mathrm{V}_2, & \mathrm{D}_{(2n)h} \approx \mathrm{D}_{2n} \otimes \mathrm{V}_2, \\
\mathrm{D}_{(2n+1)d} \approx \mathrm{D}_{(2n+1)} \otimes \mathrm{V}_2, & \mathbf{T}_h \approx \mathbf{T} \otimes \mathrm{V}_2, \\
\mathbf{O}_h \approx \mathbf{O} \otimes \mathrm{V}_2, & \mathbf{I}_h \approx \mathbf{I} \otimes \mathrm{V}_2,
\end{array} \tag{2.32}$$

where the meaning of the subscripts will be given at the end of this section.

An improper point group G of P-type does not contain the spatial inversion σ. Denote by R_k and S_j the proper rotation and improper rotation in G, respectively. σS_j is a proper rotation and is not equal to any proper rotation R_k in G. Otherwise, if $\sigma S_j = R_k$, $\sigma = R_k S_j^{-1} \in G$. It is in conflict with the assumption. Let G' be a proper point group composed of those σS_j and R_k. G' is isomorphic onto G and contains the invariant subgroup H with index 2. Conversely, from a proper group G' with an invariant subgroup H with index 2, one can construct an isomorphic group G by multiplying the elements in the coset of H in G' with σ and by remaining H invariant. G is the improper point group of P-type. According to the Schoenflies notations, some improper point groups of type P are listed as follows.

$$
\begin{array}{ll}
\mathrm{S}_{4n} \approx \mathrm{C}_{4n}, \ \text{subgroup } \mathrm{C}_{2n}, & \mathrm{C}_{(2n+1)h} \approx \mathrm{C}_{4n+2}, \ \text{subgroup } \mathrm{C}_{2n+1}, \\
\mathrm{C}_s \approx \mathrm{C}_2, \ \text{subgroup } \mathrm{C}_1, & \mathrm{C}_{Nv} \approx \mathrm{D}_N, \ \text{subgroup } \mathrm{C}_N, \\
\mathrm{D}_{(2n)d} \approx \mathrm{D}_{4n}, \ \text{subgroup } \mathrm{D}_{2n}, & \mathrm{D}_{(2n+1)h} \approx \mathrm{D}_{4n+2}, \ \text{subgroup } \mathrm{D}_{2n+1}, \\
\mathbf{T}_d \approx \mathbf{O}, \ \text{subgroup } \mathbf{T}. &
\end{array}
\tag{2.33}
$$

At last, we introduce the meaning of the subscripts in the Schoenflies notations. Choose the coordinate system such that the z-axis is along a proper or improper axis with the highest fold, called the principal axis. For the groups **T** and **I**, the z-axis is chosen to be along a 2-fold axis. G is denoted with a subscript "h" (horizontal) if and only if an improper point group G contains a reflection with respect to the $x\ y$ plane. The reflection with respect to the $x\ y$ plane is nothing but the 2-fold improper rotation around the z-axis. There is only one exception that the P-type improper point group C_{1h} is denoted by C_s, which is isomorphic onto C_2.

Under the condition that the improper point group G does not contain a reflection with respect to the $x\ y$ plane, G is denoted with a subscript "d" if G contains both the proper and improper 2-fold axes in the $x\ y$ plane, G is denoted with a subscript "v" if G contains only the improper 2-fold axes in the $x\ y$ plane, and G is denoted by S_{4n} (improper $4n$-fold axis) or C_i and $\mathrm{C}_{(2n+1)i}$ (I-type), if G contains no 2-fold axes in the $x\ y$ plane.

2.7 Exercises

1. Let E be the identical element of a group G, R and S be any two elements in the group G, R^{-1} and S^{-1} be the inverses of R and S, respectively. Try to show from Definition 2.1: (a) $RR^{-1} = E$; (b) $RE = R$; (c) if $TR = R$, then $T = E$; (d) if $TR = E$, then $T = R^{-1}$; (e) The inverse of (RS) is $S^{-1}R^{-1}$.
2. Show that a group, composed of all positive real numbers where the multiplication rule of elements is defined with the product of digits, is isomorphic onto a group, composed of all real numbers where the multiplication rule of elements is defined with the addition of digits.
3. If H_1 and H_2 are two subgroups of a group G, prove that the common elements in H_1 and H_2 also form a subgroup of G.
4. Prove that a group whose order g is a prime number must be a cyclic group C_g.

5. Show that up to isomorphism, there are only two different fourth-order groups: The cyclic group C_4 and the fourth-order inversion group.

6. Show that up to isomorphism, there are only two different sixth-order groups: The cyclic group C_6 and the symmetric group D_3 of a regular triangle.

7. Show that a group must be an Abelian group if the order of every element in the group, except for the identical element, is 2.

8. Show that all possible products generated by the Pauli matrices σ_1 and σ_2 (see Prob. 2 in Chap. 1) form a group. Draw the multiplication table of this group. Point out the order of this group, the order of each element, the classes, the invariant subgroups, and their quotient groups. Prove that this group is isomorphic onto the symmetric group D_4 of a square.

9. Show that the set of all possible products generated by $i\sigma_1$ and $i\sigma_2$ forms a group. List the multiplication table of this group. Point out the order of this group, the order of each element, the classes, the invariant subgroups, and their quotient groups in the group, respectively. Show that this group is not isomorphic onto the group D_4.

10. Up to isomorphism, prove that there are only five different eighth-order groups: The cyclic group C_8, $C_{4h} = C_4 \otimes V_2$, the symmetric group D_4 of a square, the quaternion group Q_8 (see Prob. 9), and $D_{2h} = D_2 \otimes V_2$.

11. Investigate all ninth-order groups which are not isomorphic onto each other.

12. Investigate all tenth-order groups which are not isomorphic.

13. Give an example to show that the invariant subgroup of an invariant subgroup of a group G is not necessary to be an invariant subgroup of the group G. Conversely, show that if an invariant subgroup of a group G completely belongs to a subgroup H of G, then it is also an invariant subgroup of H.

14. In a finite group G of order g, let $\mathcal{C}_\alpha = \left\{S_1, S_2, \ldots, S_{n(\alpha)}\right\}$ be a class of G containing $n(\alpha)$ elements. For any two elements S_i and S_j in the class $\mathcal{C}_\alpha$ (may be different or the same), show that the number $m(\alpha)$ of elements $P \in G$ satisfying $S_i = PS_jP^{-1}$ is $g/n(\alpha)$.

15. Prove that, being the product of two subsets, the product of two classes in a group G must be a sum aggregate of a few whole classes. Namely, the sum aggregate contains all elements conjugate to any product of two elements belonging to the two classes, respectively.

16. Calculate the multiplication table of the group $\mathbf{T}$ by extending the multiplication table of the subgroup $C_3 = \{E,\ R_1,\ R_1^2\}$ of $\mathbf{T}$.

17. The multiplication table of the finite group G is as follows.

	E	A	B	C	D	F	I	J	K	L	M	N
E	E	A	B	C	D	F	I	J	K	L	M	N
A	A	E	F	I	J	B	C	D	M	N	K	L
B	B	F	A	K	L	E	M	N	I	J	C	D
C	C	I	L	A	K	N	E	M	J	F	D	B
D	D	J	K	L	A	M	N	E	F	I	B	C
F	F	B	E	M	N	A	K	L	C	D	I	J
I	I	C	N	E	M	L	A	K	D	B	J	F
J	J	D	M	N	E	K	L	A	B	C	F	I
K	K	M	J	F	I	D	B	C	N	E	L	A
L	L	N	I	J	F	C	D	B	E	M	A	K
M	M	K	D	B	C	J	F	I	L	A	N	E
N	N	L	C	D	B	I	J	F	A	K	E	M

(a) Find the inverse of each element in G;

(b) Point out the elements which can commute with any element in G;

(c) List the period and order of each element;

(d) Find the elements in each class of G;

(e) Find all invariant subgroups in G. For each invariant subgroup, list its cosets and point out onto which group its quotient group is isomorphic;

(f) Make a judgment whether G is isomorphic onto the tetrahedral symmetric group $\mathbf{T}$, or isomorphic onto the regular six-sided polygon symmetric group D_6.

Chapter 3

THEORY OF LINEAR REPRESENTATIONS OF GROUPS

The theory of linear representations of groups is the foundation of group theory. In this chapter we will introduce the definition of a representation of a group, study the concepts and properties of inequivalent and irreducible representations, discuss the methods for finding all inequivalent and irreducible representations of a group, and demonstrate the fundamental steps for the application of group theory to physics through an example.

3.1 Linear Representations of a Group

3.1.1 *Definition of a Linear Representation*

If a given group G is isomorphic or homomorphic onto a group composed of matrices, the matrix group, which describes the property of G at least partly, is called a linear representation of G, or briefly a representation.

Definition 3.1 A matrix group $D(G)$, composed of nonsingular $m \times m$ matrices $D(R)$, is called an m-dimensional representation of a group G, or briefly a representation of G, if G is isomorphic or homomorphic onto the matrix group $D(G)$. The matrix $D(R)$, to which an element R in G maps, is called the representation matrix of R in the representation, and $\text{Tr}\, D(R) = \chi(R)$ is the character of R.

The representation matrix $D(E)$ of the identical element E is a unit matrix, $D(E) = \mathbf{1}$. The representation matrices of R and its inverse R^{-1} are mutually inverse matrices, $D(R^{-1}) = D(R)^{-1}$. The representation $D(G)$ is said to be faithful if G is isomorphic onto $D(G)$. Any group has an identical representation, or called the trivial one, where $D(R) = 1$ for every element R in G. A matrix group is its own representation, called the self-representation. A representation is called unitary (or real orthogonal) if each representation matrix $D(R)$ is unitary (or real orthogonal). In this textbook we only study the representation with a finite dimension if without special notification. An example of an infinite-dimensional unitary representation is studied in Prob. 24 of Chap. 4 of [Ma and Gu (2004)].

3.1.2 *Group Algebra and the Regular Representation*

In a group G, the multiplication of two elements has been defined, but the sum of two elements is not defined. Now, we define the sum of two elements in G, satisfying the fundamental axiom

$$\begin{aligned} &c_1R + c_2R = (c_1 + c_2)R, \\ &c_1R + c_2S = c_2S + c_1R, \\ &c_3\left(c_1R + c_2S\right) = c_3c_1R + c_3c_2S. \end{aligned} \tag{3.1}$$

The elements are assumed to be linearly independent. For a finite group G, the linear space $\mathcal{L}$ spanned by the elements R in G is called the group space of G, and the elements are its natural basis vectors. Any vector in $\mathcal{L}$ is a combination of the elements in G,

$$X = \sum_{R\in G} F(R)R, \qquad Y = \sum_{R\in G} F_1(R)R. \tag{3.2}$$

A vector X is described completely by a group function $F(R)$, which is a map of the elements R in G onto a complex number $F(R)$. For a finite group G of order g, the dimension of $\mathcal{L}$ is g and only g values can be taken for a group function $F(R)$ so that there are only g linearly independent group functions for G. The inner product of two vectors X and Y in $\mathcal{L}$ is defined as

$$\langle X|Y\rangle = \sum_{R\in G} F(R)^* F_1(R), \qquad \langle R|S\rangle = \delta_{RS}. \tag{3.3}$$

In the inner product, the group elements are orthonormal to each other. Each matrix entry $D_{\mu\nu}(R)$ of a representation $D(G)$ is a group function of G, and $D(G)$ is a matrix function. The character $\chi(R)$ is also a group function. In fact, $\chi(R)$ is a class function because $\chi(R) = \chi(TRT^{-1})$.

In a linear space $\mathcal{L}$, the sum of two vectors and the multiplication of a vector and a complex number have been defined. Also, the inner product of two vectors in $\mathcal{L}$ can be defined. A linear space $\mathcal{L}$ is called an algebra if another product of vectors in $\mathcal{L}$ can be defined to satisfy

$$XY \in \mathcal{L}, \qquad Z\left(X + Y\right) = ZX + ZY. \tag{3.4}$$

In a group space $\mathcal{L}$, the product of two vectors can be defined such that the coefficients are multiplied like two numbers and the group elements are multiplied according to the multiplication rule in the group G

$$XY = \left\{\sum_{R\in G} F(R)R\right\} \left\{\sum_{S\in G} F_1(S)S\right\} = \sum_{R\in G}\sum_{S\in G} \{F(R)F_1(S)\}(RS)$$
$$= \sum_{T\in G}\left\{\sum_{R\in G} F(R)F_1(R^{-1}T)\right\}T = \sum_{T\in G}\left\{\sum_{S\in G} F(TS^{-1})F_1(S)\right\}T.$$

Thus, the group space $\mathcal{L}$ becomes a group algebra.

In the group algebra $\mathcal{L}$, a vector is changed to another vector if an element S, or a vector in $\mathcal{L}$, left-multiplies or right-multiplies on it. Thus, the element S as well as a vector in $\mathcal{L}$ plays a role of a linear operator in $\mathcal{L}$. We first study the matrix form $D(S)$ of the left-multiplying operator S in the natural basis of $\mathcal{L}$

$$SR = \sum_{P\in G} P\, D_{PR}(S). \tag{3.5}$$

In fact, there is only one term in the sum of Eq. (3.5), namely,

$$D_{PR}(S) = \begin{cases} 1 & \text{when } P = SR, \\ 0 & \text{when } P \neq SR. \end{cases} \tag{3.6}$$

The rows and columns of the matrix $D(S)$ are enumerated by the group elements. There is only one nonvanishing entry in each column of $D(S)$ as well as in each row. Equation (3.5) gives a one-to-one correspondence between a group element S and a matrix $D(S)$, and the correspondence is invariant with respect to the multiplication rule of elements in G:

$$T(SR) = \sum_{P\in G} TPD_{PR}(S) = \sum_{Q\in G} Q\left\{\sum_{P\in G} D_{QP}(T)D_{PR}(S)\right\}$$
$$= (TS)R = \sum_{Q\in G} QD_{QR}(TS).$$

$$D(S) \longleftrightarrow S, \qquad D(T)D(S) = D(TS) \longleftrightarrow TS.$$

From Theorem 2.4, the matrices $D(S)$ form a group $D(G)$, isomorphic onto the group G. Thus, $D(G)$ is a faithful representation of G, called the regular representation. Any finite group has the regular representation. The character $\chi(S)$ in the regular representation is

$$\chi(S) = \text{Tr } D(S) = \begin{cases} g & \text{when } \ S = E, \\ 0 & \text{when } \ S \neq E. \end{cases} \tag{3.7}$$

The regular representation matrix $D(S)$ is easy to calculate from the group table. The product element $SR = T$ in the R column of the S row of the group table indicates the position (the T row) of the nonvanishing entry in the R column of $D(S)$. For example, in the order E, D, F, A, B, and C for the rows (and columns), the regular representation matrices $D(D)$ and $D(A)$ of the group D_3 are read from its group table 2.6:

$$D(D) = \begin{pmatrix} 0 & 0 & 1 & 0 & 0 & 0 \\ 1 & 0 & 0 & 0 & 0 & 0 \\ 0 & 1 & 0 & 0 & 0 & 0 \\ 0 & 0 & 0 & 0 & 0 & 1 \\ 0 & 0 & 0 & 1 & 0 & 0 \\ 0 & 0 & 0 & 0 & 1 & 0 \end{pmatrix}, \qquad D(A) = \begin{pmatrix} 0 & 0 & 0 & 1 & 0 & 0 \\ 0 & 0 & 0 & 0 & 0 & 1 \\ 0 & 0 & 0 & 0 & 1 & 0 \\ 1 & 0 & 0 & 0 & 0 & 0 \\ 0 & 0 & 1 & 0 & 0 & 0 \\ 0 & 1 & 0 & 0 & 0 & 0 \end{pmatrix}. \tag{3.8}$$

Secondly, we study the right-multiplying operator $\overline{S}$, where the bar is used to distinguish it from the left-multiplying operator S:

$$\overline{S}R \equiv RS. \tag{3.9}$$

The multiplication rule for the right-multiplying operator $\overline{S}$ is different from that for the left-multiplying operator S because the product of $\overline{T}$ and $\overline{S}$ is not equal to $\overline{(TS)}$, but $\overline{(ST)}$:

$$\overline{T}\left(\overline{S}R\right) = \overline{T}\,(RS) = (RS)\,T = R\,(ST) = \overline{(ST)}R. \tag{3.10}$$

Under this multiplication rule, the right-multiplying operators $\overline{S}$ form a group $\overline{G}$, called the intrinsic group of G. $\overline{G}$ is usually said to be anti-isomorphic onto G with the one-to-one correspondence $\overline{S} \longleftrightarrow S$. As a matter of fact, $\overline{G}$ is isomorphic onto G with the one-to-one correspondence $\overline{S} \longleftrightarrow S^{-1}$. The matrix forms of the right-multiplying operators $\overline{S}$ in the natural basis of $\mathcal{L}$ form a representation of the intrinsic group $\overline{G}$, and their transposes, denoted by $\overline{D}(S)$, form a representation of G:

$$\overline{S}R = RS = \sum_{P \in G} \overline{D}_{RP}(S)P, \tag{3.11}$$

because

$$\overline{T}(\overline{S}R) = \sum_{P\in G} \overline{D}_{RP}(S)PT = \sum_{Q\in G} \left\{\sum_{P\in G} \overline{D}_{RP}(S)\overline{D}_{PQ}(T)\right\} Q$$
$$= R(ST) = \sum_{Q\in G} \overline{D}_{RQ}(ST)\, Q,$$
$$\overline{D}(S) \longleftrightarrow S, \qquad \overline{D}(S)\overline{D}(T) = \overline{D}(ST) \longleftrightarrow ST.$$

From Eq. (3.11), one has

$$\overline{D}_{RP}(S) = \begin{cases} 1 & \text{when } P = RS, \\ 0 & \text{when } P \neq RS, \end{cases} \tag{3.12}$$

$$\overline{\chi}(S) = \text{Tr}\, \overline{D}(S) = \begin{cases} g & \text{when } S = E, \\ 0 & \text{when } S \neq E. \end{cases} \tag{3.13}$$

The matrices $\overline{D}(S)$ form another faithful representation of G. The character $\overline{\chi}(S)$ is the same as $\chi(S)$. The representation matrix $\overline{D}(S)$ is easy to calculate from the group table. The product element $RS = T$ in the R row of the S column of the group table indicates the position (the T column) of the nonvanishing entry in the R row of $\overline{D}(S)$. For example, in the order E, D, F, A, B, and C for the rows (and columns), the representation matrices $\overline{D}(D)$ and $\overline{D}(A)$ of the group D_3 are read from the group table 2.6 of D_3:

$$\overline{D}(D) = \begin{pmatrix} 0&1&0&0&0&0 \\ 0&0&1&0&0&0 \\ 1&0&0&0&0&0 \\ 0&0&0&0&0&1 \\ 0&0&0&1&0&0 \\ 0&0&0&0&1&0 \end{pmatrix}, \qquad \overline{D}(A) = \begin{pmatrix} 0&0&0&1&0&0 \\ 0&0&0&0&1&0 \\ 0&0&0&0&0&1 \\ 1&0&0&0&0&0 \\ 0&1&0&0&0&0 \\ 0&0&1&0&0&0 \end{pmatrix}. \tag{3.14}$$

3.1.3 *Class Operator and Class Space*

Assume that a finite group G of order g contains g_c classes, and a class $\mathcal{C}_\alpha$ in G consists of $n(\alpha)$ elements

$$\mathcal{C}_\alpha = \{S_1,\ S_2,\ \cdots,\ S_{n(\alpha)}\} = \{S_k | S_k = TS_jT^{-1}, T \in G\}. \tag{3.15}$$

The sum of elements S_j belonging to the class $\mathcal{C}_\alpha$ is a vector in the group algebra $\mathcal{L}$, which is also a linear operator in $\mathcal{L}$, called the class operator C_α,

$$\mathsf{C}_\alpha = \sum_{S_j\in\mathcal{C}_\alpha} S_j. \tag{3.16}$$

From Eq. (2.17) and Theorem 2.2 one has

$$\begin{aligned} &T\mathsf{C}_\alpha T^{-1} = \mathsf{C}_\alpha, \qquad [T,\ \mathsf{C}_\alpha] = 0, \qquad \forall\, T \in G, \\ &\sum_{T\in G} TS_jT^{-1} = \frac{g}{n(\alpha)}\mathsf{C}_\alpha, \qquad \forall\, S_j \in \mathcal{C}_\alpha. \end{aligned} \tag{3.17}$$

A class operator is commutable with every element in G and every class operator in $\mathcal{L}$. Conversely, if X is commutable with every element in G,

$$\sum_{\alpha=1}^{g_c} \sum_{S_j\in\mathcal{C}_\alpha} F(S_j)S_j = X = TXT^{-1} = \frac{1}{g}\sum_{T\in G} TXT^{-1}, \tag{3.18}$$

from Eq. (3.17) X has to be a linear combination of the class operators

$$\begin{aligned} &X = \frac{1}{g}\sum_{\alpha=1}^{g_c} \sum_{S_j\in\mathcal{C}_\alpha} F(S_j) \sum_{T\in G} TS_jT^{-1} = \sum_{\alpha=1}^{g_c} \mathsf{C}_\alpha F_\alpha, \\ &F_\alpha = n(\alpha)^{-1} \sum_{S_j\in\mathcal{C}_\alpha} F(S_j). \end{aligned} \tag{3.19}$$

In comparison with Eq. (3.18), $F(S_j)$ depends on the class $\mathcal{C}_\alpha$ but not on the element S_j, $F(S_j) = F_\alpha$. Since the product $\mathsf{C}_\alpha\mathsf{C}_\beta$ is commutable with every element T in G, one has

$$\mathsf{C}_\alpha\mathsf{C}_\beta = \sum_{\gamma=1}^{g_c} f(\alpha,\beta,\gamma)\mathsf{C}_\gamma, \qquad f(\alpha,\beta,\gamma) = f(\beta,\alpha,\gamma), \tag{3.20}$$

where $f(\alpha,\beta,\gamma)$ is a non-negative integer and can be calculated directly from the group table (see Prob. 15 of Chap. 2 in [Ma and Gu (2004)]).

The class operators are linearly independent and span a linear space $\mathcal{L}_c$ of dimension g_c, called the class space. Equation (3.20) shows that the class space $\mathcal{L}_c$ is closed for the multiplication of its vectors so that $\mathcal{L}_c$ is an algebra, called the class algebra.

3.2 Transformation Operators for a Scalar Function

Denote simply by x all the coordinates of degrees of freedom in a quantum system, and by $\psi(x)$ the scalar wave function. R is a linear transformation of the system, which may be either a space–time transformation such as a translation, a rotation, or an inversion etc., or an internal transformation, such as a rotation in the space of the isotopic spin, and so on. Under the transformation R, x is changed to $x' = Rx$ and the wave function $\psi(x)$ is changed to $\psi'(x')$. In order to show explicitly the dependence of the wave function on the transformation R, we introduce an operator P_R, $\psi'(x') \equiv P_R\psi(x')$. Being a scalar wave function, the value of the transformed wave function $P_R\psi$ at the point Rx should be equal to the value of the original wave function ψ at the point x, namely

$$\begin{aligned} & x \xrightarrow{R} x' = Rx \ , \qquad x = R^{-1}x', \\ & \psi(x) \xrightarrow{R} \psi'(x') = P_R\psi(Rx) = \psi(x). \end{aligned} \tag{3.21}$$

Replacing the argument (Rx) in $P_R\psi(Rx)$ with x, one has

$$P_R\psi(x) = \psi(R^{-1}x). \tag{3.22}$$

$\psi(x)$ and $P_R\psi(x)$ are two different functions of x. Equation (3.22) shows the relation between the values of two functions at different points. At the same time, the relation gives the method to calculate the transformed function $P_R\psi(x)$ from the original function $\psi(x)$. Namely, first replace the argument x in $\psi(x)$ with $R^{-1}x$, and then, regard $\psi(R^{-1}x)$ as a function of x, which is nothing but the transformed function $P_R\psi(x)$.

Obviously, P_R is a linear operator

$$P_R\left\{a\psi(x) + b\phi(x)\right\} = a\psi(R^{-1}x) + b\phi(R^{-1}x) = aP_R\psi(x) + bP_R\phi(x). \tag{3.23}$$

Equation (3.22) shows a one-to-one correspondence between the operator P_R and the transformation R. This correspondence is invariant in the product of transformations,

$$\begin{aligned} & x' \xrightarrow{S} x'' = Sx' = (SR)x, \\ & \psi'(x') \xrightarrow{S} \psi''(x'') = P_SP_R\psi(x'') = P_S\psi'(x'') = \psi'(x') = \psi(x), \\ & \psi(x) \xrightarrow{SR} P_{SR}\psi(x'') = \psi\left[(SR)^{-1}x''\right] = \psi(x). \end{aligned}$$

Namely, $P_SP_R = P_{SR}$ corresponds to SR in the same rule. If the transformations R form a group G, then P_R form a group isomorphic onto G. It is worthy to emphasize that

$$P_S\left[P_R\psi(x)\right] \neq P_S\psi(R^{-1}x) = \psi(S^{-1}R^{-1}x). \tag{3.24}$$

P_S has to act on the function $\psi' = P_R\psi$, not on the function ψ.

As an example, we discuss the translation of a system in one dimension,

$$x \xrightarrow{T(a)} x' = T(a)x = x + a, \qquad x = T(a)^{-1}x' = x' - a,$$

$$\begin{aligned} P_{T(a)}\psi(x) &= \psi[T(a)^{-1}x] = \psi(x-a) = \sum_{n=0}^{\infty} \frac{(-a)^n}{n!}\frac{d^n}{dx^n}\psi(x) \\ &= \exp\left\{-a\left(\frac{d}{dx}\right)\right\}\psi(x) = \exp\left\{-iap_x/\hbar\right\}\psi(x). \end{aligned} \tag{3.25}$$

Namely, the translation operator $P_{T(a)}$ for a scalar function is expressed as an exponential function of the momentum operator, $p_x = -i\hbar d/dx$. See another example in Prob. 2 of Chap. 3 of [Ma and Gu (2004)].

Let a linear operator $L(x)$ change a state A to another state B,

$$\psi_B(x) = L(x)\psi_A(x),$$

where $\psi_A(x)$ and $\psi_B(x)$ are the wave functions of two states, respectively. After the transformation R, the wave functions are changed,

$$\begin{aligned} \psi_A(x) &\xrightarrow{R} \psi'_A(x') = P_R\psi_A(x'), \\ \psi_B(x) &\xrightarrow{R} \psi'_B(x') = P_R\psi_B(x'). \end{aligned}$$

The operator $L(x)$ is changed to $L'(x')$ such that the operator still describes a transformation changing the state A to the state B:

$$L(x) \xrightarrow{R} L'(x'), \qquad \psi'_B(x') = L'(x')\psi'_A(x').$$

Replacing the argument x' with x, and noting that ψ_A is an arbitrary function, one has

$$L'(x)\left\{P_R\psi_A(x)\right\} = P_R\psi_B(x) = P_R\left\{L(x)\psi_A(x)\right\}.$$

$$L'(x) = P_R L(x) P_R^{-1}. \tag{3.26}$$

This is the transformation formula of a linear operator $L(x)$ in the transformation R. This formula can be understood as follows:

$$P_R\left[L(x)\psi(x)\right] = \left[P_R L(x) P_R^{-1}\right]\ \left[P_R\psi(x)\right] = L'(x)\psi'(x). \tag{3.27}$$

R is a symmetric transformation of a system if the Hamiltonian $H(x)$ of the system is invariant in R. According to Eq. (3.26), one has

$$H(x) \xrightarrow{R} P_R H(x) P_R^{-1} = H(x), \qquad [H(x),\ P_R] = 0. \tag{3.28}$$

Namely, the symmetric transformation operator P_R is commutable with the Hamiltonian $H(x)$. If the energy level E is degenerate with multiplicity m, there is m linearly independent eigenfunctions $\psi_\mu(x)$,

$$H(x)\psi_\mu(x) = E\psi_\mu(x), \qquad \mu = 1,\ 2,\ \cdots,\ m. \tag{3.29}$$

The basis functions $\psi_\mu(x)$ span an m-dimensional space $\mathcal{L}$ such that any eigenfunction of $H(x)$ with the eigenvalue E belongs to $\mathcal{L}$. From Eq. (3.28) one has

$$H(x)\left[P_R\psi_\mu(x)\right] = P_R H(x)\psi_\mu(x) = E\left[P_R\psi_\mu(x)\right]. \tag{3.30}$$

Hence, the space $\mathcal{L}$ is invariant with respect to P_R. Denote by $D(R)$ the matrix of P_R in the basis functions $\psi_\mu(x)$,

$$P_R\psi_\mu(x) = \sum_{\nu=1}^{m} \psi_\nu(x) D_{\nu\mu}(R). \tag{3.31}$$

Equation (3.31) gives a one-to-one or a many-to-one correspondence between P_R and $D(R)$, which is invariant with respect to the product of the symmetric transformations. Thus, the symmetric group G of the system is isomorphic or homomorphic onto the matrix group $D(G)$ composed of $D(R)$,

$$D(G) \sim P_G \approx G. \tag{3.32}$$

$D(G)$ is an m-dimensional representation of G, and describes the property of the static wave functions $\psi_\mu(x)$ in the symmetric transformation R. In order to study the symmetric property of a system, one has to find all representations of the symmetric group G. It is one of the main tasks of group theory to find the representations of the symmetric groups used in physics. However, the task can be simplified.

3.3 Equivalent Representations

A representation of a group G is a matrix group. The linear space on whose vectors the representation matrix $D(R)$ acts is called the representation space. The group algebra of a finite group is the representation space of the regular representation. The eigenfunctions $\psi_\mu(x)$ of $H(x)$ with the energy E span a linear space $\mathcal{L}$, which is invariant with respect to the symmetric transformation operators P_R. The matrices $D(R)$ of P_R form a representation $D(G)$ of the symmetric group G, and $\mathcal{L}$ is its representation space.

When the basis vectors in the representation space make a linear combination,

$$\phi_\mu = \sum_\nu \psi_\nu X_{\nu\mu}, \tag{3.33}$$

the matrix of P_R makes a similarity transformation

$$P_R\psi_\mu = \sum_\rho \psi_\rho D_{\rho\mu}(R), \qquad P_R\phi_\nu = \sum_\lambda \phi_\lambda \overline{D}_{\lambda\nu}(R),$$

$$\overline{D}(R) = X^{-1}D(R)X. \tag{3.34}$$

$\overline{D}(R)$ forms another representation of G. Both $D(R)$ and $\overline{D}(R)$ are the matrices of the same operator P_R in the same linear space $\mathcal{L}$, but with different basis vectors. Two representations are said to be equivalent.

Definition 3.2 A representation $\overline{D}(G)$ of a group G is called equivalent to another representation $D(G)$ of G if two representation matrices for each element R in G satisfy the same similarity transformation (3.34).

Obviously, two equivalent representations have the same dimension. Each element R in G has the same character in equivalent representations

$$\chi(R) = \overline{\chi}(R), \qquad \forall R \in G. \tag{3.35}$$

Conversely, it will be shown later that two representations of a finite group G are equivalent if Eq. (3.35) holds. From any representation $D(G)$ of G one can obtain infinite number of equivalent representations $\overline{D}(G)$ by different similarity transformations. There is no essential difference between equivalent representations. One only needs to study one representation in all equivalent ones. A representation of G is said to be real if it is equivalent to a representation composed of real representation matrices.

The representation $\overline{D}(G)$ given in Eq. (3.12) and the regular representation $D(G)$ given in Eq. (3.6) have the same character, so they are equivalent. The similarity transformation matrix X is

$$\overline{D}(R)X = XD(R), \qquad X_{RS} = \begin{cases} 1, & RS = T, \\ 0, & RS \neq T, \end{cases} \tag{3.36}$$

where T is an arbitrarily given element in G (see Prob. 7 of Chap. 3 in [Ma and Gu (2004)]).

Theorem 3.1 Any representation of a finite group is equivalent to a unitary representation, and two equivalent unitary representations can be related through a unitary similarity transformation.

Proof The first part of the Theorem means that for any representation $D(G)$ of a finite group G there is a matrix X satisfying

$$\overline{D}(R) = X^{-1}D(R)X, \qquad \overline{D}(R)^{\dagger}\overline{D}(R) = \mathbf{1}, \tag{3.37}$$

namely,

$$D(R)^{\dagger}\left(XX^{\dagger}\right)^{-1}D(R) = \left(XX^{\dagger}\right)^{-1}. \tag{3.38}$$

From Theorem 2.1 the following matrix H satisfies Eq. (3.38):

$$\begin{aligned} H &\equiv \sum_{S\in G} D(S)^{\dagger}D(S) = \sum_{S\in G} D(SR)^{\dagger}D(SR) \\ &= D(R)^{\dagger}\left(\sum_{S\in G} D(S)^{\dagger}D(S)\right)D(R) = D(R)^{\dagger}HD(R). \end{aligned} \tag{3.39}$$

The problem is how to solve X from

$$\left(XX^{\dagger}\right)^{-1} = H. \tag{3.40}$$

In fact, H is a Hermitian matrix with positive eigenvalues. H can be diagonalized by a unitary similarity transformation,

$$\Gamma = U^{-1}HU, \qquad U^{\dagger}U = \mathbf{1}, \qquad \Gamma_{\mu\mu} > 0.$$

Hence, $X = U\Gamma'U^{-1}$ with $\Gamma'_{\mu\nu} = \delta_{\mu\nu}(\Gamma_{\mu\mu})^{-1/2}$ satisfies Eq. (3.40).

Now, turn to the second part of the theorem. If two unitary representations $D(G)$ and $\overline{D}(G)$ are related by a non-unitary similarity transformation X, $X\overline{D}(R) = D(R)X$, one wants to find a matrix Y which is commutable with $\overline{D}(R)$ and (XY) is unitary. Define a Hermitian matrix $H_1 = X^{\dagger}X = \left(YY^{\dagger}\right)^{-1}$ which is commutable with $\overline{D}(R)$:

$$\overline{D}(R)^{-1}H_1\overline{D}(R) = \left[X\overline{D}(R)\right]^{\dagger}\left[X\overline{D}(R)\right] = \left[D(R)X\right]^{\dagger}\left[D(R)X\right] = H_1.$$

Y can be solved from H_1 just like X is solved from H. Diagonaling H_1, $\Gamma_1 = V^{-1}H_1V$, where the diagonal entries $(\Gamma_1)_{\mu\mu}$ are positive, one has $Y = V\Gamma_1'V^{-1}$ with $(\Gamma_1')_{\mu\nu} = \delta_{\mu\nu}(\Gamma_1)_{\mu\mu}^{-1/2}$. Thus, $V^{-1}\overline{D}(R)V$ is commutable with both Γ_1 and Γ_1'. □

The key in the proof of Theorem 3.1 is that the average $\overline{F}(G)$ of a group function $F(G)$ of a finite group G is invariant in multiplying with any group element R (see Eq. (3.39)),

$$\overline{F}(G) = \frac{1}{g}\sum_{S\in G} F(S) = \frac{1}{g}\sum_{S\in G} F(RS) = \frac{1}{g}\sum_{S\in G} F(SR). \tag{3.41}$$

Hereafter, due to Theorem 3.1, we will only discuss the unitary representations and the unitary similarity transformations of a finite group G. If the representation $D(G)$ is real, only real matrices appear in the proof of Theorem 3.1. Thus, one has the corollary.

Corollary 3.1.1 Any real representation of a finite group can be changed to a real orthogonal representation through a real similarity transformation, and the similarity transformation between two equivalent real orthogonal representations can be chosen to be real orthogonal.

3.4 Inequivalent and Irreducible Representations

3.4.1 *Irreducible Representations*

Definition 3.3 A representation $D(G)$ of a group G is said to be reducible if the representation matrix $D(R)$ of each element R of G can be changed to the same form of the echelon matrix through a common similarity transformation X,

$$X^{-1}D(R)X = \begin{pmatrix} D^{(1)}(R) & T(R) \\ 0 & D^{(2)}(R) \end{pmatrix}. \tag{3.42}$$

Otherwise, it is called an irreducible representation.

It is easy to show that both $D^{(1)}(R)$ and $D^{(2)}(R)$ form the representations of G. The character of R in the reducible representation $D(G)$ is the sum of the characters in two sub-representations

$$\chi(R) = \chi^{(1)}(R) + \chi^{(2)}(R). \tag{3.43}$$

From Definition 3.3, the representation space of $D^{(1)}(G)$ is a nontrivial invariant subspace in the representation space of a reducible representation $D(G)$. Conversely, if there is a nontrivial invariant subspace in the representation space of $D(G)$, one may choose new basis vectors respectively belonging to the subspace and its complementary subspace. In the new basis vectors the representation matrices take the form (3.42) of echelon matrices, namely, a representation of G is irreducible if and only if in the representation space there is no nontrivial invariant subspace with respect to G.

If both complementary subspaces are invariant with respect to $D(G)$, there exists a common similarity transformation X such that the representation matrix $D(R)$ of each element R of G is a direct sum of two submatrices $D^{(1)}(R)$ and $D^{(2)}(R)$,

$$X^{-1}D(R)X = \begin{pmatrix} D^{(1)}(R) & 0 \\ 0 & D^{(2)}(R) \end{pmatrix} = D^{(1)}(R) \oplus D^{(2)}(R). \tag{3.44}$$

This representation is called completely reducible. The representation $D(G)$, which is written in the direct sum of two representations $D^{(1)}(G)$ and $D^{(2)}(G)$, is called the reduced representation.

If the representation space $\mathcal{L}$ of a unitary representation $D(G)$ of a finite group G contains an m_1-dimensional subspace $\mathcal{L}_1$ invariant with respect to G, one makes a unitary similarity transformation X such that the first m_1 new basis vectors belong to $\mathcal{L}_1$ and the remaining basis vectors do not. Thus, $X^{-1}D(R)X$ is in the form (3.42). Since $X^{-1}D(R)X$ is unitary, $D^{(1)}(R)$ is unitary and $T(R) = 0$, namely, any reducible representation of a finite group is completely reducible.

A typical example for existence of the incompletely reducible representation is the following representation of $\mathcal{T}$, where $\mathcal{T}$ is the translation group in one-dimensional space and is an infinite Abelian group,

$$x \xrightarrow{T(a)} x' = x + a, \qquad T(a)T(b) = T(a+b),$$

$$D(a) = \begin{pmatrix} 1 & a \\ 0 & 1 \end{pmatrix}, \qquad D(a)D(b) = D(a+b).$$

3.4.2 *Schur Theorem*

Theorem 3.2 (The second Schur Theorem) For two inequivalent and irreducible representations $D^{(1)}(G)$ and $D^{(2)}(G)$ of a group G,

$$X = 0, \qquad \text{if } D^{(1)}(R)X = XD^{(2)}(R), \qquad \forall\, R \in G. \tag{3.45}$$

Proof The key in the proof is that there is no nontrivial invariant subspace in the representation space of an irreducible representation. Let the dimensions of $D^{(1)}(G)$ and $D^{(2)}(G)$ be m_1 and m_2, respectively. X must be an $m_1 \times m_2$ matrix. We will show in three cases that the space spanned by the column (or row) matrices of X is the null space.

(a) $m_1 > m_2$. Due to Eq. (3.45), the column matrices in X, $(\underline{X_{\cdot\mu}})_\lambda = X_{\lambda\mu}$, span an m_2-dimensional subspace invariant in the action of $D^{(1)}(R)$,

$$D^{(1)}(R)\underline{X_{\cdot\mu}} = \sum_\rho \underline{X_{\cdot\rho}} D^{(2)}(R)_{\rho\mu}. \tag{3.46}$$

Thus, the subspace is a null space because $D^{(1)}(G)$ is an irreducible representation.

(b) $m_1 = m_2$. If $\det X \neq 0$, the inverse matrix X^{-1} exists such that from Eq. (3.45), $D^{(1)}(G)$ and $D^{(2)}(G)$ are equivalent which is in contradiction to the hypothesis. If $\det X = 0$, the column matrices in X span an invariant subspace in the action of $D^{(1)}(R)$ with the dimension less than m_1. Thus, the subspace is a null space.

(c) $m_1 < m_2$. Transposing Eq. (3.45), one has $D^{(2)}(R)^T X^T = X^T D^{(1)}(R)^T$. If $X^T \neq 0$, similar to Eq. (3.46), it means that in the acting space of $D^{(2)}(R)^T$ there is an m_1-dimensional invariant subspace. Namely, there is a similarity transformation Y such that

$$Y^{-1} D^{(2)}(R)^T Y = \begin{pmatrix} D_1(R)^T & M(R) \\ 0 & D_2(R)^T \end{pmatrix}.$$

Thus, by transposing, the representation space of $D^{(2)}(R)$ contains an invariant subspace with the dimension less than m_2, which is in contradiction to the fact that $D^{(2)}(G)$ is an irreducible representation. □

Corollary 3.2.1 (The first Schur Theorem) For an irreducible representation $D(G)$ of a group G,

$$X = \lambda \mathbf{1}, \qquad \text{if } XD(R) = D(R)X, \qquad \forall R \in G. \tag{3.47}$$

Proof Letting λ be an eigenvalue of X, one defines $Y = X - \lambda\mathbf{1}$, and then, $\det Y = 0$ and $YD(R) = D(R)Y$. Thus, through the similar proof to the case (b), $Y = 0$ and $X = \lambda\mathbf{1}$. □

The Schur theorems hold for any group. Since any reducible representation of a finite group is completely reducible, there exists a nonconstant matrix commutable with its every representation matrix.

Corollary 3.2.2 A representation of a finite group is reducible if and only if there exists a nonconstant matrix commutable with its every representation matrix.

3.4.3 *Orthogonal Relation*

For a finite group G of order g, any matrix entry $D_{\mu\nu}(R)$ of a representation $D(G)$ is a group function and describes a vector in the group space with respect to the natural basis vectors. The inner product of two group functions is defined through the inner product (3.3) of two vectors in the group space,

$$\sum_{R\in G} F_1(R)^* F_2(R). \tag{3.48}$$

Two group functions are orthogonal if their inner product is vanishing. The inner self-product of one group function is the square module of the function.

Theorem 3.3 The representation matrix entries $D^i_{\mu\rho}(R)$ and $D^j_{\nu\lambda}(R)$ in two inequivalent and irreducible unitary representations of a finite group G satisfy the orthogonal relation of the group functions

$$\sum_{R\in G} D^i_{\mu\rho}(R)^* D^j_{\nu\lambda}(R) = \frac{g}{m_j}\delta_{ij}\delta_{\mu\nu}\delta_{\rho\lambda}, \tag{3.49}$$

where g is the order of G, m_j denotes the dimension of $D^j(G)$, and $D^i(R) = D^j(R)$ when $i = j$.

Proof Define two $m_i \times m_j$ matrices $Y(\mu\nu)$ and $X(\mu\nu)$, where $Y(\mu\nu)$ contains only one nonvanishing matrix entry, $Y(\mu\nu)_{\rho\lambda} = \delta_{\mu\rho}\delta_{\nu\lambda}$, and $X(\mu\nu) = \sum_R D^i(R^{-1})Y(\mu\nu)D^j(R)$:

$$X(\mu\nu)_{\rho\lambda} = \sum_{R\in G}\sum_{\tau\sigma} D^i_{\rho\tau}(R^{-1})Y(\mu\nu)_{\tau\sigma}D^j_{\sigma\lambda}(R) = \sum_{R\in G} D^i_{\mu\rho}(R)^* D^j_{\nu\lambda}(R). \tag{3.50}$$

$X(\mu\nu)_{\rho\lambda}$ is nothing but the left-hand side of Eq. (3.49). Due to Theorem 2.1, $X(\mu\nu)$ satisfies

$$X(\mu\nu)D^j(S) = \sum_{R\in G} D^i(S)D^i\left(RS\right)^{-1} Y(\mu\nu)D^j(RS) = D^i(S)X(\mu\nu).$$

From the Schur theorem, $X(\mu\nu) = 0$ when $i \neq j$, and $X(\mu\nu)_{\rho\lambda} = C(\mu\nu)\delta_{\rho\lambda}$ when $i = j$, where $C(\mu\nu)$ is a constant depending on $Y(\mu\nu)$. Substituting the expression $X(\mu\nu)_{\rho\lambda}$ into Eq. (3.50) with $i = j$ and $\rho = \lambda$ and summing over λ, one has

$$m_jC(\mu\nu) = \sum_{R\in G}\sum_{\lambda} D^j_{\lambda\mu}(R^{-1})D^j_{\nu\lambda}(R) = \sum_{R\in G} D^j_{\nu\mu}(RR^{-1}) = g\delta_{\mu\nu}.$$

Thus, Eq. (3.49) follows. □

From Theorem 3.3, all matrix entries $D^i_{\mu\rho}(R)$ and $D^j_{\nu\lambda}(R)$ with $i \neq j$ are orthogonal to each other. Therefore, the orthogonality holds if two representations are not unitary because any representation of a finite group can be changed to a unitary one by a similarity transformation. From Eq. (3.49) with $i = j$, a unitary irreducible representation $D^j(G)$ provides m_j^2 group functions $D^j_{\nu\lambda}(R)$ of G, which are orthogonal to each other and are normalized to a common number g/m_j. Since there are only g linearly independent group functions in the group space of a finite group G of order g, the square sum of the dimensions m_j of all inequivalent and irreducible representations of a finite group G is not larger than its order g.

Taking $\mu = \rho$ and $\nu = \lambda$ in Eq. (3.49) and summing over μ and ν, one has

$$\sum_{R \in G} \chi^i(R)^* \chi^j(R) = g\delta_{ij}. \tag{3.51}$$

Corollary 3.3.1 Being group functions, the characters of the inequivalent and irreducible representations of a finite group G of order g are orthogonal to each other and normalized to g.

Let a class $\mathcal{C}_\alpha$ contain $n(\alpha)$ elements. The characters $\chi^j(R)$ of the elements in $\mathcal{C}_\alpha$ are equal to each other and can be denoted by χ^j_α. Multiplied with a factor $[n(\alpha)/g]^{1/2}$, χ^j_α are orthonormal functions in the class space

$$\sum_{\alpha=1}^{g_c} [n(\alpha)/g]^{1/2} \chi^{i*}_\alpha [n(\alpha)/g]^{1/2} \chi^j_\alpha = \frac{1}{g} \sum_{\alpha=1}^{g_c} n(\alpha) \chi^{i*}_\alpha \chi^j_\alpha = \delta_{ij}. \tag{3.52}$$

Thus, the number of the inequivalent and irreducible representations of a finite group G is not larger than the number g_c of classes contained in G.

A reducible representation $D(G)$ of a finite group G can be reduced to a direct sum of some irreducible representations through a similarity transformation X

$$X^{-1}D(R)X = \bigoplus_j a_j D^j(R), \qquad \chi(R) = \sum_j a_j \chi^j(R), \tag{3.53}$$

where $\chi(R)$ is the character of R in $D(G)$. a_j is called the multiplicity of an irreducible representation $D^j(G)$ in the representation $D(G)$. From Eq. (3.51), one has

$$a_j = \frac{1}{g} \sum_{R \in G} \chi^j(R)^* \chi(R). \tag{3.54}$$

If all inequivalent and irreducible representations of a finite group G are known, the multiplicity a_j can be calculated from the character $\chi(R)$, and the similarity transformation matrix X can be calculated from Eq. (3.53). It is the main method for reducing a representation of a finite group. Two representations are equivalent if the multiplicities a_j of each irreducible representation $D^j(G)$ in them are equal to each other, respectively. From Eqs. (3.51) and (3.53), one has

$$\sum_{R\in G} |\chi(R)|^2 = g \sum_j a_j^2 \geq g. \tag{3.55}$$

The sum will be equal to g if the representation is irreducible.

Corollary 3.3.2 Two representations of a finite group are equivalent if and only if the characters of every element in them are equal to each other, respectively, i.e., $\chi(R) = \overline{\chi}(R)$.

Corollary 3.3.3 A representation of a finite group of order g is irreducible if and only if the character $\chi(R)$ in the representation satisfies

$$\sum_{R\in G} |\chi(R)|^2 = g. \tag{3.56}$$

3.4.4 *Completeness of Representations*

From the character $\chi(R)$, given in Eq. (3.6), of the regular representation of a finite group G, the multiplicity a_j of each irreducible representation $D^j(G)$ in the regular representation can be calculated to be its dimension m_j,

$$a_j = \frac{1}{g} \sum_{R\in G} \chi^j(R)^* \chi(R) = \chi^j(E)^* = m_j,$$

namely,

$$X^{-1}D(R)X = \bigoplus_j m_j D^j(R), \qquad \chi(R) = \sum_j m_j \chi^j(R). \tag{3.57}$$

Taking the character of the identical element E, one has

$$g = \chi(E) = \sum_j m_j^2. \tag{3.58}$$

Theorem 3.4 The square sum of the dimensions of all inequivalent and irreducible representations of a finite group G is equal to the order g of G.

Corollary 3.4.1 The representation matrix entries $D^j_{\nu\lambda}(R)$ of all inequivalent and irreducible representations of a finite group G construct a complete set of orthogonal basis vectors in the group space, and any group function $F(R)$ can be expanded with respect to the basis vectors

$$\begin{aligned} F(R) &= \sum_{j\mu\nu} C^j_{\mu\nu} D^j_{\mu\nu}(R), \\ C^j_{\mu\nu} &= \frac{m_j}{g} \sum_{R\in G} D^j_{\mu\nu}(R)^* F(R). \end{aligned} \tag{3.59}$$

The class function is a special group function, where the functional values for the elements in the same class are equal to each other. Expanding a class function with respect to $D^j_{\mu\nu}(R)$, one has

$$\begin{aligned} F(R) = F(SRS^{-1}) &= \frac{1}{g}\sum_{S\in G} F(SRS^{-1}) \\ &= \sum_{j\mu\nu} \frac{C^j_{\mu\nu}}{g} \sum_{S\in G} \sum_{\rho\lambda} D^j_{\mu\rho}(S) D^j_{\rho\lambda}(R) D^j_{\nu\lambda}(S)^* \\ &= \sum_j \left(\frac{1}{m_j} \sum_\mu C^j_{\mu\mu} \right) \chi^j(R). \end{aligned}$$

Thus, the characters $\chi^j(R)$ construct a complete set of basis vectors in the class space.

Corollary 3.4.2 The characters $\chi^j_{\nu\lambda}(R)$ of all inequivalent and irreducible representations of a finite group G construct a complete set of basis vectors in the class space, and any class function $F(R)$ can be expanded with respect to the basis vectors

$$\begin{aligned} F(R) = F(SRS^{-1}) &= \sum_j C_j \chi^j(R), \\ C_j &= \frac{1}{g} \sum_{R\in G} \chi^j(R)^* F(R). \end{aligned} \tag{3.60}$$

Corollary 3.4.3 The number of the inequivalent and irreducible representations of a finite group G is equal to its class number g_c,

$$\sum_j 1 = g_c. \tag{3.61}$$

Define a $g \times g$ matrix U and a $g_c \times g_c$ matrix V:

$$U_{R,j\mu\nu} = \left(\frac{m_j}{g}\right)^{1/2} D^j_{\mu\nu}(R), \qquad V_{\alpha,j} = \left(\frac{n(\alpha)}{g}\right)^{1/2} \chi^j_\alpha. \tag{3.62}$$

The orthogonal relations (3.49) and (3.52) show that U and V are both unitary. Then, one derives the complete relations of the representation matrix entries $D^j_{\mu\nu}(R)$ and the characters χ^j_α

$$\frac{1}{g}\sum_{j\mu\nu} m_j D^j_{\mu\nu}(R) D^j_{\mu\nu}(S)^* = \delta_{RS}, \tag{3.63}$$

$$\frac{n(\alpha)}{g}\sum_j \chi^j_\alpha \chi^{j*}_\beta = \delta_{\alpha\beta}. \tag{3.64}$$

3.4.5 *Character Tables of Finite Groups*

One of the main tasks of group theory is to study the symmetric group of the physical systems, including to find all inequivalent and irreducible representations and their characters, and to reduce the reducible representation through a similarity transformation.

The characters of all inequivalent and irreducible representations of a finite group G can be listed in a table, called the character table. A group has some faithful representations as well as some unfaithful ones. An unfaithful representation of G is a faithful representation of its quotient group. An important step for listing the character table of G is to find all invariant subgroups H of a finite group G and their quotient group G/H.

The characters of irreducible representations of a finite group G satisfy four conditions. Conditions (3.58) and (3.61) restrict the number and dimensions of inequivalent and irreducible representations of G. Conditions (3.52) and (3.64) show orthogonality, normalization, and completeness of characters in the character table. However, the four conditions are only the necessary ones for the characters, not sufficient ones.

$D(G)^*$, which is composed of the complex conjugate matrices $D(R)^*$, is called the conjugate representation of a representation $D(G)$. The conjugate representation of an irreducible representation is irreducible. The representation $D(G)$ is called self-conjugate if $D(G)^*$ is equivalent to $D(G)$. The characters $\chi(R)$ of a self-conjugate representation is real. The direct product of two representations of G is a representation. Especially, the direct product of an irreducible representation and a one-dimensional one is an irreducible representation of G. Those properties are helpful in listing the character table of a finite group.

The character table of a finite group with a low order is easy to find. But it needs special techniques to find the character tables of the groups with high orders. In principle, it is the task of mathematicians to find the character tables of the symmetric groups of physical systems. However, knowing some basic techniques in listing the character table is useful to a physicist to understand the results of mathematicians.

A cyclic group C_N of order N is

$$C_N = \{E,\ R,\ R^2,\ \cdots,\ R^{N-1}\}, \qquad R^N = E. \tag{3.65}$$

C_N is an Abelian group. The class number g_c is equal to its order g so that each irreducible representation is one-dimensional and satisfies $D^j(R)^N = D^j(E) = 1$:

$$D^j(R) = \exp\{-i2\pi j/N\}, \qquad 0 \le j \le (N-1). \tag{3.66}$$

In many cases, a given finite group is homomorphic onto a cyclic group. The character tables of cyclic groups are helpful in finding the one-dimensional representations of some finite groups.

Table 3.1 The character table of C_2

C_2	E	R
A	1	1
B	1	-1

Table 3.2 The character table of C_3 ($\omega = \exp\{-i2\pi/3\}$)

C_3	E	R	R^2
A	1	1	1
E	1	ω	ω^*
E'	1	ω^*	ω

Table 3.3 The character table of C_4

C_4	E	R	R^2	R^3
A	1	1	1	1
B	1	-1	1	-1
E	1	$-i$	-1	i
E'	1	i	-1	$-i$

The representation D^B of C_2 is usually called the antisymmetric one. The group C_4 contains an invariant subgroup $C_2 = \{E, R^2\}$. The representation D^B of C_4 is the antisymmetric representation of the quotient group C_4/C_2, and $D^{E'} = D^B \times D^E$. The group C_6 can be expressed as a direct product of two subgroups, $C_6 = C_2 \otimes C_3$, where $C_2 = \{E, R^3\}$ and $C_3 = \{E, R^2, R^4\}$. An irreducible representation of C_6 is the direct product of the irreducible representations of subgroups. Denote by R_μ, S_j, and T_k the elements of C_6, C_2, and C_3, respectively, where $R_\mu = S_j T_k$. Thus,

$$\begin{aligned}
&D^A(R_\mu) = D^A(S_j)D^A(T_k), && D^B(R_\mu) = D^B(S_j)D^A(T_k), \\
&D^{E_1}(R_\mu) = D^B(S_j)D^E(T_k), && D^{E_1'}(R_\mu) = D^B(S_j)D^{E'}(T_k), \\
&D^{E_2}(R_\mu) = D^A(S_j)D^{E'}(T_k), && D^{E_2'}(R_\mu) = D^A(S_j)D^E(T_k).
\end{aligned} \tag{3.67}$$

Table 3.4 The character table of C_5 ($\eta = \exp\{-i2\pi/5\}$)

C_5	E	R	R^2	R^3	R^4
A	1	1	1	1	1
E_1	1	η	η^2	η^3	η^4
E_1'	1	η^4	η^3	η^2	η
E_2	1	η^2	η^4	η	η^3
E_2'	1	η^3	η	η^4	η^2

Table 3.5 The character table of C_6 ($\omega = \exp\{-i2\pi/3\}$)

C_6	E	R	R^2	R^3	R^4	R^5
A	1	1	1	1	1	1
B	1	-1	1	-1	1	-1
E_1	1	$-\omega^*$	ω	-1	ω^*	$-\omega$
E_1'	1	$-\omega$	ω^*	-1	ω	$-\omega^*$
E_2	1	ω	ω^*	1	ω	ω^*
E_2'	1	ω^*	ω	1	ω^*	ω

In the crystal theory, a class $\mathcal{C}_\alpha$ in the character table of point groups is usually denoted by one element in $\mathcal{C}_\alpha$ and the coefficient in front of the element denotes the number $n(\alpha)$ of elements contained in $\mathcal{C}_\alpha$. The N-fold proper rotation, $N > 2$, is denoted by C_N if its axis is along z-axis, and otherwise by C_N'. The 2-fold proper rotation is denoted by C_2 if its axis is along z-axis, by C_2' if its axis is along x-axis, and otherwise by C_2''.

Table 3.6 The character table of D_2

D_2	E	C_2	C_2'	C_2''
A_1	1	1	1	1
A_2	1	1	-1	-1
B_1	1	-1	-1	1
B_2	1	-1	1	-1

The D_2 group is isomorphic onto the inverse group V_4 of order 4. The identical element E and any other element construct a subgroup C_2. In addition to the identical representation, the other three representations respectively are the antisymmetric representations of three quotient groups D_2/C_2. Remind that the character tables of C_4 and D_2 are different, but the four necessary conditions for characters are the same for the two groups.

The simplest non-Abelian group D_3 was discussed in §2.2. D_3 contains six elements and three classes. The identical element E itself is a class, two 3-fold rotations D and F construct a class denoted by C_3, and three 2-fold rotations A, B, C construct a class denoted by C_2'. The set of E, D, and F forms an invariant subgroup C_3 with index 2, and the class C_2' is its coset. Since $g = 6$, $g_c = 3$, and $1^2 + 1^2 + 2^2 = 6$, the D_3 group has two representations of one dimension, D^A and D^B, and one irreducible representation D^E of two dimensions. D^A and D^B are the representations of the quotient group D_3/C_3. There are several methods to calculate D^E. From Eq. (3.64) one has $\chi^E(C_2') = 0$. In fact, if $\chi^E(C_2') \neq 0$, the representation $D^B \times D^E$ would be a two-dimensional irreducible representation, inequivalent to D^E, which is in contradiction to Eq. (3.58). From Eq. (3.52) one obtains $\chi^E(C_3) = -1$. The character table of D_3 is listed in Table 3.7. The two-dimensional representation D^E of D_3 can also be obtained with the method of coordinate transformation (see Eq. (2.12)).

Table 3.7 The character table of D_3

D_3	E	$2C_3$	$3C_2'$
A	1	1	1
B	1	1	-1
E	2	-1	0

The C_{2n+1} group has one self-reciprocal class and one real representation. The C_{2n} has two self-reciprocal classes and two real representations. The classes in the group D_2 and in the D_3 group are all self-reciprocal and all the irreducible representations are real. Generally speaking, the number of the self-reciprocal classes in a finite group G is equal to the number of the inequivalent and irreducible self-conjugate representations of G (see Prob. 9 of Chap. 3 in [Ma and Gu (2004)]).

3.4.6 *The Character Table of the Group* **T**

The proper symmetric group **T** of a tetrahedron contains 12 elements and 4 classes. Since $1^2+1^2+1^2+3^2 = 12$, **T** has three inequivalent representations of one dimension, D^A, D^E, and $D^{E'}$, and one irreducible representation D^T of three dimensions. **T** has a subgroup D_2 composed by the identical element and three 2-fold rotations. Its index is 3. The quotient group $\mathbf{T}/D_2$ is isomorphic onto C_3, from which one obtains three inequivalent one-dimensional representations. Since $D^E \times D^T$ has to be equivalent to D^T, $\chi^T(C_3') = \chi^T(C_3'^2) = 0$. Then, one can calculate $\chi^T(C_2) = -1$ by Eq. (3.52). The character table of **T** is listed in Table 3.8.

Table 3.8 The character table of T $(\omega = e^{-i2\pi/3})$

T	E	$3C_2$	$4C_3'$	$4C_3'^2$
A	1	1	1	1
E	1	1	ω	ω^2
E'	1	1	ω^2	ω
T	3	-1	0	0

The representation matrices $D^T(R)$ can be calculated by the method of coordinate transformation (see Eq. (2.11)). The 2-fold rotation T_z^2 changes the directions of x- and y-axes. The 3-fold rotation R_1 makes a cyclic transformation of the three axes x, y, and z. Thus, the representation matrices of those two elements in the representation D^T are

$$D^T(T_z^2) = \begin{pmatrix} -1 & 0 & 0 \\ 0 & -1 & 0 \\ 0 & 0 & 1 \end{pmatrix}, \qquad D^T(R_1) = \begin{pmatrix} 0 & 0 & 1 \\ 1 & 0 & 0 \\ 0 & 1 & 0 \end{pmatrix}. \tag{3.68}$$

3.4.7 *The Character Table of the Group* **O**

Table 3.9 The character table of O

O	E	$3C_4^2$	$8C_3'$	$6C_4$	$6C_2''$
A	1	1	1	1	1
B	1	1	1	-1	-1
E	2	2	-1	0	0
T_1	3	-1	0	1	-1
T_2	3	-1	0	-1	1

The proper symmetric group **O** of a cube contains 24 elements and 5 self-reciprocal classes. Since $1^2+1^2+2^2+3^2+3^2 = 24$, **O** has two inequivalent representations of one dimension, D^A and D^B, one irreducible representation D^E of two dimensions, and two inequivalent and irreducible representations of three dimensions, D^{T_1} and D^{T_2}. **T** is an invariant subgroup of **O** with index 2, and from its quotient group one obtains the representations

D^A and D^B. The invariant subgroup D_2 of $\mathbf{T}$ is also an invariant subgroup of $\mathbf{O}$, and its quotient group $\mathbf{O}/D_2$ is isomorphic onto D_3. From the quotient group one can calculate the representation D^E of $\mathbf{O}$ (see Prob. 12 of Chap. 3 in [Ma and Gu (2004)]). One three-dimensional representation D^{T_1} can be calculated by the method of coordinate transformation,

$$D^{T_1}(T_z) = \begin{pmatrix} 0 & -1 & 0 \\ 1 & 0 & 0 \\ 0 & 0 & 1 \end{pmatrix}, \qquad D^{T_1}(R_1) = \begin{pmatrix} 0 & 0 & 1 \\ 1 & 0 & 0 \\ 0 & 1 & 0 \end{pmatrix}. \tag{3.69}$$

The other is $D^{T_2} = D^B \times D^{T_1}$.

$$D^{T_2}(T_z) = \begin{pmatrix} 0 & 1 & 0 \\ -1 & 0 & 0 \\ 0 & 0 & -1 \end{pmatrix}, \qquad D^{T_2}(R_1) = \begin{pmatrix} 0 & 0 & 1 \\ 1 & 0 & 0 \\ 0 & 1 & 0 \end{pmatrix}. \tag{3.70}$$

3.4.8 *Self-conjugate Representation*

The characters $\chi(R)$ of a self-conjugate representation is real, but a self-conjugate representation may not be real. Namely, a self-conjugate representation is not always equivalent to a representation composed of real representation matrices. The following theorem gives a criterion of a real representation.

Theorem 3.5 The similarity transformation matrix X, which relates the irreducible unitary self-conjugate representation of a finite group and its conjugate one, must be symmetric or antisymmetric. X is symmetric if and only if the representation is real.

Proof Let $D(G)$ be an irreducible unitary self-conjugate representation of a finite group G, and $X^{-1}D(R)X = D(R)^*$. Due to the Schur theorem, X is determined up to a constant factor. Without loss of generality, X can be assumed to be unitary,

$$D(R) = X^T D(R)^* X^* = \left(X^T X^{-1}\right) D(R) \left(X X^*\right).$$

From the Schur Theorem 3.2.1, $X^T X^{-1} = \tau \mathbf{1}$, $X^T = \tau X$, and $X = \tau X^T = \tau^2 X$. Thus, $\tau = \pm 1$, and X must be symmetric or antisymmetric.

τ can be expressed by the characters such that τ is independent of a similarity transformation.

$$\sum_{R\in G} \chi(R^2) = \sum_{\mu\nu} \sum_{R\in G} \left\{D_{\mu\nu}(R)^*\right\}^* D_{\nu\mu}(R)$$

$$\begin{aligned}
&= \sum_{\mu\nu} \sum_{R\in G} \left\{X^{-1}D(R)X\right\}^*_{\mu\nu} D_{\nu\mu}(R) \\
&= \sum_{\mu\nu\rho\lambda} \tau \sum_{R\in G} X_{\mu\rho} D_{\rho\lambda}(R)^* \left(X^{-1}\right)_{\nu\lambda} D_{\nu\mu}(R) \\
&= \frac{g\tau}{m} \sum_{\mu\nu} X_{\mu\nu} \left(X^{-1}\right)_{\nu\mu} = g\tau.
\end{aligned}$$

If the representation is not self-conjugate, the first line of the above equation is equal to zero owing to Theorem 3.3.

If $D(R)$ is real, X is a unit matrix and $\tau = 1$. Conversely, if X is a symmetric unitary matrix, $X^\dagger = X^* = X^{-1}$, we are going to show that X can be decomposed to Y^2, where Y is a symmetric unitary matrix. In fact, if $X\boldsymbol{a} = \lambda\boldsymbol{a}$, then $X^{-1}\boldsymbol{a}^* = \lambda^{-1}\boldsymbol{a}^*$. Both $\boldsymbol{a}$ and $\boldsymbol{a}^*$ are the eigenvectors of X with the same eigenvalue. Namely, the eigenvectors of X can be chosen to be real, and X can be diagonalized through a real orthogonal similarity transformation M. Let $M^{-1}XM = \Gamma' = \Gamma^2$. Thus, $Y = M\Gamma M^{-1}$ is unitary, $X = Y^2$, and $Y^{-1} = M\Gamma^* M^{-1} = Y^*$. Since Y is a symmetric unitary matrix and $Y^{-2}D(R)Y^2 = D(R)^*$, $Y^{-1}D(R)Y = YD(R)^*Y^{-1} = \left(Y^{-1}D(R)Y\right)^*$. $D(R)$ is a real representation. □

Corollary 3.5.1 The character in an irreducible representation of a finite group satisfies

$$\frac{1}{g}\sum_{R\in G} \chi(R^2) = \begin{cases} 1, & \text{real representation,} \\ -1, & \text{self-conjugate, but not real representation,} \\ 0, & \text{not self-conjugate representation.} \end{cases} \tag{3.71}$$

3.5 Subduced and Induced Representations

Discuss a finite group G of order g containing g_c classes. A class $\mathcal{C}_\alpha$ in G consists of $n(\alpha)$ elements. $D^j(G)$ is an m_j-dimensional irreducible representation of G. The character of $S \in \mathcal{C}_\alpha$ in $D^j(G)$ is denoted by $\chi^j(S)$ or χ^j_α. $H = \{E = T_1, T_2, \ldots, T_h\}$ is a subgroup of G with index $n = g/h$. The cosets of H in G are denoted by R_rH, $2 \le r \le n$. Assume that R_r have been chosen and $R_1 = E$ such that any element in G can be expressed as R_rT_t uniquely. The class $\overline{\mathcal{C}}_\beta$ in the subgroup H contains $\overline{n}(\beta)$ elements. $\overline{D}^k(H)$ is an $\overline{m}_k$-dimensional irreducible representation of H, and the character of $T_t \in \overline{\mathcal{C}}_\beta$ in $\overline{D}^k(H)$ is denoted by $\overline{\chi}^k(T_t)$ or $\overline{\chi}^k_\beta$.

The set of representation matrices $D^j(T_t)$, $T_t \in H$, forms a representation $D^j(H)$ of H, which is called the subduced representation from an irreducible representation $D^j(G)$ of G with respect to the subgroup H. Generally, the subduced representation is reducible with respect to the subgroup H

$$\begin{aligned} X^{-1}D^j(T_t)X &= \bigoplus_k a_{jk}\overline{D}^k(T_t), \qquad m_j = \sum_k a_{jk}\overline{m}_k, \\ a_{jk} &= \frac{1}{h}\sum_{T_t\in H} \overline{\chi}^k(T_t)^*\chi^j(T_t) = \frac{1}{h}\sum_{\beta} \overline{n}(\beta)\left(\overline{\chi}^k_\beta\right)^*\chi^j_\beta. \end{aligned} \tag{3.72}$$

Denote by ψ_μ the $\overline{m}_k$ bases in the representation space of $\overline{D}^k(H)$

$$P_{T_t}\psi_\mu = \sum_\nu \psi_\nu \overline{D}^k_{\nu\mu}(T_t).$$

Define an extended space of dimension $n\overline{m}_k$ with the bases $\psi_{r\mu} = P_{R_r}\psi_\mu$, where $\psi_{1\mu} = \psi_\mu$. The extended space is invariant with respect to the group G, and corresponds to an $n\overline{m}_k$-dimensional representation $\Delta^k(G)$ of G in the following way. For any given element S in G and for each R_r, SR_r can be expressed as R_uT_t, where u and t are completely determined by S and r. Since

$$P_S\psi_{r\mu} = P_{SR_r}\psi_\mu = P_{R_u}P_{T_t}\psi_\mu = \sum_\nu \psi_{u\nu}\overline{D}^k_{\nu\mu}(T_t),$$

one obtains

$$\Delta^k_{u\nu,r\mu}(S) = \overline{D}^k_{\nu\mu}(T_t), \qquad \chi^k(S) = \sum_{r\mu} \Delta^k_{r\mu,r\mu}(S). \tag{3.73}$$

This representation $\Delta^k(G)$ is called the induced representation from the irreducible representation $\overline{D}^k(H)$ of the subgroup H with respect to G. In general, the induced representation is reducible with respect to G:

$$\begin{aligned} Y^{-1}\Delta^k(S)Y &= \bigoplus_j b_{jk}D^j(S), \qquad (g/h)\,\overline{m}_k = \sum_j b_{jk}m_j, \\ b_{jk} &= \frac{1}{g}\sum_{S\in G} \chi^j(S)^*\chi^k(S) = \frac{1}{g}\sum_\alpha n(\alpha)\left(\chi^j_\alpha\right)^*\chi^k_\alpha, \end{aligned} \tag{3.74}$$

where the character of $S \in \mathcal{C}_\alpha$ in the representation $\Delta^k(G)$ is denoted by $\chi^k(S) = \chi^k_\alpha$. In general, some elements in the class $\mathcal{C}_\alpha$ belong to the subgroup H, and some in $\mathcal{C}_\alpha$ do not. The elements in $\mathcal{C}_\alpha$ belonging to H constitute a few whole classes of the subgroup H, denoted by $\overline{\mathcal{C}}_\beta$. It is possible that no element in $\mathcal{C}_\alpha$ belongs to the subgroup H. For this case we say that no $\overline{\mathcal{C}}_\beta$ exists. From Eq. (3.73), the diagonal element of $\Delta^k(S)$ appears only when $r = u$, i.e., $SR_r = R_rT_t$. Thus, $\chi^k(S)$ is nonvanishing only when the class $\mathcal{C}_\alpha$ contains a few elements belonging to the subgroup H. Denoting by κ_β the number of different R_r satisfying $R_r^{-1}SR_r \in \overline{\mathcal{C}}_\beta$, one has $\chi^k_\alpha = \sum_\beta \kappa_\beta \overline{\chi}^k_\beta$.

From Prob. 14 of Chap. 2, the number of elements R in G satisfying $R^{-1}SR = T_t$ is $m(\alpha) = g/n(\alpha)$. Expressing R by R_rT_x and letting $T_t \in \overline{\mathcal{C}}_\beta$, one has $R_r^{-1}SR_r = T_xT_tT_x^{-1} \in \overline{\mathcal{C}}_\beta$. On the other hand, the number of T_y in the subgroup H satisfying $T_yT_tT_y^{-1} = T_t$ is $\overline{m}(\beta) = h/\overline{n}(\beta)$. If R_rT_x satisfies $S(R_rT_x) = (R_rT_x)T_t$, then $R_rT_xT_y$ satisfies this formula too. However, the latter does not make any new contribution to the characters $\chi^k(S)$ because $R_r^{-1}SR_r = T_xT_yT_tT_y^{-1}T_x^{-1} = T_xT_tT_x^{-1}$. Therefore,

$$\kappa_\beta = \frac{m(\alpha)}{\overline{m}(\beta)} = \frac{g\,\overline{n}(\beta)}{h\,n(\alpha)}, \qquad \chi^k_\alpha = \frac{g}{h\,n(\alpha)} \sum_\beta \overline{n}(\beta)\,\overline{\chi}^k_\beta. \tag{3.75}$$

It is easy to show from Eq. (3.75) that the multiplicities b_{jk} in Eq. (3.74) is equal to the multiplicities a_{jk} in Eq. (3.72):

$$b_{jk} = \frac{1}{g}\sum_\alpha n(\alpha)\left(\chi^j_\alpha\right)^* \chi^k_\alpha = \frac{1}{h}\sum_\beta \overline{n}(\beta)\left(\chi^j_\beta\right)^* \overline{\chi}^k_\beta = a_{jk}. \tag{3.76}$$

In fact, the class $\mathcal{C}_\alpha$ in G contains a few classes $\overline{\mathcal{C}}_\beta$ of H, $\chi^j_\alpha = \chi^j_\beta$, and the different classes $\mathcal{C}_\alpha$ correspond to the different classes $\overline{\mathcal{C}}_\beta$. An element in $\mathcal{C}_\alpha$, which does not belong to H, makes no contribution to χ^k_α. Therefore, the sum over the classes $\mathcal{C}_\alpha$ in Eq. (3.76) is equivalent to the sum over the classes $\overline{\mathcal{C}}_\beta$ in H. The formula (3.76) is called the Frobenius theorem.

In terms of the method of induced representations, the character tables and the irreducible representations of the groups D_{2n+1}, D_{2n} and **I** can be calculated (see Probs. 15, 16, and 18 of Chap. 3 in [Ma and Gu (2004)]). We list the results as follows.

Table 3.10 The character table of D_5 ($p = (\sqrt{5}-1)/2$)

D_5	E	$2C_5$	$2C_5^2$	$5C_2'$
A	1	1	1	1
B	1	1	1	-1
E_1	2	p	$-p^{-1}$	0
E_2	2	$-p^{-1}$	p	0

The group D_{2n+1} contains one $(2n+1)$-fold axis, called the principal axis, and $(2n+1)$ equivalent 2-fold axes, located in the plane perpendicular to the principal axis. Two generators of D_{2n+1} are the $(2n+1)$-fold rotation C_{2n+1} and one of the 2-fold rotations C_2', $C_{2n+1}C_2' = C_2'C_{2n+1}^{-1}$. The order of D_{2n+1} is $g = 4n+2$. The number of the classes in D_{2n+1} is $g_c = n+2$. The group D_{2n+1} has two inequivalent representations of one dimension, D^A and D^B, and n inequivalent irreducible representations of two dimensions, D^{E_j}, $1 \leq j \leq n$:

Table 3.11 The character table of D_7
$(\lambda = e^{-i2\pi/7},\ q_j = \lambda^j + \lambda^{-j},\ j = 1, 2, 3)$

D_7	E	$2C_7$	$2C_7^2$	$2C_7^3$	$7C_2'$
A	1	1	1	1	1
B	1	1	1	1	−1
E_1	2	q_1	q_2	q_3	0
E_2	2	q_2	q_3	q_1	0
E_3	2	q_3	q_1	q_2	0

$$\begin{aligned} & D^A(C_{2n+1}) = D^B(C_{2n+1}) = D^A(C_2') = 1, \qquad D^B(C_2') = -1, \\ & D^{E_j}(C_{2n+1}) = \begin{pmatrix} e^{-i\frac{2j\pi}{2n+1}} & 0 \\ 0 & e^{i\frac{2j\pi}{2n+1}} \end{pmatrix}, \qquad D^{E_j}(C_2') = \begin{pmatrix} 0 & 1 \\ 1 & 0 \end{pmatrix}. \end{aligned} \tag{3.77}$$

The character table of D_3 is listed in Table 3.7. The character tables of D_5 and D_7 are listed in Tables 3.10 and 3.11.

The group D_{2n} contains one $(2n)$-fold axis, called the principal axis, and $(2n)$ 2-fold axes, located in the plane perpendicular to the principal axis. The 2-fold axes are divided into two sets, each of which contains n equivalent 2-fold axes. They form two classes, respectively. Two generators of D_{2n} are the $(2n)$-fold rotation C_{2n} and one of the 2-fold rotations C_2', $C_{2n}C_2' = C_2'C_{2n}^{-1} = C_2''$. The angle between two 2-fold axes corresponding to C_2' and C_2'' is $\pi/(2n)$. The order of D_{2n} is $g = 4n$. The number of the classes in D_{2n} is $g_c = n+3$. The group D_{2n} has four inequivalent representations of one dimension, D^{A_1}, D^{A_2}, D^{B_1}, and D^{B_2}, and $(n-1)$ inequivalent irreducible representations of two dimensions, D^{E_j}, $1 \leq j \leq n-1$.

$$\begin{aligned} & D^{A_1}(C_{2n}) = D^{A_2}(C_{2n}) = D^{A_1}(C_2') = 1, \qquad D^{A_2}(C_2') = -1, \\ & D^{B_1}(C_{2n}) = D^{B_2}(C_{2n}) = D^{B_2}(C_2') = -1, \qquad D^{B_1}(C_2') = 1, \\ & D^{E_j}(C_{2n}) = \begin{pmatrix} e^{-ij\pi/n} & 0 \\ 0 & e^{ij\pi/n} \end{pmatrix}, \qquad D^{E_j}(C_2') = \begin{pmatrix} 0 & 1 \\ 1 & 0 \end{pmatrix}. \end{aligned} \tag{3.78}$$

The character tables of D_4 and D_6 are listed in Tables 3.12 and 3.13. For the group D_{4n+2}, the rotation C_{4n+2}^{2n+1} along the principal axis through

π angle does not belong to the subgroup D_{2n+1}, and is commutable with any element in D_{4n+2}. Thus, D_{4n+2} is a direct product of two subgroups:

$$D_{4n+2} = C_2 \otimes D_{2n+1}, \qquad C_2 = \{E, C_{4n+2}^{2n+1}\}.$$

Table 3.12 The character table of D_4

D_4	E	$2C_4$	C_4^2	$2C_2'$	$2C_2''$
A_1	1	1	1	1	1
A_2	1	1	1	−1	−1
B_1	1	−1	1	1	−1
B_2	1	−1	1	−1	1
E_1	2	0	−2	0	0

Table 3.13 The character table of D_6

D_6	E	$2C_6$	$2C_6^2$	C_6^3	$3C_2'$	$3C_2''$
A_1	1	1	1	1	1	1
A_2	1	1	1	1	−1	−1
B_1	1	−1	1	−1	1	−1
B_2	1	−1	1	−1	−1	1
E_1	2	1	−1	−2	0	0
E_2	2	−1	−1	2	0	0

The proper symmetric group **I** of a regular icosahedron contains six 5-fold axes with the generators T_j, $0 \leq j \leq 5$, ten 3-fold axes with the generators R_j, $1 \leq j \leq 10$, and fifteen 2-fold axes with the generators S_j, $1 \leq j \leq 15$. All axes are nonpolar and any two axes with the same fold are equivalent to each other. The order g of **I** is 60. The number g_c of the classes in **I** is 5. **I** does not contain any nontrivial invariant subgroup. The character table of **I** is given in Table 3.14. The representation matrices of the generators of **I** are calculated in Prob. 12 of Chap. 4 of [Ma and Gu (2004)]. (Also see [Deng and Yang (1992)].)

Table 3.14 The character table of I ($p = (\sqrt{5} - 1)/2$)

I	E	$12C_5$	$12C_5^2$	$20C_3'$	$15C_2'$
A	1	1	1	1	1
T_1	3	p^{-1}	$-p$	0	−1
T_2	3	$-p$	p^{-1}	0	−1
G	4	−1	−1	1	0
H	5	0	0	−1	1

3.6 Applications in Physics

At the beginning of Chapter 2, before the study of group theory, we raised a simple example to see how to obtain some precise information of the system through analyzing its symmetry. Now, we have studied the fundamental concepts on group theory and the theory of representations. It is time to discuss the typical applications of group theory to physics.

3.6.1 *Classification of Static Wave Functions*

The first step in the application of group theory to physics is to find the symmetric transformations of a given quantum system with the Hamiltonian $H(x)$. A symmetric transformation R preserves the Hamiltonian invariant,

$$H(x) \stackrel{R}{\longrightarrow} P_R H(x) P_R = H(x), \qquad [P_R,\ H(x)] = 0. \tag{3.79}$$

P_R is the transformation operator for scalar wave functions corresponding to the symmetric transformation R. The set of the symmetric transformations is the symmetric group G of the system.

Second, find the inequivalent irreducible representations and their characters of the symmetric group of the system. This is a task of group theory. Usually, one chooses the convenient forms of the irreducible representation matrices $D^j(R)$ such that the representation matrices $D^j(A)$ of as much generators A of G as possible are diagonal. Of course, the representation matrices $D^j(B)$ of the remaining generators B are not diagonal if G is not Abelian. The representation matrices of any element in G can be calculated from those of the generators.

Third, if the energy level E is m degenerate, there are m linearly independent eigenfunctions $\psi_\mu(x)$ of $H(x)$ with the eigenvalue E:

$$H(x)\psi_\mu(x) = E\psi_\mu(x), \qquad \mu = 1, 2, \ldots, m. \tag{3.80}$$

$\psi_\mu(x)$ span an m-dimensional functional space $\mathcal{L}$. Any function $\phi(x)$ in $\mathcal{L}$ is the eigenfunction of $H(x)$ with the eigenvalue E, and any eigenfunction of $H(x)$ with the eigenvalue E belongs to $\mathcal{L}$. Due to Eq. (3.79), $P_R\psi_\mu(x)$ is an eigenfunction of $H(x)$ with the same energy E. Namely, $\mathcal{L}$ is invariant in the action of the symmetric operator P_R. We can calculate the matrix $D(R)$ of P_R in the basis function $\psi_\mu(x)$:

$$P_R\psi_\mu(x) = \psi_\mu(R^{-1}x) = \sum_{\nu=1}^{m} \psi_\nu(x) D_{\nu\mu}(R). \tag{3.81}$$

The set of $D(R)$ forms a representation of the symmetric group G of the system, called the representation corresponding to the energy E. The character $\chi(R) = \text{Tr}D(R)$ of R in $D(R)$ is easy to calculate. $D(G)$ describes the transformation rule of the eigenfunctions of $H(x)$ with E in the symmetric transformations. Generally, the representation $D(G)$ is reducible and not in the convenient form. Through a similarity transformation X, $D(G)$ can be reduced into the direct sum of irreducible representations,

$$X^{-1}D(R)X = \bigoplus_j a_j D^j(R), \qquad \chi(R) = \sum_j a_j \chi^j(R). \tag{3.82}$$

The multiplicity a_j of the irreducible representation $D^j(G)$ in the reducible representation $D(G)$ can be calculated from the orthogonal relation (3.51):

$$a_j = \frac{1}{g}\sum_{R\in G} \chi^j(R)^* \chi(R) = \frac{1}{g}\sum_\alpha n(\alpha)\chi_\alpha^{j*}\chi_\alpha. \tag{3.83}$$

Then, X can be calculated from Eq. (3.82) in the following way. When R in Eq. (3.82) is taken to be the generators A where $D^j(A)$ are diagonal, X is the similarity transformation matrix to diagonalize $D(A)$. Namely, the column matrices of X are the eigenvectors of $D(A)$ with different eigenvalues, respectively. The solution X contains some undetermined parameters. Those parameters will be partly determined in substituting X into Eq. (3.82), where R is taken to be the remaining generators B. Since the matrix on the right-hand side of Eq. (3.82) is a block matrix, X, which satisfies Eq. (3.82) for all generators, still contains some undetermined parameters whose number is $\sum_j a_j^2$. Those parameters should be chosen to make X as simple as possible.

From Eq. (3.82), the row index of X is the same as the column index of $D(R)$, denoted by μ, and the column index of X is the same as the column index of the block matrix on the right-hand side of Eq. (3.82), which is enumerated by three indices j, ρ, and r. Two indices j and ρ are denoted the irreducible representation $D^j(R)$ and its row and the additional index r is needed when $a_j > 1$ to distinguish different D^j in the reduction (3.82).

New basis functions $\Phi^j_{\rho r}(x)$ are the combinations of $\psi_\mu(x)$ by X:

$$\begin{aligned} \Phi^j_{\rho r}(x) &= \sum_\mu \psi_\mu(x) X_{\mu, j\rho r}, \\ P_R \Phi^j_{\rho r}(x) &= \sum_\lambda \Phi^j_{\lambda r}(x) D^j_{\lambda\rho}(R). \end{aligned} \tag{3.84}$$

$\Phi^j_{\rho r}(x)$ are the basis functions chosen by group theory. The function $\Phi^j_{\rho r}(x)$

is called a function belonging to the ρth row of the irreducible representation D^j of G. This is the so-called classification of the static wave functions according to the irreducible representations of the symmetric group of the system. When $a_j > 1$, there are a_j sets of basis functions $\Psi^j_{\rho r}(x)$ which belong to the representation D^j of G and are distinguished by the parameter r. Any linear combination of the functions $\Phi^j_{\rho r}(x)$ with the same j and ρ is the eigenfunction belonging to the ρth row of D^j

$$\Phi^j_{\rho s}(x) = \sum_r \Psi^j_{\rho r}(x) Y^j_{rs}, \tag{3.85}$$

where the combination coefficients Y^j_{rs} are independent of ρ. The combination matrix Y^j is related to the undetermined parameters in X.

The physical meaning of the function $\Psi^j_{\rho r}(x)$ depends on the given group G and the chosen representation D^j. Since the representation matrices $D^j(A)$ of some generators A in G are diagonal, $\Psi^j_{\rho r}(x)$ is the common eigenfunction of the operators P_A as given in Eq. (3.84).

3.6.2 *Clebsch–Gordan Series and Coefficients*

If a quantum system consists of two subsystems (see §1.6), the wave function of the system is expressed as the product of two wave functions of the subsystems or their combinations. Suppose that two functional spaces $\mathcal{L}^j$ and $\mathcal{L}^k$ of two subsystems are the representation spaces of irreducible representations $D^j(G)$ and $D^k(G)$, respectively, then

$$P_R \psi^j_\mu(x) = \sum_\rho \psi^j_\rho(x) D^j_{\rho\mu}(R), \qquad P_R \phi^k_\nu(y) = \sum_\lambda \phi^k_\lambda(y) D^k_{\lambda\nu}(R).$$

The functional space $\mathcal{L}$ of the composed system is the direct product of $\mathcal{L}^j$ and $\mathcal{L}^k$, and is spanned by $\Psi^{jk}_{\mu\nu}(x, y) = \psi^j_\mu(x)\phi^k_\nu(y)$ where $1 \leq \mu \leq m_j$ and $1 \leq \nu \leq m_k$. In the symmetric transformation, $\Psi^{jk}_{\mu\nu}(x, y)$ transforms according to the direct product representation

$$\begin{aligned} P_R \Psi^{jk}_{\mu\nu}(x, y) &= \sum_{\rho\lambda} \Psi^{jk}_{\rho\lambda}(x, y) \left[D^j(R) \times D^k(R)\right]_{\rho\lambda,\mu\nu}, \\ \left[D^j(R) \times D^k(R)\right]_{\rho\lambda,\mu\nu} &= D^j_{\rho\mu}(R) D^k_{\lambda\nu}(R). \end{aligned} \tag{3.86}$$

The representation can be reduced through a similarity transformation C^{jk}

$$\left(C^{jk}\right)^{-1} \left[D^j(R) \times D^k(R)\right] C^{jk} = \bigoplus_J a_J D^J(R). \tag{3.87}$$

The series on the right-hand side of Eq. (3.87) is called the Clebsch–Gordan series. Taking the trace of Eq. (3.87), one has

$$\chi^j(R)\chi^k(R) = \sum_J a_J \chi^J(R). \tag{3.88}$$

The multiplicity a_J as well as the matrix C^{jk} can be calculated from Eqs. (3.88) and (3.87). Similar to the discussion in the preceding subsection, the row and column indices of C^{jk} are denoted respectively by $\mu\nu$ and JMr, where the additional index r is needed when $a_J > 1$. New basis functions $\Phi^J_{Mr}(x,y)$ are combined from $\Psi^{jk}_{\mu\nu}(x,y)$ through C^{jk} such that they belong to the representation D^J of G,

$$\begin{aligned} \Phi^J_{Mr}(x,y) &= \sum_{\mu\nu} \Psi^{jk}_{\mu\nu}(x,y) C^{jk}_{\mu\nu,JMr}, \\ P_R \Phi^J_{Mr}(x,y) &= \sum_{M'} \Phi^J_{M'r}(x,y) D^J_{M'M}(R). \end{aligned} \tag{3.89}$$

The matrix entries $C^{jk}_{\mu\nu,JMr}$ are called the Clebsch–Gordan coefficients. The Clebsch–Gordan coefficients depend upon the group G and its chosen representations. In addition, as discussed in the preceding subsection, C^{jk} contains $\sum_J a_J^2$ undetermined parameters, which are chosen to make C^{jk} as simple as possible. Some examples for the calculations of the Clebsch–Gordan coefficients of point groups are given in Probs. 24–27 of Chap. 3 of [Ma and Gu (2004)].

3.6.3 *Wigner–Eckart Theorem*

Theorem 3.6 (Wigner–Eckart Theorem) The functions belonging to two inequivalent irreducible representations of a unitary transformation group P_G are orthogonal to each other. The functions belonging to different rows of a unitary irreducible representation of P_G are orthogonal to each other, and the self-inner products of those functions are independent of the row number.

Proof First of all, it is assumed that one can define an inner product of two functions such that the operators P_R are unitary:

$$\langle \phi(x)|\psi(x)\rangle = \langle P_R\phi(x)|P_R\psi(x)\rangle. \tag{3.90}$$

If P_G is the symmetric group of a physical system such as a finite group or the rotational group, P_R is unitary for the inner product usually used in physics. An important counter-example in physics is the case where P_A is the proper Lorentz transformation.

Let $D^j(G)$ and $D^k(G)$ be two inequivalent irreducible unitary representations of the group P_G. ψ^j_μ belongs to the μth row of $D^j(G)$ and ϕ^k_ν belongs to the νth row of $D^k(G)$,

$$P_R\psi^j_\mu(x) = \sum_\rho \psi^j_\rho(x)D^j_{\rho\mu}(R), \qquad P_R\phi^k_\nu(x) = \sum_\lambda \phi^k_\lambda(x)D^k_{\lambda\nu}(R).$$

Letting $\langle\phi^k_\nu(x)|\psi^j_\mu(x)\rangle = X^{kj}_{\nu\mu}$, one has

$$\begin{aligned}\langle\phi^k_\nu(x)|P_R\psi^j_\mu(x)\rangle &= \sum_\rho \langle\phi^k_\nu(x)|\psi^j_\rho(x)\rangle D^j_{\rho\mu}(R) = \sum_\rho X^{kj}_{\nu\rho}D^j_{\rho\mu}(R) \\ &= \langle P_R^{-1}\phi^k_\nu(x)|\psi^j_\mu(x)\rangle = \sum_\lambda D^k_{\lambda\nu}(R^{-1})^*\langle\phi^k_\lambda(x)|\psi^j_\mu(x)\rangle = \sum_\lambda D^k_{\nu\lambda}(R)X^{kj}_{\lambda\mu}.\end{aligned} \tag{3.91}$$

From the Schur Theorem 3.3, $X^{kj}_{\nu\mu} = \delta_{kj}\delta_{\nu\mu}\ \langle\phi^k||\psi^j\rangle$, where $\langle\phi^k||\psi^j\rangle$ is called the reduced matrix entry, which is a constant independent of the subscript μ. Thus, the Theorem is proved,

$$\langle\phi^k_\nu(x)|\psi^j_\mu(x)\rangle = \delta_{kj}\delta_{\nu\mu}\langle\phi^k||\psi^j\rangle. \tag{3.92}$$

The functions belonging to two inequivalent irreducible unitary representations of P_G are orthogonal to each other, so are their combinations. □

In quantum mechanics, most physical observables are calculated through matrix entries. When the static wave functions belong to given rows of given unitary irreducible representations, the Wigner–Eckart Theorem simplifies the problem of calculating $m_k m_j$ matrix entries $\langle\phi^k_\nu(x)|\psi^j_\mu(x)\rangle$ to a problem of calculating only one reduced matrix entry. Furthermore, the static wave functions in the real problems usually are hard to be solved such that even one matrix entry cannot be calculated because the wave functions are unknown. However, some precise information of the system can be obtained through analyzing its symmetry. Some matrix entries are known to be vanishing (selection rule), and the ratios of the matrix entries can be obtained by eliminating the reduced matrix entry as a parameter, although the parameter cannot be calculated. We will explain this method by examples later.

If a set of operators $L^k_\rho(x)$ of mechanical quantities transforms as follows in the symmetric transformations P_R,

$$P_R L^k_\rho(x) P_R^{-1} = \sum_\lambda L^k_\lambda(x) D^k_{\lambda\rho}(R), \tag{3.93}$$

$L^k_\rho(x)$ are called the irreducible tensor operators, then,

$$P_R L^k_\rho(x)\psi^j_\mu(x) = \sum_{\lambda\tau} L^k_\lambda(x)\psi^j_\tau \left[D^k(R)\times D^j(R)\right]_{\lambda\tau,\rho\mu}. \tag{3.94}$$

$L^k_\rho(x)\psi^j_\mu(x)$ can be combined by the Clebsch–Gordan coefficients to $F^J_{Mr}(x)$ which belongs to the Mth row of D^J:

$$\begin{aligned} F^J_{Mr}(x) &= \sum_{\rho\mu} L^k_\rho(x)\psi^j_\mu(x)C^{kj}_{\rho\mu,JMr}, \\ P_R F^J_{Mr}(x) &= \sum_{M'} F^J_{M'r}(x)D^J_{M'M}(R), \\ L^k_\rho(x)\psi^j_\mu(x) &= \sum_{JMr} F^J_{Mr}(x)\left[\left(C^{kj}\right)^{-1}\right]_{JMr,\rho\mu}. \end{aligned} \tag{3.95}$$

There are $(m_{j'}m_k m_j)$ matrix entries $\langle\phi^{j'}_\nu(x)|L^k_\rho(x)|\psi^j_\mu(x)\rangle$ for the mechanical quantities $L^k_\rho(x)$ between two static wave functions $\phi^{j'}_\nu(x)$ and $\psi^j_\mu(x)$. The Wigner–Eckart Theorem greatly simplifies the calculation problems,

$$\begin{aligned} \langle\phi^{j'}_\nu(x)|L^k_\rho(x)|\psi^j_\mu(x)\rangle &= \sum_{JMr} \langle\phi^{j'}_\nu(x)|F^J_{Mr}(x)\rangle \left[\left(C^{kj}\right)^{-1}\right]_{JMr,\rho\mu} \\ &= \sum_{r} \langle\phi^{j'}||L^k||\psi^j\rangle_r \left[\left(C^{kj}\right)^{-1}\right]_{j'\nu r,\rho\mu}, \end{aligned} \tag{3.96}$$

namely, the information of those matrix entries related to the symmetry of the system demonstrates itself through the Clebsch–Gordan coefficients, and the remaining information is given in a few reduced matrix entries $\langle\phi^{j'}||L^k||\psi^j\rangle_r$, which are independent of the row indices ν, ρ, and μ. The number of the reduced matrix entries is equal to the multiplicity of the irreducible representation $D^{j'}$ in the reduction of the direct product representation $D^k\times D^j$.

3.6.4 *Normal Degeneracy and Accidental Degeneracy*

If an m-degenerate energy E of the original Hamiltonian $H_0(x)$ corresponds to a representation $D(G)$ of the symmetric group G, the degeneracy is called normal if $D(G)$ is irreducible and called accidental if $D(G)$ is reducible.

We begin with the energy level of the original system to be normal degeneracy. Introduce a "symmetric perturbation" $\lambda H_1(x)$ which does not disturb the symmetric group of $H_0(x)$. Thus, $H_0(x)$ and $H_1(x)$ are both commutable with the symmetric transformation operators P_R:

$$[P_R,\ H_0(x)] = 0, \qquad [P_R,\ H_1(x)] = 0. \tag{3.97}$$

The perturbation is introduced smoothly as the parameter λ increases from 0 to 1 continuously.

Denote the eigenfunctions of $H_0(x)$ with E by $\psi^j_\mu(x)$ belonging to the μth row of an irreducible representation $D^j(G)$:

$$P_R\psi^j_\mu(x) = \sum_\nu \psi^j_\nu(x)D^j_{\nu\mu}(R). \tag{3.98}$$

For the first approximation, the energy shift ΔE^j is calculated by

$$\lambda\langle\psi^j_\nu(x)|H_1(x)|\psi^j_\mu(x)\rangle = \delta_{\nu\mu}\left(\Delta E^j\right). \tag{3.99}$$

Namely, in the first approximation, the eigenvalue does not split. In fact, it cannot split in arbitrarily high approximation. Otherwise, if it splits into, say two eigenvalues E_1 and E_2 with m_1 and m_2 eigenfunctions ($m_1 + m_2 = m$), the m_1 eigenfunctions of E_1 transform among themselves under P_R and correspond to a representation of dimension m_1. The representation cannot contain the original irreducible representation $D^j(G)$ because $m_1 < m$. Then, those m_1 eigenfunctions of E_1 would be orthogonal to all m eigenfunctions of E, and could not be obtained continuously from any combination of the original eigenfunctions. The conclusion is that a symmetric perturbation cannot split an energy with a normal degeneracy.

Now, we turn to an energy E of the original Hamiltonian with an accidental degeneracy, where the representation corresponding to E is reducible, but does not contain an irreducible representation with the multiplicity $a_j > 1$. The energy shifts under the symmetric perturbation, but splits only between eigenfunctions belonging to different irreducible representations

$$\lambda\langle\psi^k_\nu(x)|H_1(x)|\psi^j_\mu(x)\rangle = \delta_{kj}\delta_{\nu\mu}\left(\Delta E^j\right). \tag{3.100}$$

This conclusion is also non-perturbation. If the representation corresponding to E contains an irreducible representation $D^j(G)$ with the multiplicity $a_j > 1$, the sets of eigenfunctions belonging to $D^j(G)$ are combined

$$\lambda\langle\psi^k_{\nu r}(x)|H_1(x)|\psi^j_{\mu s}(x)\rangle = \delta_{kj}\delta_{\nu\mu}\left(\Delta E^j\right)_{rs}, \tag{3.101}$$

where $r,\ s = 1, 2, \ldots, a_j$. By making use of the symmetry of the system, the method of group theory greatly simplifies the calculation.

If the original Hamiltonian $H_0(x)$ and the perturbation Hamiltonian $H_1(x)$ have different symmetries, one may choose the common symmetric transformations of $H_0(x)$ and $H_1(x)$ to form the symmetric group of the system so that the perturbation Hamiltonian $H_1(x)$ is the "symmetric perturbation". It is common viewpoint that the energy level of the original system is normal degeneracy if G contains all symmetric transformations of the system. The accidental degeneracy is related to the existence of some undiscovered symmetric transformations of the system [Zou and Huang (1995)].

3.6.5 *An Example of Application*

We are going to raise a physical example to demonstrate the steps of application of group theory to physics. Discuss a quantum system with a square well potential in two dimensions. The Hamiltonian equation is ($\hbar = 2m = 1$)

$$\begin{aligned} H\psi &= -\frac{d^2\psi}{dx^2} - \frac{d^2\psi}{dy^2} + V\psi = E\psi, \\ V(x,y) &= \begin{cases} 0 & \text{when } |x| < \pi\ , \quad |y| < \pi, \\ \infty & \text{the remaining cases.} \end{cases} \end{aligned} \tag{3.102}$$

First, it is evident that the symmetric group of the system with the square well potential in two dimensions is the group D_4. The character table of D_4 is listed in Table 3.12. Two generators of D_4 are the 4-fold rotation C_4 along the z-axis and the 2-fold rotation C_2' along the x-axis. Denote C_4 by T and C_2' by S for convenience. In the two-dimensional representation D^E, their representation matrices are

$$D^E(T) = \begin{pmatrix} 0 & -1 \\ 1 & 0 \end{pmatrix}, \qquad D^E(S) = \begin{pmatrix} 1 & 0 \\ 0 & -1 \end{pmatrix}. \tag{3.103}$$

They are just the matrices in the method of coordinate transformation,

$$\begin{pmatrix} x' \\ y' \end{pmatrix} = R \begin{pmatrix} x \\ y \end{pmatrix}, \qquad R = D^E(R).$$

Second, by separation of variables, $\psi(x, y) = X(x)Y(y)$, the Hamiltonian equation becomes

$$\begin{aligned} &X'' + E_1 X = 0, \qquad Y'' + E_2 Y = 0, \qquad |x| \le \pi, \qquad |y| \le \pi, \\ &X(\pm\pi) = Y(\pm\pi) = 0, \qquad E = E_1 + E_2. \end{aligned}$$

Solving it, one has

$$X(x) = \begin{cases} \sin(mx), & E_1 = m^2, \\ \cos\left[(2m-1)x/2\right], & E_1 = (2m-1)^2/4, \end{cases}$$

$$Y(y) = \begin{cases} \sin(ny)\ , & E_2 = n^2, \\ \cos\left[(2n-1)y/2\right], & E_2 = (2n-1)^2/4, \end{cases}$$

where m and n are both positive integers. There are five kinds of energy levels with the following eigenfunctions ($m \neq n$):

(1) $E = (2m-1)^2/2$, $\psi = \cos\left[(2m-1)x/2\right]\cos\left[(2m-1)y/2\right]$.

(2) $E = 2m^2$, $\psi = \sin(mx)\sin(my)$.

(3) $E = m^2 + n^2$, $\psi_1 = \sin(mx)\sin(ny)$, $\psi_2 = \sin(nx)\sin(my)$.

(4) $E = (2m-1)^2/4 + (2n-1)^2/4$,

$$\psi_1 = \cos\left[(2m-1)x/2\right]\cos\left[(2n-1)y/2\right],$$
$$\psi_2 = \cos\left[(2n-1)x/2\right]\cos\left[(2m-1)y/2\right].$$

(5) $E = m^2 + (2n-1)^2/4$,

$$\psi_1 = \sin(mx)\cos\left[(2n-1)y/2\right], \qquad \psi_2 = \cos\left[(2n-1)x/2\right]\sin(my).$$

The matrices of T and S in the basis functions in the above five cases are calculated by Eq. (3.81), where

$$T^{-1} = \begin{pmatrix} 0 & 1 \\ -1 & 0 \end{pmatrix}, \qquad S^{-1} = \begin{pmatrix} 1 & 0 \\ 0 & -1 \end{pmatrix}. \tag{3.104}$$

Comparing the matrices with Table 3.12 and Eq. (3.103), one determines the representations in five cases.

(1) $P_T\psi = P_S\psi = \psi$, $D(T) = D(S) = 1$, and the representation is D^{A_1}.

(2) $P_T\psi = P_S\psi = -\psi$, $D(T) = D(S) = -1$, and the representation is D^{B_2}.

(3) $P_T\psi_1 = -\psi_2$, $P_T\psi_2 = -\psi_1$, $P_S\psi_1 = -\psi_1$, $P_S\psi_2 = -\psi_2$, and then

$$D(T) = \begin{pmatrix} 0 & -1 \\ -1 & 0 \end{pmatrix}, \qquad D(S) = \begin{pmatrix} -1 & 0 \\ 0 & -1 \end{pmatrix},$$
$$\chi(T) = 0, \qquad \chi(S) = -2.$$

By Eq. (3.83) one knows that the representation is equivalent to $D^{A_2} \oplus D^{B_2}$. $D(S)$ is a constant matrix and $D(T)$ can be diagonalized as

$$X^{-1}D(T)X = \begin{pmatrix} 1 & 0 \\ 0 & -1 \end{pmatrix}, \qquad X = \frac{1}{\sqrt{2}}\begin{pmatrix} 1 & 1 \\ -1 & 1 \end{pmatrix},$$
$$\Phi_1 = (\psi_1 - \psi_2)/\sqrt{2}, \qquad \Phi_2 = (\psi_1 + \psi_2)/\sqrt{2}.$$

Φ_1 belongs to D^{A_2} and Φ_2 belongs to D^{B_2}.

(4) $P_T\psi_1 = \psi_2$, $P_T\psi_2 = \psi_1$, $P_S\psi_1 = \psi_1$, $P_S\psi_2 = \psi_2$, and then

$$D(T) = \begin{pmatrix} 0 & 1 \\ 1 & 0 \end{pmatrix}, \qquad D(S) = \begin{pmatrix} 1 & 0 \\ 0 & 1 \end{pmatrix},$$
$$\chi(T) = 0, \qquad \chi(S) = 2.$$

The corresponding representation can be reduced to $D^{A_1} \oplus D^{B_1}$.

$$\Phi_1 = (\psi_1 + \psi_2)/\sqrt{2}\ , \qquad \Phi_2 = (\psi_1 - \psi_2)/\sqrt{2}\ .$$

Φ_1 belongs to D^{A_1} and Φ_2 belongs to D^{B_1}.

(5) $P_T\psi_1 = \psi_2$, $P_T\psi_2 = -\psi_1$, $P_S\psi_1 = \psi_1$, $P_S\psi_2 = -\psi_2$, and then

$$D(T) = \begin{pmatrix} 0 & -1 \\ 1 & 0 \end{pmatrix}, \qquad D(S) = \begin{pmatrix} 1 & 0 \\ 0 & -1 \end{pmatrix}.$$
$$\chi(T) = 0, \qquad \chi(S) = 0\ .$$

This is the irreducible representation D^E.

In the five kinds of energy levels, the energies (1), (2), and (5) are normal degeneracies, and the energies (3) and (4) are accidental degeneracies. For some special energies the multiplicity of degeneracy may be higher. For example, the energy $E = 65$ for $m = 1$, $n = 8$ in case (3) and for $m = 4$, $n = 7$ in case (3). $E = 50$ for $m = 1$ and $n = 7$ in case (3) and for $m = 5$ in case (2). The corresponding representation is the direct sum of representations in two cases. There is no new problem in analysis for those higher degenerate energies.

At last, we discuss the shift and split of energy in the action of a symmetric perturbation Hamiltonian $H_1 = \varepsilon x^2 y^2$. The group D_4 is the common symmetric group for both H_0 and H_1. Under the symmetric perturbation H_1, the energy of normal degeneracy shifts but does not split. For example, in case (5)

$$\langle\psi_1|H_1|\psi_1\rangle = \langle\psi_2|H_1|\psi_2\rangle = \varepsilon\left\{\frac{\pi^3}{3} - \frac{\pi}{2m^2}\right\}\left\{\frac{\pi^3}{3} - \frac{2\pi}{(2n+1)^2}\right\},$$
$$\langle\psi_1|H_1|\psi_2\rangle = \langle\psi_2|H_1|\psi_1\rangle = 0\ .$$

There is only one matrix entry that needs to be calculated.

The energy of accidental degeneracy will split under the symmetric perturbation H_1. For the cases (3) and (4), we have chosen the basis functions Φ_μ by group theory such that they belong to the irreducible representations. In those basis functions H_1 is diagonalized.

$$\begin{aligned}
\langle \Phi_1|H_1|\Phi_1\rangle &= \varepsilon \left\{\frac{\pi^3}{3} - \frac{\pi}{2m^2}\right\}\left\{\frac{\pi^3}{3} - \frac{\pi}{2n^2}\right\} \\
&\quad - \varepsilon(-1)^{m-n}\left\{\frac{2\pi}{(m-n)^2} - \frac{2\pi}{(m+n)^2}\right\}^2, \\
\langle \Phi_2|H_1|\Phi_2\rangle &= \varepsilon \left\{\frac{\pi^3}{3} - \frac{\pi}{2m^2}\right\}\left\{\frac{\pi^3}{3} - \frac{\pi}{2n^2}\right\} \\
&\quad + \varepsilon(-1)^{m-n}\left\{\frac{2\pi}{(m-n)^2} - \frac{2\pi}{(m+n)^2}\right\}^2, \\
\langle \Phi_1|H_1|\Phi_2\rangle &= \langle \Phi_2|H_1|\Phi_1\rangle = 0.
\end{aligned}$$

In comparison with the basis function ψ_μ, the calculation in the basis functions Φ_μ is simplified.

3.7 Irreducible Bases in Group Algebra

The group algebra $\mathcal{L}$ of a finite group G is invariant under left- and right-multiplication with an element of G, and corresponds to the regular representation (see Eqs. (3.6) and (3.12)). The regular representation is reducible and the multiplicity of each irreducible representation $D^j(G)$ of G in the reduction of the regular representation is its dimension m_j (see Eq. (3.57)). Reducing the regular representation, one obtains a new set of basis vectors $b^j_{\mu\nu}$ which belong to the irreducible representations $D^j(G)$:

$$\begin{gathered}
X^{-1}D(T)X = \bigoplus_j m_j D^j(T), \\
b^j_{\mu\nu} = \sum_{R\in G} R X_{R,j\mu\nu}, \qquad T b^j_{\mu\nu} = \sum_\tau b^j_{\tau\nu} D^j_{\tau\mu}(T).
\end{gathered} \tag{3.105}$$

$b^j_{\mu\nu}$ is called the irreducible basis vectors in the group algebra.

If G is the symmetric group of a quantum system, the static wave functions ψ_ρ can be combined into that belonging to the irreducible representations of G. When the irreducible basis vectors $b^j_{\mu\nu}$ in $\mathcal{L}$ are known, $b^j_{\mu\nu}\psi_\rho$, if it is not vanishing, is the function belonging to the irreducible representations $D^j(G)$:

$$T\left\{b^j_{\mu\nu}\psi_\rho\right\} = \sum_\tau \left\{b^j_{\tau\nu}\psi_\rho\right\} D^j_{\tau\mu}(T). \tag{3.106}$$

In this section we will introduce the methods for finding the irreducible basis vectors in the group algebra of a finite group.

3.7.1 *Ideal and Idempotent*

A subspace $\mathcal{L}_a$ of an algebra $\mathcal{L}$ is a sub-algebra of $\mathcal{L}$ if $yx \in \mathcal{L}_a$ for all $x \in \mathcal{L}_a$ and $y \in \mathcal{L}_a$. A subspace $\mathcal{L}_a$ of an algebra $\mathcal{L}$ is a left ideal of $\mathcal{L}$ if

$$tx \in \mathcal{L}_a, \qquad \forall\, t \in \mathcal{L} \text{ and } x \in \mathcal{L}_a. \tag{3.107}$$

A subspace $\mathcal{R}_a$ of $\mathcal{L}$ is a right ideal of $\mathcal{L}$ if

$$xt \in \mathcal{R}_a, \qquad \forall\, t \in \mathcal{L} \text{ and } x \in \mathcal{R}_a. \tag{3.108}$$

In the following, we will study the properties of the left ideals as example. Most of the properties are suitable for the right ideals. The difference of the two kinds of ideals will be noticed.

In the reduction (3.105) of the regular representation, each irreducible representation space is a left ideal of the group algebra $\mathcal{L}$. Assume that $\mathcal{L}$ is decomposed into the direct sum of n left ideals $\mathcal{L}_a$

$$\mathcal{L} = \bigoplus_{a=1}^{n} \mathcal{L}_a = \mathcal{L}_a \bigoplus \overline{\mathcal{L}}_a, \qquad \overline{\mathcal{L}}_a = \bigoplus_{b \neq a} \mathcal{L}_b, \tag{3.109}$$

where $\overline{\mathcal{L}}_a$ is the supplement of $\mathcal{L}_a$. Any vector t in $\mathcal{L}$ can be decomposed uniquely into a sum of vectors belonging to the left ideals $\mathcal{L}_a$, respectively,

$$t = \sum_{a=1}^{n} t_a, \qquad t \in \mathcal{L}, \qquad t_a \in \mathcal{L}_a. \tag{3.110}$$

Thus, the identical element E can be decomposed as

$$E = \sum_{a=1}^{n} e_a, \qquad e_a \in \mathcal{L}_a. \tag{3.111}$$

Since $t = tE = \sum_a te_a$, one has $t_a = te_a$. Then, $te_a = t$ if $t \in \mathcal{L}_a$, and $te_a = 0$ if $t \in \overline{\mathcal{L}}_a$. Replacing t with e_b, one has

$$e_b e_a = \delta_{ba} e_a. \tag{3.112}$$

The projective operator e_a, which satisfies $e_a^2 = e_a$, is called an idempotent. Those e_a satisfying Eq. (3.112) are called the mutually orthogonal idempotents. $\mathcal{L}_a$ is called the left ideal generated by e_a,

$$\mathcal{L}e_a = \mathcal{L}_a, \qquad \mathcal{L} = \bigoplus_a \mathcal{L}_a. \tag{3.113}$$

$\mathcal{R}_a = e_a\mathcal{L}$ is called the right ideal generated by e_a. Remind that the idempotent which generates a given left (or right) ideal is not unique (see the next subsection).

3.7.2 *Primitive Idempotent*

Find a complete set of basis vectors x_μ in a left ideal $\mathcal{L}_a$ generated by e_a. Since $\mathcal{L}_a$ is invariant in left-multiplication with group elements, one obtains an m-dimensional representation $D(G)$ of G,

$$Tx_\mu = \sum_{\nu=1}^{m} x_\nu D_{\nu\mu}(T), \tag{3.114}$$

where m is the dimension of $\mathcal{L}_a$. $D(G)$ is called a representation corresponding to the left ideal $\mathcal{L}_a$. For a right ideal $\mathcal{R}_a$ of dimension m, Eq. (3.114) is replaced with (see Eq. (3.11))

$$x_\mu T = \sum_{\nu=1}^{m} \overline{D}_{\mu\nu}(T) x_\nu. \tag{3.115}$$

Two left ideals are said to be equivalent if the representations corresponding to the left ideals are equivalent. Two idempotents generating two equivalent left ideals, respectively, are equivalent. By choosing suitable basis vectors in two equivalent left ideals, the corresponding representations can be the same as each other. In this case, one can define a one-to-one correspondence between two sets of basis vectors as well as their combinations, and the correspondence is invariant in left-multiplication with any element in G. Conversely, if there is a one-to-one correspondence between vectors in the two left ideals and the correspondence is invariant in left-multiplication with any element in G, the representations corresponding to the two left ideals are the same, and the two left ideals are equivalent.

If a left ideal corresponds to a reducible representation, the left ideal can be decomposed into the direct sum of two left sub-ideals,

$$\begin{array}{lll} \mathcal{L}_a = \mathcal{L}_j \bigoplus \mathcal{L}_k, & & e_a = e_j + e_k, \\ e_j^2 = e_j, & e_k^2 = e_k, & e_j e_k = e_k e_j = 0, \\ e_a e_j = e_j e_a = e_j, & & e_a e_k = e_k e_a = e_k. \end{array} \tag{3.116}$$

The decomposition can be carried over until the left sub-ideal corresponds to an irreducible representation. The left ideal is called a minimal left ideal if it corresponds to an irreducible representation of G, and the idempotent generating the minimal left ideal is called the primitive idempotent. In order to study the inequivalent irreducible representations of a finite group, one needs the criterions for equivalence, irreducibility, and completeness of representations corresponding to the left ideals.

Theorem 3.7 Two primitive idempotents e_a and e_b are equivalent if and only if there exists at least one element $S \in G$ satisfying

$$e_a S e_b \neq 0. \tag{3.117}$$

Proof If Eq. (3.117) holds, there is a map from the minimal left ideal $\mathcal{L}_a$ generated by e_a onto the minimal left ideal $\mathcal{L}_b$ generated by e_b:

$$x \in \mathcal{L}_a \quad \longrightarrow \quad y = x e_a S e_b \in \mathcal{L}_b. \tag{3.118}$$

Evidently, the map is invariant in left-multiplication with any element

$$Rx \in \mathcal{L}_a \quad \longrightarrow \quad Ry = Rx e_a S e_b \in \mathcal{L}_b. \tag{3.119}$$

We are going to show that the map is bijection, namely there is a one-to-one correspondence between vectors in the two left ideals such that the two minimal left ideals are equivalent.

First, we show that the map is surjective. If the set of y mapped from x is a subspace $\mathcal{L}_y$ of the minimal left ideal $\mathcal{L}_b$, due to Eq. (3.119), $\mathcal{L}_y$ is a left sub-ideal of $\mathcal{L}_b$. Since $\mathcal{L}_y$ contains $e_a S e_b \neq 0$, it is not a null space such that $\mathcal{L}_y = \mathcal{L}_b$.

Second, we show that $y \neq 0$ if $x \neq 0$. Otherwise, the set $\mathcal{L}_x$ of x which maps to $y = 0$ is a left sub-ideal of $\mathcal{L}_a$. Since $\mathcal{L}_x$ does not contain e_a, $\mathcal{L}_x$ is a null space.

Third, we show that the map is injective. If two vectors x and x' in $\mathcal{L}_a$ both map to the same y in $\mathcal{L}_b$, the difference $(x - x')$ maps to the null vector so that $x = x'$.

Conversely, if two minimal left ideals $\mathcal{L}_a$ and $\mathcal{L}_b$ are equivalent, and the idempotent e_a maps to $t = t e_b$ in $\mathcal{L}_b$, then, $e_a e_a$ maps to $e_a t = e_a t e_b \neq 0$. Thus, there exists at least one element S satisfying Eq. (3.117). □

Corollary 3.7.1 An idempotent e_a is primitive if and only if

$$e_a t e_a = \lambda_t e_a, \qquad \forall\ t \in \mathcal{L}, \tag{3.120}$$

where λ_t is a constant depending on t and is allowed to be vanishing.

Proof First, show the condition (3.120) to be sufficient by reduction to absurdity. If the condition (3.120) holds but e_a is not a primitive idempotent,

e_a can be decomposed into a sum given in Eq. (3.116). Thus,

$$e_j = e_a e_j e_a = \lambda_j e_a, \qquad e_j = e_j^2 = e_j \left(\lambda_j e_a\right) = \lambda_j e_j.$$

The solutions are $e_j = 0$ ($e_a = e_k$) or $\lambda_j = 1$ ($e_a = e_j$), both of which mean that e_a cannot be decomposed.

Second, show the condition (3.120) to be necessary. If $e_a t e_a = 0$, Eq. (3.120) holds. If $e_a t e_a \neq 0$, $e_a t e_a$ provides a one-to-one correspondence from the minimal left ideal $\mathcal{L}_a$ onto itself, which is invariant in the left-multiplication with any element in G. The basis x_μ in $\mathcal{L}_a$ maps to $y_\mu = x_\mu e_a t e_a = \sum_\nu x_\nu M_{\nu\mu}$. The irreducible representations in both sets of basis vectors are the same, $M^{-1}D(R)M = D(R)$. From the Schur Theorem 3.3.1, M is a constant matrix, denoted by $\lambda_t \mathbf{1}$. Thus, $y_\mu = \lambda_t x_\mu$. □

If $\lambda_t \neq 0$, another idempotent $e'_a = \lambda_t^{-1} t e_a$ generates the same left ideal as that generated by e_a. In fact, since $e'_a e_a = e'_a$ and $e_a e'_a = e_a$, one has $\mathcal{L}e_a = \mathcal{L}e_a e'_a \subset \mathcal{L}e'_a$ and $\mathcal{L}e'_a = \mathcal{L}e'_a e_a \subset \mathcal{L}e_a$. Thus, $\mathcal{L}e_a = \mathcal{L}e'_a$. Similarly, the right ideal generated by $e''_a = \lambda_t^{-1} e_a t$ is the same as that generated by e_a, $e''_a \mathcal{L} = e_a \mathcal{L}$.

Theorem 3.8 The direct sum of n left ideals $\mathcal{L}_a$ respectively generated by the orthogonal idempotents e_a (see Eq. (3.112)) is equal to the group algebra $\mathcal{L}$ if and only if the sum of e_a is equal to the identical element E:

$$\mathcal{L} = \bigoplus_{a=1}^{n} \mathcal{L}_a \iff E = \sum_{a=1}^{n} e_a. \tag{3.121}$$

Proof If E is equal to the sum of e_a, $t = tE = \sum_a t e_a$. Thus, any vector t in $\mathcal{L}$ can be decomposed uniquely into a sum of vectors $t e_a$ belonging to $\mathcal{L}_a$, respectively. Conversely, if $\mathcal{L}$ is the direct sum of $\mathcal{L}_a$, E can be decomposed uniquely into a sum of vectors belonging to $\mathcal{L}_a$, respectively, where the vector belonging to $\mathcal{L}_a$ is $E e_a = e_a$. □

3.7.3 *Two-side Ideal*

The left ideal and the right ideal, generated by the same idempotent e_a, are not the same as each other generally. If they are the same, $\mathcal{L}e_a = e_a \mathcal{L}$ is called a two-side ideal generated by an idempotent e_a,

$$\begin{gathered} \mathcal{L}e_a = e_a \mathcal{L} = \mathcal{I}_a, \\ tx \in \mathcal{I}_a, \quad xt \in \mathcal{I}_a, \quad \forall t \in \mathcal{L} \ \text{and} \ x \in \mathcal{I}_a. \end{gathered} \tag{3.122}$$

A two-side ideal is called simple if it cannot be decomposed into a direct sum of two or more two-side sub-ideals.

The group algebra $\mathcal{L}$ of a finite group G can be decomposed into a direct sum of the minimal left ideals $\mathcal{L}_\mu^{(j)}$, where the equivalent minimal left ideals are denoted by the same superscript (j). The left ideal $\mathcal{L}^{(j)}$ is the direct sum of those equivalent minimal left ideals $\mathcal{L}_\mu^{(j)}$. The identical element E is decomposed similarly.

$$\begin{aligned}
&\mathcal{L} = \bigoplus_{j=1}^{g_c} \mathcal{L}^{(j)}, \qquad && E = \sum_{j=1}^{g_c} e^{(j)}, \qquad && \mathcal{L}^{(j)} = \mathcal{L}e^{(j)}, \\
&\mathcal{L}^{(j)} = \bigoplus_{\mu=1}^{m_j} \mathcal{L}_\mu^{(j)}, && e^{(j)} = \sum_{\mu=1}^{m_j} e_\mu^{(j)}, && \mathcal{L}_\mu^{(j)} = \mathcal{L}e_\mu^{(j)}, \\
&e_\mu^{(j)} e_\nu^{(i)} = \delta_{ji}\delta_{\mu\nu} e_\mu^{(j)}, && e_\mu^{(j)} t e_\nu^{(i)} = \delta_{ji} e_\mu^{(j)} t e_\nu^{(j)}, && t \in \mathcal{L}.
\end{aligned} \tag{3.123}$$

Theorem 3.9 The left ideal $\mathcal{L}^{(j)}$ given in Eq. (3.123) is a simple two-side ideal in the group algebra $\mathcal{L}$.

Proof Letting t be an arbitrary vector in $\mathcal{L}$, one has

$$e^{(j)}t = e^{(j)}tE = \sum_{\mu i \nu} e_\mu^{(j)} t e_\nu^{(i)} = \sum_\nu \left(\sum_\mu e_\mu^{(j)} t \right) e_\nu^{(j)} \in \mathcal{L}^{(j)}. \tag{3.124}$$

Thus, $\mathcal{L}^{(j)}$ is also a right ideal. Second, we will show that the two-side ideal $\mathcal{L}^{(j)}$ is simple. Let $\mathcal{I} \subset \mathcal{L}^{(j)}$ be a two-side ideal. Denote by $\mathcal{L}_0$ an arbitrarily minimal left sub-ideal contained in $\mathcal{I}$. $\mathcal{L}_0$ has to be equivalent to each $\mathcal{L}_\mu^{(j)}$. From Theorem 3.7, there exists a group element R_μ in G such that $e_0 R_\mu e_\mu^{(j)} \neq 0$, where e_0 generates $\mathcal{L}_0$. Since $e_0 R_\mu e_\mu^{(j)}$ belongs to both $\mathcal{I}$ and $\mathcal{L}_\mu^{(j)}$, $\mathcal{L}e_0 R_\mu e_\mu^{(j)} = \mathcal{L}_\mu^{(j)} \subset \mathcal{I}$. Thus, $\mathcal{L}^{(j)} \subset \mathcal{I}$, and then, $\mathcal{L}^{(j)} = \mathcal{I}$. □

Corollary 3.9.1 The sum of all equivalent minimal left (or right) ideals in a group algebra $\mathcal{L}$ is a simple two-side ideal.

Corollary 3.9.2 The decomposition of $\mathcal{L}$ into the simple two-side ideals is unique.

3.7.4 *Standard Irreducible Basis Vectors*

Choose a complete set of basis vectors $b_{\nu\mu}^{(j)}$ in a minimal left ideal $\mathcal{L}_\mu^{(j)}$, which is generated by the primitive idempotent $e_\mu^{(j)}$ and corresponds to a unitary irreducible representation $D^j(G)$, such that

$$b_{\mu\mu}^{(j)} = e_{\mu}^{(j)}, \qquad Tb_{\rho\mu}^{(j)} = \sum_{\tau} b_{\tau\mu}^{(j)} D_{\tau\rho}^j(T). \tag{3.125}$$

Expanding an arbitrary element R of G with respect to the left ideals $\mathcal{L}_{\mu}^{(j)}$,

$$R = \sum_{j\mu} Re_{\mu}^{(j)} = \sum_{j\mu} Rb_{\mu\mu}^{(j)} = \sum_{j\mu\nu} b_{\nu\mu}^{(j)} D_{\nu\mu}^j(R). \tag{3.126}$$

The basis vectors $b_{\nu\mu}^{(j)}$ can be solved by making use of Theorem 3.3:

$$b_{\nu\mu}^{(j)} = \frac{m_j}{g} \sum_{R\in G} D_{\nu\mu}^j(R)^* R, \tag{3.127}$$

$$e_{\mu}^{(j)} = b_{\mu\mu}^{(j)} = \frac{m_j}{g} \sum_{R\in G} D_{\mu\mu}^j(R)^* R, \tag{3.128}$$

$$e^{(j)} = \sum_{\mu} e_{\mu}^{(j)} = \frac{m_j}{g} \sum_{\alpha=1}^{g_c} \chi_{\alpha}^{j*} \mathsf{C}_{\alpha}. \tag{3.129}$$

Since $\sum_j m_j \chi^j(R)$ is the character of R in the regular representation, $\sum_j e^{(j)} = E$. The basis vectors $b_{\nu\mu}^{(j)}$ given in Eq. (3.127) satisfy

$$\begin{aligned} b_{\nu\rho}^{(j)} b_{\tau\mu}^{(j)} &= \left(\frac{m_j}{g}\right)^2 \sum_{R\in G} D_{\nu\rho}^j(R)^* R \sum_{S\in G} D_{\tau\mu}^j(S)^* S \\ &= \left(\frac{m_j}{g}\right)^2 \sum_{T\in G} \sum_{\lambda} D_{\nu\lambda}^j(T)^* T \sum_{S\in G} D_{\lambda\rho}^j(S^{-1})^* D_{\tau\mu}^j(S)^* \\ &= \delta_{\rho\tau} \frac{m_j}{g} \sum_{T\in G} D_{\nu\mu}^j(T)^* T = \delta_{\rho\tau} b_{\nu\mu}^{(j)}. \end{aligned}$$

The irreducible basis vectors $b_{\nu\mu}^{(j)}$ in the group algebra $\mathcal{L}$ are called standard if they satisfy

$$b_{\mu\mu}^{(j)} = e_{\mu}^{(j)}, \qquad b_{\nu\rho}^{(j)} b_{\tau\mu}^{(j)} = \delta_{\rho\tau} b_{\nu\mu}^{(j)}. \tag{3.130}$$

Obviously, $e_{\nu}^{(j)} b_{\nu\mu}^{(j)} = b_{\nu\mu}^{(j)} = b_{\nu\mu}^{(j)} e_{\mu}^{(j)}$. Recall that the idempotent $e_{\mu}^{(j)}$ is a projective operator, but $b_{\nu\mu}^{(j)}$ is not because its square is not equal to itself. The representation matrix entry $D_{\lambda\rho}^j(T)$ can be calculated by left-multiplying Eq. (3.125) with $b_{\mu\lambda}^{(j)}$,

$$b_{\mu\lambda}^{(j)} T b_{\rho\mu}^{(j)} = e_{\mu}^{(j)} D_{\lambda\rho}^j(T), \tag{3.131}$$

where μ can be taken arbitrarily.

The basis vectors $b^{(j)}_{\nu\mu}$ with a fixed ν form a complete set of basis vectors in the right ideal $\mathcal{R}^{(j)}_{\nu}$,

$$b^{(j)}_{\nu\lambda}T = \sum_{\tau} \overline{D}_{\lambda\tau}(T) b^{(j)}_{\nu\tau}. \tag{3.132}$$

Right-multiplying Eq. (3.132) with $b^{(j)}_{\rho\nu}$, one obtains

$$b^{(j)}_{\nu\lambda}Tb^{(j)}_{\rho\nu} = e^{(j)}_{\nu}\overline{D}_{\lambda\rho}(T).$$

In comparison with Eq. (3.131), $\overline{D}_{\lambda\rho}(T) = D^{j}_{\lambda\rho}(T)$,

$$b^{(j)}_{\nu\lambda}T = \sum_{\tau} D_{\lambda\tau}(T) b^{(j)}_{\nu\tau}. \tag{3.133}$$

In summary, the standard basis vectors $b^{(j)}_{\nu\mu}$ with a fixed μ form a complete set of basis vectors in the left ideal $\mathcal{L}^{(j)}_{\mu}$, and those with a fixed ν form a complete set of basis vectors in the right ideal $\mathcal{R}^{(j)}_{\nu}$. The left ideal $\mathcal{L}^{(j)}_{\mu}$ and the right ideal $\mathcal{R}^{(j)}_{\nu}$ correspond to the same irreducible representation $D^{j}(G)$.

Theorem 3.10 In a group algebra $\mathcal{L}$, both the left ideal $\mathcal{L}e^{(j)}_{\mu}$ and the right ideal $e^{(j)}_{\mu}\mathcal{L}$ generated by the same primitive idempotent $e^{(j)}_{\mu}$ correspond to the same irreducible representation $D^{j}(G)$.

The calculation of the basis vectors $b^{j}_{\nu\mu}$ by Eq. (3.127) is not so easy when the order g of G is quite high. There is another calculation method for the basis vectors $b^{j}_{\nu\mu}$ by making use of the class operator C_{α}.

Since a class operator C_{α} is commutable with every element in G, the representation matrix of C_{α} in an irreducible representation $D^{j}(G)$ is a constant matrix (Schur Theorem). The constant is $n(\alpha)\chi^{j}_{\alpha}/m_{j}$ (see Prob. 8), where m_{j} is the dimension of the representation $D^{j}(G)$, χ^{j}_{α} is the character of the class $\mathcal{C}_{\alpha}$ in $D^{j}(G)$, and $n(\alpha)$ is the number of elements in the class $\mathcal{C}_{\alpha}$. Namely, $b^{(j)}_{\nu\mu}$ is the eigenvector of C_{α}, so is $e^{(j)}$,

$$\mathsf{C}_{\alpha}b^{(j)}_{\nu\mu} = b^{(j)}_{\nu\mu}\mathsf{C}_{\alpha} = \frac{n(\alpha)\chi^{j}_{\alpha}}{m_{j}} b^{(j)}_{\nu\mu}, \qquad \mathsf{C}_{\alpha}e^{(j)} = \frac{n(\alpha)\chi^{j}_{\alpha}}{m_{j}} e^{(j)}. \tag{3.134}$$

From Eq. (3.129) $e^{(j)}$ is a vector in the class algebra so that $e^{(j)}$ is easy to be solved by Eq. (3.20). In fact, as pointed by [Chen–Ping–Wang (2002)], $e^{(j)}$ is the common eigenvector of the matrices $M(\alpha)$ whose matrix entry $M_{\gamma\beta}(\alpha)$ is equal to the coefficient $f(\alpha,\beta,\gamma)$ in Eq. (3.20). However, we are more interested in the irreducible basis vectors $b^{(j)}_{\nu\mu}$ which satisfy Eqs. (3.125), (3.133), and (3.134). Equation (3.134) shows that $b^{(j)}_{\nu\mu}$ are the common eigenvectors of C_α with the eigenvalues $n(\alpha)\chi^j_\alpha/m_j$. In most finite groups G, one only needs to calculate the common eigenvectors of one or two C_α if the eigenvalues are not degenerate. The set of C_α is called the complete set of commuting operators I (CSCO I) by [Chen–Ping–Wang (2002)]. For example, CSCO I is the class operator C_2 (2-fold rotations) for the group D_3, C_2 and C_N (N-fold rotations) for D_N, C_3 (3-fold rotations) for **T**, C_4 (4-fold rotations) for **O**, and C_5 (5-fold rotations) for **I**.

In order to identify the row and the column of $b^{(j)}_{\nu\mu}$, one needs the set of more commuting operators, left-multiplying and right-multiplying on $b^{(j)}_{\nu\mu}$. For some finite groups with lower order, the set of commuting operators contains only a few mutually commutable group elements, denoted by A for convenience, namely, one choose elements A whose representation matrices are diagonal. The basis $b^{(j)}_{\nu\mu}$ is the common eigenvector of, in additional to CSCO I, left-multiplication and right-multiplication with A

$$Ab^{(j)}_{\nu\mu} = \eta^\nu b^{(j)}_{\nu\mu}, \qquad b^{(j)}_{\nu\mu}A = \eta^\mu b^{(j)}_{\nu\mu}, \qquad \eta = \exp\{-i2\pi/\tau\}, \tag{3.135}$$

where τ is the order of A, $A^\tau = E$. For example, A is the 2-fold rotation C'_2 for the group D_N, the 3-fold rotation R_1 for **T** and for **O**, and the 5-fold rotation T_0 for **I**. In [Chen–Ping–Wang (2002)], CSCO II is composed of CSCO I and left-multiplication with a few elements A, and CSCO III is composed of CSCO II and right-multiplication with a few elements A.

Now, we calculate the irreducible basis $b^{(j)}_{\nu\mu}$ for the group D_3 as example. CSCO-I for D_3 (see Table 2.6) is $\mathsf{C}_2 = A + B + C$. The additional operator in CSCO-II and CSCO III for D_3 is a 2-fold rotation A. The basis $b^{(j)}_{\nu\mu}$ satisfies

$$\begin{aligned}
&\mathsf{C}_2 b^{(A)}_{00} = b^{(A)}_{00}\mathsf{C}_2 = 3b^{(A)}_{00}, && \mathsf{C}_2 b^{(B)}_{11} = b^{(B)}_{11}\mathsf{C}_2 = -3b^{(B)}_{11},\\
&\mathsf{C}_2 b^{(E)}_{\nu\mu} = b^{(E)}_{\nu\mu}\mathsf{C}_2 = 0, && \nu,\ \mu = 0,1,\\
&Ab^{(A)}_{00} = b^{(A)}_{00}A = b^{(A)}_{00}, && Ab^{(B)}_{11} = b^{(B)}_{11}A = -b^{(B)}_{11},\\
&Ab^{(E)}_{\nu\mu} = (-1)^\nu b^{(E)}_{\nu\mu}, && b^{(E)}_{\nu\mu}A = (-1)^\mu b^{(E)}_{\nu\mu}.
\end{aligned} \tag{3.136}$$

In the real calculation, one first calculates the eigenfunctions $\Phi^{(r)}_{\nu\mu}$ of left- and right-multiplications of A,

$$A\Phi^{(r)}_{\nu\mu} = (-1)^\nu \Phi^{(r)}_{\nu\mu}, \qquad \Phi^{(r)}_{\nu\mu}A = (-1)^\mu \Phi^{(r)}_{\nu\mu}.$$

In fact, $\Phi^{(r)}_{\nu\mu}$ are calculated by the idempotents e^j of the cyclic group C_2 generated by A. $C_2 = \{E,\ A\}$ has two inequivalent irreducible representations D^A and D^B (see Table 3.1). The corresponding idempotents $e^{(j)}$, denoted by P_j for convenience, are

$$e^{(A)} \equiv P_0 = (E+A)/2, \qquad e^{(B)} \equiv P_1 = (E-A)/2. \tag{3.137}$$

Let $\Phi_{\nu\mu}^{(r)}$ be proportional to $P_\nu R P_\mu$. Take $R = E$ to calculate $\Phi_{\nu\mu}^{(1)}$. Then, taking R to be, say D, which does not appear in $\Phi_{\nu\mu}^{(1)}$, one obtains

$$\begin{aligned} \Phi_{\nu\mu}^{(1)} &= 4P_\nu E P_\mu = 2\delta_{\nu\mu}\left\{E + (-1)^\mu A\right\}, \\ \Phi_{\nu\mu}^{(2)} &= 4P_\nu D P_\mu = D + (-1)^{\nu+\mu} F + (-1)^\mu B + (-1)^\nu C. \end{aligned} \tag{3.138}$$

The matrix of C_2 in $\Phi_{\nu\mu}^{(r)}$ is

$$\begin{aligned} \mathsf{C}_2\,\Phi_{\mu\mu}^{(1)} &= \mathsf{C}_2\,\Phi_{\mu\mu}^{(2)} = 2(-1)^\mu\left\{E + D + F\right\} + 2\left\{A + B + C\right\} \\ &= (-1)^\mu\left[\Phi_{\mu\mu}^{(1)} + 2\Phi_{\mu\mu}^{(2)}\right], \\ \mathsf{C}_2\,\Phi_{\nu\mu}^{(2)} &= 0, \qquad \nu \neq \mu. \end{aligned}$$

When $\mu = \nu$, the matrix of C_2 in the basis functions $\Phi_{\mu\mu}^{(r)}$ is $(-1)^\mu\begin{pmatrix}1 & 1\\ 2 & 2\end{pmatrix}$. The eigenvector of C_2 with the eigenvalue $3(-1)^\mu$ is $\left[\Phi_{\mu\mu}^{(1)} + 2\Phi_{\mu\mu}^{(2)}\right]$ and that with the eigenvalue 0 is $\left[\Phi_{\rho\rho}^{(1)} - \Phi_{\rho\rho}^{(2)}\right]$. After normalization one has

$$\begin{aligned} b_{00}^{(A)} &= \frac{1}{\sqrt{6}}\left\{\Phi_{00}^{(1)} + 2\Phi_{00}^{(2)}\right\} = \frac{1}{\sqrt{6}}\left\{E + D + F + A + B + C\right\}, \\ b_{11}^{(B)} &= \frac{1}{\sqrt{6}}\left\{\Phi_{11}^{(1)} + 2\Phi_{11}^{(2)}\right\} = \frac{1}{\sqrt{6}}\left\{E + D + F - A - B - C\right\}, \\ b_{00}^{(E)} &= \frac{1}{2\sqrt{3}}\left\{\Phi_{00}^{(1)} - \Phi_{00}^{(2)}\right\} = \frac{1}{2\sqrt{3}}\left\{2E - D - F + 2A - B - C\right\}, \\ b_{11}^{(E)} &= C_{11}\left\{\Phi_{11}^{(1)} - \Phi_{11}^{(2)}\right\} = C_{11}\left\{2E - D - F - 2A + B + C\right\}. \end{aligned} \tag{3.139}$$

When $\nu \neq \mu$, $\Phi_{\nu\mu}^{(2)}$ is the eigenvector of C_2 with the eigenvalue 0,

$$\begin{aligned} b_{01}^{(E)} &= C_{01}\,\Phi_{01}^{(2)} = C_{01}\left\{D - F - B + C\right\}, \\ b_{10}^{(E)} &= C_{10}\,\Phi_{10}^{(2)} = C_{10}\left\{D - F + B - C\right\}. \end{aligned} \tag{3.140}$$

The representation D^E depends on the choice of the coefficients through Eqs. (3.125) and (3.133). Conversely, if the representation (2.12) is taken, one has

$$Db_{0\lambda}^{(E)} = -\frac{1}{2}b_{0\lambda}^{(E)} + \frac{\sqrt{3}}{2}b_{1\lambda}^{(E)}, \qquad b_{\rho 0}^{(E)}D = -\frac{1}{2}b_{\rho 0}^{(E)} - \frac{\sqrt{3}}{2}b_{\rho 1}^{(E)}. \tag{3.141}$$

Thus, the coefficients are calculated, $C_{01} = C_{10} = 1/2$ and $C_{11} = 1/\sqrt{12}$.

The calculations of the standard irreducible basis vectors $b^j_{\mu\nu}$ for the groups **T** and **O** can be found in Probs. 22 and 23 of Chap. 3 of [Ma and Gu (2004)]. The property of **I** is introduced in §2.5.4. The character table is given in Table 3.14 and calculated in Prob. 18 of Chap 3. The representation matrices of the generators T_0 and S_1 are given in Prob. 21 of Chap. 3 and calculated in Prob. 12 of Chap. 4. The CSCOs of **I** are C_5 and the left- and right-multiplications with a 5-fold rotation T_0.

$$\begin{array}{l} \mathsf{C}_5 = \sum_{k=0}^{5} \left(T_k + T_k^4\right), \qquad \mathsf{C}_5 b^j_{\mu\nu} = b^j_{\mu\nu}\mathsf{C}_5 = \alpha^j b^j_{\mu\nu}, \\ \alpha^A = 12, \quad \alpha^{T_1} = 4p^{-1}, \quad \alpha^{T_2} = -4p, \quad \alpha^G = -3, \quad \alpha^H = 0, \end{array} \tag{3.142}$$

where $p = \eta + \eta^{-1}$, $p^{-1} = -\eta^2 - \eta^{-2}$, and $\eta = \exp\{-i2\pi/5\}$.

First calculate the eigenfunctions $\Phi^{(a)}_{\mu\nu}$ of the idempotents $e^{(\mu)}$ of the cyclic group C_5 generated by T_0. $e^{(\mu)}$ is denoted by P_μ for convenience,

$$\begin{array}{lll} P_\mu = \frac{1}{5} \sum_{\rho=-2}^{2} \eta^{-\mu\rho} T_0^\rho, & \Phi^{(a)}_{\mu\nu} = P_\mu R^{(a)} P_\nu, & \\ T_0\Phi^{(a)}_{\mu\nu} = \eta^\mu \Phi^{(a)}_{\mu\nu}, & \Phi^{(a)}_{\mu\nu} T_0 = \eta^\nu \Phi^{(a)}_{\mu\nu}, & \mu, \nu \bmod 5. \end{array} \tag{3.143}$$

Take $R^{(1)} = E$ to calculate $\Phi^{(1)}_{\mu\nu}$. Then, taking $R^{(2)}$ in **I**, say S_{11}, which does not appear in $\Phi^{(1)}_{\mu\nu}$, one calculates $\Phi^{(2)}_{\mu\nu}$. In this way, one calculates $\Phi^{(3)}_{\mu\nu}$ and $\Phi^{(4)}_{\mu\nu}$ by $R^{(3)} = S_5$ and $R^{(4)} = S_{10}$. The normalized eigenfunctions are

$$\begin{aligned} \Phi^{(1)}_{\mu\mu} &= \left(E + \eta^{-\mu} T_0 + \eta^{-2\mu} T_0^2 + \eta^{2\mu} T_0^3 + \eta^\mu T_0^4\right)/\sqrt{5}, \\ \Phi^{(2)}_{\mu\overline{\mu}} &= \left(S_{11} + \eta^{-\mu} S_{14} + \eta^{-2\mu} S_{12} + \eta^{2\mu} S_{15} + \eta^\mu S_{13}\right)/\sqrt{5}, \\ \Phi^{(3)}_{\mu\nu} &= \left\{\left(S_5 + \eta^{-\mu} R_5^2 + \eta^{-2\mu} T_1^4 + \eta^{2\mu} T_4 + \eta^\mu R_4\right)\right. \\ &+ \eta^{(\mu-\nu)} \left(S_4 + \eta^{-\mu} R_4^2 + \eta^{-2\mu} T_5^4 + \eta^{2\mu} T_3 + \eta^\mu R_3\right) \\ &+ \eta^{2(\mu-\nu)} \left(S_3 + \eta^{-\mu} R_3^2 + \eta^{-2\mu} T_4^4 + \eta^{2\mu} T_2 + \eta^\mu R_2\right) \\ &+ \eta^{-2(\mu-\nu)} \left(S_2 + \eta^{-\mu} R_2^2 + \eta^{-2\mu} T_3^4 + \eta^{2\mu} T_1 + \eta^\mu R_1\right) \\ &+ \left.\eta^{-(\mu-\nu)} \left(S_1 + \eta^{-\mu} R_1^2 + \eta^{-2\mu} T_2^4 + \eta^{2\mu} T_5 + \eta^\mu R_5\right)\right\}/5, \\ \Phi^{(4)}_{\mu\nu} &= \left\{\left(S_{10} + \eta^{-\mu} T_1^3 + \eta^{-2\mu} R_6^2 + \eta^{2\mu} R_9 + \eta^\mu T_5^2\right)\right. \\ &+ \eta^{(\mu-\nu)} \left(S_9 + \eta^{-\mu} T_5^3 + \eta^{-2\mu} R_{10}^2 + \eta^{2\mu} R_8 + \eta^\mu T_4^2\right) \\ &+ \eta^{2(\mu-\nu)} \left(S_8 + \eta^{-\mu} T_4^3 + \eta^{-2\mu} R_9^2 + \eta^{2\mu} R_7 + \eta^\mu T_3^2\right) \\ &+ \eta^{-2(\mu-\nu)} \left(S_7 + \eta^{-\mu} T_3^3 + \eta^{-2\mu} R_8^2 + \eta^{2\mu} R_6 + \eta^\mu T_2^2\right) \\ &+ \left.\eta^{-(\mu-\nu)} \left(S_6 + \eta^{-\mu} T_2^3 + \eta^{-2\mu} R_7^2 + \eta^{2\mu} R_{10} + \eta^\mu T_1^2\right)\right\}/5, \end{aligned} \tag{3.144}$$

where $\Phi^{(1)}_{\mu\nu} = 0$ if $\mu \neq \nu$, $\Phi^{(2)}_{\mu\nu} = 0$ if $\mu \neq -\nu$, and $\overline{\mu}$ denotes $-\mu$.

Second, calculate the matrix of the class operator C_5 in $\Phi^{(a)}_{\mu\nu}$ and its eigenvectors. Due to Eq. (3.144), only the coefficients of the terms E, S_{11}, S_5, and S_{10} in $\mathsf{C}_5\Phi^{(a)}_{\mu\nu}$ are needed to be calculated explicitly. The group table of **I** is given in Tables 2.12 and 2.14. From Table 2.14 one has

$$\begin{aligned}
&T_1 = T_0^2R_2^2, \quad T_2 = T_0^4R_2, \quad T_3 = T_0^2R_4^2, \quad T_4 = T_0^4R_4, \quad T_5 = T_0^3S_1, \\
&T_1^4 = T_0^3R_4, \quad T_2^4 = T_0^2S_1, \quad T_3^4 = T_0R_2^2, \quad T_4^4 = T_0^3R_2, \quad T_5^4 = T_0R_4^2, \\
&S_{11} = S_{12}T_0^2 = T_0S_{12}T_0^3 = T_0^2S_{12}T_0^4 = T_0^3S_{12} = T_0^4S_{12}T_0, \\
&S_5 = R_4^2T_0^4 = T_0R_4 = T_0^2R_2^2T_0^2 = T_0^3R_2T_0^3 = T_0^4S_1T_0, \\
&S_{10} = R_6T_0^2 = T_0R_{10}T_0 = T_0^2S_8T_0^3 = T_0^3R_6^2 = T_0^4R_{10}^2T_0^4.
\end{aligned}$$

Then, say, $S_{11} = T_0^2S_{12}T_0^4 = (T_0^2R_2^2)(R_2S_{12}T_0^4) = T_1(R_{10}T_0^4) = T_1R_8^2$,

$$\begin{aligned}
E &= T_jT_j^4 = T_j^4T_j, \qquad 0 \le j \le 5, \\
S_{11} &= T_0S_{13} = T_0^4S_{14} = T_1R_8^2 = T_1^4R_6^2 = T_2T_1^2 = T_2^4T_4^2 \\
&= T_3T_1^3 = T_3^4T_4^3 = T_4R_8 = T_4^4R_6 = T_5S_8 = T_5^4S_6, \\
S_5 &= T_0R_4 = T_0^4R_5^2 = T_1T_0^2 = T_1^4T_5^3 = T_2S_4 = T_2^4S_{10} \\
&= T_3S_9 = T_3^4S_1 = T_4T_5^2 = T_4^4T_0^3 = T_5R_5 = T_5^4R_4^2, \\
S_{10} &= T_0T_5^2 = T_0^4T_1^3 = T_1R_5 = T_1^4R_{10}^2 = T_2S_5 = T_2^4S_{13} \\
&= T_3T_5^3 = T_3^4T_1^2 = T_4S_{15} = T_4^4S_1 = T_5R_{10} = T_5^4R_5^2.
\end{aligned}$$

Thus, the matrix of the class operator C_5 in $\Phi^{(a)}_{\mu\nu}$ is obtained,

$$\begin{aligned}
\mathsf{C}_5\Phi^{(1)}_{\mu\mu} &= \left[\left(\eta^{\mu} + \eta^{-\mu}\right)E + \left(\eta^{2\mu} + \eta^{-2\mu}\right)S_5 + \ldots\right]/\sqrt{5} \\
&= \left(\eta^{\mu} + \eta^{-\mu}\right)\Phi^{(1)}_{\mu\mu} + \sqrt{5}\left(\eta^{2\mu} + \eta^{-2\mu}\right)\Phi^{(3)}_{\mu\mu}, \\
\mathsf{C}_5\Phi^{(2)}_{\mu\overline{\mu}} &= \left[\left(\eta^{\mu} + \eta^{-\mu}\right)S_{11} + \left(\eta^{\mu} + \eta^{2\mu}\right)S_{10} + \ldots\right)/\sqrt{5} \\
&= \left(\eta^{\mu} + \eta^{-\mu}\right)\Phi^{(2)}_{\mu\overline{\mu}} + \sqrt{5}\left(\eta^{\mu} + \eta^{2\mu}\right)\Phi^{(4)}_{\mu\overline{\mu}}, \\
\mathsf{C}_5\Phi^{(3)}_{\mu\nu} &= \left[\delta_{\mu\nu}\left(\eta^{2\mu} + \eta^{-2\mu}\right)E + \left(\eta^{\mu} + \eta^{-\mu} + \eta^{\nu} + \eta^{-\nu}\right.\right. \\
&\quad \left.\left. + \eta^{\mu-\nu} + \eta^{\nu-\mu}\right)S_5 + \left(1 + \eta^{-\mu}\right)\left(1 + \eta^{\nu}\right)S_{10} + \ldots\right]/5 \\
&= \delta_{\mu\nu}\sqrt{5}\left(\eta^{2\mu} + \eta^{-2\mu}\right)\Phi^{(1)}_{\mu\mu} + \left(\eta^{\mu} + \eta^{-\mu} + \eta^{\nu} + \eta^{-\nu}\right. \\
&\quad \left. + \eta^{\mu-\nu} + \eta^{\nu-\mu}\right)\Phi^{(3)}_{\mu\nu} + \left(1 + \eta^{-\mu}\right)\left(1 + \eta^{\nu}\right)\Phi^{(4)}_{\mu\nu}, \\
\mathsf{C}_5\Phi^{(4)}_{\mu\nu} &= \left[\delta_{\mu(-\nu)}\left(\eta^{-\mu} + \eta^{-2\mu}\right)S_{11} + \left(1 + \eta^{\mu}\right)\left(1 + \eta^{-\nu}\right)S_5\right. \\
&\quad \left. + \left(\eta^{\mu} + \eta^{-\mu} + \eta^{\nu} + \eta^{-\nu} + \eta^{\mu+\nu} + \eta^{-\mu-\nu}\right)S_{10} + \ldots\right]/5 \\
&= \delta_{\mu(-\nu)}\sqrt{5}\left(\eta^{-\mu} + \eta^{-2\mu}\right)\Phi^{(2)}_{\mu\overline{\mu}} + \left(1 + \eta^{\mu}\right)\left(1 + \eta^{-\nu}\right)\Phi^{(3)}_{\mu\nu} \\
&\quad + \left(\eta^{\mu} + \eta^{-\mu} + \eta^{\nu} + \eta^{-\nu} + \eta^{\mu+\nu} + \eta^{-\mu-\nu}\right)\Phi^{(4)}_{\mu\nu}.
\end{aligned} \tag{3.145}$$

Its eigenvectors are the irreducible bases $b^j_{\mu\nu}$ of **I** as listed in Table 2.15.

For example, when $\mu = \nu = 0$, the matrix of C_5 in $\Phi^{(a)}_{00}$ and its eigenvectors with the eigenvalues 12, $4p^{-1}$, $-4p$, and 0, respectively, are

$$\begin{pmatrix} 2 & 0 & 2\sqrt{5} & 0 \\ 0 & 2 & 0 & 2\sqrt{5} \\ 2\sqrt{5} & 0 & 6 & 4 \\ 0 & 2\sqrt{5} & 4 & 6 \end{pmatrix}, \quad \begin{pmatrix} 1 \\ 1 \\ \sqrt{5} \\ \sqrt{5} \end{pmatrix}, \quad \begin{pmatrix} 1 \\ -1 \\ 1 \\ -1 \end{pmatrix}, \quad \begin{pmatrix} 1 \\ -1 \\ -1 \\ 1 \end{pmatrix}, \quad \begin{pmatrix} \sqrt{5} \\ \sqrt{5} \\ -1 \\ -1 \end{pmatrix}.$$

Normalizing the eigenvectors, one obtains b^A_{00}, $b^{T_1}_{00}$, $b^{T_2}_{00}$, and b^H_{00}, respectively. When $\mu = 0$ and $\nu \neq 0$ one has the matric of C_5 in $\Phi^{(3)}_{0\nu}$ and $\Phi^{(4)}_{0\nu}$ is $2\begin{pmatrix} -\eta^{2\nu} - \eta^{-2\nu} & 1+\eta^{-\nu} \\ 1+\eta^{\nu} & -\eta^{2\nu} - \eta^{-2\nu} \end{pmatrix}$. Normalizing the eigenvectors, one obtains $b^{T_1}_{0\nu}$, $b^{T_2}_{0\nu}$, and $b^H_{0\nu}$, respectively.

Table 3.15 Irreducible bases $b^j_{\mu\nu}$ in the group space of I

$$b^j_{\mu\nu} = N^{-1/2} \sum_{a=1}^{4} c_a \, \Phi^{(a)}_{\mu\nu}, \qquad \eta = \exp(-i2\pi/5), \qquad p = \eta + \eta^{-1}$$

$$b^A_{00} = \left(\Phi^{(1)}_{00} + \Phi^{(2)}_{00} + \sqrt{5}\Phi^{(3)}_{00} + \sqrt{5}\Phi^{(4)}_{00} \right) / \sqrt{12}.$$

D^{T_1}

μ	ν	c_1	c_2	c_3	c_4	N
1	1	1	0	$-p^{-1}$	$-p$	4
0	1	0	0	$-\eta^{-1}$	η^2	2
-1	1	0	η^{-2}	$-\eta^{-2}p$	$-\eta^{-1}p^{-1}$	4
1	0	0	0	$-\eta$	η^{-2}	2
0	0	1	-1	1	-1	4
-1	0	0	0	η^{-1}	$-\eta^2$	2
1	-1	0	η^2	$-\eta^2 p$	$-\eta p^{-1}$	4
0	-1	0	0	η	$-\eta^{-2}$	2
-1	-1	1	0	$-p^{-1}$	$-p$	4

D^{T_2}

μ	ν	c_1	c_2	c_3	c_4	N
2	2	1	0	$-p$	$-p^{-1}$	4
0	2	0	0	η^{-2}	$-\eta^{-1}$	2
-2	2	0	$-\eta$	ηp^{-1}	$\eta^{-2}p$	4
2	0	0	0	η^2	$-\eta$	2
0	0	1	-1	-1	1	4
-2	0	0	0	η^{-2}	$-\eta^{-1}$	2
2	-2	0	$-\eta^{-1}$	$\eta^{-1}p^{-1}$	$\eta^2 p$	4
0	-2	0	0	η^2	$-\eta$	2
-2	-2	1	0	$-p$	$-p^{-1}$	4

D^G

μ	ν	c_1	c_2	c_3	c_4	N
2	2	1	0	-1	1	3
1	2	0	0	$-\eta^{-1}p$	$-\eta^2 p^{-1}$	3
$\overline{1}$	2	0	0	$-\eta^2 p^{-1}$	$-\eta p$	3
$\overline{2}$	2	0	η	η	$-\eta^{-2}$	3
2	1	0	0	$-\eta p$	$-\eta^{-2}p^{-1}$	3
1	1	1	0	1	-1	3
$\overline{1}$	1	0	η^{-2}	$-\eta^{-2}$	η^{-1}	3
$\overline{2}$	1	0	0	$-\eta^2 p^{-1}$	$-\eta p$	3
2	$\overline{1}$	0	0	$-\eta^{-2}p^{-1}$	$-\eta^{-1}p$	3
1	$\overline{1}$	0	η^2	$-\eta^2$	η	3
$\overline{1}$	$\overline{1}$	1	0	1	-1	3
$\overline{2}$	$\overline{1}$	0	0	$-\eta^{-1}p$	$-\eta^2 p^{-1}$	3
2	$\overline{2}$	0	η^{-1}	η^{-1}	$-\eta^2$	3
1	$\overline{2}$	0	0	$-\eta^{-2}p^{-1}$	$-\eta^{-1}p$	3
$\overline{1}$	$\overline{2}$	0	0	$-\eta p$	$-\eta^{-2}p^{-1}$	3
$\overline{2}$	$\overline{2}$	1	0	-1	1	3

D^H

μ	ν	c_1	c_2	c_3	c_4	N
2	2	$\sqrt{5}$	0	p^{-2}	p^2	12
1	2	0	0	$\eta^{-1}p^{-1}$	$-\eta^2 p$	3
0	2	0	0	η^{-2}	η^{-1}	2
$\overline{1}$	2	0	0	$\eta^2 p$	$-\eta p^{-1}$	3
$\overline{2}$	2	0	$\sqrt{5}\eta$	ηp^2	$\eta^{-2}p^{-2}$	12
2	1	0	0	ηp^{-1}	$-\eta^{-2}p$	3
1	1	$\sqrt{5}$	0	p^2	p^{-2}	12
0	1	0	0	$-\eta^{-1}$	$-\eta^2$	2
$\overline{1}$	1	0	$-\sqrt{5}\eta^{-2}$	$-\eta^{-2}p^{-2}$	$-\eta^{-1}p^2$	12
$\overline{2}$	1	0	0	$-\eta^2 p$	ηp^{-1}	3
2	0	0	0	η^2	η	2
1	0	0	0	$-\eta$	$-\eta^{-2}$	2
0	0	$\sqrt{5}$	$\sqrt{5}$	-1	-1	12
$\overline{1}$	0	0	0	η^{-1}	η^2	2
$\overline{2}$	0	0	0	η^{-2}	η^{-1}	2
2	$\overline{1}$	0	0	$\eta^{-2}p$	$-\eta^{-1}p^{-1}$	3
1	$\overline{1}$	0	$-\sqrt{5}\eta^2$	$-\eta^2 p^{-2}$	$-\eta p^2$	12
0	$\overline{1}$	0	0	η	η^{-2}	2
$\overline{1}$	$\overline{1}$	$\sqrt{5}$	0	p^2	p^{-2}	12
$\overline{2}$	$\overline{1}$	0	0	$-\eta^{-1}p^{-1}$	$\eta^2 p$	3
2	$\overline{2}$	0	$\sqrt{5}\eta^{-1}$	$\eta^{-1}p^2$	$\eta^2 p^{-2}$	12
1	$\overline{2}$	0	0	$-\eta^{-2}p$	$\eta^{-1}p^{-1}$	3
0	$\overline{2}$	0	0	η^2	η	2
$\overline{1}$	$\overline{2}$	0	0	$-\eta p^{-1}$	$\eta^{-2}p$	3
$\overline{2}$	$\overline{2}$	$\sqrt{5}$	0	p^{-2}	p^2	12

3.8 Exercises

1. Let G be a non-Abelian group, $D(G)$ be a faithful representation of the group G, and $D(R)$ be the representation matrix of the element R. If the element R in G corresponds to the following matrix in the set, please decide whether the following set forms a representation of the group G. For example, in (a), if $R \longleftrightarrow D(R)^\dagger$, please decide whether the set $D(G)^\dagger$ composed of $D(R)^\dagger$ forms a representation of G.
(a) $D(R)^\dagger$; (b) $D(R)^T$; (c) $D(R^{-1})$; (d) $D(R)^*$; (e) $D(R^{-1})^\dagger$; (f) $\det D(R)$; (g) $\text{Tr}\ D(R)$.

2. The homogeneous function space of degree 2 spanned by the basis functions $\psi_1(x,y) = x^2$, $\psi_2(x,y) = xy$, and $\psi_3(x,y) = y^2$ is invariant in the following rotations R in the two-dimensional coordinate space. Calculate the matrix form $D(R)$ of the corresponding transformation operator P_R for scalar functions in the three-dimensional function space:

$$\begin{pmatrix} x' \\ y' \end{pmatrix} = R \begin{pmatrix} x \\ y \end{pmatrix}, \qquad (a)\ \ R = \begin{pmatrix} 1 & 0 \\ 0 & -1 \end{pmatrix},$$

$$(b)\ \ R = \frac{1}{2}\begin{pmatrix} -1 & -\sqrt{3} \\ \sqrt{3} & -1 \end{pmatrix}, \qquad (c)\ \ R = \begin{pmatrix} \cos\alpha & -\sin\alpha \\ \sin\alpha & \cos\alpha \end{pmatrix}.$$

By replacing the basis function $\psi_2(x,y) = xy$ with $\sqrt{2}xy$, how will the matrix form of the operator P_R change?

3. Prove that the module of any representation matrix in a one-dimensional representation of a finite group is equal to 1.

4. Prove that any irreducible representation of an infinite Abelian group is one-dimensional.

5. Prove that the similarity transformation matrix between two equivalent irreducible unitary representations of a finite group, if restricting its determinant to be 1, has to be unitary.

6. Show that for a finite group G, the sum of the characters of all elements in any irreducible representation of G, except for the identical representation, is equal to zero.

7. Calculate the similarity transformation matrix X between two equivalent regular representations $D(R)$ and $\overline{D}(R)$, given in Eqs. (3.6) and (3.12), in the group space of the D_3 group. How to generalize the result to any other finite group?

8. $\mathcal{C}_\alpha$ is a class in a finite group G, and C_α is a vector in the group space composed of the sum of all elements in $\mathcal{C}_\alpha$. Prove that the representation matrix of C_α in an irreducible representation is a constant matrix, and calculate this constant.

9. Prove that the number of self-reciprocal classes in a finite group G is equal to the number of inequivalent and irreducible self-conjugate representations of G, in other words, the number of pairs of the reciprocal classes is equal to the number of pairs of the inequivalent and irreducible non-self-conjugate representations.

10. If the group G is a direct product $H_1 \otimes H_2$ of two subgroups, show that the direct product of two irreducible representations of two subgroups is an irreducible representation of G.

11. Let each element in D_3 be the coordinate transformation in two-dimensional space:

$$\begin{pmatrix} x' \\ y' \end{pmatrix} = R \begin{pmatrix} x \\ y \end{pmatrix}, \qquad R \in \mathrm{D}_3,$$

where R is equal to the representation matrix in the two-dimensional representation $D^E(\mathrm{D}_3)$. For the generators D and A in D_3, one has

$$D = D^E(D) = \frac{1}{2}\begin{pmatrix} -1 & -\sqrt{3} \\ \sqrt{3} & -1 \end{pmatrix}, \qquad A = D^E(A) = \begin{pmatrix} 1 & 0 \\ 0 & -1 \end{pmatrix}.$$

The four-dimensional functional space spanned by the following basis functions is invariant in the group D_3:

$$\psi_1(x,y) = x^3, \quad \psi_2(x,y) = x^2 y, \quad \psi_3(x,y) = xy^2, \quad \psi_4(x,y) = y^3.$$

Calculate the representation matrices of the generators D and A of D_3 in this representation. Then, reduce this representation into the direct sum of the irreducible representations of D_3, and construct new basis functions which are the linear combinations of the original basis functions and belong to the irreducible representations.

12. Calculate the characters and the representation matrices of the proper symmetric group $\mathbf{O}$ of a cube with the method of the quotient group and the method of coordinate transformations.

13. The multiplication table for a group G of order 12 is as follows.

	E	A	B	C	D	F	I	J	K	L	M	N
E	E	A	B	C	D	F	I	J	K	L	M	N
A	A	B	E	I	L	K	N	D	M	J	F	C
B	B	E	A	N	J	M	C	L	F	D	K	I
C	C	K	L	D	E	B	M	I	N	F	J	A
D	D	N	F	E	C	L	J	M	A	B	I	K
F	F	D	N	K	M	I	E	B	L	C	A	J
I	I	M	J	L	A	E	F	N	C	K	D	B
J	J	I	M	B	N	D	L	K	E	A	C	F
K	K	L	C	M	F	N	A	E	J	I	B	D
L	L	C	K	A	I	J	D	F	B	E	N	M
M	M	J	I	F	K	C	B	A	D	N	E	L
N	N	F	D	J	B	A	K	C	I	M	L	E

(a) Find out the inverse of each element in G;

(b) Point out which elements can commute with any element in the group;

(c) List the period and the order of each element;

(d) Find out the elements in each class of G;

(e) Find out all invariant subgroups in G. For each invariant subgroup, list its cosets and point out onto which group its quotient group is isomorphic;

(f) Establish the character table of G;

(g) Decide whether G is isomorphic onto the symmetric group $\mathbf{T}$ of the tetrahedron, or the symmetric group D_6 of the regular six-sided polygon.

14. Calculate the character table of the group G given in Prob. 17 of Chap. 2.

15. Calculate all inequivalent irreducible representations of the group D_{2n+1} with the method of induced representation.

16. Calculate all inequivalent irreducible representations of the group D_{2n} with the method of induced representation.

17. Calculate all inequivalent irreducible representations of the symmetric group $\mathbf{O}$ of a cube with the method of induced representation.

18. The regular icosahedron is shown in Fig. 2.3. Calculate the character table of the proper symmetric group $\mathbf{I}$ of an icosahedron with the method of induced representation.

19. Calculate the reduction of the subduced representation from each irreducible representation of the group $\mathbf{I}$ with respect to the subgroups C_5, D_5, and $\mathbf{T}$.

20. Calculate the reduction of the subduced representation from the regular representation of the improper symmetric group $\mathbf{I}_h$ of an icosahedron with respect to the subgroup C_{5i}, D_{5d}, and $\mathbf{T}_h$.

21. Calculate the representation matrices of R_6 and S_{12} of $\mathbf{I}$ in its irreducible representations by $R_6 = S_1 T_0^2 S_1 T_0^{-1}$ and $S_{12} = T_0 S_1 T_0^3 R_6$ (see Prob. 4 in Chap. 4), where the representation matrices of the generators T_0 and S_1 are (see Prob. 12 in Chap. 4):

$$
\begin{aligned}
&D^{T_1}(T_0) = \text{diag}\left\{\eta, 1, \eta^{-1}\right\}, \\
&D^{G}(T_0) = \text{diag}\left\{\eta^2, \eta, \eta^{-1}, \eta^{-2}\right\}, \\
&D^{T_2}(T_0) = \text{diag}\left\{\eta^2, 1, \eta^{-2}\right\}, \\
&D^{H}(T_0) = \text{diag}\left\{\eta^2, \eta, 1, \eta^{-1}, \eta^{-2}\right\}, \\
&D^{T_1}(S_1) = \frac{-1}{\sqrt{5}} \begin{pmatrix} p^{-1} & \sqrt{2} & p \\ \sqrt{2} & -1 & -\sqrt{2} \\ p & -\sqrt{2} & p^{-1} \end{pmatrix}, \\
&D^{T_2}(S_1) = \frac{1}{\sqrt{5}} \begin{pmatrix} -p & \sqrt{2} & p^{-1} \\ \sqrt{2} & -1 & \sqrt{2} \\ p^{-1} & \sqrt{2} & -p \end{pmatrix}, \\
&D^{G}(S_1) = \frac{1}{\sqrt{5}} \begin{pmatrix} -1 & -p & -p^{-1} & 1 \\ -p & 1 & -1 & -p^{-1} \\ -p^{-1} & -1 & 1 & -p \\ 1 & -p^{-1} & -p & -1 \end{pmatrix}, \\
&D^{H}(S_1) = \frac{1}{5} \begin{pmatrix} p^{-2} & 2p^{-1} & \sqrt{6} & 2p & p^2 \\ 2p^{-1} & p^2 & -\sqrt{6} & -p^{-2} & -2p \\ \sqrt{6} & -\sqrt{6} & -1 & \sqrt{6} & \sqrt{6} \\ 2p & -p^{-2} & \sqrt{6} & p^2 & -2p^{-1} \\ p^2 & -2p & \sqrt{6} & -2p^{-1} & p^{-2} \end{pmatrix}.
\end{aligned}
$$

22. Please use the projection operators to reduce the regular representation of group $\mathbf{T}$.

23. Please use the projection operators to reduce the regular representation of group $\mathbf{O}$.

24. The multiplication table of the group G is given as follows.

left \ right	E	A	B	C	F	K	M	N
E	E	A	B	C	F	K	M	N
A	A	E	M	K	N	C	B	F
B	B	N	E	M	K	F	C	A
C	C	K	N	E	M	A	F	B
F	F	M	K	N	E	B	A	C
K	K	C	F	A	B	E	N	M
M	M	F	A	B	C	N	K	E
N	N	B	C	F	A	M	E	K

(a) Write the character table for the group G.
(b) If we know that the representation matrices of the generators A and B in a two-dimensional irreducible representation D are

$$D(A) = \begin{pmatrix} 1 & 0 \\ 0 & -1 \end{pmatrix}, \qquad D(B) = \begin{pmatrix} 0 & 1 \\ 1 & 0 \end{pmatrix},$$

and two sets of basis functions ψ_μ and ϕ_ν transform according to this two-dimensional representation of G, respectively:

$$P_R\psi_\mu = \sum_{\mu'} \psi_{\mu'} D_{\mu'\mu}(R), \qquad P_R\phi_\nu = \sum_{\nu'} \phi_{\nu'} D_{\nu'\nu}(R),$$

combine the product function $\psi_\mu\phi_\nu$ such that the basis function belongs to the irreducible representation of G.

25. Calculate the unitary similarity transformation matrix X for reducing the self-direct product of the three-dimensional irreducible unitary representation D^T of the group $\mathbf{T}$:

$$X^{-1}\left\{D^T(R) \times D^T(R)\right\} X = \sum_j a_j D^j(R).$$

26. Calculate the Clebsch–Gordan series and the Clebsch–Gordan coefficients for the direct product representation of each pair of two irreducible representations of the group $\mathbf{O}$.

27. Calculate the Clebsch–Gordan series and Clebsch–Gordan coefficients in the reduction of direct product representation in terms of the character table of the group $\mathbf{I}$ given in Prob. 18:

(1) $D^{T_1} \times D^{T_1}$; (2) $D^{T_1} \times D^{T_2}$; (3) $D^{T_2} \times D^{T_2}$; (4) $D^{T_1} \times D^G$; (5) $D^{T_2} \times D^G$; (6) $D^{T_1} \times D^H$; (7) $D^{T_2} \times D^H$; (8) $D^G \times D^G$; (9) $D^G \times D^H$; (10) $D^H \times D^H$.

Chapter 4

THREE-DIMENSIONAL ROTATION GROUP

A system with the spherical symmetry has a symmetric center, chosen as the origin, and the system is isotropic with respect to the origin. The system is invariant under any rotation around any axis across the origin through any angle. The symmetric group of the system is three-dimensional rotation group SO(3). The spherical symmetry is the most important symmetry studied in physics, because a system with the spherical symmetry is easier to deal with and most systems in physics have this approximative symmetry. The group SO(3) is a Lie group, where its elements can be characterized by a set of continuous parameters, and it can be dealt with by mathematical analysis. Therefore, the study on three-dimensional rotation group is very important both in physics and in mathematics.

4.1 Three-dimensional Rotations

In a real three-dimensional space, a spatial rotation keeps both the position of the origin and the distance of any two points invariant. There are two viewpoints in describing a rotation in a three-dimensional space. In one viewpoint the system rotates in a laboratory frame, and in another viewpoint the coordinate frame rotates but the system is fixed. In this book we adopt the first viewpoint. Let K be a perpendicular coordinate frame which is laboratory-fixed. An arbitrary point P is described by a position vector $\boldsymbol{r}$ which is a vector pointing from the origin O to P. The coordinate basis vectors in the laboratory frame are denoted by $\boldsymbol{e}_a$, $a = 1,\ 2,\ 3$,

$$\boldsymbol{r} = \sum_{a=1}^{3} \boldsymbol{e}_a x_a \ . \tag{4.1}$$

There is a one-to-one correspondence between the position vector $\boldsymbol{r}$ and three coordinates x_a when $\boldsymbol{e}_a$ are fixed. In a spatial rotation R the point P transforms to P' whose position vector is denoted by $\boldsymbol{r}' = \sum_a \boldsymbol{e}_a x'_a$. Since the position of the origin and the distance of any two points keep invariant in R, R can be denoted by a three-dimensional real orthogonal matrix

$$\begin{pmatrix} x'_1 \\ x'_2 \\ x'_3 \end{pmatrix} = \begin{pmatrix} R_{11} & R_{12} & R_{13} \\ R_{21} & R_{22} & R_{23} \\ R_{31} & R_{32} & R_{33} \end{pmatrix} \begin{pmatrix} x_1 \\ x_2 \\ x_3 \end{pmatrix}, \qquad \underline{x}' = R\underline{x}, \tag{4.2}$$

$$\left(\underline{x}'\right)^T \underline{x}' = \underline{x}^T \underline{x}, \qquad R^T R = \mathbf{1}, \qquad R^* = R. \tag{4.3}$$

The set of all three-dimensional real orthogonal matrices, in the multiplication rule of matrices, satisfies four axioms of a group and forms a group O(3).

Let K' be a body-fixed frame with the coordinate basis vectors $\boldsymbol{e}'_a$. The components of $\boldsymbol{r}'$ in the frame K' are invariant in the rotation R,

$$\boldsymbol{r}' = \sum_{a=1}^{3} \boldsymbol{e}_a x'_a = \sum_{b=1}^{3} \boldsymbol{e}'_b x_a. \tag{4.4}$$

Substituting Eq. (4.2) into Eq. (4.4), one obtains

$$\boldsymbol{e}'_b = \sum_{a=1}^{3} \boldsymbol{e}_a R_{ab}. \tag{4.5}$$

The chirality of a coordinate frame is determined by the mixed product of coordinate basis vectors. K is a right-handed coordinate frame,

$$\boldsymbol{e}_1 \cdot (\boldsymbol{e}_2 \times \boldsymbol{e}_3) = 1. \tag{4.6}$$

The chirality of K' depends on det R

$$\boldsymbol{e}'_1 \cdot (\boldsymbol{e}'_2 \times \boldsymbol{e}'_3) = \det R = \pm 1. \tag{4.7}$$

K' is right-handed if R is a proper rotation, det $R = 1$. K' is left-handed if R is an improper rotation, det $R = -1$. The set of all proper rotations forms an invariant subgroup of O(3), called the three-dimensional rotation group SO(3). The set of all improper rotations forms the only coset of SO(3) in O(3). The spatial inversion σ is the representative element in the coset. Every improper rotation R' is equal to the product of σ and a proper rotation R.

Study some special rotations. A rotation around the z-axis through an angle ω is denoted by $R(\boldsymbol{e}_3, \omega)$:

$$\begin{aligned} x_1' &= x_1 \cos\omega - x_2 \sin\omega, \\ x_2' &= x_1 \sin\omega + x_2 \cos\omega, \\ x_3' &= x_3, \end{aligned} \qquad R(\boldsymbol{e}_3, \omega) = \begin{pmatrix} \cos\omega & -\sin\omega & 0 \\ \sin\omega & \cos\omega & 0 \\ 0 & 0 & 1 \end{pmatrix}. \tag{4.8}$$

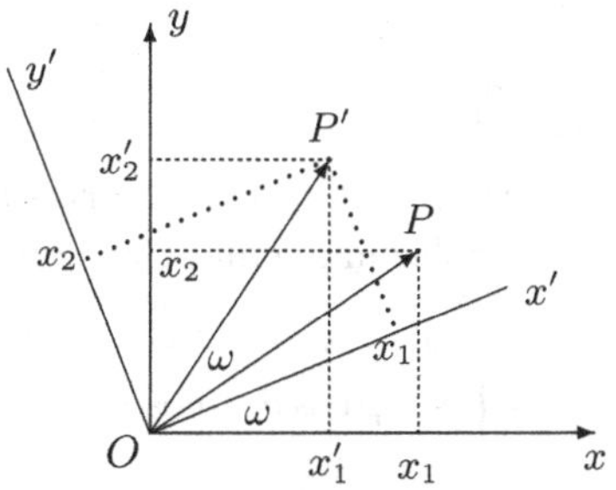

Fig. 4.1 A rotation around the z-axis.

In terms of the Pauli matrices,

$$\begin{gathered} \sigma_1 = \begin{pmatrix} 0 & 1 \\ 1 & 0 \end{pmatrix}, \qquad \sigma_2 = \begin{pmatrix} 0 & -i \\ i & 0 \end{pmatrix}, \qquad \sigma_3 = \begin{pmatrix} 1 & 0 \\ 0 & -1 \end{pmatrix}, \\ \sigma_a \sigma_b = \delta_{ab} \mathbf{1} + i \sum_{c=1}^{3} \epsilon_{abc} \sigma_c, \qquad \text{Tr}\, \sigma_a = 0, \qquad \text{Tr}\,(\sigma_a \sigma_b) = 2\delta_{ab}. \end{gathered} \tag{4.9}$$

$$\begin{aligned} \exp\{-i\omega\sigma_2\} &= \sum_n \frac{1}{n!} \omega^n (-i\sigma_2)^n \\ &= \mathbf{1} \left(1 - \frac{1}{2!}\omega^2 + \frac{1}{4!}\omega^4 - + \ldots \right) - i\sigma_2 \left(\omega - \frac{1}{3!}\omega^3 + \frac{1}{5!}\omega^5 - + \ldots \right) \\ &= \mathbf{1} \cos\omega - i\sigma_2 \sin\omega = \begin{pmatrix} \cos\omega & -\sin\omega \\ \sin\omega & \cos\omega \end{pmatrix}. \end{aligned} \tag{4.10}$$

$R(\boldsymbol{e}_3, \omega)$ can be expressed as an exponential function of matrix

$$R(\boldsymbol{e}_3, \omega) = \exp\{-i\omega T_3\} = \begin{pmatrix} \cos\omega & -\sin\omega & 0 \\ \sin\omega & \cos\omega & 0 \\ 0 & 0 & 1 \end{pmatrix}, \qquad T_3 = \begin{pmatrix} 0 & -i & 0 \\ i & 0 & 0 \\ 0 & 0 & 0 \end{pmatrix}.$$

From the cyclic of three axes, one has

$$R(\boldsymbol{e}_1, \omega) = \exp\{-i\omega T_1\} = \begin{pmatrix} 1 & 0 & 0 \\ 0 & \cos\omega & -\sin\omega \\ 0 & \sin\omega & \cos\omega \end{pmatrix}, \qquad T_1 = \begin{pmatrix} 0 & 0 & 0 \\ 0 & 0 & -i \\ 0 & i & 0 \end{pmatrix},$$

$$R(\boldsymbol{e}_2,\omega)=\exp\{-i\omega T_2\}=\begin{pmatrix}\cos\omega & 0 & \sin\omega\\ 0 & 1 & 0\\ -\sin\omega & 0 & \cos\omega\end{pmatrix},\qquad T_2=\begin{pmatrix}0 & 0 & i\\ 0 & 0 & 0\\ -i & 0 & 0\end{pmatrix}.$$

The matrix entries of T_a are

$$(T_a)_{bc}=-i\epsilon_{abc}. \tag{4.11}$$

Another useful rotation is $S(\varphi,\theta)$, which transforms $\boldsymbol{e}_3$ to the direction $\hat{\boldsymbol{n}}(\theta,\varphi)$, where θ and φ are the polar angle and the azimuthal angle of $\hat{\boldsymbol{n}}$,

$$\begin{aligned}S(\varphi,\theta)&=R(\boldsymbol{e}_3,\varphi)R(\boldsymbol{e}_2,\theta)=\begin{pmatrix}\cos\varphi\cos\theta & -\sin\varphi & \cos\varphi\sin\theta\\ \sin\varphi\cos\theta & \cos\varphi & \sin\varphi\sin\theta\\ -\sin\theta & 0 & \cos\theta\end{pmatrix},\\ S(\varphi,\theta)&\begin{pmatrix}0\\0\\1\end{pmatrix}=\begin{pmatrix}\cos\varphi\sin\theta\\ \sin\varphi\sin\theta\\ \cos\theta\end{pmatrix}=\begin{pmatrix}n_1\\n_2\\n_3\end{pmatrix}.\end{aligned} \tag{4.12}$$

It is easy to check (see Prob. 9 of Chap. 1 in [Ma and Gu (2004)])

$$\begin{aligned}S(\varphi,\theta)T_3S(\varphi,\theta)^{-1}&=n_1T_1+n_2T_2+n_3T_3=\hat{\boldsymbol{n}}\cdot\boldsymbol{T},\\ \boldsymbol{T}&=\boldsymbol{e}_1T_1+\boldsymbol{e}_2T_2+\boldsymbol{e}_3T_3.\end{aligned} \tag{4.13}$$

Thus, the rotation $R(\hat{\boldsymbol{n}},\omega)$ around the direction $\hat{\boldsymbol{n}}$ through an angle ω can be expressed as an exponential function of matrix ,

$$\begin{aligned}R(\hat{\boldsymbol{n}},\omega)&=S(\varphi,\theta)R(\boldsymbol{e}_3,\omega)S(\varphi,\theta)^{-1}=\exp\{-i\omega ST_3S^{-1}\}\\ &=\exp\{-i\omega\hat{\boldsymbol{n}}\cdot\boldsymbol{T}\}=\exp\left\{-i\sum_{a=1}^{3}\omega_aT_a\right\}.\end{aligned} \tag{4.14}$$

$$\omega_1=\omega\sin\theta\cos\varphi,\qquad \omega_2=\omega\sin\theta\sin\varphi,\qquad \omega_3=\omega\cos\theta. \tag{4.15}$$

Note that $R(\hat{\boldsymbol{n}},\omega)\hat{\boldsymbol{n}}=\hat{\boldsymbol{n}}$, $\text{Tr}R(\hat{\boldsymbol{n}},\omega)=1+2\cos\omega$, and

$$R(\hat{\boldsymbol{n}},\omega+2\pi)=R(\hat{\boldsymbol{n}},\omega)=R(-\hat{\boldsymbol{n}},2\pi-\omega),\qquad R(\hat{\boldsymbol{n}},\pi)=R(-\hat{\boldsymbol{n}},\pi). \tag{4.16}$$

On the other hand, any element R in SO(3) is a rotation around a direction $\hat{\boldsymbol{n}}$ through an angle ω where $\hat{\boldsymbol{n}}$ is the eigenvector of R to the eigenvalue 1 and ω is determined by Tr $R=1+2\cos\omega$. Thus, the element $R(\hat{\boldsymbol{n}},\omega)\in$ SO(3) can be described by $\boldsymbol{\omega}$ through its spherical coordinates (ω,θ,φ) or its rectangular coordinates $(\omega_1,\omega_2,\omega_3)$. The variation region of $\boldsymbol{\omega}$ is a spheroid with radius π, where in the sphere two end points of a diameter describe the same element (rotation). Equation (4.14) shows that the elements $R(\hat{\boldsymbol{n}},\omega)$ with the same ω form a class in SO(3).

4.2 Fundamental Concept of a Lie Group

4.2.1 *The Composition Functions of a Lie Group*

A group G is called a continuous group if its element R can be characterized by g independent and continuous real parameters r_ρ, $1 \leq \rho \leq g$, varying in a g-dimensional region. The region is called the group space of G and g is the order of G. It is required that there is a one-to-one correspondence between the group element R and the set of parameters r_ρ, at least in the region where the measure is not vanishing. For example, the group space of SO(3) is a spheroid with radius π. Each point inside the sphere corresponds to one and only one element of SO(3), but in the sphere where the measure is vanishing, two end points of a diameter correspond to one element. Hereafter, a point in the group space which corresponds to an element R of G will be said as a point R in the group space for convenience.

The parameters t_ρ of the product $T = RS$ of two elements are the functions of the parameters of the factors R and S

$$t_\rho = f_\rho(r_1 \ldots, r_g; s_1 \ldots, s_g) = f_\rho(r; s). \tag{4.17}$$

$f_\rho(r; s)$ are called the composition functions, which describe the multiplication rule of elements in G completely. A continuous group is called a Lie group if its composition functions are the analytic functions or at least piecewise continuously differentiable up to the second order. The mathematical analysis can be used in studying a Lie group so that the theory of Lie groups has been studied thoroughly.

The domain of the definition of $f_\rho(r; s)$ is a square of the group space, and the domain of function is the group space. In order to meet four axioms for a group, the composition functions have to satisfy the conditions:

$$f_\rho(f(r;s);t) = f_\rho(r;f(s;t)), \qquad f_\rho(e;r) = r_\rho, \qquad f_\rho(\overline{r};r) = e_\rho, \tag{4.18}$$

where $\overline{r}_\rho$ are the parameters of the inverse R^{-1}, and e_ρ are the parameters of the identical element E. e_ρ are usually taken to be zero for convenience. As a matter of fact, the composition functions are mainly used in the theoretical analysis, but rarely in the real calculations. Even for a very simple Lie group, such as SO(3), the composition functions are quite complicated and their forms do not write evidently.

Many fundamental concepts of a group are also suitable for a Lie group, such as the concepts of an Abelian group, a subgroup, a coset, the conjugate elements, a class, an invariant subgroup, the quotient group, the isomorphism, the homomorphism, a linear representation, the character, the equivalent representations, an irreducible representation, a self-conjugate representation, and so on. The matrix entries and the character of a representation of a Lie group are the single-valued analytic functions of the group parameters in the group space, at least in the region where the measure is not vanishing.

4.2.2 *The Local Property of a Lie Group*

The group elements in a Lie group G are said to be adjacent if their parameters differ only slightly from one another. An element $A(\alpha)$ is said to be infinitesimal if it is adjacent to the identical element E. The parameters e_ρ of E are chosen to be zero. Thus, the parameters α_ρ of an infinitesimal element $A(\alpha)$ are infinitesimal. Evidently, RA and AR are both adjacent to the element R, and the product of R^{-1} and an element adjacent to R is an infinitesimal element. To speak roughly, the point R moves continuously in the group space if R is multiplied with infinite number of infinitesimal elements continually. Conversely, if the points R and E can be connected by a continuous curve embedded completely in the group space, R is the product of infinite number of infinitesimal elements. The infinitesimal elements are related to the differential calculus of the group elements so that they describe the local property of a Lie group. The precise version of this statement will be discussed later.

The product of two infinitesimal elements $A(\alpha)$ and $B(\beta)$ is an infinitesimal element. The parameters of AB can be expanded as a Taylor series with respect to the infinitesimal parameters α_ν and β_ν

$$\begin{aligned} f_\rho(\alpha;\beta) &= f_\rho(0;0) + \sum_{\nu=1}^{g} \left(\alpha_\nu \left.\frac{\partial f_\rho(\alpha;0)}{\partial \alpha_\nu}\right|_{\alpha=0} + \beta_\nu \left.\frac{\partial f_\rho(0;\beta)}{\partial \beta_\nu}\right|_{\beta=0} \right) \\ &= \alpha_\rho + \beta_\rho, \end{aligned} \tag{4.19}$$

where $e_\rho = 0$, $AE = A$, $EB = B$, and the infinitesimal quantities of the second order have been omitted. Therefore, the product of two infinitesimal elements are commutable, and the parameters of the product are the sum of those of two elements. It does not mean that the group is Abelian. Denoting by $\overline{\alpha}_\rho$ the parameters of the inverse A^{-1} of an infinitesimal element $A(\alpha)$, one has

$$\overline{\alpha}_\rho = -\alpha_\rho. \tag{4.20}$$

4.2.3 Generators and Differential Operators

There are infinite number of infinitesimal elements in a Lie group G. How to study their property in the transformation group P_G and in a representation $D(G)$?

P_R is the transformation operator corresponding to the element R of G, $P_R\psi(x) = \psi(R^{-1}x)$, where x denotes the coordinates of all degrees of freedom in the system. Let R be an infinitesimal element $A(\alpha)$,

$$\begin{aligned}
P_A\psi(x) &= \psi(x) + \sum_{a\rho} \overline{\alpha}_\rho \left.\frac{\partial(A^{-1}x)_a}{\partial\overline{\alpha}_\rho}\right|_{\overline{\alpha}=0} \left.\frac{\partial\psi(A^{-1}x)}{\partial(A^{-1}x)_a}\right|_{\overline{\alpha}=0} \\
&= \psi(x) - i\sum_{\rho=1}^{g} \alpha_\rho I_\rho^{(0)}\psi(x), \qquad (4.21)\\
I_\rho^{(0)} &= -i\sum_a \left.\frac{\partial(Ax)_a}{\partial\alpha_\rho}\right|_{\alpha=0} \frac{\partial}{\partial x_a}.
\end{aligned}$$

The differential operators $I_\rho^{(0)}$, whose number is g, characterize the action of every infinitesimal element $A(\alpha)$ on a scalar function $\psi(x)$.

For the group SO(3), its element R is a rotation in a real three-dimensional space. If the system is a mass point and x denotes three coordinates of the mass point, one obtains from Eq. (4.11),

$$(Ax)_a = \sum_b \left\{\delta_{ab} - i\sum_d \alpha_d (T_d)_{ab}\right\} x_b = x_a - \sum_{bd} \alpha_d \epsilon_{dab} x_b. \qquad (4.22)$$

Substituting Eq. (4.22) into Eq. (4.21), one has

$$I_d^{(0)} = -i\sum_{ab} \epsilon_{dba} x_b \frac{\partial}{\partial x_a} = L_d, \qquad (4.23)$$

where the natural units are used, $\hbar = c = 1$. The differential operators $I_d^{(0)}$ of SO(3) are nothing but the orbital angular momentum operators. For an n-body system, $I_d^{(0)}$ is the total orbital angular momentum operators (see Eq. (4.220)).

If an m-dimensional functional space $\mathcal{L}$ spanned by the basis functions $\psi_\mu(x)$ is invariant with respect to the transformation group P_G and corresponds to a representation $D(G)$ of a Lie group G,

$$P_R\psi_\mu(x) = \sum_{\nu=1}^{m} \psi_\nu(x) D_{\nu\mu}(R). \qquad (4.24)$$

Expand the representation matrix $D(A)$ of an infinitesimal element,

$$D(A) = \mathbf{1} - i\sum_{\rho=1}^{g} \alpha_\rho I_\rho, \qquad I_\rho = i \left. \frac{\partial D(A)}{\partial \alpha_\rho} \right|_{\alpha=0} . \tag{4.25}$$

I_ρ is nothing but the matrix forms of the differential operators $I_\rho^{(0)}$ in the basis functions $\psi_\mu(x)$ of $\mathcal{L}$,

$$I_\rho^{(0)} \psi_\mu(x) = \sum_{\nu=1}^{m} \psi_\nu(x) \left(I_\rho\right)_{\nu\mu} . \tag{4.26}$$

I_ρ, $1 \leq \rho \leq g$, are called the generators in a representation $D(G)$ of G, which characterize the transformation of the basis functions $\psi_\mu(x)$ as well as all functions in $\mathcal{L}$ under the action of every infinitesimal element $A(\alpha)$. If $D(G)$ is a unitary representation, I_ρ are Hermitian because we introduce a factor $-i$ in front of I_ρ in Eq. (4.25).

4.2.4 *The Adjoint Representation of a Lie Group*

Let $RSR^{-1} = T$, where the parameters of T are the functions of parameters of S and R,

$$t_\rho = \psi_\rho(s_1, s_2, \ldots; r_1, r_2, \ldots) \equiv \psi_\rho(s; r). \tag{4.27}$$

For a faithful representation $D(G)$, one calculates the derivative of $D(R)D(S)D(R)^{-1} = D(T)$ with respect to the parameter s_j, and then, takes $s_j = 0$,

$$D(R) \left. \frac{\partial D(S)}{\partial s_j} D(R)^{-1} \right|_{s=0} = \sum_k \left. \frac{\partial D(T)}{\partial t_k} \right|_{t=0} \left. \frac{\partial \psi_k(s;r)}{\partial s_j} \right|_{s=0},$$

$$D(R) I_j D(R)^{-1} = \sum_k I_k D_{kj}^{\text{ad}}(R), \qquad D_{kj}^{\text{ad}}(R) = \left. \frac{\partial \psi_k(s;r)}{\partial s_j} \right|_{s=0} . \tag{4.28}$$

Similarly,

$$P_R I_j^{(0)} P_R^{-1} = \sum_k I_k^{(0)} D_{kj}^{\text{ad}}(R). \tag{4.29}$$

Equation (4.28) (or Eq. (4.29)) gives a one-to-one or many-to-one correspondence between the group element R and the matrix $D^{\text{ad}}(R)$, and the correspondence is invariant in the multiplication of group elements. Thus, $D^{\text{ad}}(R)$ is a representation of a Lie group, called the adjoint representation. The dimension of the adjoint representation is the order g of G. Every Lie group has its adjoint representation.

4.2.5 *The Global Property of a Lie Group*

In fact, the global property of a Lie group is the topological property of the group space of the Lie group. We will explain the global property of a Lie group with the language familiar to physicists.

First is the continuity of the group space. A Lie group is called mixed if its group space falls into several disjoint pieces. A mixed Lie group contains an invariant Lie subgroup G whose group space is a connected piece in which the identical element lies. G is called a connected Lie group. The set of elements related to the other connected piece is the coset of G. The property of the mixed Lie group can be characterized completely by G and the representative elements belonging to each coset, respectively. For example, the group space of O(3) falls into two pieces where det $R = 1$ and det $R = -1$, respectively. An invariant subgroup SO(3) of O(3) has a connected group space with det $R = 1$. A representative element in the coset with det $R = -1$ is usually taken to be the spatial inversion σ. The SO(3) group and σ completely characterize the group O(3) because any improper rotation in O(3) is a product of σ and a proper rotation belonging to SO(3). Hereafter, we will mainly study the connected Lie groups.

Second is the degree of continuity of the group space. The group space of a connected Lie group G can be simply or multiply connected. The Lie group G is called a multiply connected with the degree of continuity n if the connected curves of any two points in the group space are separated into n classes where any two curves in each class can be changed continuously from one to another in the group space, but two curves in different classes cannot, where "continuous change of two curves in the group space" means that there exist a set of curves $f(t)$ that depends on a continuous parameter t, $0 \leq t \leq 1$, such that each $f(t)$ embeds completely in the group space, and $f(0)$ and $f(1)$ coincide with the two curves, respectively. The Lie group with the degree of continuity $n = 1$ is called simply-connected. As proved in mathematics, for a connected Lie group G with the degree of continuity n, there exists a simply-connected Lie group G' homomorphic onto G with an n-to-one correspondence. G' is called the covering group of G.

Consider the group SO(2) composed of all rotations $R(\boldsymbol{e}_3, \varphi)$ around the z-axis. Every rotation is characterized by its rotation angle φ. φ varies in a unit circumference, which is the group space of SO(2). Any two points in the circumference can be connected by a curve in the circumference through several loops around the circumference. The loops contained in the curve is enumerated by an integer n: $+1$ for each clockwise loop and -1 for each counter-clockwise loop. The curves connecting two points are separated into classes signed by n. The degree of continuity of SO(2) is infinite.

SO(2) is an Abelian Lie group, whose representation has to satisfy the condition $D(\varphi + 2\pi) = D(\varphi)$. There are infinite number of inequivalent representations denoted by an integer m,

$$D^m(\varphi) = e^{-im\varphi}, \qquad e^{-i2m\pi} = 1. \tag{4.30}$$

The covering group of SO(2) is composed of all real numbers where the multiplication rule of elements is defined with the addition of digits. Since the elements with the parameters $\varphi + 2\pi$ and φ in the covering group are different, its representation is $e^{-i\tau\varphi}$, where τ is any number.

The group space of SO(3) is a spheroid with radius π. Two end points of a diameter in the sphere correspond to the same element. A curve connecting two points in the group space of SO(3) may contain a jump between two end points of a diameter. When the curve changes in the group space continuously, two end points change in pairs and cannot disappear. Discuss a curve connecting P to P' and containing two jumps at the sphere. Namely, the curve first goes continuously from P to A, after a jump from A to A' it goes continuously from A' to B, then after the second jump from B to B' it goes continuously from B' to P'. Now, one changes the curve in the group space such that B moves to coincide with A', and at the same time B' coincides with A. The part of the curve connecting A' to B becomes a loop which can be shrunk to the point A' continuously in the group space. In this way, two jumps contained in the curve become two successive jumps from A to A' and from A' to A, which is equal to no jump in essence. Therefore, a pair of jumps in the curve disappears continuously in the group space. The connected curves of any two points in the group space of SO(3) are separated into two classes containing even or odd number of jumps at the two end points of a diameter in the sphere, respectively. Any two curves in each class can be changed continuously from each other in the group space, but two curves in different classes cannot. The group SO(3) is doubly-connected. We will show that the two-dimensional unimodular unitary group SU(2) is the covering group of SO(3).

The third is the compactness of the group space. In an Euclidean space, a closed finite region (including the boundary) is compact. An open finite region (without the boundary) or an infinite region is not compact. A Lie group is compact if its group space is compact. The SO(3) group is compact. The Lorentz group is not compact. If one chooses the relative velocity of two inertia systems to be a parameter of the Lorentz group, its variation region is open because the velocity can tend to the velocity c of light, but cannot be equal to c. In order to generalize the formulas for the finite group given in Chap. 3 to a Lie group, the key is to define the average of a group function over the group elements. For a Lie group the average becomes an integral over the group parameters, instead of a sum over the group elements. As proved in mathematics, the integral can be defined only for a compact Lie group.

4.3 The Covering Group of SO(3)

4.3.1 *The Group SU(2)*

The set of all two-dimensional unimodular ($\det u = 1$) unitary matrices u, in the multiplication rule of matrices, forms a Lie group, called SU(2). An arbitrary element u in SU(2)

$$u = \begin{pmatrix} a & b \\ c & d \end{pmatrix} \in \mathrm{SU}(2)$$

satisfies $aa^* + cc^* = bb^* + dd^* = ad - bc = 1$ and $ab^* + cd^* = 0$. The solution is $a = d^*$, $b = -c^*$, and $|c|^2 + |d|^2 = 1$ (see Prob. 7 of Chap. 1 in [Ma and Gu (2004)]). Letting $d = \cos(\omega/2) + i\sin(\omega/2)\cos\theta$ and $c = \sin(\omega/2)\sin\theta\,(\sin\varphi - i\cos\varphi)$, one rewrites the matrix u as

$$u(\hat{\boldsymbol{n}}, \omega) = \mathbf{1}\cos\left(\frac{\omega}{2}\right) - i\,(\boldsymbol{\sigma}\cdot\hat{\boldsymbol{n}})\sin\left(\frac{\omega}{2}\right), \tag{4.31}$$

namely the generators of the self-representation of SU(2) are $\sigma_a/2$, where σ_a are the Pauli matrices given in Eq. (4.9). Let $\boldsymbol{\sigma} = \sum_a \boldsymbol{e}_a\sigma_a$,

$$\boldsymbol{\sigma}\cdot\hat{\boldsymbol{n}} = \sum_{a=1}^{3} \sigma_a n_a = \begin{pmatrix} n_3 & n_1 - in_2 \\ n_1 + in_2 & -n_3 \end{pmatrix}. \tag{4.32}$$

Taking the trace of $(\boldsymbol{\sigma}\cdot\boldsymbol{U})(\boldsymbol{\sigma}\cdot\boldsymbol{V})$ multiplying with the unit matrix $\mathbf{1}$ or σ_a, one shows

$$(\boldsymbol{\sigma}\cdot\boldsymbol{U})(\boldsymbol{\sigma}\cdot\boldsymbol{V}) = \mathbf{1}\,(\boldsymbol{U}\cdot\boldsymbol{V}) + i\boldsymbol{\sigma}\cdot(\boldsymbol{U}\times\boldsymbol{V}), \tag{4.33}$$

where $\boldsymbol{\sigma}\cdot(\boldsymbol{U}\times\boldsymbol{V}) = \sum_{abc} \epsilon_{abc}\sigma_a U_b V_c$. The direct calculation leads to

$$\begin{aligned} &u(\hat{\boldsymbol{n}}, \omega_1)u(\hat{\boldsymbol{n}}, \omega_2) = u(\hat{\boldsymbol{n}}, \omega_1 + \omega_2), \\ &u(\hat{\boldsymbol{n}}, 4\pi) = \mathbf{1}, \qquad u(\hat{\boldsymbol{n}}, 2\pi) = -\mathbf{1}, \\ &u(\hat{\boldsymbol{n}}, \omega) = u(-\hat{\boldsymbol{n}}, 4\pi - \omega) = -u(-\hat{\boldsymbol{n}}, 2\pi - \omega). \end{aligned} \tag{4.34}$$

Thus, the element $u(\hat{\boldsymbol{n}}, \omega) \in \mathrm{SU}(2)$ can be characterized by $\boldsymbol{\omega}$ through its spherical coordinates $(\omega, \theta, \varphi)$ or its rectangular coordinates $(\omega_1, \omega_2, \omega_3)$. The variation region of $\boldsymbol{\omega}$, which is the group space of SU(2), is a spheroid with radius 2π, where all points on the sphere ($\omega = 2\pi$) represent the same element, $-\mathbf{1}$. Although the curve connecting any two points in the group space of SU(2) may contain a jump on the sphere, the jump can be shrunk in the group space of SU(2) because the jump can be looked as a continuous curve on the sphere. Thus, the group space of SU(2) is a simply-connected closed region, and the SU(2) group is a simply-connected compact Lie group.

4.3.2 *Homomorphism of SU(2) onto SO(3)*

A real linear combination of three Pauli matrices is traceless and Hermitian. Conversely, any two-dimensional traceless Hermitian matrix X can be expressed as a real linear combination of the Pauli matrices. Taking the components x_a of the position vector $\boldsymbol{r}$ of an arbitrary point P in the real three-dimensional space to be the coefficients in X, one obtains a one-to-one correspondence between X and $\boldsymbol{r}$:

$$\begin{aligned} X &= \sum_{a=1}^{3} \sigma_a x_a = \boldsymbol{\sigma} \cdot \boldsymbol{r} = \begin{pmatrix} x_3 & x_1 - i x_2 \\ x_1 + i x_2 & -x_3 \end{pmatrix}, \\ x_a &= \frac{1}{2} \operatorname{Tr}(X \sigma_a), \qquad \det X = -\sum_{a=1}^{3} x_a^2 = -r^2. \end{aligned} \tag{4.35}$$

After a similarity transformation $u^{-1} \in \mathrm{SU}(2)$, X becomes another traceless Hermitian matrix X' with the same determinant,

$$X' = u X u^{-1}, \qquad \det\ X' = \det\ X. \tag{4.36}$$

X' corresponds to the position vector $\boldsymbol{r}'$ of another point P'

$$x'_b = \sum_a R_{ba} x_a, \qquad \sum_b (x'_b)^2 = \sum_a (x_a)^2, \tag{4.37}$$

where R is a real orthogonal matrix, $R \in \mathrm{O}(3)$. If u is changed from $\mathbf{1}$ continuously, R is also changed from $\mathbf{1}$ continuously. Thus, $\det R = 1$ and $R \in \mathrm{SO}(3)$. Conversely, an arbitrary element $R \in \mathrm{SO}(3)$ rotates $\boldsymbol{r}$ to $\boldsymbol{r}'$, and then, X changes to X'. Two traceless Hermitian matrices X and X' with the same determinant can be related by Eq. (4.36), where u is a unitary matrix with $\det u = 1$, namely, $u \in \mathrm{SU}(2)$. The matrix u relating X and X'

is not unique. If $u_1 X u_1^{-1} = u_2 X u_2^{-1} = X'$, $u_2^{-1} u_1$ is commutable with any matrix X, so it is a constant matrix, $u_1 = \lambda u_2$. Due to their determinants, $\lambda = \pm 1$. Since $\boldsymbol{r}$ is arbitrary, from Eq. (4.36), one obtains a two-to-one correspondence between $\pm u \in \mathrm{SU}(2)$ and $R \in \mathrm{SO}(3)$,

$$u\sigma_a u^{-1} = \sum_{b=1}^{3} \sigma_b R_{ba}. \tag{4.38}$$

Evidently, the correspondence is invariant in the multiplication of group elements. Hence, SU(2) is homomorphic onto SO(3):

$$\mathrm{SO}(3) \sim \mathrm{SU}(2). \tag{4.39}$$

To show the concrete correspondence between u and R, we calculate $u(\hat{\boldsymbol{n}},\omega)\,(\boldsymbol{\sigma}\cdot\boldsymbol{r})\,u(\hat{\boldsymbol{n}},\omega)^{-1}$. Decompose $\boldsymbol{r}$ into two components parallel and perpendicular to $\hat{\boldsymbol{n}}$, respectively, $\boldsymbol{r} = \hat{\boldsymbol{n}}a + \hat{\boldsymbol{m}}b$ where $\hat{\boldsymbol{n}}\cdot\hat{\boldsymbol{m}} = 0$. From Fig. 4.2 one sees that $R(\hat{\boldsymbol{n}},\omega)\boldsymbol{r} = \hat{\boldsymbol{n}}a + [\hat{\boldsymbol{m}}\cos\omega + (\hat{\boldsymbol{n}}\times\hat{\boldsymbol{m}})\sin\omega]\,b$.

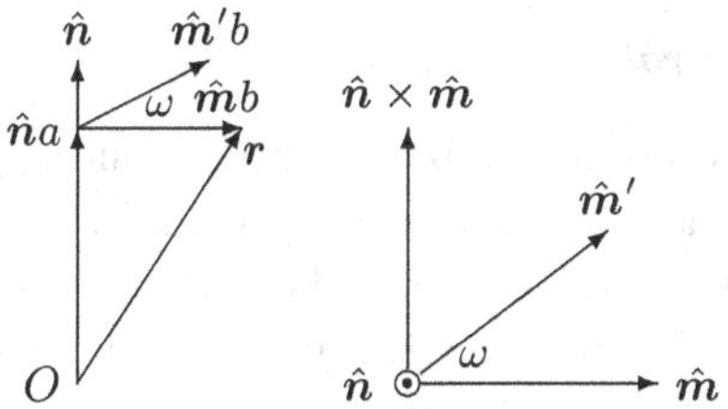

Fig. 4.2 Rotation of a vector $\boldsymbol{r}$ around $\hat{\boldsymbol{n}}$ through ω.

Due to Eq. (4.33), one has

$$\begin{aligned}
&(\boldsymbol{\sigma}\cdot\hat{\boldsymbol{n}})\,(\boldsymbol{\sigma}\cdot\hat{\boldsymbol{m}}) - (\boldsymbol{\sigma}\cdot\hat{\boldsymbol{m}})\,(\boldsymbol{\sigma}\cdot\hat{\boldsymbol{n}}) = 2i\boldsymbol{\sigma}\cdot(\hat{\boldsymbol{n}}\times\hat{\boldsymbol{m}}),\\
&(\boldsymbol{\sigma}\cdot\hat{\boldsymbol{n}})\,(\boldsymbol{\sigma}\cdot\hat{\boldsymbol{m}})\,(\boldsymbol{\sigma}\cdot\hat{\boldsymbol{n}}) = i\,\{\boldsymbol{\sigma}\cdot(\hat{\boldsymbol{n}}\times\hat{\boldsymbol{m}})\}\,(\boldsymbol{\sigma}\cdot\hat{\boldsymbol{n}}) = -\boldsymbol{\sigma}\cdot\hat{\boldsymbol{m}}.
\end{aligned}$$

Then,

$$\begin{aligned}
&u(\hat{\boldsymbol{n}},\omega)\,(\boldsymbol{\sigma}\cdot\hat{\boldsymbol{n}})\,u(\hat{\boldsymbol{n}},\omega)^{-1} = \boldsymbol{\sigma}\cdot\hat{\boldsymbol{n}},\\
&u(\hat{\boldsymbol{n}},\omega)\,(\boldsymbol{\sigma}\cdot\hat{\boldsymbol{m}})\,u(\hat{\boldsymbol{n}},\omega)^{-1} = \boldsymbol{\sigma}\cdot[\hat{\boldsymbol{m}}\cos\omega + (\hat{\boldsymbol{n}}\times\hat{\boldsymbol{m}})\sin\omega],\\
&u(\hat{\boldsymbol{n}},\omega)\,(\boldsymbol{\sigma}\cdot\boldsymbol{r})\,u(\hat{\boldsymbol{n}},\omega)^{-1} = \boldsymbol{\sigma}\cdot\boldsymbol{r}' = \boldsymbol{\sigma}\cdot[R(\hat{\boldsymbol{n}},\omega)\boldsymbol{r}].
\end{aligned} \tag{4.40}$$

$$u(\hat{\boldsymbol{n}},\omega)\sigma_a u(\hat{\boldsymbol{n}},\omega)^{-1} = \sum_{b=1}^{3} \sigma_b R_{ba}(\hat{\boldsymbol{n}},\omega). \tag{4.41}$$

The elements in SO(3) and SU(2) are both characterized by the parameter $\boldsymbol{\omega}$. The group space of SO(3), which is doubly-connected, is the spheroid with radius π. The group space of SU(2), which is simply-connected, is the spheroid with radius 2π. Inside the spheroid with radius π there is a one-to-one correspondence between elements in SO(3) and SU(2). $u(\hat{\boldsymbol{n}},\omega)$ in the ring with $\pi < \omega < 2\pi$ is equal to $-u(-\hat{\boldsymbol{n}}, 2\pi-\omega)$ owing to Eq. (4.34). The pair of $\pm u(\hat{\boldsymbol{n}},\omega)$ in SU(2) maps onto one element $R(\hat{\boldsymbol{n}},\omega)$ in SO(3). The group SU(2) is the covering group of SO(3). A faithful representation of SO(3) is simple-valued and it is an unfaithful representation of SU(2). To speak strictly, a faithful representation of SU(2) is not a representation of SO(3). However, due to physical reason, it is called a double-valued representation of SO(3). Similar to the classes in SO(3), the elements $u(\hat{\boldsymbol{n}},\omega)$ with the same ω form a class of the SU(2) group (see Prob. 13 of Chap. 4 in [Ma and Gu (2004)]).

Due to the homomorphism of SU(2) onto SO(3), it is convenient to call $u(\hat{\boldsymbol{n}},\omega) \in \mathrm{SU}(2)$ a "rotation" around the direction $\boldsymbol{n}$ through an angle ω. It will be known later that the group SU(2) is related to the spinor. Therefore, the period of a spinor is 4π in "rotation".

4.3.3 *The Group Integral*

Many properties of a finite group are based on the concept of the average of a group function which is invariant in the left- and right-multiplication with any group element. For a Lie group, if the average of a group function can be defined as an integral over the group space, those properties of a finite group will be suitable for a Lie group

$$\frac{1}{g}\sum_{R\in G} F(R) \longrightarrow \int dRF(R) = \int (dr)\ W(R)F(R), \tag{4.42}$$

$$\int dR\ F(R) = \int dR\ F(SR) = \int dR\ F(RS). \tag{4.43}$$

The group integral is linear with respect to the group function. The weight function $W(R)$ can be understood as the relative density of elements in the neighborhood of R. If $F(R) \geq 0$ and $F(R)$ is not equal to zero everywhere, the average of $F(R)$ is larger than zero. Thus, $W(R)$ has to be single valued, finite, integrable, non-negative, and not equal identically to zero in a finite region. $W(R)$ is normalized by

$$\frac{1}{g}\sum_{R\in G} 1 = 1 \longrightarrow \int dR = \int (dr)\ W(R) = 1. \tag{4.44}$$

Letting $T = SR$ and $W(E) = W_0$, one obtains from Eq. (4.43)

$$\int (dt)W(T)F(T) = \int dT\ F(T) = \int dR\ F(R)$$
$$= \int dR\ F(SR) = \int dR\ F(T) = \int (dr)W(R)F(T).$$

$$(dr)\ W(R) = (dt)\ W(T) = (d\alpha)\ W_0. \tag{4.45}$$

To speak roughly, the "number" of elements in the neighborhood of any group element is the same. $W(R)$ can be calculated through the Jacobi determinant in the replacement of integral variables. Denote by A the infinitesimal element with the parameters α_μ, and by R' the element with the parameters r'_μ in the neighborhood of a given element R. The Jacobi determinants for $R' = AR$ and $A = R'R^{-1}$ are, respectively,

$$W_0 = W(R) \left| \det \left\{ \frac{\partial f_j(\alpha; r)}{\partial \alpha_k} \right\} \right|_{\alpha=0}, \tag{4.46}$$

$$W(R) = W_0 \left| \det \left\{ \frac{\partial f_j(r'; \overline{r})}{\partial r'_k} \right\} \right|_{r'=r}. \tag{4.47}$$

Two conditions for $W(R)$ are equivalent. From any of them $W(R)$ can be calculated where W_0 is determined by the normalization condition (4.44).

For the group SU(2) the weight function is $W(\hat{\boldsymbol{n}}, \omega)$,

$$(du) = W(\hat{\boldsymbol{n}}, \omega) d\omega_1 d\omega_2 d\omega_3 = W(\hat{\boldsymbol{n}}, \omega)\omega^2 \sin\theta d\omega d\theta d\varphi. \tag{4.48}$$

Note that the elements $u(\hat{\boldsymbol{n}}, \omega)$ in SU(2) with the same ω are conjugate to each other. The conjugate elements are located on a sphere with a radius ω in the group space of SU(2) and their parameters can be related by a rotation R. The Jacobi determinant in the replacement of variables of Eq. (4.45) for two conjugate elements is nothing but $\det R = 1$. Thus, the weight function $W(\hat{\boldsymbol{n}}, \omega)$ is independent of the direction $\hat{\boldsymbol{n}}$, $W(\hat{\boldsymbol{n}}, \omega) = W(\omega)$. This is the direct result of an isotropic space.

In Eq. (4.46) the element R becomes $u(\boldsymbol{e}_3, \omega)$ and the infinitesimal element A becomes $u(A)$

$$\begin{aligned}
&u(\boldsymbol{e}_3, \omega) = \mathbf{1} \cos(\omega/2) - i\sigma_3 \sin(\omega/2), \\
&u(A) = \mathbf{1} - i\left(\sigma_1\alpha_1 + \sigma_2\alpha_2 + \sigma_3\alpha_3\right)/2, \\
&u(A)u(\boldsymbol{e}_3, \omega) = \mathbf{1} \cos\left(\omega'/2\right) - i\left(\boldsymbol{\sigma} \cdot \hat{\boldsymbol{n}}'\right) \sin\left(\omega'/2\right) \\
&\quad = \mathbf{1} \left\{\cos(\omega/2) - (\alpha_3/2)\sin(\omega/2)\right\} - i\sigma_1 \left\{\alpha_1 \cos(\omega/2) + \alpha_2 \sin(\omega/2)\right\}/2 \\
&\quad - \ i\sigma_2 \left\{\alpha_2 \cos(\omega/2) - \alpha_1 \sin(\omega/2)\right\}/2 \\
&\quad - \ i\sigma_3 \left\{\alpha_3 \cos(\omega/2) + 2\sin(\omega/2)\right\}/2,
\end{aligned}$$

where only the terms up to the first order of α_j is reserved because α_j in Eq. (4.46) is taken to be zero after the derivative. The parameters ω' and $\boldsymbol{n}'$ are calculated as follows:

$$\begin{aligned}
&\cos(\omega'/2) = \cos(\omega/2) - (\alpha_3/2)\sin(\omega/2) = \cos\{(\omega+\alpha_3)/2\},\\
&\sin(\omega'/2) = \sin\{(\omega+\alpha_3)/2\} = \sin(\omega/2) + (\alpha_3/2)\cos(\omega/2),\\
&\omega' n_1' = \omega\{\sin(\omega/2)\}^{-1}\left\{\alpha_1\cos(\omega/2)+\alpha_2\sin(\omega/2)\right\}/2,\\
&\omega' n_2' = \omega\{\sin(\omega/2)\}^{-1}\left\{\alpha_2\cos(\omega/2)-\alpha_1\sin(\omega/2)\right\}/2,\\
&\omega' n_3' = \omega'\{\sin(\omega'/2)\}^{-1}\left\{\alpha_3\cos(\omega/2)+2\sin(\omega/2)\right\}/2 = \omega' = \omega+\alpha_3.
\end{aligned}$$

Substituting them into Eq. (4.46), one has

$$\begin{aligned}
\frac{W_0}{W(\omega)} &= \left|\det\left\{\frac{\partial(\omega' n_a')}{\partial\alpha_b}\right\}\right|_{\alpha=0} = \begin{vmatrix} (\omega/2)\cot(\omega/2) & \omega/2 & 0\\ -\omega/2 & (\omega/2)\cot(\omega/2) & 0\\ 0 & 0 & 1\end{vmatrix}\\
&= \omega^2\left\{4\sin^2(\omega/2)\right\}^{-1}.
\end{aligned}$$

Thus, $W(\omega) = W_0 4\omega^{-2}\sin^2(\omega/2)$. Due to normalization

$$1 = 4W_0\int_0^{2\pi}\sin^2(\omega/2)d\omega\int_0^{\pi}\sin\theta d\theta\int_{-\pi}^{\pi}d\varphi = 16\pi^2 W_0,$$

one obtains $W_0 = \left(16\pi^2\right)^{-1}$. The group integral of a function $F(u) = F(\boldsymbol{\omega})$ of SU(2) is

$$\int(du)F(u) = \frac{1}{4\pi^2}\int_{-\pi}^{\pi}d\varphi\int_0^{\pi}\sin\theta d\theta\int_0^{2\pi}\sin^2(\omega/2)F(\boldsymbol{\omega})d\omega. \tag{4.49}$$

For the group SO(3), the variation region of ω reduces to π,

$$\int(dR)F(R) = \frac{1}{2\pi^2}\int_{-\pi}^{\pi}d\varphi\int_0^{\pi}\sin\theta d\theta\int_0^{\pi}\sin^2(\omega/2)F(\boldsymbol{\omega})d\omega. \tag{4.50}$$

For a class function, $F(\boldsymbol{\omega}) = F(\omega)$, the integral on θ and φ can be done first, namely, $\displaystyle\int\sin\theta d\theta d\varphi = 4\pi$.

Now, for a compact Lie group, such as the groups SU(2) and SO(3), the average of a group function which is invariant in the left- and right-multiplication with any group element can be defined. Thus, a compact Lie group has the following properties like those for a finite group:

(a) Any representation is equivalent to a unitary representation, and two equivalent unitary representations can be related through a unitary similarity transformation.

(b) Any real representation is equivalent to a real orthogonal representation, and two equivalent real orthogonal representations can be related through a real orthogonal similarity transformation.

(c) A representation is reducible if and only if there exists a nonconstant matrix commutable with its every representation matrix. Any reducible representation is completely reducible.

(d) The representation matrix entries and the characters of the two inequivalent irreducible unitary representations satisfy the orthogonal relations:

$$\begin{aligned}
&\int dR\ D^i_{\mu\rho}(R)^* D^j_{\nu\lambda}(R) = \frac{1}{m_j}\delta_{ij}\delta_{\mu\nu}\delta_{\rho\lambda}, \\
&\int dR\ \chi^i(R)^*\chi^j(R) = \delta_{ij}.
\end{aligned} \tag{4.51}$$

Any representation can be expanded with respect to the irreducible representations:

$$\begin{aligned}
X^{-1}D(R)X = \bigoplus_j a_j D^j(R), \qquad \chi(R) = \sum_j a_j\chi^j(R), \\
a_j = \int dR\ \chi^j(R)^*\chi(R).
\end{aligned} \tag{4.52}$$

The group integral for the characters reduces to the class integral. For the group SU(2),

$$\frac{1}{\pi}\int_0^{2\pi} d\omega\ \sin^2(\omega/2)\chi^i(\omega)^*\chi^j(\omega) = \delta_{ij}. \tag{4.53}$$

$$a_j = \frac{1}{\pi}\int_0^{2\pi} d\omega\ \sin^2(\omega/2)\chi^j(\omega)^*\chi(\omega). \tag{4.54}$$

(e) Two representations are equivalent if and only if the characters of every element in them are equal to each other, respectively, $\chi(R) = \overline{\chi}(R)$. A representation is irreducible if and only if the character $\chi(R)$ in the representation satisfies

$$\int dR\ |\chi(R)|^2 = 1. \tag{4.55}$$

For the group SU(2), Eq. (4.55) becomes

$$\frac{1}{\pi}\int_0^{2\pi} d\omega\ \sin^2(\omega/2)\,|\chi(\omega)|^2 = 1. \tag{4.56}$$

(f) The similarity transformation matrix X, which relates the irreducible unitary self-conjugate representation and its conjugate one, is symmetric or antisymmetric. X is symmetric if and only if the representation is real.

(g) Although the number of the standard irreducible basis $b^{(j)}_{\nu\mu}$ satisfying Eq. (3.130) becomes infinite for a Lie group, their formulas can still be used,

$$b^{(j)}_{\mu\mu} = e^{(j)}_{\mu}, \qquad b^{(j)}_{\nu\rho}\, b^{(j)}_{\tau\mu} = \delta_{\rho\tau}\, b^{(j)}_{\nu\mu}. \tag{4.57}$$

$$T\, b^{(j)}_{\rho\mu} = \sum_{\tau} b^{(j)}_{\tau\mu}\, D^{j}_{\tau\rho}(T), \qquad b^{(j)}_{\nu\rho}\, T = \sum_{\tau} D^{j}_{\rho\tau}(T)\, b^{(j)}_{\nu\tau}. \tag{4.58}$$

The basis $b^{(j)}_{\nu\mu}$ and the idempotent e^{j}_{μ} can be expressed as

$$\begin{aligned}
b^{(j)}_{\nu\mu} &= m_j \int dR D^{j}_{\nu\mu}(R)^* P_R, \\
e^{(j)}_{\mu} &= b^{(j)}_{\mu\mu} = m_j \int dR D^{j}_{\mu\mu}(R)^* P_R, \\
e^{(j)} &= \sum_{\mu} e^{(j)}_{\mu} = m_j \int dR\ \chi^{j}(R)^* P_R, \\
e^{(i)}_{\nu} e^{(j)}_{\mu} &= \delta_{ij}\delta_{\nu\mu} e^{(j)}_{\mu}, \qquad e^{(i)}_{\nu} t e^{(j)}_{\mu} = \delta_{ij} e^{(j)}_{\nu} t e^{(j)}_{\mu},
\end{aligned} \tag{4.59}$$

where the group element R is replaced with the transformation operator P_R which acts on the scalar wave functions, and t is a linear combination of group elements.

If the compact Lie group G is the symmetric group of a quantum system, the static wave functions $\psi_\rho(x)$ can be combined into that belonging to an irreducible representation. In fact, $b^{j}_{\nu\mu}\psi_\rho(x)$, if it is not vanishing, is just that function belonging to the νth row of the irreducible representation $D^{j}(G)$,

$$T\left\{b^{j}_{\nu\mu}\psi_\rho(x)\right\} = \sum_{\tau} \left\{b^{j}_{\tau\mu}\psi_\rho(x)\right\} D^{j}_{\tau\nu}(T). \tag{4.60}$$

4.4 Irreducible Representations of SU(2)

4.4.1 *Euler Angles*

The geometric meaning of the parameters $\boldsymbol{\omega}$ for characterizing an element $R(\hat{\boldsymbol{n}}, \omega)$ in SO(3) are evident. $R(\hat{\boldsymbol{n}}, \omega)$ is a rotation around the direction $\boldsymbol{n}$ through an angle ω. The more important merit of the parameters $\boldsymbol{\omega}$ is that, at least in the neighborhood of the identical element E, there is a one-to-one correspondence between the group element and the point in the group space. This merit makes the parameters $\boldsymbol{\omega}$ more suitable in theoretical study. However, the parameters $\boldsymbol{\omega}$ are not very convenient in the real calculation, because the parameters $\boldsymbol{\omega}$ are hard to be determined from the matrix form of R or from the relative positions of two frames before and after the rotation R. On the other hand, there is another set of parameters, called the Euler angles, which are easy to be determined from the matrix form of R and from the relative positions of two frames before and after the rotation R. There is a shortcoming for the Euler angles that the Euler angles of an element in the neighborhood of the identical element E are not unique. This shortcoming makes the Euler angles inconvenient in theoretical study.

For a given matrix $R \in \mathrm{SO}(3)$, its third column is a unit vector, to the direction of which R rotates the z-axis. Denoting by β and α the polar angle and the azimuthal angle of the unit vector, respectively, one has

$$S(\alpha,\beta)\begin{pmatrix}0\\0\\1\end{pmatrix}=\begin{pmatrix}R_{13}\\R_{23}\\R_{33}\end{pmatrix},\qquad S(\alpha,\beta)=R(\boldsymbol{e}_3,\alpha)R(\boldsymbol{e}_2,\beta),\qquad (4.61)$$

where $S(\alpha,\beta)$ is given in Eq. (4.12). Hence, $S(\alpha,\beta)^{-1}R$ preserves the z-axis invariant so that it is a rotation around z-axis, denoted by $R(\boldsymbol{e}_3,\gamma)$. Therefore, every rotation in the three-dimensional space can be expressed as a product of three rotations around the coordinate axes

$$\begin{aligned} R(\alpha,\beta,\gamma) &= R(\boldsymbol{e}_3,\alpha)R(\boldsymbol{e}_2,\beta)R(\boldsymbol{e}_3,\gamma)\\ &= \begin{pmatrix} c_\alpha c_\beta c_\gamma - s_\alpha s_\gamma & -c_\alpha c_\beta s_\gamma - s_\alpha c_\gamma & c_\alpha s_\beta \\ s_\alpha c_\beta c_\gamma + c_\alpha s_\gamma & -s_\alpha c_\beta s_\gamma + c_\alpha c_\gamma & s_\alpha s_\beta \\ -s_\beta c_\gamma & s_\beta s_\gamma & c_\beta \end{pmatrix}, \end{aligned} \qquad (4.62)$$

where $c_\alpha = \cos\alpha$, $s_\alpha = \sin\alpha$, etc. Three angles α, β, and γ are called the Euler angles. When the matrix form of R is given, β and α are respectively the polar angle and the azimuthal angle of a unit vector given in the third column of R. The third row of R is also a unit vector, whose polar angle is β and whose azimuthal angle is $(\pi-\gamma)$. If the rotation R transforms the coordinate frame K to K', then in the K coordinate frame, the polar angle and the azimuthal angle of the z'-axis of K' are β and α, respectively. In the K' coordinate frame, the polar angle and the azimuthal angle of the z-axis of K are β and $(\pi-\gamma)$. The domain of definition for the Euler angles in SO(3) are

$$-\pi \le \alpha \le \pi, \qquad 0 \le \beta \le \pi, \qquad -\pi \le \gamma \le \pi. \qquad (4.63)$$

When $\beta = 0$ or π,

$$R(\alpha, 0, \gamma) = R(\boldsymbol{e}_3, \alpha + \gamma), \qquad R(\alpha, \pi, \gamma) = S(\alpha - \gamma, \pi), \tag{4.64}$$

only one parameter in α and γ is independent. This is the shortcoming of the parameters of the Euler angles.

In Eq. (4.62) a rotation is expressed as a product of three rotations around the coordinate axes of the laboratory frame K. If one chooses the rotations around the coordinate axes of the body-fixed frame K', the product order will be changed:

$$\begin{aligned} R(\alpha, \beta, \gamma) = & \left\{ \left[R(\boldsymbol{e}_3, \alpha) R(\boldsymbol{e}_2, \beta) \right] R(\boldsymbol{e}_3, \gamma) \left[R(\boldsymbol{e}_3, \alpha) R(\boldsymbol{e}_2, \beta) \right]^{-1} \right\} \\ & \cdot \left\{ R(\boldsymbol{e}_3, \alpha) R(\boldsymbol{e}_2, \beta) R(\boldsymbol{e}_3, \alpha)^{-1} \right\} R(\boldsymbol{e}_3, \alpha). \end{aligned} \tag{4.65}$$

R is a rotation which first rotates around the z'-axis in K' through α angle, then rotates around the y'-axis in the new K' frame through β angle, and at last rotates around the z'-axis in the newer K' frame through γ angle.

The group integral of SO(3) with the parameters of the Euler angles can be calculated as follows. Let R rotate the K frame to the K' frame. Denote by P and Q the intersections of the z'-axis and the x'-axis of K' with the unit sphere in K, respectively. The position of P on the unit sphere is characterized by the polar angle β and the azimuthal angle α of the direction OP. For a given P, the position of Q is characterized by the angle γ. For the rotation in the neighborhood of R, P changes in the area $(\sin \beta d\alpha d\beta)$, and Q changes in the arc $(d\gamma)$ when P is given. Since the unit sphere is isotropic, the relative "number" of elements in the neighborhood of R is proportional to the area $(\sin \beta d\alpha d\beta)$ and to the arc $(d\gamma)$. Thus, the group integral of SO(3) with the Euler angles as parameters is

$$\int F(R) dR = \frac{1}{8\pi^2} \int_{-\pi}^{\pi} d\alpha \int_0^{\pi} \sin \beta d\beta \int_{-\pi}^{\pi} F(\alpha, \beta, \gamma) d\gamma, \tag{4.66}$$

where the coefficient is determined by the normalization

$$\int dR = \frac{1}{8\pi^2} \int_{-\pi}^{\pi} d\alpha \int_0^{\pi} \sin \beta d\beta \int_{-\pi}^{\pi} d\gamma = 1. \tag{4.67}$$

For the SU(2) group,

$$u(\alpha, \beta, \gamma) = u(\boldsymbol{e}_3, \alpha) u(\boldsymbol{e}_2, \beta) u(\boldsymbol{e}_3, \gamma), \tag{4.68}$$

where the domain of definition for the Euler angles is enlarged,

$$-\pi \le \alpha \le \pi, \qquad 0 \le \beta \le \pi, \qquad -2\pi \le \gamma \le 2\pi. \tag{4.69}$$

The group integral for SU(2) is

$$\int F(u)du = \frac{1}{16\pi^2}\int_{-\pi}^{\pi} d\alpha \int_0^{\pi} \sin\beta d\beta \int_{-2\pi}^{2\pi} F(\alpha,\beta,\gamma)d\gamma. \tag{4.70}$$

4.4.2 *Linear Representations of SU(2)*

The basic method for calculating a representation of a group G is to find a functional space $\mathcal{L}^{(j)}$ invariant with respect to G. Applying the transformation operator P_R to the basis functions ψ^j_μ in $\mathcal{L}^{(j)}$, one obtains a representation $D^j(R)$ of G:

$$P_R\psi^j_\mu(x) = \psi^j_\mu(R^{-1}x) = \sum_\nu \psi^j_\nu(x)\ D^j_{\nu\mu}(R).$$

The element $u \in$ SU(2) is a unitary transformation in a two-dimensional complex space,

$$\begin{pmatrix}\xi'\\ \eta'\end{pmatrix} = u\begin{pmatrix}\xi\\ \eta\end{pmatrix}, \qquad \begin{pmatrix}\xi''\\ \eta''\end{pmatrix} = u^{-1}\begin{pmatrix}\xi\\ \eta\end{pmatrix}.$$

The homogeneous functions of order n with respect to ξ and η construct an $(n+1)$-dimensional space $\mathcal{L}^{(n)}$ which is invariant to the group SU(2). The basis functions in $\mathcal{L}^{(n)}$ are $\xi^m\eta^{n-m}$, $m = 0,\ 1,\ \ldots,\ n$. For convenience one chooses the coefficients and enumeration of the basis functions as follows

$$\begin{aligned}&\psi^j_\mu(\xi,\eta) = \frac{(-1)^{j-\mu}}{\sqrt{(j+\mu)!(j-\mu)!}}\xi^{j-\mu}\eta^{j+\mu},\\ &j = n/2 = 0,\ 1/2,\ 1,\ 3/2,\ \ldots,\\ &\mu = j - m = j,\ (j-1),\ \ldots,\ -(j-1),\ -j.\end{aligned} \tag{4.71}$$

Calculate the matrix form $D^j_{\nu\mu}(u)$ of P_u in the basis functions $\psi^j_\mu(\xi,\eta)$,

$$P_u\psi^j_\mu(\xi,\eta) = \psi^j_\mu(\xi'',\eta'') = \sum_\nu \psi^j_\nu(\xi,\eta)D^j_{\nu\mu}(u). \tag{4.72}$$

Note that $u^{-1} = \mathbf{1}\cos(\omega/2) + i\sin(\omega/2)\,(\boldsymbol{\sigma}\cdot\boldsymbol{n})$,

$$u^{-1} = \begin{pmatrix}\cos(\omega/2) + in_3\sin(\omega/2) & \sin(\omega/2)\,(n_2 + in_1)\\ \sin(\omega/2)\,(-n_2 + in_1) & \cos(\omega/2) - in_3\sin(\omega/2)\end{pmatrix}.$$

When $\hat{\boldsymbol{n}} = \boldsymbol{e}_3$, $\xi'' = \xi \exp(i\omega/2)$ and $\eta'' = \eta \exp(-i\omega/2)$. Then,

$$\begin{aligned} P_u \psi^j_\mu(\xi,\eta) &= \frac{(-1)^{j-\mu}}{\sqrt{(j+\mu)!(j-\mu)!}} \left\{\xi e^{i\omega/2}\right\}^{j-\mu} \left\{\eta e^{-i\omega/2}\right\}^{j+\mu} \\ &= \psi^j_\mu(\xi,\eta) e^{-i\mu\omega}, \end{aligned}$$

$$D^j_{\nu\mu}(\boldsymbol{e}_3, \omega) = \delta_{\nu\mu} e^{-i\mu\omega}. \tag{4.73}$$

When $\hat{\boldsymbol{n}} = \boldsymbol{e}_2$, $\xi'' = \xi \cos(\omega/2) + \eta \sin(\omega/2)$ and $\eta'' = -\xi \sin(\omega/2) + \eta \cos(\omega/2)$. Then,

$$\begin{aligned} P_u \psi^j_\mu(\xi,\eta) &= \frac{(-1)^{j-\mu}}{\sqrt{(j+\mu)!(j-\mu)!}} \left\{\xi \cos(\omega/2) + \eta \sin(\omega/2)\right\}^{j-\mu} \\ &\quad \cdot \left\{-\xi \sin(\omega/2) + \eta \cos(\omega/2)\right\}^{j+\mu} \\ &= (-1)^{j-\mu} \sum_{n=0}^{j-\mu} \frac{\sqrt{(j-\mu)!}\left\{\xi \cos(\omega/2)\right\}^{j-\mu-n} \left\{\eta \sin(\omega/2)\right\}^n}{(j-\mu-n)!n!} \\ &\quad \cdot \sum_{m=0}^{j+\mu} \frac{\sqrt{(j+\mu)!}\left\{-\xi \sin(\omega/2)\right\}^{j+\mu-m} \left\{\eta \cos(\omega/2)\right\}^m}{(j+\mu-m)!m!}. \end{aligned}$$

In order to express the right-hand side of the equality to be a linear combination of $\psi^j_\nu(\xi,\eta)$, one makes a replacement of the summation indices from m to ν, $\nu = n + m - j$,

$$P_u \psi^j_\mu(\xi,\eta) = \sum_{\nu=-j}^{j} \frac{(-1)^{j-\nu} \xi^{j-\nu} \eta^{j+\nu}}{\{(j+\nu)!(j-\nu)!\}} D^j_{\nu\mu}(\boldsymbol{e}_2, \omega),$$

$$\begin{aligned} d^j_{\nu\mu}(\omega) \equiv\ & D^j_{\nu\mu}(\boldsymbol{e}_2, \omega) \\ =\ & \sum_n \frac{(-1)^n \left\{(j+\nu)!(j-\nu)!(j+\mu)!(j-\mu)!\right\}^{1/2}}{(j+\nu-n)!(j-\mu-n)!n!(n-\nu+\mu)!} \\ & \cdot \left\{\cos(\omega/2)\right\}^{2j+\nu-\mu-2n} \left\{\sin(\omega/2)\right\}^{2n-\nu+\mu}, \end{aligned} \tag{4.74}$$

where n runs from the maximum between 0 and $(\nu-\mu)$ to the minimum between $(j+\nu)$ and $(j-\mu)$. Hence, the representation D^j of SU(2) is

$$\begin{aligned} D^j_{\nu\mu}(\alpha,\beta,\gamma) &= \left\{D^j(\boldsymbol{e}_3,\alpha) D^j(\boldsymbol{e}_2,\beta) D^j(\boldsymbol{e}_3,\gamma)\right\}_{\nu\mu} \\ &= e^{-i\nu\alpha} d^j_{\nu\mu}(\beta) e^{-i\mu\gamma}, \end{aligned} \tag{4.75}$$

where $D^j(\boldsymbol{e}_3,\omega)$ is diagonal and $D^j(\boldsymbol{e}_2,\omega) = d^j(\omega)$ is real. As a convention, the arranging order of row (column) is j, $(j-1)$, ..., $-(j-1)$, $-j$.

The representation matrix d^j has some evident symmetry, which can be proved by replacement of summation indices:

$$\begin{aligned} d^j_{\nu\mu}(\pi) &= (-1)^{j-\mu}\delta_{\nu(-\mu)}, \qquad d^j_{\nu\mu}(2\pi) = (-1)^{2j}\delta_{\nu\mu}, \\ d^j_{\nu\mu}(\omega) &= d^j_{(-\mu)(-\nu)}(\omega) = (-1)^{\mu-\nu}d^j_{\nu\mu}(-\omega) = (-1)^{\mu-\nu}d^j_{\mu\nu}(\omega) \\ &= d^j_{\mu\nu}(-\omega) = (-1)^{\mu-\nu}d^j_{(-\nu)(-\mu)}(\omega) = (-1)^{j+\nu}d^j_{\nu(-\mu)}(\pi-\omega). \end{aligned} \tag{4.76}$$

From Eq. (4.74), one has

$$\begin{aligned} d^j_{\mu j}(\beta) &= d^j_{(-j)(-\mu)}(\beta) = (-1)^{j-\mu}d^j_{j\mu}(\beta) = (-1)^{j-\mu}d^j_{(-\mu)(-j)}(\beta) \\ &= \left\{\frac{(2j)!}{(j+\mu)!(j-\mu)!}\right\}^{1/2} \{\cos(\omega/2)\}^{j+\mu}\{\sin(\omega/2)\}^{j-\mu}, \\ d^\ell_{00}(\beta) &= \sum_{n=0}^{\ell}(-1)^n\left\{\frac{\ell!\,[\cos(\omega/2)]^{\ell-n}\,[\sin(\omega/2)]^n}{n!(\ell-n)!}\right\}^2. \end{aligned} \tag{4.77}$$

Now, we analyze the property of the representation D^j of SU(2).

(a) The dimension of D^j is $(2j+1)$, $j=0,\ 1/2,\ 1,\ 3/2,\ \ldots$. $D^0(u)=1$ and $D^{1/2}(u)=u$.

(b) When j is an integer, D^j is a single-valued representation of SO(3), but an unfaithful one of SU(2). When j is half of an odd integer, D^j is a double-valued representation of SO(3) and a faithful one of SU(2).

(c) d^j is a real orthogonal matrix and D^j is a unitary representation.

(d) The character of the class with angle ω can be calculated from the diagonal matrix $D^j(\boldsymbol{e}_3,\omega)$:

$$\chi^j(\omega) = \sum_{\mu=-j}^{j} e^{-i\mu\omega} = \sum_{\mu=-j}^{j} e^{i\mu\omega} = \frac{\sin\{(j+1/2)\omega\}}{\sin(\omega/2)}. \tag{4.78}$$

Since $\chi^j(\omega)$ satisfies the orthogonal relation (4.53), the representations D^j are inequivalent irreducible representations of SU(2).

We are going to show by reduction to absurdity that there is no other representation of SU(2) with a finite dimension which is inequivalent to all D^j. If this representation exists, its character $\chi(\omega)$ has to satisfy

$$\begin{aligned} 0 &= \frac{1}{\pi}\int_0^{2\pi} d\omega\ \sin^2(\omega/2)\chi^j(\omega)^*\chi(\omega) \\ &= \frac{1}{\pi}\int_0^{2\pi} d\omega\ \sin\{(j+1/2)\omega\}\{\chi(\omega)\sin(\omega/2)\}. \end{aligned}$$

From the theory of the Fourier series, the orthogonal basis functions

$\sin\{(j+1/2)\omega\}$, where j are half of non-negative integers, are complete in the region $[0,\ 2\pi]$, namely, any function $\{\chi(\omega)\sin(\omega/2)\}$ orthogonal to all $\sin\{(j+1/2)\omega\}$ must be zero.

When the group parameters are taken to be the Euler angles, the orthogonal relations (4.51) become

$$\begin{aligned}
&\frac{1}{16\pi^2}\int_{-\pi}^{\pi} d\alpha \int_0^{\pi} d\beta\ \sin\beta \int_{-2\pi}^{2\pi} d\gamma\ D^i_{\mu\rho}(\alpha,\beta,\gamma)^* D^j_{\nu\lambda}(\alpha,\beta,\gamma) \\
&\qquad = \frac{1}{2j+1}\ \delta_{ij}\delta_{\mu\nu}\delta_{\rho\lambda}, \\
&\frac{1}{2}\int_0^{\pi} d\beta\ \sin\beta d^i_{\mu\nu}(\beta) d^j_{\mu\nu}(\beta) = \frac{1}{2j+1}\ \delta_{ij}.
\end{aligned} \tag{4.79}$$

(e) The generators in the representation D^j are calculated by expanding the representation matrices of the rotations around the coordinate axes,

$$\begin{aligned}
D^j_{\nu\mu}(\boldsymbol{e}_3,\omega) &= \delta_{\nu\mu}\{1 - i\mu\omega + \ \dots\ \}, \\
d^j_{\nu\mu}(\omega) &= \delta_{\nu\mu} + (\omega/2)\left\{\delta_{\nu(\mu-1)}\Gamma^j_\mu - \delta_{\nu(\mu+1)}\Gamma^j_\nu\right\}, \\
D^j_{\nu\mu}(\boldsymbol{e}_1,\omega) &= \left\{D^j(\boldsymbol{e}_3,-\pi/2)d^j(\omega)D^j(\boldsymbol{e}_3,\pi/2)\right\}_{\nu\mu} \\
&= \delta_{\nu\mu} + (\omega/2)\left\{-i\delta_{\nu(\mu-1)}\Gamma^j_\mu - i\delta_{\nu(\mu+1)}\Gamma^j_\nu\right\}, \\
\left(I^j_1\right)_{\nu\mu} &= \frac{1}{2}\left[\delta_{\nu(\mu+1)}\Gamma^j_\nu + \delta_{\nu(\mu-1)}\Gamma^j_{-\nu}\right], \\
\left(I^j_2\right)_{\nu\mu} &= -\frac{i}{2}\left[\delta_{\nu(\mu+1)}\Gamma^j_\nu - \delta_{\nu(\mu-1)}\Gamma^j_{-\nu}\right], \\
\left(I^j_3\right)_{\nu\mu} &= \mu\delta_{\nu\mu},
\end{aligned} \tag{4.80}$$

where $\Gamma^j_\nu = \Gamma^j_{-\nu+1} = \{(j+\nu)(j-\nu+1)\}^{1/2}$, $\Gamma^j_{-\nu} = \Gamma^j_{\nu+1}$, and $\Gamma^j_{j-n} = \sqrt{(2j-n)(n+1)}$. In comparison with T_a given in Eq. (4.11), the representation D^1 is equivalent to the self-representation of SO(3) (see Prob. 10 of Chap. 1 in [Ma and Gu (2004)]):

$$\begin{aligned}
&M^{-1}T_aM = I^1_a, \qquad a = 1,\ 2,\ 3, \\
&M^{-1}R(\alpha,\beta,\gamma)M = D^1(\alpha,\beta,\gamma),
\end{aligned} \qquad M = \frac{1}{\sqrt{2}}\begin{pmatrix} -1 & 0 & 1 \\ -i & 0 & -i \\ 0 & \sqrt{2} & 0 \end{pmatrix}. \tag{4.81}$$

Since the generators I^j_a, including T_a, are the matrix forms of the differential operators L_a, they all satisfy the typical commutative relations

$$\left[I^j_a,\ I^j_b\right] = i\sum_{c=1}^{3} \epsilon_{abc} I^j_c. \tag{4.82}$$

In physics, the combinations of generators, $I_{\pm}^j = I_1^j \pm iI_2^j$ and $\left(I^j\right)^2 = \sum_a \left(I_a^j\right)^2$, are often used:

$$\begin{aligned}
\left(I_+^j\right)_{\nu\mu} &= \left(I_1^j + iI_2^j\right)_{\nu\mu} = \delta_{\nu(\mu+1)}\Gamma_\nu^j = \delta_{\nu(\mu+1)}\Gamma_{-\mu}^j, \\
\left(I_-^j\right)_{\nu\mu} &= \left(I_1^j - iI_2^j\right)_{\nu\mu} = \delta_{\nu(\mu-1)}\Gamma_\mu^j = \delta_{\nu(\mu-1)}\Gamma_{-\nu}^j, \\
\left(I^j\right)_{\nu\mu}^2 &= \frac{1}{2}\left\{I_+^jI_-^j + I_-^jI_+^j + 2\left(I_3^j\right)^2\right\}_{\nu\mu} = \delta_{\nu\mu}j(j+1).
\end{aligned} \tag{4.83}$$

If the basis function $\psi_\mu^j(\xi,\eta)$ belongs to the μ-row of the representation D^j (see Eq. (4.72)), it is the common eigenfunction of $\left(I^j\right)^2$ and I_3^j,

$$\begin{aligned}
\left(I^j\right)^2 \psi_\mu^j(\xi,\eta) &= j(j+1)\psi_\mu^j(\xi,\eta), \\
I_3^j\psi_\mu^j(\xi,\eta) &= \mu\psi_\mu^j(\xi,\eta), \\
I_\pm^j\psi_\mu^j(\xi,\eta) &= \Gamma_{\mp\mu}^j\psi_{\mu\pm1}^j(\xi,\eta).
\end{aligned} \tag{4.84}$$

(f) Since the character χ^j is real, D^j is self-conjugate. Due to Eq. (4.75), the similarity transformation X, transforming D^j to D^{j*}, changes the sign of ω in $D^j(\boldsymbol{e}_3,\omega)$, but preserves $D^j(\boldsymbol{e}_2,\omega)$ invariant. Thus, X is $d^j(\pi)$:

$$d^j(\pi)^{-1}D^j(\alpha,\beta,\gamma)d^j(\pi) = D^j(\alpha,\beta,\gamma)^*. \tag{4.85}$$

When j is an integer ℓ, $d^\ell(\pi)$ is symmetric [see Eq. (4.76)] and D^ℓ is real. When j is half of an odd integer, $d^j(\pi)$ is antisymmetric and D^j is self-conjugate, but not real. The explicit forms of $d_{\nu\mu}^j(\omega)$ with $1/2 \geq j \geq 3$ are listed in Prob. 9 of Chap. 4 of [Ma and Gu (2004)].

4.4.3 *Spherical Harmonics Functions*

Consider a single-body system moving in a spherically symmetric potential. Its Hamiltonian $H(\boldsymbol{r})$ is invariant in any rotation R around the center, $H(\boldsymbol{r})P_R = P_RH(\boldsymbol{r})$. The symmetry group of the system is SO(3). If the energy level E is degenerate with multiplicity n, there is n linearly independent eigenfunctions $\psi_\mu(x)$, $H(\boldsymbol{r})\psi_\mu(\boldsymbol{r}) = E\psi_\mu(x)$, where $\mu = 1,\ 2,\ \cdots,\ n$. $P_R\psi_\mu(\boldsymbol{r})$ is also an eigenfunction of $H(\boldsymbol{r})$ with the same energy E so that $P_R\psi_\mu(\boldsymbol{r})$ can be expanded with respect to the basis functions $\psi_\nu(\boldsymbol{r})$

$$P_R\psi_\mu(\boldsymbol{r}) = \sum_{\nu=1}^n \psi_\nu(\boldsymbol{r})D_{\nu\mu}(R).$$

Since $\psi_\mu(\boldsymbol{r})$ remains invariant in a rotation through an angle 2π, $D(R)$ is a single-valued representation of SO(3) corresponding to the energy E. Generally, $D(R)$ is reducible,

$$X^{-1}D(R)X = \bigoplus_{\ell} a_\ell D^\ell(R), \qquad X^{-1}I_a X = \bigoplus_{\ell} a_\ell I_a^\ell, \tag{4.86}$$

where I_a are the generators in the representation $D(R)$ and ℓ is an integer. a_ℓ can be calculated by the characters, $\chi(R) = \sum_\ell a_\ell \chi^\ell(R)$,

$$\begin{aligned} a_\ell &= \frac{2}{\pi}\int_0^\pi d\omega \ \sin^2(\omega/2)\chi^\ell(\omega)\chi(\omega) \\ &= \frac{2}{\pi}\int_0^\pi d\omega \ \sin(\omega/2)\sin\{(\ell+1/2)\omega\}\chi(\omega). \end{aligned} \tag{4.87}$$

Substituting a_ℓ into Eq. (4.86), where I_a is taken to be I_3 first, and then to be $I_\pm$, one obtains the similarity transformation matrix X. The row of X is enumerated by ρ, and the column by $\ell m\tau$, where τ is needed when $a_\ell > 1$. The new basis function $\psi^\ell_{m\tau}(\boldsymbol{r})$ is

$$\begin{aligned} \psi^\ell_{m\tau}(\boldsymbol{r}) &= \sum_\mu \psi_\mu(\boldsymbol{r}) X_{\mu,\ell m\tau} \ , \\ P_R \psi^\ell_{m\tau}(\boldsymbol{r}) &= \sum_{m'} \psi^\ell_{m'\tau}(\boldsymbol{r}) D^\ell_{m'm}(R). \end{aligned} \tag{4.88}$$

Hence, the static wave function of the single-body system with spherical symmetry can be chosen as $\psi^\ell_{m\tau}(\boldsymbol{r})$ which belongs to the m-row of the irreducible representation D^ℓ of SO(3). $\psi^\ell_{m\tau}(\boldsymbol{r})$ is the common eigenfunction of, in addition to the Hamiltonian, the orbital angular momentum L^2 and L_3 (see Eqs. (4.23), (4.26), and (4.84))

$$\begin{aligned} L^2\psi^\ell_{m\tau}(\boldsymbol{r}) &= \ell(\ell+1)\psi^\ell_{m\tau}(\boldsymbol{r}), \\ L_3\psi^\ell_{m\tau}(\boldsymbol{r}) &= m\psi^\ell_{m\tau}(\boldsymbol{r}), \\ L_\pm\psi^\ell_{m\tau}(\boldsymbol{r}) &= \Gamma^\ell_{\mp m}\psi^\ell_{(m\pm 1)\tau}(\boldsymbol{r}), \end{aligned} \tag{4.89}$$

where $L_\pm = L_1 \pm iL_2$ and $L^2 = \sum_a L_a^2$.

Now, neglect the index τ in $\psi^\ell_{m\tau}(\boldsymbol{r})$ for simplicity. Let $\boldsymbol{r} = (r,\ \theta,\ \varphi)$ and $\boldsymbol{r}_0 = (r,\ 0,\ 0)$. The rotation $T = R(\varphi,\ \theta,\ \gamma)$ changes the point $\boldsymbol{r}_0$ at the z-axis to $\boldsymbol{r}$, $\boldsymbol{r} = T\boldsymbol{r}_0$. From Eq. (4.88), one has

$$\begin{aligned} \psi^\ell_m(\boldsymbol{r}) &= \psi^\ell_m(T\boldsymbol{r}_0) = P_{T^{-1}}\psi^\ell_m(\boldsymbol{r}_0) = \sum_{m'} \psi^\ell_{m'}(\boldsymbol{r}_0) D^\ell_{mm'}(T)^* \\ &= \sum_{m'} \psi^\ell_{m'}(\boldsymbol{r}_0) e^{im\varphi} d^\ell_{mm'}(\theta) e^{im'\gamma}. \end{aligned} \tag{4.90}$$

Since the left-hand side of Eq. (4.90) is independent of γ, the right-hand side of Eq. (4.90) has to be independent of γ, namely, the terms with $m' \neq 0$ on the right-hand side of Eq. (4.90) are vanishing. Letting

$$\psi^\ell_m(\boldsymbol{r}_0) = \delta_{m0} \left(\frac{2\ell+1}{4\pi}\right)^{1/2} \phi_\ell(r), \tag{4.91}$$

from Eq. (4.90), one has

$$\psi^\ell_m(\boldsymbol{r}) = \phi_\ell(r) \left(\frac{2\ell+1}{4\pi}\right)^{1/2} D^\ell_{m0}(\varphi, \theta, 0)^*. \tag{4.92}$$

Hence, the static wave function $\psi^\ell_m(\boldsymbol{r})$ of the single-body system with the spherical symmetry can be decomposed as a product of a radial function $\phi_\ell(r)$ and an angular function $Y^\ell_m(\theta, \varphi)$,

$$Y^\ell_m(\theta, \varphi) \equiv \left(\frac{2\ell+1}{4\pi}\right)^{1/2} D^\ell_{m0}(\varphi, \theta, 0)^* = \left(\frac{2\ell+1}{4\pi}\right)^{1/2} e^{im\varphi} d^\ell_{m0}(\theta). \tag{4.93}$$

Since the radial function remains invariant in any rotation, the angular function belongs to the mth row of the irreducible representation D^ℓ of SO(3) so that it is the common eigenfunction of L^2 and L_3 as given in Eq. (4.89). In quantum mechanics, $Y^\ell_m(\theta, \varphi)$ is called the spherical harmonic function. Due to Eq. (4.79) $Y^\ell_m(\theta, \varphi)$ is orthonormal

$$\begin{aligned}
&\int_{-\pi}^{\pi} d\varphi \int_0^{\pi} d\theta \, \sin\theta Y^\ell_m(\theta, \varphi)^* Y^{\ell'}_{m'}(\theta, \varphi) \\
&= \frac{\{(2\ell+1)(2\ell'+1)\}^{1/2}}{4\pi} \int_{-\pi}^{\pi} d\varphi \int_0^{\pi} d\theta \, \sin\theta D^\ell_{m0}(\varphi, \theta, 0)^* D^{\ell'}_{m'0}(\varphi, \theta, 0) \\
&= \delta_{\ell\ell'} \delta_{mm'}.
\end{aligned} \tag{4.94}$$

From the symmetry (4.76) of $d^\ell(\theta)$, one has

$$Y^\ell_m(\theta, \varphi)^* = \left(\frac{2\ell+1}{4\pi}\right)^{1/2} e^{-im\varphi} d^\ell_{m0}(\theta) = (-1)^m Y^\ell_{-m}(\theta, \varphi). \tag{4.95}$$

The Legendre function is defined as

$$P_\ell(\cos\theta) = \left(\frac{4\pi}{2\ell+1}\right)^{1/2} Y^\ell_0(\theta, 0) = d^\ell_{00}(\theta), \tag{4.96}$$

which satisfies

$$\int_0^{\pi} d\theta \, \sin\theta P_\ell(\cos\theta) P_{\ell'}(\cos\theta) = \frac{2\delta_{\ell\ell'}}{2\ell+1}. \tag{4.97}$$

In the spatial inversion, $\boldsymbol{r} \longrightarrow -\boldsymbol{r}$, $\theta \longrightarrow \pi - \theta$, and $\varphi \longrightarrow \pi + \varphi$,

$$\begin{aligned} Y^\ell_m(\pi - \theta, \pi + \varphi) &= \left(\frac{2\ell + 1}{4\pi}\right)^{1/2} e^{-im(\pi+\varphi)} d^\ell_{m0}(\pi - \theta) \\ &= (-1)^\ell Y^\ell_m(\theta, \varphi). \end{aligned} \tag{4.98}$$

The parity of the spherical harmonic function is $(-1)^\ell$.

Multiplying $Y^\ell_m(\theta, \varphi)$ with r^ℓ, one obtains the harmonic polynomial $\mathbf{Y}^\ell_m(\boldsymbol{r})$ which is a homogeneous polynomial of order ℓ with respect to the rectangular coordinates (x_1, x_2, x_3) and satisfies the Laplace equation

$$\mathbf{Y}^\ell_m(\boldsymbol{r}) = r^\ell Y^\ell_m(\boldsymbol{r}), \qquad \nabla^2 \mathbf{Y}^\ell_m(\boldsymbol{r}) = 0. \tag{4.99}$$

$\mathbf{Y}^\ell_m(\boldsymbol{r})$ with $\ell \le 3$ are listed as follows.

$$\begin{aligned} &\mathbf{Y}^0_0(\boldsymbol{r}) = \sqrt{\frac{1}{4\pi}}, && \mathbf{Y}^1_{\pm1}(\boldsymbol{r}) = \mp\sqrt{\frac{3}{8\pi}}\left(x_1 \pm ix_2\right), \\ &\mathbf{Y}^1_0(\boldsymbol{r}) = \sqrt{\frac{3}{4\pi}}\, x_3, && \mathbf{Y}^2_{\pm2}(\boldsymbol{r}) = \sqrt{\frac{15}{32\pi}}\left(x_1 \pm ix_2\right)^2, \\ &\mathbf{Y}^2_{\pm1}(\boldsymbol{r}) = \mp\sqrt{\frac{15}{8\pi}}\left(x_1 \pm ix_2\right) x_3, && \mathbf{Y}^2_0(\boldsymbol{r}) = \sqrt{\frac{5}{16\pi}}\left(3x_3^2 - r^2\right), \\ &\mathbf{Y}^3_{\pm3}(\boldsymbol{r}) = \mp\sqrt{\frac{35}{64\pi}}\left(x_1 \pm ix_2\right)^3, && \mathbf{Y}^3_{\pm2}(\boldsymbol{r}) = \sqrt{\frac{105}{32\pi}}\left(x_1 \pm ix_2\right)^2 x_3, \\ &\mathbf{Y}^3_{\pm1}(\boldsymbol{r}) = \mp\sqrt{\frac{21}{64\pi}}\left(x_1 \pm ix_2\right)\left(5x_3^2 - r^2\right), && \mathbf{Y}^3_0(\boldsymbol{r}) = \sqrt{\frac{7}{16\pi}}\left(5x_3^2 - 3r^2\right) x_3. \end{aligned}$$

4.5 The Lie Theorems

The Lie theorems are the fundamental theorems in the theory of Lie groups which characterize the important role of generators.

Theorem 4.1 (The First Lie Theorem) The representation of a Lie group G with a connected group space is completely determined by its generators.

Proof This theorem solves the problem how the infinitesimal elements determine the property of a Lie group G with a connected group space, namely, how is the representation matrix $D(R)$ of G calculated from the generators of the representation $D(G)$?

Let $RS = T$ and $t_A = f_A(r; s)$. In the representation $D(G)$ one has $D(R) = D(T)D(S^{-1})$. Make the partial derivative on both sides of the equation with respect to the parameters r_B for the fixed S,

$$\frac{\partial D(R)}{\partial r_B}=\frac{\partial D(T)}{\partial r_B}D(S^{-1})=\sum_A \frac{\partial D(T)}{\partial t_A}\frac{\partial f_A(r;s)}{\partial r_B}D(S^{-1}).$$

Then, taking $S^{-1}=R$, one has $T=E$ and obtains from Eq. (4.25)

$$\frac{\partial D(R)}{\partial r_B}=-i\left\{\sum_A I_A S_{AB}(r)\right\}D(R), \tag{4.100}$$

where

$$S_{AB}(r)\equiv \left.\frac{\partial f_A(r;s)}{\partial r_B}\right|_{s=\overline{r}}=\left.\frac{\partial f_A(r';\overline{r})}{\partial r'_B}\right|_{r'=r}. \tag{4.101}$$

$S_{AB}(r)$ is independent of the representation $D(G)$, but depends on the choice of parameters of the Lie group G. The determinant of $S(r)$ is nothing but the Jacobi determinant in the integral transformation (see Eq. (4.47)) so that $S(r)$ is non-singular. Denote by $\overline{S}(r)$ the inverse matrix of $S(r)$:

$$\sum_D \overline{S}_{AD}(r)S_{DB}(r)=\delta_{AB}. \tag{4.102}$$

Letting $R=E$ in Eq. (4.101), one has

$$S_{AB}(E)=\delta_{AB}, \tag{4.103}$$

and Eq. (4.100) reduces to Eq. (4.25).

Equation (4.100) is solved under the m^2 boundary conditions

$$D_{\mu\nu}(R)\big|_{R=E}=\delta_{\mu\nu}, \tag{4.104}$$

where m is the dimension of the representation. Since $D(R)$ exists, one may choose a convenient path to integrate Eq. (4.100) such that only one variable changes in each segment of the path. First, let all $r_A=0$ except for r_1 which changes from 0. Equation (4.100) with the boundary condition (4.104) is a differential equation of first order and its solution $D(r_1,\ 0,\ \ldots,\ 0)$ is an exponential function. Then fix r_1 and let the remaining $r_A=0$ except for r_2 which changes from 0. Equation (4.100) with the boundary condition $D(r_1,\ 0,\ \ldots,\ 0)$ is a differential equation of first order again and its solution $D(r_1,\ r_2,\ 0,\ \ldots,\ 0)$ is the product of two exponential functions. In this way one can obtain the solution $D(R)$, which is a product of a few exponential functions. □

Corollary 4.1.1 Two representations of a Lie group with a connected group space are equivalent if and only if their generators are related by a common similarity transformation

$$I_A^{(1)} = X^{-1} I_A^{(2)} X.$$

Corollary 4.1.2 A representation of a Lie group with a connected group space is reducible if and only if its representation space contains a nontrivial invariant subspace with respect to its all generators.

Corollary 4.1.3 If $I_A^{(1)}$ and $I_A^{(2)}$ are the generators of two inequivalent irreducible representations of a Lie group with a connected group space and X satisfies

$$X I_A^{(1)} = I_A^{(2)} X$$

for all generators, then $X = 0$.

Corollary 4.1.4 If X is commutable with all generators I_A of an irreducible representation of a Lie group with a connected group space, then X is a constant matrix.

Due to Theorem 4.1, the corollaries are obvious. For the mixed Lie group, one can choose one element in each connected piece in the group space. The corollaries hold if, in addition to the generators, the representation matrices of those chosen elements also satisfy the condition.

Theorem 4.2 (The Second Lie Theorem) The generators in any representation of a Lie group G satisfy the common commutative relations

$$I_A I_B - I_B I_A = i \sum_D C_{AB}{}^D I_D, \tag{4.105}$$

where $C_{AB}{}^D$ are called the structure constants of G. The structure constants are real and independent of the representation. Conversely, if g matrices satisfy the commutative relations (4.105), they are the generators of a representation of G with order g.

Proof According to the theory of differential equations, under the boundary condition (4.104), the solutions in integrating Eq. (4.100) through two different paths which can be deformed in the group space are the same if and only if

$$\frac{\partial^2 D(R)}{\partial r_A \partial r_B} = \frac{\partial^2 D(R)}{\partial r_B \partial r_A}. \tag{4.106}$$

Differentiating Eq. (4.100) with respect to r_A, one has

$$\frac{\partial^2 D(R)}{\partial r_A \partial r_B} = -i \sum_D I_D \frac{\partial S_{DB}(r)}{\partial r_A} D(R) - \sum_{DP} I_D I_P S_{DB}(r) S_{PA}(r) D(R).$$

By interchanging A and B, it becomes

$$\frac{\partial^2 D(R)}{\partial r_B \partial r_A} = -i \sum_D I_D \frac{\partial S_{DA}(r)}{\partial r_B} D(R) - \sum_{DP} I_D I_P S_{DA}(r) S_{PB}(r) D(R).$$

Then, Eq. (4.106) becomes

$$\sum_{DP} \{I_D I_P - I_P I_D\} S_{DA}(r) S_{PB}(r) = i \sum_D I_D \left\{ \frac{\partial S_{DB}(r)}{\partial r_A} - \frac{\partial S_{DA}(r)}{\partial r_B} \right\}.$$

Right-multiplying it with $\overline{S}(r)$ (see Eq. (4.102)), one has

$$I_A I_B - I_B I_A = i \sum_D I_D \left\{ \sum_{PQ} \left(\frac{\partial S_{DQ}(r)}{\partial r_P} - \frac{\partial S_{DP}(r)}{\partial r_Q} \right) \overline{S}_{PA}(r) \overline{S}_{QB}(r) \right\}. \tag{4.107}$$

The left-hand side of Eq. (4.107) is independent of R, so is its right-hand side, namely, the quantity in the curve bracket of Eq. (4.107) is constant. Letting $R = E$, one obtains

$$C_{AB}{}^{D} = \left\{ \frac{\partial S_{DB}(r)}{r_A} - \frac{\partial S_{DA}(r)}{r_B} \right\} \Bigg|_{r=0}. \tag{4.108}$$

Thus, the structure constants are real and independent of the representation.

Conversely, if g matrices I_A satisfy the commutative relations (4.105), Eqs. (4.107) and (4.106) also hold, so that the solution $D(R)$ is independent of the paths. Due to Eq. (4.103), I_A satisfy Eq. (4.25) obviously. The next problem is whether the solution $D(R)$ constitute a representation of G, namely, whether $D(R)D(S) = D(T)$ if $RS = T$.

Integrating Eq. (4.100) with the boundary condition (4.104), first from E to S, one obtains $D(S)$. Then, integrating Eq. (4.100) with the boundary condition $D(S)$ from S to T, one has

$$\frac{\partial D(T)}{\partial t_D} = -i \sum_A I_A S_{AD}(t) D(T), \qquad D(T)|_{T=S} = D(S).$$

Thus,

$$\begin{aligned}\frac{\partial D(T)}{\partial r_B} &= \sum_D \frac{\partial D(T)}{\partial t_D}\frac{\partial f_D(r;s)}{\partial r_B}\\ &= -i\sum_A I_A\left\{\sum_D S_{AD}(t)\frac{\partial f_D(r;s)}{\partial r_B}\right\}D(T),\end{aligned}$$

$$\begin{aligned}\sum_D S_{AD}(t)\frac{\partial f_D(r;s)}{\partial r_B} &= \sum_D \left.\frac{\partial f_A(t;u)}{\partial t_D}\right|_{u=\overline{t}} \frac{\partial f_D(r;s)}{\partial r_B}\\ &= \left.\frac{\partial f_A\{f(r;s);u\}}{\partial r_B}\right|_{u=\overline{t}} = \left.\frac{\partial f_A\{r;f(s;u)\}}{\partial r_B}\right|_{f(s;u)=\overline{r}} = S_{AB}(r).\end{aligned}$$

On the other hand, right-multiplying Eqs. (4.100) and (4.104) with $D(S)$, one has

$$\begin{aligned}&\frac{\partial D(R)D(S)}{\partial r_B} = -i\sum_A I_A S_{AB}(r)D(R)D(S)\ ,\\ &D(R)D(S)|_{RS=S} = D(S).\end{aligned}$$

Since both $D(T)$ and $D(R)D(S)$ satisfy the same equation with the same boundary condition, they equal each other. □

Because the generator I_A is the matrix form of a differential operator $I_A^{(0)}$ in the representation space, $I_A^{(0)}$ satisfies the same commutative relation

$$\left[I_A^{(0)},\ I_B^{(0)}\right] = i\sum_D C_{AB}{}^D I_D^{(0)}. \tag{4.109}$$

Usually, the structure constants $C_{AB}{}^D$ are not calculated from Eq. (4.108). Instead, for a given Lie group and its parameters, $C_{AB}{}^D$ are calculated from the commutative relations of the generators in a known faithful representation, say the self-representation. The differential operators of the SO(3) group are the orbital angular momentum operators,

$$[L_a,\ L_b] = i\sum_d \epsilon_{abd}L_d, \qquad C_{ab}{}^d = \epsilon_{abd}. \tag{4.110}$$

From Eq. (4.110) one obtains that the structure constants $C_{ab}{}^d$ of both the SO(3) group and the SU(2) group are the totally antisymmetric tensor ϵ_{abd}. The generators of every representation of the SO(3) group and the SU(2) group, including their self-representations (see Eqs. (4.11) and (4.31)), have to satisfy the commutative relations (4.110). In quantum mechanics, the matrix forms (4.80) of the angular momentum operators are calculated from the commutative relations (see Prob. 14 of Chap. 4 in [Ma and Gu (2004)]).

What conditions the structure constants $C_{AB}{}^{D}$ of a Lie group should satisfy? From Eq. (4.105) and the Jacobi identity,

$$\begin{aligned}
&[[I_A,\ I_B]\,,\ I_D] + [[I_B,\ I_D]\,,\ I_A] + [[I_D,\ I_A]\,,\ I_B] \\
&\quad = I_AI_BI_D - I_BI_AI_D - I_DI_AI_B + I_DI_BI_A + I_BI_DI_A - I_DI_BI_A \\
&\qquad - I_AI_BI_D + I_AI_DI_B + I_DI_AI_B - I_AI_DI_B - I_BI_DI_A + I_BI_AI_D \\
&\quad = 0.
\end{aligned}$$

one obtains

$$\begin{gathered}
C_{AB}{}^{D} = -C_{BA}{}^{D}, \\
\sum_P \left\{C_{AB}{}^{P}C_{PD}{}^{Q} + C_{BD}{}^{P}C_{PA}{}^{Q} + C_{DA}{}^{P}C_{PB}{}^{Q}\right\} = 0.
\end{gathered} \tag{4.111}$$

Theorem 4.3 (The Third Lie Theorem) A set of constants $C_{AB}{}^{D}$ can be the structure constants of a Lie group if and only if they satisfy Eq. (4.111).

We will not prove this theorem. The Lie groups can be classified based on this theorem (see Chap. 7). Two Lie groups with the same structure constants are said to be locally isomorphic. Two locally isomorphic Lie groups are not isomorphic generally. There are two typical counterexamples. SU(2) and SO(3) have the same structure constants, but globally they are only homomorphic. The two-dimensional unitary group U(2) contains a subgroup SU(2) as well as a subgroup U(1) composed by the determinants of the elements in U(2). However, two subgroups contain two common elements $\pm\mathbf{1}$. Thus, U(2) is not isomorphic onto the group SU(2)$\otimes$ U(1), but they are locally isomorphic.

The adjoint representation of a Lie group is defined in Eq. (4.28),

$$D(R)I_AD(R)^{-1} = \sum_B I_BD^{\text{ad}}_{BA}(R).$$

When R is an infinitesimal element, from Eq. (4.25) one has

$$\begin{gathered}
D(R) = \mathbf{1} - i\sum_D r_DI_D, \qquad D^{\text{ad}}_{BA}(R) = \delta_{BA} - i\sum_D r_D\left(I^{\text{ad}}_D\right)_{BA}, \\
i\sum_B C_{DA}{}^{B}I_B = [I_D,\ I_A] = \sum_B I_B\left(I^{\text{ad}}_D\right)_{BA}.
\end{gathered} \tag{4.112}$$

Thus, the generators of the adjoint representation $D^{\text{ad}}(R)$ of a Lie group are directly related with the structure constants,

$$\left(I_A^{\rm ad}\right)_{BD} = iC_{AD}^{\ \ \ B}. \tag{4.113}$$

It is easy to check that $\left(I_A^{\rm ad}\right)_{BD}$ satisfy the commutative relation (4.105),

$$\begin{aligned}
\left[I_A^{\rm ad},\, I_B^{\rm ad}\right]_{RS} &= \sum_P \left\{\left(I_A^{\rm ad}\right)_{RP}\left(I_B^{\rm ad}\right)_{PS} - \left(I_B^{\rm ad}\right)_{RP}\left(I_A^{\rm ad}\right)_{PS}\right\} \\
&= -\sum_P \left\{C_{AP}^{\ \ \ R} C_{BS}^{\ \ \ P} - C_{BP}^{\ \ \ R} C_{AS}^{\ \ \ P}\right\} \\
&= -\sum_P \left\{C_{AP}^{\ \ \ R} C_{BS}^{\ \ \ P} + C_{BP}^{\ \ \ R} C_{SA}^{\ \ \ P}\right\} \\
&= \sum_P C_{SP}^{\ \ \ R} C_{AB}^{\ \ \ P} = i\sum_P C_{AB}^{\ \ \ P}\left(I_P^{\rm ad}\right)_{RS}.
\end{aligned}$$

Both Eqs. (4.28) and (4.113) can be used as the definitions of the adjoint representation of a Lie group. Usually, the adjoint representation of a Lie group is calculated neither by derivative in Eq. (4.28), nor by solving the differential equation (4.100) with the generators (4.113). In fact, the adjoint representation is determined by comparing the known representation and its generators with Eqs. (4.28) and (4.113). Since

$$\left(I_a^{\rm ad}\right)_{bd} = iC_{ad}^{\ \ \ b} = -i\epsilon_{abd} = \left(T_a\right)_{bd}, \tag{4.114}$$

the group SU(2) and the group SO(3) have the same adjoint representation which is nothing but the self-representation of SO(3). This conclusion can also be seen from comparing Eq. (4.41) with Eq. (4.28). Thus, for the transformation operator P_R of SO(3), Eq. (4.28) becomes

$$P_R L_a P_R^{-1} = \sum_{b=1}^{3} L_b R_{ba}. \tag{4.115}$$

Let $R = R(\varphi, \theta, 0) = S(\varphi, \theta)$. Denoting by $\hat{\boldsymbol{n}}(\theta, \varphi)$ the unit vector $R_{\cdot 3}$, one obtains the angular momentum operator at the direction $\hat{\boldsymbol{n}}$,

$$\boldsymbol{L}\cdot\hat{\boldsymbol{n}}(\theta, \varphi) = P_R L_3 P_R^{-1}, \qquad R = R(\varphi, \theta, 0). \tag{4.116}$$

Since $P_{R(\boldsymbol{e}_3,\gamma)}$ is commutable with L_3, R in Eq. (4.116) can be replaced with $R(\varphi, \theta, \gamma)$. But it is convenient to take $\gamma = 0$. If $\psi_m^\ell(x)$ belongs to the mth row of the irreducible representation $D^\ell(SO(3))$, it is the eigenfunction of L_3 with the eigenvalue m, and $P_R\psi_m^\ell(x)$ is the eigenfunction of $\boldsymbol{L}\cdot\hat{\boldsymbol{n}}$ with the eigenvalue m where $R = R(\varphi, \theta, \gamma)$ and $\hat{\boldsymbol{n}} = \hat{\boldsymbol{n}}(\theta, \varphi)$. Obviously, $P_R\psi_m^\ell(x)$ is also the eigenfunction of L^2 with the eigenvalue $\ell(\ell+1)$.

4.6 Clebsch–Gordan Coefficients of SU(2)

4.6.1 *Direct Product of Representations*

For a compound system composed by two subsystems with the spherical symmetry, the wave functions of the subsystems belong to the given irreducible representations of SU(2), respectively

$$\begin{aligned} P_u\psi^j_\mu(x^{(1)}) &= \sum_{\mu'} \psi^j_{\mu'}(x^{(1)})D^j_{\mu'\mu}(u), \\ P_u\psi^k_\nu(x^{(2)}) &= \sum_{\nu'} \psi^k_{\nu'}(x^{(2)})D^k_{\nu'\nu}(u). \end{aligned} \tag{4.117}$$

They are the eigenfunctions of the angular momentum operators, respectively. Here, we generalize the rotation of SO(3) to that of SU(2) in order to make the calculated formulas suitable for both SO(3) and SU(2). The wave function of the compound system is the product $\psi^j_\mu(x^{(1)})\psi^k_\nu(x^{(2)})$ which transforms under $u \in$ SU(2) according to the direct product of the representations

$$P_u\left\{\psi^j_\mu(x^{(1)})\psi^k_\nu(x^{(2)})\right\} = \sum_{\mu'\nu'}\left\{\psi^j_{\mu'}(x^{(1)})\psi^k_{\nu'}(x^{(2)})\right\}\left[D^j(u)\times D^k(u)\right]_{\mu'\nu',\mu\nu}. \tag{4.118}$$

Generally, the representation $D^j(u)\times D^k(u)$ is reducible and can be reduced as a direct sum of irreducible representations D^J of SU(2) by a similarity transformation C^{jk},

$$\left(C^{jk}\right)^{-1}\left\{D^j(u)\times D^k(u)\right\}C^{jk} = \bigoplus_J a_J D^J(u). \tag{4.119}$$

The series in Eq. (4.119) is called the Clebsch–Gordan series. The character of the direct product of representations is

$$\chi^j(\omega)\chi^k(\omega) = \sum_J a_J\chi^J(\omega), \tag{4.120}$$

where the multiplicity a_J can be calculated by the formula (4.87) for characters. Here, we apply the formula (4.78) directly. When $j \geq k$,

$$\chi^j(\omega) = \sum_{\mu=-j}^{j} e^{-i\mu\omega} = \sum_{\mu=-j}^{j} e^{i\mu\omega} = \frac{e^{i(j+1)\omega} - e^{-ij\omega}}{e^{i\omega}-1},$$

$$\begin{aligned}\chi^j(\omega)\chi^k(\omega) &= \sum_{\mu=-k}^{k} \frac{e^{i(j+\mu+1)\omega} - e^{-i(j+\mu)\omega}}{e^{i\omega}-1} \\ &= \sum_{J=j-k}^{j+k} \frac{e^{i(J+1)\omega} - e^{-iJ\omega}}{e^{i\omega}-1} = \sum_{J=j-k}^{j+k} \chi^J(\omega).\end{aligned} \tag{4.121}$$

When $j < k$, $j-k$ is replaced with $k-j$. Generally, one has

$$a_J = \begin{cases} 1 & \text{when } J = j+k,\ j+k-1,\ \ldots,\ |j-k|, \\ 0 & \text{the remaining cases.} \end{cases} \tag{4.122}$$

In the coupling of the angular momentums j and k of two subsystems, the total angular momentum J of the compound system can take $(2k+1)$ values (when $j \geq k$) or $(2j+1)$ values (when $j \leq k$) given in Eq. (4.122), and each value of the total angular momentum J appears once. Three angular momentums j, k, and J satisfy the rule that each one in three is not larger than the sum of the other two and not less than their difference. In addition, the three angular momentums j, k, and J are all integers, or one is integer and the other two are half of odd integers. This rule is called the rule of a triangle, denoted by $\Delta(j, k, J)$.

The row of the matrix C^{jk} is enumerated by $\mu\nu$, and its column by JM, The new basis functions are combined by the matrix entries of C^{jk},

$$\begin{aligned}\Psi^J_M(x^{(1)}, x^{(2)}) &= \sum_{\mu\nu} \left\{\psi^j_\mu(x^{(1)})\psi^k_\nu(x^{(2)})\right\} C^{jk}_{\mu\nu,JM}, \\ P_u \Psi^J_M(x^{(1)}, x^{(2)}) &= \sum_{M'} \Psi^J_{M'}(x^{(1)}, x^{(2)}) D^J_{M'M}(u),\end{aligned} \tag{4.123}$$

where $|\mu| \leq j$, $|\nu| \leq k$, and $|M| \leq J$. The entries $C^{jk}_{\mu\nu,JM}$ are called the Clebsch–Gordan coefficients, or briefly, CG coefficients. Since the coefficients are related to the addition of angular momentums, the CG coefficients are also called the vector coupling coefficients. Sometimes, the CG coefficients $C^{jk}_{\mu\nu,JM}$ are denoted by the Dirac symbols $\langle jk\mu\nu|jkJM\rangle$ or $\langle j\mu, k\nu|JM\rangle$.

Now, let us discuss the properties of the CG coefficients. Rewrite Eq. (4.119) in the form of generators. Since the generators of the direct product $D^j(u) \times D^k(u)$ of two representations are

$$I^j_a \times \mathbf{1}_{2k+1} + \mathbf{1}_{2j+1} \times I^k_a, \tag{4.124}$$

one has

$$\sum_{\rho} \left(I_a^j\right)_{\mu\rho} C^{jk}_{\rho\nu,JM} + \sum_{\lambda} \left(I_a^k\right)_{\nu\lambda} C^{jk}_{\mu\lambda,JM} = \sum_{N} C^{jk}_{\mu\nu,JN} \left(I_a^J\right)_{NM}. \quad (4.125)$$

When $a = 3$, I_3 is diagonal and Eq. (4.125) becomes

$$(\mu + \nu) C^{jk}_{\mu\nu,JM} = M C^{jk}_{\mu\nu,JM},$$

namely, $M = \mu + \nu$ in $C^{jk}_{\mu\nu,JM}$, or

$$C^{jk}_{\mu\nu,JM} = 0, \qquad \text{when} \quad M \neq \mu + \nu. \quad (4.126)$$

In the sum of two angular momentums, the component along the z-axis is summed like a scalar.

Replacing I_a in Eq. (4.125) with $I_\pm = I_1 \pm i I_2$, one has from Eq. (4.83)

$$\Gamma^j_{\pm\mu} C^{jk}_{(\mu\mp 1)\nu,JM} + \Gamma^k_{\pm\nu} C^{jk}_{\mu(\nu\mp 1),JM} = C^{jk}_{\mu\nu,J(M\pm 1)} \Gamma^J_{\mp M}. \quad (4.127)$$

This is the recurrence relation for calculating the CG coefficients of SU(2).

Since SU(2) is a compact Lie group, D^j is unitary and C^{jk} can be chosen to be unitary. As pointed out in §3.6.2, there is one undetermined phase angle for each J in the CG coefficients. Choose the phase angle such that $C^{jk}_{j(-k),J(j-k)}$ is real and positive. Taking the lower signs in Eq. (4.127) with $\mu = j$ fixed and $\nu = -k, (-k+1), \ldots, (k-1)$ one by one, one obtains that the first term on the left-hand side of Eq. (4.127) is vanishing and $C^{jk}_{j\nu,J(j+\nu)}$ are all real positive. In the same way, taking the upper sign in Eq. (4.127) with $\nu = -k$ fixed and $\mu = j, (j-1), \ldots, -j+1$ one by one, one obtains that the second term on the left-hand side of Eq. (4.127) is vanishing and $C^{jk}_{\mu(-k),J(\mu-k)}$ are all real positive. Again from Eq. (4.127), the CG coefficients for SU(2) are shown all to be real and C^{jk} is a real orthogonal matrix:

$$\begin{aligned} &\sum_{\mu\nu} C^{jk}_{\mu\nu JM} C^{jk}_{\mu\nu J'M'} = \delta_{JJ'} \delta_{MM'}, \\ &\sum_{JM} C^{jk}_{\mu\nu JM} C^{jk}_{\mu'\nu' JM} = \delta_{\mu\mu'} \delta_{\nu\nu'}. \end{aligned} \quad (4.128)$$

Due to Eq. (4.126) one has

$$\begin{aligned} &\sum_{\mu} C^{jk}_{\mu(M-\mu),JM} C^{jk}_{\mu(M-\mu),J'M} = \delta_{JJ'}, \\ &\sum_{J} C^{jk}_{\mu(M-\mu),JM} C^{jk}_{\mu'(M-\mu'),JM} = \delta_{\mu\mu'}. \end{aligned} \quad (4.129)$$

4.6.2 *Calculation of Clebsch–Gordan Coefficients*

There are two equivalent methods for the calculation of the CG coefficients of SU(2). One is based on the recurrence relation (4.127), and the other is based on the group integral for Eq. (4.119). In the first method, one takes the upper sign in Eq. (4.127) with $M = J$ fixed and $\mu = j$, $(j-1)$, $\ldots$, $-j+1$ one by one, and obtains that the term on the right-hand side of Eq. (4.127) is vanishing and $C^{jk}_{\mu(J-\mu),JJ}$ are expressed in $C^{jk}_{j(J-j),JJ}$, which is real positive and can be calculated through the normalized condition (4.129). The result is

$$C^{jk}_{\mu(J-\mu),JJ} = (-1)^{j-\mu} \cdot \left\{\frac{(2J+1)!(j+k-J)!(J+k-\mu)!(j+\mu)!}{(J+k-j)!(J+j-k)!(J+j+k+1)!(k-J+\mu)!(j-\mu)!}\right\}^{1/2}. \tag{4.130}$$

Then, $(J-M)$ times application of Eq. (4.127) with the lower sign, $C^{jk}_{\mu(M-\mu),JM}$ is expressed as a series of $C^{jk}_{m(J-m),JJ}$. Making a replacement of summation index m with $n = m - \mu$, one obtains the Racah form of the CG coefficients of SU(2)

$$\begin{aligned} C^{jk}_{\mu(M-\mu),JM} &= \left\{\frac{(2J+1)(j+k-J)!(J+M)!(J-M)!(j-\mu)!(k-M+\mu)!}{(J+j+k+1)!(J+j-k)!(J-j+k)!(j+\mu)!(k+M-\mu)!}\right\}^{1/2} \\ &\quad \cdot \sum_n \frac{(-1)^{n+j-\mu}(J+k-\mu-n)!(j+\mu+n)!}{(j-\mu-n)!(J-M-n)!n!(n+\mu+k-J)!}, \end{aligned} \tag{4.131}$$

where n runs from the maximum between 0 and $(J-k-\mu)$ to the minimum between $(j-\mu)$ and $(J-M)$. The more symmetric form, called the Van der Waerden form, of the CG coefficients can be obtained by making use of the identities on combinatorics (see Appendix A)

$$\begin{aligned} C^{jk}_{\mu(M-\mu),JM} = {} & (2J+1)^{1/2}\Delta(j,k,J)\,\{(j+\mu)!(j-\mu)! \\ & \cdot\ (k+M-\mu)!(k-M+\mu)!(J+M)!(J-M)!\}^{1/2} \\ & \cdot\ \sum_n (-1)^n\ \{(n)!(J-j-M+\mu+n)!(J-k+\mu+n)! \\ & \cdot\ (j-\mu-n)!(k+M-\mu-n)!(j+k-J-n)!\}^{-1}, \end{aligned} \tag{4.132}$$

where n runs from the maximum among 0, $(j+M-\mu-J)$, and $(k-\mu-J)$ to the minimum among $(j-\mu)$, $(k+M-\mu)$, and $(j+k-J)$, and

$$\Delta(a,b,c) = \left\{ \frac{(a+b-c)!(b+c-a)!(c+a-b)!}{(a+b+c+1)!} \right\}^{1/2}. \tag{4.133}$$

The detailed calculation can be found in §5.8 of [Ma (1993)], where the quantum parameter q is taken to be 1.

In the second method (see Chap. 17 in [Wigner (1959)]), one first rewrites Eq. (4.119) in the parameters of Euler angles. Due to Eq. (4.126) the factors on α and γ are cancelled to each other,

$$d^j_{\mu\rho}(\beta) d^k_{\nu\lambda}(\beta) = \sum_J C^{jk}_{\mu\nu,J(\mu+\nu)} d^J_{(\mu+\nu)(\rho+\lambda)}(\beta) C^{jk}_{\rho\lambda,J(\rho+\lambda)}.$$

In terms of the orthogonal relation (4.79) on the d^j function, one has

$$C^{jk}_{\mu\nu,J(\mu+\nu)} C^{jk}_{\rho\lambda,J(\rho+\lambda)} = \frac{2J+1}{2} \int_0^\pi d\beta \ \sin\beta d^J_{(\mu+\nu)(\rho+\lambda)}(\beta) d^j_{\mu\rho}(\beta) d^k_{\nu\lambda}(\beta). \tag{4.134}$$

Second, let $\rho = j$ and $\lambda = -k$ in Eq. (4.134). In terms of Eqs. (4.74), (4.77), and the integral formula

$$\frac{1}{2} \int_0^\pi d\beta \ [\cos(\beta/2)]^{2a} [\sin(\beta/2)]^{2b} = \frac{a!b!}{(a+b+1)!}, \tag{4.135}$$

one obtains

$$\begin{aligned} C^{jk}_{\mu\nu,J(\mu+\nu)} C^{jk}_{j(-k)J(j-k)} &= \frac{(2J+1)\left\{(2j)!(2k)!\right\}^{1/2}}{(J+j+k+1)!} \\ &\cdot \left\{ \frac{(J+j-k)!(J-j+k)!(J+\mu+\nu)!(J-\mu-\nu)!}{(j+\mu)!(j-\mu)!(k+\nu)!(k-\nu)!} \right\}^{1/2} \\ &\cdot \sum_m \frac{(-1)^{m+k+\nu}(J+k+\mu-m)!(m+j-\mu)!}{(J-j+k-m)!(J+\mu+\nu-m)!m!(m+j-k-\mu-\nu)!}. \end{aligned} \tag{4.136}$$

Since $C^{jk}_{j(-k)J(j-k)}$ is real and positive, it can be calculated from Eq. (4.136) with $\mu = j$ and $\nu = -k$. In the calculation Eq. (A1.3) with $u = J - j + k$, $v = J + j + k$, $r = J + j - k$, and $p = m$ is used for simplification

$$C^{jk}_{j(-k),J(j-k)} = \left\{ \frac{(2J+1)(2j)!(2k)!}{(J+j+k+1)!(j+k-J)!} \right\}^{1/2}. \tag{4.137}$$

Substituting Eq. (4.137) into Eq. (4.136) one obtains

$$C^{jk}_{\mu\nu,J(\mu+\nu)} = \left\{\frac{(j+k-J)!(J-j+k)!(J+j-k)!}{(J+j+k+1)!}\right\}^{1/2}$$
$$\cdot \left\{\frac{(2J+1)(J+\mu+\nu)!(J-\mu-\nu)!}{(j+\mu)!(j-\mu)!(k+\nu)!(k-\nu)!}\right\}^{1/2}$$
$$\cdot \sum_m \frac{(-1)^{m+k+\nu}(J+k+\mu-m)!(m+j-\mu)!}{(J-j+k-m)!(J+\mu+\nu-m)!m!(m+j-k-\mu-\nu)!}. \tag{4.138}$$

The Wigner form of the CG coefficients of SU(2) is obtained by replacing the summation index m with $n = m - k - \nu$:

$$C^{jk}_{\mu\nu,J(\mu+\nu)} = \Delta(j,k,J)\left\{\frac{(2J+1)(J+\mu+\nu)!(J-\mu-\nu)!}{(j+\mu)!(j-\mu)!(k+\nu)!(k-\nu)!}\right\}^{1/2}$$
$$\cdot \sum_n \frac{(-1)^n(J+\mu-\nu-n)!(n+j+k-\mu+\nu)!}{(J-j-\nu-n)!(J-k+\mu-n)!(n+k+\nu)!(n+j-\mu)!}, \tag{4.139}$$

where n runs from the maximum between $(-k-\nu)$ and $(-j+\mu)$ to the minimum between $(J-j-\nu)$ and $(J-k+\mu)$, and $\Delta(j,k,J)$ is given in Eq. (4.133). Three forms of the CG coefficients of SU(2) are equivalent. From them one obtains the following symmetry of the Clebsch–Gordan coefficients of SU(2), where $M = \mu + \nu$,

$$\begin{aligned}
C^{jk}_{\mu\nu,JM} &= C^{kj}_{(-\nu)(-\mu)J(-M)} = (-1)^{j+k-J}\ C^{kj}_{\nu\mu JM} \\
&= (-1)^{j+k-J}\ C^{jk}_{(-\mu)(-\nu)J(-M)} \\
&= (-1)^{k-J-\mu}\left(\frac{2J+1}{2k+1}\right)^{1/2} C^{Jj}_{(-M)\mu k(-\nu)} \\
&= (-1)^{j-J+\nu}\left(\frac{2J+1}{2j+1}\right)^{1/2} C^{kJ}_{\nu(-M)j(-\mu)}.
\end{aligned} \tag{4.140}$$

Wigner introduced the $3j$-symbols which are more symmetric

$$\begin{pmatrix} j & k & \ell \\ \mu & \nu & \rho \end{pmatrix} = (-1)^{j-k-\rho}(2\ell+1)^{-1/2} C^{jk}_{\mu\nu\ell(-\rho)}, \tag{4.141}$$

where $|j-k| \le \ell \le j+k$ and $\mu+\nu+\rho = 0$.

$$(-1)^{j+k+\ell}\begin{pmatrix} j & k & \ell \\ \mu & \nu & \rho \end{pmatrix} = \begin{pmatrix} k & j & \ell \\ \nu & \mu & \rho \end{pmatrix} = \begin{pmatrix} j & \ell & k \\ \mu & \rho & \nu \end{pmatrix} = \begin{pmatrix} j & k & \ell \\ -\mu & -\nu & -\rho \end{pmatrix}, \tag{4.142}$$

$$\sum_{\ell\rho} (2\ell+1) \begin{pmatrix} j & k & \ell \\ \mu & \nu & \rho \end{pmatrix} \begin{pmatrix} j & k & \ell \\ \mu' & \nu' & \rho \end{pmatrix} = \delta_{\mu\mu'}\, \delta_{\nu\nu'},$$
$$\sum_{\mu\nu} \begin{pmatrix} j & k & \ell \\ \mu & \nu & \rho \end{pmatrix} \begin{pmatrix} j & k & \ell' \\ \mu & \nu & \rho' \end{pmatrix} = (2\ell+1)^{-1}\, \delta_{\ell\ell'}\, \delta_{\rho\rho'}. \tag{4.143}$$

Some special CG coefficients are calculated in Probs. 16–18 of Chap. 4 of [Ma and Gu (2004)].

4.6.3 *Applications*

(a) *The permutation symmetry of wave functions in a two-body system*

The wave function of a two-body system with a given angular momentum is

$$\Psi^L_M(\boldsymbol{x}^{(1)}, \boldsymbol{x}^{(2)}) = \sum_{m_1 m_2} C^{\ell_1 \ell_2}_{m_1 m_2, LM} \psi^{\ell_1}_{m_1}(\boldsymbol{x}^{(1)}) \psi^{\ell_2}_{m_2}(\boldsymbol{x}^{(2)}).$$

In the permutation of two particles, one has

$$\begin{aligned} \Psi^L_M(\boldsymbol{x}^{(2)}, \boldsymbol{x}^{(1)}) &= \sum_{m_1 m_2} C^{\ell_1 \ell_2}_{m_1 m_2, LM} \psi^{\ell_1}_{m_1}(\boldsymbol{x}^{(2)}) \psi^{\ell_2}_{m_2}(\boldsymbol{x}^{(1)}) \\ &= \sum_{m_2 m_1} (-1)^{L-\ell_1-\ell_2} C^{\ell_2 \ell_1}_{m_2 m_1, LM} \psi^{\ell_2}_{m_2}(\boldsymbol{x}^{(1)}) \psi^{\ell_1}_{m_1}(\boldsymbol{x}^{(2)}). \end{aligned}$$

If $\ell_1 = \ell_2 = \ell$, then

$$\Psi^L_M(\boldsymbol{x}^{(2)}, \boldsymbol{x}^{(1)}) = (-1)^{L-2\ell}\, \Psi^L_M(\boldsymbol{x}^{(1)}, \boldsymbol{x}^{(2)}). \tag{4.144}$$

For example, in a system composed of two electrons with P wave ($\ell = 1$), the wave function with the total angular momentum $L = 2$ and 0 is symmetric in permutation, and that with $L = 1$ is antisymmetric.

(b) *The expansion of the Legendre function*

The product of two spherical harmonic functions of two subsystems with the same ℓ can be combined to be invariant in rotation ($L = 0$)

$$\begin{aligned} \Phi^0_0(\hat{\boldsymbol{n}}_1, \hat{\boldsymbol{n}}_2) &= \sum_m C^{\ell\ell}_{(-m)m,00} Y^\ell_{-m}(\hat{\boldsymbol{n}}_1) Y^\ell_m(\hat{\boldsymbol{n}}_2) \\ &= \sum_m \frac{(-1)^{\ell+m}}{(2\ell+1)^{1/2}} Y^\ell_{-m}(\hat{\boldsymbol{n}}_1) Y^\ell_m(\hat{\boldsymbol{n}}_2). \end{aligned}$$

Let R rotate $\hat{\boldsymbol{n}}_1$ to the z-axis and $\hat{\boldsymbol{n}}_2$ to the half $x\ z$ plane with positive x component. Denoting by θ the angle between $\hat{\boldsymbol{n}}_1$ and $\hat{\boldsymbol{n}}_2$, one has

$$Y^{\ell}_{-m}(\boldsymbol{e}_3) = Y^{\ell}_{-m}(0,0) = \delta_{m0}\left(\frac{2\ell+1}{4\pi}\right)^{1/2},$$
$$Y^{\ell}_{0}(\theta,0) = \left(\frac{2\ell+1}{4\pi}\right)^{1/2} P_\ell(\cos\theta),$$
$$Y^{\ell}_{m}(\hat{\boldsymbol{n}}_1)^* = (-1)^m Y^{\ell}_{-m}(\hat{\boldsymbol{n}}_1).$$

Hence,

$$\Phi^0_0(\hat{\boldsymbol{n}}_1,\hat{\boldsymbol{n}}_2) = \frac{(-1)^\ell(2\ell+1)^{1/2}}{4\pi}P_\ell(\cos\theta),$$

$$\begin{aligned} P_\ell(\hat{\boldsymbol{n}}_1\cdot\hat{\boldsymbol{n}}_2) = P_\ell(\cos\theta) &= \frac{4\pi}{2\ell+1}\sum_m (-1)^m Y^{\ell}_{-m}(\hat{\boldsymbol{n}}_1) Y^{\ell}_{m}(\hat{\boldsymbol{n}}_2) \\ &= \frac{4\pi}{2\ell+1}\sum_m Y^{\ell}_{m}(\hat{\boldsymbol{n}}_1)^* Y^{\ell}_{m}(\hat{\boldsymbol{n}}_2). \end{aligned} \tag{4.145}$$

In the method of partial waves in quantum mechanics, a plane wave $\exp(i\boldsymbol{k}\cdot\boldsymbol{r})$, which is a solution of the d'Alembert equation, can be expanded with respect to the Legendre function. Now, the expansion is written in the spherical harmonic functions

$$\begin{aligned} \exp(i\boldsymbol{k}\cdot\boldsymbol{r}) = \exp(ikr\cos\theta) &= \sum_{\ell=0}^{\infty} i^\ell(2\ell+1)j_\ell(kr)P_\ell(\cos\theta) \\ &= 4\pi\sum_{\ell=0}^{\infty} i^\ell j_\ell(kr)\sum_{m=-\ell}^{\ell} Y^{\ell}_{m}(\hat{\boldsymbol{k}})^* Y^{\ell}_{m}(\hat{\boldsymbol{r}}), \end{aligned} \tag{4.146}$$

where j_ℓ is the spherical Bessel function, and $\hat{\boldsymbol{k}}$ and $\hat{\boldsymbol{r}}$ are the unit vectors along the directions, respectively.

(c) *The Expansion of the product of two spherical harmonic functions*

Replacing $D^j(u)$ in Eq. (4.119) with the spherical harmonic function (see Eq. (4.93)), one has

$$\begin{aligned} Y^{\ell_1}_{m_1}(\hat{\boldsymbol{n}})Y^{\ell_2}_{m_2}(\hat{\boldsymbol{n}}) = \sum_L &\left\{\frac{(2\ell_1+1)(2\ell_2+1)}{4\pi(2L+1)}\right\}^{1/2} \\ &\cdot C^{\ell_1\ell_2}_{00,L0}C^{\ell_1\ell_2}_{m_1m_2,L(m_1+m_2)}Y^L_{m_1+m_2}(\hat{\boldsymbol{n}}). \end{aligned} \tag{4.147}$$

Since $C^{\ell_1\ell_2}_{00L0} = 0$ when $L-\ell_1-\ell_2$ is an odd integer, the parities of the wave functions on both sides of Eq. (4.147) are the same. In terms of the orthogonal relations (4.128) and (4.94), one obtains

$$C^{\ell_1\ell_2}_{00,L0}Y^L_M(\hat{\boldsymbol{n}}) = \left\{\frac{4\pi(2L+1)}{(2\ell_1+1)(2\ell_2+1)}\right\}^{1/2} \cdot \sum_m C^{\ell_1\ell_2}_{m(M-m),LM}Y^{\ell_1}_m(\hat{\boldsymbol{n}})Y^{\ell_2}_{M-m}(\hat{\boldsymbol{n}}), \tag{4.148}$$

$$\int Y^L_M(\theta,\varphi)^* Y^{\ell_1}_{m_1}(\theta,\varphi)Y^{\ell_2}_{m_2}(\theta,\varphi)\sin\theta d\theta d\varphi = \left\{\frac{(2\ell_1+1)(2\ell_2+1)}{4\pi(2L+1)}\right\}^{1/2} C^{\ell_1\ell_2}_{00,L0}C^{\ell_1\ell_2}_{m_1m_2,LM}. \tag{4.149}$$

4.6.4 *Sum of Three Angular Momentums*

If a compound system consists of three subsystems with the spherical symmetry, the wave function of the compound system with a given angular momentum can be obtained by combining first the wave functions of two subsystems, and then combining the result with the wave function of the third one. This is a reduction problem of the direct product of three irreducible representations of SU(2). In the reduction, the multiplicity a_J of an irreducible representation may be larger than one, and the result will depend upon the order of combination. For example,

$$\begin{aligned} D^1\times D^1\times D^1 &\simeq \left\{D^2\oplus D^1\oplus D^0\right\}\times D^1 \\ &\simeq \left\{D^3\oplus D^2\oplus D^1\right\}\oplus\left\{D^2\oplus D^1\oplus D^0\right\}\oplus D^1 \\ &\simeq D^3\oplus 2D^2\oplus 3D^1\oplus D^0, \end{aligned}$$

where $a_2 = 2$ and $a_1 = 3$. The functional spaces belonging to the same representation D^J can be combined arbitrarily, where the combination coefficients are independent of the row number M of the representation (the magnetic quantum number).

Two ways are commonly used in obtaining the wave function with the total angular momentum J. One is to combine the angular momentums of the first two subsystems into J_{12}, then to combine it with the angular momentum of the third one. The other is to combine the angular momentums of the last two subsystems into J_{23}, then to combine the angular momentum of the first one with J_{23}.

$$\begin{aligned} \Psi^J_M(J_{12}) &= \sum_\rho C^{J_{12}\ell}_{(M-\rho)\rho,JM}\left\{\sum_{\mu\nu} C^{jk}_{\mu\nu,J_{12}(M-\rho)}\psi^j_\mu(\boldsymbol{x}^{(1)})\psi^k_\nu(\boldsymbol{x}^{(2)})\right\}\psi^\ell_\rho(\boldsymbol{x}^{(3)}), \\ \Psi^J_M(J_{23}) &= \sum_\mu C^{jJ_{23}}_{\mu(M-\mu),JM}\psi^j_\mu(\boldsymbol{x}^{(1)})\left\{\sum_{\nu\rho} C^{k\ell}_{\nu\rho,J_{23}(M-\mu)}\psi^k_\nu(\boldsymbol{x}^{(2)})\psi^\ell_\rho(\boldsymbol{x}^{(3)})\right\}. \end{aligned} \tag{4.150}$$

Two sets of wave functions can be related by a unitary transformation X, which depends on J_{12} and J_{23}, in addition to J, j, k, and ℓ, but is independent of the magnetic quantum numbers,

$$\Psi_M^J(J_{23}) = \sum_{J_{12}} \Psi_M^J(J_{12}) X_{J_{12}J_{23}}. \tag{4.151}$$

Extracting a factor from $X_{J_{12}J_{23}}$, one uses the more symmetric coefficients, called the Racah coefficients $W[jkJ\ell; J_{12}J_{23}]$, or the $6j$-symbols,

$$\begin{aligned} X_{J_{12}J_{23}} &= \{(2J_{12}+1)(2J_{23}+1)\}^{1/2} W[jkJ\ell; J_{12}J_{23}] \\ &= (-1)^{j+k+\ell+J} \{(2J_{12}+1)(2J_{23}+1)\}^{1/2} \begin{Bmatrix} j & k & J_{12} \\ \ell & J & J_{23} \end{Bmatrix}. \end{aligned} \tag{4.152}$$

Substituting Eq. (4.150) into Eq. (4.151), one obtains

$$\begin{aligned} C^{jJ_{23}}_{\mu(M-\mu),JM}\ C^{k\ell}_{\nu\rho,J_{23}(M-\mu)} &= \sum_{J_{12}} C^{J_{12}\ell}_{(M-\rho)\rho,JM}\ C^{jk}_{\mu\nu,J_{12}(M-\rho)} \\ &\cdot\ \{(2J_{12}+1)(2J_{23}+1)\}^{1/2} W[jkJ\ell; J_{12}J_{23}]. \end{aligned} \tag{4.153}$$

In terms of the orthogonal relations (4.129), the Racah coefficients can be expressed as a product of four CG coefficients:

$$\begin{aligned} & C^{J_{12}\ell}_{(M-\rho)\rho,JM}\ \{(2J_{12}+1)(2J_{23}+1)\}^{1/2} W[jkJ\ell; J_{12}J_{23}] \\ &\quad = \sum_{\mu\nu} C^{jJ_{23}}_{\mu(M-\mu),JM}\ C^{k\ell}_{\nu\rho,J_{23}(M-\mu)} C^{jk}_{\mu\nu,J_{12}(M-\rho)}, \end{aligned} \tag{4.154}$$

$$\begin{aligned} & \{(2J_{12}+1)(2J_{23}+1)\}^{1/2} W[jkJ\ell; J_{12}J_{23}]\delta_{JJ'} \\ &\quad = \sum_{\mu\nu\rho} C^{jJ_{23}}_{\mu(M-\mu),JM}\ C^{k\ell}_{\nu\rho,J_{23}(M-\mu)} C^{J_{12}\ell}_{(M-\rho)\rho,J'M}\ C^{jk}_{\mu\nu,J_{12}(M-\rho)}. \end{aligned} \tag{4.155}$$

$$\begin{aligned} \begin{Bmatrix} j_1 & j_2 & j_3 \\ \ell_1 & \ell_2 & \ell_3 \end{Bmatrix} &= \sum_{\text{all } m \text{ and } \mu} (-1)^{\ell_1+\ell_2+\ell_3+m_1+m_2+m_3} \\ &\cdot \begin{pmatrix} j_1 & j_2 & j_3 \\ \mu_1 & \mu_2 & \mu_3 \end{pmatrix} \begin{pmatrix} j_1 & \ell_2 & \ell_3 \\ \mu_1 & m_2 & -m_3 \end{pmatrix} \begin{pmatrix} \ell_1 & j_2 & \ell_3 \\ -m_1 & \mu_2 & m_3 \end{pmatrix} \begin{pmatrix} \ell_1 & \ell_2 & j_3 \\ m_1 & -m_2 & \mu_3 \end{pmatrix}. \end{aligned} \tag{4.156}$$

Obviously, the Racah coefficients and the $6j$-symbols are all real, and X is a real orthogonal matrix. Hence, the Racah coefficients satisfy the orthogonal relations

$$\begin{aligned}
&\sum_{J_{12}} (2J_{12}+1)W[jkJ\ell; J_{12}J_{23}]W[jkJ\ell; J_{12}J'_{23}] = \frac{\delta_{J_{23}J'_{23}}}{2J_{23}+1}, \\
&\sum_{J_{23}} (2J_{23}+1)W[jkJ\ell; J_{12}J_{23}]W[jkJ\ell; J'_{12}J_{23}] = \frac{\delta_{J_{12}J'_{12}}}{2J_{12}+1}.
\end{aligned} \tag{4.157}$$

In terms of the identities on combinatorics given in Appendix A, one is able to calculate the analytic form of the Racah coefficients from Eq. (4.155):

$$\begin{aligned}
W[abcd; ef] &= (-1)^{a+b+c+d}\,\Delta(a,b,e)\Delta(d,e,c)\Delta(b,d,f)\Delta(a,f,c) \\
&\cdot \sum_z (-1)^z (z+1)!\{(z-a-b-e)!(z-c-d-e)! \\
&\cdot (z-b-d-f)!(z-a-c-f)!(a+b+c+d-z)! \\
&\cdot (a+d+e+f-z)!(b+c+e+f-z)!\}^{-1},
\end{aligned} \tag{4.158}$$

$$z = \max\begin{Bmatrix} a+b+e \\ c+d+e \\ a+c+f \\ b+d+f \end{Bmatrix}, \quad \ldots, \quad \min\begin{Bmatrix} a+b+c+d \\ a+d+e+f \\ b+c+e+f \end{Bmatrix},$$

where a, b, c, d, e, and f have to satisfy four conditions of the triangle rules given in Fig. 4.3. The detailed calculation is given in §5.4.2 of [Ma (1993)], where the quantum parameter q is taken to be 1.

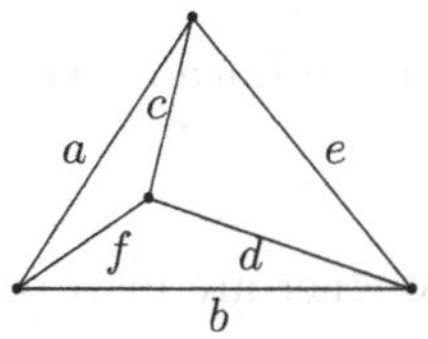

Fig. 4.3 Four triangle rules of the Racah coefficients

The formulas on the Racah coefficients are quite complicated. A graphic method may be helpful to understand them. Denote a CG coefficient by three oriented lines intersected at one point as shown in Fig. 4.4 (a). Remind that $\left(C^{jk}\right)^{-1}$ is the transpose of C^{jk}. When $J=0$ (or $j=0$, $k=0$) the corresponding oriented line can be omitted. The solid line denotes the magnetic quantum number to be summed, and the dotted line denotes the magnetic quantum number to be equal to each other, but not summed.

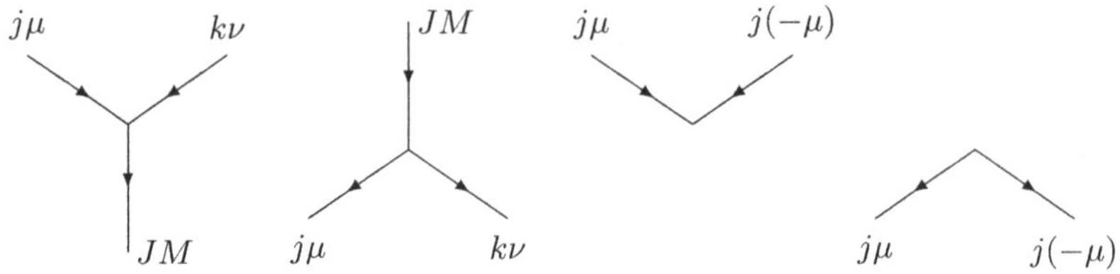

(a) $C^{jk}_{\mu\nu,JM}$ (b) $C^{jj}_{\mu(-\mu),00}$

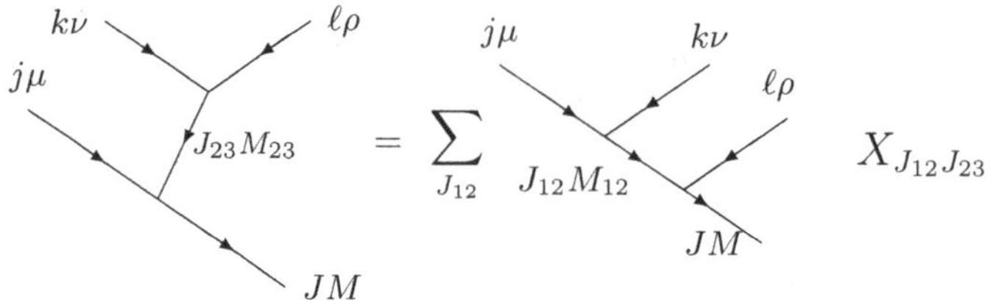

(c) The definition (4.153) for the Racah coefficients

$$\{(2J_{12}+1)(2J_{23}+1)\}^{1/2}\, W[jkJ\ell; J_{12}J_{23}] \;=$$

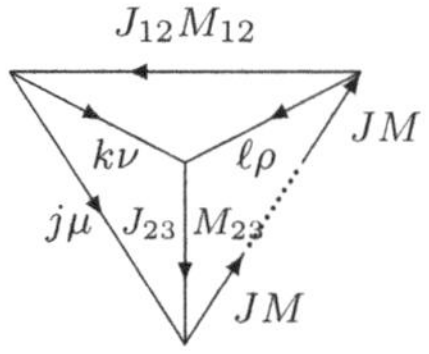

(d) The expansion (4.155) of the Racah coefficients.

Fig. 4.4 Diagram for the Racah coefficients.

The $6j$-symbols introduced by Wigner are more symmetric. There are 144 $6j$-symbols to be equal to each other:

$$\begin{aligned}
&\begin{Bmatrix} a & b & e \\ d & c & f \end{Bmatrix} = \begin{Bmatrix} b & a & e \\ c & d & f \end{Bmatrix} = \begin{Bmatrix} a & e & b \\ d & f & c \end{Bmatrix} = \begin{Bmatrix} d & c & e \\ a & b & f \end{Bmatrix} \\
&= \begin{Bmatrix} (a+b-c+d)/2 & (a+b+c-d)/2 & e \\ (a-b+c+d)/2 & (-a+b+c+d)/2 & f \end{Bmatrix} .
\end{aligned} \tag{4.159}$$

4.7 Tensors and Spinors

We have discussed the concepts of a scalar and a vector. Now, we are going to introduce the concepts of a tensor and a spinor. The definitions for a tensor and a spinor, as well as those for a scalar and a vector, are based on their transformation properties in a matrix group. In physics, if without special notification, the matrix group is the group SO(3), or sometimes the Lorentz group.

4.7.1 *Vector Fields*

Denote the position vector of an arbitrary point P in a real three-dimensional space by $\boldsymbol{r} = \sum_a \boldsymbol{e}_a x_a$, where $\boldsymbol{e}_a$ is the basis vector in the laboratory frame K and x_a is the coordinate in K. Under a rotation R in the three-dimensional space, the point P rotates to the point P', and the position vector $\boldsymbol{r}$ becomes $\boldsymbol{r}'$ with the component x'_a in K,

$$\begin{aligned} \boldsymbol{r} = \sum_a \boldsymbol{e}_a x_a &\xrightarrow{R} \boldsymbol{r}' = \sum_a \boldsymbol{e}_a x'_a, \\ x_a &\xrightarrow{R} x'_a = \sum_{b=1}^{3} R_{ab} x_b. \end{aligned} \tag{4.160}$$

A quantity ψ is called a scalar if it reserves invariant in the rotation R. The distribution of a scalar is called a scalar field. A scalar field is described by a scalar function $\psi(x)$. In a rotation R, the value $\psi'(Rx)$ of a scalar field at the point P' after the rotation is equal to the value $\psi(x)$ at the point P before the rotation. The transformation of a scalar function in a rotation R is characterized by the operator P_R,

$$\psi \xrightarrow{R} \psi' = \psi, \qquad \psi'(x) \equiv P_R \psi(x) = \psi(R^{-1}x). \tag{4.161}$$

A quantity $\boldsymbol{V}$ is called a vector if its components $\boldsymbol{V}_a$ transform in a rotation R like x_a,

$$\begin{aligned} \boldsymbol{V} = \sum_a \boldsymbol{e}_a \boldsymbol{V}_a &\xrightarrow{R} \boldsymbol{V}' = \sum_a \boldsymbol{e}_a \boldsymbol{V}'_a, \\ \boldsymbol{V}_a &\xrightarrow{R} \boldsymbol{V}'_a = \sum_{b=1}^{3} R_{ab} \boldsymbol{V}_b. \end{aligned} \tag{4.162}$$

A vector has three components, which as a whole characterize the state of the system. Denote by Q_R the transformation operator for the components,

$$\boldsymbol{V}_a' \equiv (Q_R \boldsymbol{V})_a = \sum_{b=1}^{3} R_{ab} \boldsymbol{V}_b. \tag{4.163}$$

In fact, Q_R is the R matrix.

The distribution of a vector is called a vector field. A vector field is described by three functions $\boldsymbol{V}(x)_a$ for its components. In the rotation R, the vector $\boldsymbol{V}'(Rx)_a$ at the point P' after the rotation is rotated from the vector $\boldsymbol{V}(x)_a$ at the point P before the rotation according to Eq. (4.162), $\boldsymbol{V}'(Rx)_a = \sum_b R_{ab} \boldsymbol{V}(x)_b$. Introduce the transformation operator O_R, for the rotation R,

$$\begin{aligned} &\boldsymbol{V}'(x)_a \equiv [O_R \boldsymbol{V}(x)]_a = \sum_{b=1}^{3} R_{ab} \boldsymbol{V}(R^{-1}x)_b, \\ &O_R = P_R Q_R = Q_R P_R, \\ &[Q_R \boldsymbol{V}(x)]_a = \sum_{b=1}^{3} R_{ab} \boldsymbol{V}(x)_b, \qquad [P_R \boldsymbol{V}(x)]_a = \boldsymbol{V}(R^{-1}x)_a. \end{aligned} \tag{4.164}$$

Rewrite the formulas in the vector equalities,

$$\begin{aligned} &O_R \boldsymbol{V}(x) = R\boldsymbol{V}(R^{-1}x), \\ &Q_R \boldsymbol{V}(x) = R\boldsymbol{V}(x), \qquad P_R \boldsymbol{V}(x) = \boldsymbol{V}(R^{-1}x). \end{aligned} \tag{4.165}$$

The body-fixed frame K' and its basis vectors $\boldsymbol{e}_a'$ are fixed with the system and rotate together, so that the components x_a of the position vector $\boldsymbol{r}'$ with respect to the basis vectors $\boldsymbol{e}_a'$ reserve invariant,

$$\boldsymbol{r}' = \sum_a \boldsymbol{e}_a x_a' = \sum_d \boldsymbol{e}_d' x_d, \qquad \boldsymbol{e}_d' = Q_R \boldsymbol{e}_d = \sum_{b=1}^{3} \boldsymbol{e}_b R_{bd}. \tag{4.166}$$

For the general vector field $\boldsymbol{V}(x)$, one has

$$\begin{aligned} \boldsymbol{V}(x) \xrightarrow{R} O_R \boldsymbol{V}(x) &= \sum_{a=1}^{3} \boldsymbol{e}_a [O_R \boldsymbol{V}(x)]_a = \sum_{a=1}^{3} \boldsymbol{e}_a \sum_{d=1}^{3} R_{ad} \boldsymbol{V}(R^{-1}x)_d \\ &= \sum_{d=1}^{3} \boldsymbol{e}_d' V(R^{-1}x)_d = \sum_{d=1}^{3} [Q_R \boldsymbol{e}_d] P_R V(x)_d. \end{aligned} \tag{4.167}$$

Before the rotation R, two frames K and K' coincide with each other, $\boldsymbol{e}_a = \boldsymbol{e}_a'$, and $\boldsymbol{V}(x)_a = V(x)_a$,

$$\boldsymbol{V}(x) = \sum_{a=1}^{3} \boldsymbol{e}_a \boldsymbol{V}(x)_a = \sum_{d=1}^{3} \boldsymbol{e}_d V(x)_d.$$

In the formula $\boldsymbol{V}(x) = \sum_a \boldsymbol{e}_a \boldsymbol{V}(x)_a$, $\boldsymbol{e}_a$ is fixed in the rotation and the component $\boldsymbol{V}(x)_a$, denoted by the symbols in the bold form and called the vector field, transforms according to Eq. (4.164). In the formula $\boldsymbol{V}(x) = \sum_d \boldsymbol{e}_d V(x)_d$, $\boldsymbol{e}_d$ is a basis vector in K' which coincides with the basis vector $\boldsymbol{e}_a$ in K before rotation and transforms according to Eq. (4.166), and $V(x)_d$ is only a coefficient and transforms like a scalar. In summary, under a rotation R, a vector field $\boldsymbol{V}(x)$ changes as follows

$$\begin{aligned}
&\sum_{a=1}^{3} \boldsymbol{e}_a O_R \boldsymbol{V}(x)_a = \sum_{ad} \boldsymbol{e}_a R_{ad} \boldsymbol{V}(R^{-1}x)_d = \sum_{d=1}^{3} \boldsymbol{e}'_d P_R V(x)_d, \\
&O_R \boldsymbol{V}(x)_a = \sum_{d=1}^{3} R_{ad} \boldsymbol{V}(R^{-1}x)_d, \\
&\boldsymbol{e}'_d = O_R \boldsymbol{e}_d = Q_R \boldsymbol{e}_d = \sum_{b=1}^{3} \boldsymbol{e}_b R_{bd}, \\
&O_R V(x)_d = P_R V(x)_d = V(R^{-1}x)_d.
\end{aligned} \tag{4.168}$$

The position vector field $\boldsymbol{r}(x)_a$ is a special vector field which reserves invariant in a rotation. In fact, $\boldsymbol{r}(x)_a = x_a$, and $\boldsymbol{r}(R^{-1}x)_a = (R^{-1}x)_a$,

$$\left[O_R \boldsymbol{r}(x)\right]_a = \sum_b R_{ab} \boldsymbol{r}(R^{-1}x)_b = x_a = \boldsymbol{r}(x)_a.$$

Namely,

$$\begin{aligned}
&O_R \boldsymbol{r}(x) = \boldsymbol{r}(x), \qquad \left[Q_R \boldsymbol{r}(x)\right]_a = \sum_b R_{ab} x_b, \\
&\left[P_R \boldsymbol{r}(x)\right]_a = \boldsymbol{r}(R^{-1}x)_a = \sum_b (R^{-1})_{ab} x_b.
\end{aligned} \tag{4.169}$$

The operators O_R and Q_R, as well as the operator P_R, are all the linear unitary operators. The operator $L(x)$, describing a mechanics quantity, transforms in a rotation R as follows:

$$L(x) \xrightarrow{R} O_R L(x) O_R^{-1}. \tag{4.170}$$

4.7.2 *Tensor Fields*

A tensor of order n contains n subscripts and 3^n components, which as a whole characterize the state of the system. In the rotation, each subscript transforms like a vector subscript,

$$\boldsymbol{T}_{a_1a_2\ldots a_n} \xrightarrow{R} (O_R\boldsymbol{T})_{a_1a_2\ldots a_n} = \sum_{b_1b_2\ldots b_n} R_{a_1b_1}\ldots R_{a_nb_n}\boldsymbol{T}_{b_1b_2\ldots b_n}. \tag{4.171}$$

A distribution of a tensor is called a tensor field. A tensor field of order n is described by 3^n functions for its components transforming in a rotation R as

$$\begin{aligned}
&[O_R\boldsymbol{T}(x)]_{a_1a_2\ldots a_n} = \sum_{b_1b_2\ldots b_n} R_{a_1b_1}R_{a_2b_2}\ldots R_{a_nb_n}\boldsymbol{T}(R^{-1}x)_{b_1b_2\ldots b_n},\\
&O_R = Q_RP_R, \qquad Q_R = R\times R\times\cdots\times R,\\
&[P_R\boldsymbol{T}(x)]_{a_1a_2\ldots a_n} = \boldsymbol{T}(R^{-1}x)_{a_1a_2\ldots a_n},\\
&[Q_R\boldsymbol{T}(x)]_{a_1a_2\ldots a_n} = \sum_{b_1b_2\ldots b_n} R_{a_1b_1}R_{a_2b_2}\ldots R_{a_nb_n}\boldsymbol{T}(x)_{b_1b_2\ldots b_n}.
\end{aligned} \tag{4.172}$$

The basis tensor $\boldsymbol{\theta}_{d_1d_2\ldots d_n}$ is a tensor containing only one nonvanishing component which is equal to 1

$$(\boldsymbol{\theta}_{d_1d_2\ldots d_n})_{a_1\ldots a_n} = \delta_{d_1a_1}\delta_{d_2a_2}\ \cdots\ \delta_{d_na_n}. \tag{4.173}$$

A tensor field can be expanded with respect to the basis tensor,

$$\begin{aligned}
&\boldsymbol{T}(x)_{a_1\ldots a_n} = \sum_{d_1\ldots d_n} (\boldsymbol{\theta}_{d_1\ldots d_n})_{a_1\ldots a_n}\, T(x)_{d_1\ldots d_n} = T(x)_{a_1\ldots a_n},\\
&[O_R\boldsymbol{T}(x)]_{a_1a_2\ldots a_n} = \sum_{d_1\ldots d_n} (Q_R\boldsymbol{\theta}_{d_1\ldots d_n})_{a_1\ldots a_n}\, P_RT(x)_{d_1\ldots d_n},\\
&(Q_R\boldsymbol{\theta}_{d_1d_2\ldots d_n})_{a_1a_2\ldots a_n} = \sum_{b_1\ldots b_n} R_{a_1b_1}R_{a_2b_2}\ldots R_{a_nb_n}\,(\boldsymbol{\theta}_{d_1d_2\ldots d_n})_{b_1b_2\ldots b_n}\\
&\quad = R_{a_1d_1}R_{a_2d_2}\ldots R_{a_nd_n} = \sum_{b_1\ldots b_n} (\boldsymbol{\theta}_{b_1\ldots b_n})_{a_1\ldots a_n}\, R_{b_1d_1}R_{b_2d_2}\ldots R_{b_nd_n},\\
&P_RT(x)_{d_1\ldots d_n} = T(R^{-1}x)_{d_1\ldots d_n}.
\end{aligned} \tag{4.174}$$

The scalar field is a tensor field of order 0, and a vector field is a tensor of order 1.

4.7.3 *Spinor Fields*

A spinor of rank s contains $(2s+1)$ components which as a whole characterize the state of the system. In a rotation, the spinor transforms as

$$\Psi_\sigma^{(s)} \xrightarrow{R} \left(O_R\Psi^{(s)}\right)_\sigma = \sum_\lambda D^s_{\sigma\lambda}(R)\Psi_\lambda^{(s)}. \tag{4.175}$$

A distribution of a spinor is called a spinor field. A spinor field of rank s is described by $(2s+1)$ functions for its components transforming in a rotation as

$$\begin{aligned}
&\left[O_R \Psi^{(s)}(x)\right]_\sigma = \sum_\lambda D^s_{\sigma\lambda}(R)\, \Psi^{(s)}(R^{-1}x)_\lambda, \\
&O_R = Q_R P_R, \qquad Q_R = D^s(R), \\
&\left[P_R \Psi^{(s)}(x)\right]_\sigma = \Psi^{(s)}(R^{-1}x)_\sigma, \\
&\left[Q_R \Psi^{(s)}(x)\right]_\sigma = \sum_\lambda D^s_{\sigma\lambda}(R)\, \Psi^{(s)}(x)_\lambda.
\end{aligned} \tag{4.176}$$

Usually, a spinor is denoted by a $(2s+1)\times 1$ column matrix. The spinor of rank $1/2$ is called a fundamental spinor, or briefly a spinor, where the superscript $1/2$ is often neglected. A fundamental spinor for the group SO(3) has two components and is denoted by a 2×1 column matrix.

The basis spinor $\boldsymbol{e}^{(s)}(\rho)$ is a spinor containing only one nonvanishing component which is equal to 1,

$$\boldsymbol{e}^{(s)}(\rho)_\sigma = \delta_{\rho\sigma}, \tag{4.177}$$

where the ordinal index ρ is indicated inside the round brackets. In a rotation R,

$$\begin{aligned}
\left[O_R \boldsymbol{e}^{(s)}(\rho)\right]_\sigma &= \left[Q_R \boldsymbol{e}^{(s)}(\rho)\right]_\sigma = \sum_{\lambda=-s}^{s} D^s_{\sigma\lambda}(R) \boldsymbol{e}^{(s)}(\rho)_\lambda \\
&= D^s_{\sigma\rho}(R) = \sum_{\lambda=-s}^{s} \boldsymbol{e}^{(s)}(\lambda)_\sigma D^s_{\lambda\rho}(R).
\end{aligned} \tag{4.178}$$

A spinor field can be expanded with respect to the basis spinor,

$$\begin{aligned}
&\Psi^{(s)}(x) = \sum_{\rho=-s}^{s} \boldsymbol{e}^{(s)}(\rho)\psi^{(s)}(x)_\rho, \\
&O_R \Psi^{(s)}(x) = \sum_{\rho=-s}^{s} \left\{Q_R \boldsymbol{e}^{(s)}(\rho)\right\} \left\{P_R \psi^{(s)}(x)\right\}_\rho, \\
&Q_R \boldsymbol{e}^{(s)}(\rho) = \sum_{\lambda=-s}^{s} \boldsymbol{e}^{(s)}(\lambda) D^s_{\lambda\rho}(R), \qquad P_R \psi^{(s)}(x) = \psi^{(s)}(R^{-1}x).
\end{aligned} \tag{4.179}$$

A scalar is a spinor of rank 0. A vector is a spinor of rank 1 because the self-representation of SO(3) is equivalent to D^1 (see Eq. (4.81)). The basis spinor $\boldsymbol{e}^{(1)}(\rho)$ of rank one is also called the spherical harmonic basis vector,

$$\begin{aligned}
&\boldsymbol{V}(x) = \sum_{a=1}^{3} \boldsymbol{e}_a V(x)_a = \sum_{\rho=-1}^{1} \boldsymbol{e}^{(1)}(\rho)\psi^{(1)}(x)_\rho, \\
&\boldsymbol{e}^{(1)}(\rho) = \sum_{a=1}^{3} \boldsymbol{e}_a M_{a\rho}, \qquad \psi^{(1)}(x)_\rho = \sum_{a=1}^{3} \left(M^{-1}\right)_{\rho a} V(x)_a, \\
&\begin{cases} \boldsymbol{e}^{(1)}(1) = -\left(\boldsymbol{e}_1 + i\boldsymbol{e}_2\right)/\sqrt{2}, \\ \boldsymbol{e}^{(1)}(0) = \boldsymbol{e}_3, \\ \boldsymbol{e}^{(1)}(-1) = \left(\boldsymbol{e}_1 - i\boldsymbol{e}_2\right)/\sqrt{2}, \end{cases} \\
&\begin{cases} \psi^{(1)}(x)_1 = -\left[V(x)_1 - iV(x)_2\right]/\sqrt{2}, \\ \psi^{(1)}(x)_0 = V(x)_3, \\ \psi^{(1)}(x)_{-1} = \left[V(x)_1 + iV(x)_2\right]/\sqrt{2}. \end{cases}
\end{aligned} \tag{4.180}$$

4.7.4 *Total Angular Momentum Operator*

Discuss a system characterized by a spinor field. The Hamiltonian of the system is isotropic so that the group SO(3) is the symmetric group of the system,

$$O_R H(x) = H(x) O_R, \tag{4.181}$$

where the transformation operator O_R for a spinor field is divided into two operators, $O_R = P_R Q_R$. For the infinitesimal elements,

$$\begin{aligned}
P_A &= 1 - i\sum_{a=1}^{3} \alpha_a L_a, \\
Q_A &= \mathbf{1} - i\sum_{a=1}^{3} \alpha_a S_a, \\
O_A &= \mathbf{1} - i\sum_{a=1}^{3} \alpha_a \left(L_a + S_a\right) = \mathbf{1} - i\sum_{a=1}^{3} \alpha_a J_a,
\end{aligned} \tag{4.182}$$

where S_a is the generator of D^s, and L_a is the differential operator of P_R, called the orbital angular momentum operator in physics. Their sum is denoted by J_a

$$J_a = L_a + S_a. \tag{4.183}$$

J_a, S_a, and L_a all satisfy the typical commutative relations of angular momentums.

The static wave functions with energy E construct an invariant functional space. Its basis function $\Psi_\rho(x)$ is a spinor field, transforming in a rotation R according to Eq. (4.176). But, after the transformation, it has to be a combination of the basis functions,

$$O_R \Psi_\rho(x) = D^s(R)\Psi_\rho(R^{-1}x) = \sum_\lambda \Psi_\lambda(x) D_{\lambda\rho}(R). \tag{4.184}$$

The set of the combinative coefficients $D_{\lambda\rho}(R)$ forms a representation of SO(3). Reducing the representation $D(R)$ by the method of group theory to be the direct sum of the irreducible representations of SO(3), the static wave function is combined to be $\Psi^j_\mu(x)$ belonging to the μ row of the irreducible representation D^j,

$$O_R \Psi^j_\mu(x) = D^s(R)\Psi^j_\mu(R^{-1}x) = \sum_\nu \Psi^j_\nu(x) D^j_{\nu\mu}(R). \tag{4.185}$$

Namely, $\Psi^j_\mu(x)$ is the common eigenfunction of the generators J^2 and J_3,

$$\begin{aligned}
&J^2 \Psi^j_\mu(x) = j(j+1)\Psi^j_\mu(x), \\
&J_3 \Psi^j_\mu(x) = \mu \Psi^j_\mu(x), \\
&J_\pm \Psi^j_\mu(x) = \Gamma^j_{\mp\mu} \Psi^j_{\mu\pm 1}(x), \\
&J^2 = J_1^2 + J_2^2 + J_3^2, \qquad J_\pm = J_1 \pm iJ_2.
\end{aligned} \tag{4.186}$$

Now, the system is characterized by a spinor field, and its conserved angular momentum is not the orbital angular momentum, but other mechanical quantities J^2 and J_3. J_a is the sum of the orbital angular momentum L_a and another quantity S_a related to the spinor. Both J_a and S_a satisfy the typical commutative relations of angular momentum. S_a should be a mathematical description of the spinor angular momentum, discovered and measured in experiments. Therefore, S_a is called the operator of the spinor angular momentum and J_a the operator of the total angular momentum. The total angular momentum is conserved in a spherically symmetric system characterized by a spinor field. From Eq. (4.178) the basis spinor $\boldsymbol{e}^{(s)}(\rho)$ is the common eigenfunction of the operators of the total angular momentum and the spinor angular momentum,

$$\begin{aligned}
&J_3 \boldsymbol{e}^{(s)}(\rho) = S_3 \boldsymbol{e}^{(s)}(\rho) = \rho \boldsymbol{e}^{(s)}(\rho), \\
&J_\pm \boldsymbol{e}^{(s)}(\rho) = S_\pm \boldsymbol{e}^{(s)}(\rho) = \Gamma^s_{\mp\rho} \boldsymbol{e}^{(s)}(\rho \pm 1), \\
&J^2 \boldsymbol{e}^{(s)}(\rho) = S^2 \boldsymbol{e}^{(s)}(\rho) = s(s+1)\boldsymbol{e}^{(s)}(\rho), \\
&S^2 = S_1^2 + S_2^2 + S_3^2, \qquad S_\pm = S_1 \pm iS_2.
\end{aligned} \tag{4.187}$$

There are three sets of the mutual commutable angular momentum operators, one is L^2, L_3, S^2, and S_3, the other is J^2, J_3, L^2, and S^2, and the third set is J^2, J_3, S^2, and $\boldsymbol{L}\cdot\boldsymbol{S}=\sum_a L_aS_a$. For the fundamental spinor, $s=1/2$, the common eigenfunctions of the first set are the product of the spherical harmonic functions $Y^\ell_m(\hat{\boldsymbol{n}})$ and the basis spinor $\boldsymbol{e}^{(s)}(\rho)$. Combining them by CG coefficients, one obtains the spherical spinor function, which is the common eigenfunction of the second set with the eigenvalues $j(j+1)$, μ, $\ell(\ell+1)$, and $s(s+1)$,

$$Y^{j\ell s}_\mu(\hat{\boldsymbol{n}})=\sum_\rho C^{s\ell}_{\rho(\mu-\rho)j\mu}Y^\ell_{\mu-\rho}(\hat{\boldsymbol{n}})\boldsymbol{e}^{(s)}(\rho). \tag{4.188}$$

When $s=1/2$ and $\ell=j\mp 1/2$, one has

$$\begin{aligned} Y^{j(j-1/2)(1/2)}_\mu(\hat{\boldsymbol{n}})&=\begin{pmatrix}\left(\dfrac{j+\mu}{2j}\right)^{1/2}Y^{j-1/2}_{\mu-1/2}(\hat{\boldsymbol{n}})\\ \left(\dfrac{j-\mu}{2j}\right)^{1/2}Y^{j-1/2}_{\mu+1/2}(\hat{\boldsymbol{n}})\end{pmatrix},\\ Y^{j(j+1/2)(1/2)}_\mu(\hat{\boldsymbol{n}})&=\begin{pmatrix}\left(\dfrac{j-\mu+1}{2j+2}\right)^{1/2}Y^{j+1/2}_{\mu-1/2}(\hat{\boldsymbol{n}})\\ -\left(\dfrac{j+\mu+1}{2j+2}\right)^{1/2}Y^{j+1/2}_{\mu+1/2}(\hat{\boldsymbol{n}})\end{pmatrix}. \end{aligned} \tag{4.189}$$

Since $\boldsymbol{\sigma}\cdot\hat{\boldsymbol{x}}$ is self-inverse and commutable with J_3 and J^2,

$$(\boldsymbol{\sigma}\cdot\hat{\boldsymbol{x}})\,Y^{j(j-1/2)(1/2)}_\mu(\hat{\boldsymbol{n}})=C_1Y^{j(j-1/2)(1/2)}_\mu(\hat{\boldsymbol{n}})+C_2Y^{j(j+1/2)(1/2)}_\mu(\hat{\boldsymbol{n}}),$$

where C_1 and C_2 are coefficients independent of μ. Letting $\mu=j$, one obtains $C_1=0$ and $C_2=1$. As calculated in Prob. 22 of Chap. 4 of [Ma and Gu (2004)], the common eigenfunction of the third set of the angular momentum operators with the eigenvalues $j(j+1)$, μ, $s(s+1)$, and ν is

$$\sum_{\rho=-s}^{s}\boldsymbol{e}^{(s)}(\rho)e^{i(\mu-\rho)\varphi}d^s_{\rho\nu}(\theta)d^j_{\mu\nu}(\theta). \tag{4.190}$$

4.8 Irreducible Tensor Operators and Their Application

4.8.1 *Irreducible Tensor Operators*

In quantum mechanics a physical quantity is characterized by a linear operator $L(x)$ which consists of the coordinate operators x_a, the differential operators $\partial/\partial x_a$, and the matrix operators σ_a. In a rotation R, $L(x)$ transforms as

$$L(x) \xrightarrow{R} O_R L(x) O_R^{-1}.$$

A set of $(2k+1)$ operators $L_\rho^k(x)$, $-k \leq \rho \leq k$, is called the irreducible tensor operators of rank k if those operators transform in the rotation R of SO(3) as

$$O_R L_\rho^k(x) O_R^{-1} = \sum_{\lambda=-k}^{k} L_\lambda^k(x) D_{\lambda\rho}^k(R). \tag{4.191}$$

Each operator in the set of the irreducible tensor operators characterizes a physical quantity independently, but in a rotation it relates to the other operators in the set as given in Eq. (4.191). Rewriting Eq. (4.191) in the form of generators, one has

$$\begin{aligned} &\left[J_3,\ L_\rho^k(x)\right] = \rho L_\rho^k(x), \\ &\left[J_\pm,\ L_\rho^k(x)\right] = \{(k \mp \rho)(k \pm \rho + 1)\}^{1/2} L_{\rho\pm 1}^k(x), \\ &\sum_{a=1}^{3} \left[J_a,\ \left[J_a,\ L_\rho^k(x)\right]\right] = k(k+1) L_\rho^k(x). \end{aligned} \tag{4.192}$$

This is an equivalent definition for the irreducible tensor operators. The irreducible tensor operator of rank 0 is called a symmetric operator or a scalar operator which reserves invariant in any rotation R. The irreducible tensor operators of rank 1 are called the vector operators. $L_\rho^k(x)$ are called the irreducible tensor operators of rank k with respect to the orbital space or the spinor space if replacing O_R with P_R or Q_R, respectively.

A typical example for the irreducible tensor operators is the electric multipole operators which are proportional to $Y_\rho^k(\hat{\boldsymbol{n}})$,

$$O_R Y_\rho^k(\hat{\boldsymbol{n}}) O_R^{-1} = P_R Y_\rho^k(\hat{\boldsymbol{n}}) P_R^{-1} = \sum_\lambda Y_\lambda^k(\hat{\boldsymbol{n}}) D_{\lambda\rho}^k(R). \tag{4.193}$$

They are the irreducible tensor operators of rank k with respect to the whole space and to the orbital space, but the scalar operators with respect to the spinor space. The electric dipole operators $Y_\mu^1(\hat{\boldsymbol{n}})$ are the vector operators with respect to the whole space and to the orbital space, and the scalar operators with respect to the spinor space. In fact, they are the combinations of the coordinate operators,

$$\begin{aligned}
&\left(\frac{4\pi}{3}\right)^{1/2} rY_1^1(\hat{\boldsymbol{n}}) = -\frac{1}{\sqrt{2}}(x_1 + ix_2), \\
&\left(\frac{4\pi}{3}\right)^{1/2} rY_0^1(\hat{\boldsymbol{n}}) = x_3, \\
&\left(\frac{4\pi}{3}\right)^{1/2} rY_{-1}^1(\hat{\boldsymbol{n}}) = \frac{1}{\sqrt{2}}(x_1 - ix_2), \\
&O_R Y_\rho^1(\hat{\boldsymbol{n}}) O_R^{-1} = P_R Y_\rho^1(\hat{\boldsymbol{n}}) P_R^{-1} = \sum_\lambda Y_\lambda^1(\hat{\boldsymbol{n}}) D_{\lambda\rho}^1(R), \\
&O_R x_a O_R^{-1} = P_R x_a P_R^{-1} = \sum_b x_b R_{ba}, \qquad Q_R x_a Q_R^{-1} = x_a.
\end{aligned} \tag{4.194}$$

Rewriting Eq. (4.194) in the form of generators, one has

$$[J_a, x_b] = [L_a, x_b] = i\sum_{d=1}^{3} \epsilon_{abd} x_d, \qquad [S_a, x_b] = 0. \tag{4.195}$$

In quantum mechanics there are some familiar operators having similar commutative relations,

$$\begin{aligned}
&[J_a, p_b] = [L_a, p_b] = i\sum_{d=1}^{3} \epsilon_{abd} p_d, \qquad [S_a, p_b] = 0, \\
&[J_a, L_b] = [L_a, L_b] = i\sum_{d=1}^{3} \epsilon_{abd} L_d, \qquad [S_a, L_b] = 0, \\
&[J_a, J_b] = i\sum_{d=1}^{3} \epsilon_{abd} J_d, \\
&[J_a, S_b] = [S_a, S_b] = i\sum_{d=1}^{3} \epsilon_{abd} S_d, \qquad [L_a, S_b] = 0.
\end{aligned} \tag{4.196}$$

$$\begin{aligned}
&O_R p_a O_R^{-1} = P_R p_a P_R^{-1} = \sum_b p_b R_{ba}, \qquad Q_R p_a Q_R^{-1} = p_a, \\
&O_R L_a O_R^{-1} = P_R L_a P_R^{-1} = \sum_b L_b R_{ba}, \qquad Q_R L_a Q_R^{-1} = L_a, \\
&O_R J_a O_R^{-1} = \sum_b J_b R_{ba}, \\
&O_R S_a O_R^{-1} = Q_R S_a Q_R^{-1} = \sum_b S_b R_{ba}, \qquad P_R S_a P_R^{-1} = S_a.
\end{aligned} \tag{4.197}$$

They all are the vector operators with respect to the whole space. For the vector operators, their components along a given direction $\hat{\boldsymbol{n}}$ are

$$\hat{\boldsymbol{n}} \cdot \boldsymbol{r} = O_R x_3 O_R^{-1} = P_R x_3 P_R^{-1}, \qquad \hat{\boldsymbol{n}} \cdot \boldsymbol{p} = O_R p_3 O_R^{-1} = P_R p_3 P_R^{-1},$$

$$\begin{aligned} \hat{\boldsymbol{n}} \cdot \boldsymbol{L} &= O_R L_3 O_R^{-1} = P_R L_3 P_R^{-1}, & \hat{\boldsymbol{n}} \cdot \boldsymbol{J} &= O_R J_3 O_R^{-1}, \\ \hat{\boldsymbol{n}} \cdot \boldsymbol{S} &= O_R S_3 O_R^{-1} = Q_R S_3 Q_R^{-1}, & R &= R(\varphi, \theta, 0). \end{aligned} \tag{4.198}$$

4.8.2 *Wigner–Eckart Theorem*

The static wave functions of a spherically symmetric system belong to an irreducible representation of SO(3),

$$\begin{aligned} O_R \Psi_\mu^j(x) &= \sum_\nu \Psi_\nu^j(x) D_{\nu\mu}^j(R), \\ O_R \Phi_{\mu'}^{j'}(x) &= \sum_{\nu'} \Phi_{\nu'}^{j'}(x) D_{\nu'\mu'}^{j'}(R). \end{aligned}$$

The calculation on the expectation values of the irreducible tensor operators in the static wave functions can be greatly simplified by making use of the symmetry. Since

$$\begin{aligned} O_R \left\{ L_\rho^k(x) \Psi_\mu^j(x) \right\} &= \left\{ O_R L_\rho^k(x) O_R^{-1} \right\} \left\{ O_R \Psi_\mu^j(x) \right\} \\ &= \sum_{\lambda\nu} \left\{ L_\lambda^k(x) \Psi_\nu^j(x) \right\} \left\{ D_{\lambda\rho}^k(R) D_{\nu\mu}^j(R) \right\}, \end{aligned}$$

$L_\rho^k(x) \Psi_\mu^j(x)$ transform according to the direct product of representations. Combining them by the Clebsch–Gordan coefficients, one obtains $F_M^J(x)$ belonging to the irreducible representation:

$$\begin{aligned} F_M^J(x) &= \sum_\rho L_\rho^k(x) \Psi_{M-\rho}^j(x) C_{\rho(M-\rho)JM}^{kj}, \\ L_\rho^k(x) \Psi_\mu^j(x) &= \sum_J F_{\rho+\mu}^J(x) C_{\rho\mu J(\rho+\mu)}^{kj}, \\ O_R F_M^J(x) &= \sum_{M'} F_{M'}^J(x) D_{M'M}^J(R). \end{aligned} \tag{4.199}$$

From the Wigner–Eckart theorem (see Theorem 3.6),

$$\langle \Phi_{\mu'}^{j'}(x) | F_M^J(x) \rangle = \delta_{j'J} \delta_{\mu'M} \langle \Phi^{j'} || L^k || \Psi^j \rangle, \tag{4.200}$$

where the constant $\langle \Phi^{j'} || L^k || \Psi^j \rangle$, called the reduced matrix entry, is independent of the subscripts μ', ρ, μ, and M, but is related to the explicit forms of Φ, L, and Ψ and depends on the indices j', k, and j. Hence,

$$\begin{aligned} \langle \Phi_{\mu'}^{j'}(x) | L_\rho^k(x) | \Psi_\mu^j(x) \rangle &= \sum_J C_{\rho\mu J(\rho+\mu)}^{kj} \langle \Phi_{\mu'}^{j'}(x) | F_{\rho+\mu}^J(x) \rangle \\ &= C_{\rho\mu j'\mu'}^{kj} \langle \Phi^{j'} || L^k || \Psi^j \rangle. \end{aligned} \tag{4.201}$$

In quantum mechanics, the expectation values of the irreducible tensor operators in the static wave functions are characterized in the calculation of matric entries. There are $(2j'+1)(2k+1)(2j+1)$ matrix entries in the form of $\langle \Phi^{j'}_{\mu'}(x)|L^k_\rho(x)|\Psi^j_\mu(x)\rangle$. Equation (4.201) shows that the rotational properties of the expectation values are fully demonstrated in the CG coefficients, and the detailed properties of the system and the operators are left in the reduced matrix entry. The Wigner–Eckart theorem simplifies the calculation of $(2j'+1)(2k+1)(2j+1)$ matrix entries to the calculation of only one parameter $\langle \Phi^{j'}||L^k||\Psi^j\rangle$. If there is one matrix entry with given subscripts μ', ρ, and μ which can be calculated, the remaining matrix entries all are calculable. In most cases, even one matrix entry is hard to be calculated. For example, the explicit forms of the wave functions $\Psi^j_\mu(x)$ and $\Phi^{j'}_{\mu'}(x)$ are unknown. However, through eliminating the parameter, Eq. (4.201) gives some relations among the matrix entries which sometimes can be observed in experiments. Furthermore, the CG coefficients in Eq. (4.201) have to be vanishing when some conditions are not satisfied

$$\begin{aligned} &|\mu'| = |\rho+\mu| \leq j', \qquad |\rho| \leq k, \qquad |\mu| \leq j, \\ &|k-j| \leq j' \leq k+j. \end{aligned} \tag{4.202}$$

Those conditions are called the selection rules in quantum mechanics. The above method holds when O_R is replaced with P_R or Q_R.

4.8.3 *Selection Rule and Relative Intensity of Radiation*

An isolated atom is isotropic and its symmetric group is SO(3), no matter how complicated its internal construction is. Its static wave function belongs to the irreducible representation of SO(3). In the language of quantum mechanics, the static wave function $\Psi^J_M(x)$ is the common eigenfunction of $H(x)$, J^2, and J_3. The intensity of the electric dipole radiation between two states is proportional to the square of the matrix entry of x_a

$$|\langle \Psi^{J'}_{M'}(x)|x_a|\Psi^J_M(x)\rangle|^2,$$

because the electric dipole operator is proportional to the distance of the pair of electric charges. From Eq. (4.194) the operators x_a are the combinations of the spherical harmonic functions $Y^1_\rho(\hat{\boldsymbol{n}})$, which are the vector operators with respect to the whole space and to the orbital space, and the scalar operators with respect to the spinor space. The square of the entries $|\langle \Psi^{J'}_{M'}|Y^1_\rho|\Psi^J_M\rangle|^2$ with $\rho=0$ demonstrates the intensity of the plane light of the electric dipole radiation polarized in the z-axis, and that with $\rho = \pm 1$ demonstrates the intensity of the circularly polarized light in the $x\ y$ plane.

The Wigner–Eckart theorem shows

$$\langle \Psi^{J'}_{M'}(x)|Y^1_\rho(\hat{\boldsymbol{n}})|\Psi^J_M(x)\rangle = C^{1J}_{\rho M,J'M'}\langle \Psi^{J'}||Y^1||\Psi^J\rangle. \tag{4.203}$$

The selection rules are the conditions where the CG coefficients are not vanishing,

$$\Delta M = M' - M = \rho = 0 \ \text{ or } \ \pm 1, \qquad |J-1| \leq J' \leq J+1. \tag{4.204}$$

When $J = 0$, J' cannot be equal to 0, so the section rule for J' is usually written as

$$\Delta J = J' - J = \pm 1 \text{ or } 0, \qquad 0 \not\longrightarrow 0.$$

Read as "the transition from 0 to 0 is forbidden". Considering the spatial inversion, the parities of the initial state and the final state have to be opposite because x_a has odd parity.

If the spin–orbital interaction is weak, the wave functions are first coupled into the product of two parts, the orbital part with a total orbital angular momentum L and the spinor part with a total spinor angular momentum S. Then, the static wave function is the combination of those functions,

$$\Psi^{JLS}_M = \sum_\sigma C^{LS}_{(M-\sigma)\sigma,JM}\Psi^L_{M-\sigma}W^S_\sigma\ . \tag{4.205}$$

In physics, this case is called the LS coupling. Ψ^{JLS}_M is the common eigenfunction of J^2, J_3, L^2, and S^2. In this case, there are additional selection rules obtained from the rotational properties of the wave functions and the operators in the orbital space (the action of P_R) and the spinor space (the action of Q_R) [see Eq. (4.194)],

$$\begin{aligned} &\Delta L = L' - L = \pm 1 \text{ or } 0, \qquad 0 \not\longrightarrow 0, \\ &\Delta S = S' - S = 0. \end{aligned} \tag{4.206}$$

Remind that in the complicated system of n electrons, the parity is not equal to $(-1)^L$. The transition with $\Delta L = 0$ does not violate the conservation law of parity.

The relative intensity of the electric dipole radiation is proportional to the square of the CG coefficients in Eq. (4.203),

$$\begin{aligned} &|\langle\Psi^{J'}_{M+1}|Y^1_1|\Psi^J_M\rangle|^2 \ : \ |\langle\Psi^{J'}_{M}|Y^1_0|\Psi^J_M\rangle|^2 \ : \ |\langle\Psi^{J'}_{M-1}|Y^1_{-1}|\Psi^J_M\rangle|^2 \\ &= \left(C^{1J}_{1M,J'(M+1)}\right)^2 \ : \ \left(C^{1J}_{0M,J'M}\right)^2 \ : \ \left(C^{1J}_{(-1)M,J'(M-1)}\right)^2 . \end{aligned}$$

For instance, when $J' = J - 1$, from Eq. (4.132) (or see Prob. 17), the ratio of the relative intensity is

$$(J-M)(J-M-1) \ : \ 2(J-M)(J+M) \ : \ (J+M)(J+M-1).$$

The total intensity of radiation with the given J' and J

$$\sum_{\rho=-1}^{1} \left(C^{1J}_{\rho M,J'(M+\rho)}\right)^2 = \frac{2J'+1}{2J+1} \sum_{\rho=-1}^{1} \left(C^{1J'}_{\rho(-M-\rho),J(-M)}\right)^2 = \frac{2J'+1}{2J+1}.$$

The total intensity is independent of M. Namely, the total transition probability is independent of the direction of the total angular momentum.

4.8.4 *Landé Factor and Zeeman Effects*

The Stern–Gerlach experiment discovered that a hydrogen atom in S-wave deflects as it passes through an asymmetric magnetic field. The two spectral lines on the film shows that the atom has magnetic momentum. This is the earliest experiment observing the spin of an electron. It was measured that the geromagnetic ratio for the spin angular momentum is double of that for the orbital angular momentum. Namely, the operator of total magnetic momentum of an atom $\mathcal{M}_a$ is

$$\mathcal{M}_a = -\frac{e}{2m_e}(L_a + 2S_a) = -\frac{e}{2m_e}(J_a + S_a) = -\frac{egJ_a}{2m_e},$$

where m_e is the mass of the electron, g is the Landé factor, and the natural units are used, $c = \hbar = 1$. The expectation value of $\mathcal{M}_3$ in the state Ψ^J_M is

$$\overline{\mathcal{M}_3} = \langle\Psi^J_M|\mathcal{M}_3|\Psi^J_M\rangle = \frac{-egM}{2m_e} = -\frac{eM}{2m_e} - \frac{e}{2m_e}\langle\Psi^J_M|S_3|\Psi^J_M\rangle,$$

$$(g-1)M = \langle\Psi^J_M|S_3|\Psi^J_M\rangle. \tag{4.207}$$

The matrix entries of two vector operators J_a and S_a in the same states are proportional,

$$\langle\Psi^J_M|S_a|\Psi^J_{M'}\rangle = A_J\langle\Psi^J_M|J_a|\Psi^J_{M'}\rangle, \tag{4.208}$$

where the proportional coefficient A_J is independent of the subscripts a, M, and M'. Since $J_a \Psi^J_M$ is a linear combination of $\Psi^J_{M'}$, Eq. (4.208) holds if $\Psi^J_{M'}$ is replaced with $J_a \Psi^J_M$,

$$\sum_{a=1}^{3} \langle \Psi^J_M | S_a J_a | \Psi^J_M \rangle = A_J \sum_{a=1}^{3} \langle \Psi^J_M | J_a J_a | \Psi^J_M \rangle = A_J J(J+1).$$

If Ψ^J_M is a state by LS coupling [see Eq. (4.205)], it is the common eigenfunction of J^2, J_3, L^2, and S^2. Since

$$\sum_{a=1}^{3} L_a^2 = \sum_{a=1}^{3} (J_a - S_a)^2 = \sum_{a=1}^{3} J_a^2 + \sum_{a=1}^{3} S_a^2 - 2\sum_{a=1}^{3} S_a J_a,$$

one has

$$\sum_{a=1}^{3} \langle \Psi^J_M | S_a J_a | \Psi^J_M \rangle = \{J(J+1) + S(S+1) - L(L+1)\}/2,$$

and

$$A_J = \frac{J(J+1) + S(S+1) - L(L+1)}{2J(J+1)}.$$

Taking $a = 3$ and $M' = M$ in Eq. (4.208) and comparing it with Eq. (4.207), one obtains

$$g = 1 + A_J = \frac{3J(J+1) + S(S+1) - L(L+1)}{2J(J+1)}. \tag{4.209}$$

For the system of a single electron, $S = 1/2$ and $L = J \mp 1/2$. Thus,

$$g = \begin{cases} \dfrac{2J+1}{2J} & \text{when } L = J - \dfrac{1}{2}, \\[2ex] \dfrac{2J+1}{2(J+1)} & \text{when } L = J + \dfrac{1}{2}. \end{cases} \tag{4.210}$$

If the atom is in a strong magnetic field such that the spin–orbital interaction can be neglected, the wave function can be described by $\psi^L_m W^S_\sigma$, in which the energy E is

$$E = E_0 + \frac{e}{2m_e}(m + 2\sigma)H. \tag{4.211}$$

E_0 is the energy when the magnetic field does not exist. Equation (4.211) shows that the change of energy of the system in a magnetic field is independent of L and S.

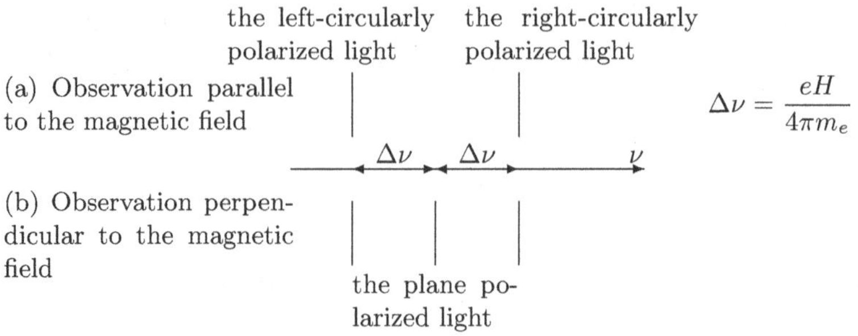

Fig. 4.5 The normal Zeeman effect

The selection rule for the electric dipole radiation is

$$\begin{aligned} &\Delta L = \pm 1 \text{ or } 0, \qquad 0 \not\longrightarrow 0. \\ &\Delta m = \pm 1,\ 0, \qquad \Delta S = 0, \qquad \Delta\sigma = 0. \end{aligned} \tag{4.212}$$

The energy of the radiative photon in the electric dipole transition is

$$\Delta E = \begin{cases} \Delta E_0 & \text{when } \Delta m = 0, \\ \Delta E_0 \mp \dfrac{eH}{2m_e} & \text{when } \Delta m = \pm 1. \end{cases} \tag{4.213}$$

The radiative light is the plane light polarized in the z-axis when $\Delta m = 0$ and is the circularly polarized light about z-axis when $\Delta m = \pm 1$. When the observation is made in the direction parallel to the magnetic field (the z-axis), the plane polarized light cannot be seen. Two spectral lines of the circularly polarized light deflect from the original position (ΔE_0) in equidistance. The red-moved line is the left-circularly polarized light ($m - m' = -\Delta m = -1$), and the other is the right-circularly polarized light ($m - m' = 1$). When the observation is made in the direction perpendicular to the magnetic field, three lines can be seen where the middle line is the plane light polarized parallel to the magnetic field. This splitting of an electron in a strong magnetic field is called the normal Zeeman effect.

For an electron in a weak magnetic field, the spin–orbital interaction cannot be neglected. The wave function has to be characterized by Ψ_M^{JLS} given in Eq. (4.205). The energy of the electron in the magnetic field is

$$E = E_0 + \frac{e}{2m_e} gMH\ .$$

The observed splitting of spectrum is related with the quantum numbers

J, L, S, and M of both the initial state and the final state:

$$\Delta E = \Delta E_0 + \frac{eH}{2m_e}\left(gM - g'M'\right) . \tag{4.214}$$

This splitting for an electron in a weak magnetic field is called the anomalous Zeeman effect.

4.9 An Isolated Quantum n-body System

An isolated quantum n-body system is invariant in the translation of space–time and the spatial rotation so that the energy, the total momentum, and total orbital angular momentum of the system are conserved. The motion of the center-of-mass and the global rotation of the system can be separated from its internal motion, and its Schrödinger equation can be reduced to the radial equation, depending only on the internal degrees of freedom.

4.9.1 *Separation of the Motion of Center-of-Mass*

Denote by $\boldsymbol{r}_k$ and by m_k, $k = 1, 2, \ldots, n$, the position vectors and the masses of n particles, respectively. The total mass of the n-body quantum system is $M = \sum_k m_k$ and its Schrödinger equation is

$$-\frac{1}{2}\sum_{k=1}^{n} m_k^{-1}\, \bigtriangledown^2_{\boldsymbol{r}_k}\, \Psi + V\Psi = E\Psi, \tag{4.215}$$

where $\bigtriangledown^2_{\boldsymbol{r}_k}$ is the Laplace operator with respect to $\boldsymbol{r}_k$, and V is a pair potential, depending upon the distance $|\boldsymbol{r}_j - \boldsymbol{r}_k|$ of each pair of particles. The natural units are used for convenience, $\hbar = c = 1$.

Replace the position vectors $\boldsymbol{r}_k$ by the Jacobi coordinate vectors $\boldsymbol{R}_j$:

$$\begin{aligned} &\boldsymbol{R}_0 = M^{-1/2}\sum_{k=1}^{n} m_k\boldsymbol{r}_k, \qquad \boldsymbol{R}_j = \mu_j^{1/2}\left(\boldsymbol{r}_j - \sum_{k=j+1}^{n}\frac{m_k\boldsymbol{r}_k}{M_{j+1}}\right), \\ &M_j = \sum_{k=j}^{n} m_k, \qquad M_1 = M, \qquad 1 \le j \le (n-1), \end{aligned} \tag{4.216}$$

where $\boldsymbol{R}_0$ describes the position of the center of mass, $\boldsymbol{R}_1$ describes the mass-weighted separation from the first particle to the center of mass of the remaining particles, $\boldsymbol{R}_2$ describes the mass-weighted separation from the second particle to the center of mass of the remaining $(n-2)$ particles, and so on. An additional factor $\sqrt{M}$ is included in the Jacobi coordinate vectors for convenience. The mass-weighted factors μ_j in the formulas for $\boldsymbol{R}_j$ are determined by the condition

$$\sum_{k=1}^{n} m_k \boldsymbol{r}_k^2 = \sum_{j=0}^{n-1} \boldsymbol{R}_j^2. \tag{4.217}$$

μ_j can be calculated one by one from the following schemes. In the center-of-mass frame, if the first $(j-1)$ particles are located at the origin and the last $(n-j)$ particles coincide with each other,

$$\begin{aligned} &\boldsymbol{r}_k = 0, \qquad 1 \leq k < j, \\ &\boldsymbol{r}_{j+1} = \boldsymbol{r}_{j+2} = \cdots = \boldsymbol{r}_n = -m_j \boldsymbol{r}_j / M_{j+1}, \\ &\boldsymbol{R}_k = 0, \qquad k \neq j, \end{aligned}$$

then,

$$\begin{aligned} \sum_{k=j}^{n} m_k \boldsymbol{r}_k^2 &= m_j \boldsymbol{r}_j^2 + M_{j+1} \left(-m_j \boldsymbol{r}_j / M_{j+1}\right)^2 = \left(m_j M_j / M_{j+1}\right) \boldsymbol{r}_j^2 \\ &= \boldsymbol{R}_j^2 = \mu_j \left[\boldsymbol{r}_j + m_j \boldsymbol{r}_j / M_{j+1}\right]^2 = \mu_j \left[M_j \boldsymbol{r}_j / M_{j+1}\right]^2, \end{aligned}$$

and

$$\mu_j = \frac{m_j M_{j+1}}{M_j}. \tag{4.218}$$

The linear relations between $\boldsymbol{r}_k$ and $\boldsymbol{R}_j$ are written in a symbolic form:

$$\boldsymbol{R}_j = \sum_{k=1}^{n} X_{jk} m_k^{1/2} \boldsymbol{r}_k, \qquad \nabla_{\boldsymbol{r}_k} = m_k^{1/2} \sum_{j=0}^{n-1} X_{jk} \nabla_{\boldsymbol{R}_j} . \tag{4.219}$$

Substituting Eq. (4.219) into Eq. (4.217), one obtains that X is a real orthogonal matrix,

$$\sum_{j=0}^{n-1} \boldsymbol{R}_j^2 = \sum_{kt} \left(m_k m_t\right)^{1/2} \boldsymbol{r}_k \boldsymbol{r}_t \sum_{j=0}^{n-1} X_{jk} X_{jt} = \sum_{k=1}^{n} m_k \boldsymbol{r}_k^2.$$

Then, $\boldsymbol{r}_k = m_k^{-1/2} \sum_j \boldsymbol{R}_j X_{jk}$, and the Laplace operator in Eq. (4.215) and the orbital angular momentum operator $\boldsymbol{L}$ are directly expressed in $\boldsymbol{R}_j$:

$$\begin{aligned} &\sum_{k=1}^{n} m_k^{-1} \nabla_{\boldsymbol{r}_k}^2 = \sum_{j=0}^{n-1} \nabla_{\boldsymbol{R}_j}^2, \\ &\boldsymbol{L} = -i \sum_{k=1}^{n} \boldsymbol{r}_k \times \nabla_{\boldsymbol{r}_k} = -i \sum_{j=0}^{n-1} \boldsymbol{R}_j \times \nabla_{\boldsymbol{R}_j}. \end{aligned} \tag{4.220}$$

By the separation of variables, one has

$$\begin{aligned} &\boldsymbol{\Psi}(\boldsymbol{R}_0, \boldsymbol{R}_1, \ldots, \boldsymbol{R}_{n-1}) = \phi(\boldsymbol{R}_0) \Psi(\boldsymbol{R}_1, \ldots, \boldsymbol{R}_{n-1}), \\ &\phi(\boldsymbol{R}_0) = e^{i\boldsymbol{P}_c \cdot \boldsymbol{R}_0/\sqrt{M}}, \qquad -\frac{1}{2} \nabla_{\boldsymbol{R}_0}^2 \phi(\boldsymbol{R}_0) = \frac{P_c^2}{2M} \phi(\boldsymbol{R}_0), \\ &-\frac{1}{2} \nabla^2 \Psi + V \Psi = \left(E - \frac{P_c^2}{2M}\right) \Psi, \qquad \nabla^2 = \sum_{j=1}^{n-1} \nabla_{\boldsymbol{R}_j}^2. \end{aligned} \tag{4.221}$$

The inverse transformation of Eq. (4.216) is

$$\begin{aligned} \boldsymbol{r}_k &= \left[\frac{M_{k+1}}{m_k M_k}\right]^{1/2} \boldsymbol{R}_k - \sum_{j=1}^{k-1} \left[\frac{m_j}{M_j M_{j+1}}\right]^{1/2} \boldsymbol{R}_j + M^{-1/2} \boldsymbol{R}_0, \\ \boldsymbol{r}_k - \boldsymbol{r}_j &= \left[\frac{M_{k+1}}{m_k M_k}\right]^{1/2} \boldsymbol{R}_k - \sum_{i=j+1}^{k-1} \left[\frac{m_i}{M_i M_{i+1}}\right]^{1/2} \boldsymbol{R}_i - \left[\frac{M_j}{m_j M_{j+1}}\right]^{1/2} \boldsymbol{R}_j. \end{aligned} \tag{4.222}$$

$|\boldsymbol{r}_k - \boldsymbol{r}_j|^2$ are the functions of $\boldsymbol{R}_i \cdot \boldsymbol{R}_t$ which are rotational invariant and independent of $\boldsymbol{R}_0$. The last equality in Eq. (4.221) is the Schrödinger equation in the center-of-mass frame, which is independent of $\boldsymbol{R}_0$ and spherically symmetric. Hereafter, we will neglect the motion of center-of-mass by simply assuming $\boldsymbol{R}_0 = 0$ and $\boldsymbol{P}_c = 0$ in Eq. (4.221) for convenience.

4.9.2 *Quantum Two-body System*

For a quantum two-body system, there is only one Jacobi coordinate vector $\boldsymbol{R}$. The eigenfunction of the angular momentum is the spherical harmonic function $Y_m^\ell(\hat{\boldsymbol{R}})$ where $\hat{\boldsymbol{R}} = \boldsymbol{R}/r$ and $r = |\boldsymbol{R}|$. Since the angular momentum is conserved, the wave function can be separated (see §4.4.3),

$$\Psi_m^\ell(\boldsymbol{R}) = \phi_\ell(r) Y_m^\ell(\hat{\boldsymbol{R}}), \qquad Y_m^\ell(\hat{\boldsymbol{R}}) = \sqrt{\frac{2\ell+1}{4\pi}}\, D_{m0}^\ell(\varphi, \theta, 0)^*, \tag{4.223}$$

where θ and φ are the polar angle and the azimuthal angle of $\hat{\boldsymbol{R}}$, respectively. Substituting Eq. (4.223) into the Schrödinger equation (4.221), one obtains the radial equation

$$\begin{aligned}
\nabla^2 \Psi^\ell_m(\boldsymbol{R}) &= Y^\ell_m(\hat{\boldsymbol{R}}) r^{-1} \frac{d^2}{dr^2} r\phi_\ell(r) - \phi_\ell(r) \frac{L^2}{r^2} Y^\ell_m(\hat{\boldsymbol{R}}) \\
&= Y^\ell_m(\hat{\boldsymbol{R}}) \left\{ \frac{d^2}{dr^2} + \frac{2}{r}\frac{d}{dr} - \frac{\ell(\ell+1)}{r^2} \right\} \phi_\ell(r) \\
&= -2\left\{E - V(r)\right\} \phi_\ell(r) Y^\ell_m(\hat{\boldsymbol{R}}),
\end{aligned} \tag{4.224}$$

where the property of the spherical harmonic function $Y^\ell_m(\hat{\boldsymbol{R}})$ is used. If $Y^\ell_m(\hat{\boldsymbol{R}})$ is replaced with the harmonic polynomial $\boldsymbol{Y}^\ell_m(\boldsymbol{R})$, the action of the Laplace operator is divided into three parts. The first part is its action on the radial functions $r^{-\ell}\phi_\ell(r)$. The second part is its action on $\boldsymbol{Y}^\ell_m(\boldsymbol{R})$ which is vanishing. The third part is its mixed action

$$\begin{aligned}
\Psi^\ell_m(\boldsymbol{R}) &= \phi_\ell(r) Y^\ell_m(\hat{\boldsymbol{R}}) = r^{-\ell}\phi_\ell(r) \boldsymbol{Y}^\ell_m(\boldsymbol{R}), \\
\nabla^2 \Psi^\ell_m(\boldsymbol{R}) &= \boldsymbol{Y}^\ell_m(\boldsymbol{R}) r^{-1} \frac{d^2}{dr^2}\left[r^{1-\ell}\phi_\ell(r)\right] + 2\frac{d}{dr}\left[r^{-\ell}\phi_\ell(r)\right] \frac{\boldsymbol{R}}{r} \cdot \nabla \boldsymbol{Y}^\ell_m(\boldsymbol{R}) \\
&= Y^\ell_m(\hat{\boldsymbol{R}}) \left\{ \frac{d^2}{dr^2} + \left[\frac{2(1-\ell)}{r} + \frac{2\ell}{r}\right]\frac{d}{dr} + \left[\frac{\ell(\ell-1)}{r^2} - \frac{2\ell^2}{r^2}\right] \right\} \phi_\ell(r).
\end{aligned}$$

The results are the same, but the spherical coordinates θ and φ do not appear in the calculation.

4.9.3 *Quantum Three-body System*

What is the generalization of the basis eigenfunctions of angular momentum to a quantum three-body system or a quantum multiple-body system? A naive idea is to generalize the spherical harmonic function $Y^\ell_m(\hat{\boldsymbol{x}})$ to $D^\ell_{mm'}(\alpha, \beta, \gamma)^*$, as was done by Wigner [Wigner (1959)]. However, the derivative of the Euler angles leads to the singularity. As shown in the preceding subsection, in terms of the harmonic polynomials $\boldsymbol{Y}^\ell_m(\boldsymbol{x})$ the angular variables are avoided in the derivation of the radial equations. What is the generalization of the harmonic polynomials to a three-body system?

For a quantum three-body system there are two Jacobi coordinate vectors $\boldsymbol{R}_1$ and $\boldsymbol{R}_2$,

$$\begin{aligned}
\boldsymbol{R}_1 &= \left(\frac{m_1(m_2+m_3)}{m_1+m_2+m_3}\right)^{1/2} \left[\boldsymbol{r}_1 - \frac{m_2\boldsymbol{r}_2 + m_3\boldsymbol{r}_3}{m_2+m_3}\right] \\
&= \left(\frac{m_1(m_1+m_2+m_3)}{m_2+m_3}\right)^{1/2} \boldsymbol{r}_1, \\
\boldsymbol{R}_2 &= \left(\frac{m_2 m_3}{m_2+m_3}\right)^{1/2} \left(\boldsymbol{r}_2 - \boldsymbol{r}_3\right).
\end{aligned} \tag{4.225}$$

The Schrödinger equation (4.221) becomes

$$\nabla^2 \Psi(\boldsymbol{R}_1, \boldsymbol{R}_2) = -2\,(E - V)\,\Psi(\boldsymbol{R}_1, \boldsymbol{R}_2), \qquad \nabla^2 = \nabla^2_{\boldsymbol{R}_1} + \nabla^2_{\boldsymbol{R}_2}, \quad (4.226)$$

where $\nabla^2_{\boldsymbol{R}_1}$ and $\nabla^2_{\boldsymbol{R}_2}$ are the Laplace operators with respect to the Jacobi coordinate vectors $\boldsymbol{R}_1$ and $\boldsymbol{R}_2$, respectively. There are nine degrees of freedom for a three-body system, where three degrees of freedom describe the motion of center-of-mass, three degrees of freedom describe the global rotation of the system, and the remaining three degrees of freedom describe the internal motion. The internal variables are denoted by

$$\xi_1 = \boldsymbol{R}_1 \cdot \boldsymbol{R}_1, \qquad \xi_2 = \boldsymbol{R}_1 \cdot \boldsymbol{R}_2, \qquad \eta_2 = \boldsymbol{R}_2 \cdot \boldsymbol{R}_2, \qquad (4.227)$$

which are invariant in the global rotation and in the space inversion of the system. The potential V is the function of the three internal variables.

On one hand, the number of linearly independent homogeneous polynomials of degree n with respect to the components of $\boldsymbol{R}_1$ and $\boldsymbol{R}_2$ is $M(n)$:

$$M(n) = \frac{1}{5!}(n+1)(n+2)(n+3)(n+4)(n+5). \qquad (4.228)$$

The independent basis eigenfunctions of angular momentum do not contain a factor of the internal variables, because the factor should be incorporated into the radial functions. The number of the homogeneous polynomials of degree n that do not contain a function of internal variables as a factor is

$$K(n) = M(n) - 3M(n-2) + 3M(n-4) - M(n-6) = 4n^2 + 2, \qquad n \geq 1.$$

On the other hand, a common eigenfunction of L^2 and L_z with the eigenvalues $\ell(\ell+1)$ and μ, which is a homogeneous polynomial of degree n, can be obtained from $\boldsymbol{Y}^q_m(\boldsymbol{R}_1)\boldsymbol{Y}^{n-q}_{m'}(\boldsymbol{R}_2)$ by the Clebsch–Gordan coefficients,

$$\boldsymbol{Y}^{q(n-q)}_{\ell\mu}(\boldsymbol{R}_1, \boldsymbol{R}_2) = \sum_m \boldsymbol{Y}^q_m(\boldsymbol{R}_1)\boldsymbol{Y}^{n-q}_{\mu-m}(\boldsymbol{R}_2)\langle q, m, n-q, \mu-m|\ell, \mu\rangle, \quad (4.229)$$

where $\ell = n,\ n-1,\ \ldots\ ,\ |n-2q|$. $\boldsymbol{Y}^{q(n-q)}_{\ell\mu}(\boldsymbol{R}_1, \boldsymbol{R}_2)$ is a homogeneous polynomial of degrees q and $n-q$ with respect to the components of $\boldsymbol{R}_1$ and $\boldsymbol{R}_2$, respectively. When $\mu = \ell = n$ and $(n-1)$, one has

$$\begin{aligned} \boldsymbol{Y}^{q(n-q)}_{\ell\ell}(\boldsymbol{R}_1, \boldsymbol{R}_2) &= (-1)^\ell \left\{ \frac{\left[(2q+1)!(2\ell-2q+1)!\right]^{1/2}}{q!(\ell-q)!2^{\ell+2}\pi} \right\} \\ &\cdot\ (R_{1x} + iR_{1y})^q (R_{2x} + iR_{2y})^{\ell-q}, \qquad 0 \leq q \leq \ell = n, \end{aligned}$$

$$\begin{aligned}
\boldsymbol{Y}_{\ell\ell}^{q(n-q)}(\boldsymbol{R}_1,\boldsymbol{R}_2) &= (-1)^{\ell}\left\{\frac{(2q+1)!(2\ell-2q+3)!}{2q(\ell+1)(\ell-q+1)}\right\}^{1/2} \\
&\cdot \left\{(q-1)!(\ell-q)!2^{\ell+2}\pi\right\}^{-1}(R_{1x}+iR_{1y})^{q-1}(R_{2x}+iR_{2y})^{\ell-q} \\
&\cdot \left\{(R_{1x}+iR_{1y})R_{2z}-R_{1z}(R_{2x}+iR_{2y})\right\}, \qquad 1\le q\le \ell=n-1.
\end{aligned} \tag{4.230}$$

It is evident that these expressions do not contain a function of the internal variables as a factor, neither do their partners with smaller μ owing to the spherical symmetry. The number of those eigenfunctions is

$$(2n+1)(n+1)+(2n-1)(n-1)=4n^2+2=K(n), \qquad n\ge 1.$$

Namely, any of the remaining eigenfunctions $\mathbf{Y}_{\ell\mu}^{q(n-q)}(\boldsymbol{R}_1,\boldsymbol{R}_2)$ with $\ell < n-1$ is a combination of polynomials, each of which contains a factor of internal variables. For example,

$$\begin{aligned}
\boldsymbol{Y}_{00}^{11}(\boldsymbol{R}_1,\boldsymbol{R}_2) &= -\frac{\sqrt{3}}{4\pi}\xi_2, \qquad \boldsymbol{Y}_{00}^{22}(\boldsymbol{R}_1,\boldsymbol{R}_2) = \frac{\sqrt{5}}{8\pi}\left\{3\xi_2^2-\xi_1\eta_2\right\}, \\
\boldsymbol{Y}_{22}^{22}(\boldsymbol{R}_1,\boldsymbol{R}_2) &= \frac{5\sqrt{21}}{56\pi}\left\{\eta_2(R_{1x}+iR_{1y})^2+\xi_1(R_{2x}+iR_{2y})^2\right. \\
&\quad \left.-3\xi_2(R_{1x}+iR_{1y})(R_{2x}+iR_{2y})\right\}.
\end{aligned}$$

The mathematical reason for this can be seen in §9.4.4.

Now, any eigenfunction of L^2 and L_z with the eigenvalues ℓ and $\mu = \ell$, which will be called the eigenfunction of angular momentum ℓ for convenience, is a combination of those homogeneous polynomials $\boldsymbol{Y}_{\ell\ell}^{q(\ell-q)}(\boldsymbol{R}_1,\boldsymbol{R}_2)$ and $\boldsymbol{Y}_{\ell\ell}^{q(\ell-q+1)}(\boldsymbol{R}_1,\boldsymbol{R}_2)$ where the combinative coefficients are functions of the internal variables. $\mathbf{Y}_{\ell\ell}^{q(\ell-q+\lambda)}(\boldsymbol{R}_1,\boldsymbol{R}_2)$ can be replaced with a simpler form $Q_q^{\ell\lambda}(\boldsymbol{R}_1,\boldsymbol{R}_2)$ by removing a constant factor which can be incorporated into the radial functions [Hsiang and Hsiang (1998)]

$$\begin{aligned}
Q_q^{\ell\lambda}(\boldsymbol{R}_1,\boldsymbol{R}_2) &= \frac{X^{q-\lambda}Y^{\ell-q}Z^{\lambda}}{(q-\lambda)!(\ell-q)!}, \qquad \lambda\le q\le \ell, \qquad \lambda=0,1, \\
X &\equiv R_{1x}+iR_{1y}, \qquad Y\equiv R_{2x}+iR_{2y}, \qquad Z\equiv XR_{2z}-R_{1z}Y.
\end{aligned} \tag{4.231}$$

Note that

$$Q_q^{\ell 1}(\boldsymbol{R}_1,\boldsymbol{R}_2) = Q_{q-1}^{(\ell-1)0}(\boldsymbol{R}_1,\boldsymbol{R}_2)Z. \tag{4.232}$$

$Q_q^{\ell\lambda}(\boldsymbol{R}_1,\boldsymbol{R}_2)$, called the generalized harmonic polynomial, is a homogeneous polynomial of degrees q and $(\ell-q+\lambda)$ with respect to the components of $\boldsymbol{R}_1$ and $\boldsymbol{R}_2$, respectively. It is the common eigenfunction of L^2, L_z, $L_{\boldsymbol{R}_1}^2$, $L_{\boldsymbol{R}_2}^2$, $\nabla_{\boldsymbol{R}_1}^2$, $\nabla_{\boldsymbol{R}_2}^2$, $\nabla_{\boldsymbol{R}_1}\cdot\nabla_{\boldsymbol{R}_2}$, and the space inversion with the eigenvalues

$\ell(\ell+1)$, ℓ, $q(q+1)$, $(\ell-q+\lambda)(\ell-q+\lambda+1)$, 0, 0, 0, and $(-1)^{\ell+\lambda}$, respectively, where $L^2_{\boldsymbol{R}_1}$ ($L^2_{\boldsymbol{R}_2}$) is the square of the partial angular momentum. Any wave function with the given angular momentum ℓ and the parity $(-1)^{\ell+\lambda}$ can be expanded with respect to $Q_q^{\ell\lambda}(\boldsymbol{R}_1, \boldsymbol{R}_2)$

$$\Psi_\ell^{\ell\lambda}(\boldsymbol{R}_1, \boldsymbol{R}_2) = \sum_{q=\lambda}^{\ell} \phi_q^{\ell\lambda}(\xi_1, \xi_2, \eta_2) Q_q^{\ell\lambda}(\boldsymbol{R}_1, \boldsymbol{R}_2), \qquad \lambda = 0, 1. \tag{4.233}$$

That is, for a three-body system, the generalized harmonic polynomials $Q_q^{\ell\lambda}(\boldsymbol{R}_1, \boldsymbol{R}_2)$ constitute a complete set of basis eigenfunctions with angular momentum ℓ and parity $(-1)^{\ell+\lambda}$. Only $(\ell+1-\lambda)$ partial angular momentum states are involved in constructing a function with angular momentum ℓ and parity $(-1)^{\ell+\lambda}$, and the contributions from the infinite number of remaining partial angular momentum states are incorporated into those of the radial functions. When substituting Eq. (4.233) into the Schrödinger equation (4.226), the action of the Laplace operator is divided into three parts. The first part is its action on the radial functions $\phi_q^{\ell\lambda}(\xi_1, \xi_2, \eta_2)$ which can be calculated by the replacement of variables directly. The second part is its action on the generalized harmonic polynomials $Q_q^{\ell\lambda}(\boldsymbol{R}_1, \boldsymbol{R}_2)$ which is vanishing. The third part is its mixed action

$$\begin{aligned}
&2\left\{\left(\partial_{\xi_1}\phi_q^{\ell\lambda}\right) 2\boldsymbol{R}_1 + \left(\partial_{\xi_2}\phi_q^{\ell\lambda}\right) \boldsymbol{R}_2\right\} \cdot \nabla_{\boldsymbol{R}_1} Q_q^{\ell\lambda} \\
&+2\left\{\left(\partial_{\xi_2}\phi_q^{\ell\lambda}\right) \boldsymbol{R}_1 + \left(\partial_{\eta_2}\phi_q^{\ell\lambda}\right) 2\boldsymbol{R}_2\right\} \cdot \nabla_{\boldsymbol{R}_2} Q_q^{\ell\tau},
\end{aligned}$$

where ∂_ξ denotes $\partial/\partial\xi$. In terms of Eqs. (4.231) one obtains

$$\begin{aligned}
&\boldsymbol{R}_1 \cdot \nabla_{\boldsymbol{R}_1} Q_q^{\ell\lambda} = q Q_q^{\ell\lambda}, && \boldsymbol{R}_2 \cdot \nabla_{\boldsymbol{R}_1} Q_q^{\ell\lambda} = (\ell - q + 1)\, Q_{q-1}^{\ell\lambda}, \\
&\boldsymbol{R}_1 \cdot \nabla_{\boldsymbol{R}_2} Q_q^{\ell\lambda} = (q - \lambda + 1)\, Q_{q+1}^{\ell\lambda}, && \boldsymbol{R}_2 \cdot \nabla_{\boldsymbol{R}_2} Q_q^{\ell\lambda} = (\ell - q + \lambda)\, Q_q^{\ell\lambda}.
\end{aligned}$$

Thus, the radial equations for the radial functions $\phi_q^{\ell\lambda}(\xi_1, \xi_2, \eta_2)$ are [Hsiang and Hsiang (1998); Gu et al. (2001b)]:

$$\begin{aligned}
&\nabla^2\phi_q^{\ell\lambda} + 4q\partial_{\xi_1}\phi_q^{\ell\lambda} + 4(\ell - q + \lambda)\partial_{\eta_2}\phi_q^{\ell\lambda} + 2(q-\lambda)\partial_{\xi_2}\phi_{q-1}^{\ell\lambda} \\
&\qquad + 2(\ell - q)\partial_{\xi_2}\phi_{q+1}^{\ell\lambda} = -2\,(E - V)\,\phi_q^{\ell\lambda}, \\
&\nabla^2\phi_q^{\ell\lambda}(\xi_1, \xi_2, \eta_2) = \Big\{4\xi_1\partial^2_{\xi_1} + 4\eta_2\partial^2_{\eta_2} + 6\left(\partial_{\xi_1} + \partial_{\eta_2}\right) \\
&\qquad + \left(\xi_1 + \eta_2\right)\partial^2_{\xi_2} + 4\xi_2\left(\partial_{\xi_1} + \partial_{\eta_2}\right)\partial_{\xi_2}\Big\}\,\psi_q^{\ell\lambda}(\xi_1, \xi_2, \eta_2), \\
&\lambda \le q \le \ell, \qquad \lambda = 0, 1.
\end{aligned} \tag{4.234}$$

4.9.4 *Quantum n-body System*

For a quantum n-body system, there are $(n-1)$ Jacobi coordinate vectors. Arbitrarily choose two Jacobi coordinate vectors, say $\boldsymbol{R}_1$ and $\boldsymbol{R}_2$. In the body-fixed frame, $\boldsymbol{R}_1$ is parallel to its z-axis, and $\boldsymbol{R}_2$ is located in its x z plane with a non-negative x-component. Among $3n$ variables of the n-body system, three variables describe the motion of center-of-mass, three variables describe the global rotation of the system, and the remaining $(3n-6)$ variables describe the internal motion. The internal variables cannot be chosen as $\boldsymbol{R}_j \cdot \boldsymbol{R}_k$, because the number of $\boldsymbol{R}_j \cdot \boldsymbol{R}_k$ is $n(n-1)/2$ which is larger than $(3n-6)$ when $n > 4$. Further, this set of internal variables is not complete because two configurations, which are related by a reflection to the plane spanned by $\boldsymbol{R}_1$ and $\boldsymbol{R}_2$, are described by the same internal variables. The complete set of internal variables are

$$\begin{array}{lll} \xi_j = \boldsymbol{R}_j \cdot \boldsymbol{R}_1, & \eta_j = \boldsymbol{R}_j \cdot \boldsymbol{R}_2, & \zeta_j = \boldsymbol{R}_j \cdot \left(\boldsymbol{R}_1 \times \boldsymbol{R}_2\right), \\ 1 \leq j \leq (n-1), & \eta_1 = \xi_2, & \zeta_1 = \zeta_2 = 0, \end{array} \tag{4.235}$$

which are invariant in the global rotation of the system. The number of the internal variables is $(3n-6)$, where ξ_j and η_j have even parity, but ζ_j have odd parity. Introduce a set of functions of internal variables,

$$\begin{aligned} \Omega_j &= (\boldsymbol{R}_1 \times \boldsymbol{R}_j) \cdot (\boldsymbol{R}_1 \times \boldsymbol{R}_2) = \xi_1 \eta_j - \xi_2 \xi_j, \\ \omega_j &= (\boldsymbol{R}_2 \times \boldsymbol{R}_j) \cdot (\boldsymbol{R}_1 \times \boldsymbol{R}_2) = \xi_2 \eta_j - \eta_2 \xi_j, \\ \Omega_1 &= \omega_2 = 0, \qquad \Omega_2 = -\omega_1 = \left(\boldsymbol{R}_1 \times \boldsymbol{R}_2\right)^2. \end{aligned} \tag{4.236}$$

In the body-fixed frame, due to Eq. (4.235), $\boldsymbol{R}_1$ is $\left(0, 0, \xi_1^{1/2}\right)$, $\boldsymbol{R}_2$ is $\left[\left(\Omega_2/\xi_1\right)^{1/2}, 0, \xi_2 \xi_1^{-1/2}\right]$, and the components R'_{jb} of $\boldsymbol{R}_j$ are

$$R'_{jx} = \Omega_j \left(\xi_1 \Omega_2\right)^{-1/2}, \qquad R'_{jy} = \zeta_j \Omega_2^{-1/2}, \qquad R'_{jz} = \xi_j \xi_1^{-1/2}, \tag{4.237}$$

where $1 \leq j \leq n-1$. The volume element of the configuration space can be calculated from the Jacobi determinant by replacement of variables:

$$\prod_{j=1}^{n-1} dR_{jx} dR_{jy} dR_{jz} = \frac{1}{4} \Omega_2^{3-n} \sin\beta d\alpha d\beta d\gamma d\xi_1 d\xi_2 d\eta_2 \prod_{j=3}^{n-1} d\xi_j d\eta_j d\zeta_j. \tag{4.238}$$

The domains of definition of the Euler angles are given in Eq. (4.63). The domains of definition of ξ_1 and η_2 are $(0, \infty)$ and the domains of definition of the remaining variables are $(-\infty, \infty)$. Because Eq. (4.222) and

$$\boldsymbol{R}_j \cdot \boldsymbol{R}_k = \Omega_2^{-1} \left(\Omega_j \eta_k - \omega_j \xi_k + \zeta_j \zeta_k \right), \tag{4.239}$$

the potential V is a function of only the internal variables.

Denoting by $R(\alpha, \beta, \gamma)$ the rotation transforming the center-of-mass frame to the body-fixed frame, one has $\boldsymbol{R}_j = R(\alpha, \beta, \gamma)\boldsymbol{R}'_j$. $R(\alpha, \beta, \gamma)$ is given in Eq. (4.62), where α, β, and γ are the Euler angles and $c_\alpha = \cos\alpha$, $s_\alpha = \sin\alpha$, etc. Through a straightforward calculation, one obtains

$$\begin{aligned}
&X = R_{1x} + iR_{1y} = \xi_1^{1/2} e^{i\alpha} s_\beta, \\
&Y = R_{2x} + iR_{2y} = \left(\Omega_2/\xi_1\right)^{1/2} e^{i\alpha} \left(c_\beta c_\gamma + i s_\gamma\right) + \xi_2 \xi_1^{-1/2} e^{i\alpha} s_\beta, \\
&R_{1z} = \xi_1^{1/2} c_\beta, \qquad R_{2z} = -\left(\Omega_2/\xi_1\right)^{1/2} s_\beta c_\gamma + \xi_2 \xi_1^{-1/2} c_\beta, \\
&Z = \left(R_{1x} + iR_{1y}\right) R_{2z} - R_{1z} \left(R_{2x} + iR_{2y}\right) = -\Omega_2^{1/2} e^{i\alpha} \left(c_\gamma + i c_\beta s_\gamma\right), \\
&R_{jx} + iR_{jy} = \Omega_2^{-1} \left\{-\omega_j X + \Omega_j Y - i\zeta_j Z\right\}, \\
&\left(R_{jx} + iR_{jy}\right) R_{kz} - R_{jz} \left(R_{kx} + iR_{ky}\right) \\
&\quad = \Omega_2^{-1} \left\{ i \left(\eta_j \zeta_k - \eta_k \zeta_j\right) X - i \left(\xi_j \zeta_k - \xi_k \zeta_j\right) Y + \left(\xi_j \eta_k - \xi_k \eta_j\right) Z \right\}.
\end{aligned} \tag{4.240}$$

Namely, the components of the Jacobi coordinate vectors $\boldsymbol{R}_j$ can be expressed as a linear combination of the components of $\boldsymbol{R}_1$ and $\boldsymbol{R}_2$, where the coefficients depend only on the internal variables. It is reasonable because $\boldsymbol{R}_1$ and $\boldsymbol{R}_2$ determine the body-fixed frame completely. Thus, each harmonic polynomial $\boldsymbol{Y}_\ell^\ell(\boldsymbol{R}_j)$ can be expressed as a combination of $Q_q^{\ell\lambda}(\boldsymbol{R}_1, \boldsymbol{R}_2)$ with the coefficients depending on the internal variables. This means that the generalized harmonic polynomials $Q_q^{\ell\lambda}(\boldsymbol{R}_1, \boldsymbol{R}_2)$ given in Eq. (4.231) do constitute a complete set of independent basis eigenfunctions with the given angular momentum ℓ for a quantum n-body system, just like they do for a quantum three-body system. A more direct proof for this can be found in [Gu et al. (2001b)].

Therefore, any function $\Psi_\ell^{\ell\lambda}(\boldsymbol{R}_1, \ldots, \boldsymbol{R}_{n-1})$ with the angular momentum ℓ and the parity $(-1)^{\ell+\lambda}$ in a quantum n-body system can be expanded with respect to the generalized harmonic polynomials $Q_q^{\ell\tau}(\boldsymbol{R}_1, \boldsymbol{R}_2)$, where the coefficients $\phi_{q\tau}^{\ell\lambda}(\xi, \eta, \zeta)$ depend on $(3n-6)$ internal variables:

$$\begin{aligned}
&\Psi_\ell^{\ell\lambda}(\boldsymbol{R}_1, \ldots, \boldsymbol{R}_{n-1}) = \sum_{\tau=0}^{1} \sum_{q=\tau}^{\ell} \phi_{q\tau}^{\ell\lambda}(\xi, \eta, \zeta) Q_q^{\ell\tau}(\boldsymbol{R}_1, \boldsymbol{R}_2), \\
&\phi_{q\tau}^{\ell\lambda}(\xi, \eta, \zeta) = \phi_{q\tau}^{\ell\lambda}(\xi_1, \ldots, \xi_{n-1}, \eta_2, \ldots, \eta_{n-1}, \zeta_3, \ldots, \zeta_{n-1}), \\
&\phi_{q\tau}^{\ell\lambda}(\xi, \eta, -\zeta) = (-1)^{\lambda-\tau} \phi_{q\tau}^{\ell\lambda}(\xi, \eta, \zeta).
\end{aligned} \tag{4.241}$$

The last equality means that the parity of $\phi_{q\tau}^{\ell\lambda}(\xi,\eta,\zeta)$ is $(-1)^{\lambda-\tau}$.

The expansion (4.241) has two important characteristics, which make it easier to derive the radial equations. One is that the generalized harmonic polynomial $Q_q^{\ell\tau}(\boldsymbol{R}_1,\boldsymbol{R}_2)$ is a homogeneous polynomial in the components of two Jacobi coordinate vectors $\boldsymbol{R}_1$ and $\boldsymbol{R}_2$, where the Euler angles do not appear explicitly. The other is the well chosen internal variables (4.235), where the internal variables ζ_j have odd parity. It is due to the existence of ζ_j that $Q_q^{\ell 0}(\boldsymbol{R}_1,\boldsymbol{R}_2)$ and $Q_q^{\ell 1}(\boldsymbol{R}_1,\boldsymbol{R}_2)$ appear together in the expansion (4.241) of the wave function.

When substituting Eq. (4.241) into the Schrödinger equation (4.221)

$$\nabla^2 \Psi_\ell^{\ell\lambda}(\boldsymbol{R}_1,\ldots,\boldsymbol{R}_{n-1}) = -2(E-V)\Psi_\ell^{\ell\lambda}(\boldsymbol{R}_1,\ldots,\boldsymbol{R}_{n-1}). \tag{4.242}$$

The action of the Laplace operator to the function $\Psi_\ell^{\ell\lambda}(\boldsymbol{R}_1,\ldots,\boldsymbol{R}_{n-1})$ is separated into three parts. The first is its action on the radial functions $\phi_{q\tau}^{\ell\lambda}(\xi,\eta,\zeta)$ which can be calculated by the replacement of variables directly,

$$\begin{aligned}
\nabla^2\phi_{q\tau}^{\ell\lambda}(\xi,\eta,\zeta) &= \Big\{4\xi_1\partial_{\xi_1}^2 + 4\eta_2\partial_{\eta_2}^2 + (\xi_1+\eta_2)\,\partial_{\xi_2}^2 \\
&+ 4\xi_2\left(\partial_{\xi_1}+\partial_{\eta_2}\right)\partial_{\xi_2} + 6\left(\partial_{\xi_1}+\partial_{\eta_2}\right) + \sum_{j=3}^{n-1}\Big[\xi_1\partial_{\xi_j}^2 + \eta_2\partial_{\eta_j}^2 \\
&+ \Omega_2\partial_{\zeta_j}^2 + 2\xi_2\partial_{\xi_j}\partial_{\eta_j} + 4\left(\xi_j\partial_{\xi_j}+\zeta_j\partial_{\zeta_j}\right)\partial_{\xi_1} \\
&+ 4\left(\eta_j\partial_{\eta_j}+\zeta_j\partial_{\zeta_j}\right)\partial_{\eta_2} + 2\left(\eta_j\partial_{\xi_j}+\xi_j\partial_{\eta_j}\right)\partial_{\xi_2}\Big] \\
&+ \Omega_2^{-1}\sum_{j,k=3}^{n-1}\Big[(\Omega_j\eta_k - \omega_j\xi_k + \zeta_j\zeta_k)\left(\partial_{\xi_j}\partial_{\xi_k}+\partial_{\eta_j}\partial_{\eta_k}\right) \\
&- 2\left(\omega_j\zeta_k - \omega_k\zeta_j\right)\partial_{\xi_j}\partial_{\zeta_k} + 2\left(\Omega_j\zeta_k - \Omega_k\zeta_j\right)\partial_{\eta_j}\partial_{\zeta_k} \\
&+ \left(\Omega_j\Omega_k + \omega_j\omega_k + \xi_1\zeta_j\zeta_k + \eta_2\zeta_j\zeta_k\right)\partial_{\zeta_j}\partial_{\zeta_k}\Big]\Big\}\phi_{q\tau}^{\ell\lambda}(\xi,\eta,\zeta).
\end{aligned} \tag{4.243}$$

The second is its action on the generalized harmonic polynomials $Q_q^{\ell\tau}(\boldsymbol{R}_1,\boldsymbol{R}_2)$ which is vanishing because $Q_q^{\ell\tau}(\boldsymbol{R}_1,\boldsymbol{R}_2)$ satisfies the Laplace equation. The third is the mixed application

$$\begin{aligned}
&2\Big\{\left(\partial_{\xi_1}\phi_{q\tau}^{\ell\lambda}\right)2\boldsymbol{R}_1 + \left(\partial_{\xi_2}\phi_{q\tau}^{\ell\lambda}\right)\boldsymbol{R}_2 + \sum_{j=3}^{n-1}\Big[\left(\partial_{\xi_j}\phi_{q\tau}^{\ell\lambda}\right)\boldsymbol{R}_j \\
&+ \left(\partial_{\zeta_j}\phi_{q\tau}^{\ell\lambda}\right)(\boldsymbol{R}_2\times\boldsymbol{R}_j)\Big]\Big\}\cdot\nabla_{\boldsymbol{R}_1}Q_q^{\ell\tau} + 2\Big\{\left(\partial_{\xi_2}\phi_{q\tau}^{\ell\lambda}\right)\boldsymbol{R}_1 + \left(\partial_{\eta_2}\phi_{q\tau}^{\ell\lambda}\right)2\boldsymbol{R}_2 \\
&+ \sum_{j=3}^{n-1}\Big[\left(\partial_{\eta_j}\phi_{q\tau}^{\ell\lambda}\right)\boldsymbol{R}_j + \left(\partial_{\zeta_j}\phi_{q\tau}^{\ell\lambda}\right)(\boldsymbol{R}_j\times\boldsymbol{R}_1)\Big]\Big\}\cdot\nabla_{\boldsymbol{R}_2}Q_q^{\ell\tau}.
\end{aligned}$$

In terms of Eqs. (4.231) and (4.240) one obtains

$$
\begin{aligned}
&\boldsymbol{R}_1 \cdot \nabla_{\boldsymbol{R}_1} Q_q^{\ell\tau} = q Q_q^{\ell\tau}, \qquad \boldsymbol{R}_2 \cdot \nabla_{\boldsymbol{R}_2} Q_q^{\ell\tau} = (\ell - q + \tau)\, Q_q^{\ell\tau}, \\
&\boldsymbol{R}_2 \cdot \nabla_{\boldsymbol{R}_1} Q_q^{\ell\tau} = (\ell - q + 1)\, Q_{q-1}^{\ell\tau}, \qquad \boldsymbol{R}_1 \cdot \nabla_{\boldsymbol{R}_2} Q_q^{\ell\tau} = (q - \tau + 1)\, Q_{q+1}^{\ell\tau}, \\
&\boldsymbol{R}_j \cdot \nabla_{\boldsymbol{R}_1} Q_q^{\ell 0} = \Omega_2^{-1} \left\{ -\omega_j q Q_q^{\ell 0} + \Omega_j (\ell - q + 1)\, Q_{q-1}^{\ell 0} - i\zeta_j Q_q^{\ell 1} \right\}, \\
&\boldsymbol{R}_j \cdot \nabla_{\boldsymbol{R}_2} Q_q^{\ell 0} = \Omega_2^{-1} \left\{ -\omega_j (q+1) Q_{q+1}^{\ell 0} + \Omega_j (\ell - q)\, Q_q^{\ell 0} - i\zeta_j Q_{q+1}^{\ell 1} \right\}, \\
&\boldsymbol{R}_j \cdot \nabla_{\boldsymbol{R}_1} Q_q^{\ell 1} = \Omega_2^{-1} \left\{ -i\eta_2 \zeta_j q^2 Q_q^{\ell 0} + i\xi_2 \zeta_j (2q-1)(\ell - q + 1) Q_{q-1}^{\ell 0} \right. \\
&\quad \left. - i\xi_1 \zeta_j (\ell - q + 2)(\ell - q + 1) Q_{q-2}^{\ell 0} - \omega_j q Q_q^{\ell 1} + \Omega_j (\ell - q + 1) Q_{q-1}^{\ell 1} \right\}, \\
&\boldsymbol{R}_j \cdot \nabla_{\boldsymbol{R}_2} Q_q^{\ell 1} = \Omega_2^{-1} \left\{ -i\eta_2 \zeta_j (q+1) q Q_{q+1}^{\ell 0} + i\xi_2 \zeta_j q (2\ell - 2q + 1) Q_q^{\ell 0} \right. \\
&\quad \left. - i\xi_1 \zeta_j (\ell - q + 1)^2 Q_{q-1}^{\ell 0} - \omega_j q Q_{q+1}^{\ell 1} + \Omega_j (\ell - q + 1) Q_q^{\ell 1} \right\}, \\
&(\boldsymbol{R}_2 \times \boldsymbol{R}_j) \cdot \nabla_{\boldsymbol{R}_1} Q_q^{\ell 0} = \Omega_2^{-1} \left\{ \eta_2 \zeta_j q Q_q^{\ell 0} - \xi_2 \zeta_j (\ell - q + 1)\, Q_{q-1}^{\ell 0} - i\omega_j Q_q^{\ell 1} \right\}, \\
&(\boldsymbol{R}_j \times \boldsymbol{R}_1) \cdot \nabla_{\boldsymbol{R}_2} Q_q^{\ell 0} = \Omega_2^{-1} \left\{ -\xi_2 \zeta_j (q+1) Q_{q+1}^{\ell 0} + \xi_1 \zeta_j (\ell - q)\, Q_q^{\ell 0} \right. \\
&\qquad \left. + i\Omega_j Q_{q+1}^{\ell 1} \right\}, \\
&(\boldsymbol{R}_2 \times \boldsymbol{R}_j) \cdot \nabla_{\boldsymbol{R}_1} Q_q^{\ell 1} = \Omega_2^{-1} \left\{ -i\eta_2 \omega_j q^2 Q_q^{\ell 0} + i\xi_2 \omega_j (2q-1)\, (\ell - q + 1)\, Q_{q-1}^{\ell 0} \right. \\
&\quad \left. - i\xi_1 \omega_j (\ell - q + 2)(\ell - q + 1) Q_{q-2}^{\ell 0} + \eta_2 \zeta_j q Q_q^{\ell 1} - \xi_2 \zeta_j (\ell - q + 1)\, Q_{q-1}^{\ell 1} \right\}, \\
&(\boldsymbol{R}_j \times \boldsymbol{R}_1) \cdot \nabla_{\boldsymbol{R}_2} Q_q^{\ell 1} = \Omega_2^{-1} \left\{ i\eta_2 \Omega_j (q+1) q Q_{q+1}^{\ell 0} - i\xi_2 \Omega_j q\, (2\ell - 2q + 1)\, Q_q^{\ell 0} \right. \\
&\quad \left. + i\xi_1 \Omega_j (\ell - q + 1)^2 Q_{q-1}^{\ell 0} - \xi_2 \zeta_j q Q_{q+1}^{\ell 1} + \xi_1 \zeta_j (\ell - q + 1)\, Q_q^{\ell 1} \right\}.
\end{aligned}
$$

Now, the radial equations are

$$
\begin{aligned}
&\triangle \phi_{q0}^{\ell\lambda} + 4 \left\{ q\partial_{\xi_1} + (\ell - q)\partial_{\eta_2} \right\} \phi_{q0}^{\ell\lambda} + 2q\partial_{\xi_2} \phi_{(q-1)0}^{\ell\lambda} + 2(\ell - q)\partial_{\xi_2} \phi_{(q+1)0}^{\ell\lambda} \\
&\quad + \sum_{j=3}^{n-1} 2\Omega_2^{-1} \Big\{ \left[-\omega_j q \partial_{\xi_j} + \Omega_j (\ell - q)\partial_{\eta_j} + \eta_2 \zeta_j q \partial_{\zeta_j} + \xi_1 \zeta_j (\ell - q)\partial_{\zeta_j} \right] \phi_{q0}^{\ell\lambda} \\
&\quad - q \left[\omega_j \partial_{\eta_j} + \xi_2 \zeta_j \partial_{\zeta_j} \right] \phi_{(q-1)0}^{\ell\lambda} + (\ell - q) \left[\Omega_j \partial_{\xi_j} - \xi_2 \zeta_j \partial_{\zeta_j} \right] \phi_{(q+1)0}^{\ell\lambda} \\
&\quad - i\eta_2 q(q-1) \left[\zeta_j \partial_{\eta_j} - \Omega_j \partial_{\zeta_j} \right] \phi_{(q-1)1}^{\ell\lambda} - iq \left[\eta_2 \zeta_j q \partial_{\xi_j} \right. \\
&\quad \left. - \xi_2 \zeta_j (2\ell - 2q + 1)\partial_{\eta_j} + \eta_2 \omega_j q \partial_{\zeta_j} + \xi_2 \Omega_j (2\ell - 2q + 1)\partial_{\zeta_j} \right] \phi_{q1}^{\ell\lambda} \\
&\quad + i(\ell - q) \left[\xi_2 \zeta_j (2q+1)\partial_{\xi_j} - \xi_1 \zeta_j (\ell - q)\partial_{\eta_j} + \xi_2 \omega_j (2q+1)\partial_{\zeta_j} \right. \\
&\quad \left. + \xi_1 \Omega_j (\ell - q)\partial_{\zeta_j} \right] \phi_{(q+1)1}^{\ell\lambda} - i\xi_1 (\ell - q)(\ell - q - 1) \left[\zeta_j \partial_{\xi_j} + \omega_j \partial_{\zeta_j} \right] \phi_{(q+2)1}^{\ell\lambda} \Big\} \\
&\quad = -2 \left[E - V \right] \phi_{q0}^{\ell\lambda},
\end{aligned}
$$

$$\begin{aligned}
&\triangle\phi^{\ell\lambda}_{q1} + 4\left\{q\partial_{\xi_1} + (\ell-q+1)\partial_{\eta_2}\right\}\phi^{\ell\lambda}_{q1} + 2(q-1)\partial_{\xi_2}\phi^{\ell\lambda}_{(q-1)1} \\
&\quad + 2(\ell-q)\partial_{\xi_2}\phi^{\ell\lambda}_{(q+1)1} + \sum_{j=3}^{n-1} 2\Omega_2^{-1}\Big\{\left[-\omega_j q\partial_{\xi_j} + \Omega_j(\ell-q+1)\partial_{\eta_j}\right. \\
&\quad \left.+\eta_2\zeta_j q\partial_{\zeta_j} + \xi_1\zeta_j(\ell-q+1)\partial_{\zeta_j}\right]\phi^{\ell\lambda}_{q1} \\
&\quad - (q-1)\left[\omega_j\partial_{\eta_j} + \xi_2\zeta_j\partial_{\zeta_j}\right]\phi^{\ell\lambda}_{(q-1)1} + (\ell-q)\left[\Omega_j\partial_{\xi_j} - \xi_2\zeta_j\partial_{\zeta_j}\right]\phi^{\ell\lambda}_{(q+1)1} \\
&\quad - i\left[\zeta_j\partial_{\eta_j} - \Omega_j\partial_{\zeta_j}\right]\phi^{\ell\lambda}_{(q-1)0} - i\left[\zeta_j\partial_{\xi_j} + \omega_j\partial_{\zeta_j}\right]\phi^{\ell\lambda}_{q0}\Big\} \\
&\quad = -2\left[E-V\right]\phi^{\ell\lambda}_{q1},
\end{aligned} \tag{4.244}$$

where $\triangle\phi^{\ell\lambda}_{q\tau}$ was given in Eq. (4.243). When $n=3$, Eq. (4.244) reduces to Eq. (4.234), where the radial functions $\phi^{\ell\lambda}_{q\tau}(\xi,\eta,\zeta)$ with $\lambda \neq \tau$ have to be vanishing because all internal variables have even parity.

4.10 Exercises

1. Prove the preliminary formula by induction:

$$\begin{aligned}
e^{\alpha}\beta e^{-\alpha} &= \beta + \frac{1}{1!}\left[\alpha,\ \beta\right] + \frac{1}{2!}\left[\alpha,\ \left[\alpha,\ \beta\right]\right] + \cdots \\
&= \sum_{n=0}^{\infty}\frac{1}{n!}\overbrace{\left[\alpha,\ \left[\alpha,\ \cdots\ \left[\alpha,\ \beta\right]\ \cdots\ \right]\right]}^{n},
\end{aligned}$$

where α and β are two matrices with the same dimension. Then, show Eq. (4.38) and prove that SU(2) is homomorphic onto SO(3).

2. Expand $R(\hat{\boldsymbol{n}},\omega) = \exp\left(-i\omega\hat{\boldsymbol{n}}\cdot\boldsymbol{T}\right)$ as a sum of matrices with the finite terms.

Hint: $\left(\hat{\boldsymbol{n}}\cdot\boldsymbol{T}\right)^3 = \hat{\boldsymbol{n}}\cdot\boldsymbol{T}$.

3. Check the following formulas in the group **O** [see the notation given in §2.5.1] in terms of the homomorphism of SU(2) onto SO(3):

$$T_zR_1 = S_3, \qquad T_zT_x = R_1, \qquad R_1R_2 = R_3^2.$$

4. Prove the following formulas for the group **I** in terms of the homomorphism of SU(2) onto SO(3) (see the notation given in §2.5.4)

$$R_6 = S_1T_0^2S_1T_0^{-1}, \qquad S_{12} = T_0S_1T_0^3R_6,$$

which were used in Prob. 21 of Chap. 3.

5. Calculate the Euler angles for the following transformation matrices R, S, and T, and write their representation matrices in D^j of SO(3):

$$\text{(a)}\quad R(\alpha,\beta,\gamma)=\frac{1}{4}\begin{pmatrix} -\sqrt{3}-2 & \sqrt{3}-2 & -\sqrt{2} \\ \sqrt{3}-2 & -\sqrt{3}-2 & \sqrt{2} \\ -\sqrt{2} & \sqrt{2} & 2\sqrt{3} \end{pmatrix},$$

$$\text{(b)}\quad S(\alpha,\beta,\gamma)=\frac{1}{8}\begin{pmatrix} \sqrt{6}+2\sqrt{3} & 3\sqrt{2}-2 & 2\sqrt{6} \\ \sqrt{2}-6 & \sqrt{6}+2\sqrt{3} & 2\sqrt{2} \\ -2\sqrt{2} & -2\sqrt{6} & 4\sqrt{2} \end{pmatrix},$$

$$\text{(c)}\quad T(\alpha,\beta,\gamma)=\frac{1}{2}\begin{pmatrix} \sqrt{3} & -1 & 0 \\ -1 & -\sqrt{3} & 0 \\ 0 & 0 & -2 \end{pmatrix}.$$

6. Calculate the Euler angles for the following rotations R, S, and T, and write their representation matrices in D^j of SO(3).
(a) R is a rotation around the direction $\hat{\boldsymbol{n}}=\boldsymbol{e}_1\sin\theta+\boldsymbol{e}_3\cos\theta$ through an acute angle θ;
(b) S is a rotation around the direction $\hat{\boldsymbol{n}}=(\boldsymbol{e}_1+\boldsymbol{e}_2+\boldsymbol{e}_3)/\sqrt{3}$ through $2\pi/3$;
(c) T is a rotation around the direction $\hat{\boldsymbol{n}}=(\boldsymbol{e}_1+\boldsymbol{e}_2)/\sqrt{2}$ through π.

7. Calculate the Euler angles for the rotations T_0, T_2, R_1, R_2, R_6, S_1, S_2, S_6, S_{11}, and S_{12} in the icosahedron group **I**, where the notations for the elements are given in §2.5.4.

8. Express the representation matrix $D^j(\hat{\boldsymbol{n}},\omega)$ of a rotation around the direction $\hat{\boldsymbol{n}}(\theta,\varphi)$ through ω in terms of $D^j(\boldsymbol{e}_3,\alpha)$ and $d^j(\beta)$.

9. Calculate all the matrix entries $d^j_{\nu\mu}(\omega)$ by Eq. (4.74), where $j=1/2$, 1, 3/2, 2, 5/2, and 3.

10. Reduce the subduced representation from the irreducible representation D^3 of SO(3) with respect to the subgroup D_3 and find the similarity transformation matrix.

11. Reduce the subduced representations from the irreducible representations D^{20} and D^{18} of SO(3) with respect to the subgroup **I** (the proper symmetry group of the icosahedron), respectively.

12. Reduce the subduced representations from the irreducible representations D^1, D^2, and D^3 of SO(3) with respect to the subgroup **I**, respectively, and calculate the similarity transformation matrices. From the results, calculate the polar angles of the axes in the icosahedron.

13. Prove that the elements $u(\hat{\boldsymbol{n}}, \omega)$ with the same ω form a class of the SU(2) group.

14. Calculate the representation matrices of the generators in an irreducible representation of SU(2) by the second Lie theorem.

15. For any one-order Lie group with the composition function $f(r; s)$, please try to find a new parameter r' such that the new composition function is the additive function, $f'(r'; s') = r' + s'$. The Lorentz transformation $A(v)$ for the boost along the z-axis with the relative velocity v is taken in the following form. The set of them forms a one-order Lie group:

$$A(v) = \begin{pmatrix} 1 & 0 & 0 & 0 \\ 0 & 1 & 0 & 0 \\ 0 & 0 & \gamma & -i\gamma v/c \\ 0 & 0 & i\gamma v/c & \gamma \end{pmatrix}, \qquad \begin{array}{l} \gamma = \left(1 - v^2/c^2\right)^{-1/2}, \\ f(v_1; v_2) = \dfrac{v_1 + v_2}{1 + v_1 v_2/c^2}. \end{array}$$

Find the new parameter with the additive composition function.

16. Directly calculate the Clebsch–Gordan coefficients for the direct product representation of two irreducible representations of SU(2) in terms of the raising and lowering operators $J_\pm$: (a) $D^{1/2} \times D^{1/2}$, (b) $D^{1/2} \times D^1$, (c) $D^1 \times D^1$, (d) $D^1 \times D^{3/2}$.

17. Prove two sets of formulas for the Clebsch–Gordan coefficients in terms of the raising and lowering operators $J_\pm$:
(a) In the reduction of $D^{1/2} \times D^j$,

$$\begin{aligned} \|(j+1/2), M\rangle = & \left(\frac{j+M+1/2}{2j+1}\right)^{1/2} |1/2, 1/2\rangle |j, M-1/2\rangle \\ & + \left(\frac{j-M+1/2}{2j+1}\right)^{1/2} |1/2, -1/2\rangle |j, M+1/2\rangle, \\ \|(j-1/2), M\rangle = & \left(\frac{j-M+1/2}{2j+1}\right)^{1/2} |1/2, 1/2\rangle |j, M-1/2\rangle \\ & - \left(\frac{j+M+1/2}{2j+1}\right)^{1/2} |1/2, -1/2\rangle |j, M+1/2\rangle. \end{aligned}$$

(b) In the reduction of $D^1 \times D^j$,

$$\begin{aligned}
\|(j+1), M\rangle =& \left\{\frac{(j+M)(j+M+1)}{2(2j+1)(j+1)}\right\}^{1/2} |1,1\rangle|j, M-1\rangle \\
&+\left\{\frac{(j-M+1)(j+M+1)}{(2j+1)(j+1)}\right\}^{1/2} |1,0\rangle|j, M\rangle \\
&+\left\{\frac{(j-M)(j-M+1)}{2(2j+1)(j+1)}\right\}^{1/2} |1,-1\rangle|j, M+1\rangle, \\
\|j, M\rangle =& \left\{\frac{(j+M)(j-M+1)}{2j(j+1)}\right\}^{1/2} |1,1\rangle|j, M-1\rangle \\
&-\frac{M}{[j(j+1)]^{1/2}}|1,0\rangle|j, M\rangle \\
&-\left\{\frac{(j-M)(j+M+1)}{2j(j+1)}\right\}^{1/2} |1,-1\rangle|j, M+1\rangle, \\
\|(j-1), M\rangle =& \left\{\frac{(j-M)(j-M+1)}{2j(2j+1)}\right\}^{1/2} |1,1\rangle|j, M-1\rangle \\
&-\left\{\frac{(j-M)(j+M)}{j(2j+1)}\right\}^{1/2} |1,0\rangle|j, M\rangle \\
&+\left\{\frac{(j+M)(j+M+1)}{2j(2j+1)}\right\}^{1/2} |1,-1\rangle|j, M+1\rangle.
\end{aligned}$$

18. Calculate the eigenfunctions of the total spinor angular momentum in a three-electron system.

19. The spherical harmonic function $Y^{\ell}_m(\hat{\boldsymbol{n}})$ belongs to the mth row of the representation D^{ℓ} of SO(3) so that it is the eigenfunction of the orbital angular momentum operator L_3 with the eigenvalue m. Calculate the eigenfunction with the eigenvalue m of the orbital angular momentum operator $\boldsymbol{L} \cdot \hat{\boldsymbol{a}}$ along the direction $\hat{\boldsymbol{a}} = (\boldsymbol{e}_1 - \boldsymbol{e}_2)/\sqrt{2}$ in terms of combining $Y^{\ell}_m(\hat{\boldsymbol{n}})$ linearly.

20. Let the function $\psi^{\ell}_m(x)$ belong to the mth row of the irreducible representation D^{ℓ} of SO(3). Calculate the eigenfunction with the eigenvalue m of the orbital angular momentum operator $\boldsymbol{L} \cdot \hat{\boldsymbol{b}}$ along the direction $\hat{\boldsymbol{b}} = (\sqrt{3}\boldsymbol{e}_2 + \boldsymbol{e}_3)/2$ in terms of combining $\psi^{\ell}_m(x)^*$ linearly.

Hint: Use the similarity transformation between the representation D^j of SO(3) and its complex conjugate representation.

21. Q_R is the rotational transformation operator in the spinor space. In the rotation Q_R, the basis spinor $e^{(s)}(\rho)$ belongs to the ρth row of the irreducible representation D^s, so that it is the common eigenfunction of the spinor angular momentum S^2 and S_3 with the eigenvalues $s(s+1)$ and ρ, respectively. Based on this property, calculate the eigenfunction of the spinor angular momentum $\boldsymbol{S} \cdot \hat{\boldsymbol{r}}$ along the radial direction, where $\hat{\boldsymbol{r}}$ is the unit vector in the radial direction.

22. There are three sets of the mutual commutable angular momentum operators. One set consists of L^2, L_3, S^2, and S_3. The other set consists of J^2, J_3, L^2, and S^2, and the third set consists of J^2, J_3, S^2, and $\boldsymbol{S} \cdot \hat{\boldsymbol{r}}$. Calculate the common eigenfunctions of the three sets of operators, respectively.

23. Calculate $\left\{d^{\ell}(\theta)\left(I_3^{\ell}\right)^2 d^{\ell}(\theta)^{-1}\right\}_{mm}$, where $d^{\ell}(\theta)$ is the representation matrix of $R(\boldsymbol{e}_2, \theta)$ in D^{ℓ} of SO(3) and I_3^{ℓ} is the third generator in the representation.

Hint: Use the property of the adjoint representation.

24. Establish the differential equation satisfied by the matrix entries $D^j_{\nu\mu}(\alpha, \beta, \gamma)$ of the representation D^j of SO(3).

25. Discuss all inequivalent and irreducible unitary representations of the SO(3) group and the SO(2,1) group.

Chapter 5

SYMMETRY OF CRYSTALS

The study on the symmetry of crystals is a typical example of the physical application of group theory. Through a systematic study by group theory only based on the translation symmetry of crystals, the crystals are classified completely that there are 11 proper crystallographic point groups, 32 crystallographic point groups, 7 crystal systems, 14 Bravais lattices, 73 symmorphic space groups, and 230 space groups. In this chapter we will study the symmetry group of crystals, their representations, and the classification of crystals.

5.1 Symmetric Group of Crystals

The fundamental character of a crystal is the spatial periodic array of the atoms composing the crystal, called the crystal lattice. By the periodic boundary condition, the crystal is invariant in the following translation (see [Ren (2006)] for the crystals of finite size):

$$\boldsymbol{r} \longrightarrow T(\boldsymbol{\ell})\boldsymbol{r} = \boldsymbol{r} + \boldsymbol{\ell}, \tag{5.1}$$

where $\boldsymbol{\ell}$ is called the vector of crystal lattice. Three fundamental periods $\boldsymbol{a}_j$ of a crystal lattice, which are not coplanar, are taken to be the basis vectors of crystal lattice, or briefly called the lattice bases. The lattice bases are said to be primitive if each vector of crystal lattice is an integral combination of the lattice bases. For simplicity, we only use the primitive lattice bases if without special notification. For three chosen lattice bases $\boldsymbol{a}_i$, a vector of crystal lattice is characterized by three integers ℓ_i:

$$\boldsymbol{\ell} = \boldsymbol{a}_1\ell_1 + \boldsymbol{a}_2\ell_2 + \boldsymbol{a}_3\ell_3 = \sum_{i=1}^{3} \boldsymbol{a}_i\ell_i, \qquad \ell_i \text{ are integers.} \tag{5.2}$$

The multiplication of two translations is defined to be a translation where two translation vectors are added. The set of all translations $T(\boldsymbol{\ell})$ which preserve the crystal invariant forms an Abelian group, called the translation group $\mathcal{T}$ of the crystal.

Usually, in addition to the translation symmetry, a crystal also reserves invariant under some other symmetric operations composed of the spatial inversion, the rotation, and the translation. A general symmetric operation is denoted by $g(R, \boldsymbol{\alpha})$,

$$\boldsymbol{r} \longrightarrow g(R, \boldsymbol{\alpha})\boldsymbol{r} = R\boldsymbol{r} + \boldsymbol{\alpha}, \tag{5.3}$$

where $R \in \mathrm{O}(3)$ is a proper or improper rotation, and $\boldsymbol{\alpha}$ is a translation vector, not necessary to be a vector of crystal lattice $\boldsymbol{\ell}$. If $\boldsymbol{\alpha} = 0$, $g(R, 0) = R$ is a proper or improper rotation which preserves the origin invariant. If $R = E$, $\boldsymbol{\alpha}$ has to be a vector of crystal lattice $\boldsymbol{\ell}$ and $g(E, \boldsymbol{\ell}) = T(\boldsymbol{\ell})$. The multiplication of two symmetric operations is defined as their successive applications,

$$g(R, \boldsymbol{\alpha})g(R', \boldsymbol{\beta})\boldsymbol{r} = g(R, \boldsymbol{\alpha})\{R'\boldsymbol{r} + \boldsymbol{\beta}\} = RR'\boldsymbol{r} + \boldsymbol{\alpha} + R\boldsymbol{\beta},$$

$$g(R, \boldsymbol{\alpha})g(R', \boldsymbol{\beta}) = g(RR', \boldsymbol{\alpha} + R\boldsymbol{\beta}). \tag{5.4}$$

The inverse of $g(R, \boldsymbol{\alpha})$ is

$$g(R, \boldsymbol{\alpha})^{-1} = g(R^{-1}, -R^{-1}\boldsymbol{\alpha}). \tag{5.5}$$

The set of all symmetric operations $g(R, \boldsymbol{\alpha})$ for a crystal with the multiplication rule (5.4) forms a group $\mathcal{S}$, called the space group of the crystal. In the multiplication (5.4) of two symmetric operations, the rotational part obeys the multiplication rule of two rotations, but the translational part is affected by the rotational part. Therefore, the set of the rotational parts R in $g(R, \boldsymbol{\alpha})$ forms a group G, called the crystallographic point group.

Removing the vector of crystal lattice $\boldsymbol{\ell}$ from $\boldsymbol{\alpha}$, the general symmetric operation can be expressed as

$$\begin{aligned} &g(R, \boldsymbol{\alpha}) = T(\boldsymbol{\ell})g(R, \boldsymbol{t}), \qquad \boldsymbol{\alpha} = \boldsymbol{\ell} + \boldsymbol{t} \\ &\boldsymbol{t} = \sum_{j=1}^{3} \boldsymbol{a}_j t_j, \qquad 0 \leq t_j < 1. \end{aligned} \tag{5.6}$$

For a given crystal with the space group $\mathcal{S}$, it is easy to show by reduction to absurdity that $\boldsymbol{t}$ in the symmetric operation $g(R, \boldsymbol{t})$ depends upon R

uniquely. In fact, if $g(R, \boldsymbol{t})$ and $g(R, \boldsymbol{t}')$ are both the symmetric operations of the crystal,

$$g(R, \boldsymbol{t})^{-1} g(R, \boldsymbol{t}') = T(-R^{-1}\boldsymbol{t} + R^{-1}\boldsymbol{t}') = T(\boldsymbol{\ell}),$$
$$\boldsymbol{t}' - \boldsymbol{t} = R\boldsymbol{\ell} = \boldsymbol{\ell}'.$$

Due to the restriction (5.6) on $\boldsymbol{t}$, $\boldsymbol{t}' = \boldsymbol{t}$.

Generally, R is not an element of $\mathcal{S}$, and G is not a subgroup of $\mathcal{S}$. The space group $\mathcal{S}$ is called the symmorphic space group if G is the subgroup of $\mathcal{S}$. Namely, in a symmorphic space group, $\boldsymbol{t}$ in each $g(R, \boldsymbol{t})$ is vanishing, and any element in $\mathcal{S}$ can be expressed as $g(R, \boldsymbol{\ell}) = T(\boldsymbol{\ell})R$.

The conjugate element of a translation $T(\boldsymbol{\ell})$ is still a translation,

$$g(R, \boldsymbol{\alpha})T(\boldsymbol{\ell})g(R, \boldsymbol{\alpha})^{-1} = g\left(E, \boldsymbol{\alpha} + R(\boldsymbol{\ell} - R^{-1}\boldsymbol{\alpha})\right) = T(R\boldsymbol{\ell}),$$
$$R\boldsymbol{\ell} = \boldsymbol{\ell}'. \tag{5.7}$$

Namely, the translation group $\mathcal{T}$ is the invariant subgroup of the space group $\mathcal{S}$. Due to Eq. (5.6), the coset of $\mathcal{T}$ is completely determined by the rotation R. Thus, the quotient group of $\mathcal{T}$ with respect to $\mathcal{S}$ is the crystallographic point group,

$$G = \mathcal{T}/\mathcal{S}. \tag{5.8}$$

It will be seen that Eq. (5.7) is a fundamental constraint for the possible crystallographic point groups, the crystal systems, and the Bravais lattices.

5.2 Crystallographic Point Groups

5.2.1 *Elements in a Crystallographic Point Group*

In the crystal theory, it is convenient to choose the lattice bases $\boldsymbol{a}_j$ to be the basis vectors. The merit for this choice is that, due to Eq. (5.7), any matrix entry $D_{ij}(R)$ of R in the bases is an integer:

$$R\boldsymbol{a}_j = \sum_{i=1}^{3} \boldsymbol{a}_i D_{ij}(R) = \boldsymbol{\ell}. \tag{5.9}$$

The shortcoming is that $\mathbf{a}_j$ are generally not orthonormal and $D(R)$ is real but not orthogonal. Let $\boldsymbol{e}_a$ be the orthonormal basis vectors in the real three-dimensional space. The matrix of the rotation R in the basis vectors $\boldsymbol{e}_a$ is denoted by $\overline{D}(R)$,

$$\boldsymbol{e}_a \cdot \boldsymbol{e}_b = \delta_{ab}, \qquad R\boldsymbol{e}_a = \sum_{b=1}^{3} \boldsymbol{e}_b \overline{D}_{ba}(R). \tag{5.10}$$

$\overline{D}(R)$ is a real orthogonal matrix, related with $D(R)$ by a real similarity transformation X,

$$\boldsymbol{a}_i = \sum_{d=1}^{3} \boldsymbol{e}_d X_{di}, \qquad D(R) = X^{-1}\overline{D}(R)X. \tag{5.11}$$

Define another set of basis vectors $\mathbf{b}_j$ satisfying

$$\boldsymbol{b}_j = \sum_{a=1}^{3} \left(X^{-1}\right)_{ja} \boldsymbol{e}_a, \qquad \boldsymbol{b}_j \cdot \boldsymbol{a}_i = \delta_{ji}. \tag{5.12}$$

$\mathbf{b}_j$ is called the basis vector of the reciprocal crystal lattice, or briefly called the reciprocal lattice basis. The matrix form of R in the basis vectors $\boldsymbol{b}_i$ is

$$\begin{aligned} R\boldsymbol{b}_i &= \sum_{d=1}^{3} \left(X^{-1}\right)_{id} R\boldsymbol{e}_d = \sum_{dc} \left(X^{-1}\right)_{id} \boldsymbol{e}_c \overline{D}_{cd}(R) \\ &= \sum_j \left\{ \sum_{dc} \left(X^{-1}\right)_{id} \left[\overline{D}(R)^{-1}\right]_{dc} X_{cj} \right\} \boldsymbol{b}_j = \sum_{j=1}^{3} \left[D(R)^{-1}\right]_{ij} \boldsymbol{b}_j. \end{aligned} \tag{5.13}$$

A rotation R in the crystal theory is usually expressed in the form of a double-vector. A double-vector can be simply understood as two merged vectors, called left-vector and right-vector, respectively. Two vectors can make the usual vector calculation independently. The double-vector form of a rotation R is

$$\begin{aligned} \vec{\vec{R}} &= \sum_{ij} D_{ij}(R)\boldsymbol{a}_i\boldsymbol{b}_j = \sum_{ij} \left[D(R)^{-1}\right]_{ij} \boldsymbol{b}_j\boldsymbol{a}_i, \\ D_{ij}(R) &= \boldsymbol{b}_i \cdot \vec{\vec{R}} \cdot \boldsymbol{a}_j = \boldsymbol{a}_j \cdot \vec{\vec{R}}^{-1} \cdot \boldsymbol{b}_i. \end{aligned} \tag{5.14}$$

Equation (5.14) shows the method how to write the double-vector of R. The coefficient of $\boldsymbol{b}_j$ in $\vec{\vec{R}}$ is $R\boldsymbol{a}_j = \sum_{ij} D_{ij}(R)\boldsymbol{a}_i$. The double-vector of the identical element is

$$\vec{\vec{1}} = \sum_j \boldsymbol{a}_j\boldsymbol{b}_j = \sum_j \boldsymbol{b}_j\boldsymbol{a}_j, \tag{5.15}$$

and the double-vector of the spatial inversion σ is that multiplied with a negative sign. The double-vector of a rotation $R(\hat{\boldsymbol{n}}, \omega)$ around the direction $\hat{\boldsymbol{n}}$ through an angle ω is (see Prob. 3 of Chap. 5 in [Ma and Gu (2004)])

$$\vec{\vec{R}}(\hat{\boldsymbol{n}}, \omega) = \hat{\boldsymbol{n}}\hat{\boldsymbol{n}} + \left(\vec{\vec{1}} - \hat{\boldsymbol{n}}\hat{\boldsymbol{n}}\right)\cos\omega + \left(\vec{\vec{1}} \times \hat{\boldsymbol{n}}\right)\sin\omega. \tag{5.16}$$

$\hat{\boldsymbol{n}}\hat{\boldsymbol{n}}$ is a projective operator along the direction $\hat{\boldsymbol{n}}$, and $\left(\vec{\vec{1}} - \hat{\boldsymbol{n}}\hat{\boldsymbol{n}}\right)$ is that to the plane perpendicular to $\hat{\boldsymbol{n}}$.

From Eq. (5.9), $D_{ij}(R)$ are all integers, so is its trace:

$$\text{Tr } D(R) = \text{Tr } \overline{D}(R) = \pm(1 + 2\cos\omega) = \text{integer}, \tag{5.17}$$

where the positive sign corresponds to the proper rotation, and the negative sign to the improper one. Thus, $\cos\omega$ is a half of integer: 0, $\pm 1/2$, and ± 1,

$$\omega = 2\pi m/N, \qquad N = 1,\ 2,\ 3,\ 4,\ \text{or}\ 6, \qquad 0 \leq m < N. \tag{5.18}$$

The element in a crystallographic point group has to be an N-fold proper or improper rotation, where N is 1, 2, 3, 4, or 6.

An integral combination $\boldsymbol{K}$ of the reciprocal lattice bases $\boldsymbol{b}_j$ is called the vector of the reciprocal crystal lattice. The inner product of $\boldsymbol{K}$ and $\boldsymbol{\ell}$ is an integer and invariant in the rotation,

$$\boldsymbol{K} \cdot \boldsymbol{\ell} = (R\boldsymbol{K}) \cdot (R\boldsymbol{\ell}) = \text{integer}, \tag{5.19}$$

namely, $R\boldsymbol{K}$ is also a vector of the reciprocal crystal lattice

$$R\boldsymbol{K} = \boldsymbol{K}'. \tag{5.20}$$

Both the crystal lattice and the reciprocal crystal lattice have the same crystallographic point group.

5.2.2 *Proper Crystallographic Point Groups*

First, if a proper crystallographic point group contains only one proper rotational axis, it is a cyclic group. According to the Schoenflies notations, they are C_N, $N = 1$, 2, 3, 4, and 6. They are denoted by N in the international notations. The generator of C_N is a rotation C_N around the rotational axis $\hat{\boldsymbol{n}}$ through an angle $2\pi/N$. The direction $\hat{\boldsymbol{n}}$ of the rotational axis in C_N is usually taken to be the z-axis. The group C_N is an Abelian group with order N. There are N inequivalent one-dimensional representations

$$D^m(C_N) = \exp(-i2\pi m/N), \qquad 0 \leq m < N. \tag{5.21}$$

Second, if a proper crystallographic point group G contains two or more than two proper rotational axes, we are going to study the restriction on the number of axes. Denote by n_N the number of the N-fold proper axes contained in G, where $N = 2$, 3, 4, and 6. Except for the identical element E, there is no common element contained in the cyclic groups with different rotational axes. Thus, n_N are related to the order g of G

$$g = 1 + n_2 + 2n_3 + 3n_4 + 5n_6. \tag{5.22}$$

For each N-fold proper axis, the double-vector of the sum of elements in the cyclic group C_N is

$$\left\{\overset{\Rightarrow}{C}_N\right\} = N\hat{\boldsymbol{n}}\hat{\boldsymbol{n}}, \qquad \{C_N\} = \sum_{m=0}^{N-1} (C_N)^m, \tag{5.23}$$

because its application to $\hat{\boldsymbol{n}}$ is $N\hat{\boldsymbol{n}}$ and its application to a vector perpendicular to $\hat{\boldsymbol{n}}$ is zero. In a set of orthonormal bases including a basis vector $\hat{\boldsymbol{n}}$, Tr $D(\{C_N\}) = N$. Removing the trace of the identical element, the trace of the sum of the remaining elements in C_N is $N - 3$. Thus, the trace of the sum of all elements in G is

$$3 + (2-3)n_2 + (3-3)n_3 + (4-3)n_4 + (6-3)n_6 = 3 - n_2 + n_4 + 3n_6.$$

On the other hand, from the rearrangement theorem (Theorem 2.1), for an arbitrary vector $\boldsymbol{r}$ one has

$$S\left\{\sum_{R\in G} R\boldsymbol{r}\right\} = \left\{\sum_{R\in G} R\boldsymbol{r}\right\}, \qquad S \in G.$$

Since G contains at least two different rotation axes, the vector in the curve bracket has to be null. Thus, the sum of all elements in G is vanishing

$$3 - n_2 + n_4 + 3n_6 = 0. \tag{5.24}$$

It is another restriction on the number of axes in G.

Since SO(3) $\sim$ SU(2), the calculation of the product of two elements in SO(3) can be replaced with that in SU(2), which is much simpler. This method can be used to calculate the product of two rotations along different axes in G where the sign in $u(\hat{\boldsymbol{n}}, \omega)$ does not matter with the calculation. The fundamental calculation formulas are

$$\begin{aligned}
&u(\hat{\boldsymbol{n}}, \omega) = \mathbf{1}\cos(\omega/2) - i\,(\boldsymbol{\sigma}\cdot\hat{\boldsymbol{n}})\sin(\omega/2), \\
&\cos(\omega/2) = \frac{1}{2}\mathrm{Tr}\,\{u(\hat{\boldsymbol{n}}, \omega)\}, \\
&(\boldsymbol{\sigma}\cdot\hat{\boldsymbol{n}}_1)\,(\boldsymbol{\sigma}\cdot\hat{\boldsymbol{n}}_2) = \mathbf{1}\,(\hat{\boldsymbol{n}}_1\cdot\hat{\boldsymbol{n}}_2) + i\boldsymbol{\sigma}\cdot(\hat{\boldsymbol{n}}_1\times\hat{\boldsymbol{n}}_2),
\end{aligned} \tag{5.25}$$

where $\cos(\omega/2)$ can take only the following values

$$\cos(\omega/2) = 0, \quad \pm\frac{1}{2}, \quad \pm\frac{1}{\sqrt{2}}, \quad \pm\frac{\sqrt{3}}{2}, \quad \pm 1. \tag{5.26}$$

Third, if G contains only 2-fold proper axes, one obtains $n_2 = 3$ from Eq. (5.24). The matrix of a 2-fold proper rotation in SU(2) is

$$u(\hat{\boldsymbol{n}}, \pi) = -i\,(\boldsymbol{\sigma} \cdot \hat{\boldsymbol{n}}) .$$

The product of two 2-fold proper rotations along different axes is

$$u(\hat{\boldsymbol{n}}_1, \pi)u(\hat{\boldsymbol{n}}_2, \pi) = -\mathbf{1}\,(\hat{\boldsymbol{n}}_1 \cdot \hat{\boldsymbol{n}}_2) - i\vec{\sigma} \cdot (\hat{\boldsymbol{n}}_1 \times \hat{\boldsymbol{n}}_2) . \tag{5.27}$$

Thus, $\hat{\boldsymbol{n}}_1 \cdot \hat{\boldsymbol{n}}_2 = 0$, and $\hat{\boldsymbol{n}}_3 = \hat{\boldsymbol{n}}_1 \times \hat{\boldsymbol{n}}_2$. The three 2-fold axes are orthogonal to each other. The group G is D_2, which is an Abelian group and isomorphic onto the inversion group V_4. Its group table is given in Table 2.4 and its character table is given in Table 3.6.

Fourth, if G contains, in addition to the 2-fold axes, only one N-fold proper axis where N is larger than 2, then the N-fold axis is called the principal axis and taken to be the z-axis. The remaining elements in G are 2-fold proper rotations with the axes perpendicular to the principal axis. Due to the restriction (5.24), $n_2 = N$. The product of two 2-fold rotations has to be an N-fold rotation, $\omega = 2m\pi/N$, and due to Eq. (5.27) $\hat{\boldsymbol{n}}_1 \cdot \hat{\boldsymbol{n}}_2 = \cos(m\pi/N)$, namely, N 2-fold axes well-distribute in the x y plane, where the angle of two neighbored axes is π/N. This group G is just D_N, where $N = 3$, 4, or 6. The D_N group contains $2N$ elements whose multiplication rule is given in Eq. (2.13). Its character table is given in Tables 3.7, 3.12, and 3.13. D_N can be expressed as a product of two subgroups, $C_N C_2'$, where C_2' is the cyclic group of 2-fold proper rotations along the x-axis. D_N is denoted by $N2'$ in the international notations.

Fifth, G contains only 2-fold and 3-fold proper axes. The matrix of 3-fold rotation in SU(2) is

$$u(\hat{\boldsymbol{n}}, \pm 2\pi/3) = \mathbf{1}/2 \mp i\,(\boldsymbol{\sigma} \cdot \hat{\boldsymbol{n}})\,\sqrt{3}/2.$$

For the product of two 3-fold rotations with different axes, the cosine of half of its rotational angle is

$$\frac{1}{2}\mathrm{Tr}\,\{u(\hat{\boldsymbol{n}}_1, \pm 2\pi/3)u(\hat{\boldsymbol{n}}_2, 2\pi/3)\} = 1/4 \mp 3\,(\hat{\boldsymbol{n}}_1 \cdot \hat{\boldsymbol{n}}_2)\,/4. \tag{5.28}$$

Due to Eq. (5.26), $\hat{\boldsymbol{n}}_1 \cdot \hat{\boldsymbol{n}}_2 = \pm 1/3$. Without loss of generality, one has

$$\hat{\boldsymbol{n}}_1 \cdot \hat{\boldsymbol{n}}_2 = \cos\theta = -1/3, \qquad \theta = 109°28'. \tag{5.29}$$

Thus, from the products of 3-fold rotations one obtains four 3-fold axes well-distributed in the three-dimensional space, $n_3 = 4$. The angle of two neighbored 3-fold axes is θ. Due to Eq. (5.24), $n_2 = 3$. Each 2-fold axis has to be along the angular bisector of two neighbored 3-fold axes. Since

$$\begin{aligned} &u(\hat{\boldsymbol{n}}_1, -2\pi/3)u(\hat{\boldsymbol{n}}_2, 2\pi/3) = i\boldsymbol{\sigma} \cdot \left(\sqrt{3}\hat{\boldsymbol{n}}_1 - \sqrt{3}\hat{\boldsymbol{n}}_2 + 3\hat{\boldsymbol{n}}_1 \times \hat{\boldsymbol{n}}_2\right)/4, \\ &\hat{\boldsymbol{n}}_1 \cdot \left(\sqrt{3}\hat{\boldsymbol{n}}_1 - \sqrt{3}\hat{\boldsymbol{n}}_2 + 3\hat{\boldsymbol{n}}_1 \times \hat{\boldsymbol{n}}_2\right)/4 = 1/\sqrt{3}, \end{aligned}$$

the cosine of the angle between the neighbored 3-fold axis and 2-fold axis is $1/\sqrt{3}$. Three 2-fold axes are equivalent to each other, and four 3-fold axes all are polar. This group G is the proper symmetric group $\mathbf{T}$ of a tetrahedron. Its group table and character table are given in Tables 2.9 and 3.8. $\mathbf{T}$ can be expressed as a product of three subgroups, $C_3'C_2C_2'$. $\mathbf{T}$ is denoted by $3'22'$ in the international notations.

At last, if G contains more than one 6-fold axes, the angle between two 6-fold axes has to be $109°28'$, because a 6-fold axis is also a 3-fold axis. On the other hand, since a 6-fold axis is also a 2-fold axis, the product of two 2-fold rotations with this angle cannot satisfy the condition (5.26). Therefore, the only remaining case is that G contains more than one 4-fold proper axes. The matrices of 4-fold rotations in SU(2) are

$$u(\hat{\boldsymbol{n}}, \pm\pi/2) = \mathbf{1}/\sqrt{2} \mp i\left(\boldsymbol{\sigma} \cdot \hat{\boldsymbol{n}}\right)/\sqrt{2}, \qquad u(\hat{\boldsymbol{n}}, \pi) = -i\left(\boldsymbol{\sigma} \cdot \hat{\boldsymbol{n}}\right).$$

For a product of two 4-fold rotations along different axes, the cosine of half of its rotational angle is

$$\begin{aligned} &\mathrm{Tr}\left\{u(\hat{\boldsymbol{n}}_1, \pm\pi/2)u(\hat{\boldsymbol{n}}_2, \pi/2)\right\}/2 = 1/2 \mp \left(\hat{\boldsymbol{n}}_1 \cdot \hat{\boldsymbol{n}}_2\right)/2, \\ &\mathrm{Tr}\left\{u(\hat{\boldsymbol{n}}_1, \pi)u(\hat{\boldsymbol{n}}_2, \pi/2)\right\}/2 = -\left(\hat{\boldsymbol{n}}_1 \cdot \hat{\boldsymbol{n}}_2\right)/\sqrt{2}. \end{aligned}$$

The only solution satisfying the condition (5.26) is $\hat{\mathbf{n}}_1 \cdot \hat{\boldsymbol{n}}_2 = 0$, namely, there are three 4-fold axes perpendicular to each other. They are nonpolar and equivalent to each other. In the products of two 4-fold rotations, there are 3-fold rotations and 2-fold rotations,

$$\begin{aligned} &u(\hat{\boldsymbol{n}}_1, \pi/2)u(\hat{\boldsymbol{n}}_2, \pi/2) = \mathbf{1}/2 - i\boldsymbol{\sigma} \cdot \left(\hat{\boldsymbol{n}}_1 + \hat{\boldsymbol{n}}_2 + \hat{\boldsymbol{n}}_1 \times \hat{\boldsymbol{n}}_2\right)/2, \\ &\hat{\boldsymbol{n}}_1 \cdot \left(\hat{\boldsymbol{n}}_1 + \hat{\boldsymbol{n}}_2 + \hat{\boldsymbol{n}}_1 \times \hat{\boldsymbol{n}}_2\right)/\sqrt{3} = 1/\sqrt{3}, \\ &u(\hat{\boldsymbol{n}}_1, \pi)u(\hat{\boldsymbol{n}}_2, \pi/2) = -i\boldsymbol{\sigma} \cdot \left(\hat{\boldsymbol{n}}_1 + \hat{\boldsymbol{n}}_1 \times \hat{\boldsymbol{n}}_2\right)/\sqrt{2}. \end{aligned}$$

The cosine of the angle between the neighbored 3-fold axis and 4-fold axis is $1/\sqrt{3}$. Four 3-fold axes well-distribute around a 4-fold axis. They are nonpolar and equivalent to each other. Since $n_3 = 4$ and $n_4 = 3$, from Eq. (5.24), one has $n_2 = 6$. Each 2-fold axis is along the angular bisector of two neighbored 4-fold axes and perpendicular to the third 4-fold axis. Six 2-fold axes are equivalent to each other. This group G is the proper symmetric group **O** of a cube. Its group table is given in Tables 2.9 and 2.11. Its character table is given in Table 3.9. **O** can be expressed as a product of three subgroups $C_3'C_4C_2''$. **O** is denoted by $3'42''$ in the international notations.

Table 5.1 The proper crystallographic point group

G	g	g_c	n_N: 2	3	4	6	No. of rep.: $1d$	$2d$	$3d$	Product of subgroups	Generators	H
C_1	1	1					1			C_1	C_1	
C_2	2	2	1				2			C_2	C_2	C_1
C_3	3	3		1			3			C_3	C_3	
C_4	4	4			1		4			C_4	C_4	C_2
C_6	6	6				1	6			C_6	C_6	C_3
D_2	4	4	3				4			C_2C_2'	C_2, C_2'	C_2
D_3	6	3	3	1			2	1		C_3C_2'	C_3, C_2'	C_3
D_4	8	5	4		1		4	1		C_4C_2'	C_4, C_2'	C_4,D_2
D_6	12	6	6			1	4	2		C_6C_2'	C_6, C_2'	C_6,D_3
T	12	4	3	4			3		1	$C_3'C_2C_2'$	C_3', C_2	
O	24	5	6	4	3		2	1	2	$C_3'C_4C_2''$	C_3', C_4	**T**

There are 11 proper crystallographic point groups G listed in Table 5.1, where g is the order of G, g_c is the number of classes in G, n_N is the number of N-fold proper axes contained in G, H is the invariant subgroup with index 2 in G, "No. of rep." denotes the number of the inequivalent irreducible representations of the given dimension, and G can be expressed as the product of subgroups, which is related to the symbol of G in the international notations. In the theory of crystals, an N-fold rotation along the z-axis is denoted by C_N, the remaining N-fold rotation with $N > 2$ by C_N', a 2-fold rotation along the x-axis is denoted by C_2', and the remaining 2-fold rotation by C_N''.

5.2.3 *Improper Crystallographic Point Group*

In §2.6, we discussed the general method for finding the improper point groups from a proper point group, and introduce the meaning of the subscripts in the Schoenflies notations. There are two types of improper point groups. An improper crystallographic point group G' of I-type is a direct product of a proper crystallographic point group G and the inversion group V_2, composed by the identical element E and the spatial inversion σ. There are 11 improper crystallographic point groups of I-type,

$$
\begin{array}{lll}
\mathrm{C}_i, & \mathrm{C}_{2h} = \mathrm{C}_2 \otimes \mathrm{C}_i, & \mathrm{C}_{3i} = \mathrm{C}_3 \otimes \mathrm{C}_i, \\
\mathrm{C}_{4h} = \mathrm{C}_4 \otimes \mathrm{C}_i, & \mathrm{C}_{6h} = \mathrm{C}_6 \otimes \mathrm{C}_i, & \mathrm{D}_{2h} = \mathrm{D}_2 \otimes \mathrm{C}_i, \\
\mathrm{D}_{3d} = \mathrm{D}_3 \otimes \mathrm{C}_i, & \mathrm{D}_{4h} = \mathrm{D}_4 \otimes \mathrm{C}_i, & \mathrm{D}_{6h} = \mathrm{D}_6 \otimes \mathrm{C}_i, \\
\mathrm{T}_h = \mathrm{T} \otimes \mathrm{C}_i, & \mathrm{O}_h = \mathrm{O} \otimes \mathrm{C}_i. &
\end{array}
\tag{5.30}
$$

If a proper crystallographic point group G contains an invariant subgroup H with index 2, one can construct an improper crystallographic point group G' of P-type by multiplying the elements in the coset of H in G with σ and remaining H invariant. G' is isomorphic onto G. From Table 5.1, there are 10 improper crystallographic point groups of P-type

$$
\begin{array}{llll}
\mathrm{C}_s \approx \mathrm{C}_2, & \mathrm{S}_4 \approx \mathrm{C}_4, & \mathrm{C}_{3h} \approx \mathrm{C}_6, & \mathrm{C}_{2v} \approx \mathrm{D}_2, \\
\mathrm{C}_{3v} \approx \mathrm{D}_3, & \mathrm{C}_{4v} \approx \mathrm{D}_{2d} \approx \mathrm{D}_4, & \mathrm{C}_{6v} \approx \mathrm{D}_{3h} \approx \mathrm{D}_6, & \mathrm{T}_d \approx \mathrm{O}.
\end{array}
\tag{5.31}
$$

The property of 32 crystallographic point groups are listed in Table 5.2.

Table 5.2 Crystallographic point groups

Proper crystallographic point groups			Improper crystallographic point groups			
	Product of		P-type		I-type	
G	subgroups	H	G'	Product	G'	Product
C_1	C_1				C_i	C_i
C_2	C_2	C_1	C_s	C_s	C_{2h}	$\mathrm{C}_i\mathrm{C}_2$
C_3	C_3				C_{3i}	C_{3i}
C_4	C_4	C_2	S_4	S_4	C_{4h}	$\mathrm{C}_i\mathrm{C}_4$
C_6	C_6	C_3	C_{3h}	C_{3h}	C_{6h}	$\mathrm{C}_i\mathrm{C}_6$
D_2	$\mathrm{C}_2\mathrm{C}_2'$	C_2	C_{2v}	$\mathrm{C}_2\mathrm{C}_s'$	D_{2h}	$\mathrm{C}_i\mathrm{C}_2\mathrm{C}_2'$
D_3	$\mathrm{C}_3\mathrm{C}_2'$	C_3	C_{3v}	$\mathrm{C}_3\mathrm{C}_s'$	D_{3d}	$\mathrm{C}_{3i}\mathrm{C}_2'$
D_4	$\mathrm{C}_4\mathrm{C}_2'$	C_4	C_{4v}	$\mathrm{C}_4\mathrm{C}_s'$	D_{4h}	$\mathrm{C}_i\mathrm{C}_4\mathrm{C}_2'$
		D_2	D_{2d}	$\mathrm{S}_4\mathrm{C}_2'$		
D_6	$\mathrm{C}_6\mathrm{C}_2'$	C_6	C_{6v}	$\mathrm{C}_6\mathrm{C}_s'$	D_{6h}	$\mathrm{C}_i\mathrm{C}_6\mathrm{C}_2'$
		D_3	D_{3h}	$\mathrm{C}_{3h}\mathrm{C}_2'$		
$\mathbf{T}$	$\mathrm{C}_3'\mathrm{C}_2\mathrm{C}_2'$				$\mathbf{T}_h$	$\mathrm{C}_{3i}'\mathrm{C}_2\mathrm{C}_2'$
$\mathbf{O}$	$\mathrm{C}_3'\mathrm{C}_4\mathrm{C}_2''$	$\mathbf{T}$	$\mathbf{T}_d$	$\mathrm{C}_3'\mathrm{S}_4\mathrm{C}_s''$	$\mathbf{O}_h$	$\mathrm{C}_{3i}'\mathrm{C}_4\mathrm{C}_2''$

An improper crystallographic point group G' of P-type is isomorphic onto the corresponding proper crystallographic point group G. They have the same character table and the irreducible representations. An improper crystallographic point group G' of I-type is a direct product of the corresponding proper crystallographic point group G and the inversion group V_2. The irreducible representation of G' is the direct product of two irreducible representations of G and V_2, respectively.

5.3 Crystal Systems and Bravais Lattice

In the previous sections we studied the translation symmetry of a crystal and obtained the constraint (5.7), which restricts the possible rotation R in the symmetric operation $g(R, \boldsymbol{\alpha})$. The conclusion is made that there are only 32 crystallographic point groups. The constraint (5.7) also restricts the choice of the basis vectors $\boldsymbol{a}_i$ of crystal lattice for each given crystallographic point group. In this section the lattice bases $\boldsymbol{a}_i$ are not assumed to be primitive. The crystals are classified into seven crystal systems based on the restriction to the choice of $\boldsymbol{a}_i$. For each crystal system we will discuss the possibility of the vectors of crystal lattice $\boldsymbol{f}$ which are the fractional combinations of the lattice bases $\boldsymbol{a}_i$. A crystal system contains different Bravais Lattices and the symmorphic space groups based on the vectors of crystal lattice $\boldsymbol{f}$. There are 14 Bravais Lattices and 73 symmorphic space groups, where the crystallographic point group is a subgroup of the space group. The general space groups will be discussed in the next section.

5.3.1 *Restrictions on Vectors of Crystal Lattice*

The conditions (5.7) and (5.20) give the restriction on the lattice basis $\boldsymbol{a}_i$ and the reciprocal lattice basis $\boldsymbol{b}_j$ when the crystallographic point group G is given. Obviously, the restriction for some crystallographic point groups may be the same. For instance, the spatial inversion σ makes no restriction on the lattice basis so that the restriction for the improper crystallographic point groups, both P-type and I-type, are the same as that for the corresponding proper crystallographic point group. It will be shown that except for $N = 2$, the restriction for C_N is the same as that for D_N, and the restriction for $\mathbf{T}$ is the same as that for $\mathbf{O}$. Based on the restrictions, the crystals are classified into seven crystal systems listed as follows, where the crystallographic point groups are also listed.

(a) Triclinic crystal system, C_1 and C_i.
(b) Monoclinic crystal system, C_2, C_s, and C_{2h}.
(c) Orthorhombic crystal system, D_2, C_{2v}, and D_{2h}.
(d) Trigonal crystal system, C_3, C_{3i}, D_3, C_{3v}, and D_{3d}.
(e) Hexagonal crystal system, C_6, C_{3h}, C_{6h}, D_6, C_{6v}, D_{3h}, and D_{6h}.
(f) Tetragonal crystal system, C_4, S_4, C_{4h}, D_4, C_{4v}, D_{2d}, and D_{4h}.
(g) Cubic crystal system, T, T_h, O, T_d, and O_h.

In the study, one finds that for some cases in the trigonal crystal system the restriction on $\boldsymbol{a}_i$ is the same as that in the hexagonal crystal system, and those cases are merged into the hexagonal crystal system. The remaining cases in the trigonal crystal system are called the rhombohedral crystal system.

For each crystal system, the lattice bases $\boldsymbol{a}_i$ are chosen in a given way according to the restriction, and then, the possibility of the vectors of crystal lattice $\boldsymbol{f}$ which are the fractional combinations of $\boldsymbol{a}_i$ is studied,

$$\boldsymbol{f} = \boldsymbol{a}_1 f_1 + \boldsymbol{a}_2 f_2 + \boldsymbol{a}_3 f_3, \qquad 0 \leq f_i < 1. \tag{5.32}$$

Denote by $T(\boldsymbol{f}) \equiv T(f_1, f_2, f_3)$ the translation operator with $\boldsymbol{f}$. The calculation shows that f_i can only be 0 or 1/2. The set of the translations where the translation vectors are the integral combinations of $\boldsymbol{a}_i$ forms an invariant subgroup $\mathcal{T}_\ell$ of the translation group $\mathcal{T}$. The cosets of $\mathcal{T}_\ell$ are generated by $T(\boldsymbol{f})$ and its powers. The crystal system is divided into a few Bravais lattices, based on the cosets as follows:

(a) Primitive translation group, denoted by P:

$$\mathcal{T} = \mathcal{T}_\ell. \tag{5.33}$$

The primitive translation group for the rhombohedral crystal system is denoted by R.

(b) Body-centered translation group, denoted by I:

$$\mathcal{T} = \mathcal{T}_\ell \otimes \left\{E,\ T\left(\frac{1}{2}, \frac{1}{2}, \frac{1}{2}\right)\right\}. \tag{5.34}$$

(c) Base-centered translation group, denoted by A, B, and C:

$$\begin{aligned}
A: \quad & \mathcal{T} = \mathcal{T}_\ell \otimes \left\{E,\ T\left(0, \frac{1}{2}, \frac{1}{2}\right)\right\}, \\
B: \quad & \mathcal{T} = \mathcal{T}_\ell \otimes \left\{E,\ T\left(\frac{1}{2}, 0, \frac{1}{2}\right)\right\}, \\
C: \quad & \mathcal{T} = \mathcal{T}_\ell \otimes \left\{E,\ T\left(\frac{1}{2}, \frac{1}{2}, 0\right)\right\}.
\end{aligned} \tag{5.35}$$

(d) Face-centered translation group, denoted by F:

$$\mathcal{T} = \mathcal{T}_\ell \otimes \left\{E,\ T\left(0, \frac{1}{2}, \frac{1}{2}\right), T\left(\frac{1}{2}, 0, \frac{1}{2}\right), T\left(\frac{1}{2}, \frac{1}{2}, 0\right)\right\}. \tag{5.36}$$

Obviously, if two of A, B, and C are the symmetric subgroups of the crystal simultaneously, its translation group is the face-centered translation group F. Combining the translation group with the crystal system, one obtains 14 Bravais lattices. A crystallographic point group headed by the symbol of the Bravais lattice denotes the symmorphic space group. In the following, we are going to count the 73 symmorphic space groups.

There are a few notation-systems for the space groups $\mathcal{S}$. The Schoenflies notations are convenient for the crystallographic point groups, but not very convenient for the space groups. Through some improvement there are international notations for the point groups, the Mauguin–Hermann notations and the international notations for the space groups. Their comparison is listed in Table 5.3. We recommend the international notations for the space groups, where a proper cyclic subgroup C_N is denoted by a number N, an improper one by a number with a bar $\overline{N}$, and the inversion group V_2 (C_i) is denoted by $\pm$. N (or $\overline{N}$) stands for an N-fold cyclic subgroup along $\boldsymbol{a}_3$. N' (or $\overline{N}'$) with $N > 2$ stands for the remaining N-fold cyclic subgroup. $2'$ (or $\overline{2}'$) stands for a 2-fold cyclic subgroup along $\boldsymbol{a}_1$. $2''$ (or $\overline{2}''$) stands for the remaining 2-fold cyclic subgroup. We will explain the phenomena later that one crystallographic point group (e.g., D_3) corresponds to two notations in Table 5.3.

Table 5.3 Comparison between notations for space groups.

Sch. = the Schoenflies notations,
INPG = the international notations for point groups,
MPN = the Mauguin–Hermann notations,
INSG = the international notations for space groups.

Sch.	INPG	MHN	INSG	Sch.	INPG	MHN	INSG
C_1	1	1	1	C_{3v}	$3m$	$31m$	$3\overline{2}''$
C_i	$\overline{1}$	$\overline{1}$	$\overline{1}$	D_{3d}	$\overline{3}m$	$\overline{3}\frac{2}{m}1$	$\overline{3}2'$
C_2	2	2	2	D_{3d}	$\overline{3}m$	$\overline{3}1\frac{2}{m}$	$\overline{3}2''$
C_s	m	m	$\overline{2}$	D_4	422	422	$42'$
C_{2h}	$2/m$	$2/m$	± 2	C_{4v}	$4mm$	$4mm$	$4\overline{2}'$
C_3	3	3	3	D_{2d}	$\overline{4}2m$	$\overline{4}2m$	$\overline{4}2'$
C_{3i}	$\overline{3}$	$\overline{3}$	$\overline{3}$	D_{2d}	$\overline{4}2m$	$\overline{4}m2$	$\overline{4}2''$
C_4	4	4	4	D_{4h}	$4/mmm$	$\frac{4}{m}\frac{2}{m}\frac{2}{m}$	$\pm 42'$
S_4	$\overline{4}$	$\overline{4}$	$\overline{4}$	D_6	622	622	$62'$
C_{4h}	$4/m$	$4/m$	± 4	C_{6v}	$6mm$	$6mm$	$6\overline{2}'$
C_6	6	6	6	D_{3h}	$\overline{6}m2$	$\overline{6}2m$	$\overline{6}2'$
C_{3h}	$\overline{6}$	$\overline{6}$	$\overline{6}$	D_{3h}	$\overline{6}m2$	$\overline{6}m2$	$\overline{6}2''$
C_{6h}	$6/m$	$6/m$	± 6	D_{6h}	$6/mmm$	$\frac{6}{m}\frac{2}{m}\frac{2}{m}$	$\pm 62'$
D_2	222	222	$22'$	T	23	23	$3'22'$
C_{2v}	$2mm$	$2mm$	$2\overline{2}'$	T_h	$m3$	$\frac{2}{m}3$	$\overline{3}'22'$
D_{2h}	mmm	$\frac{2}{m}\frac{2}{m}\frac{2}{m}$	$\pm 22'$	O	432	432	$3'42''$
D_3	32	321	$32'$	T_d	$\overline{4}3m$	$\overline{4}3m$	$3'\overline{4}\,\overline{2}''$
D_3	32	312	$32''$	O_h	$m3m$	$\frac{4}{m}3\frac{2}{m}$	$\overline{3}'42''$
C_{3v}	$3m$	$3m1$	$3\overline{2}'$				

Except for the trivial point group C_1, there are vectors of both crystal lattice and reciprocal crystal lattice along each rotational axis, and at least two noncollinear vectors of lattice bases and those of reciprocal crystal lattice at the plane perpendicular to a rotational axis. In fact, each crystallographic point group G, except for C_1, contains C_2, C_s, or C_3 as a subgroup. If G contains C_2, $\boldsymbol{\ell}+C_2\boldsymbol{\ell}$ is parallel to the 2-fold axis, and $\boldsymbol{\ell}-C_2\boldsymbol{\ell}$ is perpendicular to it. Find another vector of crystal lattice $\boldsymbol{\ell}'$ outside of the plane spanned by $\boldsymbol{\ell} \pm C_2\boldsymbol{\ell}$. $\boldsymbol{\ell}' - C_2\boldsymbol{\ell}'$ is another vector of crystal lattice at the plane perpendicular to the 2-fold axis. The situation is similar when G contains C_s. If G contains C_3, $\boldsymbol{\ell} + C_3\boldsymbol{\ell} + C_3^2\boldsymbol{\ell}$ is along the 3-fold axis, and $\boldsymbol{\ell} - C_3\boldsymbol{\ell}$ and $\boldsymbol{\ell} - C_3^2\boldsymbol{\ell}$ are two noncollinear vectors of crystal lattice at the plane perpendicular to the 3-fold axis. Since the crystallographic point group for the reciprocal crystal lattice is the same as that for the crystal lattice, the situation is also the same.

In the following the length of $\boldsymbol{a}_i$ is denoted by a_i and the angle between $\boldsymbol{a}_1$ and $\boldsymbol{a}_2$ is denoted by α_3 and its cyclic.

5.3.2 *Triclinic Crystal System*

The crystallographic point groups in the triclinic crystal system are C_1 (1) and C_i ($\overline{1}$), where the number in the bracket is the symbol of international notations. The matrices of the identical element E and the spatial inversion σ are constant ones, which make no restriction on the choice of the lattice bases. Choose three vectors of crystal lattice, which are not coplanar, to be the primitive lattice bases. Thus, there are one Bravais lattice P and two symmorphic space groups $P1$ and $P\overline{1}$ in the triclinic crystal system.

5.3.3 *Monoclinic Crystal System*

The crystallographic point groups in the monoclinic crystal system are C_2 (2), C_s ($\overline{2}$), and C_{2h} (± 2). The shortest vector of crystal lattice along the 2-fold axis is chosen to be $\boldsymbol{a}_3$. In the plane perpendicular to $\boldsymbol{a}_3$, choose two smallest vectors of crystal lattice to be $\boldsymbol{a}_1$ and $\boldsymbol{a}_2$ such that $\boldsymbol{a}_1 \times \boldsymbol{a}_2$ is along the positive direction of $\boldsymbol{a}_3$, where "two smallest vectors of crystal lattice in a plane" means that each vector of crystal lattice in the plane is an integral combination of those two smallest vectors of crystal lattice. The length a_1 and a_2 may not be the shortest among the vectors of crystal lattice in the plane. Thus, in the monoclinic crystal system the lengths of the lattice bases have no restriction, and the restrictions on the angles are

$$\alpha_1 = \alpha_2 = \pi/2. \tag{5.37}$$

From Eq. (5.14) the double-vector and the matrix of the generator C_2 in the set of lattice bases are calculated as

$$\vec{\vec{2}} = -\boldsymbol{a}_1\boldsymbol{b}_1 - \boldsymbol{a}\boldsymbol{b}_2 + \boldsymbol{a}_3\boldsymbol{b}_3, \qquad D(C_2) = \begin{pmatrix} -1 & 0 & 0 \\ 0 & -1 & 0 \\ 0 & 0 & 1 \end{pmatrix}. \tag{5.38}$$

Discuss the possibility of the vector of crystal lattice $\boldsymbol{f}$ with the fractional components

$$\boldsymbol{f} = \boldsymbol{a}_1 f_1 + \boldsymbol{a}_2 f_2 + \boldsymbol{a}_3 f_3, \qquad 0 \leq f_j < 1. \tag{5.39}$$

If $\boldsymbol{f}$ is a vector of crystal lattice, the following are ones, too

$$\boldsymbol{f} + \vec{\vec{2}} \cdot \boldsymbol{f} = \boldsymbol{a}_3 (2f_3), \qquad \boldsymbol{f} - \vec{\vec{2}} \cdot \boldsymbol{f} = \boldsymbol{a}_1 (2f_1) + \boldsymbol{a}_2 (2f_2).$$

Hence, three f_j all can be 0 or 1/2. However, when $f_3 = 0$, $f_1 = f_2 = 0$ because $\boldsymbol{a}_1$ and $\boldsymbol{a}_2$ are two smallest vectors of crystal lattice in the plane perpendicular to $\boldsymbol{a}_3$. When $f_1 = f_2 = 0$, $f_3 = 0$ because $\boldsymbol{a}_3$ is the shortest vector of crystal lattice along the 2-fold axis. Therefore, there are four possibilities for $\boldsymbol{f}$, $\boldsymbol{f} = 0$ (P-type), $\boldsymbol{f} = (\boldsymbol{a}_2 + \boldsymbol{a}_3)/2$ (A-type), $\boldsymbol{f} = (\boldsymbol{a}_1 + \boldsymbol{a}_3)/2$ (B-type), and $\boldsymbol{f} = (\boldsymbol{a}_1 + \boldsymbol{a}_2 + \boldsymbol{a}_3)/2$ (I-type). Since C-type translation group does not exist in the crystal system, those of A-type and B-type cannot exist for one crystal, neither can F-type one. The B-type translation group becomes the A-type one if $\boldsymbol{a}_2$ is chosen to be new $\boldsymbol{a}_1$ and $-\boldsymbol{a}_1$ to be new $\boldsymbol{a}_2$. Similarly, the I-type translation group becomes the A-type one if $\boldsymbol{a}_1 + \boldsymbol{a}_2$ is chosen to be new $\boldsymbol{a}_1$ and $\boldsymbol{a}_2$ remains invariant. Thus, there are two Bravais lattices P and A, and six symmorphic space groups $P2$, $P\bar{2}$, $P \pm 2$, $A2$, $A\bar{2}$, and $A \pm 2$ in the monoclinic crystal system.

5.3.4 *Orthorhombic Crystal System*

The crystallographic point groups in the orthorhombic crystal system are D_2 $(22')$, C_{2v} $(2\bar{2}')$, and D_{2h} $(\pm 22')$. The point groups all contain three perpendicular 2-fold axes (proper or improper). The shortest vector of crystal lattice along the one 2-fold proper axis is chosen to be $\boldsymbol{a}_3$, and those along the remaining two 2-fold axes to be $\boldsymbol{a}_1$ and $\boldsymbol{a}_2$ such that $\boldsymbol{a}_1 \times \boldsymbol{a}_2$ is along the positive direction of $\boldsymbol{a}_3$. Thus, in the orthorhombic crystal system the lengths of the lattice bases have no restriction, and the angles are

$$\alpha_1 = \alpha_2 = \alpha_3 = \pi/2. \tag{5.40}$$

In the set of lattice bases the double-vector and the matrix of the 2-fold rotation along $\boldsymbol{a}_3$ are given in Eq. (5.38), and those along $\boldsymbol{a}_1$ are calculated from Eq. (5.14)

$$\vec{\vec{2}}' = \boldsymbol{a}_1\boldsymbol{b}_1 - \boldsymbol{a}_2\boldsymbol{b}_2 - \boldsymbol{a}_3\boldsymbol{b}_3, \qquad D(C_2') = \begin{pmatrix} 1 & 0 & 0 \\ 0 & -1 & 0 \\ 0 & 0 & -1 \end{pmatrix}. \tag{5.41}$$

If $\boldsymbol{f}$ given in Eq. (5.39) is a vector of crystal lattice, $\boldsymbol{f} + \vec{\vec{R}} \cdot \boldsymbol{f}$ also is the one where $\vec{\vec{R}}$ is $\vec{\vec{2}}$, $\vec{\vec{2}}'$ or $\vec{\vec{2}} \cdot \vec{\vec{2}}'$. Hence, three f_j can be 0 and 1/2, respectively. Thus, for the orthorhombic crystal system, in addition to the P-type, A-type, and B-type translation groups, there are C-type and F-type translation groups.

For the point groups D_2 and D_{2h}, A-, B-, and C-type translation groups are the same. For the point group C_{2v}, the 2-fold axis along $\boldsymbol{a}_3$ is proper, but two 2-fold axes along $\boldsymbol{a}_1$ and $\boldsymbol{a}_2$ are improper so that A- and B-type translation groups are the same, but C-type is different. Those different space groups are usually classified into the same Bravais lattices. Thus, there are four Bravais lattices P, C (A), I, and F, and 13 symmorphic space groups $P22'$, $P2\overline{2}'$, $P \pm 22'$, $C22'$, $C2\overline{2}'$, $A2\overline{2}'$, $C \pm 22'$, $I22'$, $I2\overline{2}'$, $I \pm 22'$, $F22'$, $F2\overline{2}'$, and $F \pm 22'$ in the orthorhombic crystal system.

5.3.5 *Trigonal and Hexagonal Crystal System*

The crystallographic point groups in the hexagonal crystal system are C_6 (6), C_{3h} $(\overline{6})$, C_{6h} (± 6), D_6 $(62')$, C_{6v} $(6\overline{2}')$, D_{3h} $(\overline{6}2')$, and D_{6h} $(\pm 62')$. The shortest vector of crystal lattice along the 6-fold axis is chosen to be $\boldsymbol{a}_3$. For the cases with the point groups C_6, C_{3h}, and C_{6h}, the shortest vector of crystal lattice in the plane perpendicular to $\boldsymbol{a}_3$ is chosen to be $\boldsymbol{a}_1$. For the cases with the point groups D_6, C_{6v}, D_{3h}, and D_{6h}, the shortest vector of crystal lattice along the 2-fold axes in the plane perpendicular to $\boldsymbol{a}_3$ is chosen to be $\boldsymbol{a}_1$. Then, $\boldsymbol{a}_2 = C_6^2 \boldsymbol{a}_1$. Thus,

$$a_1 = a_2, \qquad \alpha_1 = \alpha_2 = \pi/2, \qquad \alpha_3 = 2\pi/3. \tag{5.42}$$

For the cases with the point group D_{3h}, two space groups whether $\boldsymbol{a}_1$ is along a proper or an improper 2-fold axis are different.

From Eq. (5.14), the double-vectors and the matrices of the generators C_6, C_2', and C_2'' in the set of lattice bases are

$$\begin{aligned}
\vec{\vec{6}} &= (\boldsymbol{a}_1 + \boldsymbol{a}_2)\,\boldsymbol{b}_1 - \boldsymbol{a}_1\boldsymbol{b}_2 + \boldsymbol{a}_3\boldsymbol{b}_3, & D(C_6) &= \begin{pmatrix} 1 & -1 & 0 \\ 1 & 0 & 0 \\ 0 & 0 & 1 \end{pmatrix}, \\
\vec{\vec{2}}' &= \boldsymbol{a}_1\boldsymbol{b}_1 - (\boldsymbol{a}_1 + \boldsymbol{a}_2)\,\boldsymbol{b}_2 - \boldsymbol{a}_3\boldsymbol{b}_3, & D(C_2') &= \begin{pmatrix} 1 & -1 & 0 \\ 0 & -1 & 0 \\ 0 & 0 & -1 \end{pmatrix}, \\
\vec{\vec{2}}'' &= -\boldsymbol{a}_2\boldsymbol{b}_1 - \boldsymbol{a}_1\boldsymbol{b}_2 - \boldsymbol{a}_3\boldsymbol{b}_3, & D(C_2') &= \begin{pmatrix} 0 & -1 & 0 \\ -1 & 0 & 0 \\ 0 & 0 & -1 \end{pmatrix}.
\end{aligned} \tag{5.43}$$

If $\boldsymbol{f}$ given in Eq. (5.39) is a vector of crystal lattice,

$$\begin{aligned}
&\boldsymbol{f} + \left(\vec{\vec{6}}\right)^3 \cdot \boldsymbol{f} = 2f_3\boldsymbol{a}_3, \\
&\boldsymbol{f} + \left(\vec{\vec{6}}\right)^2 \cdot \boldsymbol{f} + \left(\vec{\vec{6}}\right)^4 \cdot \boldsymbol{f} = 3f_3\boldsymbol{a}_3, \\
&\boldsymbol{f} - \vec{\vec{6}} \cdot \boldsymbol{f} = f_2\boldsymbol{a}_1 + (f_2 - f_1)\,\boldsymbol{a}_2,
\end{aligned} \tag{5.44}$$

are all vectors of crystal lattice. From the first two conditions in Eq. (5.44), $f_3 = 0$. For the cases with the point groups C_6, C_{3h}, and C_{6h}, $f_1 = f_2 = 0$ because the length of the vector in the third formula of Eq. (5.44) is 0 or less than a_1. For the cases with the point groups D_6, C_{6v}, D_{3h}, and D_{6h},

$$\begin{aligned}
&\boldsymbol{f} + \vec{\vec{2}}' \cdot \boldsymbol{f} = (2f_1 - f_2)\,\boldsymbol{a}_1, \\
&\boldsymbol{f} + \left(\vec{\vec{3}} \cdot \vec{\vec{2}}' \cdot \vec{\vec{3}} \cdot \vec{\vec{3}}\right) \cdot \boldsymbol{f} = (2f_2 - f_1)\,\boldsymbol{a}_2,
\end{aligned} \tag{5.45}$$

are also the vectors of crystal lattice. Thus, $2f_1 - f_2$ and $2f_2 - f_1$ are 0 or 1, respectively. They cannot be 1 at the same time because both f_1 and f_2 are less than 1. There are three solutions to Eq. (5.45):

$$f_1 = f_2 = 0, \qquad f_1 = 2f_2 = 2/3, \qquad 2f_1 = f_2 = 2/3. \tag{5.46}$$

However, both vectors $\boldsymbol{f} = (2\boldsymbol{a}_1 + \boldsymbol{a}_2)/3$ and $\boldsymbol{f}' = (\boldsymbol{a}_1 + 2\boldsymbol{a}_2)/3$ are along a 2-fold axis with the length less than a_1, so that they are ruled out. Thus, there is one Bravais lattice P and eight symmorphic space groups $P6$, $P\overline{6}$, $P{\pm}6$, $P62'$, $P6\overline{2}'$, $P\overline{6}2'$, $P\overline{6}2''$, and $P{\pm}62'$ in the hexagonal crystal system.

The crystallographic point groups in the trigonal crystal system are C_3 (3), C_{3i} ($\overline{3}$), D_3 ($32'$), C_{3v} ($3\overline{2}'$), and D_{3d} ($\overline{3}2'$). The shortest vector of crystal lattice along the 3-fold axis is chosen to be $\boldsymbol{a}_3$. For the cases with the point groups C_3 and C_{3i}, the shortest vector of crystal lattice in the plane perpendicular to $\boldsymbol{a}_3$ is chosen to be $\boldsymbol{a}_1$. For the cases with the point groups D_3, C_{3v}, and D_{3d}, the shortest vector of crystal lattice along the 2-fold axes is chosen to be $\boldsymbol{a}_1$. Then, $\boldsymbol{a}_2 = C_3\boldsymbol{a}_1$. Thus, the condition (5.42) is satisfied. The double-vectors and the matrices of the generators $C_3 = C_6^2$ and C_2' in the set of lattice bases are given in Eq. (5.43).

If $\boldsymbol{f}$ given in Eq. (5.39) is a vector of crystal lattice, $\boldsymbol{f} + \vec{\mathfrak{Z}}\cdot\boldsymbol{f} + \vec{\mathfrak{Z}}\cdot\vec{\mathfrak{Z}}\cdot\boldsymbol{f} = 3f_3\boldsymbol{a}_3$ is also a vector of crystal lattice. Thus,

$$f_3 = 0,\ 1/3,\quad \text{or}\ \ 2/3. \tag{5.47}$$

For the cases with the point groups D_3, C_{3v}, and D_{3d}, one also obtains Eqs. (5.45) and (5.46). When $f_1 = f_2 = 0$, f_3 has to be 0.

For the cases with the point groups C_3 and C_{3i}, the following vector is also a vector of crystal lattice

$$\boldsymbol{f} - \vec{\mathfrak{Z}}\cdot\boldsymbol{f} = (f_1 + f_2)\,\boldsymbol{a}_1 + (2f_2 - f_1)\,\boldsymbol{a}_2.$$

After removing the integral part, if $(f_1 + f_2)$ and $(2f_2 - f_1)$ both are not equal to 0, the length of the vector is less than a_1, which is in conflict with the assumption. If $f_1 + f_2 = 0$, one has the first solution in Eq. (5.46), and if $f_1 + f_2 = 1$ and $(2f_2 - f_1) = 0$ or 1, one has the remaining solutions of Eq. (5.46).

From Eqs. (5.46) and (5.47), there are four possibilities for $\boldsymbol{f}$ in the trigonal crystal system, where $\boldsymbol{f}'$ is calculated from $2\boldsymbol{f}$,

$$\begin{aligned}
&\text{(a)}\ \ \boldsymbol{f} = 0,\\
&\text{(b)}\ \ \boldsymbol{f} = (2\boldsymbol{a}_1 + \boldsymbol{a}_2)/3, \quad \text{and} \quad \boldsymbol{f}' = (\boldsymbol{a}_1 + 2\boldsymbol{a}_2)/3,\\
&\text{(c)}\ \ \boldsymbol{f} = (2\boldsymbol{a}_1 + \boldsymbol{a}_2 + \boldsymbol{a}_3)/3, \quad \text{and} \quad \boldsymbol{f}' = (\boldsymbol{a}_1 + 2\boldsymbol{a}_2 + 2\boldsymbol{a}_3)/3,\\
&\text{(d)}\ \ \boldsymbol{f} = (\boldsymbol{a}_1 + 2\boldsymbol{a}_2 + \boldsymbol{a}_3)/3, \quad \text{and} \quad \boldsymbol{f}' = (2\boldsymbol{a}_1 + \boldsymbol{a}_2 + 2\boldsymbol{a}_3)/3.
\end{aligned} \tag{5.48}$$

The length of the vector $\boldsymbol{f}$ in (b) is less than a_1. It is not a vector of crystal lattice for the cases with the point groups C_3 and C_{3i}. For the cases with the point groups D_3, C_{3v}, and D_{3d}, a set of new lattice bases can be defined as follows such that new lattice bases are primitive,

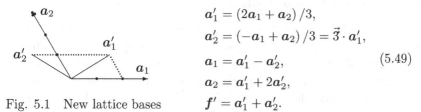

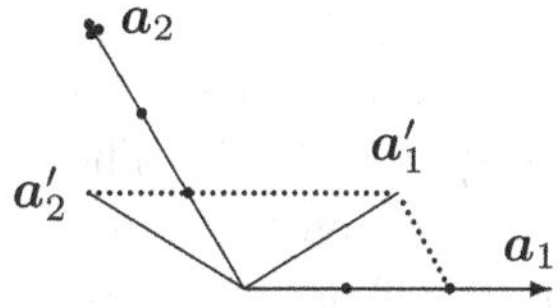

Fig. 5.1 New lattice bases

$$\begin{aligned} \boldsymbol{a}'_1 &= (2\boldsymbol{a}_1 + \boldsymbol{a}_2)/3, \\ \boldsymbol{a}'_2 &= (-\boldsymbol{a}_1 + \boldsymbol{a}_2)/3 = \vec{3} \cdot \boldsymbol{a}'_1, \\ \boldsymbol{a}_1 &= \boldsymbol{a}'_1 - \boldsymbol{a}'_2, \\ \boldsymbol{a}_2 &= \boldsymbol{a}'_1 + 2\boldsymbol{a}'_2, \\ \boldsymbol{f}' &= \boldsymbol{a}'_1 + \boldsymbol{a}'_2. \end{aligned} \tag{5.49}$$

However, in these cases new lattice bases $\boldsymbol{a}'_1$ and $\boldsymbol{a}'_2$ are not along a 2-fold axis, but along the angular bisector between two neighbored 2-fold axes (a direction with an angle $\pi/6$ to a 2-fold axis). Now, the 2-fold rotation is denoted by $\vec{2}''$ given in Eq. (5.43) instead of $\vec{2}'$. Two space groups are different, whether $\boldsymbol{a}_1$ is along a 2-fold axis or not. This Bravais lattice is merged into the P-type Bravais lattices in the hexagonal crystal system. In this case there are eight symmorphic space groups: $P3$, $P\overline{3}$, $P32'$, $P32''$, $P3\overline{2}'$, $P3\overline{2}''$, $P\overline{3}2'$, and $P\overline{3}2''$.

The length of $\boldsymbol{a}_1 + \boldsymbol{a}_2$ is the same as a_1. $\boldsymbol{a}_1 + \boldsymbol{a}_2$ is along a 2-fold axis for the cases with the point groups D_3, C_{3v}, and D_{3d}. Thus, for the cases of all five crystallographic point groups, $\boldsymbol{a}_1 + \boldsymbol{a}_2$ can also be chosen as a lattice basis instead of $\boldsymbol{a}_1$ such that the case (d) becomes the case (c). For the case (c), one defines new lattice bases, which are primitive

$$\begin{aligned} \boldsymbol{a}'_1 &= (2\boldsymbol{a}_1 + \boldsymbol{a}_2 + \boldsymbol{a}_3)/3 = \vec{3} \cdot \boldsymbol{a}'_3, & \boldsymbol{a}_1 &= \boldsymbol{a}'_1 - \boldsymbol{a}'_2, \\ \boldsymbol{a}'_2 &= (-\boldsymbol{a}_1 + \boldsymbol{a}_2 + \boldsymbol{a}_3)/3 = \vec{3} \cdot \boldsymbol{a}'_1, & \boldsymbol{a}_2 &= \boldsymbol{a}'_2 - \boldsymbol{a}'_3, \\ \boldsymbol{a}'_3 &= (-\boldsymbol{a}_1 - 2\boldsymbol{a}_2 + \boldsymbol{a}_3)/3 = \vec{3} \cdot \boldsymbol{a}'_2, & \boldsymbol{a}_3 &= \boldsymbol{a}'_1 + \boldsymbol{a}'_2 + \boldsymbol{a}'_3. \end{aligned} \tag{5.50}$$

From Eq. (5.14), the double-vectors and the matrices of the generators C_3 and C'_2 in the new lattice bases are

$$\begin{aligned} &\vec{3} = \boldsymbol{a}_2\boldsymbol{b}_1 + \boldsymbol{a}_3\boldsymbol{b}_2 + \boldsymbol{a}_1\boldsymbol{b}_3, & &\vec{2}' = -\boldsymbol{a}_2\boldsymbol{b}_1 - \boldsymbol{a}_1\boldsymbol{b}_2 - \boldsymbol{a}_3\boldsymbol{b}_3, \\ &D(C_3) = \begin{pmatrix} 0 & 0 & 1 \\ 1 & 0 & 0 \\ 0 & 1 & 0 \end{pmatrix}, & &D(C'_2) = \begin{pmatrix} 0 & -1 & 0 \\ -1 & 0 & 0 \\ 0 & 0 & -1 \end{pmatrix}. \end{aligned} \tag{5.51}$$

It is called the R-type (rhombohedral) Bravais lattice where

$$a_1 = a_2 = a_3, \qquad \alpha_1 = \alpha_2 = \alpha_3. \tag{5.52}$$

There are five symmorphic space groups belonging to the R-type Bravais lattice: $R3$, $R\overline{3}$, $R32'$, $R3\overline{2}'$, and $R\overline{3}2'$.

5.3.6 *Tetragonal Crystal System*

The crystallographic point groups in the tetragonal crystal system are C_4 (4), S_4 ($\overline{4}$), C_{4h} (± 4), D_4 ($42'$), C_{4v} ($4\overline{2}'$), D_{2d} ($\overline{4}2'$), and D_{4h} ($\pm 42'$). The shortest vector of crystal lattice along the 4-fold axis is chosen to be $\boldsymbol{a}_3$. For the cases with the point groups C_4, S_4, and C_{4h}, the shortest vector of crystal lattice in the plane perpendicular to $\boldsymbol{a}_3$ is chosen to be $\boldsymbol{a}_1$. For the cases with the point groups D_4, C_{4v}, D_{2d}, and D_{4h}, the shortest vector of crystal lattice along the 2-fold axes in the plane perpendicular to $\boldsymbol{a}_3$ is chosen to be $\boldsymbol{a}_1$. Then, $\boldsymbol{a}_2 = C_4\boldsymbol{a}_1$. Thus,

$$a_1 = a_2, \qquad \alpha_1 = \alpha_2 = \alpha_3 = \pi/2. \tag{5.53}$$

For the cases with the point group D_{2d}, two space groups whether $\boldsymbol{a}_1$ is along a proper or an improper 2-fold axis are different.

From Eq. (5.14), the double-vectors and the matrices of the generators C_4, C_2', and C_2'' in the set of lattice bases are

$$\begin{aligned}
\vec{\vec{4}} &= \boldsymbol{a}_2\boldsymbol{b}_1 - \boldsymbol{a}_1\boldsymbol{b}_2 + \boldsymbol{a}_3\boldsymbol{b}_3, & D(C_4) &= \begin{pmatrix} 0 & -1 & 0 \\ 1 & 0 & 0 \\ 0 & 0 & 1 \end{pmatrix}, \\
\vec{\vec{2}}' &= \boldsymbol{a}_1\boldsymbol{b}_1 - \boldsymbol{a}_2\boldsymbol{b}_2 - \boldsymbol{a}_3\boldsymbol{b}_3, & D(C_2') &= \begin{pmatrix} 1 & 0 & 0 \\ 0 & -1 & 0 \\ 0 & 0 & -1 \end{pmatrix}, \\
\vec{\vec{2}}'' &= \boldsymbol{a}_2\boldsymbol{b}_1 + \boldsymbol{a}_1\boldsymbol{b}_2 - \boldsymbol{a}_3\boldsymbol{b}_3, & D(C_2'') &= \begin{pmatrix} 0 & 1 & 0 \\ 1 & 0 & 0 \\ 0 & 0 & -1 \end{pmatrix}.
\end{aligned} \tag{5.54}$$

If $\boldsymbol{f}$ given in Eq. (5.39) is a vector of crystal lattice, $\boldsymbol{f} + \vec{\vec{2}} \cdot \boldsymbol{f} = 2f_3\boldsymbol{a}_3$ is also a vector of crystal lattice, so that

$$f_3 = 0, \quad \text{or} \quad 1/2. \tag{5.55}$$

For the cases with the point groups D_4, C_{4v}, D_{2d}, and D_{4h},

$$\begin{aligned}
&\boldsymbol{f} + \vec{\vec{2}}' \cdot \boldsymbol{f} = 2f_1\boldsymbol{a}_1, \\
&\boldsymbol{f} + \left\{\vec{\vec{4}} \cdot \vec{\vec{2}}' \cdot \left(\vec{\vec{4}}\right)^3\right\} \cdot \boldsymbol{f} = 2f_2\boldsymbol{a}_2, \\
&\boldsymbol{f} + \vec{\vec{2}}'' \cdot \boldsymbol{f} = (f_1 + f_2)(\boldsymbol{a}_1 + \boldsymbol{a}_2),
\end{aligned}$$

are also the vectors of crystal lattice. Thus, $f_1 = 0$ or $1/2$, $f_2 = 0$ or $1/2$, but $f_1 + f_2 \neq 1/2$ because the vector of crystal lattice $(\boldsymbol{a}_1 + \boldsymbol{a}_2)/2$ is along a 2-fold axis and with the length less than a_1, namely, $f_1 = f_2 = 0$ or $1/2$.

Note that $f_3 = 0$ when $f_1 = f_2 = 0$. Conversely, $f_1 = f_2 = 0$ when $f_3 = 0$. Therefore, there are only P-type and I-type Bravais lattices.

For the cases with the point groups C_4, S_4, and C_{4h},

$$\begin{aligned} \boldsymbol{f} - \vec{\bar{4}} \cdot \boldsymbol{f} &= \boldsymbol{a}_1 (f_1 + f_2) + \boldsymbol{a}_2 (f_2 - f_1) \equiv \boldsymbol{f}', \\ \boldsymbol{f}' - \vec{\bar{4}} \cdot \boldsymbol{f}' &= \boldsymbol{a}_1 (2f_2) + \boldsymbol{a}_2 (-2f_1), \end{aligned} \tag{5.56}$$

are also the vectors of crystal lattice. If $f_1 = f_2$, they have to be 0 or 1/2. Since $f_3 = 0$ when $f_1 = f_2 = 0$, and $f_1 = f_2 = 0$ when $f_3 = 0$, there are only P-type and I-type Bravais lattices. We are going to show that $f_1 \neq f_2$ should be ruled out, otherwise, due to Eq. (5.56), there exists a vector of crystal lattice in the plane perpendicular to $\boldsymbol{a}_3$ such that its length is less than a_1. In fact, if one of f_1 and f_2 is vanishing, the other has to be 1/2 such that the length of $\boldsymbol{f}'$ is less than a_1. If $f_1 + f_2 \geq 1$,

$$a_1^2 \left\{ (f_1 + f_2 - 1)^2 + (f_2 - f_1)^2 \right\} = a_1^2 \left\{ 1 - 2 \left(f_1 + f_2 - f_1^2 - f_2^2 \right) \right\} < a_1^2.$$

Namely, the length of $\boldsymbol{f}' - \boldsymbol{a}_1$ is less than a_1. For the same reason, $2f_1 < 1$ and $2f_2 < 1$. Then, if $f_1 + f_2 < 1$,

$$|\vec{f}'|^2 = a_1^2 \left\{ (f_1 + f_2)^2 + (f_2 - f_1)^2 \right\} = a_1^2 \left\{ 2f_1^2 + 2f_2^2 \right\} < a_1^2.$$

Thus, there are two Bravais lattices P and I, and 16 symmorphic space groups $P4$, $P\bar{4}$, $P \pm 4$, $P42'$, $P4\bar{2}'$, $P\bar{4}2'$, $P\bar{4}2''$, $P \pm 42'$, $I4$, $I\bar{4}$, $I \pm 4$, $I42'$, $I4\bar{2}'$, $I\bar{4}2'$, $I\bar{4}2''$, and $I \pm 42'$ in the tetragonal crystal system.

5.3.7 *Cubic Crystal System*

The crystallographic point groups in the cubic crystal system are **T** $(3'22')$, $\mathbf{T}_h$ $(\bar{3}'22')$, **O** $(3'42'')$, $\mathbf{T}_d$ $(3'\bar{4}\,\bar{2}'')$, and $\mathbf{O}_h$ $(\bar{3}42'')$. The shortest vectors of crystal lattice along the three 2-fold axes (for **T** and $\mathbf{T}_h$) or along the three 4-fold axes (for **O**, $\mathbf{T}_d$, and $\mathbf{O}_h$) are chosen to be the lattice bases, where $\boldsymbol{a}_1 \times \boldsymbol{a}_2$ is along the positive direction of $\boldsymbol{a}_3$. Thus,

$$a_1 = a_2 = a_3, \qquad \alpha_1 = \alpha_2 = \alpha_3 = \pi/2. \tag{5.57}$$

In the set of lattice bases, the double-vectors and the matrices of C_4 along $\boldsymbol{a}_3$, C_2' along $\boldsymbol{a}_1$, and C_2'' along $(\boldsymbol{a}_1 + \boldsymbol{a}_2)$ are given in Eq. (5.54), and those of C_3' along the direction $(\boldsymbol{a}_1 + \boldsymbol{a}_2 + \boldsymbol{a}_3)$ are given in Eq. (5.51).

If $\boldsymbol{f}$ given in Eq. (5.39) is a vector of crystal lattice, $\boldsymbol{f} + C_2\boldsymbol{f}$ is also a vector of crystal lattice, and then, $f_3 = 0$ or 1/2. Since three 4-fold axes are equivalent, f_1 and f_2 are also equal to 0 or 1/2, and the Bravais lattices A, B, and C are merged to the F-type Bravais lattice. Thus, there are three Bravais lattices P, I, and F, and 15 symmorphic space groups $P3'22'$, $P\bar{3}'22'$, $P3'42''$, $P3'\bar{4}\,\bar{2}''$, $P\bar{3}'42''$, $I3'22'$, $I\bar{3}'22'$, $I3'42''$, $I3'\bar{4}\,\bar{2}''$, $I\bar{3}'42''$, $F3'22'$, $F\bar{3}'22'$, $F3'42''$, $F3'\bar{4}\,\bar{2}''$, and $F\bar{3}'42''$ in the cubic crystal system.

Seven crystal systems, 14 Bravais lattices and 73 symmorphic space groups as well as the relations between the lattice bases and the rotational axes are listed in Table 5.4. The matrices of the generator are listed in the note.

Table 5.4 The property of symmorphic space groups

Crystal system, Bravais	Point groups		Product of subgroups	Lattice bases	Direction of rotational axes		
	Sch.	INSG					
Triclinic, P	C_1	1	C_1				
$P1$, $P\bar{1}$	C_i	$\bar{1}$	C_i				
Monoclinic,						$\boldsymbol{a}_3$	
P, A, $no.=6$	C_2	2	C_2	$\alpha_1=\alpha_2$		2	
$P2$, $P\bar{2}$, $P\pm 2$,	C_s	$\bar{2}$	C_s	$=\pi/2$		$\bar{2}$	
$A2$, $A\bar{2}$, $A\pm 2$	C_{2h}	± 2	C_iC_2			± 2	
Orthorhombic,						$\boldsymbol{a}_3$	$\boldsymbol{a}_1$
P, $C(A)$, I, F	D_2	$22'$	$C_2C'_2$	$\alpha_1=\alpha_2$		2	2
$no.=13$	C_{2v}	$2\bar{2}'$	$C_2C'_s$	$=\alpha_3$		± 2	± 2
see Note (e)	D_{2h}	$\pm 22'$	$C_iC_2C'_2$	$=\pi/2$		2	$\bar{2}$
Tetragonal,					$\boldsymbol{a}_3$	$\boldsymbol{a}_1$	$\boldsymbol{a}_1+\boldsymbol{a}_2$
P,I $no.=16$	C_4	4	C_4	$a_1=a_2$	4		
$P4$, $P\bar{4}$, $P\pm 4$,	S_4	$\bar{4}$	S_4	$\alpha_1=\alpha_2$	$\bar{4}$		
$P42'$, $P4\bar{2}'$,	C_{4h}	± 4	C_iC_4	$=\alpha_3$	± 4		
$P\bar{4}2'$, $P\bar{4}2''$,	D_4	$42'$	$C_4C'_2$	$=\pi/2$	4	2	2
$P\pm 42'$, $I4$,	C_{4v}	$4\bar{2}'$	$C_4C'_s$		4	$\bar{2}$	$\bar{2}$
$I\bar{4}$, $I\pm 4$, $I42'$,	D_{2d}	$\bar{4}2'$	$S_4C'_2$		$\bar{4}$	2	$\bar{2}$
$I4\bar{2}'$, $I\bar{4}2'$,	D_{2d}	$\bar{4}2''$	$S_4C''_2$		$\bar{4}$	$\bar{2}$	2
$I\bar{4}2''$, $I\pm 42'$	D_{4h}	$\pm 42'$	$C_iC_4C'_2$		± 4	± 2	± 2
Cubic,					$\boldsymbol{a}_3$	$\boldsymbol{a}'$	$\boldsymbol{a}_1+\boldsymbol{a}_2$
	T	$3'22'$	$C'_3C_2C'_2$	$a_1=a_2$	2	3	
P, I, F	T_h	$\bar{3}22'$	$C'_{3i}C_2C'_2$	$=a_3$	± 2	$\bar{3}$	
	O	$3'42''$	$C'_3C_4C''_2$	$\alpha_1=\alpha_2$	4	3	2
$no.=15$	T_d	$3'\bar{4}\,\bar{2}''$	$C_3S_4C''_s$	$=\alpha_3$	$\bar{4}$	3	$\bar{2}$
see Note (f)	O_h	$\bar{3}'42''$	$C'_{3i}C_4C''_2$	$=\pi/2$	± 4	$\bar{3}$	± 2
Rhombo-						$\boldsymbol{a}'$	$\boldsymbol{a}_1-\boldsymbol{a}_2$
hedral,	C_3	3	C_3	$a_1=a_2$		3	
R	C_{3i}	$\bar{3}$	C_{3i}	$=a_3$		$\bar{3}$	
$no.=5$	D_3	$32'$	$C_3C'_2$	$\alpha_1=\alpha_2$		3	2
$R3$, $R\bar{3}$, $R32'$,	C_{3v}	$3\bar{2}'$	$C_3C'_s$	$=\alpha_3$		3	$\bar{2}$
$R3\bar{2}'$, $R\bar{3}2'$	D_{3d}	$\bar{3}2'$	$C_{3i}C'_2$			$\bar{3}$	± 2

(continued)

Crystal system, Bravais	Point groups Sch.	INSG	Product of subgroups	Lattice bases	Direction of rotational axes $\boldsymbol{a}_3$	$\boldsymbol{a}_1$	$\boldsymbol{a}_1 - \boldsymbol{a}_2$
Hexagonal,							
	C_3	3	C_3	$a_1 = a_2$	3		
P	C_{3i}	$\overline{3}$	C_{3i}	$\alpha_1 = \alpha_2$	$\overline{3}$		
	D_3	$32'$	C_3C_2'	$= \pi/2$	3	2	
	D_3	$32''$	C_3C_2''	$\alpha_3 = 2\pi/3$	3		2
	C_{3v}	$3\overline{2}'$	C_3C_s'		3	$\overline{2}$	
	C_{3v}	$3\overline{2}''$	C_3C_s''		3		$\overline{2}$
$no. = 16$	D_{3d}	$\overline{3}2'$	$C_{3i}C_2'$		$\overline{3}$	± 2	
	D_{3d}	$\overline{3}2''$	$C_{3i}C_2''$		$\overline{3}$		± 2
$P3$, $P\overline{3}$,	C_6	6	C_6		6		
$P32'$, $P32''$,	C_{3h}	$\overline{6}$	C_{3h}		$\overline{6}$		
$P3\overline{2}'$, $P3\overline{2}''$,	C_{6h}	± 6	C_iC_6		± 6		
$P\overline{3}2'$, $P\overline{3}2''$	D_6	$62'$	C_6C_2'		6	2	2
$P6$, $P\overline{6}$,	C_{6v}	$6\overline{2}'$	C_6C_s'		6	$\overline{2}$	$\overline{2}$
$P\pm 6$, $P62'$,	D_{3h}	$\overline{6}2'$	$C_{3h}C_2'$		$\overline{6}$	2	$\overline{2}$
$P6\overline{2}'$, $P\overline{6}2'$,	D_{3h}	$\overline{6}2''$	$C_{3h}C_2''$		$\overline{6}$	$\overline{2}$	2
$P\overline{6}2''$, $P\pm 62'$	D_{6h}	$\pm 62'$	$C_iC_6C_2'$		± 6	± 2	± 2

Note. (a) Generators in the hexagonal crystal system:

$$C_6 = \begin{pmatrix} 1 & -1 & 0 \\ 1 & 0 & 0 \\ 0 & 0 & 1 \end{pmatrix}, \quad C_3 = \begin{pmatrix} 0 & -1 & 0 \\ 1 & -1 & 0 \\ 0 & 0 & 1 \end{pmatrix}, \quad C_2' = \begin{pmatrix} 1 & -1 & 0 \\ 0 & -1 & 0 \\ 0 & 0 & -1 \end{pmatrix}, \quad C_2'' = \begin{pmatrix} 0 & -1 & 0 \\ -1 & 0 & 0 \\ 0 & 0 & -1 \end{pmatrix}.$$

(b) Generators in the rhombohedral crystal system:

$$C_3 = \begin{pmatrix} 0 & 0 & 1 \\ 1 & 0 & 0 \\ 0 & 1 & 0 \end{pmatrix}, \quad C_2' = \begin{pmatrix} 0 & -1 & 0 \\ -1 & 0 & 0 \\ 0 & 0 & -1 \end{pmatrix}.$$

(c) Generators of other cyclic subgroups:

$$C_2 = \begin{pmatrix} -1 & 0 & 0 \\ 0 & -1 & 0 \\ 0 & 0 & 1 \end{pmatrix}, \quad C_2' = \begin{pmatrix} 1 & 0 & 0 \\ 0 & -1 & 0 \\ 0 & 0 & -1 \end{pmatrix}, \quad C_2'' = \begin{pmatrix} 0 & 1 & 0 \\ 1 & 0 & 0 \\ 0 & 0 & -1 \end{pmatrix},$$

$$C_4 = \begin{pmatrix} 0 & -1 & 0 \\ 1 & 0 & 0 \\ 0 & 0 & 1 \end{pmatrix}, \quad C_3 = \begin{pmatrix} 0 & 0 & 1 \\ 1 & 0 & 0 \\ 0 & 1 & 0 \end{pmatrix}.$$

(d) $\boldsymbol{a}' = \boldsymbol{a}_1 + \boldsymbol{a}_2 + \boldsymbol{a}_3$.

(e) There are 13 symmorphic space groups $P22'$, $P2\overline{2}'$, $P\pm 22'$, $C22'$, $C2\overline{2}'$, $A2\overline{2}'$, $C\pm 22'$, $I22'$, $I2\overline{2}'$, $I\pm 22'$, $F22'$, $F2\overline{2}'$, and $F\pm 22'$ in the orthorhombic crystal system.

(f) There are 15 symmorphic space groups $P3'22'$, $P\overline{3}'22'$, $P3'42''$, $P3'\overline{4}\,\overline{2}''$, $P\overline{3}'42''$, $I3'22'$, $I\overline{3}'22'$, $I3'42''$, $I3'\overline{4}\,\overline{2}''$, $I\overline{3}'42''$, $F3'22'$, $F\overline{3}'22'$, $F3'42''$, $F3'\overline{4}\,\overline{2}''$, and $F\overline{3}'42''$ in the cubic crystal system.

5.4 Space Group

In this section we will study the space group $\mathcal{S}$ where there is at least one symmetric operation $g(R, \boldsymbol{t})$ with nonvanishing $\boldsymbol{t}$. We will first study the constraint on $\boldsymbol{t}$ from R and the dependence of $\boldsymbol{t}$ on the choice of the origin. Then, we will briefly introduce the method of finding the inequivalent space groups. At last, we will analyze the symmetric property of a crystal from the symbol of its space group.

5.4.1 *Symmetric Elements*

For a given crystal with a crystallographic point group G, the symmetric operation in a coordinate frame is expressed generally

$$\begin{aligned} &g(R, \boldsymbol{\alpha}) = T(\boldsymbol{\ell})g(R, \boldsymbol{t}), \qquad \boldsymbol{\alpha} = \boldsymbol{\ell} + \boldsymbol{t}, \\ &\boldsymbol{t} = \sum_{j=1}^{3} \boldsymbol{a}_j t_j, \qquad 0 \leq t_j < 1, \end{aligned} \tag{5.58}$$

where $\boldsymbol{t}$ depends on R uniquely. We first study the constraint on $\boldsymbol{t}$ from R.

In this section we denote the proper rotation R by C_N and the improper rotation R by S_N for convenience, where $N = 1$, 2, 3, 4, or 6. The double-vector of the sum of the cyclic group generated by R is denoted by $\left\{\vec{\vec{R}}\right\}$. For an arbitrary vector $\boldsymbol{r}$, one has

$$\vec{\vec{R}} \cdot \left\{\vec{\vec{R}}\right\} \cdot \boldsymbol{r} = \left\{\vec{\vec{R}}\right\} \cdot \boldsymbol{r}. \tag{5.59}$$

It means that the vector $\left\{\vec{\vec{R}}\right\} \cdot \boldsymbol{r}$ does not change in the rotation R. On the other hand, C_N $(N \neq 1)$ preserves only the vector along the rotational axis $\hat{\boldsymbol{n}}$ invariant. S_2 is a reflection with respect to a plane perpendicular to the improper axis and preserves only the vector on the plane invariant. There is no vector invariant in S_N $(N \neq 2)$. Therefore,

$$\begin{aligned} &\left\{\vec{\vec{C_1}}\right\} = \vec{\vec{1}}, \qquad &&\left\{\vec{\vec{C_N}}\right\} = N\hat{\boldsymbol{n}}\hat{\boldsymbol{n}}, \qquad &&N \neq 1, \\ &\left\{\vec{\vec{S_2}}\right\} = 2\left(\vec{\vec{1}} - \hat{\boldsymbol{n}}\hat{\boldsymbol{n}}\right), \qquad &&\left\{\vec{\vec{S_N}}\right\} = 0, \qquad &&N \neq 2. \end{aligned} \tag{5.60}$$

Since $(R)^{N'} = E$, where $N' = 2N$ for $R = S_N$ with $N = 1$ or 3, and where $N' = N$ for the remaining cases, the constraint

$$\{g(R,\boldsymbol{t})\}^{N'} = T\left(\left\{\vec{\vec{R}}\right\}\cdot\boldsymbol{t}\right) = T(\boldsymbol{\ell}), \tag{5.61}$$

gives a strong restriction on $\boldsymbol{t}$:

$$\begin{aligned}
&\boldsymbol{t} = 0, && \text{when } R = C_1 = E, \\
&N\left\{\hat{\boldsymbol{n}}\left(\hat{\boldsymbol{n}}\cdot\boldsymbol{t}\right)\right\} = N\boldsymbol{t}_{\|} = m\boldsymbol{a}_{\|}, && \text{when } R = C_N, \quad N \neq 1, \\
&2\left\{\boldsymbol{t} - \hat{\boldsymbol{n}}\left(\hat{\boldsymbol{n}}\cdot\boldsymbol{t}\right)\right\} = 2\boldsymbol{t}_{\perp} = m\boldsymbol{a}_{\perp}, && \text{when } R = S_2, \\
&\boldsymbol{t} \text{ no restriction}, && \text{when } R = S_N, \quad N \neq 2,
\end{aligned} \tag{5.62}$$

where m is an integer, $\boldsymbol{t}_{\|} = \hat{\boldsymbol{n}}\left(\hat{\boldsymbol{n}}\cdot\boldsymbol{t}\right)$ is the parallel component of $\boldsymbol{t}$, and $\boldsymbol{t}_{\perp} = \boldsymbol{t} - \boldsymbol{t}_{\|}$ the perpendicular component of $\boldsymbol{t}$. When $R = C_N$, $N \neq 1$, $\boldsymbol{a}_{\|}$ is the shortest vector of the crystal lattice along the rotational axis $\hat{\boldsymbol{n}}$ of C_N. When $R = S_2$, $\boldsymbol{a}_{\perp}$ is the shortest vector of the crystal lattice along the direction of $\boldsymbol{t}_{\perp}$. We are going to show that the part of $\boldsymbol{t}$ restricted by the constraint is independent of the choice of the origin, and the remaining part can be removed by a suitable choice of the origin.

In the laboratory frame K,

$$g(R,\boldsymbol{t})\boldsymbol{r} = R\boldsymbol{r} + \boldsymbol{t},$$

and a symmetric operation $g(R,\boldsymbol{t})$ moves the origin O with a translation vector $\boldsymbol{t}$. Denote by $\boldsymbol{r}_0$ the position vector of O' in K. The operation $g(R,\boldsymbol{t})$ moves O' to the position $R\boldsymbol{r}_0+\boldsymbol{t}$, namely, the translation of O' in $g(R,\boldsymbol{t})$ is $(R-E)\boldsymbol{r}_0+\boldsymbol{t}$. Move in parallel the coordinate frame K to the coordinate frame K' with the origin O', where the symmetric operation $g(R,\boldsymbol{t})$ becomes g',

$$g' = g\left[R,(R-E)\boldsymbol{r}_0 + \boldsymbol{t}\right] = T(-\boldsymbol{r}_0)g(R,\boldsymbol{t})T(\boldsymbol{r}_0). \tag{5.63}$$

Hence, the operator forms of the same symmetric operation depend upon the position of the origin of the coordinate frame. The suitable choice of the position of the origin may simplify the operator forms of the symmetric operations. Obviously, only one origin can be chosen for a given crystal.

For $R = C_N$, $N \neq 1$, the shift $\boldsymbol{r}_0$ of the origin changes $\boldsymbol{t}$ with $(C_N - E)\boldsymbol{r}_0$, which is perpendicular to the rotational axis of C_N. In the perpendicular plane the eigenvalue of C_N is not equal to one such that the equation $(C_N - E)\boldsymbol{r}_0 = -\boldsymbol{t}_{\perp}$ has a unique solution $\boldsymbol{r}_0$. Thus, $\boldsymbol{t}_{\perp} = \left(\vec{\mathbf{1}} - \hat{\boldsymbol{n}}\hat{\boldsymbol{n}}\right)\cdot\boldsymbol{t}$ in $g(C_N,\boldsymbol{t})$ can be removed by the choice of the origin.

For $R = S_2$, which is a reflection with respect to the plane perpendicular to the improper rotational axis of S_2, the shift $\boldsymbol{r}_0$ of the origin changes $\boldsymbol{t}$ with $(S_2 - E)\boldsymbol{r}_0$. The equation $(S_2 - E)\boldsymbol{r}_0 = -\boldsymbol{t}_{\parallel}$ has a solution, $\boldsymbol{r}_0 = \boldsymbol{t}_{\parallel}/2$. Thus, $\boldsymbol{t}_{\parallel} = \hat{\boldsymbol{n}}\hat{\boldsymbol{n}} \cdot \boldsymbol{t}$ in $g(S_2, \boldsymbol{t})$ can be removed by the choice of the origin.

For $R = S_N$, $N \neq 2$, the shift $\boldsymbol{r}_0$ of the origin changes $\boldsymbol{t}$ with $(S_N - E)\boldsymbol{r}_0$. Since the eigenvalue of S_N is not equal to one such that the equation $(S_N - E)\boldsymbol{r}_0 = -\boldsymbol{t}$ has a unique solution $\boldsymbol{r}_0$, the vector $\boldsymbol{t}$ in $g(S_N, \boldsymbol{t})$ can be removed by the choice of the origin.

In summary, the part of $\boldsymbol{t}$ which is not restricted by the constraint can be removed by a suitable choice of the origin, but the part of $\boldsymbol{t}$ restricted by the constraint is independent of the choice of the origin. Remind that the position of the origin can be further chosen in the condition that the simplified $\boldsymbol{t}$ does not change. We will discuss this with examples later.

A point is called the symmetric center of a symmetric operation $g(R, \boldsymbol{t})$ if the point does not change in $g(R, \boldsymbol{t})$. A symmetric operation $g(R, \boldsymbol{t})$ is called closed if it has a symmetric center, otherwise, it is called an open operation. The position vector $\boldsymbol{r}_0$ of a symmetric center satisfies

$$[g(R, \boldsymbol{t}) - E]\, \boldsymbol{r}_0 = (R - E)\, \boldsymbol{r}_0 + \boldsymbol{t} = 0. \tag{5.64}$$

In comparison with Eq. (5.63), the operator form of the symmetric operation $g(R, \boldsymbol{t})$ becomes R if the symmetric center is chosen to be the origin of a new coordinate frame. Since $R^{N'} = E$ and the operator form of E is independent of the choice of the origin, one has

$$[g(R, \boldsymbol{t})]^{N'} = \begin{cases} E, & \text{closed operation,} \\ T(\boldsymbol{\ell}) \neq E, & \text{open operation.} \end{cases} \tag{5.65}$$

This is another definition for a closed or an open operation. A symmetric operation $g(R, \mathbf{t})$ is called closed if its power is equal to E, otherwise, it is called an open operation. Due to Eq. (5.61), the vector $\boldsymbol{t}$ in a closed operation $g(R, \boldsymbol{t})$ satisfies

$$\left\{ \vec{\vec{R}} \right\} \cdot \boldsymbol{t} = 0. \tag{5.66}$$

Hence, for a closed operation, m in Eq. (5.62) has to be 0 such that the vector $\boldsymbol{t}$ in a closed operation $g(R, \boldsymbol{t})$ can be removed by a suitable choice of the origin.

The position vector $\boldsymbol{r}_0$ of the symmetric center of a closed symmetric operation can be calculated from Eq. (5.64). For a closed symmetric operation $g(C_N, \boldsymbol{t})$, $N \neq 1$, every point on a straight line, which is parallel to the rotational axis of C_N and across the symmetric center of $g(C_N, \boldsymbol{t})$, is a symmetric center. For a closed symmetric operation $g(S_2, \boldsymbol{t})$, every point on a plane, which is perpendicular to the improper rotational axis of S_2 and across the symmetric center of $g(S_2, \boldsymbol{t})$, is a symmetric center.

There are only two kinds of open operations where m in Eq. (5.62) is not vanishing. In a suitable choice of the origin, two open operations can be expressed as

$$\begin{aligned} &g\left(C_N, \boldsymbol{t}_{\parallel}\right), \qquad \boldsymbol{t}_{\parallel} = m\boldsymbol{a}_{\parallel}/N \neq 0, \qquad 0 < m < N \neq 1, \\ &g\left(S_2, \boldsymbol{t}_{\perp}\right), \qquad \boldsymbol{t}_{\perp} = \boldsymbol{a}_{\perp}/2 \neq 0. \end{aligned} \tag{5.67}$$

In the first kind of open operations, $\boldsymbol{t}_{\parallel}$ is a multiple of the shortest vector of the crystal lattice along the rotational axis $\hat{\boldsymbol{n}}$ of C_N divided by N. This axis is called a screw axis, which is parallel to $\hat{\boldsymbol{n}}$ and across the point $\boldsymbol{r}_0$, where $\boldsymbol{r}_0$ is the solution of the following equation:

$$(E - C_N)\boldsymbol{r}_0 = \boldsymbol{t} - \boldsymbol{t}_{\parallel} = \boldsymbol{t}_{\perp}. \tag{5.68}$$

$\mathbf{t}_{\parallel}$ is called the gliding vector of the screw axis. In the second kind of open operations, $\boldsymbol{t}_{\perp}$ is on the reflection plane of S_2 and equal to half of the shortest vector of the crystal lattice along the direction of $\boldsymbol{t}_{\perp}$. This reflection plane is called a gliding plane, which is perpendicular to the improper rotational axis of S_2 and across the point $\boldsymbol{r}_0$, where $\boldsymbol{r}_0$ satisfies

$$(E - S_2)\boldsymbol{r}_0 = \boldsymbol{t} - \boldsymbol{t}_{\perp} = \boldsymbol{t}_{\parallel}, \qquad \boldsymbol{r}_0 = \boldsymbol{t}_{\parallel}/2. \tag{5.69}$$

$\boldsymbol{t}_{\perp}$ is called the gliding vector of the gliding plane. The gliding vectors of both the screw axis and the gliding plane have to satisfy the restriction (5.62).

A straight line is called a symmetric straight line of a symmetric operation $g(R, \boldsymbol{t})$ if the straight line does not change in $g(R, \boldsymbol{t})$. A screw axis is a symmetric straight line although there is no symmetric center on it. A plane is called a symmetric plane of a symmetric operation $g(R, \boldsymbol{t})$ if the plane does not change in $g(R, \boldsymbol{t})$. A gliding plane is a symmetric plane although there is no symmetric center on it.

5.4.2 *Symbols of a Space Group*

For a given crystal, its symmetric operations $g(R, \boldsymbol{t})$ and the distribution of the vectors of the crystal lattice determine its crystallographic point group, and then, determine its crystal system and its Bravais lattice. Neglecting the translation vectors $\boldsymbol{t}$ in $g(R, \boldsymbol{t})$ temporarily, one has the symbol of the symmorphic space group. The symbol of the space group $\mathcal{S}$ of the crystal is obtained by attaching some subscripts to the symbol of the symmorphic space group.

The basis vectors of a crystal lattice $\boldsymbol{a}_j$ and the possible vector of crystal lattice $\boldsymbol{f}$ which is a fractional combination of $\boldsymbol{a}_j$ are determined from its crystal system and its Bravais lattice. A symmetric operation of a crystal is generally written as

$$g(R, \boldsymbol{\alpha}) = T(\boldsymbol{L})g(R, \boldsymbol{t}), \qquad \boldsymbol{L} = \boldsymbol{\ell} \text{ or } \boldsymbol{\ell} + \boldsymbol{f}, \tag{5.70}$$

where $\boldsymbol{\ell}$ is an integral combination of $\boldsymbol{a}_j$. When the crystallographic point group is C_1, $\boldsymbol{t} = 0$, and there are only the symmorphic space groups. The remaining 31 crystallographic point groups G, from Table 5.3, can be expressed as a product of one, two, or three cyclic subgroups, respectively. For a given space group, each generator R of the cyclic subgroups has its corresponding translation vector $\boldsymbol{t}$. Thus, the general form of the element in a space group $\mathcal{S}$ is

$$T(\boldsymbol{L}) \left\{g(R, \boldsymbol{t})\right\}^n, \tag{5.71a}$$

$$T(\boldsymbol{L}) \left\{g(R, \boldsymbol{t})\right\}^n \left\{g(R_1, \boldsymbol{p})\right\}^{n_1}, \tag{5.71b}$$

$$T(\boldsymbol{L}) \left\{g(R, \boldsymbol{t})\right\}^n \left\{g(R_2, \boldsymbol{q})\right\}^{n_2} \left\{g(R_1, \boldsymbol{p})\right\}^{n_1}, \tag{5.71c}$$

where n, n_1, and n_2 are integers, and $\boldsymbol{t}$, $\boldsymbol{p}$, and $\boldsymbol{q}$ are the fractional combinations of $\boldsymbol{a}_j$ with the coefficients:

$$0 \leq t_j < 1, \qquad 0 \leq p_j < 1, \qquad 0 \leq q_j < 1. \tag{5.72}$$

In the international notations for the space groups (INSG), the symbol for a space group $\mathcal{S}$ is obtained from the symbol of its corresponding symmorphic space group by attaching the components of $\boldsymbol{t}$, $\boldsymbol{p}$, and $\boldsymbol{q}$ as subscripts to the numbers of the cyclic subgroups. If one of $\boldsymbol{t}$, $\boldsymbol{p}$, and $\boldsymbol{q}$ is vanishing, the corresponding 0 subscripts can be omitted.

The translation vectors $\boldsymbol{t}$, $\boldsymbol{p}$, and $\boldsymbol{q}$ have to satisfy three conditions. The first condition comes from the constraint (5.61) which was discussed in the preceding subsection. The second condition comes from the group property. The product of two elements of $\mathcal{S}$ given in Eq. (5.71) has to be an element of $\mathcal{S}$ and can also be written in the form (5.71). Especially, the translation vectors $\boldsymbol{t}$, $\boldsymbol{p}$, and $\boldsymbol{q}$ do not change in the product. The third condition comes from the choice of the origin of the coordinate frame. The choice of the origin of the coordinate frame affects the translation vectors $\boldsymbol{t}$, $\boldsymbol{p}$, and $\boldsymbol{q}$, as well as the symbol of the space group. The different symbols due to the different choices of the origin represent the same space group. They are called the equivalent symbols. For a given crystal, one has to choose one common origin of the coordinate frame. The task of group theory is to take the simplest symbol among the equivalent ones by the suitable choice of the origin and to find out all inequivalent symbols for the space groups. There are 230 space groups listed in Appendix C, where the Schoenflies notations and the international notations for the space groups are given for comparison.

5.4.3 *Method for Determining the Space Groups*

First, the origin of the coordinate frame is chosen to make the translation vector $\boldsymbol{t}$ in the factor $g(R, \boldsymbol{t})$ of Eq. (5.71) as simple as possible. According to $g(R, \boldsymbol{t})$, the space groups are classified into three types.

Type A: $R = S_N$, $N \neq 2$. There is a symmetric center for $g(S_N, \boldsymbol{t})$. Choosing the symmetric center to be the origin of the coordinate frame, one has $\boldsymbol{t} = 0$. Under the condition $\boldsymbol{t} = 0$, the origin can be further chosen to simplify $\boldsymbol{p}$ and $\boldsymbol{q}$, where the shift $\boldsymbol{r}_0$ of the origin satisfies

$$(S_N - E)\,\boldsymbol{r}_0 = \boldsymbol{L}. \tag{5.73}$$

There are 15 crystallographic point groups belonging to type A: C_i, C_{3i}, S_4, C_{3h}, C_{2h}, C_{4h}, C_{6h}, D_{2d}, D_{3d}, D_{3h}, D_{2h}, D_{4h}, D_{6h}, T_h, and O_h.

Type B: $R = C_N$, $N \neq 1$. There is a symmetric straight line for $g(C_N, \boldsymbol{t})$. Choosing the origin of the coordinate frame at the symmetric straight line, one has $\boldsymbol{t} = \boldsymbol{t}_{\parallel} = m\boldsymbol{a}_{\parallel}/N$, $0 \leq m < N$, where $\boldsymbol{a}_{\parallel}$ is the shortest vector of the crystal lattice along the rotational axis $\hat{\boldsymbol{n}}$ of C_N. The origin can be further chosen with a shift $\boldsymbol{r}_0$ satisfying

$$(C_N - E)\,\boldsymbol{r}_0 = \boldsymbol{L} + m'\boldsymbol{a}_{\parallel}/N, \qquad 0 \leq m' < N. \tag{5.74}$$

Note that the left-hand side of Eq. (5.74) is perpendicular to the rotational axis $\hat{\boldsymbol{n}}$. The second term in the right-hand side of Eq. (5.74) appears in the case where $\boldsymbol{L}$ contains a component along $\hat{\boldsymbol{n}}$. This additional term with $m' \neq 0$ in Eq. (5.74) can cancel $\boldsymbol{t}_{\parallel}$ or partly. Furthermore, the origin can still be chosen to simplify $\boldsymbol{p}$ and $\boldsymbol{q}$ under the condition that the simplified $\boldsymbol{t}$ does not change. We will discuss this simplification with example later.

There are 15 crystallographic point groups belonging to type B: C_2, C_3, C_4, C_6, D_2, C_{2v}, D_3, C_{3v}, D_4, C_{4v}, D_6, C_{6v}, T, O, and T_d.

Type C: $R = S_2$ and the crystallographic point group is C_s. There is a symmetric plane for $g(S_2, \boldsymbol{t})$. Choosing the origin of the coordinate frame at the symmetric plane, one has $\boldsymbol{t} = 0$ or $\boldsymbol{a}_\perp/2$. For the Bravais lattice P, if $\boldsymbol{t} \neq 0$, one may choose $\boldsymbol{a}_\perp$ to be the lattice basis $\boldsymbol{a}_1$. For the Bravais lattice A, if $\boldsymbol{t}$ contains a component $\boldsymbol{a}_2/2$, this component can be removed by a suitable choice of the origin because $\boldsymbol{a}_2/2 = \boldsymbol{f} - \boldsymbol{a}_3/2$. Therefore, in addition to the symmorphic space groups, there are two space groups in type C: $P\overline{2}_{\frac{1}{2}00}$ and $A\overline{2}_{\frac{1}{2}00}$.

For the space groups in type A and type B, the restrictions on the translation vectors $\boldsymbol{p}$ and $\boldsymbol{q}$ need to be studied further. It is found that the condition $R_1R^{-1} = RR_1$ holds for the space groups in the form of Eq. (5.71b) because in those space groups either $R = C_i$ or R_1 is a proper or improper 2-fold rotation with the rotational axis perpendicular to the rotational axis of R (see Table 5.3). Thus, $\boldsymbol{p}$ satisfies

$$g\left(R_1, \boldsymbol{p} + \boldsymbol{L}\right) = g\left(R, \boldsymbol{t}\right) g\left(R_1, \boldsymbol{p}\right) g\left(R, \boldsymbol{t}\right). \tag{5.75}$$

Under the restrictions (5.62) and (5.75), the origin of the coordinate frame is chosen according to Eq. (5.73) or Eq. (5.74) to make $\boldsymbol{p}$ as simple as possible.

For the space groups in the form of Eq. (5.71c), the restriction like Eq. (5.75) becomes more complicated. Through a direct calculation one obtains

$$\begin{array}{lll} R_1R_2^{-1} = R_2R_1, & R_1R = R^iR_2^jR_1^k, & R_2R = R^{i'}R_2^{j'}R_1^{k'}, \\ i = k = i' = j' = 1, \quad j = k' = 0, & & \text{for } D_{2h},\ D_{4h}, \text{ and } D_{6h}, \\ i = j = i' = j' = k' = 1, \quad k = 0, & & \text{for T and } T_h, \\ i = i' = j' = -1, \quad j = 1, \quad k = k' = 0, & & \text{for O},\ T_d, \text{ and } O_h. \end{array} \tag{5.76}$$

Thus, $\boldsymbol{p}$ and $\boldsymbol{q}$ satisfy

$$\begin{aligned} g\left(R_1, \boldsymbol{p} + \boldsymbol{L}_1\right) &= g\left(R_2, \boldsymbol{q}\right) g\left(R_1, \boldsymbol{p}\right) g\left(R_2, \boldsymbol{q}\right), \\ g\left(R_1, \boldsymbol{p} + \boldsymbol{L}_2\right) &= g\left(R, \boldsymbol{t}\right)^i g\left(R_2, \boldsymbol{q}\right)^j g\left(R_1, \boldsymbol{p}\right)^k g\left(R, \boldsymbol{t}\right)^{-1}, \\ g\left(R_2, \boldsymbol{q} + \boldsymbol{L}_3\right) &= g\left(R, \boldsymbol{t}\right)^{i'} g\left(R_2, \boldsymbol{q}\right)^{j'} g\left(R_1, \boldsymbol{p}\right)^{k'} g\left(R, \boldsymbol{t}\right)^{-1}. \end{aligned} \tag{5.77}$$

Under the restrictions (5.62) and (5.77), the origin of the coordinate frame is chosen according to Eq. (5.73) or Eq. (5.74) to make $\boldsymbol{p}$ and $\boldsymbol{q}$ as simple as possible.

5.4.4 *Example for the Space Groups in Type A*

Discuss a crystal with the crystallographic point group C_{4h} (± 4), which can be expressed as the product of subgroups C_iC_4. The crystal belongs to the tetragonal crystal system, $a_1 = a_2$ and $\alpha_1 = \alpha_2 = \alpha_3 = \pi/2$. The lattice basis $\boldsymbol{a}_3$ is along a 4-fold axis. The double-vector of C_4 is

$$\vec{\bar{4}} = \boldsymbol{a}_2\boldsymbol{b}_1 - \boldsymbol{a}_1\boldsymbol{b}_2 + \boldsymbol{a}_3\boldsymbol{b}_3.$$

There are two Bravais lattices P and I. $\boldsymbol{f} = 0$ for the Bravais lattices P, and $\boldsymbol{f} = (\boldsymbol{a}_1 + \boldsymbol{a}_2 + \boldsymbol{a}_3)/2$ for the Bravais lattices I.

The origin of the coordinate frame is taken at the symmetric center of $g(S_1, \boldsymbol{t})$, then, $\boldsymbol{t} = 0$. The origin is allowed to make further shift $\boldsymbol{r}_0$ satisfying $(S_1 - E)\,\boldsymbol{r}_0 = -2\boldsymbol{r}_0 = \boldsymbol{L}$.

$$-\boldsymbol{r}_0 = \frac{\boldsymbol{L}}{2} = \begin{cases} \boldsymbol{\ell}/2, & P \text{ lattice,} \\ \boldsymbol{\ell}/2 \ \text{ or } \ (\boldsymbol{a}_1 + \boldsymbol{a}_2 + \boldsymbol{a}_3)/4, & I \text{ lattice.} \end{cases}$$

Because

$$(C_4 - E)\boldsymbol{\ell}/2 = -(\ell_2 + \ell_1)\,\boldsymbol{a}_1/2 + (\ell_1 - \ell_2)\,\boldsymbol{a}_2/2,$$
$$(C_4 - E)\,(\boldsymbol{a}_1 + \boldsymbol{a}_2 + \boldsymbol{a}_3)/4 = -\boldsymbol{a}_1/2,$$

$\boldsymbol{p}$ changes in the shift $\boldsymbol{r}_0$ of the origin

$$(C_4 - E)\boldsymbol{r}_0 = \begin{cases} (\boldsymbol{a}_1 - \boldsymbol{a}_2)/2 \ \text{ or } \ (\boldsymbol{a}_1 + \boldsymbol{a}_2)/2, & P \text{ lattice,} \\ (\boldsymbol{a}_1 - \boldsymbol{a}_2)/2, \ (\boldsymbol{a}_1 + \boldsymbol{a}_2)/2, \text{ or } \boldsymbol{a}_1/2, & I \text{ lattice.} \end{cases} \tag{5.78}$$

From Eq. (5.75) one has

$$S_1 g(C_4, \boldsymbol{p}) S_1 = g(C_4, -\boldsymbol{p}) = g(C_4, \boldsymbol{p} + \boldsymbol{L}'),$$

$$\boldsymbol{p} = -\frac{\boldsymbol{L}'}{2} = \begin{cases} \boldsymbol{\ell}/2, & P \text{ lattice,} \\ \boldsymbol{\ell}/2 \ \text{ or } \ (\boldsymbol{a}_1 + \boldsymbol{a}_2 + \boldsymbol{a}_3)/4, & I \text{ lattice.} \end{cases}$$

Thus, in the Bravais lattice P, $\boldsymbol{p}$ can be

$$\begin{array}{llll} 0, & \boldsymbol{a}_1/2, & (\boldsymbol{a}_1 + \boldsymbol{a}_2)/2, & (\boldsymbol{a}_1 + \boldsymbol{a}_3)/2, \\ \boldsymbol{a}_2/2, & \boldsymbol{a}_3/2, & (\boldsymbol{a}_2 + \boldsymbol{a}_3)/2, & (\boldsymbol{a}_1 + \boldsymbol{a}_2 + \boldsymbol{a}_3)/2. \end{array}$$

Due to Eq. (5.78), $\boldsymbol{a}_1/2$ and $\boldsymbol{a}_2/2$ are equivalent, $(\boldsymbol{a}_1 + \boldsymbol{a}_3)/2$ and $(\boldsymbol{a}_2 + \boldsymbol{a}_3)/2$ are equivalent, $(\boldsymbol{a}_1 + \boldsymbol{a}_2 + \boldsymbol{a}_3)/2$ and $\boldsymbol{a}_3/2$ are equivalent, and $(\boldsymbol{a}_1 + \boldsymbol{a}_2)/2$ can be removed, namely, $\boldsymbol{p}$ can be 0, $\boldsymbol{a}_1/2$, $\boldsymbol{a}_3/2$, and $(\boldsymbol{a}_2 + \boldsymbol{a}_3)/2$ in the Bravais lattice P. In the Bravais lattice I, because $(\boldsymbol{a}_1 + \boldsymbol{a}_2 + \boldsymbol{a}_3)/2$ is a vector of crystal lattice, from Eq. (5.78) $\boldsymbol{a}_1/2$, $\boldsymbol{a}_3/2$, and $(\boldsymbol{a}_2 + \boldsymbol{a}_3)/2$ can be removed, but there is another $\boldsymbol{p} = (\boldsymbol{a}_1 + \boldsymbol{a}_2 + \boldsymbol{a}_3)/4$. In summary, the space groups with the crystallographic point group C_{4h} (± 4) are

$$P \pm 4, \quad P \pm 4_{\frac{1}{2}00}, \quad P \pm 4_{00\frac{1}{2}}, \quad P \pm 4_{0\frac{1}{2}\frac{1}{2}}, \quad I \pm 4, \quad I \pm 4_{\frac{1}{4}\frac{1}{4}\frac{1}{4}}. \tag{5.79}$$

5.4.5 *Example for the Space Groups in Type B*

Discuss a crystal with the crystallographic point group D_3, which can be expressed as the product of subgroups C_3C_2' or C_3C_2'', depending on whether the lattice basis $\boldsymbol{a}_1$ is along a 2-fold axis or not. Correspondingly, its international symbol is $32'$ or $32''$. The crystal belongs to the hexagonal crystal system or the rhombohedral crystal system. In both cases, $\boldsymbol{f} = 0$.

(a) *The Bravais lattice P in the hexagonal system.*

$\boldsymbol{a}_3$ is along the 3-fold axis, $a_1 = a_2$, $\alpha_1 = \alpha_2 = \pi/2$, and $\alpha_3 = 2\pi/3$. The double-vectors of the generators are

$$\begin{aligned}
\vec{3} &= \boldsymbol{a}_2\boldsymbol{b}_1 - (\boldsymbol{a}_1 + \boldsymbol{a}_2)\boldsymbol{b}_2 + \boldsymbol{a}_3\boldsymbol{b}_3, && \text{for } 32' \text{ or } 32'', \\
\vec{2}' &= \boldsymbol{a}_1\boldsymbol{b}_1 - (\boldsymbol{a}_1 + \boldsymbol{a}_2)\boldsymbol{b}_2 - \boldsymbol{a}_3\boldsymbol{b}_3, && \text{for } 32', \\
\vec{2}'' &= -\boldsymbol{a}_2\boldsymbol{b}_1 - \boldsymbol{a}_1\boldsymbol{b}_2 - \boldsymbol{a}_3\boldsymbol{b}_3, && \text{for } 32''.
\end{aligned}$$

The origin of the coordinate frame is taken at the symmetric straight line of $g(C_3, \boldsymbol{t})$ such that $\boldsymbol{t} = 0$, $\boldsymbol{a}_3/3$, or $2\boldsymbol{a}_3/3$. The origin is allowed to make further shift $\boldsymbol{r}_0$ satisfying

$$\begin{gathered}
(C_3 - E)\,\boldsymbol{r}_0 = -\boldsymbol{a}_1(r_{01} + r_{02}) + \boldsymbol{a}_2(r_{01} - 2r_{02}) = \boldsymbol{\ell}, \\
r_{01} = (-2\ell_1 + \ell_2)/3, \qquad r_{02} = -(\ell_1 + \ell_2)/3, \qquad r_{03} \ \text{arbitrary},
\end{gathered}$$

where ℓ_1 and ℓ_2 are integers. For the crystallographic point group $32'$, $\boldsymbol{p}$ changes in the shift $\boldsymbol{r}_0$ of the origin

$$\begin{aligned}
(C_2' - E)\boldsymbol{r}_0 &= -\boldsymbol{a}_1 r_{02} - \boldsymbol{a}_2 2r_{02} - \boldsymbol{a}_3 2r_{03} \\
&= (\boldsymbol{a}_1 + 2\boldsymbol{a}_2)(\ell_1 + \ell_2)/3 - \boldsymbol{a}_3 2r_{03}.
\end{aligned} \tag{5.80}$$

For the crystallographic point group $32''$, $\boldsymbol{p}$ changes in the shift $\boldsymbol{r}_0$ of the origin

$$\begin{aligned}
(C_2'' - E)\boldsymbol{r}_0 &= -(\boldsymbol{a}_1 + \boldsymbol{a}_2)(r_{01} + r_{02}) - \boldsymbol{a}_3 2r_{03} \\
&= (\boldsymbol{a}_1 + \boldsymbol{a}_2)\ell_1 - \boldsymbol{a}_3 2r_{03}.
\end{aligned} \tag{5.81}$$

Note that $(\boldsymbol{a}_1 + 2\boldsymbol{a}_2)$ is perpendicular to the rotational axis of C_2', and $(\boldsymbol{a}_1 + \boldsymbol{a}_2)$ is perpendicular to the rotational axis of C_2''.

On the other hand, for the crystallographic point group $32'$, from Eq. (5.62), $p_1 = 0$ or $1/2$. Due to Eq. (5.75) $\boldsymbol{p}$ has to satisfy

$$\begin{aligned} g(C_2', \boldsymbol{p} + \boldsymbol{\ell}') &= g(C_3, \boldsymbol{t}) g(C_2', \boldsymbol{p}) g(C_3, \boldsymbol{t}) \\ &= g(C_2', \boldsymbol{t} + C_3 \boldsymbol{p} + C_3 C_2' \boldsymbol{t}) = g(C_2', C_3 \boldsymbol{p}), \end{aligned}$$

where $\boldsymbol{t} + C_3 C_2' \boldsymbol{t} = 0$ because $\boldsymbol{t}$ is along $\boldsymbol{a}_3$. Hence,

$$\boldsymbol{\ell}' = (C_3 - E)\boldsymbol{p} = -\boldsymbol{a}_1(p_1 + p_2) + \boldsymbol{a}_2(p_1 - 2p_2).$$

The solution is that p_3 is arbitrary, $p_1 = (-2\ell_1' + \ell_2')/3$, and $p_2 = -(\ell_1' + \ell_2')/3$. Since ℓ_1' and ℓ_2' are integers, $p_1 = 1/2$ is ruled out, and then, $p_1 = p_2 = 0$. p_3 can be removed by Eq. (5.80).

For the crystallographic point group $32''$, the shortest vector of crystal lattice along the rotational axis of C_2'' is $(\boldsymbol{a}_1 - \boldsymbol{a}_2)$. From the constraint (5.61) the following vector is a vector of crystal lattice

$$(E + C_2'')\boldsymbol{p} = (\boldsymbol{a}_1 - \boldsymbol{a}_2)(p_1 - p_2).$$

Thus, $p_1 = p_2$. Due to Eq. (5.75), $\boldsymbol{p}$ has to satisfy

$$\begin{aligned} g(C_2'', \boldsymbol{p} + \boldsymbol{\ell}') &= g(C_3, \boldsymbol{t}) g(C_2'', \boldsymbol{p}) g(C_3, \boldsymbol{t}) \\ &= g(C_2'', \boldsymbol{t} + C_3 \boldsymbol{p} + C_3 C_2'' \boldsymbol{t}) = g(C_2'', C_3 \boldsymbol{p}), \end{aligned}$$

where $\boldsymbol{t} + C_3 C_2'' \boldsymbol{t} = 0$, because $\boldsymbol{t}$ is along $\boldsymbol{a}_3$. Hence,

$$\boldsymbol{\ell}' = (C_3 - E)\boldsymbol{p} = -\boldsymbol{a}_1(p_1 + p_2) + \boldsymbol{a}_2(p_1 - 2p_2) = -2p_1\boldsymbol{a}_1 - p_1\boldsymbol{a}_2,$$

namely, $p_1 = p_2 = 0$. p_3 can be removed by Eq. (5.81).

In summary, the space groups with the crystallographic point group D_3 in the hexagonal crystal system are

$$P32', \quad P3_{00\frac{1}{3}}2', \quad P3_{00\frac{2}{3}}2', \quad P32'', \quad P3_{00\frac{1}{3}}2'', \quad P3_{00\frac{2}{3}}2''. \tag{5.82}$$

(b) *The Bravais lattice R in the rhombohedral crystal system.*

Three lattice bases $\boldsymbol{a}_j$ are well-distributed around the 3-fold axis, $a_1 = a_2 = a_3$ and $\alpha_1 = \alpha_2 = \alpha_3$. The double-vectors of the generators are

$$\vec{3} = \boldsymbol{a}_2\boldsymbol{b}_1 + \boldsymbol{a}_3\boldsymbol{b}_2 + \boldsymbol{a}_1\boldsymbol{b}_3, \qquad \vec{2}' = -\boldsymbol{a}_2\boldsymbol{b}_1 - \boldsymbol{a}_1\boldsymbol{b}_2 - \boldsymbol{a}_3\boldsymbol{b}_3.$$

The origin of the coordinate frame is taken at the symmetric straight line of $g(C_3, \boldsymbol{t})$, then,

$$\boldsymbol{t} = \boldsymbol{t}_{\parallel} = m\boldsymbol{a}_{\parallel}/3, \qquad \boldsymbol{a}_{\parallel} = \boldsymbol{a}_1 + \boldsymbol{a}_2 + \boldsymbol{a}_3, \qquad m = 0,\ 1,\ 2.$$

The origin makes a further shift $\boldsymbol{r}_0$ according to Eq. (5.74)

$$\begin{aligned}\boldsymbol{\ell}+(\boldsymbol{a}_1+\boldsymbol{a}_2+\boldsymbol{a}_3)m'/3 &= (C_3-E)\,\boldsymbol{r}_0 \\ &= \boldsymbol{a}_1(-r_{01}+r_{03})+\boldsymbol{a}_2(r_{01}-r_{02})+\boldsymbol{a}_3(r_{02}-r_{03}).\end{aligned} \tag{5.83}$$

Since the sum of the three coefficients with respect to the lattice bases $\boldsymbol{a}_j$ on the right-hand side of Eq. (5.83) is 0, that on the left-hand side has to be zero, $m' = -\sum_j \ell_j$, namely, $\boldsymbol{t}$ in $g(C_3, \boldsymbol{t})$ can be removed by a suitable choice of the origin. The reason is that the component of $\boldsymbol{a}_j$ along the 3-fold axis is equal to $\boldsymbol{a}_{\|}/3$.

Under the condition $\boldsymbol{t}=0$, the origin is allowed to make further shift,

$$(C_3-E)\,\boldsymbol{r}_0 = \boldsymbol{a}_1(-r_{01}+r_{03})+\boldsymbol{a}_2(r_{01}-r_{02})+\boldsymbol{a}_3(r_{02}-r_{03}) = \boldsymbol{\ell}.$$

The solution is that r_{03} is arbitrary, and both $r_{01}-r_{03}$ and $r_{02}-r_{03}$ are integers. $\boldsymbol{p}$ changes in the shift of the origin

$$\begin{aligned}(C_2'-E)\boldsymbol{r}_0 &= -(\boldsymbol{a}_1+\boldsymbol{a}_2)(r_{01}+r_{02})-2\boldsymbol{a}_3 r_{03} \\ &= -(\boldsymbol{a}_1+\boldsymbol{a}_2+\boldsymbol{a}_3)2r_{03}-(\boldsymbol{a}_1+\boldsymbol{a}_2)(r_{01}+r_{02}-2r_{03}),\end{aligned} \tag{5.84}$$

namely, $\boldsymbol{p}$ changes an arbitrary multiple of $(\boldsymbol{a}_1+\boldsymbol{a}_2+\boldsymbol{a}_3)$.

On the other hand, from Eq. (5.61), $\boldsymbol{p}$ satisfies

$$(C_2'+E)\boldsymbol{p} = (\boldsymbol{a}_1-\boldsymbol{a}_2)(p_1-p_2) = \boldsymbol{\ell}.$$

Thus, $p_1 = p_2$. From Eq. (5.75) one has

$$g(C_2', \boldsymbol{p}+\boldsymbol{\ell}') = C_3 g(C_2', \boldsymbol{p}) C_3 = g(C_2', C_3\boldsymbol{p}).$$

Hence,

$$\boldsymbol{\ell}' = (C_3-E)\boldsymbol{p} = \boldsymbol{a}_1(-p_1+p_3)+\boldsymbol{a}_2(p_1-p_2)+\boldsymbol{a}_3(p_2-p_3).$$

The solution is $p_1 = p_2 = p_3$ which is just removed by a suitable choice of the origin [see Eq. (5.84)]. Therefore, the space groups with the crystallographic point group D_3 in the rhombohedral crystal system is only the symmorphic space group $R32$.

5.4.6 *Analysis of the Symmetry of a Crystal*

We have introduced the method for finding the space groups of crystals by group theory. Perhaps, the important problem for most readers is how to analyze the symmetry of a crystal from its international symbol of the space group. We will demonstrate this problem through an example.

We study a crystal with the space group

$$I \pm 4_{\frac{1}{4}\frac{1}{4}\frac{1}{4}} 2'_{\frac{1}{2}00}.$$

From the international symbol one knows that the crystal belongs to the Bravais lattice I in the tetragonal crystal system with the crystallographic point group D_{4h}. $\boldsymbol{a}_3$ is along the 4-fold axis, $\boldsymbol{a}_1$ is along a 2-fold axis, and $\boldsymbol{a}_2 = C_4\boldsymbol{a}_1$. $\alpha_1 = \alpha_2 = \alpha_3 = \pi/2$ and $a_1 = a_2$. The shortest vector of crystal lattice along an inequivalent 2-fold axis is $\boldsymbol{a}_1 \pm \boldsymbol{a}_2$ with the length $\sqrt{2}a_1$. The lattice bases are not primitive. There is a vector of crystal lattice $\boldsymbol{f}$ which is a fractional combination of the lattice bases $\boldsymbol{a}_j$

$$\boldsymbol{L} = \boldsymbol{\ell} + \boldsymbol{f}, \qquad \boldsymbol{f} = (\boldsymbol{a}_1 + \boldsymbol{a}_2 + \boldsymbol{a}_3)/2.$$

The double-vectors of the generators in the space group are

$$C_4 = \boldsymbol{a}_2\boldsymbol{b}_1 - \boldsymbol{a}_1\boldsymbol{b}_2 + \boldsymbol{a}_3\boldsymbol{b}_3,$$
$$C'_2 = \boldsymbol{a}_1\boldsymbol{b}_1 - \boldsymbol{a}_2\boldsymbol{b}_2 - \boldsymbol{a}_3\boldsymbol{b}_3.$$

The general form of an element in the space group is

$$T(\boldsymbol{\ell})T(\boldsymbol{f})^{n_1} S_1^{n_2} g\left[C_4, (\boldsymbol{a}_1 + \boldsymbol{a}_2 + \boldsymbol{a}_3)/4\right]^m g\left(C'_2, \boldsymbol{a}_1/2\right)^{n_3},$$

where S_1 is the spatial inversion with respect to the origin of the coordinate system. n_1, n_2, and n_3 are equal to 0 or 1, respectively. m is taken to be 0, 1, 2, or 3.

The rotational axis of $g(C_4, \boldsymbol{q})$ is parallel to $\boldsymbol{a}_3$, and $\boldsymbol{q}_{\parallel} = \boldsymbol{a}_3/4$, $\boldsymbol{q}_{\perp} = (\boldsymbol{a}_1 + \boldsymbol{a}_2)/4$. Its screw axis is parallel to $\boldsymbol{a}_3$ and across the point $\boldsymbol{r}_0$

$$(C_4 - E)\,\boldsymbol{r}_0 + \boldsymbol{q}_{\perp} = 0, \qquad r_{01} + r_{02} = -r_{01} + r_{02} = 1/4.$$

The solution is $r_{01} = 0$, $r_{02} = 1/4$, namely, the screw axis of $g(C_4, \boldsymbol{q})$ is parallel to $\boldsymbol{a}_3$ and across the point $\boldsymbol{a}_2/4$ with the gliding vector $\boldsymbol{a}_3/4$.

The rotational axis of $g(C'_2, \boldsymbol{p})$ is parallel to $\boldsymbol{a}_1$, and $\boldsymbol{p} = \boldsymbol{p}_{\parallel} = \boldsymbol{a}_1/2$. The screw axis of $g(C'_2, \boldsymbol{p})$ is parallel to $\boldsymbol{a}_1$ and across the origin with the gliding vector $\boldsymbol{a}_1/2$. Since $S_1 g\left(C'_2, \boldsymbol{a}_1/2\right) = T(-\boldsymbol{a}_1) g\left(S'_2, \boldsymbol{a}_1/2\right)$, where S'_2 is an improper 2-fold rotation along $\boldsymbol{a}_1$, $\boldsymbol{p} = \boldsymbol{p}_{\parallel} = \boldsymbol{a}_1/2$, and the symmetric plane of $g\left(S'_2, \boldsymbol{a}_1/2\right)$ is perpendicular to $\boldsymbol{a}_1$ and across the point $\boldsymbol{a}_1/4$.

From an arbitrary point $\boldsymbol{r} = \boldsymbol{a}_1 x_1 + \boldsymbol{a}_2 x_2 + \boldsymbol{a}_3 x_3$, one can obtain eight equivalent points in a crystal cell through the symmetric operations

$g\left[C_4, (\boldsymbol{a}_1+\boldsymbol{a}_2+\boldsymbol{a}_3)/4\right]$ and $g\left(C_2', \boldsymbol{a}_1/2\right)$:

$$\begin{aligned}
\boldsymbol{r}_1 &= \boldsymbol{a}_1 x_1 + \boldsymbol{a}_2 x_2 + \boldsymbol{a}_3 x_3, \\
\boldsymbol{r}_2 &= \boldsymbol{a}_1\left(-x_2+1/4\right) + \boldsymbol{a}_2\left(x_1+1/4\right) + \boldsymbol{a}_3\left(x_3+1/4\right), \\
\boldsymbol{r}_3 &= -\boldsymbol{a}_1 x_1 + \boldsymbol{a}_2\left(-x_2+1/2\right) + \boldsymbol{a}_3\left(x_3+1/2\right), \\
\boldsymbol{r}_4 &= \boldsymbol{a}_1\left(x_2-1/4\right) + \boldsymbol{a}_2\left(-x_1+1/4\right) + \boldsymbol{a}_3\left(x_3+3/4\right), \\
\boldsymbol{r}_5 &= \boldsymbol{a}_1\left(x_1+1/2\right) - \boldsymbol{a}_2 x_2 - \boldsymbol{a}_3 x_3, \\
\boldsymbol{r}_6 &= \boldsymbol{a}_1\left(x_2+1/4\right) + \boldsymbol{a}_2\left(x_1+3/4\right) + \boldsymbol{a}_3\left(-x_3+1/4\right), \\
\boldsymbol{r}_7 &= \boldsymbol{a}_1\left(-x_1+1/2\right) + \boldsymbol{a}_2\left(x_2+1/2\right) + \boldsymbol{a}_3\left(-x_3+1/2\right), \\
\boldsymbol{r}_8 &= \boldsymbol{a}_1\left(-x_2-1/4\right) + \boldsymbol{a}_2\left(-x_1+3/4\right) + \boldsymbol{a}_3\left(-x_3+3/4\right).
\end{aligned}$$

Then, the remaining 24 equivalent points can be obtained through the symmetric operations $T(\boldsymbol{f})$ and S_1.

5.5 Linear Representations of Space Groups

The translation group $\mathcal{T}$ is an Abelian invariant subgroup of the space group $\mathcal{S}$. The irreducible representations of $\mathcal{T}$ are one-dimensional. The irreducible representations of $\mathcal{S}$ will be established by two steps. First, find the subduced representation $D(\boldsymbol{\ell})$ of an irreducible representation of $\mathcal{S}$ with respect to $\mathcal{T}$ which is taken to be diagonal. Then, calculate the representation matrices of the elements $g(R, \boldsymbol{t})$ in the irreducible representation of $\mathcal{S}$. In this section, the lattice bases $\boldsymbol{a}_j$ are primitive for convenience.

5.5.1 *Irreducible Representations of* $\mathcal{T}$

Denote by $\boldsymbol{a}_j$ the primitive basis vectors of crystal lattice and by $\boldsymbol{b}_i$ the basis vectors of reciprocal crystal lattice. The inner product of a vector of crystal lattice $\boldsymbol{\ell} = \sum_j \boldsymbol{a}_j \ell_j$ and a vector of reciprocal crystal lattice $\boldsymbol{K} = \sum_i \boldsymbol{b}_i K_i$, where ℓ_j and K_i are all integers, is an integer:

$$\boldsymbol{K} \cdot \boldsymbol{\ell} = \sum_j K_j \ell_j = \text{integer}. \tag{5.85}$$

If R is an element of the crystallographic point group G,

$$(R\boldsymbol{K}) \cdot \boldsymbol{\ell} = \boldsymbol{K} \cdot \left(R^{-1}\boldsymbol{\ell}\right) = \boldsymbol{K} \cdot \boldsymbol{\ell}' = \text{integer}, \tag{5.86}$$

namely, $R\boldsymbol{K} = \boldsymbol{K}'$. The crystallographic point group G is the same for both the crystal lattice and the reciprocal crystal lattice.

The real crystal is finite so that there is no rigorous translational symmetry for it (see [Ren (2006)]). Usually, the periodic boundary condition is assumed to restore the translational symmetry of the crystal such that the translation group $\mathcal{T}$ is a finite Abelian group.

A parallelepiped spanned by three lattice bases is called a crystal cell. Assume that along three lattice bases $\boldsymbol{a}_j$ there are N_j crystal cells, respectively, where N_j are very large natural numbers. Thus, the translation group $\mathcal{T}$ is a direct product of three translation subgroups

$$\mathcal{T} = \mathcal{T}^{(1)} \otimes \mathcal{T}^{(2)} \otimes \mathcal{T}^{(3)}. \tag{5.87}$$

The order of $\mathcal{T}^{(j)}$ is N_j, and the order of $\mathcal{T}$ is $N_1N_2N_3$ if there is no common factor among N_1, N_2, and N_3 for simplicity. When N_j goes to infinity, the finite crystal becomes an ideal crystal.

$\mathcal{T}^{(1)}$, for example, is a cyclic group and has N_1 one-dimensional inequivalent representations. The representation matrix of its element $T(\boldsymbol{a}_1\ell_1)$ is

$$D^{k_1}(\boldsymbol{a}_1\ell_1) = \exp\left(-i2\pi p_1\ell_1/N_1\right) = \exp\left(-i2\pi k_1\ell_1\right). \tag{5.88}$$

The one-dimensional representation D^{k_1} of $\mathcal{T}^{(1)}$ is denoted by $k_1 = p_1/N_1$, where p_1 is an integer, $0 \leq p_1 < N_1$. Thus, there are $(N_1N_2N_3)$ one-dimensional inequivalent representations characterized by a vector $\boldsymbol{k}$ in the space of the reciprocal crystal lattice

$$\boldsymbol{k} = \sum_{j=1}^{3} \boldsymbol{b}_j k_j, \qquad k_j = p_j/N_j. \tag{5.89}$$

The representation matrix of an element of $\mathcal{T}$ in the representation is

$$D^{\boldsymbol{k}}(\boldsymbol{\ell}) = e^{-i2\pi(\boldsymbol{k}\cdot\boldsymbol{\ell})} = \exp\left(-i2\pi\sum_{j=1}^{3} k_j\ell_j\right), \qquad 0 \leq k_j < 1. \tag{5.90}$$

$\boldsymbol{k}$ is called the wave vector, and the space of reciprocal crystal lattice is called the space of wave vectors, or $\boldsymbol{k}$ space. In the region, $0 \leq k_j < 1$, there are $(N_1N_2N_3)$ different wave vectors $\boldsymbol{k}$ denoting $(N_1N_2N_3)$ inequivalent irreducible representations of $\mathcal{T}$. When N_j goes to infinity, $\boldsymbol{k}$ varies continuously in that region.

If the difference $\boldsymbol{k} - \boldsymbol{k}'$ is a vector of the reciprocal crystal lattice $\boldsymbol{K}$, both $\boldsymbol{k}$ and $\boldsymbol{k}'$ characterize the same representation of $\mathcal{T}$, and are said to be the equivalent wave vectors. In this meaning, there is the translation symmetry in the $\boldsymbol{k}$ space. The region $0 \leq k_j < 1$ plays the role of a reciprocal crystal cell in the $\boldsymbol{k}$ space, just like a crystal cell in the crystal lattice. However, this region is not very convenient in practice because the point in the region may move out of the region in the rotation R of the crystallographic point group. In solid-state physics the Brillouin zone is introduced as a reciprocal crystal cell. The vertical bisector plane of the line from the origin to a point of reciprocal crystal lattice is called the Bragg plane. The region surrounded by the nearest Bragg planes to the origin is called the first Brillouin zone. The region from the first Brillouin zone to the next Bragg planes is called the second Brillouin zone, and so on. A Brillouin zone, except for the first one, is not connected, but each Brillouin zone has the same volume and contains $(N_1N_2N_3)$ wave vectors equivalent to that given in Eq. (5.89), respectively. Under a rotation R, a wave vector inside a Brillouin zone moves to another wave vector inside it, and a wave vector on its boundary moves to another wave vector on the boundary. There is a one-to-one correspondence between a wave vector inside a Brillouin zone and an inequivalent irreducible representation of $\mathcal{T}$. But the wave vectors on the boundary of a Brillouin zone may be equivalent, namely they may characterize the same representation.

5.5.2 *Star of Wave Vectors and Group of Wave Vectors*

Denote by $D(\mathcal{S})$ an m-dimensional irreducible representation of the space group $\mathcal{S}$, whose subduced representation with respect to the translation group $\mathcal{T}$ is reduced,

$$D(E, \boldsymbol{\ell}) = \text{diag}\left\{e^{-i2\pi \boldsymbol{k}_1 \cdot \boldsymbol{\ell}},\ e^{-i2\pi \boldsymbol{k}_2 \cdot \boldsymbol{\ell}},\ \cdots,\ e^{-i2\pi \boldsymbol{k}_m \cdot \boldsymbol{\ell}}\right\}. \tag{5.91}$$

Note that the subscript in $\boldsymbol{k}_\rho$ is not the component index, but the row (column) index of the matrix. The wave vectors $\boldsymbol{k}_\rho$ with a different subscript ρ is not necessary different.

Since $\mathcal{T}$ is an invariant subgroup of $\mathcal{S}$,

$$\begin{aligned} g(E, \boldsymbol{\ell})g(R, \boldsymbol{\alpha}) &= g(R, \boldsymbol{\alpha})g(E, R^{-1}\boldsymbol{\ell}), \\ D(E, \boldsymbol{\ell})D(R, \boldsymbol{\alpha}) &= D(R, \boldsymbol{\alpha})D(E, R^{-1}\boldsymbol{\ell}). \end{aligned} \tag{5.92}$$

This conjugate transformation changes $\boldsymbol{k}_\rho$ in the exponential function into $R\boldsymbol{k}_\rho$. On the other hand, the eigenvalues do not change in a similarity transformation, so $R\boldsymbol{k}_\rho$ has to be equivalent to one $\boldsymbol{k}_\tau$ in Eq. (5.91):

$$R\boldsymbol{k}_\rho = \boldsymbol{k}_\tau + \boldsymbol{K}. \tag{5.93}$$

The wave vectors in the first Brillouin zone is called mutual conjugate if they can be related by a rotation R in the crystallographic point group G through Eq. (5.93). The set of conjugate vectors is called the star of wave vectors. The number of inequivalent wave vectors in a star of wave vectors is called the index of the star, denoted by q.

Since $D(E, \boldsymbol{\ell})$ is diagonal, one obtains from Eq. (5.92)

$$\exp\left\{-i2\pi \boldsymbol{k}_\tau \cdot \boldsymbol{\ell}\right\} \ D_{\tau\rho}(R, \boldsymbol{\alpha}) = D_{\tau\rho}(R, \boldsymbol{\alpha}) \ \exp\left\{-i2\pi \left(R\boldsymbol{k}_\rho\right) \cdot \boldsymbol{\ell}\right\}. \quad (5.94)$$

Thus, $D_{\tau\rho}(R, \boldsymbol{\alpha}) \neq 0$ only if Eq. (5.93) holds. Since the representation $D(\mathcal{S})$ is irreducible, any two wave vectors $\boldsymbol{k}_\rho$ and $\boldsymbol{k}_\tau$ in Eq. (5.91) are conjugate, namely, there is an element R in G such that Eq. (5.93) holds. Otherwise, the wave vectors in Eq. (5.91) have to be divided into at least two classes, where $\boldsymbol{k}_\rho$ and $\boldsymbol{k}_\tau$ in different classes are not conjugate, so that the representation matrix entry $D_{\tau\rho}(R, \boldsymbol{\alpha})$ of every element $g(R, \boldsymbol{\alpha})$ in $\mathcal{S}$ is vanishing and the representation is reducible.

One is able to collect the same diagonal entries in Eq. (5.91) together by a simple similarity transformation,

$$D(E, \boldsymbol{\ell}) = \text{diag}\left\{e^{-i2\pi \boldsymbol{k}_1 \cdot \boldsymbol{\ell}}, \cdots, e^{-i2\pi \boldsymbol{k}_1 \cdot \boldsymbol{\ell}}, \cdots, e^{-i2\pi \boldsymbol{k}_q \cdot \boldsymbol{\ell}}, \cdots, e^{-i2\pi \boldsymbol{k}_q \cdot \boldsymbol{\ell}}\right\},$$

where $\boldsymbol{k}_\mu \neq \boldsymbol{k}_\nu$ if $\mu \neq \nu$. The multiplicity of each $\boldsymbol{k}_\mu$ is the same $d = m/q$. Each irreducible representation of $\mathcal{S}$ corresponds to a star of wave vectors, where the dimension m of the representation is a multiple of the index q of the star.

Arbitrarily choose one wave vector in the star, say $\boldsymbol{k}_1$. There are some elements P in G satisfying

$$P\boldsymbol{k}_1 = \boldsymbol{k}_1 \quad \text{or} \quad \boldsymbol{k}_1 + \boldsymbol{K}_P. \quad (5.95)$$

$\boldsymbol{k}_1$ is said to be invariant in P. The set of P forms a subgroup $H(\boldsymbol{k}_1)$ of G. The left coset of $H(\boldsymbol{k}_1)$ is denoted by $R_\mu H(\boldsymbol{k}_1)$,

$$R_\mu \boldsymbol{k}_1 = \boldsymbol{k}_\mu, \quad \text{or} \quad \boldsymbol{k}_\mu + \boldsymbol{K}_\mu. \quad (5.96)$$

The nonvanishing vector of reciprocal crystal lattice appears in Eqs. (5.95) and (5.96) only when the star of wave vectors is located at the boundary of the Brillouin zone. The index of $H(\boldsymbol{k}_1)$ is equal to the index q of the star of wave vectors. Although the choice of R_μ is not unique, R_μ is assumed to be fixed, where $R_1 = E$. Any element in G can be expressed as $R_\mu P$.

The set of $g(P, \boldsymbol{\alpha})$, where $P \in H(\boldsymbol{k}_1)$, forms a subgroup $\mathcal{S}(\boldsymbol{k}_1)$ of $\mathcal{S}$, called the wave vectors group with respect to $\boldsymbol{k}_1$. The index of $\mathcal{S}(\boldsymbol{k}_1)$ in $\mathcal{S}$

is q. $\mathcal{T}$ is also an invariant subgroup of $\mathcal{S}(\boldsymbol{k}_1)$, where the quotient group $\mathcal{S}(\boldsymbol{k}_1)/\mathcal{T}$ is isomorphic onto $H(\boldsymbol{k}_1)$.

5.5.3 *Representation Matrices of Elements in* $\mathcal{S}$

We discuss an arbitrary m-dimensional unitary irreducible representation $D(\mathcal{S})$ of $\mathcal{S}$ which corresponds to a star of wave vectors with index q. Divide the representation matrix $D(R, \boldsymbol{\alpha})$ of an arbitrary element $g(R, \boldsymbol{\alpha})$ of $\mathcal{S}$ into q^2 square submatrices with the dimension $d = m/q$,

$$D(R,\boldsymbol{\alpha}) = \begin{pmatrix} D_{11}(R,\boldsymbol{\alpha}) & D_{12}(R,\boldsymbol{\alpha}) & \cdots & D_{1q}(R,\boldsymbol{\alpha}) \\ D_{21}(R,\boldsymbol{\alpha}) & D_{22}(R,\boldsymbol{\alpha}) & \cdots & D_{2q}(R,\boldsymbol{\alpha}) \\ \cdots & \cdots & \cdots & \cdots \\ D_{q1}(R,\boldsymbol{\alpha}) & D_{q2}(R,\boldsymbol{\alpha}) & \cdots & D_{qq}(R,\boldsymbol{\alpha}) \end{pmatrix}. \tag{5.97}$$

It will be shown that there is only one nonvanishing submatrix in each column (or row) of $D(R, \boldsymbol{\alpha})$. First, the representation matrix of a translation $T(\boldsymbol{\ell})$ is diagonal,

$$D_{\mu\nu}(E,\boldsymbol{\ell}) = \mathbf{1}_d \delta_{\mu\nu} \exp\left(-i2\pi \boldsymbol{k}_\mu \cdot \boldsymbol{\ell}\right), \tag{5.98}$$

where $\mathbf{1}_d$ is a d-dimensional unit matrix.

Second, discuss the representation matrix of $g(R_\mu, \boldsymbol{t}_\mu)$. Substituting Eq. (5.98) into Eq. (5.92), one has

$$\exp\left\{-i2\pi \boldsymbol{k}_\nu \cdot \boldsymbol{\ell}\right\} \; D_{\nu 1}(R_\mu, \boldsymbol{t}_\mu) = D_{\nu 1}(R_\mu, \boldsymbol{t}_\mu) \; \exp\left\{-i2\pi \left(R_\mu \boldsymbol{k}_1\right) \cdot \boldsymbol{\ell}\right\}.$$

Thus, in the first column only the submatrix in the μth row is nonvanishing,

$$D_{\nu 1}(R_\mu, \boldsymbol{t}_\mu) = \delta_{\nu\mu} D_{\mu 1}(R_\mu, \boldsymbol{t}_\mu), \tag{5.99}$$

where $D_{\mu 1}(R_\mu, \boldsymbol{t}_\mu)$ is unitary.

Up to now, only $D(E, \boldsymbol{\ell})$ has been determined. The representation $D(\mathcal{S})$ is allowed to be made a unitary similarity transformation X, which is commutable with $D(E, \boldsymbol{\ell})$. In order to simplify $D(R_\mu, \boldsymbol{t}_\mu)$, one chooses:

$$X = \bigoplus_\mu X_\mu, \qquad X_\mu = \begin{cases} \mathbf{1}_d & \text{when } \mu = 1, \\ D_{\mu 1}(R_\mu, \boldsymbol{t}_\mu) & \text{when } \mu \neq 1. \end{cases} \tag{5.100}$$

The representation after the similarity transformation is still denoted by the same symbol D. $D(E, \boldsymbol{\ell})$ remains the form given in Eq. (5.98), but

$$D_{\nu 1}(R_\mu, \boldsymbol{t}_\mu) = \mathbf{1}_d \delta_{\mu\nu}. \tag{5.101}$$

Since the representation is unitary, the submatrices in the μth row are vanishing except for that in the first column,

$$D_{\mu\nu}(R_\mu, \boldsymbol{t}_\mu) = \mathbf{1}_d \delta_{1\nu}. \tag{5.102}$$

Third, through similar derivation, from Eq. (5.94), one obtains the representation matrix of $g(P, \alpha)$

$$D_{\nu 1}(P, \boldsymbol{\alpha}) = \delta_{\nu 1} D_{11}(P, \boldsymbol{\alpha}), \qquad D_{1\nu}(P, \boldsymbol{\alpha}) = \delta_{1\nu} D_{11}(P, \boldsymbol{\alpha}). \tag{5.103}$$

At last, calculate the submatrix $D_{\rho\mu}(R, \boldsymbol{t})$ in the μth column of the representation matrix of $g(R, \boldsymbol{t})$. Since the product RR_μ is an element of G, it can be expressed as $R_\nu P$,

$$RR_\mu = R_\nu P, \qquad g(R, \boldsymbol{t}) = g(R_\nu, \boldsymbol{t}_\nu) g(P, \boldsymbol{\alpha}) g(R_\mu, \boldsymbol{t}_\mu)^{-1}. \tag{5.104}$$

For the given R and R_μ, $\boldsymbol{t}$, $\boldsymbol{t}_\mu$, R_ν, $\boldsymbol{t}_\nu$, P, and $\boldsymbol{\alpha}$ in the space group $\mathcal{S}$ are all determined. The νth row of the μth column of the representation matrix of $g(R, \boldsymbol{t})$ is

$$D_{\nu\mu}(R, \boldsymbol{t}) = \sum_{\tau\lambda} D_{\nu\tau}(R_\nu, \boldsymbol{t}_\nu) D_{\tau\lambda}(P, \boldsymbol{\alpha}) D_{\mu\lambda}(R_\mu, \boldsymbol{t}_\mu)^* = D_{11}(P, \boldsymbol{\alpha}). \tag{5.105}$$

Since the representation is unitary,

$$D_{\rho\mu}(R, \boldsymbol{t}) = \delta_{\rho\nu} D_{11}(P, \boldsymbol{\alpha}), \qquad D_{\nu\rho}(R, \boldsymbol{t}) = \delta_{\rho\mu} D_{11}(P, \boldsymbol{\alpha}). \tag{5.106}$$

Thus, in each column (or row) of $D(R, \boldsymbol{t})$ there is only one nonvanishing submatrix, which is expressed as $D_{11}(P, \boldsymbol{\alpha})$. The representation matrix of any element of the space group $\mathcal{S}$ is completely determined by the irreducible representation $D_{11}(P, \boldsymbol{\alpha})$ of the wave vector group.

5.5.4 *Irreducible Representations of $\mathcal{S}(\boldsymbol{k}_1)$*

We first prove that the representation $D(\mathcal{S})$ of the space group $\mathcal{S}$ is irreducible if and only if the representation $D_{11}(P, \boldsymbol{\alpha})$ of the wave vector group $\mathcal{S}(\boldsymbol{k}_1)$ is irreducible. In fact, if D_{11} is reducible, there is a nonconstant matrix Y commutable with the matrices $D_{11}(P, \boldsymbol{\alpha})$. Then, $X = \mathbf{1}_q \times Y$ is commutable with all matrices $D(R, \boldsymbol{t})$ and $D(E, \boldsymbol{\ell})$, so that $D(\mathcal{S})$ is reducible. Conversely, if $D(\mathcal{S})$ is reducible, there is a nonconstant matrix X commutable with all representation matrices of $D(\mathcal{S})$. Since X is commutable with $D(E, \boldsymbol{\ell})$, $X = \bigoplus_\mu Y_\mu$. Since X is commutable with $D(R_\mu, \boldsymbol{t}_\mu)$,

Y_μ is independent of μ and commutable with $D_{11}(P,\boldsymbol{\alpha})$. Thus, $D_{11}(P,\boldsymbol{\alpha})$ is reducible because Y_1 is not a constant matrix.

Second, make a correspondence from an element $g(P,\boldsymbol{\alpha})$ of the wave vector group $\mathcal{S}(\boldsymbol{k}_1)$ to the matrix

$$D_{11}(P,\boldsymbol{\alpha}) = \exp\left(-i2\pi\boldsymbol{k}_1\cdot\boldsymbol{\alpha}\right)\Gamma(P), \tag{5.107}$$

where $\Gamma(H)$ is an irreducible representation of the point group $H(\boldsymbol{k}_1)$. One wants to show whether this correspondence remains invariant in the multiplication of the two elements of $\mathcal{S}(\boldsymbol{k}_1)$,

$$g(P,\boldsymbol{\alpha})g(P',\boldsymbol{\alpha}') = g(PP',\boldsymbol{\alpha}+P\boldsymbol{\alpha}'),$$

namely, whether the following equation holds,

$$D_{11}(P,\boldsymbol{\alpha})D_{11}(P',\boldsymbol{\alpha}') = D_{11}(PP',\boldsymbol{\alpha}+P\boldsymbol{\alpha}'). \tag{5.108}$$

Note that $\Gamma(P)\Gamma(P') = \Gamma(PP')$. The exponential function on the left-hand side of Eq. (5.108) is

$$\exp\left[-\mathrm{i}2\pi\boldsymbol{k}_1\cdot(\boldsymbol{\alpha}+\boldsymbol{\alpha}')\right],$$

and that on the right-hand side of Eq. (5.108) is

$$\exp\left[-\mathrm{i}2\pi\boldsymbol{k}_1\cdot(\boldsymbol{\alpha}+P\boldsymbol{\alpha}')\right] = \exp\left(-\mathrm{i}2\pi\boldsymbol{k}_1\cdot\boldsymbol{\alpha}\right)\exp\left[-\mathrm{i}2\pi(P^{-1}\boldsymbol{k}_1)\cdot\boldsymbol{\alpha}'\right].$$

If the star of wave vectors is located inside the first Brillouin zone, $P^{-1}\boldsymbol{k}_1 = \boldsymbol{k}_1$, Eq. (5.108) holds. If $\mathcal{S}$ is a symmorphic space group, $\boldsymbol{\alpha}' = \boldsymbol{\ell}$, even though the star of wave vectors is located on the boundary of the first Brillouin zone, $P^{-1}\boldsymbol{k}_1 = \boldsymbol{k}_1 + \boldsymbol{K}$, one has $\exp\left(-\mathrm{i}2\pi\boldsymbol{K}\cdot\boldsymbol{\ell}\right) = 1$, and Eq. (5.108) still holds. Therefore, for the two special cases, the set of $D_{11}(P,\boldsymbol{\alpha})$ given in Eq. (5.107) forms an irreducible representation of the wave vector group. Remind that the number of the irreducible representations in the form (5.107) is equal to the number of classes of $H(\boldsymbol{k}_1)$. We will not discuss the remaining cases where the star of wave vectors is located on the boundary of the first Brillouin zone but the space group $\mathcal{S}$ is not symmorphic.

In summary, an irreducible representation $D(\mathcal{S})$ of a space group $\mathcal{S}$ is characterized by the star of wave vectors and the irreducible representation $\Gamma(H)$ of the point group $H(\boldsymbol{k}_1)$. Its dimension is $m = qd$, where q is the index of the star of wave vectors and d is the dimension of the representation Γ. The representation matrix of an element of the translation subgroup $\mathcal{T}$ is diagonal and given in Eq. (5.98). The representation matrix of an element $g(R,\boldsymbol{t})$ in the coset of $\mathcal{T}$ is composed of $q\times q$ submatrices with the dimension d as given in Eq. (5.97). In each row (or column) there is only one nonvanishing submatrix, which is related with the irreducible representation $D_{11}(P, \boldsymbol{\alpha})$ of the wave vector group $\mathcal{S}(\boldsymbol{k}_1)$ (see Eqs. (5.106) and (5.107)). It can be proved that for a given star of wave vectors, the representation $D(\mathcal{S})$ is independent of the choice of $\boldsymbol{k}_1$.

5.5.5 *The Bloch Theorem*

Let $\psi_{1m}(\boldsymbol{r})$ be the basis functions belonging to the irreducible representation $D_{11}(P, \boldsymbol{\alpha})$ of the wave vector group $\mathcal{S}(\boldsymbol{k}_1)$

$$P_{g(P,\boldsymbol{\alpha})}\psi_{1m}(\boldsymbol{r}) = \sum_{m'} \psi_{1m'}(\boldsymbol{r}) \left\{D_{11}(P, \boldsymbol{\alpha})\right\}_{m'm}. \tag{5.109}$$

For a translation $T(\boldsymbol{\ell})$, due to Eq. (5.98),

$$P_{g(E,\boldsymbol{\ell})}\psi_{1m}(\boldsymbol{r}) = \psi_{1m}(\boldsymbol{r} - \boldsymbol{\ell}) = \psi_{1m}(\boldsymbol{r}) \exp\left(-\mathrm{i}2\pi \boldsymbol{k}_1 \cdot \boldsymbol{\ell}\right). \tag{5.110}$$

Thus,

$$\psi_{1m}(\boldsymbol{r}) = \exp\left(\mathrm{i}2\pi \boldsymbol{k}_1 \cdot \boldsymbol{r}\right) u_m(\boldsymbol{r}), \qquad u_m(\boldsymbol{r} - \boldsymbol{\ell}) = u_m(\boldsymbol{r}). \tag{5.111}$$

Extending the basis functions, one has

$$\begin{aligned}
\psi_{\mu m}(\boldsymbol{r}) &= P_{R_\mu}\psi_{1m}(\boldsymbol{r}) = \psi_{1m}(R_\mu^{-1}\boldsymbol{r}) \\
&= \exp\left\{\mathrm{i}2\pi \boldsymbol{k}_1 \cdot \left(R_\mu^{-1}\boldsymbol{r}\right)\right\} u_m(R_\mu^{-1}\boldsymbol{r}) \\
&= \exp\left\{\mathrm{i}2\pi \left(R_\mu \boldsymbol{k}_1\right) \cdot \boldsymbol{r}\right\} P_{R_\mu} u_m(\boldsymbol{r}),
\end{aligned} \tag{5.112}$$

$$\begin{aligned}
P_{R_\mu} u_m(\boldsymbol{r} - \boldsymbol{\ell}) &= u_m(R_\mu^{-1}\boldsymbol{r} - R_\mu^{-1}\boldsymbol{\ell}) = u_m(R_\mu^{-1}\boldsymbol{r} - \boldsymbol{\ell}') \\
&= u_m(R_\mu^{-1}\boldsymbol{r}) = P_{R_\mu} u_m(\boldsymbol{r}).
\end{aligned}$$

The basis functions $\psi_{\mu m}(\boldsymbol{r})$ belong to the irreducible representation $D(\mathcal{S})$ characterized by the star of wave vectors and the irreducible representation $\Gamma(H)$ of the point group $H(\boldsymbol{k}_1)$. In fact, the definition (5.112) for $\psi_{\mu m}(\boldsymbol{r})$ shows that Eq. (5.101) holds, so does Eq. (5.102). For the translation $T(\boldsymbol{\ell})$,

$$\begin{aligned}
P_{g(E,\boldsymbol{\ell})}\psi_{\mu m}(\boldsymbol{r}) &= \psi_{\mu m}(\boldsymbol{r} - \boldsymbol{\ell}) = \exp\left\{-\mathrm{i}2\pi \left(R_\mu \boldsymbol{k}_1\right) \cdot \boldsymbol{\ell}\right\} \psi_{\mu m}(\boldsymbol{r}) \\
&= \exp\left\{-\mathrm{i}2\pi \boldsymbol{k}_\mu \cdot \boldsymbol{\ell}\right\} \psi_{\mu m}(\boldsymbol{r}).
\end{aligned} \tag{5.113}$$

Thus, Eq. (5.98) as well as Eqs. (5.105) and (5.106) hold.

The static wave function of an electron moving in the crystal belongs to an irreducible representation of the space group $\mathcal{S}$ of the crystal. If $\psi_{1m}(\boldsymbol{r})$

is an eigenfunction of the Hamiltonian, its partners $\psi_{\mu m}(\boldsymbol{r})$ in an irreducible representation of $\mathcal{S}$ are the degenerate eigenfunctions of the Hamiltonian with the same energy.

The Bloch theorem, which is a fundamental theorem in solid-state physics (see p.211 in [Bradley and Cracknell (1972)]), says that the static wave function of an electron moving in a potential with the translation symmetry satisfies

$$\psi_{\boldsymbol{k}}(\boldsymbol{r}) = u_{\boldsymbol{k}}(\boldsymbol{r})e^{\mathrm{i}\boldsymbol{k}\cdot\boldsymbol{r}}, \qquad u_{\boldsymbol{k}}(\boldsymbol{r}-\boldsymbol{\ell}) = u_{\boldsymbol{k}}(\boldsymbol{r}). \tag{5.114}$$

Equation (5.112) gives an explanation of the Bloch theorem from group theory.

5.5.6 *Energy Band in a Crystal*

In this subsection we will discuss the dependence of the energy of an electron in a crystal on the wave vector $\boldsymbol{k}$ by the approximation of free electrons. An electron in a crystal is moving in a potential $V(\boldsymbol{r})$ with the symmetry of the space group $\mathcal{S}$. $V(\boldsymbol{r})$ varies around its average field V_0, and the difference $V_1(\boldsymbol{r}) = V(\boldsymbol{r}) - V_0$ plays the role of perturbation. The Hamiltonian of the electron is

$$H(\boldsymbol{r}) = H_0(\boldsymbol{r}) + V_1(\boldsymbol{r}), \qquad H_0(\boldsymbol{r}) = -\frac{\hbar^2}{2m_e}\nabla^2 + V_0, \tag{5.115}$$

where m_e is the mass of the electron. If the energy level is normal degenerate, the eigenfunction of the energy belongs to an irreducible representation of the space group $\mathcal{S}$, namely, $\psi_{\mu m}(\boldsymbol{r})$ given in Eq. (5.112) has the same energy, independent of μ and m. For simplicity we denote $\boldsymbol{k}_1$ by $\boldsymbol{k}$ and omit the subscripts μ and m in the wave function,

$$\psi(\boldsymbol{r}) = \exp\left(\mathrm{i}2\pi\boldsymbol{k}\cdot\boldsymbol{r}\right)u(\boldsymbol{r}), \qquad u(\boldsymbol{r}-\boldsymbol{\ell}) = u(\boldsymbol{r}). \tag{5.116}$$

u can be made a Fourier expansion with respect to the vector of reciprocal crystal lattice $\boldsymbol{K}_n$

$$\begin{aligned} u(\boldsymbol{r}) &= \sum_n u_n \exp\left(-\mathrm{i}2\pi\boldsymbol{K}_n\cdot\boldsymbol{r}\right), \\ \psi(\boldsymbol{r}) &= \sum_n u_n \exp\left\{\mathrm{i}2\pi\left(\boldsymbol{k}-\boldsymbol{K}_n\right)\cdot\boldsymbol{r}\right\}, \end{aligned} \tag{5.117}$$

where u_n is a constant to be determined. Each term in the expansion

(5.117) is an eigenfunction of H_0 with the eigenvalue E_n

$$E_n = \frac{h^2}{2m_e}(\boldsymbol{k} - \boldsymbol{K}_n)^2 + V_0. \tag{5.118}$$

E_n varies continuously as the wave vector $\boldsymbol{k} - \boldsymbol{K}_n$ changes.

Calculate the energy correction to the ground state with $n = 0$. Since the first approximation is vanishing, the energy correction comes from the second approximation of the energy. When $\boldsymbol{k}$ is near the center of the first Brillouin zone, E_0 depends on $\boldsymbol{k}^2$ continuously. $E_n - E_0$ is large and u_n with $n \neq 0$ is much smaller than u_0. When $\boldsymbol{k}$ is in the boundary of the first Brillouin zone, $(\boldsymbol{k} - \boldsymbol{K}_n)^2$ may be closed to $\boldsymbol{k}^2$, and the energy interference occurs. The result is that one energy level enhances and the other depresses, so that an energy gap appears.

The condition for the boundary of the Brillouin zones is

$$\boldsymbol{k}^2 = (\boldsymbol{k} - \boldsymbol{K})^2 .$$

The solution is

$$2\boldsymbol{k} \cdot \boldsymbol{K} = \boldsymbol{K}^2. \tag{5.119}$$

It is the equation for the perpendicular bisector plane of the line from the origin to a point of reciprocal crystal lattice, called the Bragg plane in solid-state physics. The Brillouin zones are divided by the Bragg planes.

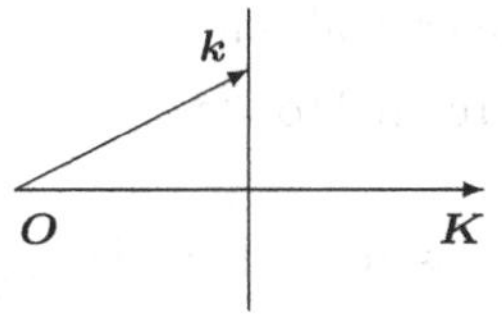

Fig. 5.2 The Bragg plane.

5.6 Exercises

1. In the rectangular coordinate frame, write the double-vector forms and the matrix forms of the proper and improper six-fold rotations around the z-axis.

2. Express the directions of the proper and improper rotational axes in the groups $\mathbf{T}_d$ and $\mathbf{O}_h$ by the basis vectors $\boldsymbol{e}_j$ of the rectangular coordinate frame. Express the double-vector forms of the generators of the proper rotational axes in $\mathbf{O}_h$ by $\boldsymbol{e}_j$.

3. Write the double-vector form of a rotation $R(\hat{\boldsymbol{n}}, \omega)$ around the direction $\hat{\boldsymbol{n}}$ through the angle ω by the identity $\overleftrightarrow{\mathbf{1}}$ and the unit vector $\hat{\boldsymbol{n}}$.

4. Let R be a rotation around the direction $\hat{\boldsymbol{n}} = (\boldsymbol{e}_1 + \boldsymbol{e}_2)/\sqrt{2}$ through $2\pi/3$. Find the symmetric straight line for $g(R, \mathbf{t})$ where (a) $\mathbf{t} = \boldsymbol{e}_3$, (b) $\mathbf{t} = \boldsymbol{e}_1 + \boldsymbol{e}_3$. If the symmetric straight line is a screw line, find its gliding vector. Simplify $g(R, \mathbf{t})$ by moving the origin to the symmetric straight line for checking your result.

5. Let R be a reflection with respect to the $x\ y$ plane. Find the symmetric plane of $g(R, \mathbf{t})$ where (a) $\mathbf{t} = \boldsymbol{e}_3$, (b) $\mathbf{t} = \boldsymbol{e}_1 + \boldsymbol{e}_3$. If the symmetric plane is a gliding plane, find its gliding vector. Simplify $g(R, \mathbf{t})$ by moving the origin to the symmetric plane for checking your result.

6. Analyze the symmetry property of the crystal with the following space group: No. 52 [D_{2h}^6, $P \pm 2(\frac{1}{2}\frac{1}{2}0)2'(\frac{1}{2}\frac{1}{2}\frac{1}{2})$], No. 161 [$\mathrm{C}_{3v}^6$, $R3\overline{2}'(\frac{1}{2}\frac{1}{2}\frac{1}{2})$], and No. 199 [$\mathrm{T}^5$, $I3'2(\frac{1}{2}0\frac{1}{2})2'(\frac{1}{2}\frac{1}{2}0)$]. Point (a) the general form of the symmetric operation; (b) the relations of directions and lengths among three lattice bases; (c) the symmetric straight line, symmetric plane, and the gliding vector of the generator in each cyclic subgroup of the space group, if they exist; (d) the equivalent point to an arbitrary point $\mathbf{r} = \mathbf{a}_1 x_1 + \mathbf{a}_2 x_2 + \mathbf{a}_3 x_3$ in the crystal cell.

7. Calculate the 12 space groups related to the crystallographic point group D_{2d}.

8. Calculate the nine space groups related to the crystallographic point group D_2.

Chapter 6

PERMUTATION GROUPS

A permutation group describes the permutation symmetry of a system of identical particles. We will establish the representation theory of the permutation groups by the Young operator method that is an elegant theory and is widely used in modern physics. From the viewpoint of mathematics, the theory of permutation groups is very interesting because each row in the group table of a finite group shows a permutation of group elements such that every finite group is a subgroup of a permutation group.

6.1 Multiplication of Permutations

6.1.1 *Permutations*

A rearrangement of n objects, denoted by digits 1, 2, $\cdots$, n, is called a permutation. A permutation R which moves the jth object into the r_jth position is characterized by a $2 \times n$ matrix,

$$R = \begin{pmatrix} 1 & 2 & \ldots & n \\ r_1 & r_2 & \ldots & r_n \end{pmatrix}. \tag{6.1}$$

In expression (6.1), the order of the columns does not matter, but the corresponding relation between two digits in each column is essential, for example,

$$R = \begin{pmatrix} 1 & 2 & 3 & 4 & 5 \\ 3 & 4 & 5 & 2 & 1 \end{pmatrix} = \begin{pmatrix} 5 & 4 & 1 & 2 & 3 \\ 1 & 2 & 3 & 4 & 5 \end{pmatrix}. \tag{6.2}$$

A digit j describes an object located in the jth position. Hereafter, a permutation of objects is said as a permutation of digits for simplicity. Usually, this simplification does not cause any confusion.

The multiplication of two permutations does not satisfy the multiplication rule of matrices, although a permutation is expressed by a matrix. The multiplication SR of two permutations is defined as successive applications of R and then S. In other words, the jth object is transformed into the r_jth position by R, and then, into the s_{r_j}th position by S. Note that in the second permutation S, the jth object in the new order after the permutation R is transformed into the s_jth position, no matter in which position it was located before the first permutation R. From the definition of multiplication, SR can be calculated as follows. First, change the order of columns in S or R such that the first row of S is the same as the second row of R. Then, combine the first row of R and the second row of S to a new $2 \times n$ matrix, which is the matrix form of SR. For example,

$$S = \begin{pmatrix} 1 & 2 & \ldots & n \\ s_1 & s_2 & \ldots & s_n \end{pmatrix} = \begin{pmatrix} r_1 & r_2 & \ldots & r_n \\ s_{r_1} & s_{r_2} & \ldots & s_{r_n} \end{pmatrix},$$

$$SR = \begin{pmatrix} r_1 & r_2 & \ldots & r_n \\ s_{r_1} & s_{r_2} & \ldots & s_{r_n} \end{pmatrix} \begin{pmatrix} 1 & 2 & \ldots & n \\ r_1 & r_2 & \ldots & r_n \end{pmatrix} = \begin{pmatrix} 1 & 2 & \ldots & n \\ s_{r_1} & s_{r_2} & \ldots & s_{r_n} \end{pmatrix}. \tag{6.3}$$

The multiplication of two permutations is a permutation. Being a multiplication of transformations, the multiplication of permutations satisfies the associative law, $(RS)T = R(ST)$, but it is generally not commutable. The identical permutation E is characterized by a matrix whose two rows are exactly the same,

$$E = \begin{pmatrix} 1 & 2 & 3 \ldots n \\ 1 & 2 & 3 \ldots n \end{pmatrix}. \tag{6.4}$$

From the multiplication rule (6.3), $ER = R$ and E is the identical element. The matrix of R^{-1} is obtained from that of R by switching two rows,

$$R^{-1} = \begin{pmatrix} r_1 & r_2 & \ldots & r_n \\ 1 & 2 & \ldots & n \end{pmatrix}, \tag{6.5}$$

$$R^{-1}R = \begin{pmatrix} r_1 & r_2 & \ldots & r_n \\ 1 & 2 & \ldots & n \end{pmatrix} \begin{pmatrix} 1 & 2 & \ldots & n \\ r_1 & r_2 & \ldots & r_n \end{pmatrix} = E.$$

There are $n!$ different permutations among n objects. In the multiplication rule (6.3) the set of $n!$ permutations satisfies the four axioms for a group, and forms the permutation group of n objects, denoted by S_n. The order of S_n is $g = n!$. Choosing arbitrarily m objects among n objects, one obtains a permutation group S_m from the permutations of m objects. S_m is a subgroup of S_n.

In the literature, there are different definitions for a permutation. For example, a permutation R given in Eq. (6.1) may be defined as a transformation moving the r_jth object to the jth position. In fact, two definitions are mutually inverse. Attention must be paid to the definition when one begins to read a new book.

6.1.2 *Cycles*

If a permutation S preserves $(n-\ell)$ objects invariant and changes the remaining ℓ objects in order, S is called a cycle with length ℓ, characterized by a one-row matrix:

$$\begin{aligned} S &= \begin{pmatrix} a_1 & a_2 & \ldots & a_{\ell-1} & a_\ell & b_1 & \ldots & b_{n-\ell} \\ a_2 & a_3 & \ldots & a_\ell & a_1 & b_1 & \ldots & b_{n-\ell} \end{pmatrix} \\ &= (a_1 \ a_2 \ \ldots \ a_{\ell-1} \ a_\ell) = (a_2 \ a_3 \ \ldots \ a_\ell \ a_1) . \end{aligned} \tag{6.6}$$

In the one-row matrix for a cycle, the order of digits is essential, while the transformation of digits in sequence is permitted. A cycle with length 1 is the identical element E. A cycle with length 2 is called a transposition,

$$(a \ b) = (b \ a) , \qquad (a \ b)\,(a \ b) = E. \tag{6.7}$$

The order of a cycle S with length ℓ is ℓ (see Prob. 3 of Chap. 6 in [Ma and Gu (2004)]), namely, ℓ is the smallest power with $S^\ell = E$.

Two cycles are called independent if they do not contain any common digit (object). The multiplication of two independent cycles is commutable. Any permutation can be decomposed as a product of independent cycles uniquely up to the order. In fact, checking an arbitrary digit a_1 in a given permutation R, one finds that after R, a_1 changes to a_2, a_2 changes to a_3, and so on. In this chain there must exist an object, say a_ℓ, such that a_ℓ changes to a_1. Thus, R contains a cycle $(a_1, \ a_2, \ \ldots, \ a_\ell)$ whose length is ℓ. In the remaining objects one checks another object, say b_1, and finds another chain of objects which gives another cycle independent of the first one. This process can be done until all objects are contained in the cycles. Thus, the permutation is decomposed into a product of independent cycles. For example,

$$\begin{aligned} R &= \begin{pmatrix} 1 & 2 & 3 & 4 & 5 \\ 3 & 4 & 5 & 2 & 1 \end{pmatrix} = (1 \ 3 \ 5)\,(2 \ 4) = (2 \ 4)\,(1 \ 3 \ 5) , \\ S &= \begin{pmatrix} 1 & 2 & 3 & 4 & 5 \\ 3 & 1 & 2 & 4 & 5 \end{pmatrix} = (1 \ 3 \ 2)\,(4)\,(5) = (1 \ 3 \ 2) . \end{aligned} \tag{6.8}$$

The set of lengths of the independent cycles in a permutation R is called its cycle structure. For example, the cycle structure of R in Eq. (6.8) is (3 2), and that of S is $(3,1,1) \equiv (3,1^2)$. The order of the cycle lengths ℓ_i in the cycle structure of a permutation is arbitrary and the repetitive cycle lengths can be expressed as an exponent. It is a common sense that the simplest form of a permutation is its decomposition into a product of independent cycles.

In the calculation of the multiplication of two permutations one has to deal with the multiplication of two cycles with a few common objects. The multiplication of two cycles with one common object can be calculated as

$$\begin{aligned}(a\ b\ c\ d)\,(d\ e\ f) &= \begin{pmatrix} a\ b\ c\ e\ f\ d \\ b\ c\ d\ e\ f\ a \end{pmatrix} \begin{pmatrix} a\ b\ c\ d\ e\ f \\ a\ b\ c\ e\ f\ d \end{pmatrix} = \begin{pmatrix} a\ b\ c\ d\ e\ f \\ b\ c\ d\ e\ f\ a \end{pmatrix} \\ &= (a\ b\ c\ d\ e\ f)\,.\end{aligned}$$

Generally, two cycles with one common object are connected to be one,

$$(a\ \cdots\ b\ c)(c\ d\ \cdots\ f) = (a\ \cdots\ b\ c\ d\ \cdots\ f). \tag{6.9}$$

Conversely, a cycle can be decomposed into a product of two cycles by cutting it in any digit and repeating the digit. Generalizing this method, one can calculate the multiplication of two cycles with more than one common objects by the above "cutting" and "connecting", say

$$\begin{aligned}&(a_1\ \ldots a_i\ c\ a_{i+1}\ \ldots\ a_j\ d)\,(d\ b_1\ \ldots b_r\ c\ b_{r+1}\ \ldots\ b_s) \\ &= (a_1\ \ldots a_i\ c)\,(c\ a_{i+1}\ \ldots\ a_j\ d)\,(d\ b_1\ \ldots b_r\ c)\,(c\ b_{r+1}\ \ldots\ b_s) \\ &= (a_1\ \ldots a_i\ c)\,(a_{i+1}\ \ldots\ a_j\ d\ c)\,(c\ d\ b_1\ \ldots b_r)\,(c\ b_{r+1}\ \ldots\ b_s) \\ &= (a_1\ \ldots a_i\ c)\,(a_{i+1}\ \ldots\ a_j\ d)\,(d\ c)\,(c\ d)\,(d\ b_1\ \ldots b_r)\,(c\ b_{r+1}\ \ldots\ b_s) \\ &= (a_1\ \ldots a_i\ c)\,(a_{i+1}\ \ldots\ a_j\ d)\,(d\ b_1\ \ldots b_r)\,(c\ b_{r+1}\ \ldots\ b_s) \\ &= (a_1\ \ldots a_i\ c\ b_{r+1}\ \ldots\ b_s)\,(a_{i+1}\ \ldots\ a_j\ d\ b_1\ \ldots b_r)\,.\end{aligned}$$

6.1.3 *Classes in a Permutation Group*

The multiplication rule (6.3) of permutations can be understood from other viewpoints. When a permutation R is left-multiplied by a permutation S, the digits (objects) in the second row of R make a S permutation, and when R is right-multiplied by S, the digits (objects) in the first row of R make an S^{-1} permutation. Therefore, the conjugate permutation SRS^{-1} of R can be calculated by transforming all digits in both rows of R by the permutation S. Especially, when R is a cycle, one has

$$S\,(a\;b\;c\;\ldots\;d)\,S^{-1} = (s_a\;s_b\;s_c\;\ldots\;s_d)\,. \tag{6.10}$$

Formula (6.10) for the multiplication of permutations is called the interchanging rule. From the rule, two permutations are conjugate to each other if and only if their cycle structures are the same. The class in a permutation group is characterized by the cycle structure of the permutations in the class.

Equation (6.10) can be rewritten in another way,

$$S\,(a\;b\;c\;\ldots\;d) = (s_a\;s_b\;s_c\;\ldots\;s_d)\,S. \tag{6.11}$$

When a permutation S is moved from the left-side of a permutation R to its right-side, the digits (objects) in R are made an S permutation. Conversely, when S is moved from the right-side of R to its left-side, the digits in R are made an S^{-1} permutation.

A set of positive integers ℓ_j, whose sum is equal to a positive integer n, is called a partition of n. The cycle structure of a permutation R as well as a class is characterized by a partition of n,

$$(\ell_1, \ell_2, \ldots, \ell_m)\,, \qquad \sum_{i=1}^{m} \ell_i = n. \tag{6.12}$$

The number $g_c(n)$ of classes in S_n is equal to the number of partitions of n. There is no analytic formula for $g_c(n)$ (see Chap. 1 in [Angrews (1976)]), but all partitions for a given positive integer n can be listed without dropping (see the next section). Some $g_c(n)$ are given as follows.

$$\begin{array}{llll}
g_c(1) = 1, & g_c(2) = 2, & g_c(3) = 3, & g_c(4) = 5, \\
g_c(5) = 7, & g_c(6) = 11, & g_c(7) = 15, & g_c(8) = 22, \\
g_c(9) = 30, & g_c(10) = 42, & g_c(20) = 627, & g_c(50) = 204226, \\
g_c(100) = 190569292, & & g_c(200) = 3972999029388. &
\end{array}$$

On the other hand, the number $n(\alpha)$ of the elements in a class $\mathcal{C}_\alpha$ of S_n can be expressed analytically. Denote by $N[n, \ell, m]$ the number of elements in S_n which contain m cycles with length ℓ and the length of the remaining cycles is 1,

$$N[n, \ell, m] = [m!]^{-1}\,[(\ell-1)!]^m \prod_{a=0}^{m-1} \binom{n - a\ell}{\ell} = \frac{n!}{\ell^m m!(n - m\ell)!}.$$

If the number of integers ℓ in the partition of $\mathcal{C}_\alpha$ is m_ℓ, $1 \leq \ell \leq k$, one has

$$n(\alpha) = \prod_{\ell=1}^{k} N[n - T_\ell, \ell, m_\ell] = n! \prod_{\ell=1}^{k} (m_\ell! \ell^{m_\ell})^{-1}, \quad T_\ell = \sum_{a=1}^{\ell-1} a m_a, \quad (6.13)$$

where $T_1 = 0$ and $T_{k+1} = n$.

6.1.4 *Alternating Subgroups*

Each cycle can be decomposed into a product of transpositions by cutting the digits in the cycle, so can each permutation:

$$(a\ b\ c\ \ldots\ p\ q) = (a\ b)\,(b\ c) \ldots (p\ q)\,. \qquad (6.14)$$

This decomposition is not unique. However, it is fixed for a given permutation whether the number of the transpositions in its decomposition is even or odd. This conclusion can be proved by making use of the Vandermonde determinant. The Vandermonde determinant is defined as

$$D(x_1, x_2, \ldots, x_n) = \begin{vmatrix} 1 & 1 & \ldots & 1 \\ x_1 & x_2 & \ldots & x_n \\ x_1^2 & x_2^2 & \ldots & x_n^2 \\ \ldots & \ldots & \ldots & \ldots \\ x_1^{n-1} & x_2^{n-1} & \ldots & x_n^{n-1} \end{vmatrix}.$$

The determinant changes its sign when the variables x_j are made a transposition. When the variables x_j are made a permutation R, it is completely determined by R whether the determinant changes its sign or not. On the other hand, R can be decomposed into a product of transpositions. Whether the determinant changes its sign or not depends on whether the number of transpositions in the decomposition is odd or even. Thus, the conclusion is proved.

A permutation is called even (or odd) if it is decomposed into a product of even (or odd) transpositions. The product of two even (or two odd) permutations is an even permutation. The product of an even permutation and an odd permutation is an odd permutation. The identical element E is an even permutation. The set of all even permutations in a permutation group S_n forms an invariant subgroup of S_n with index 2, called the alternating subgroup of S_n, and the set of all odd permutations in S_n forms its coset. The quotient group is isomorphic onto the inversion group V_2. Thus, S_n, $n > 1$, has an antisymmetric representation, where the representation matrix of R is called the permutation parity of R, denoted by

$$\delta(R) = \begin{cases} 1 & \text{when } R \text{ is even,} \\ -1 & \text{when } R \text{ is odd.} \end{cases} \tag{6.15}$$

The permutation parity of a cycle with length ℓ is $(-1)^{\ell-1}$.

6.1.5 *Transposition of Two Neighbored Objects*

Denote by $P_a = (a\ a+1)$ the transposition of two neighbored objects a and $(a+1)$. From the interchanging rule P_a satisfies

$$\begin{aligned} &P_a^2 = E, \qquad P_a P_{a+1} P_a = P_{a+1} P_a P_{a+1}, \\ &P_a P_b = P_b P_a, \quad \text{when} \ \ |a-b| \geq 2. \end{aligned} \tag{6.16}$$

Each transposition can be expressed as a product of the transpositions of neighbored objects, so can a permutation,

$$\begin{aligned} (a\ \ d) = (d\ \ a) &= P_{a-1} P_{a-2} \ldots P_{d+1} P_d P_{d+1} \ldots P_{a-2} P_{a-1} \\ &= P_d P_{d+1} \ldots P_{a-2} P_{a-1} P_{a-2} \ldots P_{d+1} P_d, \qquad d < a. \end{aligned} \tag{6.17}$$

Denote by W a cycle of length n,

$$W = (1\ 2\ \ldots\ n)\,, \qquad W^{-1} = W^{n-1}. \tag{6.18}$$

As shown in Prob. 4 of Chap. 6 of [Ma and Gu (2004)], W and P_1 are the generators of a permutation group S_n. The rank of S_n is 2.

6.2 Young Patterns, Young Tableaux, and Young Operators

6.2.1 *Young Patterns*

A class in the permutation group S_n is characterized by a partition of n, $(\ell) = (\ell_1, \ell_2, \ldots, \ell_m)$ with $\sum_j \ell_j = n$. Since the number of the inequivalent irreducible representations of a finite group is equal to the number of its classes, therefore, the irreducible representation of S_n can also be described by a partition of n, denoted by

$$\begin{aligned} &[\lambda] = [\lambda_1,\ \lambda_2,\ \ldots,\ \lambda_m], \\ &\lambda_1 \geq \lambda_2 \geq \ldots \geq \lambda_m > 0, \end{aligned} \qquad \sum_{j=1}^{m} \lambda_j = n. \tag{6.19}$$

Note that the irreducible representation $[\lambda]$ has no relation with the class (ℓ), no matter whether the partitions are the same or not. Based on a partition $[\lambda]$, a Young pattern $[\lambda]$ (or called a Young diagram) can be defined. We will show later that a Young pattern $[\lambda]$ characterizes an irreducible representation of S_n.

A Young pattern $[\lambda]$ consists of n boxes lined up on the top and on the left, where the jth row contains λ_j boxes. For example, the Young pattern $[3, 2]$ is

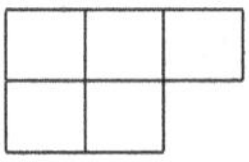

Note that in a Young pattern, the number of boxes in the upper row is not less than that in the lower row, and the number of boxes in the left column is not less than that in the right column. In order to emphasize the above rules for a Young pattern, a Young pattern is sometimes called a regular Young pattern in the literature. In fact, we will not be interested in an irregular Young pattern.

A Young pattern $[\lambda]$ is said to be larger than a Young pattern $[\lambda']$ if there is a row number j such that $\lambda_i = \lambda'_i$, $1 \leq i < j$, and $\lambda_j > \lambda'_j$. There is no analytic formula for the number of different Young patterns with n boxes. However, one can list all different Young patterns with n boxes from the largest to the smallest. Namely, one lists the Young patterns in the order that λ_1 decreases from n one by one, then for a given λ_1, λ_2 decreases from the minimum of λ_1 and $(n - \lambda_1)$ one by one, third for the given λ_1 and λ_2, λ_3 decreases from the minimum of λ_2 and $(n - \lambda_1 - \lambda_2)$ one by one, and so on. For example, the Young patterns for $n = 7$ are listed as follows:

$$\begin{array}{llll}
[7], & [6,1], & [5,2], & [5,1,1], \\
[4,3], & [4,2,1], & [4,1,1,1], & [3,3,1], \\
[3,2,2], & [3,2,1,1], & [3,1,1,1,1], & [2,2,2,1], \\
[2,2,1,1,1], & [2,1,1,1,1,1], & [1,1,1,1,1,1,1]. &
\end{array}$$

6.2.2 *Young Tableaux*

Filling n digits 1, 2, ..., n arbitrarily into the Young pattern $[\lambda]$ with n boxes, one obtains a Young tableau. There are $n!$ different Young tableaux for a given Young pattern with n boxes. A Young tableau is said to be standard if the digit on the left is smaller than the digit on the right in the same row, and the upper digit is smaller than the lower digit in the same column. It is proved that the number $d_{[\lambda]}$ of the standard Young tableaux for the Young pattern $[\lambda]$ with m rows is

$$d_{[\lambda]}(\mathrm{S}_n) = \frac{n!}{r_1!r_2!\ldots r_m!} \prod_{j<k}^{m} (r_j - r_k), \tag{6.20}$$

where $r_j = \lambda_j + m - j$. The square sum of the numbers $d_{[\lambda]}$ is $n!$

$$\sum_{[\lambda]} d_{[\lambda]}(\mathrm{S}_n)^2 = n!. \tag{6.21}$$

The proof can be found in Chap. IV in [Boerner (1963)].

The formula (6.20) for calculating $d_{[\lambda]}$ is not the simplest because there are some common factors between the numerator and the denominator. For example,

$$d_{[3,2,1,1]}(\mathrm{S}_7) = 7! \times \frac{5 \times 4 \times 2}{6!} \times \frac{3 \times 2}{4!} \times \frac{1}{2!} \times \frac{1}{1!} = 35.$$

A simpler method for calculating $d_{[\lambda]}$ is called the hook rule. The hook number h_{ij} of the box at the jth column of the ith row in a Young pattern $[\lambda]$ is equal to the number of boxes at its right in the ith row, plus the number of boxes below it in the jth column, and plus 1. $Y_h^{[\lambda]}$ is the product of the digits filled in a tableau of the Young pattern $[\lambda]$ where the box in the jth column of the ith row is filled with its hook number h_{ij}. The number $d_{[\lambda]}$ of standard Young tableaux for the Young pattern $[\lambda]$ is

$$d_{[\lambda]}(\mathrm{S}_n) = \frac{n!}{Y_h^{[\lambda]}} = \frac{n!}{\prod_{ij} h_{ij}}. \tag{6.22}$$

For comparison, $d_{[3,2,1,1]}$ is re-calculated from Eq. (6.22),

$$Y_h^{[3,2,1,1]} = \begin{array}{|c|c|c|} \hline 6 & 3 & 1 \\ \hline 4 & 1 & \\ \cline{1-2} 2 & & \\ \cline{1-1} 1 & & \\ \cline{1-1} \end{array}, \qquad d_{[3,2,1,1]}(\mathrm{S}_7) = \frac{7!}{6 \times 3 \times 4 \times 2} = 35.$$

From the calculation one finds that two formulas are the same because for each row i,

$$\prod_{j=1}^{\lambda_i} h_{ij} = m_i! \left\{ \prod_{j=i+1}^{m} (r_i - r_j) \right\}^{-1}.$$

Compare the filled digits in two standard Young tableaux of a given Young pattern from left to right of the first row, and then those of the second row, and so on. For the first different filled digits, a smaller filled digit corresponds to a smaller standard Young tableau. Enumerate the standard Young tableaux of a given Young pattern $[\lambda]$ by an integer μ from 1 to $d_{[\lambda]}$. This increasing order is the so-called dictionary order. The readers are suggested to learn how to arrange the standard Young tableaux of a given Young pattern in the dictionary order. For example, the order of the standard Young tableaux of the Young pattern $[3, 2]$ is

1	2	3
4	5	

1	2	4
3	5	

1	2	5
3	4	

1	3	4
2	5	

1	3	5
2	4	

Two Young patterns related by a transpose are called the associated Young patterns. The corresponding standard Young tableaux of two associated Young patterns are also related by a transpose, but the larger Young tableau of one Young pattern becomes the smaller of its associated Young pattern. The numbers of the standard Young tableaux of two associated Young patterns are the same.

6.2.3 *Young Operators*

A permutation of the digits in the jth row of a given Young tableau is called a horizontal permutation P_j of the Young tableau, and a permutation of the digits in the kth column is called its vertical permutation Q_k. The product of horizontal permutations is also a horizontal permutation, denoted by P. The product of vertical permutations is a vertical permutation, denoted by Q. The sum of all horizontal permutations of a given Young tableau is called its horizontal operator, and the sum of all vertical permutations multiplied by their permutation parities is called its vertical operator. The Young operator of a given Young tableau is the product of $\mathcal{P}$ and $\mathcal{Q}$:

$$\begin{aligned}
&\mathcal{P} = \sum P = \sum \prod_j P_j = \prod_j \left(\sum P_j\right), \\
&\mathcal{Q} = \sum \delta(Q)Q = \sum \prod_k \delta(Q_k)Q_k = \prod_k \left[\sum \delta(Q_k)Q_k\right], \qquad (6.23) \\
&\mathcal{Y} = \mathcal{P}\mathcal{Q}.
\end{aligned}$$

P and Q are also said to be the horizontal permutation and the vertical permutation of the Young operator $\mathcal{Y}$, respectively. A Young operator is said to be standard if its Young tableau is standard. Since for a given

Young operator $\mathcal{Y}$, its Young tableau and its Young pattern are fixed, the Young tableau and the Young pattern are usually respectively called the Young pattern $\mathcal{Y}$ and the Young tableau $\mathcal{Y}$ for convenience. The symbol $\mathcal{Y}$ denotes the Young operator itself. Except for the identical element E, no permutation can be both the horizontal permutation and the vertical permutation belonging to the same Young tableau. The readers are suggested to be familiar with writing the expansion of a Young operator of a given Young tableau. The following is an example:

1	2	3
4	5	

$$\begin{aligned}
\mathcal{Y} &= \{E + (1\ 2) + (1\ 3) + (2\ 3) + (1\ 2\ 3) + (3\ 2\ 1)\} \\
&\quad \cdot \{E + (4\ 5)\}\{E - (1\ 4)\}\{E - (2\ 5)\} \\
&= \{E + (1\ 2) + (1\ 3) + (2\ 3) + (1\ 2\ 3) + (3\ 2\ 1) + (4\ 5) + (1\ 2)(4\ 5) \\
&\quad + (1\ 3)(4\ 5) + (2\ 3)(4\ 5) + (1\ 2\ 3)(4\ 5) + (3\ 2\ 1)(4\ 5)\} \\
&\quad \cdot \{E - (1\ 4) - (2\ 5) + (1\ 4)(2\ 5)\} \\
&= \{E + (1\ 2) + (1\ 3) + (2\ 3) + (1\ 2\ 3) + (3\ 2\ 1) + (4\ 5) + (1\ 2)(4\ 5) \\
&\quad + (1\ 3)(4\ 5) + (2\ 3)(4\ 5) + (1\ 2\ 3)(4\ 5) + (3\ 2\ 1)(4\ 5)\} \\
&\quad - \{(1\ 4) + (2\ 1\ 4) + (3\ 1\ 4) + (2\ 3)(1\ 4) + (2\ 3\ 1\ 4) + (3\ 2\ 1\ 4) \\
&\quad + (5\ 4\ 1) + (2\ 1\ 5\ 4) + (3\ 1\ 5\ 4) + (2\ 3)(5\ 4\ 1) + (2\ 3\ 1\ 5\ 4) \\
&\quad + (3\ 2\ 1\ 5\ 4)\} - \{(2\ 5) + (1\ 2\ 5) + (1\ 3)(2\ 5) + (3\ 2\ 5) + (3\ 1\ 2\ 5) \\
&\quad + (1\ 3\ 2\ 5) + (4\ 5\ 2) + (1\ 2\ 4\ 5) + (1\ 3)(4\ 5\ 2) + (3\ 2\ 4\ 5) \\
&\quad + (3\ 1\ 2\ 4\ 5) + (1\ 3\ 2\ 4\ 5)\} + \{(1\ 4)(2\ 5) + (1\ 4\ 2\ 5) + (3\ 1\ 4)(2\ 5) \\
&\quad + (1\ 4)(3\ 2\ 5) + (3\ 1\ 4\ 2\ 5) + (1\ 4\ 3\ 2\ 5) + (4\ 1\ 5\ 2) + (4\ 2)(1\ 5) \\
&\quad + (4\ 3\ 1\ 5\ 2) + (3\ 2\ 4\ 1\ 5) + (4\ 2)(3\ 1\ 5) + (4\ 3\ 2)(1\ 5)\}.
\end{aligned}$$

From the definition of a Young operator or from the above example one sees that a Young operator is a vector in the group algebra of the permutation group S_n,

$$\mathcal{Y} = \sum_{R \in \mathrm{S}_n} F(R)R, \tag{6.24}$$

$$F(E) = F(P) = \delta(Q)F(Q) = \delta(Q)F(PQ) = 1. \tag{6.25}$$

The coefficients $F(R)$ are taken to be 1, -1, or 0. The permutation R with nonvanishing coefficient $F(R)$ in Eq. (6.24) is called a permutation belonging to the Young operator $\mathcal{Y}$ (and the Young tableau $\mathcal{Y}$), and the remaining permutations do not belong to $\mathcal{Y}$. We will discuss the criterion whether a permutation belongs to the Young operator $\mathcal{Y}$ or not.

It is easy to determine a permutation S uniquely which transforms a Young tableau $\mathcal{Y}$ to another Young tableau $\mathcal{Y}'$ with the same Young pattern. In fact, the first row of S is filled with the digits of the Young tableau $\mathcal{Y}$ and the second row of S is filled with the corresponding digits of the Young tableau $\mathcal{Y}'$ in the same order. For example,

$$\text{Young tableau } \mathcal{Y} = \begin{array}{|c|c|c|}\hline 1 & 3 & 5 \\ \hline 2 & 4 & \\ \cline{1-2}\end{array} \qquad \text{Young tableau } \mathcal{Y}' = \begin{array}{|c|c|c|}\hline 1 & 2 & 3 \\ \hline 4 & 5 & \\ \cline{1-2}\end{array}$$

$$S = \begin{pmatrix} 1 & 3 & 5 & 2 & 4 \\ 1 & 2 & 3 & 4 & 5 \end{pmatrix}.$$

Then, one has from the interchanging rule (6.10)

$$S\mathcal{Y}S^{-1} = \mathcal{Y}', \qquad S\mathcal{P}S^{-1} = \mathcal{P}', \qquad S\mathcal{Q}S^{-1} = \mathcal{Q}'. \tag{6.26}$$

6.2.4 *Fundamental Property of Young Operators*

For a given Young operator $\mathcal{Y}$, the set of its horizontal permutations forms a group, and the horizontal operator $\mathcal{P}$ is the sum of elements in the group. Similarly, the set of its vertical permutations forms a group, and the vertical operator $\mathcal{Q}$ is the algebraic sum of elements in the group with their permutation parities. Thus, from the rearrangement theorem (Theorem 2.1), one has

$$P\mathcal{P} = \mathcal{P}P = \mathcal{P}, \qquad Q\mathcal{Q} = \mathcal{Q}Q = \delta(Q)\mathcal{Q}. \tag{6.27}$$

The Young operator $\mathcal{Y}$ satisfies

$$P\mathcal{Y} = \delta(Q)\mathcal{Y}Q = \mathcal{Y}. \tag{6.28}$$

It is the fundamental property of a Young operator. Except for the normalization condition $F(E) = 1$, Eq. (6.25) can be derived from this property. In fact, from Eq. (6.24) one has

$$\begin{aligned} P^{-1}\mathcal{Y} &= \sum_{R\in \mathrm{S}_n} F(R)P^{-1}R = \sum_{S\in \mathrm{S}_n} F(PS)S = \mathcal{Y} = \sum_{S\in \mathrm{S}_n} F(S)S, \\ \mathcal{Y}Q^{-1} &= \sum_{R\in \mathrm{S}_n} F(R)RQ^{-1} = \sum_{S\in \mathrm{S}_n} F(SQ)S = \delta(Q)\mathcal{Y} = \sum_{S\in \mathrm{S}_n} \delta(Q)F(S)S. \end{aligned}$$

Thus,

$$F(S) = F(PS) = \delta(Q)F(SQ) = \delta(Q)F(PSQ). \tag{6.29}$$

Letting $S = E$ and assuming $F(E) = 1$, one obtains Eq. (6.25). Fock found another important property of a Young operator.

Theorem 6.1 (Fock conditions) Let λ and λ' be the numbers of boxes in the jth row and in the j'th row of a Young tableau $\mathcal{Y}$, respectively. Denote by a_μ the digits filled in the jth row and by b_ν those in the j'th row. If $\lambda \geq \lambda'$, then

$$\left\{ E + \sum_{\mu=1}^{\lambda} (a_\mu \, b_\nu) \right\} \mathcal{Y} = 0. \tag{6.30}$$

Let τ and τ' be the numbers of boxes in the kth column and in the k'th column of a Young tableau $\mathcal{Y}$, respectively. Denote by c_μ the digits filled in the kth column and by d_ν those in the k'th column. If $\tau \geq \tau'$, then

$$\mathcal{Y} \left\{ E - \sum_{\mu=1}^{\tau} (c_\mu \, d_\nu) \right\} = 0. \tag{6.31}$$

The following example shows the filling positions of a_μ, b_ν, c_μ, and d_ν in a Young tableau, where $j = k = 2$ and $j' = k' = 3$:

a_1	a_2	a_3	a_4	a_5
	b_ν			

c_1			
c_2	d_ν		
c_3			
c_4			
c_5			

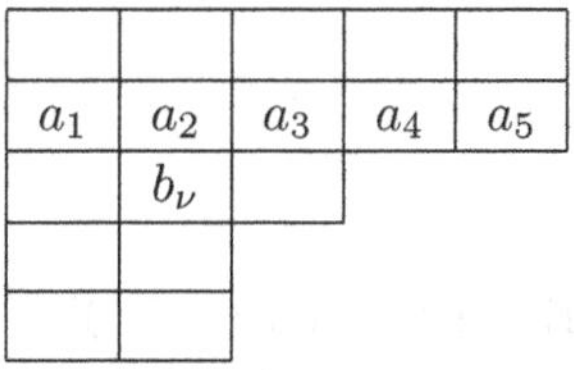

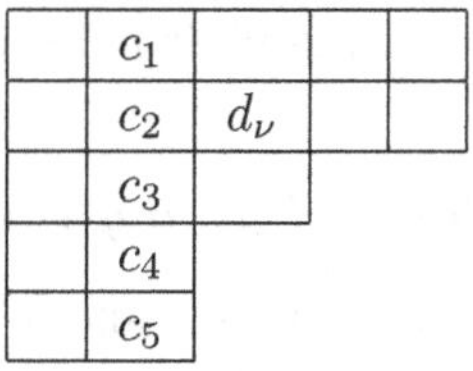

Proof The proofs for the two Fock conditions are similar. In the following we will prove the condition (6.30) as example. Denote by $(\sum P_j)$ and $(\sum P_{j'})$ the sums of horizontal permutations in the jth row and in the j'th row, respectively. $\mathcal{P}'$ is the sum of horizontal permutations not containing the digits in the jth row and the j'th row. Hence,

$$\mathcal{Y} = \mathcal{P}\mathcal{Q} = \left(\sum P_j \right) \left(\sum P_{j'} \right) \mathcal{P}' \mathcal{Q}.$$

$(\sum P_j)$ is the totally symmetric operator for the λ objects in the jth row, because it is the sum of $\lambda!$ different permutations of those objects. Left-multiplying it with $\left\{ E + \sum_\mu (a_\mu \, b_\nu) \right\}$, one obtains a sum of $(\lambda+1)!$ different permutations of the object b_ν and λ objects in the jth row. Hence, the sum is the totally symmetric operator for those $(\lambda + 1)$ objects, denoted by

$\sum P_j(b_\nu)$. When each term $P_{j'}$ in the sum $(\sum P_{j'})$ moves from the right-side of $\sum P_j(b_\nu)$ to its left-side, the only change of $\sum P_j(b_\nu)$ is that b_ν may be replaced with another digit b_ρ in the j' row,

$$\left[\sum P_j(b_\nu)\right] P_{j'} = P_{j'} \left[\sum P_j(b_\rho)\right]. \tag{6.32}$$

$\sum P_j(b_\rho)$ is commutable with $\mathcal{P}'$. Since $\lambda \geq \lambda'$, there exists an object a_ρ in the jth row which is located at the same column as b_ρ. The transposition $(a_\rho \; b_\rho)$ preserves $[\sum P_j(b_\rho)]$ invariant but changes the sign of $\mathcal{Q}$, so that

$$\left[\sum P_j(b_\rho)\right] \mathcal{Q} = \left[\sum P_j(b_\rho)\right] (a_\rho \; b_\rho)\, \mathcal{Q} = -\left[\sum P_j(b_\rho)\right] \mathcal{Q} = 0. \tag{6.33}$$

Thus, Eq. (6.30) is proved. □

The key in the proof is that if the box number λ of the jth row is not less than the box number λ' of the j'th row, $\mathcal{Y}$ is annihilated from the left by symmetrizing b_ν in the j'th row with all a_μ in the jth row. Note that Eq. (6.30) holds if $j < j'$, but when $j > j'$, it holds only if $\lambda = \lambda'$. Similarly, if the box number τ of the kth column is not less than the box number τ' of the k'th column, $\mathcal{Y}$ is annihilated from the right by antisymmetrizing d_ν in the k'th column with all c_μ in the kth column. Equation (6.31) holds if $k < k'$, but when $k > k'$, it holds only if $\tau = \tau'$.

6.2.5 *Products of Young Operators*

For two Young operators $\mathcal{Y}$ and $\mathcal{Y}'$, one cannot derive $\mathcal{Y}\mathcal{Y}' = 0$ from $\mathcal{Y}'\mathcal{Y} = 0$, and vice versa. Two Young operators are called orthogonal only when the two products vanish simultaneously. The method used in Eq. (6.33) is the typical one for proving that the product of two Young operators vanishes.

Theorem 6.2 If there exist two digits a and b in one row of a Young tableau $\mathcal{Y}$ which also occur in one column of a Young tableau $\mathcal{Y}'$, or equivalently, if $T_0 = (a \; b)$ is both the horizontal transposition of a Young operator $\mathcal{Y} = \mathcal{P}\mathcal{Q}$ and the vertical transposition of a Young operator $\mathcal{Y}' = \mathcal{P}'\mathcal{Q}'$, then

$$\mathcal{Q}'\mathcal{P} = 0 \;\; \text{and} \;\; \mathcal{Y}'\mathcal{Y} = 0. \tag{6.34}$$

Proof $\mathcal{Q}'\mathcal{P} = \mathcal{Q}'T_0\mathcal{P} = -\mathcal{Q}'\mathcal{P} = 0$. □

The subscript 0 indicates T_0 to be a transposition. In proving the following corollaries, the key is to find the pair of digits.

Corollary 6.2.1 If a Young pattern $\mathcal{Y}'$ is less than a Young pattern $\mathcal{Y}$, then $\mathcal{Y}'\mathcal{Y} = 0$.

Proof Denote by $[\lambda']$ and $[\lambda]$ the partitions of the Young patterns $\mathcal{Y}'$ and $\mathcal{Y}$, respectively. Since the Young pattern $\mathcal{Y}'$ is less than the Young pattern $\mathcal{Y}$, there is a row number j such that $\lambda'_i = \lambda_i$ when $i < j$ and $\lambda'_j < \lambda_j$.

Check the digits filled in the first $(j-1)$ rows of the Young tableau $\mathcal{Y}$ to see whether there exist two digits in one row of the Young tableau $\mathcal{Y}$ which also occur in one column of the Young tableau $\mathcal{Y}'$. If yes, from Theorem 6.2, $\mathcal{Y}'\mathcal{Y} = 0$. If no, there is a vertical permutation Q' of the Young tableau $\mathcal{Y}'$ which transforms the Young tableau $\mathcal{Y}'$ to the Young tableau $\mathcal{Y}''$ such that each row among the first $(j-1)$ rows of both the Young tableau $\mathcal{Y}$ and the Young tableau $\mathcal{Y}''$ contains the same digits. Note that

$$\mathcal{Y}'' = Q'\mathcal{Y}'Q'^{-1} = \delta(Q')Q'\mathcal{Y}', \qquad \mathcal{Y}' = \delta(Q')Q'^{-1}\mathcal{Y}''.$$

Since $\lambda'_j < \lambda_j$, there exist at least two digits in the jth row of the Young tableau $\mathcal{Y}$, which occur in one column of the Young tableau $\mathcal{Y}''$. Thus, $\mathcal{Y}''\mathcal{Y} = 0$, and then, $\mathcal{Y}'\mathcal{Y} = 0$. In this corollary the Young operators are not necessary to be standard. □

Corollary 6.2.2 For a given Young pattern, if a standard Young tableau $\mathcal{Y}'$ is larger than a standard Young tableau $\mathcal{Y}$, then $\mathcal{Y}'\mathcal{Y} = 0$.

Proof Compare the filled digits in two standard Young tableaux $\mathcal{Y}$ and $\mathcal{Y}'$ from left to right of the first row, and then those of the second row, and so on. If the first different filled digits occur at the jth column of the ith row where the digit in the Young tableau $\mathcal{Y}'$ is a and the digit in the Young tableau $\mathcal{Y}$ is b, then $a > b$. Thus, b has to be filled in the j'th column of the i'th row of the standard Young tableau $\mathcal{Y}'$, where $j' < j$ and $i' > i$. The digits filled in the j'th column of the ith row of both the Young tableaux $\mathcal{Y}$ and $\mathcal{Y}'$ are the same, say c. Thus, the pair of digits b and c occurs both in the same row of the Young tableau $\mathcal{Y}$ and in the same column of the Young tableau $\mathcal{Y}'$, so that from Theorem 6.2, $\mathcal{Y}'\mathcal{Y} = 0$. □

Corollary 6.2.3 For a given Young pattern, if the digits which occur in one column of the Young tableau $\mathcal{Y}'$ never occur in the same row of the Young tableau $\mathcal{Y}$, the permutation R transforming the Young tableau $\mathcal{Y}$ to the Young tableau $\mathcal{Y}'$ belongs to both the Young tableaux $\mathcal{Y}$ and $\mathcal{Y}'$.

Proof Remind that the condition of Corollary 6.2.3 is equivalent to that the digits which occur in one row of the Young tableau $\mathcal{Y}$ never occur in the same column of the Young tableau $\mathcal{Y}$.

Let P be a horizontal permutation of the Young tableau $\mathcal{Y}$ which transforms the Young tableau $\mathcal{Y}$ to the Young tableau $\mathcal{Y}''$ such that each column of both the Young tableau $\mathcal{Y}'$ and the Young tableau $\mathcal{Y}''$ contains the same digits. Thus, the permutation transforming the Young tableau $\mathcal{Y}''$ to the Young tableau $\mathcal{Y}'$ is a vertical permutation Q'' of the Young tableau $\mathcal{Y}''$, $Q'' = PQP^{-1}$, where Q is a vertical permutation of the Young tableau $\mathcal{Y}$. Therefore, $R = Q''P = PQ$ and $R = RRR^{-1} = (RPR^{-1})(RQR^{-1})$. □

The converse and negative theorem of Corollary 6.2.3 says that if R transforming the Young tableau $\mathcal{Y}$ to the Young tableau $\mathcal{Y}'$ does not belong to $\mathcal{Y}$, there at least exists a horizontal transposition P_0 of $\mathcal{Y}$ which is also a vertical one of $\mathcal{Y}'$, $P_0 = Q_0' = RQ_0R^{-1}$. Namely,

$$R = P_0 R Q_0. \tag{6.35}$$

By the way, the converse theorem of Corollary 6.2.3 holds obviously. In fact, if R transforming the Young tableau $\mathcal{Y}$ to the Young tableau $\mathcal{Y}'$ belongs to $\mathcal{Y}$, then, $R\mathcal{Y}R^{-1} = \mathcal{Y}'$, $R = PQ = (RQR^{-1})P$, and the Young tableau $\mathcal{Y}''$ is transformed both from the Young tableau $\mathcal{Y}$ by a horizontal permutation P of $\mathcal{Y}$ and from the Young tableau $\mathcal{Y}'$ by a vertical permutation $(RQR^{-1})^{-1}$ of $\mathcal{Y}'$. Thus, the digits which occur in one column of the Young tableau $\mathcal{Y}'$ never occur in the same row of the Young tableau $\mathcal{Y}$.

Corollary 6.2.4 A permutation R does not belong to $\mathcal{Y}$ if and only if there is a horizontal transposition P_0 and a vertical transposition Q_0 of $\mathcal{Y}$ such that Eq. (6.35) holds.

Proof From the converse and negative theorem of Corollary 6.2.3, if R does not belong to the Young tableau $\mathcal{Y}$, Eq. (6.35) holds. Conversely, if $R = P_0 R Q_0$, from Eq. (6.29), $F(R) = F(P_0 R Q_0) = -F(R) = 0$. □

6.3 Irreducible Representations of $\mathbf{S}_n$

6.3.1 *Primitive Idempotents in the Group Algebra of S_n*

In §3.7 we introduced the method of finding the standard irreducible bases $b^j_{\mu\nu}$ in the group algebra $\mathcal{L}$ of a finite group G in terms of idempotents. The idempotent e_a is a projective operator in $\mathcal{L}$, satisfying $e_a^2 = e_a$. The idempotents e_a are called mutually orthogonal if $e_b e_a = \delta_{ba} e_a$. There are three main theorems for idempotents. Corollary 3.7.1 says that an idempotent e_a is primitive if and only if

$$e_a t e_a = \lambda_t e_a, \qquad \forall t \in \mathcal{L}, \tag{6.36}$$

where λ_t is a constant depending on t and is allowed to be 0. Theorem 3.7 says that two primitive idempotents e_a and e_b are equivalent if and only if there exists at least one element $S \in G$ satisfying

$$e_a S e_b \neq 0. \tag{6.37}$$

Theorem 3.8 says that the direct sum of n left ideals $\mathcal{L}_a$ generated by the orthogonal idempotents e_a, respectively, is equal to the group algebra $\mathcal{L}$ if and only if the sum of e_a is equal to the identical element E:

$$\mathcal{L} = \bigoplus_{a=1}^{n} \mathcal{L}_a \iff E = \sum_{a=1}^{n} e_a. \tag{6.38}$$

We are going to show that the Young operator $\mathcal{Y}$ is proportional to the primitive idempotent of the permutation group S_n, and then, to calculate the irreducible representations of S_n.

Theorem 6.3 If a vector $\mathcal{X}$ in the group algebra $\mathcal{L}$ of S_n satisfies

$$P\mathcal{X} = \delta(Q)\mathcal{X}Q = \mathcal{X} \tag{6.39}$$

for all horizontal permutations P and vertical permutations Q of $\mathcal{Y}$, then $\mathcal{X}$ is proportional to the Young operator $\mathcal{Y}$

$$\mathcal{X} = \lambda \mathcal{Y}. \tag{6.40}$$

Proof Let $\mathcal{X} = \sum_{R \in S_n} F_1(R) R$. Similar to the proof of Eq. (6.29), from Eq. (6.39) one has

$$F_1(S) = F_1(PS) = \delta(Q) F_1(SQ) = \delta(Q) F_1(PSQ).$$

Taking $S = E$ and $F_1(E) = \lambda$, one obtains

$$\lambda = F_1(E) = F_1(P) = \delta(Q) F_1(Q) = \delta(Q) F_1(PQ).$$

If a permutation R does not belong to $\mathcal{Y}$, then from Corollary 6.2.4

$$F_1(R) = F_1(P_0 R Q_0) = -F_1(R) = 0.$$

In comparison with (6.25), Eq. (6.40) is proved. □

Since $\mathcal{Y}t\mathcal{Y}$ satisfies Eq. (6.39), Corollary 6.3.1 follows directly.

Corollary 6.3.1 For any vector t in the group algebra $\mathcal{L}$ of S_n, one has

$$\mathcal{Y}t\mathcal{Y} = \lambda_t \mathcal{Y}, \tag{6.41}$$

where λ_t is a constant depending upon t and is allowed to be 0.

Corollary 6.3.2 The square of a Young operator $\mathcal{Y}$ is not vanishing:

$$\mathcal{Y}\mathcal{Y} = \lambda \mathcal{Y} \neq 0. \tag{6.42}$$

Proof We are going to calculate the constant λ explicitly. The right ideal generated by $\mathcal{Y}$ is denoted by $\mathcal{R} = \mathcal{Y}\mathcal{L}$. $\mathcal{R}$ is not empty because it contains at least a nonvanishing vector $\mathcal{Y}$. Let $f \neq 0$ be the dimension of $\mathcal{R}$. Take a complete set of basis vectors x_μ in the group algebra $\mathcal{L}$ of S_n such that the first f basis vectors x_μ, $\mu \leq f$, constitute a set of bases in $\mathcal{R}$, and the last $(n! - f)$ basis vectors x_μ, $\mu > f$, do not belong to $\mathcal{R}$. Since $\mathcal{R}$ is generated by $\mathcal{Y}$, any vector in $\mathcal{R}$, including its basis vector x_μ, can be expressed as a product of $\mathcal{Y}$ and another vector in $\mathcal{R}$:

$$x_\mu = \mathcal{Y} y_\mu, \qquad 1 \leq \mu \leq f. \tag{6.43}$$

Now, calculate the product $\mathcal{Y} x_\mu$ from two viewpoints. On one hand, x_μ is a basis vector in $\mathcal{L}$, and $\mathcal{Y}$ is an operator applying to x_μ. Thus, the matrix form $\overline{D}(\mathcal{Y})$ of $\mathcal{Y}$ in the basis vectors x_μ,

$$\mathcal{Y} x_\mu = \sum_{\nu=1}^{n!} x_\nu \overline{D}_{\nu\mu}(\mathcal{Y}), \tag{6.44}$$

is the representation matrix of $\mathcal{Y}$ in the representation $\overline{D}(S_n)$, which is equivalent to the regular representation of S_n. Then,

$$\text{Tr}\,\overline{D}(\mathcal{Y}) = \text{Tr}\,\overline{D}(E) = n!. \tag{6.45}$$

On the other hand, since $\mathcal{Y} x_\mu \in \mathcal{Y}\mathcal{L} = \mathcal{R}$, the summation in Eq. (6.44) only contains the terms with $\nu \leq f$,

$$\overline{D}_{\nu\mu}(\mathcal{Y}) = 0 \qquad \text{when} \quad \nu > f. \tag{6.46}$$

When $\mu \leq f$, from Eq. (6.43) one has $\mathcal{Y} x_\mu = \mathcal{Y}\mathcal{Y} y_\mu = \lambda \mathcal{Y} y_\mu = \lambda x_\mu$. Thus, $\overline{D}_{\nu\mu}(\mathcal{Y}) = \delta_{\nu\mu}\lambda$ when $\mu \leq f$, and $\text{Tr}\,\overline{D}(\mathcal{Y}) = f\lambda$. In comparison with Eq. (6.45), one obtains

$$\lambda = n!/f \neq 0, \tag{6.47}$$

where $f \neq 0$. □

Corollary 6.3.3 $a = (f/n!)\mathcal{Y}$ is a primitive idempotent of the permutation group S_n.

Corollary 6.3.4 If the digits in one column of the Young tableau $\mathcal{Y}'$ never occur in the same row of the Young tableau $\mathcal{Y}$, then $\mathcal{Y}'\mathcal{Y} \neq 0$.

Proof From Corollary 6.2.3, the permutation R transforming the Young tableau $\mathcal{Y}$ to the Young tableau $\mathcal{Y}'$ belongs to the Young tableau $\mathcal{Y}$. Thus, $\mathcal{Y}' = R\mathcal{Y}R^{-1}$, $R = PQ$, and $\mathcal{Y}'\mathcal{Y} = R\mathcal{Y}Q^{-1}P^{-1}\mathcal{Y} = \delta(Q)R\mathcal{Y}\mathcal{Y} = \delta(Q)\lambda R\mathcal{Y} \neq 0$. □

Corollary 6.3.5 Two minimal left ideals generated by the Young operators $\mathcal{Y}$ and $\mathcal{Y}'$, respectively, are equivalent if and only if their Young patterns are the same as each other.

Proof If the Young patterns $\mathcal{Y}$ and $\mathcal{Y}'$ are the same, there exist a permutation R transforming the Young tableau $\mathcal{Y}$ to the Young tableau $\mathcal{Y}'$ such that $\mathcal{Y}' = R\mathcal{Y}R^{-1}$, and $\mathcal{Y}'R\mathcal{Y} = R\mathcal{Y}\mathcal{Y} \neq 0$. From Theorem 3.7, two left ideals are equivalent.

Conversely, if two Young patterns $\mathcal{Y}$ and $\mathcal{Y}'$ are different, without loss of generality, the Young pattern $\mathcal{Y}$ is assumed to be larger than the Young pattern $\mathcal{Y}'$. For any permutation R, the Young pattern $\mathcal{Y}''$, where $\mathcal{Y}'' = R\mathcal{Y}R^{-1}$, is the same as the Young pattern $\mathcal{Y}$. Then, due to Corollary 6.2.1, $\mathcal{Y}'\mathcal{Y}'' = 0$ and $\mathcal{Y}'R\mathcal{Y} = \mathcal{Y}'\mathcal{Y}''R = 0$. □

Therefore, an irreducible representation of S_n can be characterized by a Young pattern $[\lambda]$. Two representations denoted by different Young patterns are not equivalent to each other. Thus, the following Corollary follows Theorem 3.7.

Corollary 6.3.6. Two Young operators corresponding to different Young patterns $\mathcal{Y}$ and $\mathcal{Y}'$ are orthogonal to each other, $\mathcal{Y}\mathcal{Y}' = \mathcal{Y}'\mathcal{Y} = 0$.

The number of different Young patterns is equal to the number of partitions of n, which is equal to the number $g_c(n)$ of classes in S_n. Hence, the irreducible representations denoted by all different Young patterns with n boxes constitute a complete set of the inequivalent and irreducible representations of S_n.

6.3.2 *Orthogonal Primitive Idempotents of S_n*

Two Young operators corresponding to different Young patterns are orthogonal to each other. However, two standard Young operators corresponding to the same Young pattern are not necessary to be orthogonal. The non-orthogonal standard Young operators occur only for S_n with $n \geq 5$. For $n = 5$ there are two Young patterns, $[3, 2]$ and $[2, 2, 1]$, where some standard Young operators are not orthogonal to each other. For example, list the standard Young tableaux for $[3, 2]$ from the smallest to the largest:

$$\begin{array}{ccccc}
\mathcal{Y}_1 & \mathcal{Y}_2 & \mathcal{Y}_3 & \mathcal{Y}_4 & \mathcal{Y}_5 \\[4pt]
\begin{array}{|c|c|c|}\hline 1 & 2 & 3 \\ \hline 4 & 5 \\ \cline{1-2}\end{array} &
\begin{array}{|c|c|c|}\hline 1 & 2 & 4 \\ \hline 3 & 5 \\ \cline{1-2}\end{array} &
\begin{array}{|c|c|c|}\hline 1 & 2 & 5 \\ \hline 3 & 4 \\ \cline{1-2}\end{array} &
\begin{array}{|c|c|c|}\hline 1 & 3 & 4 \\ \hline 2 & 5 \\ \cline{1-2}\end{array} &
\begin{array}{|c|c|c|}\hline 1 & 3 & 5 \\ \hline 2 & 4 \\ \cline{1-2}\end{array}
\end{array}$$

Due to Corollary 6.2.2, $\mathcal{Y}_\mu \mathcal{Y}_\nu = 0$ when $\mu > \nu$. Check the product $\mathcal{Y}_\nu \mathcal{Y}_\mu$ with $\mu > \nu$ one by one whether two digits in one row of the Young tableau $\mathcal{Y}_\mu$ occur in the same column of the Young tableau $\mathcal{Y}_\nu$. If no, $\mathcal{Y}_\nu \mathcal{Y}_\mu \neq 0$. The result is that only

$$\mathcal{Y}_1 \mathcal{Y}_5 \neq 0. \tag{6.48}$$

The permutation R_{15} transforming the Young tableau $\mathcal{Y}_5$ to the Young tableau $\mathcal{Y}_1$ is

$$\begin{aligned}
R_{15} = \begin{pmatrix} 1\ 3\ 5\ 2\ 4 \\ 1\ 2\ 3\ 4\ 5 \end{pmatrix} &= (3\ 2\ 4\ 5) = (2\ 4)\,(4\ 5\ 3) \\
&= (2\ 4)\,(5\ 3)\ \ (3\ 4) = P_5\ Q_5 \\
&= (4\ 5)\,(3\ 2)\ \ (2\ 5) = P_1\ Q_1,
\end{aligned} \tag{6.49}$$

where $P_5 = (2\ 4)\,(5\ 3)$, $Q_5 = (3\ 4)$, $P_1 = (4\ 5)(3\ 2)$, and $Q_1 = (2\ 5)$. The decomposition of R_{15} in Eq. (6.49) is a typical technique for the decomposition of the permutation between two non-orthogonal Young operators.

For a given Young pattern, we want to orthogonalize the standard Young operators by left-multiplying or right-multiplying them with some vectors y_μ in $\mathcal{L}$. In the above example, there are two sets of orthogonal Young operators. One set is

$$\mathcal{Y}_1' = \mathcal{Y}_1\,[E - P_5]\,, \qquad \mathcal{Y}_\nu' = \mathcal{Y}_\nu, \qquad \nu > 1. \tag{6.50}$$

Since $\mathcal{Y}_1 P_5 = \mathcal{Y}_1 R_{15} Q_5^{-1} = \delta(Q_5) R_{15} \mathcal{Y}_5$, $\mathcal{Y}_1' \mathcal{Y}_\mu = \mathcal{Y}_1 \mathcal{Y}_\mu$ when $\mu < 5$. $\mathcal{Y}_1' \mathcal{Y}_5 = \mathcal{Y}_1 (E - P_5) \mathcal{Y}_5 = 0$. The other set is

$$\mathcal{Y}_5'' = [E + Q_1]\,\mathcal{Y}_5, \qquad \mathcal{Y}_\mu'' = \mathcal{Y}_\mu, \qquad \mu < 5. \tag{6.51}$$

Since $Q_1 \mathcal{Y}_5 = P_1^{-1} R_{15} \mathcal{Y}_5 = \mathcal{Y}_1 R_{15}$, $\mathcal{Y}_\nu \mathcal{Y}_5'' = \mathcal{Y}_\nu \mathcal{Y}_5$ when $\nu > 1$. $\mathcal{Y}_1 \mathcal{Y}_5'' = \mathcal{Y}_1 (E + Q_1) \mathcal{Y}_5 = 0$.

Generally, for a given Young pattern $[\lambda]$ where the standard Young operators $\mathcal{Y}_\mu^{[\lambda]}$ are not orthogonal completely, we want to choose some vectors

$y_\mu^{[\lambda]}$ or $\overline{y}_\mu^{[\lambda]}$ such that the new set of $\mathcal{Y}_\mu^{[\lambda]} y_\mu^{[\lambda]}$ or the new set of $\overline{y}_\mu^{[\lambda]} \mathcal{Y}_\mu^{[\lambda]}$ are mutually orthogonal. We will discuss the first set in detail and give the result for the second set. Since the Young pattern $[\lambda]$ is fixed, in the following we will omit the superscript $[\lambda]$ for simplicity.

The problem is to find y_μ such that

$$\mathcal{Q}_\mu y_\mu \mathcal{P}_\nu = \delta_{\mu\nu} \mathcal{Q}_\mu \mathcal{P}_\mu, \qquad 1 \leq \mu \leq d, \qquad 1 \leq \nu \leq d. \tag{6.52}$$

Then, $\mathcal{Y}_\mu y_\mu \mathcal{Y}_\nu = \delta_{\mu\nu} \mathcal{Y}_\mu \mathcal{Y}_\mu$. Denote by $R_{\mu\nu}$ the permutation transforming the standard Young tableau $\mathcal{Y}_\nu$ to the standard Young tableau $\mathcal{Y}_\mu$,

$$\begin{aligned} &R_{\mu\nu} \mathcal{Y}_\nu = \mathcal{Y}_\mu R_{\mu\nu}, \qquad R_{\mu\nu} \mathcal{P}_\nu = \mathcal{P}_\mu R_{\mu\nu}, \qquad R_{\mu\nu} \mathcal{Q}_\nu = \mathcal{Q}_\mu R_{\mu\nu}, \\ &R_{\mu\rho} R_{\rho\nu} = R_{\mu\nu}, \qquad R_{\mu\mu} = E. \end{aligned} \tag{6.53}$$

Due to Corollary 6.2.3,

$$R_{\mu\nu} = P_\nu^{(\mu)} Q_\nu^{(\mu)} = \overline{P}_\mu^{(\nu)} \overline{Q}_\mu^{(\nu)} \qquad \text{when } \mathcal{Y}_\mu \mathcal{Y}_\nu \neq 0, \tag{6.54}$$

where $P_\nu^{(\mu)}$ and $Q_\nu^{(\mu)}$ are the horizontal permutation and the vertical permutation of the Young tableau $\mathcal{Y}_\nu$, respectively, and $\overline{P}_\mu^{(\nu)}$ and $\overline{Q}_\mu^{(\nu)}$ are those of $\mathcal{Y}_\mu$. Let

$$P_{\mu\nu} = \begin{cases} P_\nu^{(\mu)} & \text{when } \mathcal{Y}_\mu \mathcal{Y}_\nu \neq 0, \\ 0 & \text{when } \mathcal{Y}_\mu \mathcal{Y}_\nu = 0. \end{cases} \tag{6.55}$$

Obviously, when $\mathcal{Y}_\mu \mathcal{Y}_\nu \neq 0$,

$$\begin{aligned} P_{\mu\nu} \mathcal{Q}_\nu &= R_{\mu\nu} \mathcal{Q}_\nu \left(Q_\nu^{(\mu)}\right)^{-1} = \mathcal{Q}_\mu P_{\mu\nu}, \\ \mathcal{P}_\nu P_{\mu\nu} &= P_{\mu\nu} \mathcal{P}_\nu = \mathcal{P}_\nu, \\ \mathcal{Y}_\mu P_{\mu\nu} &= R_{\mu\nu} \mathcal{Y}_\nu \left(Q_\nu^{(\mu)}\right)^{-1} = \delta(Q_\nu^{(\mu)}) R_{\mu\nu} \mathcal{Y}_\nu. \end{aligned} \tag{6.56}$$

Define y_μ one by one from $\mu = d$ to $\mu = 1$,

$$y_\mu = E - \sum_{\rho=\mu+1}^{d} P_{\mu\rho} y_\rho, \qquad y_d = E, \qquad 1 \leq \mu \leq d. \tag{6.57}$$

It is easy to show by induction that Eq. (6.52) holds. As a matter of fact, Eq. (6.52) holds when $\mu = d$ owing to Corollary 6.2.2. Suppose that Eq. (6.52) holds for $\mu > \tau$. For $\mu = \tau$, Eq. (6.52) also holds because

$$\begin{aligned}\mathcal{Q}_\tau y_\tau \mathcal{P}_\nu &= \mathcal{Q}_\tau \mathcal{P}_\nu - \sum_{\rho=\tau+1}^{d} \mathcal{Q}_\tau P_{\tau\rho} y_\rho \mathcal{P}_\nu = \mathcal{Q}_\tau \mathcal{P}_\nu - \sum_{\rho=\tau+1}^{d} P_{\tau\rho} \mathcal{Q}_\rho y_\rho \mathcal{P}_\nu \\ &= \begin{cases} 0 & \text{when } \nu < \tau, \\ \mathcal{Q}_\tau \mathcal{P}_\tau & \text{when } \nu = \tau, \\ \mathcal{Q}_\tau \mathcal{P}_\nu - P_{\tau\nu} \mathcal{Q}_\nu \mathcal{P}_\nu = 0 & \text{when } \nu > \tau. \end{cases}\end{aligned}$$

Note that y_μ is the algebraic sum of elements of S_n with the coefficients ± 1, and due to Eq. (6.56)

$$\mathcal{Y}_\mu y_\mu = \mathcal{Y}_\mu - \sum_{\mathcal{Y}_\mu \mathcal{Y}_\nu \neq 0,\ \nu=\mu+1}^{d} \delta(Q_\nu^{(\mu)}) R_{\mu\nu} \mathcal{Y}_\nu y_\nu = \sum_{\rho=\mu}^{d} t_\rho \mathcal{Y}_\rho, \tag{6.58}$$

where the sum index ν in the middle expression runs over from $\mu + 1$ to d and in the condition $\mathcal{Y}_\mu \mathcal{Y}_\nu \neq 0$, and t_ρ in the last expression is a vector in $\mathcal{L}$ which is allowed to be zero.

Similarly, let

$$\overline{Q}_{\mu\nu} = \begin{cases} \delta(\overline{Q}_\mu^{(\nu)}) \overline{Q}_\mu^{(\nu)} & \text{when } \mathcal{Y}_\mu \mathcal{Y}_\nu \neq 0, \\ 0 & \text{when } \mathcal{Y}_\mu \mathcal{Y}_\nu = 0. \end{cases} \tag{6.59}$$

$\overline{y}_\nu$ are defined one by one from $\nu = 1$ to $\nu = d$,

$$\overline{y}_\nu = E - \sum_{\rho=1}^{\nu-1} \overline{y}_\rho \overline{Q}_{\rho\nu}, \qquad \overline{y}_1 = E, \qquad 1 \leq \nu \leq d, \tag{6.60}$$

such that $\mathcal{Y}_\mu \overline{y}_\nu \mathcal{Y}_\nu = \delta_{\mu\nu} \mathcal{Y}_\nu \mathcal{Y}_\nu$.

Theorem 6.4 The following $e_\mu^{[\lambda]}$ constitute a complete set of orthogonal primitive idempotents:

$$e_\mu^{[\lambda]} = \frac{d_{[\lambda]}}{n!} \mathcal{Y}_\mu^{[\lambda]} y_\mu^{[\lambda]}, \qquad 1 \leq \mu \leq d_{[\lambda]}, \tag{6.61}$$

and the identical element E can be decomposed as

$$E = \frac{1}{n!} \sum_{[\lambda]} d_{[\lambda]} \sum_{\mu=1}^{d_{[\lambda]}} \mathcal{Y}_\mu^{[\lambda]} y_\mu^{[\lambda]}. \tag{6.62}$$

Proof For a given Young pattern $[\lambda]$, there are $d_{[\lambda]}$ orthogonal primitive idempotents $e_\mu^{[\lambda]}$, given in (6.61) but replacing $d_{[\lambda]}$ with $f_{[\lambda]}$, where $f_{[\lambda]}$ is the dimension of the left ideal $\mathcal{L}_\mu^{[\lambda]}$ generated by $e_\mu^{[\lambda]}$ (see Theorem 3.10). Since the multiplicity of each irreducible representation in the reduction of the regular representation is equal to the dimension of the representation, one has $d_{[\lambda]} \leq f_{[\lambda]}$. On the other hand, the square sum of the dimensions of the inequivalent and irreducible representations of a finite group G is equal to its order,

$$\sum_{[\lambda]} f_{[\lambda]}^2 = n!. \tag{6.63}$$

In comparison with (6.21) one obtains

$$d_{[\lambda]} = f_{[\lambda]}. \tag{6.64}$$

Therefore, the direct sum of the left ideals $\mathcal{L}_\mu^{[\lambda]}$ generated by $e_\mu^{[\lambda]}$ is equal to the group algebra $\mathcal{L}$, and Eq. (6.62) follows Theorem 3.8. □

In the same reason, $\overline{e}_\mu^{[\lambda]}$ also constitute a complete set of orthogonal primitive idempotents:

$$\overline{e}_\mu^{[\lambda]} = \frac{d_{[\lambda]}}{n!} \overline{y}_\mu^{[\lambda]} \mathcal{Y}_\mu^{[\lambda]}, \qquad 1 \leq \mu \leq d_{[\lambda]}, \tag{6.65}$$

and the identical element E can be decomposed as

$$E = \frac{1}{n!} \sum_{[\lambda]} d_{[\lambda]} \sum_{\mu=1}^{d_{[\lambda]}} \overline{y}_\mu^{[\lambda]} \mathcal{Y}_\mu^{[\lambda]}. \tag{6.66}$$

6.3.3 *Calculation of Representation Matrices for S_n*

For a given Young pattern $[\lambda]$, we are going to choose a set of standard bases $b_{\mu\nu}^{[\lambda]}$ and calculate the representation matrices. Since the Young pattern $[\lambda]$ is fixed, we omit the superscript $[\lambda]$ for simplicity.

In terms of the permutations $R_{\mu\nu}$ given in Eq. (6.53), where $R_{\mu\nu}$ transforms the standard Young tableau $\mathcal{Y}_\nu$ to the standard Young tableau $\mathcal{Y}_\mu$, d^2 basis vectors $b_{\mu\nu}$ can be defined:

$$\begin{aligned} b_{\mu\nu} = e_\mu R_{\mu\nu} e_\nu &= (d/n!)^2 \, \mathcal{Y}_\mu y_\mu R_{\mu\nu} \mathcal{Y}_\nu y_\nu = (d/n!)^2 \, \mathcal{Y}_\mu y_\mu \mathcal{Y}_\mu R_{\mu\nu} y_\nu \\ &= (d/n!) \, \mathcal{Y}_\mu R_{\mu\nu} y_\nu = (d/n!) \, R_{\mu\nu} \mathcal{Y}_\nu y_\nu \\ &= R_{\mu\nu} e_\nu. \end{aligned} \tag{6.67}$$

Those basis vectors $b_{\mu\nu}$ are standard because they satisfy the conditions (3.130),

$$b_{\mu\rho} b_{\lambda\nu} = \delta_{\rho\lambda} b_{\mu\nu}, \qquad b_{\mu\mu} = e_\mu = (d/n!) \, \mathcal{Y}_\mu y_\mu. \tag{6.68}$$

For a given ν, d basis vectors $b_{\mu\nu}$ are the complete bases in the left ideal $\mathcal{L}_\nu$, and for a given μ, d basis vectors $b_{\mu\nu}$ are that in the right ideal $\mathcal{R}_\mu$. In the standard bases, the representations of both the left ideal $\mathcal{L}_\nu$ and the right ideal $\mathcal{R}_\mu$ are the same,

$$Sb_{\mu\nu} = \sum_{\rho=1}^{d} b_{\rho\nu} D_{\rho\mu}(S), \qquad b_{\mu\nu}S = \sum_{\rho=1}^{d} D_{\nu\rho}(S) b_{\mu\rho}. \tag{6.69}$$

Replacing ν with τ in the first equality of Eq. (6.69) and left-multiplying it with $b_{\tau\nu}$, one obtains

$$D_{\nu\mu}(S)e_\tau = b_{\tau\nu}Sb_{\mu\tau} = (d/n!)^2 R_{\tau\nu}\mathcal{Y}_\nu y_\nu S\mathcal{Y}_\mu R_{\mu\tau}y_\tau, \tag{6.70}$$

where τ is arbitrary, $1 \leq \tau \leq d$. Due to Corollary 3.7.1, the right-hand side of Eq. (6.70) is proportional to e_τ. The representation matrix of any element S of S_n in the representation $[\lambda]$ is calculated from Eq. (6.70).

In calculating $D_{\nu\mu}(S)$, one has to move out the quantity $y_\nu S$ between two Young operators in Eq. (6.70) such that two Young operators reduce to one Young operator. y_ν is an algebraic sum of group elements with the coefficients ± 1 and can be expressed as follows formally,

$$y_\nu = \sum_k \delta_k T_k, \tag{6.71}$$

where T_k is a permutation and $\delta_k = \pm 1$. Denote by $\mathcal{Y}_{\nu k}$ the Young tableau transformed from the Young tableau $\mathcal{Y}_\nu$ by $(T_k)^{-1}$, and by $\mathcal{Y}_\mu(S)$ the Young tableau transformed from the Young tableau $\mathcal{Y}_\mu$ by S. Hence, Eq. (6.70) with $\tau = 1$ becomes

$$D_{\nu\mu}(S)e_1 = \sum_k \delta_k (d/n!)^2 R_{1\nu}T_k\mathcal{Y}_{\nu k}\mathcal{Y}_\mu(S)SR_{\mu 1}y_1. \tag{6.72}$$

Now, we calculate the product of two Young operators. If two digits in one row of the Young tableau $\mathcal{Y}_\mu(S)$ occur in the same column of the Young tableau $\mathcal{Y}_{\nu k}$, the product $\mathcal{Y}_{\nu k}\mathcal{Y}_\mu(S)$ is vanishing. If the digits in one row of the Young tableau $\mathcal{Y}_\mu(S)$ never occur in the same column of the Young tableau $\mathcal{Y}_{\nu k}$, from Corollary 6.2.3, the permutation transforming the Young tableau $\mathcal{Y}_\mu(S)$ to the Young tableau $\mathcal{Y}_{\nu k}$ belongs to $\mathcal{Y}_\mu(S)$,

$$P_\mu(S)Q_\mu(S) = \left[P_\mu(S)Q_\mu(S)Q_\mu(S)Q_\mu(S)^{-1}P_\mu(S)^{-1}\right]P_\mu(S) = Q_{\nu k}P_\mu(S).$$

The quantity in the bracket, denoted by $Q_{\nu k}$, is a vertical permutation of the Young tableau $\mathcal{Y}_{\nu k}$. $(Q_{\nu k})^{-1}$ transforms the Young tableau $\mathcal{Y}_{\nu k}$ to the

Young tableau $\mathcal{Y}'$ such that the digits in each row of the Young tableau $\mathcal{Y}'$ also occur in the same row of the Young tableau $\mathcal{Y}_\mu(S)$. Hence,

$$\begin{aligned}(d/n!)\,\mathcal{Y}_{\nu k}\mathcal{Y}_\mu(S) &= (d/n!)\,\mathcal{Y}_{\nu k}\delta(Q_{\nu k})Q_{\nu k}P_\mu(S)\mathcal{Y}_\mu(S)\\ &= (d/n!)\,\delta(Q_{\nu k})Q_{\nu k}P_\mu(S)\mathcal{Y}_\mu(S)\mathcal{Y}_\mu(S)\\ &= \delta(Q_{\nu k})Q_{\nu k}P_\mu(S)\mathcal{Y}_\mu(S).\end{aligned}$$

Substituting it into Eq. (6.72), one obtains

$$D_{\nu\mu}(S)e_1 = \sum_k \delta_k\delta(Q_{\nu k})\,(d/n!)\,\{R_{1\nu}T_kQ_{\nu k}P_\mu(S)SR_{\mu 1}\}\,\mathcal{Y}_1 y_1. \tag{6.73}$$

The product of permutations in the curve bracket of Eq. (6.73) has to be equal to the identical element E because the right-hand side of Eq. (6.73) is proportional to $e_1 = (d/n!)\mathcal{Y}_1 y_1$. In fact, $R_{\mu 1}$ first transforms the Young tableau $\mathcal{Y}_1$ to the Young tableau $\mathcal{Y}_\mu$. Then, S transforms the Young tableau $\mathcal{Y}_\mu$ to the Young tableau $\mathcal{Y}_\mu(S)$. Third, $Q_{\nu k}P_\mu(S)$ transforms the Young tableau $\mathcal{Y}_\mu(S)$ to the Young tableau $\mathcal{Y}_{\nu k}$. Fourth, T_k transforms the Young tableau $\mathcal{Y}_{\nu k}$ to the Young tableau $\mathcal{Y}_\nu$. At last $R_{1\nu}$ transforms the Young tableau $\mathcal{Y}_\nu$ to the original Young tableau $\mathcal{Y}_1$. Namely, the product of permutations preserves the Young tableau $\mathcal{Y}_1$ invariant so that it is equal to the identical element E. Hence,

$$D_{\nu\mu}(S) = \sum_k \delta_k\delta(Q_{\nu k}). \tag{6.74}$$

It means that in the standard basis vectors $b_{\mu\nu}$, the representation matrix entry $D_{\nu\mu}(S)$ is always an integer, so that each irreducible representation of S_n is real and its characters are integers.

Equation (6.74) provides a method for calculating $D_{\nu\mu}(S)$. Due to Eq. (6.71) δ_k is known. $\delta(Q_{\nu k})$ can be calculated by comparing two Young tableaux $\mathcal{Y}_{\nu k}$ and $\mathcal{Y}_\mu(S)$. If two digits in one row of the Young tableau $\mathcal{Y}_\mu(S)$ occur in the same column of the Young tableau $\mathcal{Y}_{\nu k}$, $\delta(Q_{\nu k}) = 0$. Otherwise, from Corollary 6.2.3, there is a vertical permutation $Q_{\nu k}^{-1}$ of $\mathcal{Y}_{\nu k}$ that transforms the Young tableau $\mathcal{Y}_{\nu k}$ into the Young tableau $\mathcal{Y}'$ such that the digits in each row of the Young tableau $\mathcal{Y}'$ also occur in the same row of the Young tableau $\mathcal{Y}_\mu(S)$. $\delta(Q_{\nu k})$ is the permutation parity of $Q_{\nu k}^{-1}$.

The matrix entry $D_{\nu\mu}^{[\lambda]}(S)$ of S in the irreducible representation $[\lambda]$ of S_n can be calculated by the tabular method as follows. Denote by $\mathcal{Y}_\mu(S)$ the Young tableau transformed from the standard Young tableau $\mathcal{Y}_\mu^{[\lambda]}$ by the permutation S. List $\mathcal{Y}_\mu(S)$, $1 \leq \mu \leq d_{[\lambda]}$, on the first row of the table to designate its columns. Let $y_\nu^{[\lambda]} = \sum_k \delta_k T_k$. Denote by $\mathcal{Y}_{\nu k}$ the Young tableau transformed from the standard Young tableau $\mathcal{Y}_\nu^{[\lambda]}$ by the permutation T_k^{-1}. List the sum of the Young tableaux $\sum_k \delta_k \mathcal{Y}_{\nu k}$, $1 \leq \nu \leq d_{[\lambda]}$, on the first column of the table to designate its rows. The representation matrix entry $D_{\nu\mu}^{[\lambda]}(S)$ is equal to $\sum_k \delta_k A_{\nu k}^{\mu}(S)$, which is filled in the μth column of the νth row of the table. $A_{\nu k}^{\mu}(S)$ is calculated by comparing two Young tableaux $\mathcal{Y}_{\nu k}$ and $\mathcal{Y}_\mu(S)$. $A_{\nu k}^{\mu}(S) = 0$ if there are two digits in one row of the Young tableau $\mathcal{Y}_\mu(S)$ which also occur in the same column of the Young tableau $\mathcal{Y}_{\nu k}$. Otherwise, $A_{\nu k}^{\mu}(S)$ is the permutation parity of the vertical permutation of the Young tableau $\mathcal{Y}_{\nu k}$, which transforms the Young tableau $\mathcal{Y}_{\nu k}$ to the Young tableau $\mathcal{Y}'$ such that the digits in each row of the Young tableau $\mathcal{Y}'$ also occur in the same row of the Young tableau $\mathcal{Y}_\mu(S)$. The sum of the diagonal entries in the table is the character $\chi^{[\lambda]}(S)$. The calculated irreducible representation $D^{[\lambda]}(S)$ is generally not unitary.

Table 6.1 Tabular method for calculating the irreducible representation matrix $D_{\nu\mu}^{[\lambda]}(S)$ of $\mathbf{S}_n$

$$[\lambda] = [3,2], \qquad S = (1\ 2\ 3\ 4\ 5), \qquad y_\nu = \sum_k \delta_k T_k,$$

T_k^{-1} transforms the Young tableau $\mathcal{Y}_\nu$ to the Young tableau $\mathcal{Y}_{\nu k}$,
S transforms the Young tableau $\mathcal{Y}_\mu$ to the Young tableau $\mathcal{Y}_\mu(S)$.

$\sum_k \delta_k$ {Young tableau $\mathcal{Y}_{\nu k}$}	Young tableau $\mathcal{Y}_\mu(S)$				
	2 3 4 5 1	2 3 5 4 1	2 3 1 4 5	2 4 5 3 1	2 4 1 3 5
1 2 3 / 4 5 − 1 4 5 / 2 3	−1 − 0	0 − 1	1 − 0	0 + 1	0 − 0
1 2 4 3 5	−1	0	0	0	1
1 2 5 3 4	0	−1	0	0	0
1 3 4 2 5	−1	0	0	1	0
1 3 5 2 4	0	−1	0	1	0

As example, in Table 6.1 the representation matrix $D^{[3,2]}(S)$ of S_5, where $S = (1\ 2\ 3\ 4\ 5)$, is calculated by the tabular method:

$$D^{[3,2]}[(1\ 2\ 3\ 4\ 5)] = \begin{pmatrix} -1 & -1 & 1 & 1 & 0 \\ -1 & 0 & 0 & 0 & 1 \\ 0 & -1 & 0 & 0 & 0 \\ -1 & 0 & 0 & 1 & 0 \\ 0 & -1 & 0 & 1 & 0 \end{pmatrix}.$$

$D^{[3,2]}[(1\ 2)]$ and $D^{[3,2]}[(5\ 4\ 3\ 2\ 1)]$ can be similarly calculated

$$\begin{pmatrix} 1 & 0 & 0 & -1 & -1 \\ 0 & 1 & 0 & -1 & 0 \\ 0 & 0 & 1 & 0 & -1 \\ 0 & 0 & 0 & -1 & 0 \\ 0 & 0 & 0 & 0 & -1 \end{pmatrix} \quad \text{and} \quad \begin{pmatrix} 0 & 0 & -1 & -1 & 1 \\ 0 & 0 & -1 & 0 & 0 \\ 1 & 0 & -1 & -1 & 0 \\ 0 & 0 & -1 & 0 & 1 \\ 0 & 1 & -1 & -1 & 1 \end{pmatrix}.$$

Then, the representation matrices for the elements in each class are calculated by the products

$$\begin{array}{ll} (2\ 3) = (1\ 2\ 3\ 4\ 5)(1\ 2)(5\ 4\ 3\ 2\ 1), & (1\ 2\ 3) = (1\ 2)(2\ 3), \\ (3\ 4) = (1\ 2\ 3\ 4\ 5)(2\ 3)(5\ 4\ 3\ 2\ 1), & (2\ 3\ 4\ 5) = (1\ 2)(1\ 2\ 3\ 4\ 5), \\ (4\ 5) = (1\ 2\ 3\ 4\ 5)(3\ 4)(5\ 4\ 3\ 2\ 1), & (1\ 2)(3\ 4), \quad (1\ 2\ 3)(4\ 5). \end{array}$$

The characters for the representation are listed in Table 6.2.

Table 6.2. Character table for representation $[3,2]$ of S_5

Class (ℓ)	(1^5)	$(2,1^3)$	$(2^2,1)$	$(3,1^2)$	$(3,2)$	$(4,1)$	(5)
Character	5	1	1	−1	1	−1	0

6.3.4 *Calculation of Characters by Graphic Method*

The representation matrices of S_n, as well as the characters in the representation, can be calculated by the tabular method. However, there is a graphic method for calculating the character of a class (ℓ) in the irreducible representation $[\lambda]$. The integers ℓ_j of the partition (ℓ) can be arranged in any order, but in the increasing order, $\ell_1 \leq \ell_2 \leq \cdots \leq \ell_m$, will simplify the calculation. According to the following rule, we first fill ℓ_1 digits 1 into the Young pattern $[\lambda]$, then fill ℓ_2 digits 2, and so on, until filling ℓ_m digits m. The filling rule is as follows:

(a) The boxes filled with each digit, say j, are connected such that from the lowest and the leftmost box one can go through all the boxes filled with j only upward and rightward.

(b) Each time when ℓ_j digits j are filled, all the boxes filled with digits $i \leq j$ form a standard Young pattern, namely, the boxes are lined up on the top and on the left such that the number of boxes in the upper row is not less than that in the lower row and there is no unfilled box embedding between two filled boxes.

It is said to be one regular application if all digits are filled into the Young pattern according to the rule. The filling parity for the digit j is defined to be 1 if the number of rows of the boxes filled with j is odd, and to be -1 if that is even. The filling parity of a regular application is defined to be the product of the filling parities of m digits. The character $\chi^{[\lambda]}[(\ell)]$ of the class (ℓ) in the representation $[\lambda]$ is equal to the sum of the filling parities of all regular applications. $\chi^{[\lambda]}[(\ell)] = 0$ if there is no regular application, namely, if m digits cannot be filled in the Young pattern, all according to the above rule. For the class (1^n) composed of only the identical element E, each regular application is just a standard Young tableau, so its character is nothing but the dimension of the representation. Usually, the character of the class (1^n) is calculated by the hook rule instead of the graphic method. In Table 6.3 all the regular applications of each class for the Young pattern $[3, 2]$ are listed, and their characters are calculated. The five regular applications of the class (1^5) are omitted.

Table 6.3 The character table for the representation [3,2] of S$_5$ calculated by the graphic method

Class	(1^5)	$(1^3, 2)$	$(1, 2^2)$	$(1^2, 3)$	$(2, 3)$	$(1, 4)$	(5)
Regular application		1 2 3 4 4	1 2 2 3 3	1 3 3 2 3	1 2 2 1 2	1 2 2 2 2	
Filling parity		1	1	-1	1	-1	
$\chi^{[3,2]}[(\ell)]$	5	1	1	-1	1	-1	0

If one changes the order of ℓ_j for (ℓ), there may be more regular applications, but the calculated results of the characters are the same. For example, if one changes the order for the class $(1, 2, 2)$ to be $(2, 2, 1)$, there are three regular applications in the Young pattern $[3, 2]$:

$$\begin{array}{lccc} & 1\,1\,3 & 1\,2\,2 & 1\,2\,3 \\ & 2\,2 & 1\,3 & 1\,2 \\ \text{Filling parity} = & 1 & ,\ (-1)\ , & 1 \end{array} \qquad \chi(1, 2^2) = 1 - 1 + 1 = 1.$$

The Young pattern $[n]$ with one row has one standard Young tableau, and the corresponding Young operator is the sum of all elements in S_n. Thus, $[n]$ characterizes the identical representation, where the representation matrix of each element in S_n is 1. The Young pattern $[1^n]$ with one column also has one standard Young tableau, and the corresponding Young operator is the sum of all elements in S_n, multiplied with their permutation parities. Thus, $[1^n]$ characterizes the antisymmetric representation, where the representation matrix of each element R is its permutation parity $\delta(R)$. Denote by $[\tilde{\lambda}]$ the associate Young pattern of $[\lambda]$. According to the graphic method, the characters of a class (ℓ) in two associate Young patterns differ only with a permutation parity $\delta[(\ell)]$ of the elements in the class,

$$\chi^{[\tilde{\lambda}]}[(\ell)] = \delta[(\ell)]\chi^{[\lambda]}[(\ell)]. \tag{6.75}$$

If $[\tilde{\lambda}] = [\lambda]$, the Young pattern $[\lambda]$ is called self-associate, and the character of a class (ℓ) with odd permutation parity is 0. In fact, the transpose of each regular application of the Young pattern $[\lambda]$ is a regular application of the associate Young pattern $[\tilde{\lambda}]$, where the positions of each digit, say j, are the same as each other except for the interchanging of rows and columns. The sum of the row number and the column number of boxes filled with the digits j in the Young patterns is $(\ell_j + 1)$, so that the product of two filling parities for the digit j in the two Young patterns is equal to the permutation parity $(-1)^{\ell_j+1}$ of a cycle with length ℓ_j.

6.3.5 *The Permutation Group* S_3

As an example, we calculate the standard bases and the inequivalent irreducible representations of S_3 by the Young operator method. S_3 is isomorphic onto the symmetric group D_3 of a regular triangle. There are six elements and three classes in S_3. The class (1^3) contains only the identical element E. The class $(2,1)$ contains three elements, $A = (2\ 3)$, $B = (3\ 1)$, and $C = (1\ 2)$. The class (3) contains two elements, $D = (3\ 2\ 1)$ and $F = (1\ 2\ 3)$. There are three Young patterns for S_3.

The Young pattern $[3]$ characterizes the identical representation, where the representation matrix of any group element is equal to 1.

$\mathcal{Y}^{[3]}$ | 1 | 2 | 3 |

$$b^{[3]} = e^{[3]} = \{E + (1\ 2) + (2\ 3) + (3\ 1) + (1\ 2\ 3) + (3\ 2\ 1)\}/6.$$

For the Young pattern $[2,1]$, there are two standard Young tableaux.

The representation $[2,1]$ is two-dimensional.

$$\mathcal{Y}_1^{[2,1]} \quad \begin{array}{|c|c|}\hline 1 & 2 \\ \hline 3 \\ \cline{1-1}\end{array} \qquad \text{and} \qquad \mathcal{Y}_2^{[2,1]} \quad \begin{array}{|c|c|}\hline 1 & 3 \\ \hline 2 \\ \cline{1-1}\end{array}$$

The idempotents and the standard basis vectors are

$$\begin{aligned}
b_{11}^{[2,1]} &= e_1^{[2,1]} = \{E + (1\ 2) - (1\ 3) - (2\ 1\ 3)\}/3, \\
b_{21}^{[2,1]} &= (2\ 3)e_1^{[2,1]} = \{(2\ 3) + (3\ 2\ 1) - (2\ 3\ 1) - (2\ 1)\}/3, \\
b_{12}^{[2,1]} &= (2\ 3)e_2^{[2,1]} = \{(2\ 3) + (2\ 3\ 1) - (3\ 2\ 1) - (3\ 1)\}/3, \\
b_{22}^{[2,1]} &= e_2^{[2,1]} = \{E + (1\ 3) - (1\ 2) - (3\ 1\ 2)\}/3.
\end{aligned}$$

Table 6.4. The representation matrices of generators in $[2,1]$ of $\mathbf{S}_3$

$\mathcal{Y}_\nu$	$S=(1\ 2)$: 2 1 / 3	$S=(1\ 2)$: 2 3 / 1	$S'=(1\ 2\ 3)$: 2 3 / 1	$S'=(1\ 2\ 3)$: 2 1 / 3
1 2 / 3	1	−1	−1	1
1 3 / 2	0	−1	−1	0

The representation matrices of the generators (1 2) and (1 2 3) are calculated by the tabular method (see Table 6.4):

$$D^{[2,1]}[(1\ 2)] = \begin{pmatrix} 1 & -1 \\ 0 & -1 \end{pmatrix}, \qquad D^{[2,1]}[(1\ 2\ 3)] = \begin{pmatrix} -1 & 1 \\ -1 & 0 \end{pmatrix}.$$

It is not a real orthogonal representation because the standard basis vectors are not orthonormal in the group algebra. Through a similarity transformation X, the new basis vectors are orthonormal and the representation becomes real orthogonal (see Eq. (2.12)).

$$X = \frac{1}{2}\begin{pmatrix} 3 & -\sqrt{3} \\ 3 & \sqrt{3} \end{pmatrix}, \qquad X^{-1}D^{[2,1]}(R)X = \overline{D}(R),$$

$$\overline{D}[(1\ 2)] = \frac{1}{2}\begin{pmatrix} -1 & -\sqrt{3} \\ -\sqrt{3} & 1 \end{pmatrix}, \qquad \overline{D}[(1\ 2\ 3)] = \frac{1}{2}\begin{pmatrix} -1 & \sqrt{3} \\ -\sqrt{3} & -1 \end{pmatrix},$$

$$\begin{aligned}
\phi_1 &= (3/2)(b_{11} + b_{21}) = \{E + (2\ 3) - (3\ 1) - (1\ 2\ 3)\}/2, \\
\phi_2 &= (\sqrt{3}/2)(-b_{11} + b_{21}) \\
&= \{-E + (2\ 3) + (3\ 1) - 2(1\ 2) - (1\ 2\ 3) + 2(3\ 2\ 1)\}/(2\sqrt{3}).
\end{aligned}$$

The Young pattern $[1,1,1]$ characterizes the antisymmetric representation, where the representation matrix of any group element R is equal to its permutation parity $\delta(R)$.

$$\mathcal{Y}^{[1^3]} \quad \begin{array}{|c|}\hline 1 \\ \hline 2 \\ \hline 3 \\ \hline \end{array}$$

$$b^{[1,1,1]} = e^{[1,1,1]} = \{E - (1\ 2) - (2\ 3) - (3\ 1) + (1\ 2\ 3) + (3\ 2\ 1)\}/6.$$

6.3.6 *Inner Product of Irreducible Representations of S_n*

The direct product of two irreducible representations of S_n is specially called the inner product, because there is another product called their outer product (see the last section in this chapter). The inner product is usually reducible and can be reduced to the Clebsch–Gordan series by the character formula (3.54).

$$\begin{aligned} & C^{-1}\left[D^{[\lambda]}(R) \times D^{[\mu]}(R)\right] C = \bigoplus_{\nu} a_{\lambda\mu\nu} D^{[\nu]}(R), \\ & \chi^{[\lambda]}(R)\chi^{[\mu]}(R) = \sum_{\nu} a_{\lambda\mu\nu}\chi^{[\nu]}(R), \\ & a_{\lambda\mu\nu} = \frac{1}{n!}\sum_{R\in S_n} \chi^{[\lambda]}(R)\chi^{[\mu]}(R)\chi^{[\nu]}(R). \end{aligned} \tag{6.76}$$

Since the characters in the irreducible representations of S_n are real, $a_{\lambda\mu\nu}$ is totally symmetric with respect to three subscripts. This property can be used to simplify the calculation of the Clebsch–Gordan series. Noting Eq. (6.75) , one has

$$[n] \times [\lambda] \simeq [\lambda], \qquad [1^n] \times [\lambda] \simeq [\widetilde{\lambda}], \qquad [\lambda] \times [\mu] \simeq [\widetilde{\lambda}] \times [\widetilde{\mu}]. \tag{6.77}$$

Due to Eq. (6.76) one concludes that there is one identical representation $[n]$ in the reduction of $[\lambda] \times [\mu]$ if and only if $[\lambda] = [\mu]$, and there is one antisymmetric representation $[1^n]$ in the reduction of $[\lambda] \times [\mu]$ if and only if $[\lambda] = [\widetilde{\mu}]$. For the group S_3 one has

$$\begin{aligned} & [3] \times [3] \simeq [1^3] \times [1^3] \simeq [3], \qquad [3] \times [1^3] \simeq [1^3], \\ & [3] \times [2,1] \simeq [1^3] \times [2,1] \simeq [2,1], \\ & [2,1] \times [2,1] \simeq [3] \oplus [1^3] \oplus [2,1]. \end{aligned} \tag{6.78}$$

Some results for the Clebsch–Gordan series can be found in Prob. 31 of Chap. 6 of [Ma and Gu (2004)].

6.4 Real Orthogonal Representation of S_n

The merit of the tabular method for calculating the representations of S_n is that the basis vectors are well known and the representation matrix entries are integers. Its shortcoming is that the calculated representation is not real orthogonal. In this section we are going to show a method to combine the basis vectors such that the new representation $[\lambda]$ is real orthogonal [Tong et al. (1992)]. In this section we neglect the superscript $[\lambda]$ for simplicity, because the representation $[\lambda]$ is fixed,

In the group algebra $\mathcal{L}$ of S_n, a transposition $(a\ d)$ is unitary and Hermitian. Introduce a set of Hermitian operators M_a in $\mathcal{L}$:

$$\begin{gathered} M_a = \sum_{d=1}^{a-1} (a\ d) = \sum_{d=1}^{a-1} P_{a-1}P_{a-2} \cdots P_{d+1}P_dP_{d+1} \cdots P_{a-2}P_{a-1}, \\ 2 \leq a \leq n, \qquad M_1 = 0, \end{gathered} \tag{6.79}$$

where $P_a = (a\ a+1)$ is the transposition of two neighboring objects. From the definition one has

$$M_{a+1} = P_a + P_aM_aP_a, \qquad P_aM_{a+1} = E + M_aP_a. \tag{6.80}$$

It is easy to show from (6.15) that

$$P_aM_b = M_bP_a, \qquad \text{if} \ \ b < a \ \ \text{or} \ \ b > a+1. \tag{6.81}$$

Then, the Hermitian operators M_a are commutable with each other

$$[M_a,\ M_b] = 0. \tag{6.82}$$

Theorem 6.5 In the standard bases $b_{\nu\rho}$ (6.67), the matrix form $D(M_a)$ of M_a is an upper triangular matrix with the known diagonal entries,

$$\begin{aligned} & M_ab_{\nu\rho} = \sum_{\mu} b_{\mu\rho}D_{\mu\nu}(M_a), \qquad D_{\nu\nu}(M_a) = m_\nu(a), \\ & D_{\mu\nu}(M_a) = 0 \qquad \text{when} \ \ \mu > \nu, \end{aligned} \tag{6.83}$$

where the rows and columns are enumerated by the standard Young tableaux $\mathcal{Y}_\nu$ in the increasing order. If a is filled in the $c_\nu(a)$th column of the $r_\nu(a)$th row of the Young tableau $\mathcal{Y}_\nu$, $m_\nu(a) = c_\nu(a) - r_\nu(a)$ is called the content of the digit a in the standard Young tableau $\mathcal{Y}_\nu$.

Proof Decompose M_a as $M_a = M_a^{(1)} + M_a^{(2)} + M_a^{(3)} + M_a^{(4)}$,

$$M_a^{(1)} = \sum_i \left(a \ \ a_i\right), \qquad M_a^{(2)} = \sum_j \left(a \ \ b_j\right) + \sum_k \left(a \ \ d_k\right),$$
$$M_a^{(3)} = -\sum_k \left(a \ \ d_k\right), \qquad M_a^{(4)} = \sum_\ell \left(a \ \ t_\ell\right),$$

where a_i denotes the digits filled in the boxes on the left of the box a at the $r_\nu(a)$th row of the standard Young tableau $\mathcal{Y}_\nu$, b_j and d_k denote the digits, respectively smaller and larger than a, filled in the boxes at the first $[r_\nu(a)-1]$ rows, and t_ℓ denote the digits less than a and filled in the lower rows than the r_νth row. From the symmetric property (6.28) and the Fock condition (6.30) one has

$$M_a^{(1)} b_{\nu\rho} = \left\{c_\nu(a) - 1\right\} b_{\nu\rho}, \qquad M_a^{(2)} b_{\nu\rho} = \left\{1 - r_\nu(a)\right\} b_{\nu\rho}.$$

When applying each transposition in $M_a^{(3)}$ and $M_a^{(4)}$ to the Young tableau $\mathcal{Y}_\nu$, a smaller digit in a lower row is interchanged with a larger digit in the upper row. Although the transformed Young tableau is generally no longer standard, it can be proved in terms of the similar method used in the proof for Corollary 6.3.2 that

$$\mathcal{Y}_\mu M_a^{(3)} \mathcal{Y}_\nu = \mathcal{Y}_\mu M_a^{(4)} \mathcal{Y}_\nu = 0, \qquad \text{when} \ \ \mu \geq \nu.$$

Noting (6.58) one has

$$\begin{aligned} &\mathcal{Y}_\mu y_\mu M_a^{(3)} \mathcal{Y}_\nu = \mathcal{Y}_\mu y_\mu M_a^{(4)} \mathcal{Y}_\nu = 0, \\ &b_{\rho\mu} M_a^{(3)} b_{\nu\rho} = b_{\rho\mu} M_a^{(4)} b_{\nu\rho} = 0, \end{aligned} \qquad \text{when} \ \ \mu \geq \nu.$$

□

Letting X be a similarity transformation which changes the representation $D(\mathrm{S}_n)$ to a real orthogonal representation $\overline{D}(\mathrm{S}_n)$ such that the representation matrices of M_a are diagonal,

$$\begin{aligned} &\overline{D}_{\mu\nu}(M_a) = \left[X^{-1} D(M_a) X\right]_{\mu\nu} = \delta_{\mu\nu} m_\nu(a), \\ &\overline{D}(P_a) = X^{-1} D(P_a) X = \overline{D}(P_a)^* = \overline{D}(P_a)^T. \end{aligned} \tag{6.84}$$

Because there are no two different standard Young tableaux $\mathcal{Y}_\mu$ and $\mathcal{Y}_\nu$ satisfying $m_\mu(a) = m_\nu(a)$ for every a, the eigenvalues $m_\nu(a)$ of M_a are not degenerate. In other words, the set of M_a, $2 \leq a \leq n$, is a complete set of the Hermitian operators in the group algebra $\mathcal{L}$ of S_n. Since $D(M_a)$ are real upper triangular matrices and $\overline{D}(M_a)$ are real diagonal, X has to be a real upper triangular matrix. The new basis vectors, called orthogonal bases, are calculated in terms of X matrix

$$\begin{aligned} \phi_{\mu\nu} &= \sum_{\rho=1}^{d} b_{\rho\nu} X_{\rho\mu} \in \mathcal{L}_{\nu}, & R\phi_{\mu\nu} &= \sum_{\rho=1}^{d} \phi_{\rho\nu} \overline{D}_{\rho\mu}(R), \\ \overline{\phi}_{\mu\nu} &= \sum_{\rho=1}^{d} \left(X^{-1}\right)_{\nu\rho} b_{\mu\rho} \in \mathcal{R}_{\mu}, & \overline{\phi}_{\mu\nu} R &= \sum_{\rho=1}^{d} \overline{D}_{\nu\rho}(R) \overline{\phi}_{\mu\rho}, \end{aligned} \tag{6.85}$$

or

$$\begin{aligned} \Phi_{\mu\nu} &= \sum_{\rho=1}^{d} \sum_{\tau=1}^{d} \left(X^{-1}\right)_{\nu\tau} b_{\rho\tau} X_{\rho\mu}, \\ R\Phi_{\mu\nu} = \sum_{\rho} \Phi_{\rho\nu} \overline{D}_{\rho\mu}(R), &\qquad \Phi_{\mu\nu} R = \sum_{\rho} \overline{D}_{\nu\rho}(R) \Phi_{\mu\rho}. \end{aligned} \tag{6.86}$$

$\overline{D}(P_a)$ as well as X can be calculated from Eqs. (6.80), (6.81), and (6.84). In fact, substituting Eq. (6.84) into Eq. (6.81), one has

$$\overline{D}_{\mu\nu}(P_a) \left\{ m_{\mu}(b) - m_{\nu}(b) \right\} = 0, \qquad \text{if } \ b < a \ \text{ or } \ b > a+1, \tag{6.87}$$

namely, $\overline{D}_{\mu\nu}(P_a) \neq 0$ only if $m_{\mu}(b) = m_{\nu}(b)$ for every b except for $b = a$ and $b = a+1$. If a and $(a+1)$ do not occur in the same row and in the same column of the standard Young Tableau $\mathcal{Y}_{\nu}$, another standard Young tableau, denoted by $\mathcal{Y}_{\nu_a}$, can be obtained from the Young tableau $\mathcal{Y}_{\nu}$ by interchanging a and $(a+1)$,

$$\begin{gathered} m_{\nu}(a) = m_{\nu_a}(a+1), \qquad m_{\nu}(a+1) = m_{\nu_a}(a), \\ \left| m_{\nu}(a) - m_{\nu}(a+1) \right| > 1. \end{gathered} \tag{6.88}$$

The relation between ν and ν_a is mutual. On the other hand, if a and $(a+1)$ occur in the same row or in the same column of the standard Young tableau $\mathcal{Y}_{\nu}$, the Young tableau obtained by the interchanging is no longer standard. In these cases we will say that ν_a does not exist for the standard Young tableau $\mathcal{Y}_{\nu}$.

Therefore, $\overline{D}_{\mu\nu}(P_a) \neq 0$ only if $\mu = \nu$ or $\mu = \nu_a$, so that $\overline{D}(P_a)$ is a block matrix. The submatrix $\overline{D}_{\nu\nu}(P_a)$ is one-dimensional if ν_a does not exist and is two-dimensional if ν_a exists,

$$\begin{pmatrix} \overline{D}_{\nu\nu}(P_a) & \overline{D}_{\nu\nu_a}(P_a) \\ \overline{D}_{\nu_a\nu}(P_a) & \overline{D}_{\nu_a\nu_a}(P_a) \end{pmatrix}, \qquad \overline{D}_{\nu\nu_a}(P_a) = \overline{D}_{\nu_a\nu}(P_a), \tag{6.89}$$

where, without loss of generality, we assume $\nu < \nu_a$, namely, a occurs at the right of and upper than $(a+1)$ in the standard Young tableau $\mathcal{Y}_\nu$.

Substituting Eq. (6.84) into Eq. (6.80), one obtains

$$-\overline{D}_{\nu\nu}(P_a) = \overline{D}_{\nu_a\nu_a}(P_a) = m^{-1}, \qquad m \equiv m_\nu(a) - m_\nu(a+1).$$

The non-diagonal entries of the two-dimensional submatrix can be calculated from $P_a^2 = E$,

$$\overline{D}_{\nu\nu_a}(P_a) = \overline{D}_{\nu_a\nu}(P_a) = \frac{\left(m^2-1\right)^{1/2}}{|m|}.$$

It is proved that the square root can be positive by choosing the phase angles of the basis vectors and noting the condition (6.16). Thus, the one-dimensional submatrix is

$$\overline{D}_{\nu\nu}(P_a) = \begin{cases} 1, & a \text{ and } (a+1) \text{ occur in the same row}, \\ -1, & a \text{ and } (a+1) \text{ occur in the same column}, \end{cases} \tag{6.90}$$

and the two-dimensional submatrix (6.89) becomes

$$\frac{1}{m}\begin{pmatrix} -1 & (m^2-1)^{1/2} \\ (m^2-1)^{1/2} & 1 \end{pmatrix}, \qquad m = m_\nu(a) - m_\nu(a+1) > 1, \tag{6.91}$$

where m is equal to the steps of going from a to $(a+1)$ in the Young tableau $\mathcal{Y}_\nu$ downward or leftward. Equations (6.90) and (6.91) give the calculation method for the real orthogonal representation matrix $\overline{D}(P_a)$ in $[\lambda]$. It is easy to calculate the similarity transformation matrix X from $D(P_a)$ and $\overline{D}(P_a)$. In the following we calculate the similarity transformation matrix X for the representation $[3,2]$ of S_5 as an example.

The standard Young tableaux with the Young pattern $[3,2]$ are listed from $\mathcal{Y}_1^{[3,2]}$ to $\mathcal{Y}_5^{[3,2]}$ as follows:

1 2 3	1 2 4	1 2 5	1 3 4	1 3 5
4 5	3 5	3 4	2 5	2 4

The representation matrices $D(P_a)$ can be calculated by the tabular method (see §6.3)

$$D(P_1) = \begin{pmatrix} 1 & 0 & 0 & -1 & -1 \\ 0 & 1 & 0 & -1 & 0 \\ 0 & 0 & 1 & 0 & -1 \\ 0 & 0 & 0 & -1 & 0 \\ 0 & 0 & 0 & 0 & -1 \end{pmatrix}, \quad D(P_2) = \begin{pmatrix} 1 & 0 & 0 & 0 & 0 \\ 0 & 0 & 0 & 1 & 0 \\ 0 & 0 & 0 & 0 & 1 \\ 0 & 1 & 0 & 0 & 0 \\ 0 & 0 & 1 & 0 & 0 \end{pmatrix},$$

$$D(P_3) = \begin{pmatrix} 0 & 1 & 0 & 0 & -1 \\ 1 & 0 & 0 & 0 & -1 \\ 0 & 0 & 1 & 0 & -1 \\ 0 & 0 & 0 & 1 & -1 \\ 0 & 0 & 0 & 0 & -1 \end{pmatrix}, \qquad D(P_4) = \begin{pmatrix} 1 & 0 & 0 & 0 & 0 \\ 0 & 0 & 1 & 0 & 0 \\ 0 & 1 & 0 & 0 & 0 \\ 0 & 0 & 0 & 0 & 1 \\ 0 & 0 & 0 & 1 & 0 \end{pmatrix}.$$

The orthogonal representation matrices $\overline{D}(P_a)$ are calculated from Eqs. (6.90) and (6.91). For example, in the calculation of $\overline{D}(P_2)$ one needs to check the positions of 2 and 3 in the Young tableaux $\mathcal{Y}_\nu$. In the interchanging of 2 and 3, the Young tableau $\mathcal{Y}_2$ is changed to the Young tableau $\mathcal{Y}_4$ with $m = 2$, and the Young tableau $\mathcal{Y}_3$ is changed to the Young tableau $\mathcal{Y}_5$ with $m = 2$, so that $\overline{D}(P_2)$ is a block matrix with one 1×1 submatrix and two 2×2 submatrices.

$$\overline{D}(P_1) = \begin{pmatrix} 1 & 0 & 0 & 0 & 0 \\ 0 & 1 & 0 & 0 & 0 \\ 0 & 0 & 1 & 0 & 0 \\ 0 & 0 & 0 & -1 & 0 \\ 0 & 0 & 0 & 0 & -1 \end{pmatrix}, \qquad \overline{D}(P_2) = \frac{1}{2}\begin{pmatrix} 2 & 0 & 0 & 0 & 0 \\ 0 & -1 & 0 & \sqrt{3} & 0 \\ 0 & 0 & -1 & 0 & \sqrt{3} \\ 0 & \sqrt{3} & 0 & 1 & 0 \\ 0 & 0 & \sqrt{3} & 0 & 1 \end{pmatrix},$$

$$\overline{D}(P_3) = \frac{1}{3}\begin{pmatrix} -1 & \sqrt{8} & 0 & 0 & 0 \\ \sqrt{8} & 1 & 0 & 0 & 0 \\ 0 & 0 & 3 & 0 & 0 \\ 0 & 0 & 0 & 3 & 0 \\ 0 & 0 & 0 & 0 & -3 \end{pmatrix}, \qquad \overline{D}(P_4) = \frac{1}{2}\begin{pmatrix} 2 & 0 & 0 & 0 & 0 \\ 0 & -1 & \sqrt{3} & 0 & 0 \\ 0 & \sqrt{3} & 1 & 0 & 0 \\ 0 & 0 & 0 & -1 & \sqrt{3} \\ 0 & 0 & 0 & \sqrt{3} & 1 \end{pmatrix}.$$

The similarity transformation matrix X, $D(P_a)X = X\overline{D}(P_a)$, is an upper triangular one, whose column matrices X_μ are denoted by

$$X_1 = \begin{pmatrix} 1 \\ 0 \\ 0 \\ 0 \\ 0 \end{pmatrix}, \quad X_2 = \begin{pmatrix} a_1 \\ a_2 \\ 0 \\ 0 \\ 0 \end{pmatrix}, \quad X_3 = \begin{pmatrix} b_1 \\ b_2 \\ b_3 \\ 0 \\ 0 \end{pmatrix}, \quad X_4 = \begin{pmatrix} c_1 \\ c_2 \\ c_3 \\ c_4 \\ 0 \end{pmatrix}, \quad X_5 = \begin{pmatrix} d_1 \\ d_2 \\ d_3 \\ d_4 \\ d_5 \end{pmatrix}.$$

From

$$D(P_3)X_1 = -\frac{1}{3}X_1 + \frac{\sqrt{8}}{3}X_2, \qquad \begin{pmatrix} 0 \\ 1 \end{pmatrix} = -\frac{1}{3}\begin{pmatrix} 1 \\ 0 \end{pmatrix} + \frac{\sqrt{8}}{3}\begin{pmatrix} a_1 \\ a_2 \end{pmatrix},$$

one obtains $a_1 = 1/\sqrt{8}$ and $a_2 = 3/\sqrt{8}$. From

$$D(P_4)X_2 = -\frac{1}{2}X_2 + \frac{\sqrt{3}}{2}X_3, \qquad \frac{1}{\sqrt{8}}\begin{pmatrix}1\\0\\3\end{pmatrix} = -\frac{1}{2\sqrt{8}}\begin{pmatrix}1\\3\\0\end{pmatrix} + \frac{\sqrt{3}}{2}\begin{pmatrix}b_1\\b_2\\b_3\end{pmatrix},$$

one gets $b_1 = b_2 = \sqrt{3/8}$ and $b_3 = \sqrt{3/2}$. From

$$D(P_2)X_2 = -\frac{1}{2}X_2 + \frac{\sqrt{3}}{2}X_4, \qquad \frac{1}{\sqrt{8}}\begin{pmatrix}1\\0\\0\\3\end{pmatrix} = -\frac{1}{2\sqrt{8}}\begin{pmatrix}1\\3\\0\\0\end{pmatrix} + \frac{\sqrt{3}}{2}\begin{pmatrix}c_1\\c_2\\c_3\\c_4\end{pmatrix},$$

one has $c_1 = c_2 = \sqrt{3/8}$, $c_3 = 0$, and $c_4 = \sqrt{3/2}$. From

$$D(P_4)X_4 = -\frac{1}{2}X_4 + \frac{\sqrt{3}}{2}X_5, \qquad \sqrt{\frac{3}{8}}\begin{pmatrix}1\\0\\1\\0\\2\end{pmatrix} = -\frac{1}{2}\sqrt{\frac{3}{8}}\begin{pmatrix}1\\1\\0\\2\\0\end{pmatrix} + \frac{\sqrt{3}}{2}\begin{pmatrix}d_1\\d_2\\d_3\\d_4\\d_5\end{pmatrix},$$

one obtains $d_1 = 3/\sqrt{8}$, $d_2 = 1/\sqrt{8}$, $d_3 = d_4 = 1/\sqrt{2}$, and $d_5 = \sqrt{2}$. At last, the similarity transformation matrix X is

$$X = \frac{1}{\sqrt{8}}\begin{pmatrix}\sqrt{8} & 1 & \sqrt{3} & \sqrt{3} & 3\\ 0 & 3 & \sqrt{3} & \sqrt{3} & 1\\ 0 & 0 & 2\sqrt{3} & 0 & 2\\ 0 & 0 & 0 & 2\sqrt{3} & 2\\ 0 & 0 & 0 & 0 & 4\end{pmatrix}. \tag{6.92}$$

It is easy to check that X satisfies (6.84). The orthogonal bases $\phi_{\nu 1}$, removed a factor $1/\sqrt{8}$, are

$$\begin{aligned}
\phi_{11} &= \sqrt{8}e_1,\\
\phi_{21} &= \{E + 3(3\ 4)\}\, e_1,\\
\phi_{31} &= \sqrt{3}\,\{E + (3\ 4) + 2(3\ 5\ 4)\}\, e_1,\\
\phi_{41} &= \sqrt{3}\,\{E + (3\ 4) + 2(2\ 3\ 4)\}\, e_1,\\
\phi_{51} &= (3E + (3\ 4) + 2(3\ 5\ 4) + 2(2\ 3\ 4) + 4(2\ 3\ 5\ 4))\, e_1.
\end{aligned}$$

The similarity transformation matrices $X_{[\lambda]}$, $D^{[\lambda]}(P_a)X_{[\lambda]} = X_{[\lambda]}\overline{D}^{[\lambda]}(P_a)$, for the representations $[3,1]$, $[2,1,1]$, $[2,1]$, and $[2,2]$ are calculated in Probs. 22–24 of Chap. 6 of [Ma and Gu (2004)],

$$X_{[2,1]} = X_{[2,2]} = \begin{pmatrix}1 & \sqrt{1/3}\\ 0 & 2/\sqrt{3}\end{pmatrix},$$

$$X_{[3,1]} = \begin{pmatrix} 1 & \sqrt{1/8} & \sqrt{3/8} \\ 0 & 3/\sqrt{8} & \sqrt{3/8} \\ 0 & 0 & \sqrt{3/2} \end{pmatrix}, \quad X_{[2,1,1]} = \begin{pmatrix} 1 & \sqrt{1/3} & -\sqrt{1/6} \\ 0 & 2/\sqrt{3} & \sqrt{1/6} \\ 0 & 0 & \sqrt{3/2} \end{pmatrix}.$$

6.5 Outer Product of Irreducible Representations of S_n

The outer product of representations of the permutation groups is related to the induced representation from a representation of the subgroup $\mathrm{S}_n \otimes \mathrm{S}_m$ with respect to the representations of the group S_{n+m}. In this section the graphic method for calculating the outer product of representations will be given.

6.5.1 *Representations of S_{n+m} and Its Subgroup $S_n \otimes S_m$*

First of all, we introduce our notations for this problem. The group algebra $\mathcal{L}$ of the permutation group S_{n+m} among $(n+m)$ objects is $(n+m)!$ dimensional. The primitive idempotent $e^{[\omega]}$ in $\mathcal{L}$ generates the minimal left ideal $\mathcal{L}^{[\omega]} = \mathcal{L}e^{[\omega]}$, which corresponds to the $d_{[\omega]}$-dimensional irreducible representation $[\omega]$, where the Young pattern $[\omega]$ contains $(n+m)$ boxes.

Denote by S_n the permutation subgroup among the first n objects in the $(n+m)$ objects, and by S_m the subgroup among the last m objects. The product of elements belonging to two subgroups S_n and S_m, respectively, is commutable because they permute different objects. The only common element in two subgroups is the identical element E. Therefore, we are able to define the direct product $\mathrm{S}_n \otimes \mathrm{S}_m$ of two subgroups, which is a subgroup of S_{n+m} with the index N,

$$N = \binom{n+m}{n} = \frac{(n+m)!}{n!m!}. \tag{6.93}$$

Its left-coset is denoted by

$$T_\alpha \left(\mathrm{S}_n \otimes \mathrm{S}_m \right), \qquad 2 \leq \alpha \leq N.$$

Each T_α moves the first n objects to n new positions, which are different for different T_α. Although those T_α are not unique, it is assumed that they have been chosen. The group algebra of the subgroup is denoted by $\mathcal{L}^{nm}$. $\mathcal{L}$ is decomposed as follows:

$$\mathcal{L} = \bigoplus_{\alpha=1}^{N} T_\alpha \mathcal{L}^{nm}, \qquad T_1 = E. \tag{6.94}$$

Denote by $e^{[\lambda][\mu]}$ the primitive idempotent in $\mathcal{L}^{nm}$. $e^{[\lambda][\mu]}$ generates the minimal left ideal of $\mathcal{L}^{nm}$, $\mathcal{L}^{[\lambda][\mu]} = \mathcal{L}^{nm}e^{[\lambda][\mu]}$, corresponding to the irreducible representation $[\lambda] \times [\mu]$ of $S_n \otimes S_m$ with the dimension $d_{[\lambda]}d_{[\mu]}$. The Young patterns $[\lambda]$ and $[\mu]$ contain n and m boxes, respectively. Note that $e^{[\lambda][\mu]}$ is an idempotent of $\mathcal{L}$ but generally not primitive, and $\mathcal{L}^{[\lambda][\mu]}$ is a subalgebra of $\mathcal{L}$ but not its left ideal.

Now, we discuss two related problems. One problem is the subduced representation from an irreducible representation $[\omega]$ of S_{n+m} with respect to the subgroup $S_n \otimes S_m$, which is generally reducible and can be reduced as a direct sum of the irreducible representations $[\lambda] \times [\mu]$ of $S_n \otimes S_m$

$$[\omega] \simeq \bigoplus a^{\omega}_{\lambda\mu} \, \{[\lambda] \times [\mu]\}, \qquad d_{[\omega]} = \sum a^{\omega}_{\lambda\mu} \, d_{[\lambda]}d_{[\mu]}. \tag{6.95}$$

$a^{\omega}_{\lambda\mu}$ is the multiplicity of $[\lambda] \times [\mu]$ in the subduced representation $[\omega]$ and can be calculated by the character formula in the subgroup

$$a^{\omega}_{\lambda\mu} = \frac{1}{n!m!} \sum_{H \in S_n \otimes S_m} \chi^{[\lambda][\mu]}(H)\chi^{[\omega]}(H), \tag{6.96}$$

where $\chi^{[\lambda][\mu]}(H)$ and $\chi^{[\omega]}(H)$ are the characters of H in the representations $[\lambda] \times [\mu]$ and $[\omega]$, respectively.

From the viewpoint of the group algebra, if a vector t_j in $\mathcal{L}$ satisfies

$$e^{[\lambda][\mu]}t_j e^{[\omega]} \neq 0, \tag{6.97}$$

$e^{[\lambda][\mu]}t_j e^{[\omega]}$ provides a map from the minimal left ideal $\mathcal{L}^{[\lambda][\mu]}$ in $\mathcal{L}^{nm}$ onto the left ideal $\mathcal{L}^{[\omega]}$ in $\mathcal{L}$:

$$\mathcal{L}^{nm}e^{[\lambda][\mu]}t_j e^{[\omega]} \subset \mathcal{L}e^{[\omega]} = \mathcal{L}^{[\omega]}, \tag{6.98}$$

namely, $[\lambda] \times [\mu]$ is contained in the subduced representation $[\omega]$. From Eq. (6.95), the number of the linearly independent maps (6.97) is equal to $a^{\omega}_{\lambda\mu}$. In other words, there are $a^{\omega}_{\lambda\mu}$ vectors t_j, $1 \leq j \leq a^{\omega}_{\lambda\mu}$ such that $e^{[\lambda][\mu]}t_j e^{[\omega]}$ are linearly independent of each other. Usually, the standard bases in $\mathcal{L}^{[\omega]}$ are chosen to be $t_j e^{[\omega]}$ one by one until $a^{\omega}_{\lambda\mu}$-independent $e^{[\lambda][\mu]}t_j e^{[\omega]}$ have been found. The Clebsch–Gordan coefficients for the decomposition can also be calculated from the map (6.98) (see Prob. 35).

Theorem 3.10 says that the left ideal and the right ideal generated by the same primitive idempotent in a group algebra correspond to the same irreducible representation. Therefore, there are also $a^{\omega}_{\lambda\mu}$ vectors t'_j such that $e^{\omega}t'_j e^{[\lambda][\mu]}$ are linearly independent of each other,

$$e^{[\omega]}t'_j e^{[\lambda][\mu]} \neq 0, \qquad 1 \leq j \leq a^{\omega}_{\lambda\mu}. \tag{6.99}$$

The other problem is the induced representation from an irreducible representation $[\lambda] \times [\mu]$ of the subgroup $S_n \otimes S_m$ with respect to the group S_{n+m}. The induced representation is called the outer product of representations in the permutation group, denoted by $[\lambda] \otimes [\mu]$. The outer product is generally reducible and can be decomposed as a direct sum of the irreducible representations $[\omega]$ of S_{n+m},

$$[\lambda] \otimes [\mu] \simeq \sum b^{\omega}_{\lambda\mu}[\omega], \qquad \frac{(n+m)!}{n!m!} d_{[\lambda]}d_{[\mu]} = \sum b^{\omega}_{\lambda\mu}\, d_{[\omega]}. \tag{6.100}$$

Due to the Frobenius theorem (see Eq. (3.76)), $b^{\omega}_{\lambda\mu} = a^{\omega}_{\lambda\mu}$,

$$a^{\omega}_{\lambda\mu} = b^{\omega}_{\lambda\mu} = \frac{1}{(n+m)!} \sum_{R \in S_{n+m}} \chi^{[\lambda]\otimes[\mu]}(R)\chi^{[\omega]}(R), \tag{6.101}$$

where $\chi^{[\lambda]\otimes[\mu]}(R)$ is the character of the element R of S_{n+m} in the outer product representation $[\lambda] \otimes [\mu]$.

From the viewpoint of the group algebra, $e^{[\lambda][\mu]}$ is also an idempotent in $\mathcal{L}$, but generally not primitive. $e^{[\lambda][\mu]}$ generates a left ideal $\mathcal{L}_{\lambda\mu}$ in $\mathcal{L}$:

$$\mathcal{L}_{\lambda\mu} = \mathcal{L}e^{[\lambda][\mu]}. \tag{6.102}$$

$\mathcal{L}_{\lambda\mu}$ corresponds a representation of S_{n+m}, which is $[\lambda] \otimes [\mu]$.

Substituting (6.94) into (6.102), one has

$$\mathcal{L}_{\lambda\mu} = \bigoplus_{\alpha=1}^{N} T_\alpha \mathcal{L}^{nm} e^{[\lambda][\mu]} = \bigoplus_{\alpha=1}^{N} T_\alpha \mathcal{L}^{[\lambda][\mu]}, \qquad T_1 = E. \tag{6.103}$$

Each term in the sum is a $d_{[\lambda]}d_{[\mu]}$ dimensional subspace in $\mathcal{L}$, and there is no common vector between any two subspaces. Hence, the dimension of the representation $[\lambda] \otimes [\mu]$ is

$$d_{[\lambda]\otimes[\mu]} = \frac{(n+m)!}{n!m!} d_{[\lambda]}d_{[\mu]}. \tag{6.104}$$

Since

$$\mathcal{L}e^{[\omega]}t'_j e^{[\lambda][\mu]} \subset \mathcal{L}e^{[\lambda][\mu]} = \mathcal{L}_{\lambda\mu}, \tag{6.105}$$

$e^{[\omega]}t'_j e^{[\lambda][\mu]}$ provides a map from the minimal left ideal $\mathcal{L}^{[\omega]}$ in $\mathcal{L}$ onto the left ideal $\mathcal{L}_{\lambda\mu}$ in $\mathcal{L}$. This is just the Frobenius theorem that the multiplicity of $[\omega]$ in $[\lambda] \otimes [\mu]$ is equal to $a^{\omega}_{\lambda\mu}$. The basis vectors in $\mathcal{L}_{\lambda\mu}$ are calculated by left-multiplying T_α to the standard bases in $\mathcal{L}^{[\lambda][\mu]}$. Then, from the map (6.105) the basis vectors of $\mathcal{L}e^{[\omega]}$ are expressed by the basis vectors of $\mathcal{L}_{\lambda\mu}$ where the coefficients are the Clebsch–Gordan coefficients for the decomposition (see Prob. 37).

6.5.2 *Littlewood–Richardson Rule*

There is a simple graphic method for calculating $b^{\omega}_{\lambda\mu}$, called the Littlewood–Richardson rule. Choose one representation between $[\lambda]$ and $[\mu]$, say $[\lambda]$. Usually, choose $[\lambda]$ whose box number is not less than that of $[\mu]$ for convenience. Fill the boxes in the jth row of the Young pattern $[\mu]$ with the digit j. Attach the boxes of the Young pattern $[\mu]$ row by row, beginning with the first row, into the Young pattern $[\lambda]$ in the following way.

(a) Each time when the boxes in one row of the Young pattern $\mathcal{Y}_\mu$ have been attached, the resultant diagram constitutes a Young pattern. Namely, it is lined up on the top and on the left, where the box number of the higher row is not less than that of the lower row.

(b) The boxes filled with the same digit never occur in the same column of the resultant diagram.

(c) After all the boxes of the Young pattern $\mathcal{Y}_\mu$ are attached, read the digits filled in the boxes of the resultant diagram from right to left, row by row beginning with the first row, such that at every step of the reading process the number of boxes filled by a smaller digit is never less than that filled by a larger digit.

If the resultant diagram, when all the boxes of the Young pattern $\mathcal{Y}_\mu$ were attached according to the above rule, is the Young pattern $[\omega]$, the representation $[\omega]$ of S_{n+m} appears in the reduction of the outer product $[\lambda] \otimes [\mu]$. The times of appearing of a Young pattern $[\omega]$ in the different resultant diagrams is the multiplicity of $[\omega]$ in the reduction. Let us give two examples to show how to use the Littlewood–Richardson rule.

Example 1. Calculate the reduction of $[2, 1] \otimes [2, 1]$.

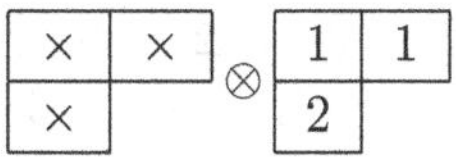

Attaching the boxes filled by the digit 1 to the first Young pattern according to the Littlewood–Richardson rule, one obtains

$$
\begin{array}{llll} \times & \times & 1 & 1 \\ \times \end{array} \qquad
\begin{array}{lll} \times & \times & 1 \\ \times & 1 \end{array} \qquad
\begin{array}{lll} \times & \times & 1 \\ \times \\ 1 \end{array} \qquad
\begin{array}{ll} \times & \times \\ \times & 1 \\ 1 \end{array}
$$

Then, attach the box filled by the digit 2 to the above diagrams, respectively. Note that the box filled by the digit 2 cannot be attached in the first row due to the rule (c). It also cannot be attached to the second row of the fourth diagram. The allowed diagrams are as follows:

$$
\begin{array}{llll} \times & \times & 1 & 1 \\ \times & 2 \end{array} \qquad
\begin{array}{llll} \times & \times & 1 & 1 \\ \times \\ 2 \end{array} \qquad
\begin{array}{lll} \times & \times & 1 \\ \times & 1 & 2 \end{array} \qquad
\begin{array}{lll} \times & \times & 1 \\ \times & 1 \\ 2 \end{array}
$$

$$
\begin{array}{lll} \times & \times & 1 \\ \times & 2 \\ 1 \end{array} \qquad
\begin{array}{lll} \times & \times & 1 \\ \times \\ 1 \\ 2 \end{array} \qquad
\begin{array}{ll} \times & \times \\ \times & 1 \\ 1 & 2 \end{array} \qquad
\begin{array}{ll} \times & \times \\ \times & 1 \\ 1 \\ 2 \end{array}
$$

The reduction of $[2,1] \otimes [2,1]$ and their dimensions are given as follows

$$
\begin{aligned}
[2,1] \otimes [2,1] \simeq & \ [4,2] \oplus [4,1,1] \oplus [3,3] \oplus 2\ [3,2,1] \\
& \oplus [3,1,1,1] \oplus [2,2,2] \oplus [2,2,1,1], \\
20 \times 2 \times 2 = & \ 9 + 10 + 5 + 2 \times 16 + 10 + 5 + 9.
\end{aligned}
$$

Due to the Frobenius theorem, the Littlewood–Richardson rule can also be used to calculate the reduction of the subduced representation $[\omega]$ with respect to the subgroup $S_n \otimes S_m$. In this reduction, $[\lambda]$ and $[\mu]$ run over all representations of S_n and S_m, respectively, but the resultant diagram has to be the given $[\omega]$.

Example 2. Reduction of the subduced representation $[3,2,1]$ of S_6 with respect to the subgroup $S_3 \otimes S_3$.

$$
\begin{array}{lll} \times & \times & \times \\ 1 & 1 \\ 2 \end{array}
\oplus
\begin{array}{lll} \times & \times & 1 \\ \times & 1 \\ 1 \end{array}
\oplus
\begin{array}{lll} \times & \times & 1 \\ \times & 1 \\ 2 \end{array}
\oplus
\begin{array}{lll} \times & \times & 1 \\ \times & 2 \\ 1 \end{array}
\oplus
\begin{array}{lll} \times & \times & 1 \\ \times & 2 \\ 3 \end{array}
\oplus
\begin{array}{lll} \times & 1 & 1 \\ \times & 2 \\ \times \end{array}
$$

Hence,

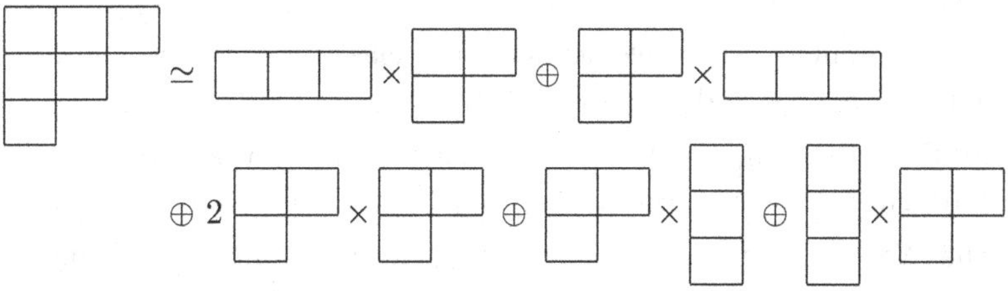

Check the dimensions in the reduction,

$$16 = 1 \times 2 + 2 \times 1 + 2 \times 2 \times 2 + 2 \times 1 + 1 \times 2.$$

6.6 Exercises

1. Simplify the following permutations into the product of cycles without any common object:

   (*a*) $(1\ 2)(2\ 3)(1\ 2)$, (*b*) $(1\ 2\ 3)(1\ 3\ 4)(3\ 2\ 1)$,

   (*c*) $(1\ 2\ 3\ 4)^{-1}$, (*d*) $(1\ 2\ 4\ 5)(4\ 3\ 2\ 6)$,

   (*e*) $(1\ 2\ 3)(4\ 2\ 6)(3\ 4\ 5\ 6)$.

2. Please show that every permutation can be decomposed as a product of the transpositions P_a of the neighbored objects.
3. Prove that the order of a cycle R with length ℓ is ℓ, namely, $R^{\ell} = E$.
4. Prove that $P_1 = (1\ 2)$ and $W = (1\ 2\ 3\ \ldots\ n)$ are two generators of the permutation group S_n.
5. There are 52 pieces of playing cards in a set of poker. The order of cards is changed in each shuffle, namely, the cards are made a permutation. If the shuffle is "strictly" done in the following rule: first separate the cards into two parts in equal number, then pick up one card from each part in order. After the strict shuffle the first and the last cards do not change their positions, while the remaining cards are rearranged. Try to find this permutation, to decompose it into a product of cycles without any common object, to write the cycle structure of the permutation, and to explain at least how many times of the strict shuffles will make the order of cards into its original one.

6. Write all Young patterns of the permutation groups S_5, S_6, and S_7 from the largest to the smallest, respectively.

7. Calculate the number $n(\ell)$ of elements in each class (ℓ) of the permutation groups S_4, S_5, S_6, and S_7, respectively.

8. Calculate the number $d_{[\lambda]}(S_n)$ of the standard Young tableaux for each Young pattern $[\lambda]$ of the permutation groups S_n, $5 \leq n \leq 9$.

9. Write the Young operators corresponding to the following Young tableaux:

(*a*)

1	2	3
4		

(*b*)

1	2
3	4

(*c*)

1	2	3	4
5			

10. Write five standard Young tableaux $\mathcal{Y}_\mu$ corresponding to the Young pattern [3,2] of S_5 from the smallest to the largest, and calculate the permutations $R_{\mu\nu}$ transforming the Young tableau $\mathcal{Y}_\nu$ into Young tableau $\mathcal{Y}_\mu$.

11. Calculate the permutation R_{12} transforming the following Young tableau $\mathcal{Y}_2$ to $\mathcal{Y}_1$, and check the formulas $\mathcal{P}_1 R_{12} = R_{12}\mathcal{P}_2$, $\mathcal{Q}_1 R_{12} = R_{12}\mathcal{Q}_2$, and $\mathcal{Y}_1 R_{12} = R_{12}\mathcal{Y}_2$, where

$\mathcal{Y}_1$:

1	2	3
4		

, $\mathcal{Y}_2$:

1	2	4
3		

.

12. It can be seen that there is no pair of digits filled in the same row of the Young tableau $\mathcal{Y}$ and in the same column of the Young tableau $\mathcal{Y}'$. Calculate the permutation R transforming the Young tableau $\mathcal{Y}$ into the Young tableau $\mathcal{Y}'$, and express R as PQ belonging to the Young tableau $\mathcal{Y}$, and as $P'Q'$ belonging to the Young tableau $\mathcal{Y}'$:

The Young tableau $\mathcal{Y}$

1	2	4	7
3	5	9	
6	8		

The Young tableau $\mathcal{Y}'$

1	2	3	4
5	6	7	
8	9		

13. Try to expand all nonstandard Young operators $\mathcal{Y}$ of the Young pattern [2,1] with respect to the standard Young operators $\mathcal{Y}_\mu$, $\mathcal{Y} = \sum_\mu t_\mu \mathcal{Y}_\mu$, where t_μ is the vector in the group space of S_3.

14. Expand explicitly the identity of the S_4 group with respect to the Young operators.

15. Calculate the orthogonal primitive idempotents for the Young pattern $[2, 2, 1]$ of the permutation group S_5.

16. Calculate the orthogonal primitive idempotents for the Young pattern $[4, 2]$ of the permutation group S_6.

17. Calculate the orthogonal primitive idempotents for the Young pattern $[3, 2, 1]$ of the permutation group S_6.

18. Calculate the orthogonal primitive idempotents for the Young patterns $[3, 3]$ and $[4,1,1]$ of the permutation group S_6, respectively.

19. Give an example to demonstrate that the primitive idempotent generating a minimum left ideal is not unique.

20. Calculate the Clebsch–Gordan coefficients for the reduction of the self-product of the irreducible representation $[2, 1]$ of the S_3 group in the standard bases and in the orthogonal bases, respectively.

21. Calculate the standard bases for the irreducible representation $[3, 1]$ in the group space of S_4, and calculate the representation matrix of the transposition P_a for the neighbored objects in the standard bases by the tabular method.

22. Calculate the real orthogonal representation matrix of the transpositions P_a for the neighbored objects in the irreducible representation $[3, 1]$ of the S_4 group in the orthogonal bases by the formulas (6.90) and (6.91). Calculate the similarity transformation matrix between two representations in the standard bases (Prob. 21) and in the orthogonal bases and write explicitly the orthogonal bases $\Phi_{\mu\nu}$ in the group space of S_4.

23. Calculate the representation matrices of the transpositions P_a for the neighbored objects both in the standard bases and in the orthogonal bases of the irreducible representation $[2, 1, 1]$ of the S_4 group. Calculate explicitly the orthogonal bases $\Phi_{\mu\nu}$ of the irreducible representation $[2, 1, 1]$ in the group space of S_4.

24. Calculate the standard bases and the orthogonal bases in the group space of S_4 for the irreducible representation $[2, 2]$, and calculate the representation matrices of the transpositions P_a for the neighbored objects in these two sets of bases.

25. Calculate the symmetric bases of the oscillatory wave function of a molecule with the symmetry of a regular tetrahedron, for example, the methane CH_4.

26. Calculate the representation matrices of the generators (1 2) and (1 2 3 4 5) of the S_5 group in the irreducible representation $[2,2,1]$ by the tabular method.

27. Respectively calculate the real orthogonal representation matrices of the transpositions P_a of the neighbored objects in two equivalent irreducible representations $[2^3] \simeq [1^6] \times [3,3]$ of S_6, and find the similarity transformation matrix X between them.

28. Calculate the character of each class of the permutation group S_5 in the representation $[2,2,1]$ by the graphic method.

29. Calculate the characters of each class of the permutation group S_6 in the representations $[3,2,1]$, $[3,3]$, and $[2^3]$ by the graphic method, respectively.

30. Calculate and fill in the character tables of S_3, S_4, S_5, S_6, and S_7 by the graphic method.

31. Calculate the Clebsch–Gordan series of the inner product of each pair of the irreducible representations for the permutation groups S_3, S_4, S_5, S_6, and S_7 by the character method.

32. Calculate the Clebsch–Gordan coefficients for the reduction of the self-product of the irreducible representation $[3,2]$ of S_5.

33. Calculate the reduction of the following outer products of the representations in the permutation groups by the Littlewood–Richardson rule:

(1) $[3,2,1] \otimes [3]$, (2) $[3,2] \otimes [2,1]$, (3) $[2,1] \otimes [4,2^3]$.

34. Calculate the reduction of the subduced representations from the following irreducible representations of S_6 with respect to the subgroup $S_3 \otimes S_3$:

$$(1)\ [4,2]\ , \qquad (2)\ [2,2,1,1]\ , \qquad (3)\ [3,3]\ .$$

35. Calculate the similarity transformation matrix in the reduction of the subduced representation from the irreducible representation $[3,3]$ of S_6 with respect to the subgroup $S_3 \otimes S_3$.

36. Calculate the representation matrices of the generators of S_4 in the induced representation from two-dimensional irreducible representation $[2,1]$ of S_3 with respect to S_4.

37. Calculate the similarity transformation matrix in the reduction of the outer product representation $[2,1] \otimes [2]$ of the permutation group.

Chapter 7

LIE GROUPS AND LIE ALGEBRAS

In Chap. 4 the fundamental concepts on Lie groups have been introduced through the SO(3) group and its covering group SU(2). In this chapter we will study the general theory on Lie groups and Lie algebras, such as the property of simple and semisimple Lie algebras, the regular commutative relations of the Cartan–Weyl bases of a semisimple Lie algebra, the classification of the simple Lie algebras, the Chevalley bases of generators, the weights, the irreducible representations, and the decomposition of direct product representations. The SU(N) groups, the SO(N) groups, and the USp(2ℓ) groups are studied related to the classical Lie algebras in this chapter and further studied in the succeeding chapters because they are widely used in physics.

7.1 Lie Algebras and its Structure Constants

7.1.1 *The Global Property of a Lie Group*

We will begin with a review of the concepts of Lie groups. An element R in a Lie group G can be characterized by g-independent real parameters r_A, $1 \leq A \leq g$, varying continuously in a g-dimensional region. The parameters e_A of the identical element E are usually taken to be zero for convenience. g is called the order of the Lie group G and the region is called the group space of G. It is required that there is a one-to-one correspondence between the set of parameters r_A and the group element R at least in the region where the measure is not vanishing. The parameters t_A of the product element are the real analytical functions of the parameters of the factor elements

$$T = RS, \qquad t_A = f_A(r_1 \ldots, r_g; s_1 \ldots, s_g) = f_A(r; s). \tag{7.1}$$

$f_A(r;s)$ are called the composition functions, characterizing the multiplication rule of elements in G completely. The composition functions have to satisfy some conditions to meet four axioms for a group. The topological property of the group space describes the global property of G.

If the group space of G falls into several disjoint pieces, G is called a mixed Lie group. G contains an invariant Lie subgroup H whose group space is a connected piece in which the identical element lies. The set of elements related to the other connected piece is the coset of H. The property of the mixed Lie group G is characterized completely by H and the representative elements, respectively belonging to each coset. Hereafter, we will mainly study the connected Lie groups.

A connected Lie group G is called multiply connected with the degree of continuity n if the connected curves of any two points in the group space are separated into n classes where any two curves in each class can be changed continuously from one to another in the group space, but two curves in different classes cannot. The Lie group with the degree of continuity $n = 1$ is called simply-connected. For a Lie group G with the degree of continuity n, there exists a covering group G' which is simply-connected and homomorphic onto G with an n-to-one correspondence. A faithful representation of G' is an n-valued representation of G.

A Lie group is compact if its group space is compact. If the group space is an Euclidean space, a closed finite region (including the boundary) is compact, and an open finite region (without the boundary) or an infinite region is not compact. The group integral can be defined only for a compact Lie group so that the most properties of a finite group can be generalized to a compact Lie group, such as Theorems 3.1, 3.3, 3.5, and 3.6. The representations of a noncompact Lie group G can be obtained through those of a compact Lie group which is "close" to G (see §7.4.3 and §9.5).

A Lie group is called simple if it does not contain any nontrivial invariant Lie subgroup. A Lie group is called semisimple if it does not contain any Abelian invariant Lie subgroup (including the whole Lie group). A Lie group of order 1 is Abelian so that it is simple but not semisimple.

7.1.2 *The Local Property of a Lie Group*

An infinitesimal element R which is located at the neighborhood of the identical element E in the group space is characterized by infinitesimal parameters $r_A = \alpha_A$. In a representation $D(G)$ of a Lie group G, the representation matrix of the identical element E is the unit matrix, $D(E) =$

1, and that of an infinitesimal element $R(\alpha)$ can be expanded as a Taylor series. Up to the first order one has

$$D(R) = \mathbf{1} - i\sum_A r_A I_A + \ldots, \qquad I_A = i\left.\frac{\partial D(R)}{\partial r_A}\right|_{R=E}. \tag{7.2}$$

I_A are called the generators in a representation $D(G)$. If the representation is faithful, the generators are linearly independent of each other. The generators characterize the property of the infinitesimal elements in a representation of a Lie group G, so that, as shown by the Lie Theorems, they can characterize the local property of the Lie group.

The First Lie Theorem says that the representation of a Lie group G with a connected group space is completely determined by its generators. $D(R)$ satisfies a differential equation with a given boundary condition

$$\frac{\partial D(R)}{\partial r_A} = -i\sum_B I_B S_{BA}(r) D(R), \qquad D(R)|_{R=E} = \mathbf{1}, \tag{7.3}$$

where the real functions

$$S_{BA}(r) = \left.\frac{\partial f_B(r;s)}{\partial r_A}\right|_{S=R^{-1}} \tag{7.4}$$

depend upon the choice of the parameters of the Lie group. They are independent of the given representation.

The Second Lie Theorem says that the generators in any representation of a Lie group G satisfy the common commutative relations

$$[I_A,\ I_B] = i\sum_D C_{AB}{}^D I_D, \tag{7.5}$$

where the real numbers $C_{AB}{}^D$ are called the structure constants of G. Conversely, if g matrices satisfy the commutative relations (7.5), they are the generators in a representation of G with order g.

The structure constants are calculated from the generators in a known representation. Although the structure constants are independent of the representation, they depend on the choice of the group parameters. If the parameters are re-chosen, from Eqs. (7.2) and (7.5), both the generators and the structure constants are changed,

$$\begin{gathered} r'_A = \sum_B r_B \left(X^{-1}\right)_{BA}, \qquad I'_A = \sum_B X_{AB} I_B, \qquad \det X \neq 0, \\ C'_{AB}{}^D = \sum_{PQR} X_{AP} X_{BQ} C_{PQ}{}^R \left(X^{-1}\right)_{RD}. \end{gathered} \tag{7.6}$$

In fact,

$$\begin{aligned}[I'_A,\, I'_B] &= \sum_{PQ} X_{AP}X_{BQ}\,[I_P,\, I_Q] = i\sum_{PQR} X_{AP}X_{BQ}C_{PQ}{}^{R}I_R \\ &= i\sum_{D}\left\{\sum_{PQR} X_{AP}X_{BQ}C_{PQ}{}^{R}\left(X^{-1}\right)_{RD}\right\}I'_D.\end{aligned}$$

The Third Lie Theorem says that a set of constants $C_{AB}{}^{D}$ can be the structure constants of a Lie group if and only if they satisfy

$$\begin{aligned}C_{AB}{}^{D} &= -C_{BA}{}^{D}, \\ \sum_{P}\left\{C_{AB}{}^{P}C_{PD}{}^{Q} + C_{BD}{}^{P}C_{PA}{}^{Q} + C_{DA}{}^{P}C_{PB}{}^{Q}\right\} &= 0.\end{aligned} \tag{7.7}$$

Based on this theorem the Lie groups can be classified by their structure constants. We will study the classification later in this chapter.

The adjoint representation is very important for a Lie group. If I_A are the generators in a representation $D(G)$ of a Lie group G, the adjoint representation $D^{\mathrm{ad}}(G)$ satisfies

$$D(R)I_BD(R)^{-1} = \sum_{D} I_D D^{\mathrm{ad}}_{DB}(R). \tag{7.8}$$

When R is an infinitesimal element, one has

$$[I_A,\, I_B] = \sum_{D} I_D\left(I^{\mathrm{ad}}_A\right)_{DB}, \qquad \left(I^{\mathrm{ad}}_A\right)_{DB} = iC_{AB}{}^{D}. \tag{7.9}$$

The structure constants characterize some important properties of a Lie group G. A Lie group is Abelian if and only if its structure constants are all vanishing. A Lie group is simple if and only if its adjoint representation is irreducible. A Lie group is semisimple if and only if its adjoint representation is completely reducible and does not contain the identical representation.

We are going to show that if a Lie group G is compact, its parameters can be chosen such that each structure constant of G is totally antisymmetric with respect to its three indices. The adjoint representation of a Lie group is real for any set of real group parameters owing to its definition (4.28). For a compact Lie group G, a real representation is equivalent to a real orthogonal one. Let Y^T be the real similarity transformation changing the adjoint representation to be real orthogonal so that its generators $\overline{I}_A$ becomes antisymmetric

$$\begin{aligned}\left(\overline{I}_A\right)_{BD} &= -\left(\overline{I}_A\right)_{DB} = \left\{\left(Y^T\right)^{-1} I_A^{\rm ad} Y^T\right\}_{BD} \\ &= \sum_{PQ} \left(Y^{-1}\right)_{PB} \left(I_A^{\rm ad}\right)_{PQ} Y_{DQ}.\end{aligned} \tag{7.10}$$

Now, change the parameters in G such that the combination matrix X in Eq. (7.6) is equal to Y in Eq. (7.10). Thus, each new structure constant of G is totally antisymmetric with respect to its three indices

$$\begin{aligned}C'_{AB}{}^{D} &= \sum_{PQR} Y_{AP} Y_{BQ} C_{PQ}{}^{R} \left(Y^{-1}\right)_{RD} \\ &= -i \sum_{P} Y_{AP} \left\{\sum_{QR} \left(Y^{-1}\right)_{RD} \left(I_P\right)_{RQ} Y_{BQ}\right\} \\ &= -i \sum_{P} Y_{AP} \left(\overline{I}_P\right)_{DB} = -C'_{AD}{}^{B}.\end{aligned}$$

7.1.3 *The Lie Algebra*

Let I_A be the generators in a faithful representation of a Lie group G. I_A satisfy the commutative relations (7.5), or equivalently,

$$\left[(-iI_A),\ (-iI_B)\right] = \sum_{D} C_{AB}{}^{D}(-iI_D). \tag{7.11}$$

For the given structure constants $C_{AB}{}^{D}$, the map (7.11) of two generators $(-iI_A)$ and $(-iI_B)$ onto a combination of generators $(-iI_D)$ is called the Lie product. The real linear space spanned by the basis vectors $(-iI_A)$ is closed for the Lie product of vectors in the space

$$\begin{aligned}&X = \sum_{A} (-iI_A)\, x_A, \qquad Y = \sum_{B} (-iI_B)\, y_B, \\ &[X,\ Y] = \sum_{AB} x_A y_B \left[(-iI_A),\ (-iI_B)\right] = \sum_{D} (-iI_D) \left\{\sum_{AB} x_A y_B C_{AB}{}^{D}\right\}.\end{aligned}$$

This real space is called the real Lie algebra of a Lie group G, denoted by $\mathcal{L}_R$. Generalizing the real space $\mathcal{L}_R$ to a complex space $\mathcal{L}$ under the same product rule of vectors, one obtains a complex Lie algebra, briefly called a Lie algebra. Evidently, the Lie algebra is not an associate algebra because the Lie product satisfies the Jacobi identity

$$\left[I_A,\ [I_B,\ I_C]\right] - \left[[I_A,\ I_B],\ I_C\right] = \left[I_B,\ [I_A,\ I_C]\right]. \tag{7.12}$$

${C_{AB}}^D$ are the common structure constants of the Lie group G and the corresponding Lie algebra $\mathcal{L}$. A representation of a Lie group is also a representation (or called "module" in mathematics) of the corresponding Lie algebra. The Lie algebra $\mathcal{L}$ is the complexification of the real Lie algebra $\mathcal{L}_R$, and $\mathcal{L}_R$ is a real form of $\mathcal{L}$. In this textbook we only discuss the Lie algebra $\mathcal{L}$ which has its Lie group G and its real form $\mathcal{L}_R$.

The real Lie algebra is important for manifesting the compactness of a Lie group. For example, the SO(4) group is a real orthogonal group in the four-dimensional Euclidean space. If the fourth coordinate is changed to be imaginary, $x_4 = ict$, the orthogonal group becomes the proper Lorentz group L_p. Both SO(4) and L_p have the same Lie algebra, but different real Lie algebras. If the imaginary parameters of a group are allowed to be used, the generators in the corresponding representations of SO(4) and L_p are the same but some parameters are different by a factor i. A representation of SO(4) becomes a representation of L_p if some real parameters of SO(4) are changed to be imaginary. This is the standard method to find the irreducible representation of a noncompact Lie group from that of a compact Lie group. We will discuss it in Chap. 9 in detail. The real Lie algebra of a compact Lie group is called the compact real Lie algebra.

Two Lie algebras are isomorphic if their structure constants are the same, but two Lie groups with the same structure constants are only locally isomorphic generally. There are two typical examples where two locally isomorphic Lie groups are not isomorphic. The SU(2) group is homomorphic onto the SO(3) group, but they have the same structure constants. The U(2) group contains two subgroups SU(2) and U(1), but does not contain a subgroup SU(2)$\otimes$U(1) because two subgroups SU(2) and U(1) have a common element $-\mathbf{1}$ other than the identical element $\mathbf{1}$. Therefore, the U(2) group is not isomorphic onto the group SU(2)$\otimes$ U(1), although they are locally isomorphic onto each other.

The commonly used concepts for an algebra can also be used for a Lie algebra and for a real Lie algebra, such as subalgebras, ideals, Abelian ideals, the direct sum, the semi-direct sum, the isomorphism, the homomorphism, and so on. For a Lie algebra, one also can introduce some concepts relating to a Lie group, such as a simple Lie algebra, a semisimple Lie algebra, representations, and so on.

A non-null subset $\mathcal{L}_1$ of a Lie algebra $\mathcal{L}$, $\mathcal{L}_1 \subset \mathcal{L}$, is a Lie subalgebra of $\mathcal{L}$ if it is closed with respect to the sum and the Lie product of the vectors in the subset

$$c_1X + c_2Y \in \mathcal{L}_1 \qquad [X,\ Y] \in \mathcal{L}_1, \tag{7.13}$$

where $X \in \mathcal{L}_1$, $Y \in \mathcal{L}_1$, and c_1 and c_2 are two arbitrary complex numbers. Furthermore, a Lie subalgebra $\mathcal{L}_1$ of a Lie algebra $\mathcal{L}$ is an ideal of $\mathcal{L}$ if

$$[X,\ Y] \in \mathcal{L}_1, \qquad \forall\ X \in \mathcal{L}_1 \ \text{ and } \ Y \in \mathcal{L}. \tag{7.14}$$

There is no difference between a left-ideal and a right-ideal of a Lie algebra $\mathcal{L}$. A Lie algebra is called simple if it does not contain any ideal except for the whole algebra. A Lie algebra is called semisimple if it does not contain any Abelian ideal. A one-dimensional Lie algebra is Abelian, so it is simple, but not semisimple. A simple Lie algebra with dimension higher than 1 must be semisimple. The Lie algebra of a simple or a semisimple Lie group is simple or semisimple, respectively.

A Lie algebra $\mathcal{L}$ is called the direct sum of two Lie subalgebras, $\mathcal{L} = \mathcal{L}_1 \oplus \mathcal{L}_2$, if

$$\mathcal{L}_1 + \mathcal{L}_2 = \mathcal{L}, \qquad \mathcal{L}_1 \bigcap \mathcal{L}_2 = \emptyset, \qquad [\mathcal{L}_1,\ \mathcal{L}_2] = 0. \tag{7.15}$$

Evidently, both $\mathcal{L}_1$ and $\mathcal{L}_2$ are ideals of the direct sum $\mathcal{L}$. A Lie algebra $\mathcal{L}$ is called the semidirect sum of two Lie subalgebras, $\mathcal{L} = \mathcal{L}_1 \oplus_s \mathcal{L}_2$, if

$$\mathcal{L}_1 + \mathcal{L}_2 = \mathcal{L}, \qquad \mathcal{L}_1 \bigcap \mathcal{L}_2 = \emptyset, \qquad [\mathcal{L}_1,\ \mathcal{L}_2] \subset \mathcal{L}_1. \tag{7.16}$$

Now, $\mathcal{L}_1$ is an ideal of $\mathcal{L}$, but $\mathcal{L}_2$ is not.

If a Lie algebra $\mathcal{L}$ is homomorphic onto a Lie algebra $\mathcal{L}'$, $\mathcal{L}$ can be decomposed into a semidirect sum $\mathcal{L} = \mathcal{L}_1 \oplus_s \mathcal{L}_2$ such that $\mathcal{L}_2$ is isomorphic onto $\mathcal{L}'$. $\mathcal{L}_1$ is an ideal of $\mathcal{L}$ and maps to null vector in $\mathcal{L}'$. $\mathcal{L}_1$ is called the kernel of the homomorphism between $\mathcal{L}$ and $\mathcal{L}'$.

Define a series of Lie subalgebras $\mathcal{L}^{(n)}$, $n = 1, 2, \ldots$, from a Lie algebra $\mathcal{L}$

$$\mathcal{L}^{(1)} = [\mathcal{L},\ \mathcal{L}], \qquad \mathcal{L}^{(2)} = \left[\mathcal{L}^{(1)},\ \mathcal{L}^{(1)}\right], \quad \ldots . \tag{7.17}$$

A Lie algebra $\mathcal{L}$ is called solvable if there exists an integer m in the series of subalgebras $\mathcal{L}^{(n)}$ such that $\mathcal{L}^{(m)} = \emptyset$. It is proved that any Lie algebra can be decomposed into a semidirect sum of a solvable Lie algebra $\mathcal{L}_1$ and a semisimple Lie algebra $\mathcal{L}_2$, $\mathcal{L} = \mathcal{L}_1 \oplus_s \mathcal{L}_2$. Any irreducible representation with a finite dimension of a solvable Lie algebra is one-dimensional.

7.1.4 *The Killing Form and the Cartan Criteria*

Based on the Third Lie Theorem, the Lie groups as well as the Lie algebras can be classified by their structure constants. However, the structure constants do depend on the choice of the group parameters. We hope to find what property of the structure constants reflects the essence of a Lie algebra.

The Killing form g_{AB} of a Lie algebra is defined as

$$g_{AB} = \sum_{PQ} C_{AP}{}^{Q} C_{BQ}{}^{P} = -\text{Tr}\left(I_A^{\text{ad}} I_B^{\text{ad}}\right) = g_{BA}. \tag{7.18}$$

g_{AB} is a symmetric matrix. Defined the structure constants C_{ABD} with three subscripts which is totally antisymmetric

$$\begin{aligned} C_{ABD} &= \sum_{P} C_{AB}{}^{P} g_{PD} = \sum_{PQR} C_{AB}{}^{P} C_{PQ}{}^{R} C_{DR}{}^{Q} \\ &= - \sum_{PQR} \left\{C_{BQ}{}^{P} C_{PA}{}^{R} + C_{QA}{}^{P} C_{PB}{}^{R}\right\} C_{DR}{}^{Q} \\ &= \sum_{PQR} \left\{C_{BQ}{}^{P} C_{AP}{}^{R} C_{DR}{}^{Q} - C_{AQ}{}^{P} C_{BP}{}^{R} C_{DR}{}^{Q}\right\} \\ &= -C_{BAD} = -C_{ADB}. \end{aligned} \tag{7.19}$$

In the nomenclature of tensors (see Appendix B), with respect to the re-choice (7.6) of the group parameters, the generators I_A is a covariant vector, the structure constant $C_{AB}{}^{D}$ is a mixed tensor of rank $(2, 1)$, C_{ABD} is a covariant antisymmetric tensor of rank 3, and the Killing form is a covariant symmetric tensor of rank 2,

$$g'_{AB} = \sum_{PQ} X_{AP} X_{BQ} g_{PQ}, \qquad g' = X g X^T. \tag{7.20}$$

The Killing form is a real symmetric matrix. A real symmetric matrix can be diagonalized through a real orthogonal similarity transformation. Then, the diagonal entries can be changed to be ± 1 or 0 through the real scale transformation. Namely, for a real Lie algebra, through the re-choice of the real group parameters, the Killing form can be transformed into a diagonal matrix with the diagonal entries ± 1 and 0. For a Lie algebra, the sign of the diagonal entry can be removed by a scale transformation. However, the number of the zero eigenvalues of g_{AB} is essential for a Lie algebra and cannot be changed by the re-choice of the parameters.

Theorem 7.1 (The Cartan Criteria) A Lie algebra is semisimple if and only if its Killing form is nonsingular,

$$\det\ g \neq 0, \tag{7.21}$$

and a real semisimple Lie algebra is compact if and only if its Killing form is negative definite.

We will not prove this theorem here. Since the Killing form g_{AB} of a semisimple Lie algebra is nonsingular, g_{AB} has its inverse matrix g^{AB}

$$g^{AB} = g^{BA}, \qquad \sum_{D} g^{AD} g_{DB} = \delta_B^A. \tag{7.22}$$

g_{AB} and g^{AB} can be used as the metric tensors. Define an operator C_n which is a homogeneous polynomial of order n with respect to generators,

$$\begin{aligned} C_n = \sum_{(D)} \sum_{(A)} \sum_{(B)} & C_{A_1 D_1}{}^{D_2} C_{A_2 D_2}{}^{D_3} \cdots C_{A_n D_n}{}^{D_1} \\ & \times g^{A_1 B_1} \cdots g^{A_n B_n} I_{B_1} I_{B_2} \cdots I_{B_n}, \end{aligned} \tag{7.23}$$

which is commutable with any generators I_A in the Lie algebra,

$$[C_n,\ I_A] = 0, \qquad \forall\ I_A \in \mathcal{L}. \tag{7.24}$$

The reader is encouraged to prove Eq. (7.24). C_n is called the Casimir operator of order n. Remind that the Casimir operator of order 2 is

$$C_2 = \sum_{AB} g^{AB} I_A I_B. \tag{7.25}$$

7.2 The Regular Form of a Semisimple Lie Algebra

7.2.1 *The Inner Product in a Semisimple Lie Algebra*

Express the basis vector I_A in a semisimple Lie algebra $\mathcal{L}$ by the Dirac symbol which is commonly used in quantum mechanics [Georgi (1982)]

$$|I_A\rangle \equiv |A\rangle, \qquad 1 \leq A \leq g. \tag{7.26}$$

The inner product of two vectors in $\mathcal{L}$ is defined with its Killing form,

$$\langle A|B\rangle = -g_{AB} = \mathrm{Tr}\left(I_A^{\mathrm{ad}} I_B^{\mathrm{ad}}\right), \tag{7.27}$$

which is bilinear with respect to two vectors $\langle A|$ and $|B\rangle$

$$\begin{aligned} &\langle A|(c_1 B + c_2 D)\rangle = c_1 \langle A|B\rangle + c_2 \langle A|D\rangle, \\ &\langle (c_1 A + c_2 D)|B\rangle = c_1 \langle A|B\rangle + c_2 \langle D|B\rangle, \\ &\langle A|B\rangle = \langle B|A\rangle, \\ &\langle A|D|B\rangle \equiv \langle A|\,[D,\ B]\rangle = \langle [A,\ D]\,|B\rangle. \end{aligned} \tag{7.28}$$

When the group parameters are changed, the inner product will be related to the new Killing form in the same way,

$$\begin{aligned} |X_\mu\rangle &= \left|\sum_A X_{\mu A} I_A\right\rangle = \sum_A X_{\mu A}|A\rangle, \\ \langle X_\mu|X_\nu\rangle &= -\sum_{AB} X_{\mu A} X_{\nu B} g_{AB} = -g_{\mu\nu}. \end{aligned} \tag{7.29}$$

7.2.2 *The Cartan Subalgebra*

In a semisimple Lie algebra $\mathcal{L}$, any vector is also a linear operator which transforms a vector to another by the Lie product

$$X|Y\rangle = |\,[X,\ Y]\rangle. \tag{7.30}$$

Then, one is able to calculate the eigenvalue and the eigenvector of a vector X in $\mathcal{L}$. Evidently, any operator is its own eigenvector with zero eigenvalue,

$$X|X\rangle = |\,[X,\ X]\rangle = 0.$$

Denote by ℓ_X the multiplicity of zero eigenvalue of X. Among all vectors in $\mathcal{L}$ the minimal ℓ_X is denoted by ℓ

$$\ell = \min\ \ell_X > 0. \tag{7.31}$$

ℓ is called the rank of the Lie algebra $\mathcal{L}$ as well as the rank of the Lie group G. A vector X in $\mathcal{L}$ is called regular if $\ell_X = \ell$.

Theorem 7.2 There are ℓ linearly independent eigenvectors H_j of a regular vector X of a semisimple Lie algebra $\mathcal{L}$ with rank ℓ. H_j are commutable with each other

$$[H_j,\ H_k] = 0, \qquad 1 \le j \le \ell, \qquad 1 \le k \le \ell, \tag{7.32}$$

and X is a combination of H_j. In the remaining subspace of $\mathcal{L}$, the $(g-\ell)$ common eigenvectors $E_{\boldsymbol{\alpha}}$ of ℓ operators H_j are nondegenerate,

$$H_j|E_{\boldsymbol{\alpha}}\rangle = \alpha_j|E_{\boldsymbol{\alpha}}\rangle, \qquad [H_j,\ E_{\boldsymbol{\alpha}}] = \alpha_j E_{\boldsymbol{\alpha}}. \tag{7.33}$$

We will not prove this theorem here. The Abelian subalgebra $\mathcal{H}$ spanned by ℓ generators H_j is called the Cartan subalgebra of $\mathcal{L}$. $\mathcal{H}$ is not an ideal of $\mathcal{L}$. The set of H_j is the largest set of the mutually commutable generators in $\mathcal{L}$. The ℓ-dimensional vector $\boldsymbol{\alpha}$ is called a root, and the ℓ-dimensional space is called the root space. There is a one-to-one correspondence between the root $\boldsymbol{\alpha}$ and the generator $E_{\boldsymbol{\alpha}}$. Sometimes, H_j is also called the generator corresponding to the zero root. Note that the "zero root" is degenerate with the multiplicity ℓ.

The choice of the Cartan subalgebra is not unique. In fact, for any group element $R \in G$, the set of $H'_j = RH_jR^{-1}$ also spans a Cartan subalgebra, conjugate to the original one.

7.2.3 *Regular Commutative Relations of Generators*

The basis vectors H_j and $E_{\boldsymbol{\alpha}}$ in a semisimple Lie algebra $\mathcal{L}$ are called the Cartan–Weyl bases, or the regular bases. The regular bases satisfy the commutative relations (7.32) and (7.33). We are going to study their remaining commutative relations.

From Eq. (7.28) one has

$$\langle E_{\boldsymbol{\beta}}|H_j|E_{\boldsymbol{\alpha}}\rangle = \alpha_j\langle E_{\boldsymbol{\beta}}|E_{\boldsymbol{\alpha}}\rangle = -\beta_j\langle E_{\boldsymbol{\beta}}|E_{\boldsymbol{\alpha}}\rangle ,$$

namely, $(\alpha_j + \beta_j)\langle E_{\boldsymbol{\beta}}|E_{\boldsymbol{\alpha}}\rangle = 0$,

$$\langle E_{\boldsymbol{\beta}}|E_{\boldsymbol{\alpha}}\rangle = 0 \qquad \text{if} \;\; \boldsymbol{\alpha} \neq -\boldsymbol{\beta}. \tag{7.34}$$

For the same reason,

$$\langle H_j|E_{\boldsymbol{\alpha}}\rangle = -g_{j\boldsymbol{\alpha}} = 0. \tag{7.35}$$

For a given root $\boldsymbol{\alpha}$, if $-\boldsymbol{\alpha}$ is not a root, $g_{\boldsymbol{\alpha}B}$ are all vanishing, which conflicts to that $\mathcal{L}$ is semisimple. Thus, in a semisimple Lie algebra, the roots $\pm\boldsymbol{\alpha}$ have to appear in pairs. Choose the factor $b_{\boldsymbol{\alpha}}$ in $E_{\boldsymbol{\alpha}}$ such that

$$\langle E_{-\boldsymbol{\alpha}}|E_{\boldsymbol{\alpha}}\rangle = -g_{(-\boldsymbol{\alpha})\boldsymbol{\alpha}} = 1. \tag{7.36}$$

The factor $b_{\boldsymbol{\alpha}}$ in $E_{\boldsymbol{\alpha}}$ can still be chosen in the condition $b_{\boldsymbol{\alpha}}b_{-\boldsymbol{\alpha}} = 1$.

The part g_{jk} of the Killing form related to the Cartan subalgebra $\mathcal{H}$ can be expressed by the components of roots:

$$\begin{aligned} &\langle H_j|H_k\rangle = -g_{jk}, \\ &g_{jk} = \sum_{AB} C_{jA}{}^{B} C_{kB}{}^{A} = \sum_{\boldsymbol{\alpha}\in\Delta} C_{j\boldsymbol{\alpha}}{}^{\boldsymbol{\alpha}} C_{k\boldsymbol{\alpha}}{}^{\boldsymbol{\alpha}} = -\sum_{\boldsymbol{\alpha}\in\Delta} \alpha_j\alpha_k, \end{aligned} \tag{7.37}$$

where Δ is the set of all roots in $\mathcal{L}$. Due to Eq. (7.35) g_{jk} is also nonsingular, det $(g_{jk}) \neq 0$. Define $-g_{jk}$ and its inverse $-g^{jk}$ to be the metric tensor in the root space such that they can be used to raise a subscript and to lower a superscript of the vector in the root space. The inner product of two vectors in the root space is defined as

$$\begin{aligned}\boldsymbol{V}\cdot\boldsymbol{U} &= -\sum_{jk} g^{jk}V_jU_k = \sum_k V^kU_k = -\sum_{jk} g_{jk}V^jU^k \\ &= \sum_{jk}\sum_{\boldsymbol{\alpha}\in\Delta}\left(\alpha_jV^j\right)\left(\alpha_kU^k\right) = \sum_{\boldsymbol{\alpha}\in\Delta}(\boldsymbol{\alpha}\cdot\boldsymbol{V})(\boldsymbol{\alpha}\cdot\boldsymbol{U}).\end{aligned} \tag{7.38}$$

From the Jacobi identity,

$$\begin{aligned}0 = &\ [H_j,\ [E_{\boldsymbol{\alpha}},\ E_{\boldsymbol{\beta}}]] + [E_{\boldsymbol{\alpha}},\ [E_{\boldsymbol{\beta}},\ H_j]] + [E_{\boldsymbol{\beta}},\ [H_j,\ E_{\boldsymbol{\alpha}}]] \\ = &\ [H_j,\ [E_{\boldsymbol{\alpha}},\ E_{\boldsymbol{\beta}}]] - \beta_j[E_{\boldsymbol{\alpha}},\ E_{\boldsymbol{\beta}}] + \alpha_j[E_{\boldsymbol{\beta}},\ E_{\boldsymbol{\alpha}}],\end{aligned}$$

one has

$$\begin{aligned}&[H_j,\ [E_{\boldsymbol{\alpha}},\ E_{\boldsymbol{\beta}}]] = (\alpha_j+\beta_j)[E_{\boldsymbol{\alpha}},\ E_{\boldsymbol{\beta}}], \\ &H_j\,|[E_{\boldsymbol{\alpha}},\ E_{\boldsymbol{\beta}}]\rangle = (\alpha_j+\beta_j)\,|[E_{\boldsymbol{\alpha}},\ E_{\boldsymbol{\beta}}]\rangle.\end{aligned} \tag{7.39}$$

Since the non-zero root is nondegenerate, one concludes that $[E_{\boldsymbol{\alpha}},\ E_{\boldsymbol{\beta}}]$ is proportional to $E_{\boldsymbol{\alpha}+\boldsymbol{\beta}}$ if $\boldsymbol{\alpha}+\boldsymbol{\beta}$ is a root, and $[E_{\boldsymbol{\alpha}},\ E_{\boldsymbol{\beta}}]=0$ if $\boldsymbol{\alpha}+\boldsymbol{\beta}\neq 0$ is not a root. When $\boldsymbol{\beta}=-\boldsymbol{\alpha}$, $[E_{\boldsymbol{\alpha}},\ E_{-\boldsymbol{\alpha}}]$ belongs to the Cartan subalgebra and can be expressed as $\sum_j \lambda^j H_j$, where

$$\begin{aligned}-\sum_j g_{kj}\lambda^j &= \sum_j \langle H_k|H_j\rangle\,\lambda^j = \langle H_k|\,[E_{\boldsymbol{\alpha}},\ E_{-\boldsymbol{\alpha}}]\rangle \\ &= \langle [H_k,\ E_{\boldsymbol{\alpha}}]\,|E_{-\boldsymbol{\alpha}}\rangle = \alpha_k\langle E_{\boldsymbol{\alpha}}|E_{-\boldsymbol{\alpha}}\rangle = \alpha_k. \\ \lambda^j &= -\sum_k g^{jk}\alpha_k = \alpha^j.\end{aligned} \tag{7.40}$$

λ^j is nothing but the contravariant component of the root $\boldsymbol{\alpha}$.

In summary, the regular commutative relations among the Cartan–Weyl bases in a semisimple Lie algebra $\mathcal{L}$ are

$$\begin{aligned}&[H_j,\ H_k]=0, \qquad [H_j,\ E_{\boldsymbol{\alpha}}]=\alpha_jE_{\boldsymbol{\alpha}}, \\ &[E_{\boldsymbol{\alpha}},\ E_{\boldsymbol{\beta}}] = \begin{cases} N_{\boldsymbol{\alpha},\boldsymbol{\beta}}E_{\boldsymbol{\alpha}+\boldsymbol{\beta}} & \text{when } \boldsymbol{\alpha}+\boldsymbol{\beta} \text{ is a root,} \\ \sum_j \alpha^jH_j = \boldsymbol{\alpha}\cdot\boldsymbol{H}\equiv H_{\boldsymbol{\alpha}} & \text{when } \boldsymbol{\beta}=-\boldsymbol{\alpha}, \\ 0, & \text{the remaining cases,}\end{cases}\end{aligned} \tag{7.41}$$

where the antisymmetric coefficients $N_{\boldsymbol{\alpha},\boldsymbol{\beta}}=-N_{\boldsymbol{\beta},\boldsymbol{\alpha}}$ are to be determined. $E_{\boldsymbol{\alpha}}$ can be multiplied with a factor $b_{\boldsymbol{\alpha}}$ satisfying $b_{\boldsymbol{\alpha}}b_{-\boldsymbol{\alpha}}=1$, namely, the ratio $b_{\boldsymbol{\alpha}}/b_{-\boldsymbol{\alpha}}$ can be chosen arbitrarily. H_j as well as the components α_j of a root $\boldsymbol{\alpha}$ can be made an arbitrary nonsingular combination.

7.2.4 *The Inner Product of Roots*

Theorem 7.3 The inner product of any two non-zero roots $\boldsymbol{\alpha}$ and $\boldsymbol{\beta}$ in a semisimple Lie algebra $\mathcal{L}$ satisfies

$$\Gamma\left(\boldsymbol{\alpha}/\boldsymbol{\beta}\right) \equiv \frac{2\boldsymbol{\alpha}\cdot\boldsymbol{\beta}}{\boldsymbol{\beta}\cdot\boldsymbol{\beta}} = \text{integer}, \tag{7.42}$$

and $\boldsymbol{\alpha} - \Gamma\left(\boldsymbol{\alpha}/\boldsymbol{\beta}\right)\boldsymbol{\beta}$ is a root in $\mathcal{L}$.

Proof Construct a root chain from a root $\boldsymbol{\alpha}$ by adding and subtracting another root $\boldsymbol{\beta}$ successively

$$\ldots,\ (\boldsymbol{\alpha}-2\boldsymbol{\beta}),\ (\boldsymbol{\alpha}-\boldsymbol{\beta}),\ \boldsymbol{\alpha},\ (\boldsymbol{\alpha}+\boldsymbol{\beta}),\ (\boldsymbol{\alpha}+2\boldsymbol{\beta}),\ \ldots$$

Since the number of roots in $\mathcal{L}$ is finite, the root chain has to break off at two sides after a finite number of terms. Without loss of generality, one has

$$\begin{array}{lll} \boldsymbol{\alpha}+n\boldsymbol{\beta}, & -q \leq n \leq p, & \text{are all the roots,} \\ \boldsymbol{\alpha}-(q+1)\boldsymbol{\beta} \ \text{ and } & \boldsymbol{\alpha}+(p+1)\boldsymbol{\beta} & \text{are not the roots,} \end{array}$$

where p and q are both non-negative integers. From Eq. (7.41) one has

$$N_{\boldsymbol{\alpha},\boldsymbol{\beta}} = -N_{\boldsymbol{\beta},\boldsymbol{\alpha}}, \qquad N_{(\boldsymbol{\alpha}+p\boldsymbol{\beta}),\boldsymbol{\beta}} = N_{(\boldsymbol{\alpha}-q\boldsymbol{\beta}),-\boldsymbol{\beta}} = 0. \tag{7.43}$$

Letting

$$F_n = -N_{(\boldsymbol{\alpha}+n\boldsymbol{\beta}),\boldsymbol{\beta}}N_{(\boldsymbol{\alpha}+(n+1)\boldsymbol{\beta}),-\boldsymbol{\beta}}, \qquad F_p = F_{-q-1} = 0, \tag{7.44}$$

one obtains from the Jacobi identity

$$\begin{aligned} 0 &= \left[E_{\boldsymbol{\alpha}+n\boldsymbol{\beta}},\ \left[E_{\boldsymbol{\beta}},\ E_{-\boldsymbol{\beta}}\right]\right] + \left[E_{\boldsymbol{\beta}},\ \left[E_{-\boldsymbol{\beta}},\ E_{\boldsymbol{\alpha}+n\boldsymbol{\beta}}\right]\right] \\ &\quad + \left[E_{-\boldsymbol{\beta}},\ \left[E_{\boldsymbol{\alpha}+n\boldsymbol{\beta}},\ E_{\boldsymbol{\beta}}\right]\right] \\ &= \sum_j \beta^j \left[E_{\boldsymbol{\alpha}+n\boldsymbol{\beta}},\ H_j\right] - N_{(\boldsymbol{\alpha}+n\boldsymbol{\beta}),-\boldsymbol{\beta}}\left[E_{\boldsymbol{\beta}},\ E_{\boldsymbol{\alpha}+(n-1)\boldsymbol{\beta}}\right] \\ &\quad + N_{(\boldsymbol{\alpha}+n\boldsymbol{\beta}),\boldsymbol{\beta}}\left[E_{-\boldsymbol{\beta}},\ E_{\boldsymbol{\alpha}+(n+1)\boldsymbol{\beta}}\right] \\ &= \left\{-\boldsymbol{\beta}\cdot(\boldsymbol{\alpha}+n\boldsymbol{\beta}) - F_{n-1} + F_n\right\} E_{\boldsymbol{\alpha}+n\boldsymbol{\beta}}. \end{aligned}$$

Thus, F_n satisfies the recursive relation

$$\begin{aligned} F_n &= F_{n-1} + \boldsymbol{\beta} \cdot \{\boldsymbol{\alpha} + n\boldsymbol{\beta}\} \\ &= F_{n-2} + \boldsymbol{\beta} \cdot \{2\boldsymbol{\alpha} + (n + n - 1)\,\boldsymbol{\beta}\} = \ldots \\ &= F_{n-(n+q+1)} + \boldsymbol{\beta} \cdot \left\{(n + q + 1)\boldsymbol{\alpha} + \frac{1}{2}\,(n - q)\,(n + q + 1)\,\boldsymbol{\beta}\right\} \\ &= \frac{1}{2}(n + q + 1)\boldsymbol{\beta} \cdot \{2\boldsymbol{\alpha} + (n - q)\,\boldsymbol{\beta}\}\,. \end{aligned} \tag{7.45}$$

When $n = p$, one has

$$2\boldsymbol{\alpha} \cdot \boldsymbol{\beta} = (q - p)\,(\boldsymbol{\beta} \cdot \boldsymbol{\beta})\,. \tag{7.46}$$

If $\boldsymbol{\beta} \cdot \boldsymbol{\beta} = 0$, $\boldsymbol{\beta}$ is orthogonal to each root $\boldsymbol{\alpha}$ in $\mathcal{L}$ so that $H_{\boldsymbol{\beta}}$ is commutable with each generator in $\mathcal{L}$,

$$[H_{\boldsymbol{\beta}},\ H_k] = 0, \qquad [H_{\boldsymbol{\beta}},\ E_{\boldsymbol{\alpha}}] = (\boldsymbol{\beta} \cdot \boldsymbol{\alpha})\,E_{\boldsymbol{\alpha}} = 0. \tag{7.47}$$

Thus, $H_{\boldsymbol{\beta}}$ spans an Abelian ideal in $\mathcal{L}$ which is in conflict with the fact that $\mathcal{L}$ is semisimple. Since $\boldsymbol{\beta} \cdot \boldsymbol{\beta} \neq 0$, from Eq. (7.46) one has

$$\Gamma\,(\boldsymbol{\alpha}/\boldsymbol{\beta}) = \frac{2\boldsymbol{\alpha} \cdot \boldsymbol{\beta}}{\boldsymbol{\beta} \cdot \boldsymbol{\beta}} = q - p. \tag{7.48}$$

It gives Eq. (7.42). Due to $-q \le (p - q) \le p$, $\boldsymbol{\alpha} - \Gamma(\boldsymbol{\alpha}/\boldsymbol{\beta})\boldsymbol{\beta}$ is a root.

At last, the theorem also holds if the root chain $\boldsymbol{\alpha} + n\boldsymbol{\beta}$ contains a zero root. Without loss of generality, let $\boldsymbol{\alpha} = m\boldsymbol{\beta}$. Defining the generator $E_{\boldsymbol{\alpha}-m\boldsymbol{\beta}} = E_0 = H_{\boldsymbol{\beta}}$ corresponding to the zero root, one has

$$\begin{aligned} &\left[E_{\boldsymbol{\alpha}-(m-1)\boldsymbol{\beta}},\ E_{-\boldsymbol{\beta}}\right] = H_{\boldsymbol{\beta}}, && N_{(\boldsymbol{\alpha}-(m-1)\boldsymbol{\beta}),-\boldsymbol{\beta}} = 1, \\ &[E_{\boldsymbol{\alpha}-m\boldsymbol{\beta}},\ E_{\pm\boldsymbol{\beta}}] = \pm\,(\boldsymbol{\beta} \cdot \boldsymbol{\beta})\,E_{\pm\boldsymbol{\beta}}, && N_{(\boldsymbol{\alpha}-m\boldsymbol{\beta}),\pm\boldsymbol{\beta}} = \pm\,(\boldsymbol{\beta} \cdot \boldsymbol{\beta})\,, \\ &\left[E_{\boldsymbol{\alpha}-(m+1)\boldsymbol{\beta}},\ E_{\boldsymbol{\beta}}\right] = -H_{\boldsymbol{\beta}}, && N_{(\boldsymbol{\alpha}-(m+1)\boldsymbol{\beta}),\boldsymbol{\beta}} = -1. \end{aligned}$$

Thus, the above proof holds for this case. □

Corollary 7.3.1 The number of linearly independent roots in a semisimple Lie algebra $\mathcal{L}$ with rank ℓ is ℓ.

Proof Prove the corollary by reduction to absurdity. If the number is less than ℓ, there is at least a non-zero vector $\boldsymbol{V}$ orthogonal to each roots in $\mathcal{L}$ so that $\boldsymbol{V} \cdot \boldsymbol{H}$ is commutable with each generator in $\mathcal{L}$. This contradicts that $\mathcal{L}$ is semisimple. □

Corollary 7.3.2 In the root space of a semisimple Lie algebra, the inner product of any two vectors which are the real combinations of roots is real, and the inner self-product of a nonvanishing real combination of roots is positive real.

Proof First, due to Eqs. (7.38) and (7.46) one has

$$\boldsymbol{\beta}\cdot\boldsymbol{\beta} = \sum_{\boldsymbol{\alpha}\in\Delta} (\boldsymbol{\alpha}\cdot\boldsymbol{\beta})^2 = \frac{1}{4}\left\{\sum_{\boldsymbol{\alpha}\in\Delta} (q_{\boldsymbol{\alpha}} - p_{\boldsymbol{\alpha}})^2\right\}(\boldsymbol{\beta}\cdot\boldsymbol{\beta})^2,$$

where $q_{\boldsymbol{\alpha}}$ and $p_{\boldsymbol{\alpha}}$ are the integral parameters in the root chain constructed from $\boldsymbol{\alpha}$ by adding and subtracting a root $\boldsymbol{\beta}$ successively. Since $\boldsymbol{\beta}\cdot\boldsymbol{\beta} \neq 0$,

$$\boldsymbol{\beta}\cdot\boldsymbol{\beta} = 4\left\{\sum_{\boldsymbol{\alpha}\in\Delta} (q_{\boldsymbol{\alpha}} - p_{\boldsymbol{\alpha}})^2\right\}^{-1} = \text{positive real.} \tag{7.49}$$

Then, due to Eq. (7.46), the inner product $\boldsymbol{\alpha}\cdot\boldsymbol{\beta}$ of any two roots is real.

Second, introducing two real combinations of roots

$$\boldsymbol{V} = \sum_{\boldsymbol{\beta}\in\Delta} b_{\boldsymbol{\beta}}\boldsymbol{\beta}, \qquad \boldsymbol{U} = \sum_{\boldsymbol{\gamma}\in\Delta} c_{\boldsymbol{\gamma}}\boldsymbol{\gamma},$$

one has

$$\boldsymbol{V}\cdot\boldsymbol{U} = \sum_{\boldsymbol{\beta}\in\Delta}\sum_{\boldsymbol{\gamma}\in\Delta} b_{\boldsymbol{\beta}}c_{\boldsymbol{\gamma}}(\boldsymbol{\beta}\cdot\boldsymbol{\gamma}) = \text{real.}$$

Due to Eq. (7.38),

$$\boldsymbol{V}\cdot\boldsymbol{V} = \sum_{\boldsymbol{\alpha}\in\Delta} (\boldsymbol{\alpha}\cdot\boldsymbol{V})^2 \geq 0.$$

$\boldsymbol{V}\cdot\boldsymbol{V} > 0$ only if $\boldsymbol{V} \neq 0$. □

Corollary 7.3.3

$$N_{\boldsymbol{\alpha},\boldsymbol{\beta}}N_{-\boldsymbol{\alpha},-\boldsymbol{\beta}} = -\frac{1}{2}p(q+1)(\boldsymbol{\beta}\cdot\boldsymbol{\beta}). \tag{7.50}$$

Proof Letting $n = 0$ in Eq. (7.45) one has

$$-N_{\boldsymbol{\alpha},\boldsymbol{\beta}}N_{(\boldsymbol{\alpha}+\boldsymbol{\beta}),-\boldsymbol{\beta}} = F_0 = \frac{1}{2}(q+1)(2\boldsymbol{\alpha}\cdot\boldsymbol{\beta} - q\boldsymbol{\beta}\cdot\boldsymbol{\beta}) = -\frac{1}{2}p(q+1)(\boldsymbol{\beta}\cdot\boldsymbol{\beta}). \tag{7.51}$$

If $\boldsymbol{\gamma} = \boldsymbol{\alpha} + \boldsymbol{\beta}$, and three roots are all non-zero,

$$\begin{aligned}\langle E_{-\boldsymbol{\alpha}}|E_{-\boldsymbol{\beta}}|E_{\boldsymbol{\gamma}}\rangle &= \langle E_{-\boldsymbol{\alpha}}\,|[E_{-\boldsymbol{\beta}},\ E_{\boldsymbol{\gamma}}]\rangle = N_{-\boldsymbol{\beta},\boldsymbol{\gamma}}\langle E_{-\boldsymbol{\alpha}}|E_{\boldsymbol{\alpha}}\rangle = N_{-\boldsymbol{\beta},\boldsymbol{\gamma}}\\ &= \langle[E_{-\boldsymbol{\alpha}},\ E_{-\boldsymbol{\beta}}]|\,E_{\boldsymbol{\gamma}}\rangle = N_{-\boldsymbol{\alpha},-\boldsymbol{\beta}}\langle E_{-\boldsymbol{\gamma}}|E_{\boldsymbol{\gamma}}\rangle = N_{-\boldsymbol{\alpha},-\boldsymbol{\beta}}.\end{aligned}$$

Due to Eq. (7.43),

$$N_{-\boldsymbol{\alpha},-\boldsymbol{\beta}} = N_{-\boldsymbol{\beta},\boldsymbol{\gamma}} = -N_{(\boldsymbol{\alpha}+\boldsymbol{\beta}),-\boldsymbol{\beta}}. \tag{7.52}$$

Equation (7.50) follows Eq. (7.51). □

Corollary 7.3.4 $N_{\boldsymbol{\alpha},\boldsymbol{\beta}} \neq 0$ if $\boldsymbol{\alpha}$, $\boldsymbol{\beta}$, and $(\boldsymbol{\alpha}+\boldsymbol{\beta})$ are all non-zero roots.

Corollary 7.3.5 Except for zero roots, there are only two roots $\pm\boldsymbol{\alpha}$ along the direction of $\boldsymbol{\alpha}$.

Proof Let $t\boldsymbol{\alpha}$ be a non-zero root, then $\Gamma(t\boldsymbol{\alpha}/\boldsymbol{\alpha}) = 2t$ and $\Gamma\left(\boldsymbol{\alpha}/(t\boldsymbol{\alpha})\right) = 2/t$. From the conditions that $2t$ and $2/t$ are both integers, one obtains $t = \pm 1$, ± 2, or $\pm 1/2$. Since $N_{\boldsymbol{\alpha},\boldsymbol{\alpha}} = 0$, $\pm 2\boldsymbol{\alpha}$ is not a root. Then, $\pm\boldsymbol{\alpha}/2$ is not a root, otherwise $\boldsymbol{\alpha} = 2(\boldsymbol{\alpha}/2)$ is not a root. □

7.2.5 *Positive Roots and Simple Roots*

Theorem 7.4 For a semisimple Lie algebra $\mathcal{L}$, the bases H_j in the Cartan subalgebra $\mathcal{H}$ of $\mathcal{L}$ can be chosen such that the roots are all real and the root space is real Euclidean.

Proof First, there exist ℓ linearly independent roots in $\mathcal{L}$, say $\boldsymbol{\beta}^{(r)}$, $1 \leq r \leq \ell$, where ℓ is the rank of $\mathcal{L}$. Any root $\boldsymbol{\alpha}$ can be expanded with respect to $\boldsymbol{\beta}^{(r)}$, $\boldsymbol{\alpha} = \sum_r x_r \boldsymbol{\beta}^{(r)}$. Taking the inner product of each terms of the equation with $2\boldsymbol{\beta}^{(s)}/\left(\boldsymbol{\beta}^{(s)}\cdot\boldsymbol{\beta}^{(s)}\right)$, one obtains

$$\Gamma\left(\boldsymbol{\alpha}/\boldsymbol{\beta}^{(s)}\right) = \sum_{r=1}^{\ell} x_r \Gamma\left(\boldsymbol{\beta}^{(r)}/\boldsymbol{\beta}^{(s)}\right).$$

This is the coupled linear algebraic equation with respect to the variables x_r. Since all the coefficients in the equation are integers and the solutions x_r do exist, x_r have to be all real. Namely, each root in $\mathcal{L}$ is a real combination of $\boldsymbol{\beta}^{(r)}$.

Second, for the set of new basis vectors H_r in $\mathcal{H}$, $H_r = \boldsymbol{\beta}^{(r)}\cdot\boldsymbol{H}$, g_{jk} becomes real,

$$g_{rs} = \sum_{jk} \beta^{(r)j}\beta^{(s)k} g_{jk} = -\sum_{\boldsymbol{\alpha}\in\Delta}\left(\boldsymbol{\alpha}\cdot\boldsymbol{\beta}^{(r)}\right)\left(\boldsymbol{\alpha}\cdot\boldsymbol{\beta}^{(s)}\right) = \text{real}.$$

The new g_{rs} can be diagonalized through a real orthogonal transformation of H_r, and then become $-\delta_{rs}$ through a scale transformation.

Third, denote by H_j the transformed bases in $\mathcal{H}$, where $g_{jk} = -\delta_{jk}$. In the condition $g_{jk} = -\delta_{jk}$, H_r as well as the components of roots can still be made an orthogonal transformation. Due to Corollary 7.3.2, the root space is real Euclidean only if the basis roots $\boldsymbol{\beta}^{(r)}$ can be transformed into real through the orthogonal transformation. Assume that the real and imaginary parts of the jth component of $\boldsymbol{\beta}^{(r)}$ are $a_j^{(r)}$ and $b_j^{(r)}$, respectively, $\beta_j^{(r)} = a_j^{(r)} + ib_j^{(r)}$. Make a real orthogonal transformation on $a_j^{(1)}$ such that $a_1^{(1)}$ is non-negative and the remaining components are vanishing. Since $\boldsymbol{\beta}^{(1)} \cdot \boldsymbol{\beta}^{(1)}$ is real positive, $a_1^{(1)} > 0$, and $b_1^{(1)} = 0$. Then, make a real orthogonal transformation on $b_j^{(1)}$, where $j > 1$, such that $b_2^{(1)}$ is non-negative and the remaining components are vanishing. The components of the transformed root $\boldsymbol{\beta}^{(1)}$ become $\beta_1^{(1)} = a_1^{(1)} = a$ and $\beta_2^{(1)} = ib_2^{(1)} = ib$, and the remaining components are vanishing. Since $\boldsymbol{\beta}^{(1)} \cdot \boldsymbol{\beta}^{(1)} > 0$, $a > b \geq 0$. Making an orthogonal transformation on the first two components,

$$\left(a^2 - b^2\right)^{-1/2} \begin{pmatrix} a & ib \\ -ib & a \end{pmatrix} \begin{pmatrix} a \\ ib \end{pmatrix} = \begin{pmatrix} \left(a^2 - b^2\right)^{1/2} \\ 0 \end{pmatrix},$$

one has that only the first component $\beta_1^{(1)}$ of the transformed root $\boldsymbol{\beta}^{(1)}$ is positive real, and its remaining components are all vanishing. Since $\boldsymbol{\beta}^{(1)} \cdot \boldsymbol{\beta}^{(r)}$ is real, the first component $\beta_1^{(r)}$ of each basis root $\boldsymbol{\beta}^{(r)}$ is real. Define new basis vectors in $\mathcal{L}$ by the real combinations, $\boldsymbol{\gamma}^{(r)} = \boldsymbol{\beta}^{(r)} - \left(\beta_1^{(r)}/\beta_1^{(1)}\right)\boldsymbol{\beta}^{(1)}$, where $r > 1$. Thus, the first components of $\boldsymbol{\gamma}^{(r)}$ with $r > 1$ are vanishing, and its remaining components are equal to those of $\boldsymbol{\beta}^{(r)}$. Each $\boldsymbol{\gamma}^{(r)}$ is not a null vector.

The rest can be deduced by analogy. Preserving the first one-dimensional subspace invariant, one makes the orthogonal transformation on the remaining $(\ell - 1)$-dimensional subspace such that the first components $\beta_1^{(r)}$ remain invariant, the second component $a_2^{(2)}$ of the transformed root $\boldsymbol{\beta}^{(2)}$ is positive real, and its remaining components are all vanishing. Corollary 7.3.2 is used in the proof. Thus, the first and the second components of all basis roots $\boldsymbol{\beta}^{(r)}$ are real. In the same way, one can prove that all the basis roots are real. Therefore, all roots in $\mathcal{L}$ are real, $g_{jk} = -\delta_{jk}$, and the root space is real Euclidean. □

For a given order of H_j in the Cartan subalgebra, a root $\boldsymbol{\alpha}$ is called positive if its first nonvanishing component is positive, and the root is negative if the component is negative. A positive root is called a simple root if it cannot be expressed as a non-negative integral combination of other positive roots. Therefore, any positive root is equal to a non-negative integral combination of the simple roots, and the sum of the coefficients in the combination is called the level of the positive root. The negative roots are similar. Obviously, the number of the simple roots in a semisimple Lie algebra with rank ℓ is not less than ℓ. It is equal to ℓ if all simple roots are linearly independent of each other.

Theorem 7.5 The difference of two simple roots is not a root, the inner product of two simple roots is not larger than 0, and the number of the simple roots in a semisimple Lie algebra with rank ℓ is equal to ℓ.

Proof First, denote by γ the difference of two simple roots $\boldsymbol{\alpha}$ and $\boldsymbol{\beta}$, $\gamma = \boldsymbol{\alpha} - \boldsymbol{\beta}$. γ is not a positive root, otherwise $\boldsymbol{\alpha}$ is a sum of two positive roots $\boldsymbol{\beta}$ and γ. γ is not a negative root, otherwise $\boldsymbol{\beta}$ is a sum of two positive roots $\boldsymbol{\alpha}$ and $-\gamma$. Thus, the difference of two simple roots is not a root.

Second, from Theorem 7.3, the inner product of two simple roots is not larger than 0,

$$\Gamma\left(\boldsymbol{\alpha}/\boldsymbol{\beta}\right) = \frac{2\boldsymbol{\alpha}\cdot\boldsymbol{\beta}}{\boldsymbol{\beta}\cdot\boldsymbol{\beta}} = q - p = -p \leq 0. \tag{7.53}$$

At last, prove by reduction to absurdity that simple roots are linearly independent. Assume that there is a real linear relation among simple roots,

$$\sum_j c_j \boldsymbol{\alpha}^{(j)} - \sum_k d_k \boldsymbol{\beta}^{(k)} = 0, \qquad c_j > 0, \qquad d_k > 0.$$

Since $\boldsymbol{\alpha}^{(j)} \neq \boldsymbol{\beta}^{(k)}$, $\boldsymbol{\alpha}^{(j)} \cdot \boldsymbol{\beta}^{(k)} \leq 0$. Then, it is in contradiction that the inner self-product of the nonvanishing vector $\boldsymbol{V}$ is not larger than zero,

$$\boldsymbol{V} = \sum_j c_j \boldsymbol{\alpha}^{(j)} = \sum_k d_k \boldsymbol{\beta}^{(k)} \neq 0,$$

$$\boldsymbol{V}\cdot\boldsymbol{V} = \sum_{jk} c_j d_k \boldsymbol{\alpha}^{(j)} \cdot \boldsymbol{\beta}^{(k)} \leq 0. \qquad \square$$

Theorem 7.6 Up to isomorphism, any semisimple Lie algebra $\mathcal{L}$ has one and only one compact real form.

Proof We only sketch the proof. Since $E_{\boldsymbol{\alpha}}$ can still be multiplied with a factor $b_{\boldsymbol{\alpha}}$ satisfying $b_{\boldsymbol{\alpha}} b_{-\boldsymbol{\alpha}} = 1$, the ratios $b_{\boldsymbol{\alpha}}/b_{-\boldsymbol{\alpha}}$ can be chosen such that

$$N_{\boldsymbol{\alpha},\boldsymbol{\beta}} = -N_{-\boldsymbol{\alpha},-\boldsymbol{\beta}}. \tag{7.54}$$

In fact, from Eq. (7.41) one has

$$\frac{N'_{\boldsymbol{\alpha},\boldsymbol{\beta}}}{N'_{-\boldsymbol{\alpha},-\boldsymbol{\beta}}} = \frac{b_{\boldsymbol{\alpha}}}{b_{-\boldsymbol{\alpha}}}\,\frac{b_{\boldsymbol{\beta}}}{b_{-\boldsymbol{\beta}}}\,\frac{b_{-\boldsymbol{\alpha}-\boldsymbol{\beta}}}{b_{\boldsymbol{\alpha}+\boldsymbol{\beta}}}\,\frac{N_{\boldsymbol{\alpha},\boldsymbol{\beta}}}{N_{-\boldsymbol{\alpha},-\boldsymbol{\beta}}}.$$

One is able to choose the ratios $b_{\alpha}/b_{-\alpha}$ such that Eq. (7.54) for the positive roots are satisfied one by one as their levels increase. Under the condition (7.54) the Killing form g_{AB} becomes $-\delta_{AB}$ if $E_{\pm\alpha}$ are replaced with

$$E_{\alpha 1} = \left(E_{\alpha} + E_{-\alpha}\right)/\sqrt{2}, \qquad E_{\alpha 2} = -i\left(E_{\alpha} - E_{-\alpha}\right)/\sqrt{2}. \qquad (7.55)$$

It can be shown straightforwardly that the structure constants $C_{AB}{}^{D}$ are all real. The real Lie algebra with the basis vectors $(-iH_j)$, $(-iE_{\alpha 1})$, and $(-iE_{\alpha 2})$ is the compact real form of $\mathcal{L}$. □

Corollary 7.6.1 Any representation with finite dimension of a semisimple Lie algebra is completely reducible.

Since the adjoint representation of a semisimple Lie algebra is completely reducible, the following Corollary follows.

Corollary 7.6.2 Any semisimple Lie algebra can be decomposed to a direct sum of some non-Abelian simple Lie algebras .

Now, the classification problem of semisimple Lie algebras reduces to the classification of simple Lie algebras with dimension larger than 1. In the classification of simple Lie algebras one will find that in each simple Lie algebra $\mathcal{L}$, there is only one root $\boldsymbol{\omega}$ whose level is the highest among all roots in $\mathcal{L}$. $\boldsymbol{\omega}$ is called the largest root of $\mathcal{L}$, which is related to the adjoint representation of $\mathcal{L}$.

7.3 Classification of Simple Lie Algebras

A simple Lie algebra of one dimension is Abelian. In this section we are going to study the classification of simple Lie algebras with the dimension larger than 1. In the compact real form of a simple Lie algebra $\mathcal{L}$, the root space is real Euclidean. Hereafter, denote by $\boldsymbol{r}_{\mu}$ the simple roots of $\mathcal{L}$.

7.3.1 *Angle between Two Simple Roots*

The inner product of two simple roots $\boldsymbol{r}_{\mu}$ is not larger than 0 so that their angle θ is not less than $\pi/2$. The cosine square of θ is

$$\begin{aligned} 4\cos^2\theta &= 4\,\frac{\left(\boldsymbol{r}_{\mu}\cdot\boldsymbol{r}_{\nu}\right)^2}{|\boldsymbol{r}_{\mu}|^2\;|\boldsymbol{r}_{\nu}|^2} = \frac{2\boldsymbol{r}_{\mu}\cdot\boldsymbol{r}_{\nu}}{|\boldsymbol{r}_{\nu}|^2}\;\frac{2\boldsymbol{r}_{\nu}\cdot\boldsymbol{r}_{\mu}}{|\boldsymbol{r}_{\mu}|^2} \\ &= \Gamma\left(\boldsymbol{r}_{\mu}/\boldsymbol{r}_{\nu}\right)\;\Gamma\left(\boldsymbol{r}_{\nu}/\boldsymbol{r}_{\mu}\right) = \text{integer}. \end{aligned} \qquad (7.56)$$

Without loss of generality, let the length of $\boldsymbol{r}_{\mu}$ be not less than that of $\boldsymbol{r}_{\nu}$,

$$\frac{|\boldsymbol{r}_\mu|^2}{|\boldsymbol{r}_\nu|^2} = \frac{\Gamma\left(\boldsymbol{r}_\mu/\boldsymbol{r}_\nu\right)}{\Gamma\left(\boldsymbol{r}_\nu/\boldsymbol{r}_\mu\right)} \geq 1. \tag{7.57}$$

There are only four solutions for $4\cos^2\theta$ listed in Table 7.1.

Table 7.1 The angles and the lengths of simple roots

θ	$\cos^2\theta$	$\Gamma\left(\boldsymbol{r}_\mu/\boldsymbol{r}_\nu\right)$	$\Gamma\left(\boldsymbol{r}_\nu/\boldsymbol{r}_\mu\right)$	$\|\boldsymbol{r}_\mu\|/\|\boldsymbol{r}_\nu\|$
$5\pi/6$ (150°)	$3/4$	-3	-1	$\sqrt{3}$
$3\pi/4$ (135°)	$1/2$	-2	-1	$\sqrt{2}$
$2\pi/3$ (120°)	$1/4$	-1	-1	1
$\pi/2$ (90°)	0	0	0	arbitrary

7.3.2 *Dynkin Diagrams*

The Dynkin diagram for a simple Lie algebra is drawn by the following rule. It will be shown that there are one or two different lengths among the simple roots in any simple Lie algebra $\mathcal{L}$. Denote each longer simple root by a white circle, and denote each shorter simple root, if it exists, by a black circle. Two circles denoting two simple roots are connected by a single link, a double link, or a triple link depending upon their angle to be $2\pi/3$, $3\pi/4$, or $5\pi/6$, respectively. The ratio of their square lengths is 1, 2, or 3, respectively. Two circles are not connected if two simple roots are orthogonal and the ratio of their lengths is not restricted. Now, we are going to study what kinds of Dynkin diagrams of $\mathcal{L}$ are allowed based on the property of a simple Lie algebra.

1. *The Dynkin diagram of a simple Lie algebra is connected.*

If a Dynkin diagram is divided into two unconnected parts, two simple roots belonging to different parts, say $\boldsymbol{r}$ and $\boldsymbol{r}'$, are orthogonal to each other. From Eq. (7.53) both the sum and the difference of those two simple roots are not roots. Thus, the generators are divided into two classes, respectively related to the roots in two parts. The generators $E_{\boldsymbol{r}}$ and $\boldsymbol{r}\cdot\boldsymbol{H}$ belonging to the first class are commutable with the generators $E_{\boldsymbol{r}'}$ and $\boldsymbol{r}'\cdot\boldsymbol{H}$ belonging to the second one. Therefore, $\mathcal{L}$ is decomposed into the direct sum of two ideals, which contradicts to that $\mathcal{L}$ is simple.

2. *The Dynkin diagram contains no loop.*

If a Dynkin diagram contains a smallest loop composed of n circles, the neighbored circles denoting the simple roots $\boldsymbol{u}_j$ and $\boldsymbol{u}_{j+1}$ are connected by links and there is no link inside the loop. Assume $\boldsymbol{u}_{n+1} = \boldsymbol{u}_1$ for convenience. The sum $\boldsymbol{\alpha}$ of the simple roots is not vanishing.

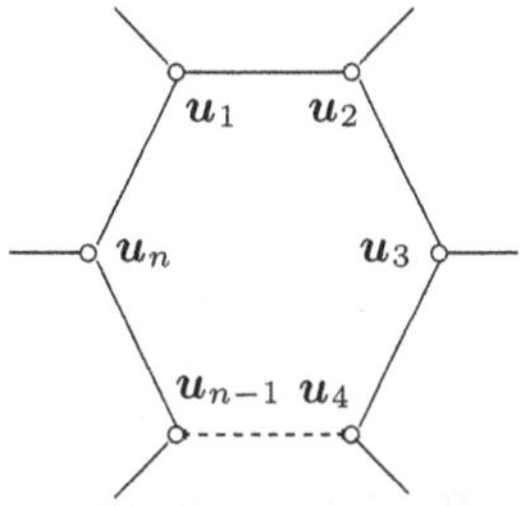

$$\begin{aligned}
\boldsymbol{\alpha} &= \sum_{j=1}^{n} \boldsymbol{u}_j \neq 0, \\
|\boldsymbol{\alpha}|^2 &= \sum_{j=1}^{n} |\boldsymbol{u}_j|^2 + 2\sum_{j=1}^{n} \boldsymbol{u}_j \cdot \boldsymbol{u}_{j+1} \\
&= \sum_{j=1}^{n} |\boldsymbol{u}_j|^2 \left\{1 + \Gamma(\boldsymbol{u}_{j+1}/\boldsymbol{u}_j)\right\} \leq 0.
\end{aligned}$$

It contradicts to that $\boldsymbol{\alpha}$ is nonvanishing.

3. *The number of links fetching out from one circle is less than four.*

Let a simple root $\boldsymbol{r}$ connect with n simple roots $\boldsymbol{u}_j$, $1 \leq j \leq n$. The number of links fetching out from $\boldsymbol{r}$ is equal to $\sum_j \Gamma\left(\boldsymbol{r}/\boldsymbol{u}_j\right) \Gamma\left(\boldsymbol{u}_j/\boldsymbol{r}\right)$. Since there is no loop, any two simple roots $\boldsymbol{u}_j$ are orthogonal to each other, and $\boldsymbol{r}$ is linearly independent of n simple roots $\boldsymbol{r}_j$. Thus,

$$|\boldsymbol{r}|^2 > \sum_{j=1}^{n} \left(\boldsymbol{r} \cdot \boldsymbol{u}_j\right)^2 / |\boldsymbol{u}_j|^2 = \frac{|\boldsymbol{r}|^2}{4} \sum_{j=1}^{n} \Gamma\left(\boldsymbol{r}/\boldsymbol{u}_j\right) \Gamma\left(\boldsymbol{u}_j/\boldsymbol{r}\right). \tag{7.58}$$

The conclusion follows Eq. (7.58) by removing a factor $\boldsymbol{r}^2$.

From this property, one concludes that there is only one Dynkin diagram with a triple link:

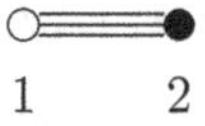

Its algebra is called the Lie algebra G_2. The following diagrams as well as those by interchanging the white and black circles are not allowed:

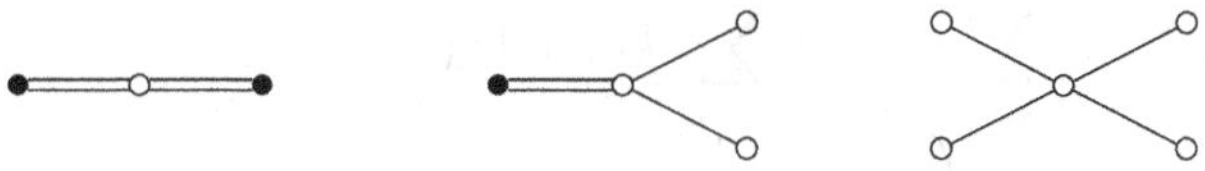

It will be shown that the diagrams by replacing the circle on the center of the above diagrams with a circle chain connected by single links are also not allowed:

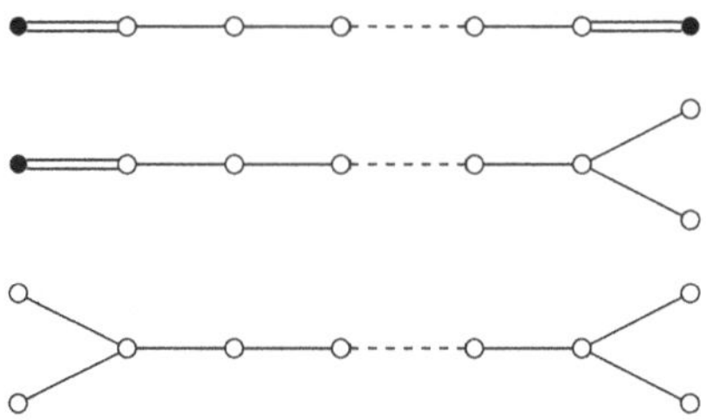

Let the circle chain contain m circles, denoting m simple roots $\boldsymbol{v}_j$ with the same length v, $1 \leq j \leq m$. The length of their sum, $\boldsymbol{v} = \sum_j \boldsymbol{v}_j$, is also equal to v,

$$|\boldsymbol{v}|^2 = \sum_{j=1}^{m} |\boldsymbol{v}_j|^2 + 2 \sum_{j=1}^{m-1} \boldsymbol{v}_j \cdot \boldsymbol{v}_{j+1} = mv^2 - (m-1)v^2 = v^2.$$

Replacing $\boldsymbol{r}$ with $\boldsymbol{v}$ in the above proof, one also shows that the number of links fetching out from the circle chain is less than four.

4. *The Dynkin diagrams with a double link*

The general form of the Dynkin diagrams with a double link is

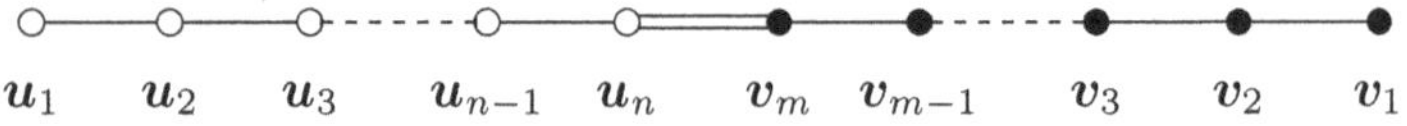

where the length of $\boldsymbol{v}_k$ is v and that of $\boldsymbol{u}_j$ is $\sqrt{2}v$. Letting

$$\boldsymbol{u} = \sum_{j=1}^{n} j\boldsymbol{u}_j, \qquad \boldsymbol{v} = \sum_{k=1}^{m} k\boldsymbol{u}_k,$$

one has

$$\begin{aligned} |\boldsymbol{v}|^2 &= \sum_{k=1}^{m} k^2 v^2 + \sum_{k=1}^{m-1} k(k+1)\left(-v^2\right) \\ &= v^2 \left(m^2 - \sum_{k=1}^{m-1} k\right) = \frac{1}{2}m(m+1)v^2 \, . \\ |\boldsymbol{u}|^2 &= n(n+1)v^2. \end{aligned} \tag{7.59}$$

Since $\boldsymbol{u}$ and $\boldsymbol{v}$ are not collinear, one obtains

$$\begin{aligned} 0 < \ & |\boldsymbol{u}|^2|\boldsymbol{v}|^2 - (\boldsymbol{u}\cdot\boldsymbol{v})^2 \\ = \ & \frac{1}{2}n(n+1)m(m+1)v^4 - (mn)^2\left(\boldsymbol{u}_n\cdot\boldsymbol{v}_m\right)^2 \\ = \ & \frac{1}{2}nm\,(n+m+1-mn)\,v^4. \end{aligned}$$

Then, $(m-1)(n-1) < 2$. If $m = 1$, n is an arbitrarily positive integer, denoted by $\ell - 1$. The Dynkin diagram is

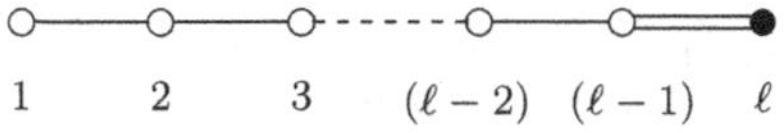

Its algebra is called the Lie algebra B_ℓ. If $n = 1$, m is an arbitrarily positive integer $\ell - 1$, and the Dynkin diagram is

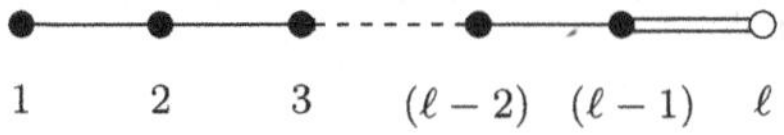

Its algebra is called the Lie algebra C_ℓ. If $n = m = 2$, the Dynkin diagram is

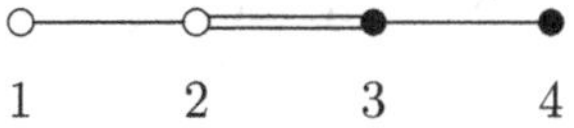

Its algebra is called the Lie algebra F_4.

5. *The Dynkin diagrams with a bifurcation*

The general form of the Dynkin diagrams with a bifurcation is

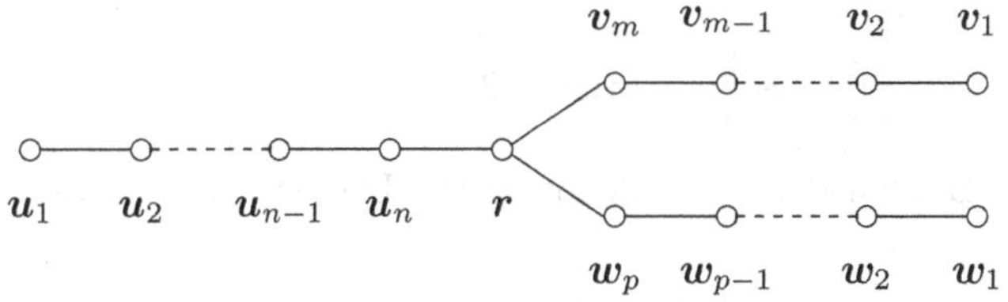

The lengths of all simple roots are the same and denoted by v. Let

$$\boldsymbol{u} = \sum_{j=1}^{n} j\boldsymbol{u}_j, \qquad \boldsymbol{v} = \sum_{k=1}^{m} k\boldsymbol{v}_k, \qquad \boldsymbol{w} = \sum_{s=1}^{p} s\boldsymbol{w}_s.$$

They are orthogonal to each other and linearly independent of $\boldsymbol{r}$, so that

$$v^2 = |\boldsymbol{r}|^2 > \frac{(\boldsymbol{r}\cdot\boldsymbol{u})^2}{|\boldsymbol{u}|^2} + \frac{(\boldsymbol{r}\cdot\boldsymbol{v})^2}{|\boldsymbol{v}|^2} + \frac{(\boldsymbol{r}\cdot\boldsymbol{w})^2}{|\boldsymbol{w}|^2}.$$

Due to Eq. (7.59),

$$\frac{(\boldsymbol{r}\cdot\boldsymbol{u})^2}{|\boldsymbol{u}|^2} = \frac{n^2v^4/4}{n(n+1)v^2/2} = \frac{nv^2}{2(n+1)} = \frac{v^2}{2} - \frac{v^2}{2(n+1)},$$
$$\frac{3}{2} - \frac{1}{2}\left(\frac{1}{n+1} + \frac{1}{m+1} + \frac{1}{p+1}\right) < 1,$$
$$\frac{1}{n+1} + \frac{1}{m+1} + \frac{1}{p+1} > 1.$$

Without loss of generality, letting $p \le m \le n$, and replacing n and m with p, one has

$$\frac{3}{p+1} > 1, \qquad \text{then,}\ \ p = 1.$$

Replacing n with m, one has

$$\frac{2}{m+1} > \frac{1}{2}, \qquad \text{then,}\ \ m = 1\ \ \text{or}\ \ 2.$$

If $m = p = 1$, n can be chosen arbitrarily. Letting $n = \ell - 3$, one obtains the Dynkin diagram

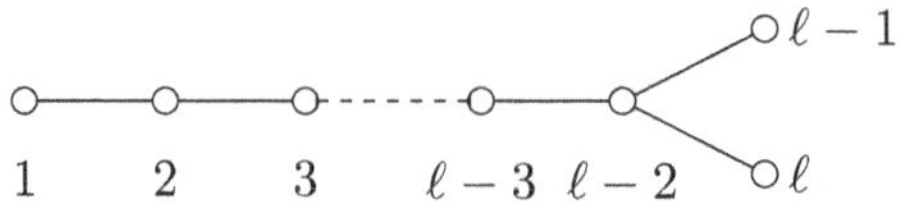

This algebra is called the Lie algebra D_ℓ. If $p = 1$ and $m = 2$, one has

$$\frac{1}{n+1} > \frac{1}{6}, \qquad \text{then,}\ \ 2 \le n < 5.$$

There are three Dynkin diagrams, denoting the Lie algebras E_6, E_7, and E_8,

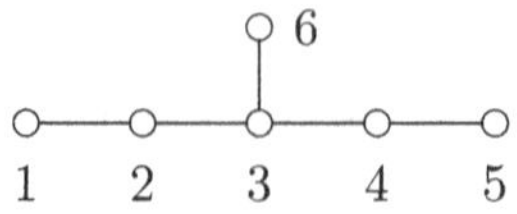

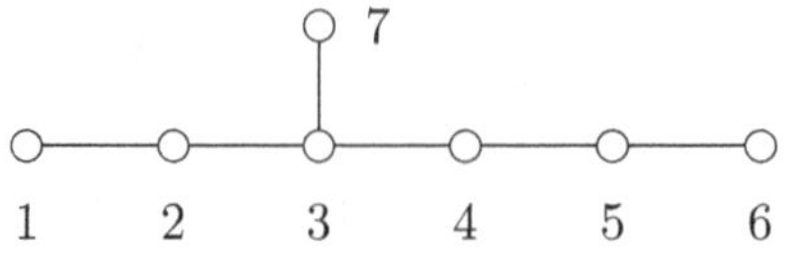

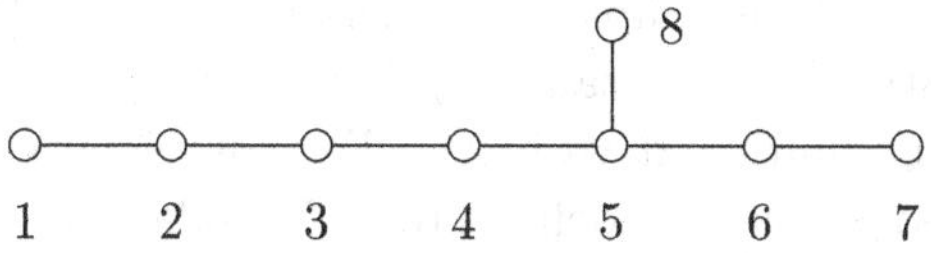

6. *The Dynkin diagrams with only the single links denote the Lie algebras A_ℓ.*

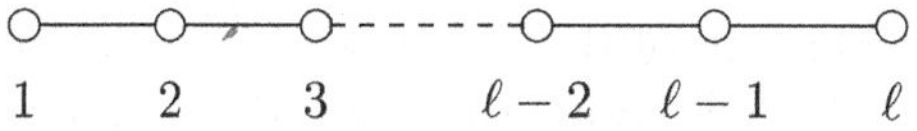

In summary, for the simple Lie algebras with dimension larger than 1, there are four sets of the classical Lie algebras A_ℓ, B_ℓ, C_ℓ, and D_ℓ and five exceptional Lie algebras G_2, F_4, E_6, E_7, and E_8. Their Dynkin diagrams are listed in Fig. 7.1.

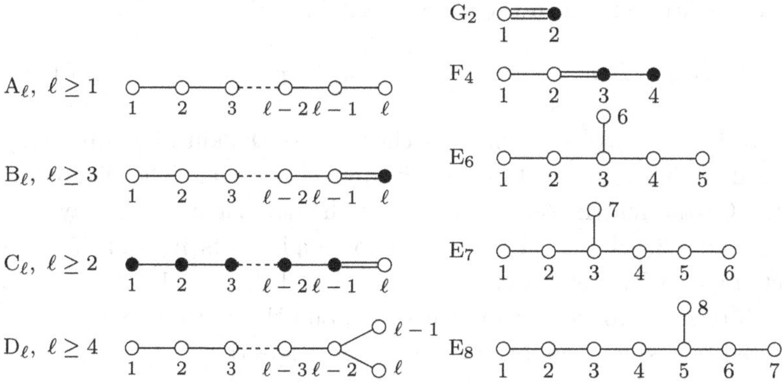

Fig. 7.1 The Dynkin diagrams of simple Lie algebras

In the next section we are going to show that the Lie algebra of the SU($\ell+1$) group is A_ℓ, that of SO($2\ell+1$) is B_ℓ, that of SO(2ℓ) is D_ℓ, and that of USp(2ℓ) is C_ℓ. From Fig. 7.1 one finds that some Dynkin diagrams are the same such that the corresponding Lie groups are locally isomorphic,

$$\begin{array}{ll} B_1 \approx A_1 \approx C_1, & SO(3) \sim SU(2) \approx USp(2), \\ B_2 \approx C_2, & SO(5) \sim USp(4), \\ D_2 \approx A_1 \oplus A_1, & SO(4) \sim SU(2) \otimes SU(2)', \\ D_3 \approx A_3 & SO(6) \sim SU(4). \end{array} \tag{7.60}$$

If all generators in a given simple Lie algebra are multiplied with a common factor λ, the structure constants ${C_{AB}}^D$ are multiplied with λ and the Killing form g_{AB} is multiplied with λ^2. For convenience, the inner product (7.38) in the root space is re-defined to be Euclidean again, namely, the metric tensor in the root space is δ_{AB} instead of $-g_{AB}$. Denote by d_μ the half of square length of a simple root $\boldsymbol{r}_\mu$,

$$d_\mu = \frac{1}{2}\, \boldsymbol{r}_\mu \cdot \boldsymbol{r}_\mu. \tag{7.61}$$

Usually in mathematics, the half of the square length of the longer simple root is normalized to be $d_\mu = 1$. But in physics, the length of each root in A_ℓ is normalized to be $d_\mu = 1/2$, namely the generators are multiplied with a factor $1/\sqrt{2}$ (see next section).

7.3.3 *The Cartan Matrix*

For a simple Lie algebra $\mathcal{L}$ of rank ℓ, there are ℓ simple roots $\boldsymbol{r}_\mu$. Define an ℓ-dimensional matrix A, called the Cartan matrix of $\mathcal{L}$,

$$A_{\mu\nu} = \Gamma\left(\boldsymbol{r}_\nu/\boldsymbol{r}_\mu\right) = \frac{2\boldsymbol{r}_\nu \cdot \boldsymbol{r}_\mu}{|\boldsymbol{r}_\mu|^2} = d_\mu^{-1}\, \left(\boldsymbol{r}_\nu \cdot \boldsymbol{r}_\mu\right). \tag{7.62}$$

When the order of simple roots in $\mathcal{L}$ is chosen, the Dynkin diagram of $\mathcal{L}$ completely determines its Cartan matrix, and vice versa. The diagonal entry in the Cartan matrix A is always 2 and the nondiagonal one may be 0, -1, -2, and -3. $A_{\mu\nu} = A_{\nu\mu} = 0$ if two simple roots $\boldsymbol{r}_\mu$ and $\boldsymbol{r}_\nu$ are disconnected in the Dynkin diagram. $A_{\mu\nu} = -1$ and $A_{\nu\mu} = -1, -2$, or -3, if the length of $\boldsymbol{r}_\mu$ is not less than that of $\boldsymbol{r}_\nu$, and those two simple roots are connected by a single, double, or triple link, respectively.

7.4 Classical Simple Lie Algebras

7.4.1 *The SU(N) Group and its Lie Algebra*

The set of all $N \times N$ unimodular unitary matrices u,

$$u^\dagger u = \mathbf{1}, \qquad \det u = 1, \tag{7.63}$$

in the multiplication rule of matrices, constitutes a group, called the N-dimensional special unitary matrix group, denoted by SU(N). An N-dimensional complex matrix contains $2N^2$ real parameters. The column matrices of a unitary matrix are normal and orthogonal to each other.

There are N real constraints for the normalization and $N(N-1)$ real constraints for the orthogonality. One constraint comes from the determinant. Thus, the number of independent real parameters needed for characterizing the elements of SU(N) is $g = 2N^2 - N - N(N-1) - 1 = N^2 - 1$. The group space is a connected closed region so that SU(N) is a simply-connected compact Lie group with order $g = N^2 - 1$.

Any element u of SU(N) can be diagonalized through a unitary similarity transformation X,

$$\begin{aligned} & u = X \exp\{-i\Phi\} X^{-1} = \exp(-iH), \qquad H = X\Phi X^{-1}, \\ & \Phi = \text{diag}\{\varphi_1, \ \varphi_2, \ \ldots, \ \varphi_N\}, \end{aligned} \tag{7.64}$$

where the diagonal entry of $X^{-1}uX$ is written as $\exp(-i\varphi_a)$. The phase angle φ_a is determined up to a multiple of 2π,

$$\varphi_N = -\sum_{a=1}^{N-1} \varphi_a, \qquad -\pi \leq \varphi_a \leq \pi, \qquad 1 \leq a \leq (N-1), \tag{7.65}$$

where the sum of the phase angles is chosen to be 0 owing to det $u = 1$.

Due to $X \in$ SU(N), Eq. (7.64) shows that u is conjugate to $\exp(-i\Phi)$ which is diagonal. Thus, the classes of SU(N) is characterized by $(N-1)$ parameters φ_a given in Eq. (7.65). The integral on the classes of SU(N) is proved to be

$$\begin{aligned} & \int (d\varphi) W(\varphi) F(\varphi) = \int_{-\pi}^{\pi} d\varphi_1 \ldots \int_{-\pi}^{\pi} d\varphi_{N-1} \int d\varphi_N \delta\left(\sum_{a=1}^{N} \varphi_a\right) W(\varphi) F(\varphi), \\ & W(\varphi) = \frac{1}{\Omega} \prod_{a<b}^{N} \sin^2\left(\frac{\varphi_a - \varphi_b}{2}\right), \\ & \Omega = \int_{-\pi}^{\pi} d\varphi_1 \ldots \int_{-\pi}^{\pi} d\varphi_{N-1} \int d\varphi_N \delta\left(\sum_{a=1}^{N} \varphi_a\right) W(\varphi). \end{aligned} \tag{7.66}$$

The Hermitian traceless matrix H in Eq. (7.64) characterizes the element u of SU(N) completely. H can be expanded with respect to the Hermitian traceless basis matrices. In physics, the basis matrices of N-dimensions are divided into three types, generalized from $\sigma_a/2$, respectively,

$$\begin{aligned} & \left(T_{ab}^{(1)}\right)_{cd} = \frac{1}{2}\left(\delta_{ac}\delta_{bd} + \delta_{ad}\delta_{bc}\right), \\ & \left(T_{ab}^{(2)}\right)_{cd} = -\frac{i}{2}\left(\delta_{ac}\delta_{bd} - \delta_{ad}\delta_{bc}\right), \end{aligned} \qquad 1 \leq a < b \leq N,$$

$$\left(T_a^{(3)}\right)_{cd} = \begin{cases} \delta_{cd}\left[2a(a-1)\right]^{-1/2}, & \text{when } c < a, \\ -\delta_{cd}\left[(a-1)/(2a)\right]^{1/2}, & \text{when } c = a, \\ 0, & \text{when } c > a, \end{cases} \qquad 2 \le a \le N, \tag{7.67}$$

where the subscripts a and b are the ordinal indices of the generator, while c and d are the row and column indices of the matrix. The matrix $T_{ab}^{(1)}$ is symmetric with respect to both ab and cd, but $T_{ab}^{(2)}$ is antisymmetric. $T_a^{(3)}$ is diagonal.

$$\begin{aligned} \left(T_{ab}^{(1)}\right)_{cd} &= \left(T_{ba}^{(1)}\right)_{cd} = \left(T_{ab}^{(1)}\right)_{dc}, \\ \left(T_{ab}^{(2)}\right)_{cd} &= -\left(T_{ba}^{(2)}\right)_{cd} = -\left(T_{ab}^{(1)}\right)_{dc}, \end{aligned} \tag{7.68}$$

$$\begin{aligned} T_2^{(3)} &= \text{diag}\left\{1,\ -1,\ 0,\ \ldots, 0\right\}/2, \\ T_3^{(3)} &= \text{diag}\left\{1,\ 1,\ -2,\ 0,\ \ldots, 0\right\}/(2\sqrt{3}), \\ T_4^{(3)} &= \text{diag}\left\{1,\ 1,\ 1,\ -3,\ 0,\ \ldots, 0\right\}/(2\sqrt{6}), \\ T_a^{(3)} &= \text{diag}\left\{1,\ \ldots,\ 1,\ -(a-1),\ 0,\ \ldots, 0\right\}/\sqrt{2a(a-1)}. \end{aligned} \tag{7.69}$$

There are $N(N-1)/2$ basis matrices $T_{ab}^{(1)}$, $N(N-1)/2$ basis matrices $T_{ab}^{(2)}$, and $(N-1)$ basis matrices $T_a^{(3)}$. Altogether, the number of basis matrices is (N^2-1).

Any Hermitian traceless matrix of N dimensions can be expanded with respect to three types of basis matrices where the coefficients are real. The expanded coefficients of H are denoted by $\omega_{ab}^{(1)}$, $\omega_{ab}^{(2)}$, and $\omega_a^{(3)}$, which are the parameters of the SU(N) group,

$$H = \sum_{a<b} \left\{\omega_{ab}^{(1)} T_{ab}^{(1)} + \omega_{ab}^{(2)} T_{ab}^{(2)}\right\} + \sum_{a=2}^{N} \omega_a^{(3)} T_a^{(3)}, \qquad u = \exp\left(-iH\right). \tag{7.70}$$

Three types of basis matrices $T_{ab}^{(1)}$, $T_{ab}^{(2)}$, and $T_a^{(3)}$ are the generators in the self-representation of SU(N). Usually, the generators in three types are enumerated uniformly in the following order:

$$\begin{array}{llll} T_1 = T_{12}^{(1)}, & T_2 = T_{12}^{(2)}, & T_3 = T_2^{(3)}, & T_4 = T_{13}^{(1)}, \\ T_5 = T_{13}^{(2)}, & T_6 = T_{23}^{(1)}, & T_7 = T_{23}^{(2)}, & T_8 = T_3^{(3)}, \\ T_9 = T_{14}^{(1)}, & T_{10} = T_{14}^{(2)}, & T_{11} = T_{24}^{(1)}, & T_{12} = T_{24}^{(2)}, \\ T_{13} = T_{34}^{(1)}, & T_{14} = T_{34}^{(2)}, & T_{15} = T_4^{(3)}, & \ldots \end{array} \tag{7.71}$$

The generators T_A satisfy the orthonormal condition:

$$\text{Tr}\,(T_A T_B) = \frac{1}{2}\,\delta_{AB}, \qquad A,\ B \leq (N^2-1). \tag{7.72}$$

The orthonormal condition (7.72) guarantees that the structure constant ${C_{AB}}^D$ is totally antisymmetric with respect to its three indices. In fact, multiplying the commutative relation (7.5) with T_C and taking the trace, one obtains

$${C_{AB}}^C = -2i\text{Tr}\left\{T_A T_B T_C - T_B T_A T_C\right\}. \tag{7.73}$$

It shows that the SU((N) group is a compact Lie group. In physical literatures, the antisymmetric structure constants ${C_{AB}}^D$ are usually denoted by f_{ABD}. In order to write the commutative relations of generators in the self-representation in a unified form, one introduces the diagonal matrices

$$\begin{aligned}
&\left(T^{(1)}_{aa}\right)_{cd} = \delta_{ac}\delta_{cd},\\
&T^{(3)}_a = \left(\frac{1}{2a(a-1)}\right)^{1/2}\left\{\sum_{b=1}^{a-1} T^{(1)}_{bb} - (a-1)T^{(1)}_{aa}\right\},\\
&T^{(1)}_{11} - T^{(1)}_{aa} = \sum_{b=2}^{a-1}\left(\frac{2}{b(b-1)}\right)^{1/2} T^{(3)}_b + \left(\frac{2a}{a-1}\right)^{1/2} T^{(3)}_a.
\end{aligned} \tag{7.74}$$

Through a straightforward calculation, one obtains the commutative relations of generators in the self-representation as follows,

$$\begin{aligned}
&\left[T^{(1)}_{ab},\ T^{(1)}_{cd}\right] = \frac{i}{2}\left(\delta_{bc}T^{(2)}_{ad} + \delta_{ad}T^{(2)}_{bc} + \delta_{ac}T^{(2)}_{bd} + \delta_{bd}T^{(2)}_{ac}\right),\\
&\left[T^{(2)}_{ab},\ T^{(2)}_{cd}\right] = \frac{-i}{2}\left(\delta_{bc}T^{(2)}_{ad} + \delta_{ad}T^{(2)}_{bc} - \delta_{ac}T^{(2)}_{bd} - \delta_{bd}T^{(2)}_{ac}\right),\\
&\left[T^{(1)}_{ab},\ T^{(2)}_{cd}\right] = \frac{-i}{2}\left(\delta_{bc}T^{(1)}_{ad} - \delta_{ad}T^{(1)}_{bc} + \delta_{ac}T^{(1)}_{bd} - \delta_{bd}T^{(1)}_{ac}\right),\\
&\left[T^{(3)}_a,\ T^{(1)}_{ab}\right] = -i\left\{(a-1)/(2a)\right\}^{1/2} T^{(2)}_{ab},\\
&\left[T^{(3)}_a,\ T^{(2)}_{ab}\right] = i\left\{(a-1)/(2a)\right\}^{1/2} T^{(1)}_{ab},\\
&\left[T^{(3)}_c,\ T^{(1)}_{ab}\right] = i\left\{2c(c-1)\right\}^{-1/2} T^{(2)}_{ab},\\
&\left[T^{(3)}_c,\ T^{(2)}_{ab}\right] = -i\left\{2c(c-1)\right\}^{-1/2} T^{(1)}_{ab}, \qquad\qquad a < c < b.\\
&\left[T^{(3)}_b,\ T^{(1)}_{ab}\right] = i\left\{b/(2b-2)\right\}^{1/2} T^{(2)}_{ab},\\
&\left[T^{(3)}_b,\ T^{(2)}_{ab}\right] = -i\left\{b/(2b-2)\right\}^{1/2} T^{(1)}_{ab},\\
&\left[T^{(3)}_a,\ T^{(3)}_b\right] = \left[T^{(3)}_b,\ T^{(1)}_{ac}\right] = \left[T^{(3)}_a,\ T^{(1)}_{cb}\right]\\
&\qquad = \left[T^{(3)}_b,\ T^{(2)}_{ac}\right] = \left[T^{(3)}_a,\ T^{(2)}_{cb}\right] = 0,
\end{aligned} \tag{7.75}$$

Note that from the second formula in Eq. (7.75), the set of generators $T_{ab}^{(2)}$ spans a subalgebra in the Lie algebra of SU(N).

There are N constant matrices in the SU(N) group,

$$T_m = \omega^m \mathbf{1}, \qquad \omega = \exp\{-i2\pi/N\}, \qquad 0 \leq m \leq (N-1). \tag{7.76}$$

They are commutable with any element in SU(N) and forms the center of SU(N), denoted by Z_N. Z_N is an Abelian invariant subgroup of SU(N). The group space of the quotient group SU(N)$/Z_N$ is connected with the degree of continuity N, and SU(N) is its covering group.

In order to combine the generators in the self-representation of SU(N) to be the Cartan–Weyl bases, one chooses $(N-1)$ commutable generators $T_a^{(3)}$ to span the Cartan subalgebra. In the mathematics convention,

$$H_j = \sqrt{2}\, T_{N-j+1}^{(3)}, \qquad j = 1,\ 2,\ \ldots,\ \ell, \qquad \ell \equiv N-1. \tag{7.77}$$

The factor $\sqrt{2}$ is introduced to make $d_\mu = 1$ for the simple roots. The normalized common eigenvectors of H_j are

$$E_{\pm\boldsymbol{\alpha}_{ab}} = T_{ab}^{(1)} \pm iT_{ab}^{(2)}, \qquad a < b,$$

$$[H_j,\ E_{\boldsymbol{\alpha}_{ab}}] = \begin{cases} -\left[(N-j)/(N-j+1)\right]^{1/2} E_{\boldsymbol{\alpha}_{ab}}, & N-j+1 = a, \\ \left[(N-j+1)(N-j)\right]^{-1/2} E_{\boldsymbol{\alpha}_{ab}}, & a < N-j+1 < b, \\ \left[(N-j+1)/(N-j)\right]^{1/2} E_{\boldsymbol{\alpha}_{ab}}, & N-j+1 = b, \\ 0, & \text{the remaining cases.} \end{cases} \tag{7.78}$$

Introduce $\ell+1$ vectors $\boldsymbol{V}_a$, distributing equally in an ℓ-dimensional space,

$$\begin{aligned} &(\boldsymbol{V}_a)_j = \left(T_{N-j+1}^{(3)}\right)_{aa} = \sqrt{1/2}\,(H_j)_{aa}, \qquad 1 \leq j \leq \ell, \\ &\sum_{a=1}^{\ell+1} \boldsymbol{V}_a = 0, \qquad \boldsymbol{V}_a \cdot \boldsymbol{V}_b = \frac{\delta_{ab}}{2} - \frac{1}{2(\ell+1)}, \qquad 1 \leq a \leq \ell+1, \\ &\sqrt{2}\,(\boldsymbol{V}_a)_j = \begin{cases} \left[(N-j+1)(N-j)\right]^{-1/2}, & a < N-j+1, \\ -\left[(N-j)/(N-j+1)\right]^{1/2}, & a = N-j+1, \\ 0, & a > N-j+1, \end{cases} \end{aligned} \tag{7.79}$$

$$\begin{aligned} \sqrt{2}\boldsymbol{V}_1 &= \left\{\sqrt{\frac{1}{(\ell+1)\ell}}, \sqrt{\frac{1}{\ell(\ell-1)}}, \ldots, \sqrt{\frac{1}{6}}, \sqrt{\frac{1}{2}}\right\}, \\ \sqrt{2}\boldsymbol{V}_a &= \left\{\sqrt{\frac{1}{(\ell+1)\ell}}, \sqrt{\frac{1}{\ell(\ell-1)}}, \ldots, \sqrt{\frac{1}{(a+1)a}}, -\sqrt{\frac{a-1}{a}},\ 0, \ldots,\ 0\right\}. \end{aligned}$$

Thus, the positive root $\boldsymbol{\alpha}_{ab}$ for the generator $E_{\boldsymbol{\alpha}_{ab}}$ is

$$\boldsymbol{\alpha}_{ab} = \sqrt{2}\left[\boldsymbol{V}_a - \boldsymbol{V}_b\right] = \sum_{\mu=a}^{b-1} \boldsymbol{r}_\mu, \qquad a < b, \tag{7.80}$$

where $\boldsymbol{r}_\mu$ are the simple roots of SU(N)

$$\boldsymbol{r}_\mu = \sqrt{2}\left[\boldsymbol{V}_\mu - \boldsymbol{V}_{\mu+1}\right], \qquad 1 \leq \mu \leq \ell. \tag{7.81}$$

From Eq. (7.79), the inner product of two simple roots is

$$\boldsymbol{r}_\mu \cdot \boldsymbol{r}_\nu = 2\delta_{\mu\nu} - \delta_{\mu(\nu-1)} - \delta_{\mu(\nu+1)}, \qquad d_\mu = 1. \tag{7.82}$$

Therefore, the lengths of the simple roots of SU(N) are the same and the angle of the neighbored simple roots is $2\pi/3$, namely, the Lie algebra of SU($\ell + 1$) is A_ℓ. The largest root in A_ℓ is

$$\boldsymbol{\omega} = \sqrt{2}\left[\boldsymbol{V}_1 - \boldsymbol{V}_{\ell+1}\right] = \sum_{\mu=1}^{\ell} \boldsymbol{r}_\mu. \tag{7.83}$$

7.4.2 *The SO(N) Group and its Lie Algebra*

The set of all $N \times N$ real orthogonal matrices R,

$$R^T R = \mathbf{1}, \qquad R^* = R, \tag{7.84}$$

in the multiplication rule of matrices, constitutes a group, called the N-dimensional real orthogonal matrix group, denoted by O(N). From Eq. (7.84) one has det $R = \pm 1$. Thus, O(N) is a mixed Lie group. The subset of elements with det $R = 1$ forms an invariant subgroup of O(N), denoted by SO(N). An N-dimensional real matrix contains N^2 real parameters. The column matrices of a real orthogonal matrix are normal and orthogonal to each other. There are N real constraints for the normalization and $N(N-1)/2$ real constraints for the orthogonality. The constraint on the determinant is a discontinuous condition, which does not decrease the parameters. Thus, the number of independent real parameters needed for characterizing the elements of SO(N) is $g = N^2 - N - N(N-1)/2 = N(N-1)/2$. The group space is a doubly-connected closed region so that SO(N) is a doubly-connected compact Lie group with order $g = N(N-1)/2$.

SO(N) is a subgroup of SU(N). The elements of the subgroup SO(N) can be obtained by taking the parameters $\omega_{ab}^{(1)}$ and $\omega_a^{(3)}$ of SU(N) to be vanishing. In the convention, one takes $\omega_{ab} = \omega_{ab}^{(2)}/2$ and $T_{ab} = 2T_{ab}^{(2)}$ for SO(N),

$$\begin{aligned} &R = \exp\left\{-i \sum_{a<b=2}^{N} \omega_{ab} T_{ab}\right\}, \\ &(T_{ab})_{cd} = -i\left(\delta_{ac}\delta_{bd} - \delta_{ad}\delta_{bc}\right), \\ &[T_{ab},\ T_{cd}] = -i\left\{\delta_{bc}T_{ad} + \delta_{ad}T_{bc} - \delta_{bd}T_{ac} - \delta_{ac}T_{bd}\right\}. \end{aligned} \tag{7.85}$$

The generators in the self-representation of SO(N) satisfy the orthonormal condition so that the structure constant is totally antisymmetric with respect to three indices,

$$\mathrm{Tr}\left(T_{ab}T_{cd}\right) = 2\left(\delta_{ac}\delta_{bd} - \delta_{ad}\delta_{bc}\right). \tag{7.86}$$

Two generators T_{ab} and T_{cd} of SO(N) are commutable with each other if their subscripts are all different. The commutable generators

$$H_j = T_{(2j-1)(2j)}, \qquad 1 \leq j \leq \ell, \tag{7.87}$$

span the Cartan subalgebra of both the groups SO(2ℓ) and SO($2\ell+1$). Combining the remaining generators to be the common eigenvectors $E_{\boldsymbol{\alpha}}$ of H_j, $[H_j,\ E_{\boldsymbol{\alpha}}] = \alpha_j E_{\boldsymbol{\alpha}}$, one obtains four types of generators $E_{\boldsymbol{\alpha}}$ for SO(2ℓ)

$$\begin{aligned} E_{ab}^{(1)} &= \frac{1}{2}\left[T_{(2a)(2b-1)} - iT_{(2a-1)(2b-1)} - iT_{(2a)(2b)} - T_{(2a-1)(2b)}\right], \\ E_{ab}^{(2)} &= \frac{1}{2}\left[T_{(2a)(2b-1)} + iT_{(2a-1)(2b-1)} + iT_{(2a)(2b)} - T_{(2a-1)(2b)}\right], \\ E_{ab}^{(3)} &= \frac{1}{2}\left[T_{(2a)(2b-1)} - iT_{(2a-1)(2b-1)} + iT_{(2a)(2b)} + T_{(2a-1)(2b)}\right], \\ E_{ab}^{(4)} &= \frac{1}{2}\left[T_{(2a)(2b-1)} + iT_{(2a-1)(2b-1)} - iT_{(2a)(2b)} + T_{(2a-1)(2b)}\right], \end{aligned} \qquad a < b, \tag{7.88}$$

with the eigenvalues (roots) $\{\boldsymbol{e}_a - \boldsymbol{e}_b\}_j$, $\{-\boldsymbol{e}_a + \boldsymbol{e}_b\}_j$, $\{\boldsymbol{e}_a + \boldsymbol{e}_b\}_j$, and $\{-\boldsymbol{e}_a - \boldsymbol{e}_b\}_j$, respectively, where $\{\boldsymbol{e}_a\}_j = \delta_{aj}$. Two more types of generators for SO($2\ell+1$) are

$$\begin{aligned} E_a^{(5)} &= \sqrt{\frac{1}{2}}\left[T_{(2a)(2\ell+1)} - iT_{(2a-1)(2\ell+1)}\right], \\ E_a^{(6)} &= \sqrt{\frac{1}{2}}\left[T_{(2a)(2\ell+1)} + iT_{(2a-1)(2\ell+1)}\right], \end{aligned} \tag{7.89}$$

with the eigenvalues $\{\boldsymbol{e}_a\}_j$ and $-\{\boldsymbol{e}_a\}_j$. The generators labelled by (1), (3), and (5) correspond to the positive roots.

The simple roots $\boldsymbol{r}_\mu$ and the largest root $\boldsymbol{\omega}$ for SO(2ℓ) are

$$\begin{aligned} &\boldsymbol{r}_\mu = \boldsymbol{e}_\mu - \boldsymbol{e}_{\mu+1}, \qquad \boldsymbol{r}_\ell = \boldsymbol{e}_{\ell-1} + \boldsymbol{e}_\ell, \qquad 1 \leq \mu \leq \ell - 1, \\ &\boldsymbol{e}_a - \boldsymbol{e}_b = \sum_{\mu=a}^{b-1} \boldsymbol{r}_\mu, \qquad \boldsymbol{e}_a + \boldsymbol{e}_b = \sum_{\mu=a}^{b-1} \boldsymbol{r}_\mu + 2\sum_{\nu=b}^{\ell-2} \boldsymbol{r}_\nu + \boldsymbol{r}_{\ell-1} + \boldsymbol{r}_\ell, \end{aligned} \tag{7.90}$$

$$\boldsymbol{\omega} = \boldsymbol{e}_1 + \boldsymbol{e}_2 = \boldsymbol{r}_1 + 2\sum_{\nu=2}^{\ell-2} \boldsymbol{r}_\nu + \boldsymbol{r}_{\ell-1} + \boldsymbol{r}_\ell. \tag{7.91}$$

The angle of two neighbored roots is $2\pi/3$ except for $\boldsymbol{r}_\ell$, which is orthogonal to $\boldsymbol{r}_{\ell-1}$, but with an angle $2\pi/3$ to $\boldsymbol{r}_{\ell-2}$. $d_\mu = 1$ for all simple roots. Thus, the algebra of SO(2ℓ) is D_ℓ.

The simple roots $\boldsymbol{r}_\mu$ and the largest root $\boldsymbol{\omega}$ for SO($2\ell+1$) are

$$\begin{aligned} &\boldsymbol{r}_\mu = \boldsymbol{e}_\mu - \boldsymbol{e}_{\mu+1}, \qquad \boldsymbol{r}_\ell = \boldsymbol{e}_\ell, \qquad 1 \leq \mu \leq \ell - 1, \\ &\boldsymbol{e}_a - \boldsymbol{e}_b = \sum_{\mu=a}^{b-1} \boldsymbol{r}_\mu, \quad \boldsymbol{e}_a + \boldsymbol{e}_b = \sum_{\mu=a}^{b-1} \boldsymbol{r}_\mu + 2\sum_{\nu=b}^{\ell} \boldsymbol{r}_\nu, \quad \boldsymbol{e}_a = \sum_{\mu=a}^{\ell} \boldsymbol{r}_\mu, \end{aligned} \tag{7.92}$$

$$\boldsymbol{\omega} = \boldsymbol{e}_1 + \boldsymbol{e}_2 = \boldsymbol{r}_1 + 2\sum_{\nu=2}^{\ell} \boldsymbol{r}_\nu. \tag{7.93}$$

Except for $\boldsymbol{r}_\ell$, the angle of two neighbored roots is $2\pi/3$ and $d_\mu = 1$. But, $d_\ell = 1/2$ and the angle between $\boldsymbol{r}_\ell$ and $\boldsymbol{r}_{\ell-1}$ is $3\pi/4$. Thus, the algebra of SO($2\ell+1$) is B_ℓ.

7.4.3 *The* USp(2ℓ) *Group and its Lie Algebra*

In a (2ℓ)-dimensional space, the vector index a is taken to be j or $\overline{j}$, $1 \leq j \leq \ell$, in the following order:

$$a = 1,\ \overline{1},\ 2,\ \overline{2},\ \ldots,\ \ell,\ \overline{\ell}. \tag{7.94}$$

Generalizing the real orthogonal matrix by replacing the unit matrix with a real antisymmetric matrix J, one obtains the real pseudo-orthogonal matrix R,

$$R^T J R = J, \qquad R^* = R, \tag{7.95}$$

where

$$J_{ab} = \begin{cases} 1 & \text{when } a = j, \quad b = \overline{j}, \\ -1 & \text{when } a = \overline{j}, \quad b = j, \\ 0 & \text{the remaining cases}, \end{cases} \tag{7.96}$$

$$J = \mathbf{1}_\ell \times (i\sigma_2) = -J^{-1} = -J^T, \qquad \det\ J = 1.$$

Study the set of all (2ℓ)-dimensional real pseudo-orthogonal matrices R in the multiplication rule of matrices. The product of two real pseudo-orthogonal matrices is a real pseudo-orthogonal matrix. The matrix product satisfies the associative law. The unit matrix is also a real pseudo-orthogonal matrix and plays the role of the identical element. The inverse $R^{-1} = -JR^TJ$ and the transpose R^T of R are also real pseudo-orthogonal,

$$RJR^T = J, \qquad \left(R^{-1}\right)^T JR^{-1} = (-JRJ)JR^{-1} = J. \tag{7.97}$$

Thus, the set of all (2ℓ)-dimensional real pseudo-orthogonal matrices R, in the multiplication rule of matrices, constitutes a group, called the (2ℓ)-dimensional real symplectic group, denoted by $\mathrm{Sp}(2\ell, R)$.

We are going to show that $\det\ R = 1$ so that $\mathrm{Sp}(2\ell, R)$ is a connected Lie group. Let R be a transformation matrix in a (2ℓ)-dimensional real space

$$x_a \xrightarrow{R} x'_a = \sum_b R_{ab}x_b, \qquad R \in \mathrm{Sp}(2\ell, R). \tag{7.98}$$

The pseudo-inner product of two vectors x and y is defined as

$$\{x,\ y\}_J \equiv \sum_{ab} x_a J_{ab} y_b = \sum_{j=1}^{\ell} \left(x_j y_{\overline{j}} - x_{\overline{j}} y_j\right), \tag{7.99}$$

which is invariant in a real pseudo-orthogonal transformation R

$$\{x,\ y\}_J = \{Rx,\ Ry\}_J\,. \tag{7.100}$$

The pseudo-inner self-product of a vector is vanishing.

In terms of the totally antisymmetric tensor one has

$$\sum_{a_1 \ldots a_{2\ell}} \epsilon_{a_1 \ldots a_{2\ell}} J_{a_1 a_2} \cdots J_{a_{2\ell-1} a_{2\ell}} = 2^\ell \ell!. \tag{7.101}$$

In fact, in the sum there is only $2^\ell \ell!$ nonvanishing terms, which come from the transpose of each J matrix and the permutations among ℓ different J.

Denote by X_a^b the column matrix at the bth column of a $(2\ell) \times (2\ell)$ matrix X. The determinant of X is

$$\epsilon_{a_1 \ldots a_{2\ell}} \det X = \sum_{b_1 \ldots b_{2\ell}} \epsilon_{b_1 \ldots b_{2\ell}} X_{a_1}^{b_1} \ldots X_{a_{2\ell}}^{b_{2\ell}}. \tag{7.102}$$

Thus,

$$\begin{aligned}\det X &= \left(2^\ell \ell!\right)^{-1} \sum_{a_1 \ldots a_{2\ell}} (\det X)\, \epsilon_{a_1 \ldots a_{2\ell}} J_{a_1 a_2} \cdots J_{a_{2\ell-1} a_{2\ell}} \\ &= \left(2^\ell \ell!\right)^{-1} \sum_{a_1 \ldots a_{2\ell}} J_{a_1 a_2} \cdots J_{a_{2\ell-1} a_{2\ell}} \sum_{b_1 \ldots b_{2\ell}} \epsilon_{b_1 \ldots b_{2\ell}} X_{a_1}^{b_1} \ldots X_{a_{2\ell}}^{b_{2\ell}} \\ &= \left(2^\ell \ell!\right)^{-1} \sum_{b_1 \ldots b_{2\ell}} \epsilon_{b_1 \ldots b_{2\ell}} \left\{X^{b_1},\, X^{b_2}\right\}_J \cdots \left\{X^{b_{2\ell-1}},\, X^{b_{2\ell}}\right\}_J .\end{aligned}$$

Since the column matrix in RX is obtained from the column matrix in X by a real pseudo-orthogonal transformation R, one has from Eq. (7.100) that det $(RX) = \det X$ so that det $R = 1$.

From the definition (7.95), R in Sp$(2\ell, R)$ is given as follows if it is a diagonal matrix

$$R = \text{diag}\left\{e^{\omega_1},\, e^{-\omega_1},\, e^{\omega_2},\, e^{-\omega_2},\, \ldots,\, e^{\omega_\ell},\, e^{-\omega_\ell}\right\}, \tag{7.103}$$

where the real parameters ω_j can be taken to be infinitely large such that Sp$(2\ell, R)$ is not a compact Lie group.

Replace the real matrix R in Eq. (7.95) with the unitary matrix u,

$$u^T J u = J, \qquad u^\dagger = u^{-1}. \tag{7.104}$$

The set of all (2ℓ)-dimensional unitary pseudo-orthogonal matrices u, in the multiplication rule of matrices, forms a group, called the (2ℓ)-dimensional unitary symplectic group, denoted by USp(2ℓ). It can be similarly shown that det $u = 1$. Since the module of any entry in a unitary matrix is finite, USp(2ℓ) is a connected compact Lie group. When $\ell = 1$, $J = i\sigma_2$ and $(\boldsymbol{\sigma} \cdot \hat{\boldsymbol{n}})^T (i\sigma_2) = -(i\sigma_2)(\boldsymbol{\sigma} \cdot \hat{\boldsymbol{n}})$. Thus, the elements in SU(2) satisfy the definition (7.104) so that SU(2) = USp(2).

Discuss the infinitesimal elements $R \in$ Sp$(2\ell, R)$ and $u \in$USp(2ℓ)

$$\begin{aligned} &R = \mathbf{1} - i\alpha X, \qquad X^T = JXJ, \qquad X^* = -X, \\ &u = \mathbf{1} - i\beta Y, \qquad Y^T = JYJ, \qquad Y^\dagger = Y. \end{aligned} \tag{7.105}$$

Both the imaginary matrix X and the Hermitian matrix Y contain $(4\ell^2)$ real parameters. The component forms of $X^T = JXJ$ and $Y^T = JYJ$ are

$$X_{kj} = -X_{\overline{j}\overline{k}}, \qquad X_{\overline{k}j} = X_{\overline{j}k}, \qquad X_{k\overline{j}} = X_{j\overline{k}},$$
$$Y_{kj} = -Y_{\overline{j}\overline{k}}, \qquad Y_{\overline{k}j} = Y_{\overline{j}k}.$$

$X^T = JXJ$ gives $\ell^2 + \ell(\ell-1)/2 + \ell(\ell-1)/2 = \ell(2\ell-1)$ real constraints. $Y^T = JYJ$ gives $\ell^2 + \ell(\ell-1) = \ell(2\ell-1)$ real constraints. Thus, the orders of both Sp$(2\ell, R)$ and USp(2ℓ) are $g = 4\ell^2 - \ell(2\ell-1) = \ell(2\ell+1)$.

In the self-representation of USp(2ℓ) the generators can be expressed in terms of the generators $T^{(r)}_{jk}$ $(r = 1, 2)$ in SU(ℓ) and the Pauli matrices σ_a. From Eq. (7.105) they are

$$\begin{aligned} &T^{(2)}_{jk} \times \mathbf{1}_2, \qquad T^{(1)}_{jk} \times \sigma_d, \qquad T^{(1)}_{jj} \times \sigma_d/\sqrt{2}, \\ &1 \le d \le 3, \qquad 1 \le j < k \le \ell. \end{aligned} \tag{7.106}$$

They satisfied the orthonormal condition $\text{Tr}(T_A T_B) = \delta_{AB}$ so that the structure constant $C_{AB}{}^D$ of USp(2ℓ) is totally antisymmetric and the Killing form g_{AB} is a negative constant matrix. These coincide with the fact that USp(2ℓ) is compact. The generators in the self-representation of Sp$(2\ell, R)$ are pure imaginary and can be obtained from Eq. (7.106) by replacing σ_1 and σ_3 with $\tau_1 = i\sigma_1$ and $\tau_3 = i\sigma_3$, respectively. If one moves the additional factor i from the generators of Sp$(2\ell, R)$ to its parameters, the generators in the corresponding representations of Sp$(2\ell, R)$ and USp(2ℓ) are the same. This is the standard method for calculating the irreducible representations of a noncompact Lie group. We will calculate the representations of the Lorentz group by this method in Chap. 9.

The diagonal matrices in the generators (7.106) span the Cartan subalgebra of the USp(2ℓ) group,

$$H_j = T^{(1)}_{jj} \times \sigma_3/\sqrt{2}, \qquad 1 \le j \le \ell. \tag{7.107}$$

The remaining generators are combined to be the eigenvectors of H_j, $[H_j,\ E_{\boldsymbol{\alpha}}] = \alpha_j E_{\boldsymbol{\alpha}}$:

$$\begin{aligned} E^{(1)}_{jk} &= \left\{T^{(1)}_{jk} \times \sigma_3 + iT^{(2)}_{jk} \times \mathbf{1}_2\right\}/\sqrt{2}, \\ E^{(2)}_{jk} &= \left\{T^{(1)}_{jk} \times \sigma_3 - iT^{(2)}_{jk} \times \mathbf{1}_2\right\}/\sqrt{2}, \qquad 1 \le j < k \le \ell, \\ E^{(3)}_{jk} &= T^{(1)}_{jk} \times (\sigma_1 + i\sigma_2)/\sqrt{2}, \\ E^{(4)}_{jk} &= T^{(1)}_{jk} \times (\sigma_1 - i\sigma_2)/\sqrt{2}, \end{aligned}$$

$$\begin{aligned} E_j^{(5)} &= T_{jj}^{(1)} \times (\sigma_1 + i\sigma_2)/2, \\ E_j^{(6)} &= T_{jj}^{(1)} \times (\sigma_1 - i\sigma_2)/2, \end{aligned} \tag{7.108}$$

with the eigenvalues $\sqrt{1/2}\,(\boldsymbol{e}_j - \boldsymbol{e}_k)$, $-\sqrt{1/2}\,(\boldsymbol{e}_j - \boldsymbol{e}_k)$, $\sqrt{1/2}\,(\boldsymbol{e}_j + \boldsymbol{e}_k)$, $-\sqrt{1/2}(\boldsymbol{e}_j + \boldsymbol{e}_k)$, $\sqrt{2}\boldsymbol{e}_j$, and $-\sqrt{2}\boldsymbol{e}_j$, respectively. The positive roots are

$$\begin{aligned} \sqrt{1/2}\,(\boldsymbol{e}_j - \boldsymbol{e}_k) &= \sum_{\mu=j}^{k-1} \boldsymbol{r}_\mu, \\ \sqrt{1/2}\,(\boldsymbol{e}_j + \boldsymbol{e}_k) &= \sum_{\mu=j}^{k-1} \boldsymbol{r}_\mu + 2\sum_{\mu=k}^{\ell-1} \boldsymbol{r}_\mu + \boldsymbol{r}_\ell, \\ \sqrt{2}\boldsymbol{e}_j &= 2\sum_{\mu=j}^{\ell-1} \boldsymbol{r}_\mu + \boldsymbol{r}_\ell, \end{aligned}$$

where $\boldsymbol{r}_\mu$ are the simple roots

$$\begin{aligned} &\boldsymbol{r}_\mu = \sqrt{1/2}\,(\boldsymbol{e}_\mu - \boldsymbol{e}_{\mu+1}), \qquad d_\mu = 1/2, \qquad 1 \leq \mu \leq \ell - 1, \\ &\boldsymbol{r}_\ell = \sqrt{2}\boldsymbol{e}_\ell, \qquad d_\ell = 1. \end{aligned} \tag{7.109}$$

The angle of two neighbored simple roots is $2\pi/3$ except for the angle of $\boldsymbol{r}_{\ell-1}$ and $\boldsymbol{r}_\ell$ which is $3\pi/4$. Thus, the algebra of USp(2ℓ) is C_ℓ. The largest root $\boldsymbol{\omega}$ in C_ℓ is

$$\boldsymbol{\omega} = \sqrt{2}\boldsymbol{e}_1 = 2\sum_{\mu=1}^{\ell-1} \boldsymbol{r}_\mu + \boldsymbol{r}_\ell. \tag{7.110}$$

7.5 Representations of a Simple Lie Algebra

A representation of a Lie group G is also a representation of the Lie algebra $\mathcal{L}$ of G. In a representation of a Lie algebra, one only studies its generators. A simple Lie algebra has its compact real form where its representation is equivalent to a unitary one.

7.5.1 *Representations and Weights*

In a unitary representation of a simple Lie algebra, the Cartan–Weyl bases of generators, denoted by $D(H_j)$ and $D(E_{\boldsymbol{\alpha}})$ for convenience, satisfy

$$D(H_j)^\dagger = D(H_j), \qquad D(E_{\boldsymbol{\alpha}})^\dagger = D(E_{-\boldsymbol{\alpha}}).$$

The basis vectors in the representation space, called the basis states and denoted by $|\boldsymbol{m}\rangle$, are usually chosen to be the common eigenvectors of the commutable Hermitian operators H_j,

$$H_j \; |\boldsymbol{m}\rangle = m_j \; |\boldsymbol{m}\rangle. \tag{7.111}$$

The ℓ-dimensional vector $\boldsymbol{m} = (m_1, \; \ldots, \; m_\ell)$, whose components are the eigenvalues m_j, is called the weight of the basis state $|\boldsymbol{m}\rangle$. The ℓ-dimensional space is called the weight space. A weight $\boldsymbol{m}$ is said to be multiple if the number n of the linearly independent basis states $|\boldsymbol{m}\rangle$ with the weight $\boldsymbol{m}$ in the representation space is larger than 1, and n is called the multiplicity of the weight $\boldsymbol{m}$. A weight is called single if its multiplicity is 1. For the adjoint representation, the representation space is the Lie algebra itself, a weight is a root, and the weight space coincides with the root space.

Except for the identical representation, any irreducible representation of a simple Lie algebra with dimension larger than 1 is faithful and the generators are linearly independent of each other. Thus, there are ℓ linearly independent weights in a representation of $\mathcal{L}$ with rank ℓ. From the regular commutative relations, $[H_j, \; E_{\boldsymbol{\alpha}}] = \alpha_j E_{\boldsymbol{\alpha}}$ and $[E_{\boldsymbol{\alpha}}, \; E_{-\boldsymbol{\alpha}}] = \boldsymbol{\alpha} \cdot \boldsymbol{H}$, one has

$$\text{Tr}\; D(E_{\boldsymbol{\alpha}}) = 0, \qquad \text{Tr}\; D(H_{\boldsymbol{\alpha}}) = 0, \qquad \text{Tr}\; D(H_j) = 0, \tag{7.112}$$

namely, the sum of weights of all basis states in an irreducible representation is vanishing,

$$\sum \boldsymbol{m} = 0. \tag{7.113}$$

A weight $\boldsymbol{m}$ is said to be higher than a weight $\boldsymbol{m}'$ if the first nonvanishing component of $\boldsymbol{m} - \boldsymbol{m}'$ is positive. For a positive root $\boldsymbol{\alpha}$, one has

$$\begin{aligned} H_j \left(E_{\pm\boldsymbol{\alpha}} \; |\boldsymbol{m}\rangle \right) &= \; [H_j, \; E_{\pm\boldsymbol{\alpha}}] \; |\boldsymbol{m}\rangle + E_{\pm\boldsymbol{\alpha}} H_j \; |\boldsymbol{m}\rangle \\ &= \; (m_j \pm \alpha_j) \; \left(E_{\pm\boldsymbol{\alpha}} \; |\boldsymbol{m}\rangle \right). \end{aligned} \tag{7.114}$$

Thus, the action of $E_{\boldsymbol{\alpha}}$ on $|\boldsymbol{m}\rangle$ raises its weight $\boldsymbol{m}$ by a positive root $\boldsymbol{\alpha}$, and that of $E_{-\boldsymbol{\alpha}}$ lowers by $\boldsymbol{\alpha}$, so that $E_{\boldsymbol{\alpha}}$ is called the raising operator and $E_{-\boldsymbol{\alpha}}$ the lowering operator. In a representation space with a finite dimension, there is a basis state $|\boldsymbol{M}\rangle$ whose weight $\boldsymbol{M}$ is higher than the weight of any other basis state $|\boldsymbol{m}\rangle$. $\boldsymbol{M}$ is called the highest weight of the representation and $|\boldsymbol{M}\rangle$ is the highest weight state. Any generators $E_{\boldsymbol{\alpha}}$ with the positive root $\boldsymbol{\alpha}$ annihilates the highest weight state $|\boldsymbol{M}\rangle$,

$$E_{\boldsymbol{\alpha}} \; |\boldsymbol{M}\rangle = 0, \qquad \forall \; \text{positive root} \;\; \boldsymbol{\alpha}. \tag{7.115}$$

This is the main method to calculate the highest weight in a representation.

Theorem 7.7 The highest weight $\boldsymbol{M}$ in an irreducible representation of a simple Lie algebra $\mathcal{L}$ is single. Two irreducible representations of $\mathcal{L}$ are equivalent if and only if their highest weights are the same.

Proof Let $|\boldsymbol{M}\rangle$ be the highest weight state. We are going to prove that if there is another state $|\boldsymbol{M}\rangle'$ with the highest weight $\boldsymbol{M}$, it has to be proportional to $|\boldsymbol{M}\rangle$. Since the representation is irreducible, $|\boldsymbol{M}\rangle'$ can be expressed as a combination of the following states each of which has the highest weight $\boldsymbol{M}$,

$$E_{\boldsymbol{\lambda}} \ldots E_{\boldsymbol{\beta}} E_{\boldsymbol{\alpha}} \, |\boldsymbol{M}\rangle. \tag{7.116}$$

If a raising operator, say $E_{\boldsymbol{\rho}}$, is contained in Eq. (7.116), move $E_{\boldsymbol{\rho}}$ rightward according to $E_{\boldsymbol{\rho}} E_{\boldsymbol{\tau}} = E_{\boldsymbol{\tau}} E_{\boldsymbol{\rho}} + [E_{\boldsymbol{\rho}}, \ E_{\boldsymbol{\tau}}]$. The additional term in the move is $N_{\boldsymbol{\rho},\boldsymbol{\tau}} E_{\boldsymbol{\rho}+\boldsymbol{\tau}}$ if $\boldsymbol{\rho}+\boldsymbol{\tau}$ is a root, $\boldsymbol{\rho} \cdot \boldsymbol{H}$ if $\boldsymbol{\rho}+\boldsymbol{\tau} = 0$, and 0 if $\boldsymbol{\rho}+\boldsymbol{\tau}$ is not a root. In each case, the additional term contains less operators than the original one. When the raising operator $E_{\boldsymbol{\rho}}$ moves to the rightmost position, the term is annihilated owing to Eq. (7.115), and the number of operators contained in each remaining term decreases. Making those moves of the raising operators in each term until it does not contain any raising operator, one finds that in the same time the lowering operators also disappear in the result terms because the weight of the term is $\boldsymbol{M}$. Namely, each term is proportional to $|\boldsymbol{M}\rangle$.

The highest weights of two equivalent representations are obviously the same. Conversely, if the highest weights of two representations are the same, define a correspondence between the basis states of two representations

$$\begin{aligned} |\boldsymbol{M}\rangle \ &\longleftrightarrow \ |\boldsymbol{M}\rangle' \\ |j\rangle \equiv E_{\boldsymbol{\lambda}} \ldots E_{\boldsymbol{\beta}} E_{\boldsymbol{\alpha}} \, |\boldsymbol{M}\rangle \ &\longleftrightarrow \ |j\rangle' \equiv E_{\boldsymbol{\lambda}} \ldots E_{\boldsymbol{\beta}} E_{\boldsymbol{\alpha}} \, |\boldsymbol{M}\rangle'. \end{aligned}$$

We are going to show that the correspondence is one-to-one, namely, any linear relation which holds in one representation space also holds in another representation space, and vice versa,

$$\forall \ \sum_j c_j \, |j\rangle = 0, \qquad \exists \ \sum_j c_j \, |j\rangle' \equiv |w\rangle' = 0.$$

By reduction to absurdity, if $|w\rangle' \neq 0$, it corresponds to the null state in the first representation space, so does the state $|w\rangle'$ applied by the generators. Then, the set of those states constitutes an invariant subspace in the second representation space, which is not equal to the whole space because it does not contain at least $|\boldsymbol{M}\rangle'$. It is in contradiction to that the representation is irreducible. □

Hereafter, an irreducible representation of a simple Lie algebra $\mathcal{L}$ is also called the highest weight representation. The basis states in the adjoint representation space of a simple Lie algebra are the generators, and the weights are the roots, so that the highest weight of the adjoint representation is the largest root $\boldsymbol{\omega}$ of $\mathcal{L}$.

7.5.2 *Weight Chain and Weyl Reflections*

Theorem 7.8 If $\boldsymbol{\alpha}$ is a nonvanishing root in a simple Lie algebra $\mathcal{L}$ and $\boldsymbol{m}$ is a weight in an irreducible representation space of $\mathcal{L}$, then,

$$\frac{2\boldsymbol{m}\cdot\boldsymbol{\alpha}}{\boldsymbol{\alpha}\cdot\boldsymbol{\alpha}} \equiv \Gamma\left(\boldsymbol{m}/\boldsymbol{\alpha}\right) = \text{ integer}, \tag{7.117}$$

and $\boldsymbol{m} - \Gamma\left(\boldsymbol{m}/\boldsymbol{\alpha}\right)\boldsymbol{\alpha}$ is also a weight with the same multiplicity as $\boldsymbol{m}$.

Proof Although Theorem 7.8 looks similar to Theorem 7.3, there is an important difference between them that the weight $\boldsymbol{m}$ is always on the numerator and the multiplicity of a weight may be larger than 1. For a multiple weight, any combination of the basis states with the same weight is also a state in the representation space. Before proving the theorem one has to find a convention for choosing the basis states.

Without loss of generality, let

$$\boldsymbol{m}\cdot\boldsymbol{\alpha} \geq 0. \tag{7.118}$$

Arbitrarily choose a basis state $|\boldsymbol{m}\rangle$ with the weight $\boldsymbol{m}$ and apply to it with $E_{\boldsymbol{\alpha}}$ successively until the state is annihilated,

$$\begin{aligned} &E_{\boldsymbol{\alpha}}^{n}\ |\boldsymbol{m}\rangle \neq 0, && 0 \leq n \leq p, \\ &|\boldsymbol{m}+p\boldsymbol{\alpha}\rangle \equiv E_{\boldsymbol{\alpha}}^{p}\ |\boldsymbol{m}\rangle, && E_{\boldsymbol{\alpha}}^{p+1}\ |\boldsymbol{m}\rangle = 0. \end{aligned}$$

Forget the choice of $|\boldsymbol{m}\rangle$. One defines a chain of basis states by applying $E_{-\boldsymbol{\alpha}}$ to $|\boldsymbol{m}+p\boldsymbol{\alpha}\rangle$ successively,

$$\begin{aligned} &|\boldsymbol{m}+n\boldsymbol{\alpha}\rangle \equiv E_{-\boldsymbol{\alpha}}^{p-n}\ |\boldsymbol{m}+p\boldsymbol{\alpha}\rangle, && -q \leq n \leq p, \\ &E_{-\boldsymbol{\alpha}}^{p+q+1}\ |\boldsymbol{m}+p\boldsymbol{\alpha}\rangle = 0, && p+q+1 > 0, \end{aligned} \tag{7.119}$$

where q is an integer larger than $-(p+1)$. We do not care whether or not the state $E_{-\boldsymbol{\alpha}}^{p}\ |\boldsymbol{m}+p\boldsymbol{\alpha}\rangle$ coincides with the original state $|\boldsymbol{m}\rangle$. The problem is whether the subspace spanned by the basis states $|\boldsymbol{m}+n\boldsymbol{\alpha}\rangle$ is closed in the application of $E_{\pm\boldsymbol{\alpha}}$, namely to show

$$E_{\boldsymbol{\alpha}}\ |\boldsymbol{m}+n\boldsymbol{\alpha}\rangle = B_n\ |\boldsymbol{m}+(n+1)\boldsymbol{\alpha}\rangle, \qquad -q \leq n \leq p. \tag{7.120}$$

Prove it by induction. Equation (7.120) holds for $n=p$ and $n=p-1$,

$$\begin{aligned}
&E_{\boldsymbol{\alpha}}\ |\boldsymbol{m}+p\boldsymbol{\alpha}\rangle = 0,\\
&E_{\boldsymbol{\alpha}}\ |\boldsymbol{m}+(p-1)\boldsymbol{\alpha}\rangle = [E_{\boldsymbol{\alpha}},\ E_{-\boldsymbol{\alpha}}]\ |\boldsymbol{m}+p\boldsymbol{\alpha}\rangle\\
&\qquad = \boldsymbol{\alpha}\cdot\boldsymbol{H}\ |\boldsymbol{m}+p\boldsymbol{\alpha}\rangle = \left(\boldsymbol{m}\cdot\boldsymbol{\alpha}+p|\boldsymbol{\alpha}|^2\right)\ |\boldsymbol{m}+p\boldsymbol{\alpha}\rangle,\\
&B_p = 0, \qquad B_{p-1} = \boldsymbol{m}\cdot\boldsymbol{\alpha}+p|\boldsymbol{\alpha}|^2.
\end{aligned} \tag{7.121}$$

If Eq. (7.120) holds for $n>k$, when $n=k$,

$$\begin{aligned}
E_{\boldsymbol{\alpha}}\ |\boldsymbol{m}+k\boldsymbol{\alpha}\rangle &= E_{\boldsymbol{\alpha}}E_{-\boldsymbol{\alpha}}\ |\boldsymbol{m}+(k+1)\boldsymbol{\alpha}\rangle\\
&= [E_{\boldsymbol{\alpha}},\ E_{-\boldsymbol{\alpha}}]\ |\boldsymbol{m}+(k+1)\boldsymbol{\alpha}\rangle + E_{-\boldsymbol{\alpha}}E_{\boldsymbol{\alpha}}\ |\boldsymbol{m}+(k+1)\boldsymbol{\alpha}\rangle\\
&= \left\{\boldsymbol{m}\cdot\boldsymbol{\alpha}+(k+1)|\boldsymbol{\alpha}|^2+B_{k+1}\right\}\ |\boldsymbol{m}+(k+1)\boldsymbol{\alpha}\rangle.
\end{aligned}$$

Thus, Eq. (7.120) is proved, and B_n satisfies the recursive relation

$$\begin{aligned}
B_n &= B_{n+1}+\boldsymbol{m}\cdot\boldsymbol{\alpha}+(n+1)|\boldsymbol{\alpha}|^2\\
&= B_{n+2}+2\boldsymbol{m}\cdot\boldsymbol{\alpha}+\{(n+1)+(n+2)\}\,|\boldsymbol{\alpha}|^2\\
&= \ldots\\
&= B_{n+(p-n)}+(p-n)\boldsymbol{m}\cdot\boldsymbol{\alpha}+\frac{1}{2}(p-n)(n+p+1)|\boldsymbol{\alpha}|^2\\
&= \frac{1}{2}(p-n)\left\{2\boldsymbol{m}\cdot\boldsymbol{\alpha}+(n+p+1)|\boldsymbol{\alpha}|^2\right\},\\
B_{-q} &= \frac{1}{2}(p+q)\left\{2\boldsymbol{m}\cdot\boldsymbol{\alpha}+(-q+p+1)|\boldsymbol{\alpha}|^2\right\}.
\end{aligned}$$

On the other hand,

$$\begin{aligned}
0 &= E_{\boldsymbol{\alpha}}E_{-\boldsymbol{\alpha}}\ |\boldsymbol{m}-q\boldsymbol{\alpha}\rangle\\
&= \{[E_{\boldsymbol{\alpha}},\ E_{-\boldsymbol{\alpha}}]+E_{-\boldsymbol{\alpha}}E_{\boldsymbol{\alpha}}\}\ |\boldsymbol{m}-q\boldsymbol{\alpha}\rangle\\
&= \left\{\boldsymbol{m}\cdot\boldsymbol{\alpha}-q|\boldsymbol{\alpha}|^2+B_{-q}\right\}\ |\boldsymbol{m}-q\boldsymbol{\alpha}\rangle,\\
B_{-q} &= -\boldsymbol{m}\cdot\boldsymbol{\alpha}+q|\boldsymbol{\alpha}|^2.
\end{aligned}$$

In comparison one obtains

$$(p+q+1)\left\{2\boldsymbol{m}\cdot\boldsymbol{\alpha}-(q-p)|\boldsymbol{\alpha}|^2\right\} = 0.$$

Since $(p+q+1)>0$ and $|\boldsymbol{\alpha}|^2>0$, Eq. (7.117) is proved and the integer is calculated to be $q-p$,

$$\Gamma\left(\boldsymbol{m}/\boldsymbol{\alpha}\right)=\frac{2\boldsymbol{m}\cdot\boldsymbol{\alpha}}{|\boldsymbol{\alpha}|^2}=q-p=\ \text{integer}. \tag{7.122}$$

Due to our convention (7.118), $q\geq p\geq 0$. In the state chain $|\boldsymbol{m}+n\boldsymbol{\alpha}\rangle$ given in Eq. (7.119), there are both states $|\boldsymbol{m}\rangle$ and $|\boldsymbol{m}'\rangle$ with the weight $\boldsymbol{m}$ and $\boldsymbol{m}'$, respectively, where

$$\boldsymbol{m}'\equiv\boldsymbol{m}-\Gamma\left(\boldsymbol{m}/\boldsymbol{\alpha}\right)\boldsymbol{\alpha}=\boldsymbol{m}-(q-p)\boldsymbol{\alpha}. \tag{7.123}$$

In the subspace orthogonal to the states given in Eq. (7.119) one finds another state with the weight $\boldsymbol{m}$ and repeats the steps to obtain another state chain containing two states with the weight $\boldsymbol{m}$ and $\boldsymbol{m}'$, respectively. If the multiplicity of the weight $\boldsymbol{m}$ is d, one is able to find d state chains so that the multiplicity d' of the weight $\boldsymbol{m}'$ is not less than d. Conversely, if the state chain is calculated from the state $|\boldsymbol{m}'\rangle$, one obtains $d\geq d'$. Thus, two weights $\boldsymbol{m}$ and $\boldsymbol{m}'$ have the same multiplicity and are called the equivalent weight □

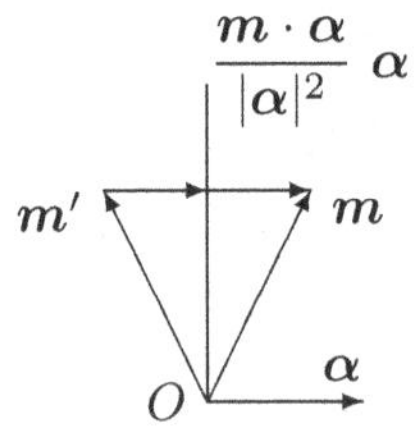

Fig. 7.2 A Weyl reflection

Two weights $\boldsymbol{m}$ and $\boldsymbol{m}'$ are the mirror images with respect to the plane perpendicular to $\boldsymbol{\alpha}$ and across the origin. This reflection is called a Weyl reflection in the weight space. The product of two Weyl reflections is defined as their successive applications. The set of all Weyl reflections and their products for a representation forms the Weyl group W. The weights related by the elements of the Weyl group are equivalent, and the number of equivalent weights is called the size of the Weyl orbit of the weights.

A weight $\boldsymbol{M}$ satisfying

$$\Gamma\left(\boldsymbol{M}/\boldsymbol{r}_\mu\right)=\ \text{non-negative integer}, \qquad \forall\ \text{simple root}\ \ \boldsymbol{r}_\mu, \tag{7.124}$$

is called a dominant weight. In an irreducible representation of a simple Lie algebra $\mathcal{L}$ there are a few dominant weights, single or multiple. Each weight m in the representation is equivalent to one dominant weight. The dimension of the representation is equal to the sum of products of the multiplicity of each dominant weight and its size of Weyl orbit. The highest weight of an irreducible representation is a dominant weight and simple because of Eqs. (7.122), (7.115), and Theorem 7.7. Dynkin proved that a space constructed by applying the lowering operators E_{r_μ} successively to a state $|\boldsymbol{M}\rangle$ with a dominant weight $\boldsymbol{M}$ is finite and corresponds to an irreducible representation of $\mathcal{L}$. This is the foundation for the method of the block weight diagram, which will be discussed in the next section.

Theorem 7.9 Any dominant weight $\boldsymbol{M}$ is the highest weight of one irreducible representation of a simple Lie algebra $\mathcal{L}$ with a finite dimension.

7.5.3 *Mathematical Property of Representations*

The highest weight $\boldsymbol{M}$ gives the full property of an irreducible representation of a simple Lie algebra $\mathcal{L}$. In this subsection we only quote some mathematical results of a highest weight representation.

Let G be a compact simple Lie group with the Lie algebra $\mathcal{L}$, and H be the Abelian Lie subgroup produced from the Cartan subalgebra $\mathcal{H}$. The elements in H are characterized by ℓ parameters $\boldsymbol{\varphi} \equiv (\varphi_1, \varphi_2, \ldots, \varphi_\ell)$,

$$R = \exp\left\{-i\boldsymbol{\varphi} \cdot \boldsymbol{H}\right\} = \exp\left\{-i\sum_{j=1}^{\ell} \varphi_j H_j\right\} \in H. \tag{7.125}$$

Each element in G is conjugate to an element in H so that the class in G is characterized by the parameters $\boldsymbol{\varphi}$. In an irreducible representation $D^{\boldsymbol{M}}$ the character $\chi(\boldsymbol{M}, \boldsymbol{\varphi})$ of the class $\boldsymbol{\varphi}$ is

$$\chi(\boldsymbol{M}, \boldsymbol{\varphi}) = \text{Tr}\ \exp\left\{-i\sum_{j=1}^{\ell} \varphi_j D(H_j)\right\} = \sum_{\boldsymbol{m}} b(\boldsymbol{m}) \exp\left\{-i\boldsymbol{\varphi} \cdot \boldsymbol{m}\right\}. \tag{7.126}$$

$b(\boldsymbol{m})$ is the multiplicity of the weight $\boldsymbol{m}$, and in fact, the sum in Eq. (7.126) runs over all basis states in the representation space. The character $\chi(\boldsymbol{M}, \boldsymbol{\varphi})$ can be calculated by the girdle $\xi(\boldsymbol{K}, \boldsymbol{\varphi})$ introduced by Weyl,

$$\xi(\boldsymbol{K}, \boldsymbol{\varphi}) = \sum_{S \in W} \delta_S \exp\left\{-i\sum_{j=1}^{\ell} (S\boldsymbol{K})_j \varphi_j\right\}, \tag{7.127}$$

where S is the element in the Weyl group and the reflection parity δ_S is 1 if S is a product of even Weyl reflections, and -1 if S is a product of odd Weyl reflections. $\boldsymbol{K} = \boldsymbol{M} + \boldsymbol{\rho}$, where $\boldsymbol{\rho}$ is the half sum of all positive roots in $\mathcal{L}$. Then,

$$\chi(\boldsymbol{M}, \boldsymbol{\varphi}) = \frac{\xi(\boldsymbol{K}, \boldsymbol{\varphi})}{\xi(\boldsymbol{\rho}, \boldsymbol{\varphi})}. \tag{7.128}$$

$\xi(\boldsymbol{\rho}, \boldsymbol{\varphi})$ is independent of the representation and is related to the class integral of the Lie group. The orthonormal relations of the characters of two irreducible representations are expressed as

$$\int \chi(\boldsymbol{M}, \boldsymbol{\varphi})^* \chi(\boldsymbol{M}', \boldsymbol{\varphi}) |\xi(\boldsymbol{\rho}, \boldsymbol{\varphi})|^2 (d\boldsymbol{\varphi}) = \int \xi(\boldsymbol{K}, \boldsymbol{\varphi})^* \xi(\boldsymbol{K}', \boldsymbol{\varphi}) (d\boldsymbol{\varphi}) = \delta_{\boldsymbol{K}\boldsymbol{K}'}. \tag{7.129}$$

When $\boldsymbol{\varphi}$ goes to zero, $\chi(\boldsymbol{M}, \boldsymbol{\varphi})$ is an indefinite form and tends to the dimension $d(\boldsymbol{M})$,

$$d(\boldsymbol{M}) = \prod_{\boldsymbol{\alpha} \in \Delta_+} \left\{ 1 + \frac{\boldsymbol{M} \cdot \boldsymbol{\alpha}}{\boldsymbol{\rho} \cdot \boldsymbol{\alpha}} \right\}, \tag{7.130}$$

where Δ_+ is the set of the positive roots in $\mathcal{L}$.

7.5.4 *Fundamental Dominant Weights*

In the weight space of a simple Lie algebra $\mathcal{L}$ with rank ℓ, there are ℓ dominant weights $\boldsymbol{w}_\mu$ satisfying

$$\Gamma\left(\boldsymbol{w}_\mu / \boldsymbol{r}_\nu\right) = d_\nu^{-1} \left(\boldsymbol{w}_\mu \cdot \boldsymbol{r}_\nu\right) = \delta_{\mu\nu}, \qquad d_\nu = \boldsymbol{r}_\nu \cdot \boldsymbol{r}_\nu / 2. \tag{7.131}$$

$\boldsymbol{w}_\mu$ is called the fundamental dominant weight. An irreducible representation whose highest weight is $\boldsymbol{w}_\mu$ is called the fundamental representation of $\mathcal{L}$. Comparing Eq. (7.131) with Eq. (7.62), one has

$$\boldsymbol{r}_\mu = \sum_{\nu=1}^{\ell} \boldsymbol{w}_\nu A_{\nu\mu}, \qquad \boldsymbol{w}_\nu = \sum_{\mu=1}^{\ell} \boldsymbol{r}_\mu \left(A^{-1}\right)_{\mu\nu}. \tag{7.132}$$

$A_{\nu\mu}$ is the component of the simple root $\boldsymbol{r}_\mu$ with respect to $\boldsymbol{w}_\nu$. $A_{\nu\mu}$ is equal to an integer 2, -1, -2, -3, or 0. The fundamental dominant weight are usually chosen to be the basis vectors in the weight space and in the root space such that the components of both the weight and the root are integers. In fact, $\boldsymbol{M} = \sum_\mu M_\mu \boldsymbol{w}_\mu$ where $M_\mu = \Gamma\left(\boldsymbol{M}/\boldsymbol{r}_\mu\right)$ is a non-negative integer. $\boldsymbol{m} = \sum_\mu m_\mu \boldsymbol{w}_\mu$ where $m_\mu = \Gamma\left(\boldsymbol{m}/\boldsymbol{r}_\mu\right)$ is an integer. The Weyl reflection relation of two equivalent weights also becomes simpler,

$$\boldsymbol{m}' \xrightarrow{\boldsymbol{r}_\mu} \boldsymbol{m} - m_\mu \boldsymbol{r}_\mu, \qquad \boldsymbol{m} = \sum_\mu m_\mu \boldsymbol{w}_\mu. \tag{7.133}$$

By making use of the symbol d_μ, one has from Eq. (7.131)

$$(\boldsymbol{r}_\mu \cdot \boldsymbol{r}_\nu) = d_\mu A_{\mu\nu}, \qquad (\boldsymbol{r}_\mu \cdot \boldsymbol{w}_\nu) = d_\mu \delta_{\mu\nu}. \tag{7.134}$$

$d_\mu A_{\mu\nu}$ is usually called the symmetrized Cartan matrix. Furthermore,

$$(\boldsymbol{w}_\mu \cdot \boldsymbol{w}_\nu) = d_\mu \left(A^{-1}\right)_{\mu\nu}. \tag{7.135}$$

The shortage of the fundamental dominant weights is that they are not orthonormal.

7.5.5 *The Casimir Operator of Order 2*

The Casimir operator C_2 of order 2 is given in Eq. (7.25) which is commutable with any generator I_B. When the Killing form is a constant matrix, $C_2 = \sum_A I_A I_A$. Due to the Schur theorem, C_2 is a constant matrix $\mathbf{1}C_2(\boldsymbol{M})$ in an irreducible representation $D^{\boldsymbol{M}}$ of a simple Lie algebra. The constant $C_2(\boldsymbol{M})$ is called the Casimir invariant of order 2 in the representation $D^{\boldsymbol{M}}$, or briefly the Casimir invariant.

In the Cartan–Weyl bases of generators,

$$C_2 = \sum_A I_A I_A = \sum_{j=1}^{\ell} H_j H_j + \sum_{\boldsymbol{\alpha}\in\Delta_+} \left\{E_{\boldsymbol{\alpha}} E_{-\boldsymbol{\alpha}} + E_{-\boldsymbol{\alpha}} E_{\boldsymbol{\alpha}}\right\}. \tag{7.136}$$

Applying C_2 to the highest weight state $|\boldsymbol{M}\rangle$, one has from Eq. (7.115),

$$\begin{aligned} C_2 \left|\boldsymbol{M}\right\rangle &= \boldsymbol{M}^2 \left|\boldsymbol{M}\right\rangle + \sum_{\boldsymbol{\alpha}\in\Delta_+} \left[E_{\boldsymbol{\alpha}} E_{-\boldsymbol{\alpha}} + E_{-\boldsymbol{\alpha}} E_{\boldsymbol{\alpha}}\right] \left|\boldsymbol{M}\right\rangle \\ &= \boldsymbol{M}^2 \left|\boldsymbol{M}\right\rangle + \sum_{\boldsymbol{\alpha}\in\Delta_+} \left[E_{\boldsymbol{\alpha}},\ E_{-\boldsymbol{\alpha}}\right] \left|\boldsymbol{M}\right\rangle \\ &= \boldsymbol{M}\cdot(\boldsymbol{M}+2\boldsymbol{\rho}) = \sum_{\mu,\nu=1}^{\ell} M_\mu d_\mu \left(A^{-1}\right)_{\mu\nu} (M_\nu + 2), \end{aligned} \tag{7.137}$$

where $\boldsymbol{\rho}$ is the half sum of the positive roots in the simple Lie algebra $\mathcal{L}$, and it is equal to the sum of the fundamental dominant weights

$$2\boldsymbol{\rho} = \sum_{\boldsymbol{\alpha}\in\Delta_+} \boldsymbol{\alpha} = 2\sum_{\mu=1}^{\ell} \boldsymbol{w}_\mu. \tag{7.138}$$

In fact, it can be shown as follows that the inner product of $2\boldsymbol{\rho}$ and any simple root $\boldsymbol{r}_\mu$ is $2d_\mu$. From any positive root $\boldsymbol{\alpha}$, except for the simple root $\boldsymbol{r}_\mu$, one is able to construct a root chain in Δ_+,

$$(\boldsymbol{\alpha}-q\boldsymbol{r}_\mu),\ \ldots,\ (\boldsymbol{\alpha}-\boldsymbol{r}_\mu),\ \boldsymbol{\alpha},\ (\boldsymbol{\alpha}+\boldsymbol{r}_\mu),\ \ldots,\ (\boldsymbol{\alpha}+p\boldsymbol{r}_\mu).$$

The sum of the roots in the root chain is $(p+q+1)\left[\boldsymbol{\alpha}+\boldsymbol{r}_\mu(p-q)/2\right]$, and due to Eq. (7.46) it is orthogonal to $\boldsymbol{r}_\mu$. The inner self-product of $\boldsymbol{r}_\mu$ is $2d_\mu$. Thus, Eq. (7.138) is proved.

There is another way to calculate the Casimir invariant. The matrix form of the generator I_A in the representation $D^{\boldsymbol{M}}$ is denoted by $I_A^{\boldsymbol{M}}$. Define a g-dimensional matrix $T(\boldsymbol{M})$, whose entry is $T_{AB}(\boldsymbol{M})=\text{Tr}\left[I_A^{\boldsymbol{M}}I_B^{\boldsymbol{M}}\right]$. For a compact simple Lie algebra, the adjoint representation is irreducible and the structure constant $C_{AB}{}^D$ is totally antisymmetric with respect to its three indices. Then, it can be shown from Eq. (7.9) that $T(\boldsymbol{M})$ is commutable with any generator I_A^{ad} in the adjoint representation,

$$\begin{aligned}
\left[T(\boldsymbol{M}),\ I_A^{\text{ad}}\right]_{BD} &= \sum_P \left\{\text{Tr}\left(I_B^{\boldsymbol{M}}I_P^{\boldsymbol{M}}\right)\left(I_A^{\text{ad}}\right)_{PD}-\left(I_A^{\text{ad}}\right)_{BP}\text{Tr}\left(I_P^{\boldsymbol{M}}I_D^{\boldsymbol{M}}\right)\right\}\\
&= i\text{Tr}\sum_P\left\{I_B^{\boldsymbol{M}}I_P^{\boldsymbol{M}}C_{AD}{}^P+C_{AB}{}^PI_P^{\boldsymbol{M}}I_D^{\boldsymbol{M}}\right\}\\
&= \text{Tr}\left\{I_B^{\boldsymbol{M}}\left(I_A^{\boldsymbol{M}}I_D^{\boldsymbol{M}}-I_D^{\boldsymbol{M}}I_A^{\boldsymbol{M}}\right)+\left(I_A^{\boldsymbol{M}}I_B^{\boldsymbol{M}}-I_B^{\boldsymbol{M}}I_A^{\boldsymbol{M}}\right)I_D^{\boldsymbol{M}}\right\}=0.
\end{aligned}$$

Thus, $T(\boldsymbol{M})$ is a constant matrix, denoted by $T_{AB}(\boldsymbol{M}) = \delta_{AB}T_2(\boldsymbol{M})$. Taking the trace, one has

$$g\ T_2(\boldsymbol{M})=\sum_A T_{AA}(\boldsymbol{M})=\text{Tr}\left\{\sum_A I_A^{\boldsymbol{M}}I_A^{\boldsymbol{M}}\right\}=d(\boldsymbol{M})C_2(\boldsymbol{M}), \tag{7.139}$$

where $d(\boldsymbol{M})$ is the dimension of the representation $D^{\boldsymbol{M}}$. The highest weight $\boldsymbol{M}$ of the adjoint representation is the largest root $\boldsymbol{\omega}$,

$$g_{AB}=-\delta_{AB}T_2(\boldsymbol{\omega})=-\delta_{AB}C_2(\boldsymbol{\omega}). \tag{7.140}$$

7.6 Main Data of Simple Lie Algebras

In this section we list the Dynkin diagram, the Cartan matrix A and its inverse A^{-1}, the simple roots $\boldsymbol{r}_\mu$, the positive roots, the largest root $\boldsymbol{\omega}$, the fundamental dominant weights $\boldsymbol{w}_\mu$, some Casimir invariants $C_2(\boldsymbol{M})$, and the formulas for the dimensions $d(\boldsymbol{M})$ of most simple Lie algebras. $\boldsymbol{M}_0$ is the self-representation. M_μ is the component of the highest weight $\boldsymbol{M}$

in the basis vectors $\boldsymbol{w}_\mu$. The Killing form is $g_{AB} = -\delta_{AB}C_2(\boldsymbol{\omega})$. $\boldsymbol{e}_\mu$ are the Euclidean basis vectors. The basis vectors $\boldsymbol{V}_j$ are given in Eq. (7.79). Due to the mathematical conventions, the half of square length of a longer simple root is $d_\mu = 1$. For the Lie algebra A_ℓ, $d_\mu = 1/2$ in the physical convention such that $C_2(\boldsymbol{M})$ as well as the Killing form shortens by a factor 2.

7.6.1 *Lie Algebra* A_ℓ *and Lie Group* $\mathrm{SU}(\ell+1)$

The order of A_ℓ is $\ell(\ell+2)$. There are $\ell(\ell+1)$ roots, denoted by $\sqrt{2}\,(\boldsymbol{V}_a - \boldsymbol{V}_b)$. The Dynkin diagram of A_ℓ is

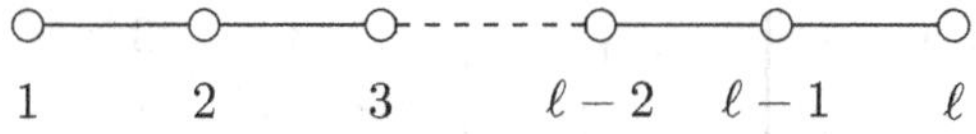

$$A = \begin{pmatrix} 2 & -1 & 0 & 0 & \dots & 0 & 0 \\ -1 & 2 & -1 & 0 & \dots & 0 & 0 \\ 0 & -1 & 2 & -1 & \dots & 0 & 0 \\ \cdot & \cdot & \cdot & \cdot & \dots & \cdot & \cdot \\ 0 & 0 & 0 & 0 & \dots & 2 & -1 \\ 0 & 0 & 0 & 0 & \dots & -1 & 2 \end{pmatrix},$$

$$A^{-1} = \frac{1}{\ell+1}\begin{pmatrix} 1\cdot\ell & 1\cdot(\ell-1) & 1\cdot(\ell-2) & \dots & 1\cdot 2 & 1\cdot 1 \\ 1\cdot(\ell-1) & 2\cdot(\ell-1) & 2\cdot(\ell-2) & \dots & 2\cdot 2 & 2\cdot 1 \\ 1\cdot(\ell-2) & 2\cdot(\ell-2) & 3\cdot(\ell-2) & \dots & 3\cdot 2 & 3\cdot 1 \\ \cdot & \cdot & \cdot & \dots & \cdot & \cdot \\ 1\cdot 2 & 2\cdot 2 & 3\cdot 2 & \dots & (\ell-1)\cdot 2 & (\ell-1)\cdot 1 \\ 1\cdot 1 & 2\cdot 1 & 3\cdot 1 & \dots & (\ell-1)\cdot 1 & \ell\cdot 1 \end{pmatrix}$$

$$\begin{aligned} &d_\mu = 1, && 1 \le \mu \le \ell, \\ &\boldsymbol{M}_0 = \boldsymbol{w}_1\ , && \boldsymbol{M}_{\text{adj}} = \boldsymbol{\omega} = \boldsymbol{w}_1 + \boldsymbol{w}_\ell, \\ &d(\boldsymbol{M}_0) = \ell + 1\ , && d(\boldsymbol{M}_{\text{adj}}) = \ell(\ell+2), \\ &C_2(\boldsymbol{M}_0) = \frac{\ell^2 + 2\ell}{\ell+1}, && C_2(\boldsymbol{M}_{\text{adj}}) = 2(\ell+1). \end{aligned}$$

$$\boldsymbol{r}_\mu = \sqrt{2}\,(\boldsymbol{V}_\mu - \boldsymbol{V}_{\mu+1})\,, \qquad \boldsymbol{w}_\mu = \sqrt{2}\sum_{\nu=1}^{\mu} \boldsymbol{V}_\nu, \qquad 1 \le \mu \le \ell\,.$$

$$d(\boldsymbol{M}) = \prod_{a<b}^{\ell+1} \left\{1 + \sum_{\mu=a}^{b-1} \frac{M_\mu}{b-a}\right\}.$$

7.6.2 *Lie Algebra* B_ℓ *and Lie Group* $\mathrm{SO}(2\ell+1)$

Being a Lie algebra, $\mathrm{B}_1 \approx \mathrm{A}_1$ and $\mathrm{B}_2 \approx \mathrm{C}_2$. Here we list the data of Lie algebra B_ℓ with $\ell \geq 3$. The order of B_ℓ is $\ell(2\ell+1)$. There are ℓ^2 positive roots, denoted by $\boldsymbol{e}_\mu$ and $\boldsymbol{e}_\mu \pm \boldsymbol{e}_\nu$. The Dynkin diagram of B_ℓ is

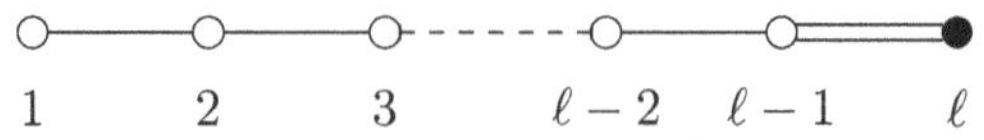

$$A = \begin{pmatrix} 2 & -1 & 0 & \ldots & 0 & 0 \\ -1 & 2 & -1 & \ldots & 0 & 0 \\ 0 & -1 & 2 & \ldots & 0 & 0 \\ \cdot & \cdot & \cdot & \ldots & \cdot & \cdot \\ 0 & 0 & 0 & \ldots & 2 & -1 \\ 0 & 0 & 0 & \ldots & -2 & 2 \end{pmatrix}, \quad A^{-1} = \begin{pmatrix} 1 & 1 & 1 & \ldots & 1 & 1/2 \\ 1 & 2 & 2 & \ldots & 2 & 1 \\ 1 & 2 & 3 & \ldots & 3 & 3/2 \\ \cdot & \cdot & \cdot & \ldots & \cdot & \cdot \\ 1 & 2 & 3 & \ldots & \ell-1 & (\ell-1)/2 \\ 1 & 2 & 3 & \ldots & \ell-1 & \ell/2 \end{pmatrix},$$

$$\begin{array}{lll} d_\mu = 1, & 1 \leq \mu \leq \ell - 1, & d_\ell = 1/2, \\ \boldsymbol{M}_0 = \boldsymbol{w}_1\ , & \boldsymbol{M}_{\mathrm{adj}} = \boldsymbol{\omega} = \boldsymbol{w}_2, & \boldsymbol{M}_S = \boldsymbol{w}_\ell, \\ d(\boldsymbol{M}_0) = 2\ell + 1, & d(\boldsymbol{M}_{\mathrm{adj}}) = \ell(2\ell+1), & d(\boldsymbol{M}_S) = 2^\ell, \\ C_2(\boldsymbol{M}_0) = 2\ell, & C_2(\boldsymbol{M}_{\mathrm{adj}}) = 2(2\ell-1), & C_2(\boldsymbol{M}_S) = \ell(2\ell+1)/4. \end{array}$$

$$\begin{aligned} \boldsymbol{r}_\mu &= \boldsymbol{e}_\mu - \boldsymbol{e}_{\mu+1}, \qquad 1 \leq \mu \leq (\ell-1), \qquad \boldsymbol{r}_\ell = \boldsymbol{e}_\ell, \\ \boldsymbol{w}_\mu &= \sum_{\nu=1}^{\mu} \boldsymbol{e}_\nu, \qquad 1 \leq \mu < \ell, \qquad \boldsymbol{w}_\ell = \frac{1}{2}\sum_{\nu=1}^{\ell} \boldsymbol{e}_\nu. \end{aligned}$$

$$\begin{aligned} d(\boldsymbol{M}) = \prod_{\lambda=1}^{\ell} & \left\{ 1 + \frac{M_\ell + 2\displaystyle\sum_{\rho=\lambda}^{\ell-1} M_\rho}{1 + 2(\ell - \lambda)} \right\} \\ \cdot \prod_{\mu<\nu}^{\ell} & \left\{ \left(1 + \frac{M_\ell + \displaystyle\sum_{\rho=\mu}^{\nu-1} M_\rho + \sum_{\rho=\nu}^{\ell-1} 2M_\rho}{1 + 2\ell - \mu - \nu} \right) \left(1 + \frac{\displaystyle\sum_{\rho=\mu}^{\nu-1} M_\rho}{\nu - \mu} \right) \right\}. \end{aligned}$$

7.6.3 *Lie Algebra* C_ℓ *and Lie Group* USp(2ℓ)

Being a Lie algebra, $C_1 \approx A_1$. Here we list the data of Lie algebra C_ℓ with $\ell \geq 2$. The order of C_ℓ is $\ell(2\ell+1)$. There are ℓ^2 positive roots, denoted by $\sqrt{1/2}\,(\boldsymbol{e}_\mu \pm \boldsymbol{e}_\nu)$ and $\sqrt{2}\boldsymbol{e}_\mu$. The Dynkin diagram of C_ℓ is

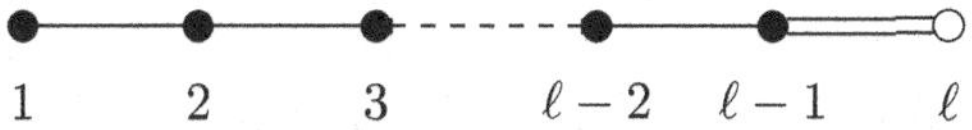

$$A = \begin{pmatrix} 2 & -1 & 0 & \ldots & 0 & 0 \\ -1 & 2 & -1 & \ldots & 0 & 0 \\ 0 & -1 & 2 & \ldots & 0 & 0 \\ \cdot & \cdot & \cdot & \ldots & \cdot & \cdot \\ 0 & 0 & 0 & \ldots & 2 & -2 \\ 0 & 0 & 0 & \ldots & -1 & 2 \end{pmatrix}, \quad A^{-1} = \begin{pmatrix} 1 & 1 & 1 & \ldots & 1 & 1 \\ 1 & 2 & 2 & \ldots & 2 & 2 \\ 1 & 2 & 3 & \ldots & 3 & 3 \\ \cdot & \cdot & \cdot & \ldots & \cdot & \cdot \\ 1 & 2 & 3 & \ldots & \ell-1 & \ell-1 \\ 1/2 & 1 & 3/2 & \ldots & (\ell-1)/2 & \ell/2 \end{pmatrix}$$

$$\begin{aligned} & d_\mu = 1/2, \quad 1 \leq \mu \leq (\ell-1), && d_\ell = 1, \\ & \boldsymbol{M}_0 = \boldsymbol{w}_1, && \boldsymbol{M}_{\text{adj}} = \boldsymbol{\omega} = 2\boldsymbol{w}_1, \\ & d(\boldsymbol{M}_0) = 2\ell, && d(\boldsymbol{M}_{\text{adj}}) = \ell(2\ell+1), \\ & C_2(\boldsymbol{M}_0) = \ell + 1/2, && C_2(\boldsymbol{M}_{\text{adj}}) = 2(\ell+1). \end{aligned}$$

$$\boldsymbol{r}_\mu = \sqrt{1/2}\,\{\boldsymbol{e}_\mu - \boldsymbol{e}_{\mu+1}\}, \quad 1 \leq \mu \leq (\ell-1), \quad \boldsymbol{r}_\ell = \sqrt{2}\boldsymbol{e}_\ell,$$

$$\boldsymbol{w}_\mu = \sqrt{1/2}\sum_{\nu=1}^{\mu} \boldsymbol{e}_\nu, \qquad 1 \leq \mu \leq \ell.$$

$$d(\boldsymbol{M}) = \prod_{\lambda=1}^{\ell} \left\{ 1 + \frac{\sum_{\rho=\lambda}^{\ell} M_\rho}{\ell + 1 - \lambda} \right\}$$

$$\cdot \prod_{\mu<\nu}^{\ell} \left\{ \left(1 + \frac{\sum_{\rho=\mu}^{\nu-1} M_\rho + \sum_{\rho=\nu}^{\ell} 2M_\rho}{2\ell + 2 - \mu - \nu} \right) \left(1 + \frac{\sum_{\rho=\mu}^{\nu-1} M_\rho}{\nu - \mu} \right) \right\}.$$

7.6.4 *Lie Algebra* D_ℓ *and Lie Group* SO(2ℓ)

Being a Lie algebra, $D_2 \approx A_1 \oplus A_1$ and $D_3 \approx A_3$. Here we list the data of Lie algebra D_ℓ with $\ell \geq 4$. The order of D_ℓ is $\ell(2\ell-1)$. There are $\ell(\ell-1)$

positive roots, denoted by $\boldsymbol{e}_\mu \pm \boldsymbol{e}_\nu$. The Dynkin diagram of D_ℓ is

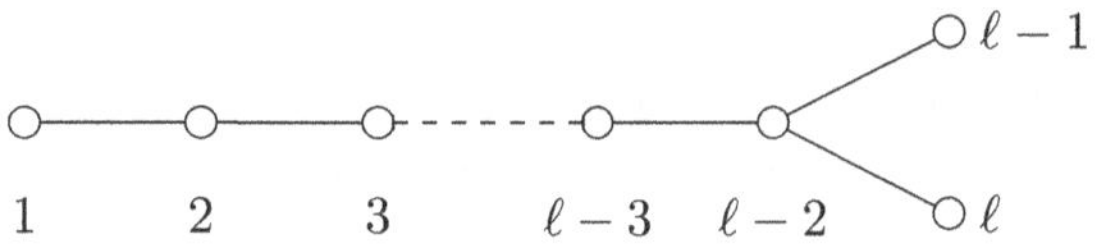

$$A = \begin{pmatrix} 2 & -1 & 0 & 0 & \ldots & 0 & 0 & 0 \\ -1 & 2 & -1 & 0 & \ldots & 0 & 0 & 0 \\ 0 & -1 & 2 & -1 & \ldots & 0 & 0 & 0 \\ \cdot & \cdot & \cdot & \cdot & \ldots & \cdot & \cdot & \cdot \\ 0 & 0 & 0 & 0 & \ldots & 2 & -1 & -1 \\ 0 & 0 & 0 & 0 & \ldots & -1 & 2 & 0 \\ 0 & 0 & 0 & 0 & \ldots & -1 & 0 & 2 \end{pmatrix},$$

$$A^{-1} = \frac{1}{2}\begin{pmatrix} 2 & 2 & 2 & \ldots & 2 & 1 & 1 \\ 2 & 4 & 4 & \ldots & 4 & 2 & 2 \\ 2 & 4 & 6 & \ldots & 6 & 3 & 3 \\ \cdot & \cdot & \cdot & \ldots & \cdot & \cdot & \cdot \\ 2 & 4 & 6 & \ldots & 2(\ell-2) & \ell-2 & \ell-2 \\ 1 & 2 & 3 & \ldots & \ell-2 & \ell/2 & (\ell-2)/2 \\ 1 & 2 & 3 & \ldots & \ell-2 & (\ell-2)/2 & \ell/2 \end{pmatrix},$$

$$\begin{aligned} &d_\mu = 1, && 1 \le \mu \le \ell, \\ &\boldsymbol{M}_0 = \boldsymbol{w}_1, && \boldsymbol{M}_{\text{adj}} = \boldsymbol{\omega} = \boldsymbol{w}_2, \\ &\boldsymbol{M}_{S1} = \boldsymbol{w}_{\ell-1}, && \boldsymbol{M}_{S2} = \boldsymbol{w}_\ell, \\ &d(\boldsymbol{M}_0) = 2\ell, && d(\boldsymbol{M}_{\text{adj}}) = \ell(2\ell-1), \\ &d(\boldsymbol{M}_{S1}) = d(\boldsymbol{M}_{S2}) = 2^{\ell-1}, && C_2(\boldsymbol{M}_0) = 2\ell - 1, \\ &C_2(\boldsymbol{M}_{\text{adj}}) = 4(\ell-1), && C_2(\boldsymbol{M}_{S1}) = C_2(\boldsymbol{M}_{S2}) = \ell(2\ell-1)/4. \end{aligned}$$

$$\begin{aligned} &\boldsymbol{r}_\mu = \boldsymbol{e}_\mu - \boldsymbol{e}_{\mu+1}\ , \quad 1 \le \mu \le (\ell-1), \qquad \boldsymbol{r}_\ell = \boldsymbol{e}_{\ell-1} + \boldsymbol{e}_\ell, \\ &\boldsymbol{w}_\nu = \sum_{\rho=1}^{\nu} \boldsymbol{e}_\rho, \qquad 1 \le \nu \le \ell - 2, \\ &\boldsymbol{w}_{\ell-1} = \frac{1}{2}\sum_{\rho=1}^{\ell-1} \boldsymbol{e}_\rho - \frac{1}{2}\boldsymbol{e}_\ell, \qquad \boldsymbol{w}_\ell = \frac{1}{2}\sum_{\rho=1}^{\ell} \boldsymbol{e}_\rho. \end{aligned}$$

$$d(\boldsymbol{M}) = \prod_{\mu<\nu}^{\ell} \left\{ \left(1 + \frac{\sum_{\rho=\mu}^{\nu-1} M_\rho + \sum_{\rho=\nu}^{\ell} 2M_\rho - M_{\ell-1} - M_\ell}{2\ell - \mu - \nu} \right) \left(1 + \frac{\sum_{\rho=\mu}^{\nu-1} M_\rho}{\nu - \mu} \right) \right\}.$$

7.6.5 *Lie Algebra* G_2

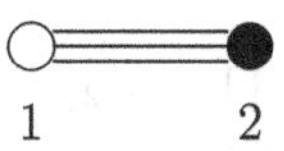

$$A = \begin{pmatrix} 2 & -1 \\ -3 & 2 \end{pmatrix}, \qquad A^{-1} = \begin{pmatrix} 2 & 1 \\ 3 & 2 \end{pmatrix},$$

$$\begin{array}{llll} d_1 = 1, & d_2 = 1/3\ , & \boldsymbol{M}_0 = \boldsymbol{w}_2, & \boldsymbol{M}_{\text{adj}} = \boldsymbol{\omega} = \boldsymbol{w}_1, \\ d(\boldsymbol{M}_0) = 7, & d(\boldsymbol{M}_{\text{adj}}) = 14, & C_2(\boldsymbol{M}_0) = 4, & C_2(\boldsymbol{M}_{\text{adj}}) = 8. \end{array}$$

The order of G_2 is 14. There are six positive roots,

$$\begin{array}{ll} \boldsymbol{r}_1 = \sqrt{2}\,\boldsymbol{e}_2, & \boldsymbol{r}_2 = \sqrt{1/6}\,\boldsymbol{e}_1 - \sqrt{1/2}\,\boldsymbol{e}_2, \\ \boldsymbol{r}_1 + \boldsymbol{r}_2 = \sqrt{1/6}\,\boldsymbol{e}_1 + \sqrt{1/2}\,\boldsymbol{e}_2, & \boldsymbol{r}_1 + 2\boldsymbol{r}_2 = \sqrt{2/3}\,\boldsymbol{e}_1, \\ \boldsymbol{r}_1 + 3\boldsymbol{r}_2 = \sqrt{3/2}\,\boldsymbol{e}_1 - \sqrt{1/2}\,\boldsymbol{e}_2\ , & 2\boldsymbol{r}_1 + 3\boldsymbol{r}_2 = \sqrt{3/2}\,\boldsymbol{e}_1 + \sqrt{1/2}\,\boldsymbol{e}_2, \\ \boldsymbol{w}_1 = \sqrt{3/2}\,\boldsymbol{e}_1 + \sqrt{1/2}\,\boldsymbol{e}_2, & \boldsymbol{w}_2 = \sqrt{2/3}\,\boldsymbol{e}_1. \end{array}$$

$$\begin{aligned} d(\boldsymbol{M}) = &\ (1+M_1)(1+M_2)(4+3M_1+M_2)(5+3M_1+2M_2) \\ &\cdot (2+M_1+M_2)(3+2M_1+M_2)/120\ . \end{aligned}$$

7.6.6 *Lie Algebra* F_4

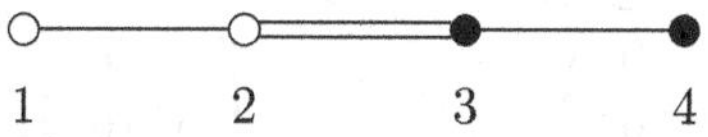

$$A = \begin{pmatrix} 2 & -1 & 0 & 0 \\ -1 & 2 & -1 & 0 \\ 0 & -2 & 2 & -1 \\ 0 & 0 & -1 & 2 \end{pmatrix}, \quad A^{-1} = \begin{pmatrix} 2 & 3 & 2 & 1 \\ 3 & 6 & 4 & 2 \\ 4 & 8 & 6 & 3 \\ 2 & 4 & 3 & 2 \end{pmatrix},$$

$$\begin{array}{lll} d_1 = d_2 = 1, & d_3 = d_4 = 1/2, & \\ \boldsymbol{M}_0 = \boldsymbol{w}_4, & d(\boldsymbol{M}_0) = 26, & C_2(\boldsymbol{M}_0) = 12, \\ \boldsymbol{M}_{\text{adj}} = \boldsymbol{\omega} = \boldsymbol{w}_1, & d(\boldsymbol{M}_{\text{adj}}) = 52, & C_2(\boldsymbol{M}_{\text{adj}}) = 18. \end{array}$$

$$\begin{aligned} &\boldsymbol{r}_1 = \boldsymbol{e}_2 - \boldsymbol{e}_3, && \boldsymbol{r}_2 = \boldsymbol{e}_3 - \boldsymbol{e}_4, \\ &\boldsymbol{r}_3 = \boldsymbol{e}_4, && \boldsymbol{r}_4 = (\boldsymbol{e}_1 - \boldsymbol{e}_2 - \boldsymbol{e}_3 - \boldsymbol{e}_4)/2, \\ &\boldsymbol{w}_1 = \boldsymbol{e}_1 + \boldsymbol{e}_2, && \boldsymbol{w}_2 = 2\boldsymbol{e}_1 + \boldsymbol{e}_2 + \boldsymbol{e}_3, \\ &\boldsymbol{w}_3 = (3\boldsymbol{e}_1 + \boldsymbol{e}_2 + \boldsymbol{e}_3 + \boldsymbol{e}_4)/2\ , && \boldsymbol{w}_4 = \boldsymbol{e}_1. \end{aligned}$$

The order of F_4 is 52. B_4 is a subalgebra of F_4. F_4 contains 48 roots, including 32 roots in B_4 and 16 other roots denoted by $a\boldsymbol{e}_1+b\boldsymbol{e}_2+c\boldsymbol{e}_3+d\boldsymbol{e}_4$, where four parameters are taken to be ± 1 independently.

$$\begin{aligned} d(\boldsymbol{M}) = &\{1 + (2M_1 + 4M_2 + 3M_3 + 2M_4)/11\}\{1 + M_1\}\{1 + M_2\} \\ &\cdot\ \{1 + M_3\}\{1 + M_4\}\{1 + (2M_1 + 2M_2 + M_3)/5\} \\ &\cdot\ \{1 + (2M_2 + M_3)/3\}\{1 + (2M_1 + 3M_2 + 2M_3 + M_4)/8\} \\ &\cdot\ \{1 + (M_2 + M_3 + M_4)/3\}\{1 + (M_1 + 3M_2 + 2M_3 + M_4)/7\} \\ &\cdot\ \{1 + (M_1 + M_2 + M_3 + M_4)/4\}\{1 + (M_1 + 2M_2 + 2M_3 + M_4)/6\} \\ &\cdot\ \{1 + (M_1 + 2M_2 + M_3 + M_4)/5\}\{1 + (M_1 + 2M_2 + M_3)/4\} \\ &\cdot\ \{1 + (M_1 + M_2 + M_3)/3\}\{1 + (M_1 + M_2)/2\}\{1 + (M_2 + M_3)/2\} \\ &\cdot\ \{1 + (2M_1 + 4M_2 + 3M_3 + M_4)/10\} \\ &\cdot\ \{1 + (2M_2 + 2M_3 + M_4)/5\}\{1 + (2M_1 + 2M_2 + 2M_3 + M_4)/7\} \\ &\cdot\ \{1 + (2M_1 + 4M_2 + 2M_3 + M_4)/9\}\{1 + (M_3 + M_4)/2\} \\ &\cdot\ \{1 + (2M_2 + M_3 + M_4)/4\}\{1 + (2M_1 + 2M_2 + M_3 + M_4)/6\}\ . \end{aligned}$$

7.6.7 *Lie Algebra* E_6

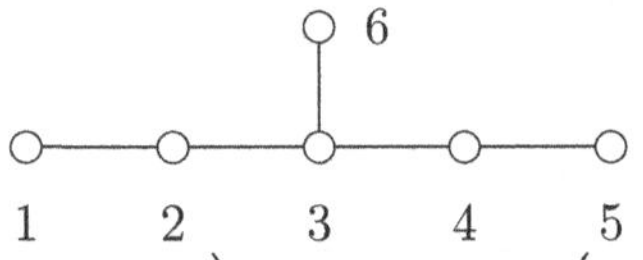

$$A = \begin{pmatrix} 2 & -1 & 0 & 0 & 0 & 0 \\ -1 & 2 & -1 & 0 & 0 & 0 \\ 0 & -1 & 2 & -1 & 0 & -1 \\ 0 & 0 & -1 & 2 & -1 & 0 \\ 0 & 0 & 0 & -1 & 2 & 0 \\ 0 & 0 & -1 & 0 & 0 & 2 \end{pmatrix}, \qquad A^{-1} = \frac{1}{3}\begin{pmatrix} 4 & 5 & 6 & 4 & 2 & 3 \\ 5 & 10 & 12 & 8 & 4 & 6 \\ 6 & 12 & 18 & 12 & 6 & 9 \\ 4 & 8 & 12 & 10 & 5 & 6 \\ 2 & 4 & 6 & 5 & 4 & 3 \\ 3 & 6 & 9 & 6 & 3 & 6 \end{pmatrix},$$

$$\begin{aligned} &d_\mu = 1, && 1 \leq \mu \leq 6, && \boldsymbol{M}_0 = \boldsymbol{w}_1, && \boldsymbol{M}_{\text{adj}} = \boldsymbol{\omega} = \boldsymbol{w}_6, \\ &d(\boldsymbol{M}_0) = 27\ , && d(\boldsymbol{M}_{\text{adj}}) = 78, && C_2(\boldsymbol{M}_0) = 52/3\ , && C_2(\boldsymbol{M}_{\text{adj}}) = 24. \end{aligned}$$

The order of E_6 is 78. A_5 is a subalgebra of E_6. E_6 contains 72 roots, including all 30 roots in A_5, $\sqrt{2}\,(\boldsymbol{U}_a - \boldsymbol{U}_b)$, and 42 other roots denoted by $\pm\boldsymbol{e}_6$ and $\sqrt{2}\,(\boldsymbol{U}_a + \boldsymbol{U}_b + \boldsymbol{U}_c) \pm \boldsymbol{e}_6/\sqrt{2}$, where $1 \leq a < b < c \leq 6$. $\boldsymbol{V}_a$ are the basis vectors for A_5 (see Eq. (7.79)). The first five components of $\boldsymbol{U}_a$ are the components of $\boldsymbol{V}_a$ but the sixth component is 0.

$$\begin{aligned}
&\boldsymbol{r}_1 = \sqrt{2}\boldsymbol{e}_6, && \boldsymbol{r}_2 = \sqrt{2}\left(\boldsymbol{U}_3 + \boldsymbol{U}_4 + \boldsymbol{U}_5\right) - \boldsymbol{e}_6/\sqrt{2}, \\
&\boldsymbol{r}_3 = \sqrt{2}\left(\boldsymbol{U}_2 - \boldsymbol{U}_3\right), && \boldsymbol{r}_4 = \sqrt{2}\left(\boldsymbol{U}_3 - \boldsymbol{U}_4\right), \\
&\boldsymbol{r}_5 = \sqrt{2}\left(\boldsymbol{U}_4 - \boldsymbol{U}_5\right), && \boldsymbol{r}_6 = \sqrt{2}\left(\boldsymbol{U}_1 - \boldsymbol{U}_2\right), \\
&\boldsymbol{w}_1 = -\sqrt{2}\boldsymbol{U}_6 + \boldsymbol{e}_6/\sqrt{2}, && \boldsymbol{w}_2 = -2\sqrt{2}\boldsymbol{U}_6, \\
&\boldsymbol{w}_3 = \sqrt{2}\left(\boldsymbol{U}_1 + \boldsymbol{U}_2 - 2\boldsymbol{U}_6\right), && \boldsymbol{w}_4 = \sqrt{2}\left(\boldsymbol{U}_1 + \boldsymbol{U}_2 + \boldsymbol{U}_3 - \boldsymbol{U}_6\right), \\
&\boldsymbol{w}_5 = -\sqrt{2}\left(\boldsymbol{U}_5 + \boldsymbol{U}_6\right), && \boldsymbol{w}_6 = \sqrt{2}\left(\boldsymbol{U}_1 - \boldsymbol{U}_6\right).
\end{aligned}$$

7.6.8 *Lie Algebra* E_7

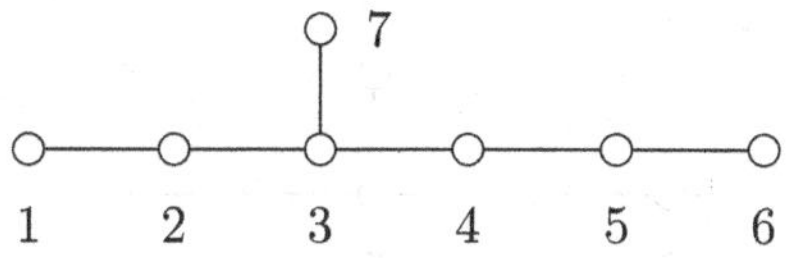

$$A = \begin{pmatrix} 2 & -1 & 0 & 0 & 0 & 0 & 0 \\ -1 & 2 & -1 & 0 & 0 & 0 & 0 \\ 0 & -1 & 2 & -1 & 0 & 0 & -1 \\ 0 & 0 & -1 & 2 & -1 & 0 & 0 \\ 0 & 0 & 0 & -1 & 2 & -1 & 0 \\ 0 & 0 & 0 & 0 & -1 & 2 & 0 \\ 0 & 0 & -1 & 0 & 0 & 0 & 2 \end{pmatrix},$$

$$A^{-1} = \frac{1}{2} \begin{pmatrix} 4 & 6 & 8 & 6 & 4 & 2 & 4 \\ 6 & 12 & 16 & 12 & 8 & 4 & 8 \\ 8 & 16 & 24 & 18 & 12 & 6 & 12 \\ 6 & 12 & 18 & 15 & 10 & 5 & 9 \\ 4 & 8 & 12 & 10 & 8 & 4 & 6 \\ 2 & 4 & 6 & 5 & 4 & 3 & 3 \\ 4 & 8 & 12 & 9 & 6 & 3 & 7 \end{pmatrix},$$

$$\begin{array}{llll}
d_\mu = 1, & 1 \leq \mu \leq 7, & \boldsymbol{M}_0 = \boldsymbol{w}_6, & \boldsymbol{M}_{\rm adj} = \boldsymbol{\omega} = \boldsymbol{w}_1, \\
d(\boldsymbol{M}_0) = 56, & d(\boldsymbol{M}_{\rm adj}) = 133, & C_2(\boldsymbol{M}_0) = 57/2, & C_2(\boldsymbol{M}_{\rm adj}) = 36.
\end{array}$$

$$\begin{aligned}
&\boldsymbol{r}_1 = \sqrt{2}\left(\boldsymbol{V}_1 - \boldsymbol{V}_2\right), && \boldsymbol{r}_2 = \sqrt{2}\left(\boldsymbol{V}_2 - \boldsymbol{V}_3\right), \\
&\boldsymbol{r}_3 = \sqrt{2}\left(\boldsymbol{V}_3 - \boldsymbol{V}_4\right), && \boldsymbol{r}_4 = \sqrt{2}\left(\boldsymbol{V}_4 - \boldsymbol{V}_5\right), \\
&\boldsymbol{r}_5 = \sqrt{2}\left(\boldsymbol{V}_5 - \boldsymbol{V}_6\right), && \boldsymbol{r}_6 = \sqrt{2}\left(\boldsymbol{V}_6 - \boldsymbol{V}_7\right),
\end{aligned}$$

$$\begin{aligned}
&\boldsymbol{r}_7 = \sqrt{2}\,(\boldsymbol{V}_4 + \boldsymbol{V}_5 + \boldsymbol{V}_6 + \boldsymbol{V}_7)\,, && \boldsymbol{w}_1 = \sqrt{2}\,(\boldsymbol{V}_1 - \boldsymbol{V}_8)\,,\\
&\boldsymbol{w}_2 = \sqrt{2}\,(\boldsymbol{V}_1 + \boldsymbol{V}_2 - 2\boldsymbol{V}_8)\,, && \boldsymbol{w}_3 = \sqrt{2}\,(\boldsymbol{V}_1 + \boldsymbol{V}_2 + \boldsymbol{V}_3 - 3\boldsymbol{V}_8)\,,\\
&\boldsymbol{w}_4 = -\sqrt{2}\,(\boldsymbol{V}_5 + \boldsymbol{V}_6 + \boldsymbol{V}_7 + 3\boldsymbol{V}_8)\,, && \boldsymbol{w}_5 = -\sqrt{2}\,(\boldsymbol{V}_6 + \boldsymbol{V}_7 + 2\boldsymbol{V}_8)\,,\\
&\boldsymbol{w}_6 = -\sqrt{2}\,(\boldsymbol{V}_7 + \boldsymbol{V}_8)\,, && \boldsymbol{w}_7 = -2\sqrt{2}\boldsymbol{V}_8,
\end{aligned}$$

where $\boldsymbol{V}_a$ are the basis vectors for A_7 (see Eq. (7.79)). The order of E_7 is 133. A_7 is a subalgebra of E_7. E_7 contains 126 roots, including all 56 roots in A_7, $\sqrt{2}\,(\boldsymbol{V}_a - \boldsymbol{V}_b)$, and 70 other roots denoted by $\sqrt{2}\,(\boldsymbol{V}_a + \boldsymbol{V}_b + \boldsymbol{V}_c + \boldsymbol{V}_d)$, where $1 \leq a < b < c < d \leq 8$.

7.6.9 *Lie Algebra* E_8

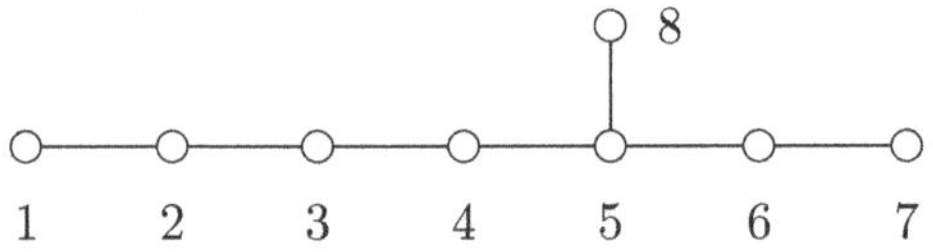

$$A = \begin{pmatrix}
2 & -1 & 0 & 0 & 0 & 0 & 0 & 0\\
-1 & 2 & -1 & 0 & 0 & 0 & 0 & 0\\
0 & -1 & 2 & -1 & 0 & 0 & 0 & 0\\
0 & 0 & -1 & 2 & -1 & 0 & 0 & 0\\
0 & 0 & 0 & -1 & 2 & -1 & 0 & -1\\
0 & 0 & 0 & 0 & -1 & 2 & -1 & 0\\
0 & 0 & 0 & 0 & 0 & -1 & 2 & 0\\
0 & 0 & 0 & 0 & -1 & 0 & 0 & 2
\end{pmatrix},$$

$$A^{-1} = \begin{pmatrix}
2 & 3 & 4 & 5 & 6 & 4 & 2 & 3\\
3 & 6 & 8 & 10 & 12 & 8 & 4 & 6\\
4 & 8 & 12 & 15 & 18 & 12 & 6 & 9\\
5 & 10 & 15 & 20 & 24 & 16 & 8 & 12\\
6 & 12 & 18 & 24 & 30 & 20 & 10 & 15\\
4 & 8 & 12 & 16 & 20 & 14 & 7 & 10\\
2 & 4 & 6 & 8 & 10 & 7 & 4 & 5\\
3 & 6 & 9 & 12 & 15 & 10 & 5 & 8
\end{pmatrix},$$

$$\begin{aligned}
&d_\mu = 1, \quad 1 \leq \mu \leq 8, && \boldsymbol{M}_0 = \boldsymbol{M}_{\text{adj}} = \boldsymbol{\omega} = \boldsymbol{w}_1,\\
&d(\boldsymbol{M}_0) = 248, && C_2(\boldsymbol{M}_{\text{adj}}) = 60.
\end{aligned}$$

$$\begin{aligned}
&\boldsymbol{r}_1 = \boldsymbol{e}_2 - \boldsymbol{e}_3, && \boldsymbol{r}_2 = \boldsymbol{e}_3 - \boldsymbol{e}_4, && \boldsymbol{r}_3 = \boldsymbol{e}_4 - \boldsymbol{e}_5,\\
&\boldsymbol{r}_4 = \boldsymbol{e}_5 - \boldsymbol{e}_6, && \boldsymbol{r}_5 = \boldsymbol{e}_6 - \boldsymbol{e}_7, && \boldsymbol{r}_6 = \boldsymbol{e}_7 - \boldsymbol{e}_8,
\end{aligned}$$

$$r_7 = 2^{-1}\left\{ e_1 + e_8 - \sum_{j=2}^{7} e_j \right\}, \qquad r_8 = e_7 + e_8,$$

$$w_1 = e_1 + e_2, \qquad w_2 = 2e_1 + e_2 + e_3,$$

$$w_3 = 3e_1 + \sum_{j=2}^{4} e_j, \qquad w_4 = 4e_1 + \sum_{j=2}^{5} e_j,$$

$$w_5 = 5e_1 + \sum_{j=2}^{6} e_j\ , \qquad w_6 = \frac{7}{2}e_1 + \frac{1}{2}\sum_{j=2}^{7} e_j - \frac{1}{2}e_8,$$

$$w_7 = 2e_1\ , \qquad w_8 = \frac{5}{2}e_1 + \frac{1}{2}\sum_{j=2}^{8} e_j.$$

The order of E_8 is 248. D_8 is a subalgebra of E_8. E_8 contains 240 roots, including all 112 roots in D_8, $\pm(e_\mu + e_\nu)$ and $(e_\mu - e_\nu)$, and 128 other roots denoted by $(\pm e_1 \pm e_2 \pm e_3 \pm e_4 \pm e_5 \pm e_6 \pm e_7 \pm e_8)/2$, where even signs are plus and the remaining signs are minus.

7.7 Block Weight Diagrams

In this section we will introduce a method for calculating the matrix forms of generators in a highest weight representation.

7.7.1 *Chevalley Bases*

Due to Eq. (7.41), $E_{\alpha+\beta}$ can be expressed as the commutative relation of E_α and E_β, so that only the generators related to the simple roots are needed to be calculated. Chevalley introduced 3ℓ bases E_μ, F_μ, and H_μ for generators, called the Chevalley bases,

$$\begin{aligned} & d_\mu^{-1/2} E_{r_\mu} \longrightarrow E_\mu, \qquad d_\mu^{-1/2} E_{-r_\mu} \longrightarrow F_\mu, \\ & d_\mu^{-1} \sum_{j=1}^{\ell} \left(r_\mu\right)_j H_j \equiv d_\mu^{-1} r_\mu \cdot H \longrightarrow H_\mu. \end{aligned} \tag{7.141}$$

Thus, the regular commutative relations (7.41) become

$$\begin{aligned} & [H_\mu,\ H_\nu] = 0, \qquad [H_\mu,\ E_\nu] = A_{\mu\nu} E_\nu, \\ & [H_\mu,\ F_\nu] = -A_{\mu\nu} F_\nu, \qquad [E_\mu,\ F_\nu] = \delta_{\mu\nu} H_\mu, \end{aligned} \tag{7.142}$$

where $A_{\mu\nu} = (\boldsymbol{r}_\nu)_\mu$ is the Cartan matrix. Recall that the difference of two simple roots is not a root, and the root chain $\boldsymbol{r}_\mu + n\boldsymbol{r}_\nu$, $0 \leq n \leq p$, breaks off at $p = -A_{\nu\mu}$. Therefore, in addition to Eq. (7.142) the Chevalley bases satisfy the Serre relations

$$\begin{aligned} &\underbrace{[E_\nu,\ [E_\nu,\ \ldots\ [E_\nu,}_{1-A_{\nu\mu}}\ E_\mu]\ \ldots\ ] = 0, \\ &\underbrace{[F_\nu,\ [F_\nu,\ \ldots\ [F_\nu,}_{1-A_{\nu\mu}}\ F_\mu]\ \ldots\ ] = 0. \end{aligned} \tag{7.143}$$

As discussed in Chap. 4, the basis states in an irreducible representation space of the SU(2) group (A_1 algebra) are denoted by $|j,m\rangle$, $-j \leq m \leq j$. The actions of the generators J_a on the basis state are

$$J_3|j,m\rangle = m|j,m\rangle, \quad (J_1 - iJ_-)\,|j,m\rangle = \sqrt{(j+m)(j-m+1)}|j,m-1\rangle.$$

The Chevalley bases for the A_1 algebra are

$$\begin{aligned} &H = 2J_3, \qquad &&E = J_1 + iJ_2, \qquad &&F = J_1 - iJ_2, \\ &[H,\ E] = 2E, &&[H,\ F] = -2F, &&[E,\ F] = H. \end{aligned} \tag{7.144}$$

Changing the notations for the basis states, one has

$$\begin{aligned} &|j,j\rangle = |M,M\rangle, \qquad |j,m\rangle = |M,M-2n\rangle, \\ &H\ |M,M-2n\rangle = (M-2n)\ |M,M-2n\rangle, \\ &F\ |M,M-2n\rangle = \sqrt{(M-n)(n+1)}\ |M,M-2n-2\rangle, \\ &E\ |M,M-2n-2\rangle = \sqrt{(M-n)(n+1)}\ |M,M-2n\rangle, \end{aligned} \tag{7.145}$$

where $0 \leq n \leq M = 2j$ and $m = j - n$. The matrix form of H is diagonal and the form of F is the transpose of that of E.

Now, three generators of Chevalley bases with a given subscript μ satisfy the same commutative relations as Eq. (7.144),

$$[H_\mu,\ E_\mu] = 2E_\mu, \qquad [H_\mu,\ F_\mu] = -2F_\mu, \qquad [E_\mu,\ F_\mu] = H_\mu. \tag{7.146}$$

Thus, in the simple Lie algebra $\mathcal{L}$, three generators E_μ, F_μ, and H_μ span an A_1 subalgebra, denoted by $\mathcal{A}_\mu$. In an irreducible representation space of $\mathcal{L}$, some states span an irreducible representation space of $\mathcal{A}_\mu$, called $\mathcal{A}_\mu$-multiplet. Those $\mathcal{A}_\mu$-multiplets embed in the representation space of $\mathcal{L}$ in a complicated way. The method of the block weight diagram will analyze the complicated overlapping of the $\mathcal{A}_\mu$-multiplets and calculate the matrix forms of the generators.

7.7.2 *Orthonormal Basis States*

In an irreducible representation space of $\mathcal{L}$, the common eigenstates of H_μ are chosen to be the basis states

$$H_\mu \, |\boldsymbol{M}, \boldsymbol{m}\rangle = m_\mu \, |\boldsymbol{M}, \boldsymbol{m}\rangle. \qquad (7.147)$$

The weight $\boldsymbol{m}$ is expanded with respect to the fundamental dominant weight $\boldsymbol{w}_\mu$ so that the components m_μ all are integers. Due to Eq. (7.142) the action of E_μ raises the weight $\boldsymbol{m}$ of the state $|\boldsymbol{M}, \boldsymbol{m}\rangle$ by a simple root $\boldsymbol{r}_\mu$ and the action of F_μ lowers $\boldsymbol{m}$ by $\boldsymbol{r}_\mu$. The states with different weights are orthogonal to each other. For convenience, the basis states are required to be orthonormal. In addition, when the weight is n-multiple, the basis states can be made any arbitrary unitary transformation. When $n = 1$ only the phase of the basis state can be chosen. The phases and the unitary transformations should be chosen carefully as will be discussed later.

In a block weight diagram of an irreducible representation of $\mathcal{L}$, each orthonormal basis state with weight $\boldsymbol{m}$ is denoted by a block filled with its components m_μ. When a component is negative, it is denoted by a number with a bar. For example, a state with a single weight $\boldsymbol{m} = -2\boldsymbol{w}_1 + \boldsymbol{w}_2$ is denoted by $\boxed{\overline{2}, 1}$. When the weight is multiple, an ordinal number is added as a subscript such as $\boxed{(\overline{2}, 1)_2}$.

A suitable combination of states $|\boldsymbol{M}, \boldsymbol{m}\rangle$ with a given weight $\boldsymbol{m}$ in an irreducible representation $D^{\boldsymbol{M}}$ can be obtained by applications of a few lowering operators F_μ to the highest weight state $|\boldsymbol{M}, \boldsymbol{M}\rangle$, so that the difference $\boldsymbol{M} - \boldsymbol{m}$ is equal to the integral sum of the simple roots $\boldsymbol{r}_\mu$,

$$\boldsymbol{M} - \boldsymbol{m} = \sum_{\mu=1}^{\ell} C_\mu \boldsymbol{r}_\mu, \qquad h(\boldsymbol{M}, \boldsymbol{m}) = \sum_{\mu=1}^{\ell} C_\mu. \qquad (7.148)$$

$h(\boldsymbol{M}, \boldsymbol{m})$ is called the height of the weight $\boldsymbol{m}$ in the representation $D^{\boldsymbol{M}}$. The height of the highest weight is 0.

In the block weight diagram, the block for the highest weight is located in the first row. The remaining blocks are arranged downward as the height increases. The blocks for the weights with the same height are put in the same row. Two blocks in the neighbored rows are connected by a special line for the index μ if their states are related by the lowering operator F_μ. The line is indicated by the corresponding matrix entry of F_μ except for the entry 1 which may be neglected for convenience. The line is removed when the matrix entry is 0.

There are two main problems in the calculations of the basis states and the generators of the highest weight representation $D^{\boldsymbol{M}}$. One is how to pick up the $\mathcal{A}_\mu$-multiplets embedding in the representation space. The other is how to determine the multiplicities of the weights.

First, pick up the $\mathcal{A}_\mu$-multiplets. For a normalized basis state $|\boldsymbol{M}, \boldsymbol{m}\rangle$ which has a positive component m_μ of $\boldsymbol{m}$ and is annihilated by E_μ,

$$E_\mu \, |\boldsymbol{M}, \boldsymbol{m}\rangle = 0, \qquad m_\mu > 0, \tag{7.149}$$

one is able to construct an $\mathcal{A}_\mu$-multiplet, which contains $(m_\mu + 1)$ orthonormal basis states, by applications of F_μ successively,

$$\begin{aligned} &F_\mu \, |\boldsymbol{M}, \boldsymbol{m} - n\boldsymbol{r}_\mu\rangle = \sqrt{(m_\mu - n)(n+1)} \; |\boldsymbol{M}, \boldsymbol{m} - (n+1)\boldsymbol{r}_\mu\rangle, \\ &0 \leq n \leq m_\mu - 1, \qquad F_\mu \, |\boldsymbol{M}, \boldsymbol{m} - m_\mu \boldsymbol{r}_\mu\rangle = 0. \end{aligned} \tag{7.150}$$

It can be shown by induction that

$$E_\mu \, |\boldsymbol{M}, \boldsymbol{m} - (n+1)\boldsymbol{r}_\mu\rangle = \sqrt{(m_\mu - n)(n+1)} \; |\boldsymbol{M}, \boldsymbol{m} - n\boldsymbol{r}_\mu\rangle, \tag{7.151}$$

namely, the matrix form of E_μ is the transpose of that of F_μ.

The highest weight state $|\boldsymbol{M}, \boldsymbol{M}\rangle$ satisfies the condition (7.149) so that one first constructs the $\mathcal{A}_\mu$-multiplets from the highest weight state by Eq. (7.150). If there is a basis state in the $\mathcal{A}_\mu$-multiplet again satisfying the condition (7.149) but with another positive component, say m_ν, one is able to construct an $\mathcal{A}_\nu$-multiplet by Eq. (7.150). New $\mathcal{A}_\rho$-multiplets can be constructed one by one from the basis states satisfying the condition (7.149). The connected multiplets beginning from the highest weight state are called a path. The state on the path is said to have a path connecting the highest weight.

If the weights in an $\mathcal{A}_\mu$-multiplet all are single, $|\boldsymbol{M}, \boldsymbol{m} - n\boldsymbol{r}_\mu\rangle$, including their phases, have been chosen to be the orthonormal basis states in the representation space. The difficulty comes from the intersection of, say, an $\mathcal{A}_\mu$-multiplet and an $\mathcal{A}_\nu$-multiplet at a weight $\boldsymbol{m}'$. Generally, those two states with the weight $\boldsymbol{m}'$ are not orthogonal to each other. One can define that one basis state $|\boldsymbol{M}, (\boldsymbol{m}')_1\rangle$ belongs to, say, the $\mathcal{A}_\mu$-multiplet, and the other basis state $|\boldsymbol{M}, (\boldsymbol{m}')_2\rangle$ is orthogonal to the first one. The state with the weight $\boldsymbol{m}'$, belonging to the $\mathcal{A}_\nu$-multiplet, is a combination of above two basis states, and the matrix entries can be determined by the formula

$$[E_\mu, \; F_\nu] = \delta_{\mu\nu} H_\mu. \tag{7.152}$$

The detail will be explained through examples later.

Second, determine the multiplicities of the weights. It is well known that the multiplicities of equivalent weights are the same. Any weight is equivalent to a dominant weight, so that the key is to determine the multiplicities of the dominant weights in the representation.

The highest weight $\boldsymbol{M}$ is a dominant weight and simple. A weight equivalent to $\boldsymbol{M}$ is simple. A weight belongs to the same $\mathcal{A}_\mu$-multiplet as $\boldsymbol{M}$ is simple. A weight connecting to the highest weight through only one path is simple. Beginning with the highest weight state, one constructs $\mathcal{A}_\rho$-multiplets one by one from the basis states satisfying the condition (7.149) until a dominant weight $\boldsymbol{M}_1$ appears. Assume that the multiplicity of the dominant weight $\boldsymbol{M}_1$ is equal to the number n of paths through which $\boldsymbol{M}_1$ connects to the highest weight $\boldsymbol{M}$. Define n orthonormal basis states $|\boldsymbol{M}, (\boldsymbol{M}_1)_a\rangle$, $1 \leq a \leq n$. According to the formula (7.152), one alculates the matrix entry of each F_μ which lowers a state to the states $|\boldsymbol{M}, (\boldsymbol{M}_1)_a\rangle$. If there is a state $|\boldsymbol{M}, (\boldsymbol{M}_1)_a\rangle$ to which the matrix entry of each F_μ is vanishing, it is a null state and the multiplicity of the dominant weight $\boldsymbol{M}_1$ decreases by 1.

7.7.3 *Method of Block Weight Diagram*

For a given highest weight representation $D^{\boldsymbol{M}}$ of a simple Lie algebra $\mathcal{L}$, draw its block weight diagram according to the following steps:

(a) Write the expansions of $\boldsymbol{r}_\nu = \sum_\mu \boldsymbol{w}_\mu A_{\mu\nu}$ (see Eq. (7.132)) where $A_{\mu\nu}$ is the Cartan matrix of $\mathcal{L}$.

(b) From Eq. (7.133), calculate the weights equivalent to the highest weight. Those weights are all single.

(c) In the first row of the block weight diagram, draw a block filled with the components of the highest weight $\boldsymbol{M}$. For each positive component M_μ, one constructs an $\mathcal{A}_\mu$-multiplet by Eq. (7.150). Draw the blocks for the states in the multiplet downward one by one and connect the neighbored blocks with a special line attached with the matrix entry of F_μ as given in Eq. (7.150). Use different line (solid line, dotted line, and double-line) for different μ. The matrix entry is usually neglected if it is equal to 1. The line is removed if the matrix entry is 0.

(d) In the multiplets, find the weight $\boldsymbol{m}$ satisfying the condition (7.149) and construct new $\mathcal{A}_\mu$-multiplet from the state with the weight $\boldsymbol{m}$ until an intersection between multiplets appears. If the weight $\boldsymbol{m}'$ at the intersection is not a dominant weight, its multiplicity is known. If $\boldsymbol{m}'$ is a dominant weight, check the number n of paths through which $\boldsymbol{m}'$ connects to the highest weight $\boldsymbol{M}$, and assume the multiplicity of $\boldsymbol{m}'$ to be n.

Let us explain the calculation for the case where the multiplicity of $\boldsymbol{m}'$ is 2 for simplicity. Assume that the intersecting multiplets are $\mathcal{A}_\mu$-multiplet and $\mathcal{A}_\nu$-multiplet. Define one state $|\boldsymbol{M},(\boldsymbol{m}')_1\rangle$ belonging to the $\mathcal{A}_\mu$-multiplet and the other $|\boldsymbol{M},(\boldsymbol{m}')_2\rangle$ as the highest weight state of a new $\mathcal{A}_\mu$-multiplet, so that $E_\mu|\boldsymbol{M},(\boldsymbol{m}')_1\rangle$ is known and $E_\mu|\boldsymbol{M},(\boldsymbol{m}')_2\rangle = 0$. The matrix entries of F_ν as well as E_ν can be calculated by Eq. (7.152). If $E_\nu|\boldsymbol{M},(\boldsymbol{m}')_2\rangle$ is also equal to 0, $|\boldsymbol{M},(\boldsymbol{m}')_2\rangle = 0$ and $\boldsymbol{m}'$ is single.

(e) If there is a dominant weight appearing in an above multiplet and its multiplicity has been calculated, calculate its equivalent weights by Eq. (7.133). They have the same multiplicity.

(f) Continue this method until a state with a weight $-\boldsymbol{M}'$ appears, where $-\boldsymbol{M}'$ has no positive component (see Theorem 7.9). $-\boldsymbol{M}'$ is the lowest weight in the irreducible representation space $D^{\boldsymbol{M}}$. If $\boldsymbol{M}' = \boldsymbol{M}$, the representation $D^{\boldsymbol{M}}$ is self-conjugate. Otherwise, two representations $D^{\boldsymbol{M}}$ and $D^{\boldsymbol{M}'}$ are conjugate to each other.

Denote by $D^{\boldsymbol{M}'}$ the conjugate representation of $D^{\boldsymbol{M}}$

$$\begin{gathered}
D^{\boldsymbol{M}}(R) = \mathbf{1} - i\sum_A \omega_A D^{\boldsymbol{M}}(I_A), \\
D^{\boldsymbol{M}'}(R) = \mathbf{1} - i\sum_A \omega_A D^{\boldsymbol{M}'}(I_A) = \mathbf{1} - i\sum_A \omega_A \left\{-D^{\boldsymbol{M}}(I_A)^*\right\}, \\
D^{\boldsymbol{M}'}(I_A) = -D^{\boldsymbol{M}}(I_A)^* = -D^{\boldsymbol{M}}(I_A)^T.
\end{gathered}$$

The representation matrix of the Chevalley bases in $D^{\boldsymbol{M}'}$ are

$$\begin{aligned}
D^{\boldsymbol{M}'}(H_\mu) &= -D^{\boldsymbol{M}}(H_\mu), \\
D^{\boldsymbol{M}'}(E_\mu) &= -D^{\boldsymbol{M}}(F_\mu), \\
D^{\boldsymbol{M}'}(F_\mu) &= -D^{\boldsymbol{M}}(E_\mu).
\end{aligned}$$

The minus sign can be attracted into the basis states in the conjugate representation $D^{\boldsymbol{M}'}$

$$|\boldsymbol{M}', -\boldsymbol{m}\rangle = (-1)^{h(\boldsymbol{M},\boldsymbol{m})}\, |\boldsymbol{M},\boldsymbol{m}\rangle^*. \tag{7.153}$$

Thus,

$$\begin{aligned}
D^{\boldsymbol{M}'}_{-\boldsymbol{m}+\boldsymbol{r}_\mu,-\boldsymbol{m}}(E_\mu) &= D^{\boldsymbol{M}}_{\boldsymbol{m}-\boldsymbol{r}_\mu,\boldsymbol{m}}(F_\mu), \\
D^{\boldsymbol{M}'}_{-\boldsymbol{m}-\boldsymbol{r}_\mu,-\boldsymbol{m}}(F_\mu) &= D^{\boldsymbol{M}}_{\boldsymbol{m}+\boldsymbol{r}_\mu,\boldsymbol{m}}(E_\mu).
\end{aligned} \tag{7.154}$$

The block weight diagrams of two conjugate representations are upside down of each other, and the corresponding states are related by Eq. (7.153).

In drawing a block weight diagram one should check step by step. Check whether the difference of two weights related by F_μ is $\boldsymbol{r}_\mu$, whether Eq. (7.152) holds when it applies to each basis state, and whether the multiplicities of the equivalent weights are the same. In the completed block weight diagram, the number of blocks with the same height first increases and then decreases as the height increases, symmetric up and down like a spindle. An excellent table book [Bremner et al. (1985)] is recommended where the useful data of the highest weight representations of all simple Lie algebras with the rank less than 13 are listed such as their dimensions, the number of different heights, and the multiplicities and orbit sizes of the dominant weights in the representation. In the succeeding subsections some examples will be given to explain the method of the block weight diagrams.

7.7.4 *Some Representations of A_2*

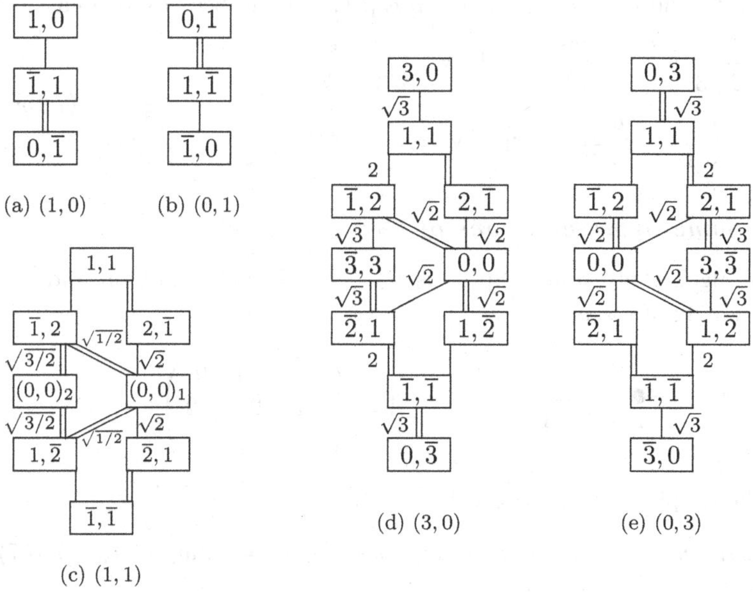

Fig. 7.3 The block weight diagrams of some representations of A_2.

The Lie algebra of the SU(3) group is A_2. It has wide application in particle physics. The representations of A_2 are easy to study by the method of block weight diagrams. The calculations are left to be exercises for readers (see Prob. 9 of Chap. 7 in [Ma and Gu (2004)]). The results are given in Fig. 7.3. From Fig. 7.3 one sees that the representation $(1,0)$ is the self-representation of SU(3), and the representation $(0,1)$ is its conjugate one. The representation $(0,3)$ is conjugate to the representation $(3,0)$. The representation $(1,1)$ is the adjoint representation of SU(3). The matrix entries of generators are indicated. From Fig. 7.3 (c) one sees that there is a double weight $(0,0)$ in the representation $(1,1)$. The matrix entries of generators related to the double weight are listed as follows:

$$\begin{aligned}
&F_1|(1,1),(2,\overline{1})\rangle = \sqrt{2}|(1,1),(0,0)_1\rangle, \\
&F_1|(1,1),(0,0)_1\rangle = \sqrt{2}|(1,1),(\overline{2},1)\rangle, \\
&F_2|(1,1),(\overline{1},2)\rangle = \sqrt{1/2}|(1,1),(0,0)_1\rangle + \sqrt{3/2}|(1,1),(0,0)_2\rangle, \\
&F_2|(1,1),(0,0)_1\rangle = \sqrt{1/2}|(1,1),(1,\overline{2})\rangle, \\
&F_2|(1,1),(0,0)_2\rangle = \sqrt{3/2}|(1,1),(1,\overline{2})\rangle.
\end{aligned} \tag{7.155}$$

Three positive roots can be read from Fig. 7.3 (c), $\boldsymbol{r}_1 = 2\boldsymbol{w}_1 - \boldsymbol{w}_2$, $\boldsymbol{r}_2 = -\boldsymbol{w}_1 + 2\boldsymbol{w}_2$, and $\boldsymbol{\alpha} = \boldsymbol{w}_1 + \boldsymbol{w}_2$. From Eq. (7.79) and Sec. 7.6 one has

$$\begin{aligned}
&\boldsymbol{V}_1 = \frac{\boldsymbol{e}_1}{2\sqrt{3}} + \frac{\boldsymbol{e}_2}{2} = \frac{\boldsymbol{w}_1}{\sqrt{2}}, \qquad \boldsymbol{V}_2 = \frac{\boldsymbol{e}_1}{2\sqrt{3}} - \frac{\boldsymbol{e}_2}{2}, \\
&\boldsymbol{V}_3 = -\frac{\boldsymbol{e}_1}{\sqrt{3}} = -\frac{\boldsymbol{w}_2}{\sqrt{2}}, \qquad \boldsymbol{r}_1 = \sqrt{2}\boldsymbol{e}_2, \qquad \boldsymbol{r}_2 = \sqrt{\frac{3}{2}}\boldsymbol{e}_1 - \sqrt{\frac{1}{2}}\boldsymbol{e}_2.
\end{aligned} \tag{7.156}$$

7.7.5 *Some Representations of C_3*

The Lie algebra of group USp(6) is C_3. The Dynkin diagram and the Cartan matrix of C_3 are

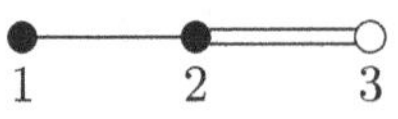

$$A = \begin{pmatrix} 2 & -1 & 0 \\ -1 & 2 & -2 \\ 0 & -1 & 2 \end{pmatrix}.$$

Thus, the simple roots of C_3 are

$$\boldsymbol{r}_1 = 2\boldsymbol{w}_1 - \boldsymbol{w}_2, \qquad \boldsymbol{r}_2 = -\boldsymbol{w}_1 + 2\boldsymbol{w}_2 - \boldsymbol{w}_3, \qquad \boldsymbol{r}_3 = -2\boldsymbol{w}_2 + 2\boldsymbol{w}_3. \tag{7.157}$$

We are going to study three fundamental representations of C_3. First, draw the block weight diagram of $\boldsymbol{M}^{(1)} = (1,0,0)$. The weights equivalent to $\boldsymbol{M}^{(1)}$ are

$$(1,0,0) \xrightarrow{r_1} (\overline{1},1,0) \xrightarrow{r_2} (0,\overline{1},1) \xrightarrow{r_3} (0,1,\overline{1}) \xrightarrow{r_2} (1,\overline{1},0) \xrightarrow{r_1} (\overline{1},0,0).$$

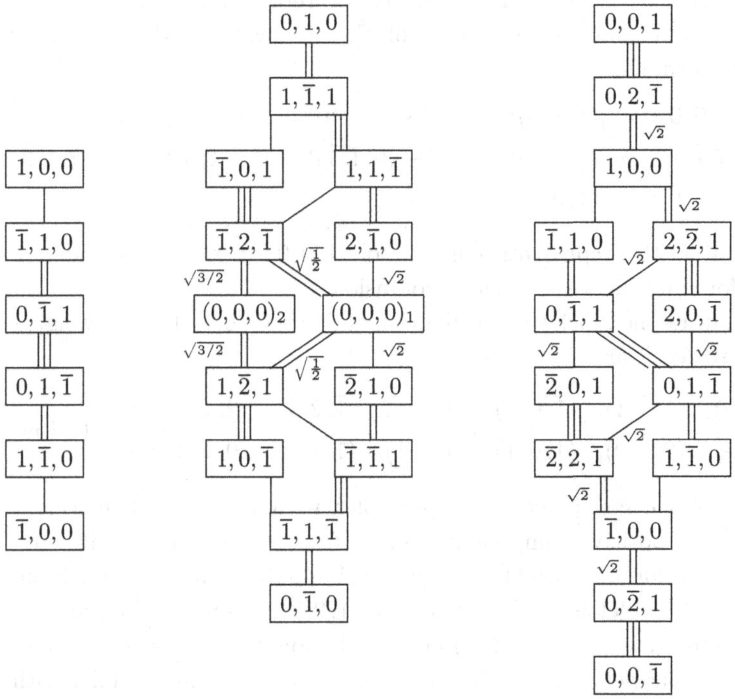

(a) (1,0,0) (b) (0,1,0) (c) (0,0,1)

Fig. 7.4 The block weight diagrams of some representations of C_3.

Since the first component of $\boldsymbol{M}^{(1)}$ is 1, one constructs an $\mathcal{A}_1$-doublet, $(1,0,0)$ and $(\overline{1},1,0)$, where the matrix entry of F_1 is 1 and the difference between the two weights is $\boldsymbol{r}_1 = (2,\overline{1},0)$. Then, from $(\overline{1},1,0)$, one constructs an $\mathcal{A}_2$-doublet with $(0,\overline{1},1)$, where the matrix entry of F_2 is 1 and the difference between the two weights is $\boldsymbol{r}_2 = (\overline{1},2,\overline{1})$. From $(0,\overline{1},1)$, an $\mathcal{A}_3$-doublet is obtained with $(0,1,\overline{1})$, where the matrix entry of F_3 is 1 and the difference between the two weights is $\boldsymbol{r}_3 = (0,\overline{2},2)$. From $(0,1,\overline{1})$, one has an $\mathcal{A}_2$-doublet with $(1,\overline{1},0)$, and the matrix entry of F_2 is 1. At last, from $(1,\overline{1},0)$, one has an $\mathcal{A}_1$-doublet with $(\overline{1},0,0)$, and the matrix entry of F_1 is 1. $(\overline{1},0,0)$ is the lowest weight because it contains no positive component. Since $(\overline{1},0,0) = -\boldsymbol{M}^{(1)}$, the representation $(1,0,0)$ is a self-conjugate representation. In fact, it is the self-representation of USp(6). The block weight diagram is given in Fig. 7.4 (a). In the representation $(1,0,0)$, there is only one dominant weight, and six weights are all single and equivalent to each other. Since the $\mathcal{A}_\mu$-multiplets in the representation are all doublets, the nonvanishing matrix entries of F_μ are always 1, which can also be shown in formulas

$$\begin{aligned} &F_1 \ |(1,0,0)\rangle = \ |(\overline{1},1,0)\rangle, \qquad &&F_2 \ |(\overline{1},1,0)\rangle = \ |(0,\overline{1},1)\rangle, \\ &F_3 \ |(0,\overline{1},1)\rangle = \ |(0,1,\overline{1})\rangle, \qquad &&F_2 \ |(0,1,\overline{1})\rangle = \ |(1,\overline{1},0)\rangle, \\ &F_1 \ |(1,\overline{1},0)\rangle = \ |(\overline{1},0,0)\rangle. \end{aligned} \tag{7.158}$$

Here and after, the representation symbol $(1,0,0)$ in the basis states is neglected for convenience if without confusion.

Second, draw the block weight diagram of $\boldsymbol{M}^{(2)} = (0,1,0)$. The weights equivalent to $\boldsymbol{M}^{(2)}$ are

$$\begin{array}{llllll} (0,1,0), & (1,\overline{1},1), & (\overline{1},0,1), & (1,1,\overline{1}), & (\overline{1},2,\overline{1}), & (2,\overline{1},0), \\ (1,\overline{2},1), & (\overline{2},1,0), & (1,0,\overline{1}), & (\overline{1},\overline{1},1), & (\overline{1},1,\overline{1}), & (0,\overline{1},0). \end{array} \tag{7.159}$$

From $(0,1,0)$ one constructs an $\mathcal{A}_2$-doublet with $(1,\overline{1},1)$. The weight $(1,\overline{1},1)$ has two positive components so that from it two doublets are constructed, an $\mathcal{A}_1$-doublet with $(\overline{1},0,1)$ and an $\mathcal{A}_3$-doublet with $(1,1,\overline{1})$. From $(\overline{1},0,1)$ an $\mathcal{A}_3$-doublet is constructed with $(\overline{1},2,\overline{1})$. From $(1,1,\overline{1})$, an $\mathcal{A}_1$-doublet is constructed with $(\overline{1},2,\overline{1})$ and an $\mathcal{A}_2$-doublet is constructed with $(2,\overline{1},0)$. Now, the weight $(\overline{1},2,\overline{1})$ appears twice, one in an $\mathcal{A}_3$-doublet with $(\overline{1},0,1)$ and the other in an $\mathcal{A}_1$-doublet with $(1,1,\overline{1})$. Since $(\overline{1},2,\overline{1})$ is equivalent to the highest weight $\boldsymbol{M}^{(2)}$, it is single and the two states in two doublets have to coincide with each other. In fact, if $F_3|(\overline{1},0,1)\rangle = |(\overline{1},2,\overline{1})\rangle$, one has from Eq. (7.152)

$$|(1,1,\overline{1})\rangle = F_3|(1,\overline{1},1)\rangle = F_3E_1|(\overline{1},0,1)\rangle = E_1F_3|(\overline{1},0,1)\rangle = E_1|(\overline{1},2,\overline{1})\rangle.$$

From $(\overline{1},2,\overline{1})$, an $\mathcal{A}_2$-triplet is constructed with $(0,0,0)$ and $(1,\overline{2},1)$. From $(2,\overline{1},0)$, an $\mathcal{A}_1$-triplet is constructed with $(0,0,0)$ and $(\overline{2},1,0)$. $(0,0,0)$ is a dominant weight and has two paths connecting to the highest weight $\boldsymbol{M}^{(2)}$ so that one has to assume temporarily that its multiplicity may be 2. Define two basis states $|(0,0,0)_1\rangle$ and $|(0,0,0)_2\rangle$ such that $|(0,0,0)_1\rangle$ belongs to the $\mathcal{A}_1$-triplet with $(2,\overline{1},0)$ and $(\overline{2},1,0)$, and $|(0,0,0)_2\rangle$ belongs to an $\mathcal{A}_1$-singlet,

$$\begin{aligned} &F_1|(2,\overline{1},0)\rangle = \sqrt{2}\ |(0,0,0)_1\rangle, \qquad F_1|(0,0,0)_1\rangle = \sqrt{2}\ |(\overline{2},1,0)\rangle, \\ &E_1|(0,0,0)_2\rangle = F_1|(0,0,0)_2\rangle = 0, \\ &F_2|(\overline{1},2,\overline{1})\rangle = a_1\ |(0,0,0)_1\rangle + a_2\ |(0,0,0)_2\rangle, \qquad a_1^2 + a_2^2 = 2, \\ &F_2|(0,0,0)_1\rangle = a_3\ |(1,\overline{2},1)\rangle, \qquad F_2|(0,0,0)_2\rangle = a_4\ |(1,\overline{2},1)\rangle. \end{aligned}$$

Applying Eq. (7.152) to $|(\overline{1},2,\overline{1})\rangle$, one has

$$E_1F_2|(\overline{1},2,\overline{1})\rangle = a_1\sqrt{2}\ |(2,\overline{1},0)\rangle = F_2E_1|(\overline{1},2,\overline{1})\rangle = \ |(2,\overline{1},0)\rangle.$$

Thus, $a_1 = \sqrt{1/2}$. Choosing the phase of $|(0,0,0)_2\rangle$ such that a_2 is a positive number, one has $a_2 = \sqrt{2-a_1^2} = \sqrt{3/2}$. Applying Eq. (7.152) to $|(0,0,0)_1\rangle$, one has

$$\begin{aligned}&E_2F_2|(0,0,0)_1\rangle = a_3^2\ |(0,0,0)_1\rangle + a_3a_4\ |(0,0,0)_2\rangle \\ &= (F_2E_2+H_2)\,|(0,0,0)_1\rangle = (1/2)\ |(0,0,0)_1\rangle + \sqrt{3/4}\ |(0,0,0)_2\rangle.\end{aligned}$$

Choosing the phase of $|(1,\overline{2},1)\rangle$ such that a_3 is a positive number, one has $a_3 = \sqrt{1/2}$, and then, $a_4 = \sqrt{3/2}$. The lower half of the diagram is symmetric to the upper half. Similarly, one obtains

$$\begin{array}{ll} F_1|(1,\overline{2},1)\rangle = |(\overline{1},\overline{1},1)\rangle, & F_3|(1,\overline{2},1)\rangle = |(1,0,\overline{1})\rangle, \\ F_2|(\overline{2},1,0)\rangle = |(\overline{1},\overline{1},1)\rangle, & F_3|(\overline{1},\overline{1},1)\rangle = |(\overline{1},1,\overline{1})\rangle, \\ F_1|(1,0,\overline{1})\rangle = |(\overline{1},1,\overline{1})\rangle, & F_2|(\overline{1},1,\overline{1})\rangle = |(0,\overline{1},0)\rangle. \end{array}$$

The representation $(0,1,0)$ is self-conjugate with dimension 14. In addition to $\boldsymbol{M}^{(2)}$, it contains another dominant weight $(0,0,0)$ with multiplicity 2.

At last, draw the block weight diagram of $\boldsymbol{M}^{(3)} = (0,0,1)$. The weights equivalent to $\boldsymbol{M}^{(3)}$ are

$$\begin{array}{llll} (0,0,1), & (0,2,\overline{1}), & (2,\overline{2},1), & (2,0,\overline{1}), \\ (\overline{2},0,1), & (\overline{2},2,\overline{1}), & (0,\overline{2},1), & (0,0,\overline{1}). \end{array}$$

From $(0,0,1)$, an $\mathcal{A}_3$-doublet is constructed with $(0,2,\overline{1})$. From $(0,2,\overline{1})$, an $\mathcal{A}_2$-triplet is constructed with $(1,0,0)$ and $(2,\overline{2},1)$,

$$F_3|(0,2,\overline{1})\rangle = \sqrt{2}\ |(1,0,0)\rangle, \qquad F_3|(1,0,0)\rangle = \sqrt{2}\ |(2,\overline{2},1)\rangle.$$

$|(1,0,0)\rangle$ is a dominant weight but has only one path connecting to the highest weight $\boldsymbol{M}^{(3)}$ so that it is single. The weights equivalent to $(1,0,0)$ have been calculated. From $(1,0,0)$ an $\mathcal{A}_1$-doublet is constructed with $(\overline{1},1,0)$. From $(\overline{1},1,0)$ an $\mathcal{A}_2$-doublet is constructed with $(0,\overline{1},1)$. From $(2,\overline{2},1)$ an $\mathcal{A}_1$-triplet is constructed with $(0,\overline{1},1)$ and $(\overline{2},0,1)$. Since $(0,\overline{1},1)$ is equivalent to $(1,0,0)$, it is single and two states in the $\mathcal{A}_2$-doublet with $(\overline{1},1,0)$ and in the $\mathcal{A}_1$-triplet with $(2,\overline{2},1)$ and $(\overline{2},0,1)$ have to coincide. In fact, if $F_2|(\overline{1},1,0)\rangle = |(0,\overline{1},1)\rangle$, one has

$$\sqrt{2}|(2,\overline{2},1)\rangle = F_2|(1,0,0)\rangle = F_2E_1|(\overline{1},1,0)\rangle = E_1F_2|(\overline{1},1,0)\rangle = E_1|(0,\overline{1},1)\rangle\,.$$

The remaining part of the block weight diagram for $(0,0,1)$ can be drawn similarly. The matrix entries of F_μ are listed as follows:

$$\begin{array}{ll}
F_3\ |2,\overline{2},1\rangle = \ |2,0,\overline{1}\rangle, & F_1\ |2,0,\overline{1}\rangle = \sqrt{2}\ |0,1,\overline{1}\rangle, \\
F_1\ |0,1,\overline{1}\rangle = \sqrt{2}\ |\overline{2},2,\overline{1}\rangle, & F_3\ |0,\overline{1},1\rangle = \ |0,1,\overline{1}\rangle, \\
F_3\ |\overline{2},0,1\rangle = \ |\overline{2},2,\overline{1}\rangle, & F_2\ |0,1,\overline{1}\rangle = \ |1,\overline{1},0\rangle, \\
F_1\ |1,\overline{1},0\rangle = \ |\overline{1},0,0\rangle, & F_2\ |\overline{2},2,\overline{1}\rangle = \sqrt{2}\ |\overline{1},0,0\rangle, \\
F_2\ |\overline{1},0,0\rangle = \sqrt{2}\ |0,\overline{2},1\rangle, & F_3\ |0,\overline{2},1\rangle = \ |0,0,\overline{1}\rangle.
\end{array}$$

The representation $(0,0,1)$ is self-conjugate with dimension 14. It contains two dominant weights $(0,0,1)$ and $(1,0,0)$ which are both single.

7.7.6 *Planar Weight Diagrams*

For a two-rank Lie algebra, each weight has two components so that the weights and their multiplicities in a representation D^M can be drawn evidently in a planar rectangular coordinate system, called the planar weight diagram. Since the fundamental dominant weights $\boldsymbol{w}_\mu$ are generally not orthonormal and not necessary to be along the coordinate axes, it is more convenient to express the weights and roots in the unit vectors $\boldsymbol{e}_a$ along the coordinate axes. The planar weight diagram of an irreducible representation is the inversion of that of its conjugate representation with respect to the origin. The planar weight diagram of a self-conjugate irreducible representation is symmetric with respect to the inversion.

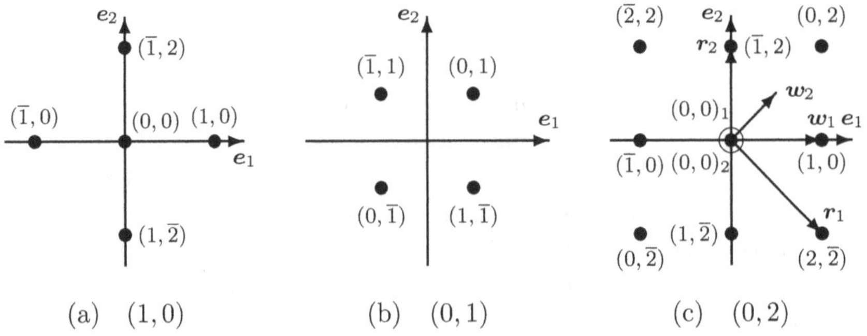

Fig. 7.5 The planar weight diagrams of some representations of SO(5).

For the Lie group SU(3) (the Lie algebra A_2), the expressions of the fundamental dominant weights $\boldsymbol{w}_\mu$ and the simple roots $\boldsymbol{r}_\nu$ with respect to $\boldsymbol{e}_a$ are given in Eq. (7.156). The planar weight diagrams of some representations in A_2 are shown in Fig. 7.3 of [Ma and Gu (2004)]. For the Lie group SO(5) (the Lie algebra B_2), which is locally isomorphic to the group USp(4) (the Lie algebra C_2), the expressions of $\boldsymbol{w}_\mu$ and $\boldsymbol{r}_\nu$ are

$$\boldsymbol{w}_1 = \boldsymbol{e}_1, \qquad \boldsymbol{w}_2 = \frac{1}{2}\left(\boldsymbol{e}_1 + \boldsymbol{e}_2\right), \qquad \boldsymbol{r}_1 = \boldsymbol{e}_1 - \boldsymbol{e}_2, \qquad \boldsymbol{r}_2 = \boldsymbol{e}_2. \quad (7.160)$$

The planar weight diagrams of some representations of B_2 are given in Fig. 7.5. The readers are encouraged to compare the planar weight diagrams of B_2 with those of C_2 which are given in Prob. 10 of Chap. 7 of [Ma and Gu (2004)]. The planar weight diagrams of some representations of G_2 are shown in Prob. 11 of Chap. 7 of [Ma and Gu (2004)].

7.8 Clebsch–Gordan Coefficients

The direct product $D^{\boldsymbol{M}^{(1)}} \times D^{\boldsymbol{M}^{(2)}}$ of two irreducible representations $D^{\boldsymbol{M}^{(1)}}$ and $D^{\boldsymbol{M}^{(2)}}$ of a simple Lie algebra $\mathcal{L}$ is generally reducible and can be reduced by a unitary similarity transformation $C^{\boldsymbol{M}^{(1)}\boldsymbol{M}^{(2)}}$,

$$\left(C^{\boldsymbol{M}^{(1)}\boldsymbol{M}^{(2)}}\right)^{-1}\left(D^{\boldsymbol{M}^{(1)}} \times D^{\boldsymbol{M}^{(2)}}\right) C^{\boldsymbol{M}^{(1)}\boldsymbol{M}^{(2)}} = \bigoplus_{\boldsymbol{M}} a_{\boldsymbol{M}} D^{\boldsymbol{M}}. \quad (7.161)$$

$a_{\boldsymbol{M}}$ is the multiplicity of the representation $D^{\boldsymbol{M}}$ in the reduction. Denote by $d(\boldsymbol{M})$ the dimension of $D^{\boldsymbol{M}}$. The dimension of the space of direct product of two representations is $d(\boldsymbol{M}^{(1)})d(\boldsymbol{M}^{(2)})$, so is the dimension of $C^{\boldsymbol{M}^{(1)}\boldsymbol{M}^{(2)}}$. There are two sets of basis vectors in the product space. Before transformation, the basis states are the products of two basis states of the representations $D^{\boldsymbol{M}^{(1)}}$ and $D^{\boldsymbol{M}^{(2)}}$, denoted by $|\boldsymbol{M}^{(1)}, \boldsymbol{m}^{(1)}\rangle|\boldsymbol{M}^{(2)}, \boldsymbol{m}^{(2)}\rangle$, or briefly by $|\boldsymbol{m}^{(1)}\rangle|\boldsymbol{m}^{(2)}\rangle$. After transformation, the basis states belong to the representation $D^{\boldsymbol{M}}$, denoted by $||\boldsymbol{M}, (r), \boldsymbol{m}\rangle$, where the ordinal index r is used when the multiplicity $a_{\boldsymbol{M}}$ of $D^{\boldsymbol{M}}$ is larger than 1. Two sets of basis states are related by the Clebsch–Gordan (CG) matrix $C^{\boldsymbol{M}^{(1)}\boldsymbol{M}^{(2)}}$,

$$||\boldsymbol{M}, (r), \boldsymbol{m}\rangle = \sum_{\boldsymbol{m}^{(1)}, \boldsymbol{m}^{(2)}} |\boldsymbol{m}^{(1)}\rangle|\boldsymbol{m}^{(2)}\rangle\, C^{\boldsymbol{M}^{(1)}\boldsymbol{M}^{(2)}}_{\boldsymbol{m}^{(1)}, \boldsymbol{m}^{(2)}, \boldsymbol{M}, (r), \boldsymbol{m}}, \quad (7.162)$$

where the sum runs over all states, especially when the weights $\boldsymbol{m}^{(1)}$ and $\boldsymbol{m}^{(2)}$ are multiple. The series on the right-hand side in Eq. (7.161) is called the CG series, and the matrix entries of $C^{\boldsymbol{M}^{(1)}\boldsymbol{M}^{(2)}}$ are the CG coefficients. Applying the generator I_A to the two sides of Eq. (7.162), one has

$$I_A||\boldsymbol{M},\boldsymbol{m}\rangle = \sum_{\boldsymbol{m}^{(1)},\boldsymbol{m}^{(2)}} \left\{\left(I_A|\boldsymbol{m}^{(1)}\rangle\right)|\boldsymbol{m}^{(2)}\rangle + |\boldsymbol{m}^{(1)}\rangle\left(I_A|\boldsymbol{m}^{(2)}\rangle\right)\right\} \cdot C^{\boldsymbol{M}^{(1)}\boldsymbol{M}^{(2)}}_{\boldsymbol{m}^{(1)},\boldsymbol{m}^{(2)},\boldsymbol{M},(r),\boldsymbol{m}} . \tag{7.163}$$

When I_A is taken to be H_μ, one obtains

$$\boldsymbol{m}\, C^{\boldsymbol{M}^{(1)}\boldsymbol{M}^{(2)}}_{\boldsymbol{m}^{(1)},\boldsymbol{m}^{(2)},\boldsymbol{M},(r),\boldsymbol{m}} = \left(\boldsymbol{m}^{(1)} + \boldsymbol{m}^{(2)}\right) C^{\boldsymbol{M}^{(1)}\boldsymbol{M}^{(2)}}_{\boldsymbol{m}^{(1)},\boldsymbol{m}^{(2)},\boldsymbol{M},(r),\boldsymbol{m}} ,$$

namely,

$$C^{\boldsymbol{M}^{(1)}\boldsymbol{M}^{(2)}}_{\boldsymbol{m}^{(1)},\boldsymbol{m}^{(2)},\boldsymbol{M},(r),\boldsymbol{m}} = 0, \qquad \text{when } \boldsymbol{m} \neq \boldsymbol{m}^{(1)} + \boldsymbol{m}^{(2)}. \tag{7.164}$$

The weights before and after the transformation are the same.

There are two main tasks in reducing the direct product of two representations. One is to determine the CG series. The other is to find the expansions (7.162) for the highest weight states $||\boldsymbol{M},(r),\boldsymbol{M}\rangle$

$$||\boldsymbol{M},(r),\boldsymbol{M}\rangle = \sum_{\boldsymbol{m}^{(1)}} |\boldsymbol{m}^{(1)}\rangle|\boldsymbol{M}-\boldsymbol{m}^{(1)}\rangle\, C^{\boldsymbol{M}^{(1)}\boldsymbol{M}^{(2)}}_{\boldsymbol{m}^{(1)},(\boldsymbol{M}-\boldsymbol{m}^{(1)}),\boldsymbol{M},(r),\boldsymbol{M}} . \tag{7.165}$$

The expansions for the remaining states can be calculated by applying the lowering operators F_μ to Eq. (7.165) (see Eq. (7.163)). In this section the method of the dominant weight diagrams will be introduced to calculate the CG series and CG coefficients.

7.8.1 *Representations in the CG Series*

What kind of representations $\boldsymbol{M}$ will appear in the CG series? If $\boldsymbol{m}^{(1)}$ in Eq. (7.165) is not the highest weight $\boldsymbol{M}^{(1)}$, there must exist a raising operator E_μ such that

$$E_\mu\ |\boldsymbol{m}^{(1)}\rangle = a\ |\boldsymbol{m}^{(1)} + \boldsymbol{r}_\mu\rangle + \dots .$$

The suspension points denote the possible states with the multiple weight. Since E_μ annihilates the highest weight state $||\boldsymbol{M},(r),\boldsymbol{M}\rangle$ (see Eq. (7.115)), the expansion (7.165) has to contain another term $|\boldsymbol{m}^{(1)}+\boldsymbol{r}_\mu\rangle|\boldsymbol{M}-\boldsymbol{m}^{(1)}-\boldsymbol{r}_\mu\rangle$ such that $E_\mu|\boldsymbol{M}-\boldsymbol{m}^{(1)}-\boldsymbol{r}_\mu\rangle = b|\boldsymbol{M}-\boldsymbol{m}^{(1)}\rangle + \dots$, and

$$aC^{\boldsymbol{M}^{(1)}\boldsymbol{M}^{(2)}}_{\boldsymbol{m}^{(1)},(\boldsymbol{M}-\boldsymbol{m}^{(1)}),\boldsymbol{M},(r),\boldsymbol{M}} + bC^{\boldsymbol{M}^{(1)}\boldsymbol{M}^{(2)}}_{(\boldsymbol{m}^{(1)}+\boldsymbol{r}_\mu),(\boldsymbol{M}-\boldsymbol{m}^{(1)}-\boldsymbol{r}_\mu),\boldsymbol{M},(r),\boldsymbol{M}} = 0.$$

Thus, if the expansion (7.165) contains a term with $\boldsymbol{m}^{(1)} \neq \boldsymbol{M}^{(1)}$, it has to contain another term with a higher weight, say $\boldsymbol{m}^{(1)} + \boldsymbol{r}_\mu$. In other words, the expansion (7.165) has to contain a term $|\boldsymbol{M}^{(1)}\rangle|\boldsymbol{M} - \boldsymbol{M}^{(1)}\rangle$ and, for the same reason, contain another term $|\boldsymbol{M} - \boldsymbol{M}^{(2)}\rangle|\boldsymbol{M}^{(2)}\rangle$,

$$\boldsymbol{M} = \boldsymbol{M}^{(1)} + \boldsymbol{m}^{(2)} = \boldsymbol{m}^{(1)} + \boldsymbol{M}^{(2)}. \tag{7.166}$$

This is the necessary condition for a highest weight representation $D^{\boldsymbol{M}}$ appearing in the CG series (7.161), namely, it is easy to write the CG series (7.161) from the block weight diagram of $D^{\boldsymbol{M}^{(1)}}$ or $D^{\boldsymbol{M}^{(2)}}$ based on the condition (7.166). The problem is how to calculate the multiplicity $a_{\boldsymbol{M}}$. If $a_{\boldsymbol{M}} = 0$, the representation $D^{\boldsymbol{M}}$ disappears in the CG series.

In the space of direct product $D^{\boldsymbol{M}^{(1)}} \times D^{\boldsymbol{M}^{(2)}}$, the highest weight is single and its basis state is

$$||\boldsymbol{M}_0, \boldsymbol{M}_0\rangle = |\boldsymbol{M}^{(1)}\rangle|\boldsymbol{M}^{(2)}\rangle, \qquad \boldsymbol{M}_0 = \boldsymbol{M}^{(1)} + \boldsymbol{M}^{(2)}. \tag{7.167}$$

Due to the condition (7.166), the highest weight $\boldsymbol{M}$ of any representation in the CG series (7.161) satisfies

$$\boldsymbol{M} = \boldsymbol{M}^{(1)} + \boldsymbol{m}^{(2)} = \boldsymbol{M}_0 - \left[\boldsymbol{M}^{(2)} - \boldsymbol{m}^{(2)}\right] = \boldsymbol{M}_0 - \sum_\mu c_\mu \boldsymbol{r}_\mu. \tag{7.168}$$

$c = \sum_\mu c_\mu$ is called the level of $\boldsymbol{M}$ in the CG series. Hereafter, $\boldsymbol{M}$ with level c is denoted by $\boldsymbol{M}_c$. Sometimes, there are a few highest weights with the same level c, then, they can be denoted by $\boldsymbol{M}_c$, $\boldsymbol{M}'_c$, and so on.

7.8.2 *Method of Dominant Weight Diagram*

There are two sets of basis vectors in the space of the direct product of two representations, $|\boldsymbol{M}^{(1)}, \boldsymbol{m}^{(1)}\rangle|\boldsymbol{M}^{(2)}, \boldsymbol{m}^{(2)}\rangle$ and $||\boldsymbol{M}_c, (r), \boldsymbol{m}\rangle$. $\boldsymbol{M}_c = \boldsymbol{M}^{(1)} + \boldsymbol{m}^{(2)}$ can be found from the block weight diagram of $D^{\boldsymbol{M}^{(2)}}$. The size $OS(\boldsymbol{M}_c)$ of Weyl orbit of $\boldsymbol{M}_c$ is calculated through Eq. (7.133). The number n_c of the linearly independent basis states $|\boldsymbol{M}^{(1)}, \boldsymbol{m}^{(1)}\rangle|\boldsymbol{M}^{(2)}, \boldsymbol{m}^{(2)}\rangle$ with $\boldsymbol{m}^{(1)} + \boldsymbol{m}^{(2)} = \boldsymbol{M}_c$ in the product space can be counted.

The Clebsch–Gordan coefficients are shown through the expansions (7.162). The expansion for the highest weight state of $D^{\boldsymbol{M}_0}$ is given in Eq. (7.167). Applying the lowering operators F_μ to Eq. (7.167), one obtains the expansions of the basis states in $D^{\boldsymbol{M}_0}$, especially those expansions with the weight $\boldsymbol{M}_c$, if exist. Denote by n_{0c} the multiplicity of $\boldsymbol{M}_c$ in $D^{\boldsymbol{M}_0}$. Thus, the multiplicity a_1 of $D^{\boldsymbol{M}_1}$ is equal to the difference $n_1 - n_{01}$.

There are two ways to calculate the expansion of the highest weight state in $D^{\boldsymbol{M}_1}$. One is based on that the highest weight state is orthogonal to those states with the same weight in the representation $D^{\boldsymbol{M}_0}$. The other is based on the property (7.115) that every raising operator E_μ annihilates the highest weight state. Then, by the lowering operators F_μ, calculate the expansions of the basis states with the weight $\boldsymbol{M}_c$ in $D^{\boldsymbol{M}_1}$ as well as its multiplicity n_{1c}. The multiplicity a_2 of $D^{\boldsymbol{M}_2}$ is equal to the difference $n_2 - n_{02} - a_1 n_{12}$. Generally, the multiplicity a_c of $D^{\boldsymbol{M}_c}$ is equal to

$$a_c = n_c - \sum_{c'=0}^{c-1} a_{c'} n_{c'c}. \tag{7.169}$$

The dimension of the representation space $D^{\boldsymbol{M}^{(1)}} \times D^{\boldsymbol{M}^{(2)}}$ is

$$\begin{aligned} d(\boldsymbol{M}^{(1)})d(\boldsymbol{M}^{(2)}) &= \sum_c OS(\boldsymbol{M}_c) n_c = \sum_c a_c d(\boldsymbol{M}_c), \\ d(\boldsymbol{M}_c) &= \sum_{c'} OS(\boldsymbol{M}_{c'}) n_{cc'}. \end{aligned} \tag{7.170}$$

When $\boldsymbol{M}^{(1)} = \boldsymbol{M}^{(2)}$, the expansion (7.162) of the basis state in each representation $D^{\boldsymbol{M}_c}$ is symmetric or antisymmetric with respect to the interchange between $|\boldsymbol{m}^{(1)}\rangle$ and $|\boldsymbol{m}^{(2)}\rangle$ such that the representations $\boldsymbol{M}_c$ in the CG series are divided into symmetric ones and antisymmetric ones. The sum of dimensions of the symmetric representations and antisymmetric ones is $d(\boldsymbol{M}^{(1)})^2$, and their difference is $d(\boldsymbol{M}^{(1)})$.

The method of dominant weight diagram consists of two tables and the calculations of the expansions of the highest weight states of $\boldsymbol{M}_c$. One table lists the dominant weights $\boldsymbol{M}_c$ based on the condition (7.166), their multiplicities n_c, and the linearly independent basis states $|\boldsymbol{m}^{(1)}\rangle|\boldsymbol{M}_c - \boldsymbol{m}^{(1)}\rangle$. The other table consists of a few columns. In the leftmost column the products $OS(\boldsymbol{M}_c) \times n_c$, where $OS(\boldsymbol{M}_c)$ is the size of Weyl orbit of $\boldsymbol{M}_c$, are listed and their sum is the dimension $d(\boldsymbol{M}^{(1)})d(\boldsymbol{M}^{(2)})$ of the product space. Then, on the right of the table, each representation $D^{\boldsymbol{M}_c}$ corresponds to one column and is arranged column by column in the increasing order of c. For each representation $D^{\boldsymbol{M}_c}$, one calculates its multiplicity a_c in the CG series, the expansions of its highest weight states, the multiplicities $n_{cc'}$ of $\boldsymbol{M}_{c'}$ contained in $D^{\boldsymbol{M}_c}$, and its dimension $d(\boldsymbol{M}_c) = \sum_{c'} OS(\boldsymbol{M}_{c'}) n_{cc'}$. In the bottom of the table, the dimension of the product space is calculated to be the sum of $a_c d(\boldsymbol{M}_c)$, and the Clebsch–Gordan series is listed. The method will be explained in detail in the next subsection.

7.8.3 *Reductions of Direct Product Representations in* A_2

The algebra of SU(3) is A_2. The block weight diagrams of some representations of A_2 are listed in Fig. 7.3. We first calculate the reduction of the direct product of two fundamental representations, $(1,0)\times(0,1)$, where the highest weight directly denotes the representation for simplification.

From Fig. 7.3 (b), in the product space there are two dominant weights $\boldsymbol{M}_c$, $(1,1)=(1,0)+(0,1)$ and $(0,0)=(1,0)+(\overline{1},0)$, with the multiplicities 1 and 3, respectively (see the upper table of Fig. 7.6). The expansion for the highest weight state $(1,1)$ is given in Eq. (7.167). The expansions of other basis states in the representation $(1,1)$ are calculated by applying the lowering operators F_μ to the highest weight state where Eq. (7.163) and the block weight diagrams in Fig. 7.3 are used.

$$\begin{aligned}
&||(1,1),(1,1)\rangle = |(1,0)\rangle|(0,1)\rangle,\\
&||(1,1),(\overline{1},2)\rangle = F_1||(1,1),(1,1)\rangle = |(\overline{1},1)\rangle|(0,1)\rangle,\\
&||(1,1),(2,\overline{1})\rangle = F_2||(1,1),(1,1)\rangle = |(1,0)\rangle|(1,\overline{1})\rangle,\\
&||(1,1),(0,0)_1\rangle = \sqrt{1/2}F_1||(1,1),(2,\overline{1})\rangle\\
&\quad = \sqrt{1/2}\left\{|(1,0)\rangle|(\overline{1},0)\rangle + |(\overline{1},1)\rangle|(1,\overline{1})\rangle\right\},\\
&||(1,1),(0,0)_2\rangle = \sqrt{2/3}\left\{F_2||(1,1),(\overline{1},2)\rangle - \sqrt{1/2}||(1,1),(0,0)_1\rangle\right\}\\
&\quad = \sqrt{1/6}\left\{2\left[|(\overline{1},1)\rangle|(1,\overline{1})\rangle + |(0,\overline{1})\rangle|(0,1)\rangle\right]\right.\\
&\qquad \left. - \left[|(1,0)\rangle|(\overline{1},0)\rangle + |(\overline{1},1)\rangle|(1,\overline{1})\rangle\right]\right\}\\
&\quad = \sqrt{1/6}\left\{-|(1,0)\rangle|(\overline{1},0)\rangle + |(\overline{1},1)\rangle|(1,\overline{1})\rangle + 2\,|(0,\overline{1})\rangle|(0,1)\rangle\right\}\\
&||(1,1),(\overline{2},1)\rangle = \sqrt{1/2}F_1||(1,1),(0,0)_1\rangle = |(\overline{1},1)\rangle|(\overline{1},0)\rangle,\\
&||(1,1),(1,\overline{2})\rangle = \sqrt{2}F_2||(1,1),(0,0)_1\rangle = |(0,\overline{1})\rangle|(1,\overline{1})\rangle,\\
&||(1,1),(\overline{1},\overline{1})\rangle = F_2||(1,1),(\overline{2},1)\rangle = |(0,\overline{1})\rangle|(\overline{1},0)\rangle.
\end{aligned} \tag{7.171}$$

The representation $(1,1)$ contains a single dominant weight $(1,1)$ and a double dominant weight $(0,0)$. Thus, the multiplicity of the representation $(0,0)$ is $a_1 = n_1 - n_{01} = 3-2 = 1$. Since the highest weight state $||(0,0),(0,0)\rangle$ is orthogonal to $||(1,1),(0,0)_1\rangle$ and $||(1,1),(0,0)_2\rangle$, one has

$$||(0,0),(0,0)\rangle = \sqrt{1/3}\left\{|(1,0)\rangle|(\overline{1},0)\rangle - |(\overline{1},1)\rangle|(1,\overline{1})\rangle + |(0,\overline{1})\rangle|(0,1)\rangle\right\}. \tag{7.172}$$

The dominant weight diagram for the reduction of $(1,0)\times(0,1)$ of A_2 is given in Fig. 7.6.

$\boldsymbol{M}_c$	n_c	Independent basis states
$(1,1)$	1	$\lvert(1,0)\rangle\lvert(0,1)\rangle$
$(0,0)$	3	$\lvert(1,0)\rangle\lvert(\overline{1},0)\rangle,\ \lvert(\overline{1},1)\rangle\lvert(1,\overline{1})\rangle,\ \lvert(0,\overline{1})\rangle\lvert(0,1)\rangle$

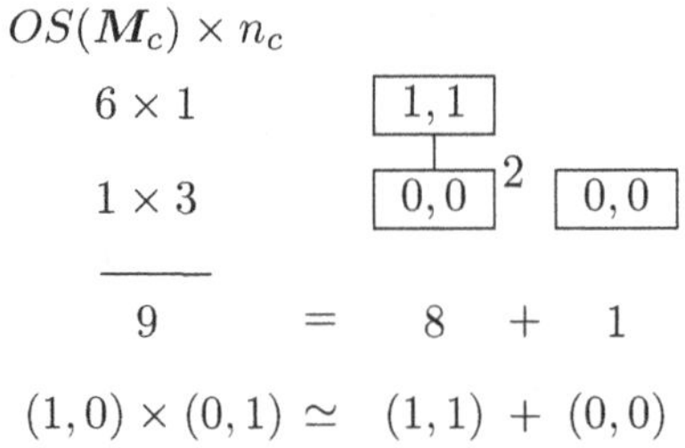

Fig. 7.6 The dominant weight diagram for $(1,0)\times(0,1)$ of A_2.

Now, we calculate the reduction of the direct product of two adjoint representations, $(1,1)\times(1,1)$. From Fig. 7.3 (c), in the product space there are five dominant weights, (2,2), (3,0), (0,3), (1,1), and (0,0). Their multiplicities and the sizes of Weyl orbits are 1, 2, 2, 6, 10 and 6, 3, 3, 6, 1, respectively. The independent basis states are listed in the upper table of Fig. 7.7 where the states by interchanging $|\boldsymbol{m}^{(1)}\rangle$ and $|\boldsymbol{m}^{(2)}\rangle$ are neglected. Except for (2,2), the block weight diagrams of the representations $\boldsymbol{M}_c$ are given in Fig. 7.3. The block weight diagram of (2,2) is left to readers as exercise. The representation $(2,2)$ contains the dominant weights (2,2), (3,0), (0,3), (1,1), and (0,0) with the multiplicities 1, 1, 1, 2, and 3, respectively. The results can also be calculated by the method of the Young operators (see Chap. 8). Based on those data, one calculates the multiplicities of representations in the CG series to be $a_{(2,2)}=a_{(3,0)}=a_{(0,3)}=a_{(0,0)}=1$ and $a_{(1,1)}=2$. In Fig. 7.7, the number on the right shoulder of the block denotes the multiplicity of the dominant weight in the representation, and the number above the the highest weight denotes the multiplicity of the representation in the CG series. The number is neglected if it is equal to 1. The expansion for the highest weight state $(2,2)$ is given in Eq. (7.167), and those of the remaining representations contain the coefficients to be determined:

$$\begin{aligned}
\lvert\lvert(2,2),(2,2)\rangle &= \lvert(1,1)\rangle\lvert(1,1)\rangle,\\
\lvert\lvert(3,0),(3,0)\rangle &= a_1\ \lvert(1,1)\rangle\lvert(2,\overline{1})\rangle + a_2\ \lvert(2,\overline{1})\rangle\lvert(1,1)\rangle,\\
\lvert\lvert(0,3),(0,3)\rangle &= b_1\ \lvert(1,1)\rangle\lvert(\overline{1},2)\rangle + b_2\ \lvert(\overline{1},2)\rangle\lvert(1,1)\rangle,
\end{aligned}$$

$$
\begin{aligned}
&||(1,1),(1,1)\rangle_S = c_1 \left\{|(1,1)\rangle|(0,0)_1\rangle + |(0,0)_1\rangle|(1,1)\rangle\right\} \\
&\quad + c_2 \left\{|(1,1)\rangle|(0,0)_2\rangle + |(0,0)_2\rangle|(1,1)\rangle\right\} \\
&\quad + c_3 \left\{|(\overline{1},2)\rangle|(2,\overline{1})\rangle + |(2,\overline{1})\rangle|(\overline{1},2)\rangle\right\}, \\
&||(1,1),(1,1)\rangle_A = d_1 \left\{|(1,1)\rangle|(0,0)_1\rangle - |(0,0)_1\rangle|(1,1)\rangle\right\} \\
&\quad + d_2 \left\{|(1,1)\rangle|(0,0)_2\rangle - |(0,0)_2\rangle|(1,1)\rangle\right\} \\
&\quad + d_3 \left\{|(\overline{1},2)\rangle|(2,\overline{1})\rangle - |(2,\overline{1})\rangle|(\overline{1},2)\rangle\right\}, \qquad (7.173) \\
&||(0,0),(0,0)\rangle = e_1\ |(1,1)\rangle|(\overline{1},\overline{1})\rangle + e_2\ |(\overline{1},2)\rangle|(1,\overline{2})\rangle \\
&\quad + e_3\ |(2,\overline{1})\rangle|(\overline{2},1)\rangle + e_4\ |(0,0)_1\rangle|(0,0)_1\rangle \\
&\quad + e_5\ |(0,0)_2\rangle|(0,0)_2\rangle + e_6\ |(0,0)_1\rangle|(0,0)_2\rangle \\
&\quad + e_7\ |(0,0)_2\rangle|(0,0)_1\rangle + e_8\ |(\overline{2},1)\rangle|(2,\overline{1})\rangle \\
&\quad + e_9\ |(1,\overline{2})\rangle|(\overline{1},2)\rangle + e_{10}\ |(\overline{1},\overline{1})\rangle|(1,1)\rangle.
\end{aligned}
$$

M_c	n_c	Independent basis states
$(2,2)$	1	$\vert(1,1)\rangle\vert(1,1)\rangle$,
$(3,0)$	2	$\vert(1,1)\rangle\vert(2,\overline{1})\rangle$,
$(0,3)$	2	$\vert(1,1)\rangle\vert(\overline{1},2)\rangle$,
$(1,1)$	6	$\vert(1,1)\rangle\vert(0,0)_1\rangle$, $\vert(1,1)\rangle\vert(0,0)_2\rangle$, $\vert(\overline{1},2)\rangle\vert(2,\overline{1})\rangle$,
$(0,0)$	10	$\vert(1,1)\rangle\vert(\overline{1},\overline{1})\rangle$, $\vert(2,\overline{1})\rangle\vert(\overline{2},1)\rangle$, $\vert(\overline{1},2)\rangle\vert(1,\overline{2})\rangle$,
		$\vert(0,0)_a\rangle\vert(0,0)_b\rangle$, $a,b=1,2$.

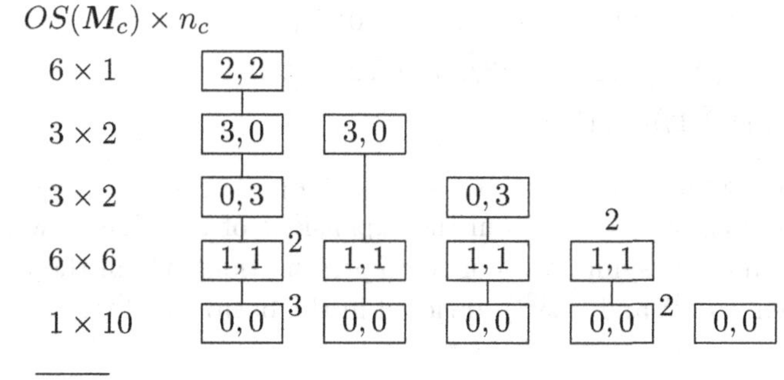

$64 \quad = \ 27 \ + \ 10 \ + \ 10 \ + \ 2\times 8 + \ 1$

$(1,1)\times(1,1) \simeq (2,2)_S + (3,0)_A + (0,3)_A + (1,1)_S + (1,1)_A + (0,0)_S$

Fig. 7.7 The dominant weight diagram for $(1,1)\times(1,1)$ of A_2

Since E_2 annihilates $||(3,0),(3,0)\rangle$, one has $a_1 = -a_2$. Due to normalization, $a_1 = \sqrt{1/2}$. Similarly, $b_1 = -b_2 = \sqrt{1/2}$. Applying E_1 and E_2 to $||(1,1),(1,1)\rangle_S$, one obtains

$$\begin{aligned}
&E_1||(1,1),(1,1)\rangle_S = \sqrt{2}c_1 \left\{|(1,1)\rangle|(2,\overline{1})\rangle + |(2,\overline{1})\rangle|(1,1)\rangle\right\} \\
&\quad + c_3 \left\{|(1,1)\rangle|(2,\overline{1})\rangle + |(2,\overline{1})\rangle|(1,1)\rangle\right\} = 0, \\
&E_2||(1,1),(1,1)\rangle_S = \sqrt{1/2}c_1 \left\{|(1,1)\rangle|(\overline{1},2)\rangle + |(\overline{1},2)\rangle|(1,1)\rangle\right\} \\
&\quad + \sqrt{3/2}c_2 \left\{|(1,1)\rangle|(\overline{1},2)\rangle + |(\overline{1},2)\rangle|(1,1)\rangle\right\} \\
&\quad + c_3 \left\{|(\overline{1},2)\rangle|(1,1)\rangle + |(1,1)\rangle|(\overline{1},2)\rangle\right\} = 0.
\end{aligned}$$

Thus, $\sqrt{2}c_1 + c_3 = 0$ and $c_1 + \sqrt{3}c_2 + \sqrt{2}c_3 = 0$. The normalized solution is $c_1 = \sqrt{3/20}$, $c_2 = \sqrt{1/20}$, and $c_3 = -\sqrt{3/10}$. The formulas for applications of E_1 and E_2 to $||(1,1),(1,1)\rangle_A$ are similar except for replacing c_α with d_α and changing a sign in each curve bracket. From the formulas one obtains $\sqrt{2}d_1 + d_3 = 0$ and $d_1 + \sqrt{3}d_2 - \sqrt{2}d_3 = 0$. The normalized solution is $d_1 = \sqrt{1/12}$, $d_2 = -1/2$, and $d_3 = -\sqrt{1/6}$. Applying E_1 and E_2 to $||(0,0),(0,0)\rangle$, one obtains

$$\begin{aligned}
&E_1||(0,0),(0,0)\rangle = (e_1 + e_2)\,|(1,1)\rangle|(1,\overline{2})\rangle + \sqrt{2}\,(e_3 + e_4)\,|(2,\overline{1})\rangle|(0,0)_1\rangle \\
&\quad + \sqrt{2}\,(e_4 + e_8)\,|(0,0)_1\rangle|(2,\overline{1})\rangle + \sqrt{2}\; e_6|(2,\overline{1})\rangle|(0,0)_2\rangle \\
&\quad + \sqrt{2}\; e_7|(0,0)_2\rangle|(2,\overline{1})\rangle + (e_9 + e_{10})\,|(1,\overline{2})\rangle|(1,1)\rangle = 0, \\
&E_2||(0,0),(0,0)\rangle = (e_1 + e_3)\,|(1,1)\rangle|(\overline{2},1)\rangle \\
&\quad + \sqrt{1/2}\,(e_2 + e_4)\,|(\overline{1},2)\rangle|(0,0)_1\rangle + \sqrt{3/2}\,(e_2 + e_5)\,|(\overline{1},2)\rangle|(0,0)_2\rangle \\
&\quad + \sqrt{1/2}\,(e_4 + e_9)\,|(0,0)_1\rangle|(\overline{1},2)\rangle + \sqrt{3/2}\,(e_5 + e_9)\,|(0,0)_2\rangle|(\overline{1},2)\rangle \\
&\quad + \sqrt{1/2}\; e_6|(\overline{1},2)\rangle|(0,0)_2\rangle + \sqrt{3/2}\; e_6|(0,0)_1\rangle|(\overline{1},2)\rangle \\
&\quad + \sqrt{1/2}\; e_7|(0,0)_2\rangle|(\overline{1},2)\rangle + \sqrt{3/2}\; e_7|(\overline{1},2)\rangle|(0,0)_1\rangle \\
&\quad + (e_8 + e_{10})\,|(\overline{2},1)\rangle|(1,1)\rangle = 0.
\end{aligned}$$

The normalized solution is $e_1 = -e_2 = -e_3 = e_4 = e_5 = -e_8 = -e_9 = e_{10} = \sqrt{1/8}$ and $e_6 = e_7 = 0$. From the expansions of the highest weight states, one obtains the symmetries of the representations $\boldsymbol{M}_c$ in the interchange between $|\boldsymbol{m}^{(1)}\rangle$ and $|\boldsymbol{m}^{(2)}\rangle$, denoted in the bottom of Fig. 7.7.

7.9 Exercises

1. Prove that the number $p_{\alpha\beta} + q_{\alpha\beta} + 1$ of roots in a root chain $\boldsymbol{\alpha} + n\boldsymbol{\beta}$, $-q_{\alpha\beta} \leq n \leq p_{\alpha\beta}$, of a simple Lie algebra is less than five.

2. Calculate the Cartan matrix of the Lie algebra E_6.

3. Draw the Dynkin diagram of a simple Lie algebra where its Cartan matrix is as follows, and indicate the enumeration for the simple roots.

$$A = \begin{pmatrix} 2 & -1 & 0 & 0 \\ -1 & 2 & -1 & 0 \\ 0 & -2 & 2 & -1 \\ 0 & 0 & -1 & 2 \end{pmatrix},$$

$$A_{\mu\nu} = \Gamma\left[\boldsymbol{r}_\nu/\boldsymbol{r}_\mu\right] = \frac{2\boldsymbol{r}_\nu \cdot \boldsymbol{r}_\mu}{\boldsymbol{r}_\mu \cdot \boldsymbol{r}_\mu} .$$

4. Calculate all simple roots and positive roots in the G_2 Lie algebra.

5. Calculate all simple roots and positive roots in the F_4 Lie algebra.

6. Calculate all simple roots and positive roots in the C_2 Lie algebra.

7. Calculate all simple roots and positive roots in the B_3 Lie algebra.

8. Calculate all simple roots and positive roots in the D_4 Lie algebra.

9. Draw the block weight diagrams and the planar weight diagrams of the representations $(1,0)$, $(0,1)$, $(1,1)$, $(3,0)$, and $(0,3)$ of the A_2 Lie algebra (the SU(3) group).

10. Draw the block weight diagrams and the planar weight diagrams of two fundamental representations and the adjoint representation $(2,0)$ of the C_2 Lie algebra.

11. Draw the block weight diagrams and the planar weight diagrams of three representations $(0,1)$, $(1,0)$, and $(0,2)$ of the exceptional Lie algebra G_2.

12. Calculate the Clebsch–Gordan series and the Clebsch–Gordan coefficients for the direct product representation $(1,0) \times (1,0)$ of the C_2 Lie algebra.

13. Calculate the Clebsch–Gordan series and the Clebsch–Gordan coefficients for the direct product representation $(0,1) \times (0,1)$ of the C_2 Lie algebra.

14. Calculate the Clebsch–Gordan series for the direct product representation $(1,0)\times(0,1)$ of the C_2 Lie algebra and the expansion for the highest weight state of each irreducible representation in the Clebsch–Gordan series.

15. Calculate the Clebsch–Gordan series for the following direct product representations of the G_2 Lie algebra and the expansion for the highest weight state of each irreducible representation in the CG series:

$$\text{(a)}\ \ (0,1)\times(0,1),\qquad \text{(b)}\ \ (0,1)\times(1,0),\qquad \text{(c)}\ \ (1,0)\times(1,0),$$

where the dimensions $d(\boldsymbol{M})$ of some representations $\boldsymbol{M}$, the Weyl orbital sizes $OS(\boldsymbol{M})$ of $\boldsymbol{M}$, and the multiplicities of the dominant weights in the representation $\boldsymbol{M}$ are listed in the following table.

			The multiplicity of the dominant weight						
$\boldsymbol{M}$	$d(\boldsymbol{M})$	$OS(\boldsymbol{M})$	$(0,0)$	$(0,1)$	$(1,0)$	$(0,2)$	$(1,1)$	$(0,3)$	$(2,0)$
$(0,0)$	1	1	1						
$(0,1)$	7	6	1	1					
$(1,0)$	14	6	2	1	1				
$(0,2)$	27	6	3	2	1	1			
$(1,1)$	64	12	4	4	2	2	1		
$(0,3)$	77	6	5	4	3	2	1	1	
$(2,0)$	77	6	5	3	3	2	1	1	1

16. Calculate the Clebsch–Gordan series for the direct product representation $(0,0,0,1)\times(0,0,0,1)$ of the F_4 Lie algebra and the expansion for the highest weight state of each irreducible representation in the Clebsch–Gordan series, where the dimensions $d(\boldsymbol{M})$ of some representations $\boldsymbol{M}$, the Weyl orbital sizes $OS(\boldsymbol{M})$ of $\boldsymbol{M}$, and the multiplicities of the dominant weights in the representation $\boldsymbol{M}$ are listed in the following table.

			The multiplicity of the dominant weight				
$\boldsymbol{M}$	$d(\boldsymbol{M})$	OS	$(0,0,0,0)$	$(0,0,0,1)$	$(1,0,0,0)$	$(0,0,1,0)$	$(0,0,0,2)$
$(0,0,0,0)$	1	1	1				
$(0,0,0,1)$	26	24	2	1			
$(1,0,0,0)$	52	24	4	1	1		
$(0,0,1,0)$	273	96	9	5	2	1	
$(0,0,0,2)$	324	24	12	5	3	1	1

Chapter 8

UNITARY GROUPS

In this chapter the reduction of the tensor space is studied by the method of Young operators. The symmetry of the tensor subspaces is characterized by the Young pattern. The independent and complete irreducible basis tensors in the tensor subspace are described by the tensor Young tableaux. Combining with the method of block weight diagrams, one orthogonalizes the irreducible basis tensors which are the orthonormal basis functions belonging to the given irreducible representation. Some applications of the SU(3) group to the particle physics are briefly introduced.

8.1 Irreducible Representations of SU(N)

The concept of tensor is related to the transformation group (see Appendix B). In this chapter, we study the tensor with respect to the SU(N) group. The element of SU(N) is an $N \times N$ unitary matrix, which transforms a vector $\boldsymbol{V}$ in a complex space of dimension N,

$$\boldsymbol{V}_a \xrightarrow{u} \boldsymbol{V}_a' \equiv (O_u \boldsymbol{V})_a = \sum_{b=1}^{N} u_{ab} \boldsymbol{V}_b, \qquad 1 \leq a \leq N. \tag{8.1}$$

A tensor $\boldsymbol{T}_{a_1,\ldots,a_n}$ of rank n contains n indices and N^n components. In the SU(N) transformation u, each index plays the role of a vector index,

$$\boldsymbol{T}_{a_1\ldots a_n} \xrightarrow{u} (O_u \boldsymbol{T})_{a_1\ldots a_n} = \sum_{b_1\ldots b_n} u_{a_1 b_1} \ldots u_{a_n b_n} \boldsymbol{T}_{b_1\ldots b_n}, \tag{8.2}$$

namely, the transformation matrix of a tensor of rank n is the direct product of n matrices u. The direct product of n self-representations is generally reducible. For example, the direct product self-representations of SU(2) are

$$D^{1/2} \times D^{1/2} \times D^{1/2} \simeq D^{3/2} \oplus 2\ D^{1/2},$$
$$D^{1/2} \times D^{1/2} \times D^{1/2} \times D^{1/2} \simeq D^{2} \oplus 3\ D^{1} \oplus 2\ D^{0}.$$

We are going to reduce the direct product representation of SU(N) and find the irreducible basis tensors.

8.1.1 *Reduction of a Tensor Space*

The tensor space $\mathcal{T}$ is an N^n-dimensional linear space which is invariant in the SU(N) transformations, namely, the transformed tensor still belongs to $\mathcal{T}$. The u matrices appear in the transformation (8.2) in the form of the matrix entries u_{ab}. The product of u_{ab} is commutable. Therefore, the permutation symmetry of the tensor indices is invariant in the SU(N) transformation O_u. The subset of tensors with the same permutation symmetry of indices constitutes an invariant subspace of $\mathcal{T}$ in O_u. It is easy to define the totally symmetric tensors and the totally antisymmetric tensors. How to describe the invariant subspaces with other permutation symmetries? For example, one cannot define a tensor of rank 3 whose first two indices are symmetric and the last two indices are antisymmetric.

We begin with the definition of a permutation R on a tensor, $R\boldsymbol{T} = \boldsymbol{T}_R$, which has to satisfy the group property. The definition for a transposition $(j\ k)$ contains no confusion because it is self-inverse,

$$[(j\ k)\boldsymbol{T}]_{a_1 \ldots a_j \ldots a_k \ldots a_n} = \boldsymbol{T}'_{a_1 \ldots a_j \ldots a_k \ldots a_n} = \boldsymbol{T}_{a_1 \ldots a_k \ldots a_j \ldots a_n}. \tag{8.3}$$

A permutation R on a tensor can be defined based on the property that a permutation R can be decomposed into a product of transpositions. For a simple example, one has

$$\begin{gathered} R = \begin{pmatrix} 1 & 2 & 3 \\ 2 & 3 & 1 \end{pmatrix} = \begin{pmatrix} 3 & 1 & 2 \\ 1 & 2 & 3 \end{pmatrix} = (1\ 2\ 3) = (1\ 2)(2\ 3)\ , \\ [(2\ 3)\boldsymbol{T}]_{a_1 a_2 a_3} = \boldsymbol{T}'_{a_1 a_2 a_3} = \boldsymbol{T}_{a_1 a_3 a_2}, \\ (R\boldsymbol{T})_{a_1 a_2 a_3} = [(1\ 2)\boldsymbol{T}']_{a_1 a_2 a_3} = \boldsymbol{T}'_{a_2 a_1 a_3} = \boldsymbol{T}_{a_2 a_3 a_1} \neq T_{a_3 a_1 a_2}. \end{gathered} \tag{8.4}$$

Generally,

$$R = \begin{pmatrix} 1 & 2 & \ldots & n \\ r_1 & r_2 & \ldots & r_n \end{pmatrix} = \begin{pmatrix} \overline{r}_1 & \overline{r}_2 & \ldots & \overline{r}_n \\ 1 & 2 & \ldots & n \end{pmatrix}, \tag{8.5}$$

$$(R\boldsymbol{T})_{a_1 \ldots a_n} \equiv (\boldsymbol{T}_R)_{a_1 \ldots a_n} = \boldsymbol{T}_{a_{r_1} \ldots a_{r_n}} \neq \boldsymbol{T}_{a_{\overline{r}_1} \ldots a_{\overline{r}_n}}. \tag{8.6}$$

The permutation R moves the r_jth index a_{r_j}, NOT the $\overline{r}_j$th index $a_{\overline{r}_j}$, to the jth position. Equivalently, R moves the jth index a_j to the $\overline{r}_j$th position, NOT to the r_jth position. The set of permutations of tensors forms the permutation group S_n and the tensor space $\mathcal{T}$ is invariant in S_n.

It is well known that a tensor of rank 2 can be decomposed into the sum of symmetric and antisymmetric tensors,

$$\boldsymbol{T}_{ab} = \frac{1}{2}\left\{\boldsymbol{T}_{ab} + \boldsymbol{T}_{ba}\right\} + \frac{1}{2}\left\{\boldsymbol{T}_{ab} - \boldsymbol{T}_{ba}\right\}.$$

The decomposition can be written in terms of the Young operators,

$$\begin{aligned}\boldsymbol{T}_{ab} &= \frac{1}{2}\left\{E + (1\ 2)\right\}\boldsymbol{T}_{ab} + \frac{1}{2}\left\{E - (1\ 2)\right\}\boldsymbol{T}_{ab} \\ &= \frac{1}{2}\left\{\mathcal{Y}^{[2]} + \mathcal{Y}^{[1,1]}\right\}\boldsymbol{T}_{ab} = E\boldsymbol{T}_{ab},\end{aligned} \tag{8.7}$$

namely, the decomposition can be achieved by the expansion (6.62) of the identical element with respect to the Young operators which are the projective operators,

$$\boldsymbol{T}_{a_1\ldots a_n} = E\boldsymbol{T}_{a_1\ldots a_n} = \frac{1}{n!}\sum_{[\lambda]} d_{[\lambda]} \sum_{\mu} \mathcal{Y}_\mu^{[\lambda]} y_\mu^{[\lambda]} \boldsymbol{T}_{a_1\ldots a_n}. \tag{8.8}$$

For example, a tensor of rank 3 is decomposed as

$$\boldsymbol{T}_{abc} = \frac{1}{6}\mathcal{Y}^{[3]}\boldsymbol{T}_{abc} + \frac{1}{3}\mathcal{Y}_1^{[2,1]}\boldsymbol{T}_{abc} + \frac{1}{3}\mathcal{Y}_2^{[2,1]}\boldsymbol{T}_{abc} + \frac{1}{6}\mathcal{Y}^{[1,1,1]}\boldsymbol{T}_{abc}. \tag{8.9}$$

The first term is a totally symmetric tensor, the last term is a totally antisymmetric tensor, and the remaining terms are tensors with mixed symmetry. The Young operators decompose the tensor space $\mathcal{T}$ into the direct sum of tensor subspaces $\mathcal{T}_\mu^{[\lambda]}$,

$$\mathcal{T} = E\mathcal{T} = \frac{1}{n!}\bigoplus_{[\lambda]} d_{[\lambda]} \bigoplus_{\mu} \mathcal{Y}_\mu^{[\lambda]} y_\mu^{[\lambda]} \mathcal{T} = \bigoplus_{[\lambda]}\bigoplus_{\mu} \mathcal{T}_\mu^{[\lambda]}. \tag{8.10}$$

Let us study the property of the tensor subspace $\mathcal{T}_\mu^{[\lambda]}$. First, there is no common tensor between two subspaces $\mathcal{T}_\mu^{[\lambda]}$ and $\mathcal{T}_\nu^{[\omega]}$ because the Young operators are orthogonal to each other. Thus, the decomposition (8.10) is in the form of direct sum. Second, the constant factor $d_{[\lambda]}/n!$ and the operator $y_\mu^{[\lambda]}$ do not make any change with the subspace $\mathcal{T}_\mu^{[\lambda]}$. In fact, due to $y_\mu^{[\lambda]}\mathcal{T} \subset \mathcal{T}$ and $\mathcal{Y}_\mu^{[\lambda]}\mathcal{T} \subset \mathcal{T}$, one has

$$\mathcal{Y}_\mu^{[\lambda]}\left\{y_\mu^{[\lambda]}\mathcal{T}\right\} \subset \mathcal{Y}_\mu^{[\lambda]}\mathcal{T}, \qquad \mathcal{Y}_\mu^{[\lambda]}\mathcal{T} = \mathcal{Y}_\mu^{[\lambda]} y_\mu^{[\lambda]}\left\{\frac{d_{[\lambda]}}{n!}\mathcal{Y}_\mu^{[\lambda]}\mathcal{T}\right\} \subset \mathcal{Y}_\mu^{[\lambda]} y_\mu^{[\lambda]}\mathcal{T}.$$

Thus,

$$\mathcal{T}_\mu^{[\lambda]} = \frac{d_{[\lambda]}}{n!}\,\mathcal{Y}_\mu^{[\lambda]} y_\mu^{[\lambda]} \mathcal{T} = \mathcal{Y}_\mu^{[\lambda]} \mathcal{T}. \tag{8.11}$$

For the same reason,

$$R\mathcal{T} = \mathcal{T}, \qquad \mathcal{Y}_\mu^{[\lambda]} R\mathcal{T} = \mathcal{T}_\mu^{[\lambda]}, \qquad \forall\ R \in \mathrm{S}_n. \tag{8.12}$$

Third, as shown in Theorem 8.1, the subspace $\mathcal{T}_\mu^{[\lambda]}$ is invariant in O_u.

Theorem 8.1 (Weyl reciprocity) The permutation R and the SU(N) transformation O_u for a tensor is commutable with each other .

Proof The key of the proof is that the matrix entries u_{ab} in Eq. (8.2) are commutable:

$$\begin{aligned}(O_u R\boldsymbol{T})_{a_1\ldots a_n} &= (O_u \boldsymbol{T}_R)_{a_1\ldots a_n} = \sum_{b_1\ldots b_n} u_{a_1 b_1}\ldots u_{a_n b_n}\,(\boldsymbol{T}_R)_{b_1\ldots b_n} \\ &= \sum_{b_1\ldots b_n} u_{a_{r_1} b_{r_1}}\ldots u_{a_{r_n} b_{r_n}} \boldsymbol{T}_{b_{r_1}\ldots b_{r_n}} = (O_u\boldsymbol{T})_{a_{r_1}\ldots a_{r_n}} = (RO_u\boldsymbol{T})_{a_1\ldots a_n}.\end{aligned} \tag{8.13}$$

Thus, $O_u\left\{\mathcal{Y}_\mu^{[\lambda]}\boldsymbol{T}\right\} = \mathcal{Y}_\mu^{[\lambda]}\left\{O_u\boldsymbol{T}\right\} \subset \mathcal{Y}_\mu^{[\lambda]}\mathcal{T} = \mathcal{T}_\mu^{[\lambda]}$. □

Remind that R on the left of $\mathcal{Y}_\mu^{[\lambda]}\mathcal{T}$ may change the subspace

$$R_{\nu\mu}\mathcal{Y}_\mu^{[\lambda]}\mathcal{T} = \mathcal{Y}_\nu^{[\lambda]} R_{\nu\mu}\mathcal{T} = \mathcal{T}_\nu^{[\lambda]}, \tag{8.14}$$

where $R_{\nu\mu}$ is the permutation transforming the standard Young tableau $\mathcal{Y}_\mu^{[\lambda]}$ to the standard Young tableau $\mathcal{Y}_\nu^{[\lambda]}$.

8.1.2 *Basis Tensors in the Tensor Subspace*

First of all, we review the property of the basis vectors (see §4.7). A basis vector $\boldsymbol{\theta}_d$ is a special vector with only one nonvanishing component, which is equal to 1, $(\boldsymbol{\theta}_d)_a = \delta_{da}$. Any vector $\boldsymbol{V}$ can be expanded with respect to the basis vectors

$$\boldsymbol{V} = \sum_{d=1}^{N} V_d\boldsymbol{\theta}_d, \qquad (\boldsymbol{V})_a = \sum_{d=1}^{N} V_d\,(\boldsymbol{\theta}_d)_a = V_a. \tag{8.15}$$

Note that $(\boldsymbol{V})_a$ and V_a are different in the SU(N) transformation although they are equal in value.

A basis tensor $\boldsymbol{\theta}_{d_1\ldots d_n}$ is a special tensor with only one nonvanishing component which is equal to 1,

$$(\boldsymbol{\theta}_{d_1\ldots d_n})_{a_1\ldots a_n} = \delta_{d_1a_1}\delta_{d_2a_2}\ldots\delta_{d_na_n} = (\boldsymbol{\theta}_{d_1})_{a_1}(\boldsymbol{\theta}_{d_2})_{a_2}\ldots(\boldsymbol{\theta}_{d_n})_{a_n}. \tag{8.16}$$

Being a tensor, the basis tensor transforms in O_u and in R as follows:

$$\begin{aligned}
(O_u\boldsymbol{\theta}_{d_1\ldots d_n})_{a_1\ldots a_n} &= \sum_{b_1\ldots b_n} u_{a_1b_1}\cdots u_{a_nb_n}(\boldsymbol{\theta}_{d_1\ldots d_n})_{b_1\ldots b_n} \\
&= u_{a_1d_1}\cdots u_{a_nd_n} = \sum_{b_1\ldots b_n} (\boldsymbol{\theta}_{b_1\ldots b_n})_{a_1\ldots a_n} u_{b_1d_1}\cdots u_{b_nd_n}, \\
(R\boldsymbol{\theta}_{d_1\ldots d_n})_{a_1\ldots a_n} &= (\boldsymbol{\theta}_{d_1\ldots d_n})_{a_{r_1}\ldots a_{r_n}} = \delta_{d_1a_{r_1}}\delta_{d_2a_{r_2}}\cdots\delta_{d_na_{r_n}} \\
&= \delta_{d_{\overline{r}_1}a_1}\delta_{d_{\overline{r}_2}a_2}\cdots\delta_{d_{\overline{r}_n}a_n} = \left(\boldsymbol{\theta}_{d_{\overline{r}_1}\ldots d_{\overline{r}_n}}\right)_{a_1\ldots a_n},
\end{aligned}$$

namely,

$$O_u\boldsymbol{\theta}_{d_1\ldots d_n} = \sum_{b_1\ldots b_n} \boldsymbol{\theta}_{b_1\ldots b_n} u_{b_1d_1}\cdots u_{b_nd_n}, \tag{8.17}$$

$$R\boldsymbol{\theta}_{d_1\ldots d_n} = \boldsymbol{\theta}_{d_{\overline{r}_1}\ldots d_{\overline{r}_n}} \neq \boldsymbol{\theta}_{d_{r_1}\ldots d_{r_n}}. \tag{8.18}$$

R transforms a basis tensor $\boldsymbol{\theta}_{d_1\ldots d_n}$ to another basis tensor $\boldsymbol{\theta}_{d_{\overline{r}_1}\ldots d_{\overline{r}_n}}$, where the jth index d_j moves to the r_jth position, NOT to the $\overline{r}_j$th position. Equivalently, the $\overline{r}_j$th index $d_{\overline{r}_j}$, NOT the r_jth index d_{r_j}, moves to the jth position. In a simple example $R = (1\ 2\ 3) = (1\ 2)(2\ 3)$,

$$(2\ 3)\boldsymbol{\theta}_{d_1d_2d_3} = \boldsymbol{\Theta}_{d_1d_3d_2}, \qquad R\boldsymbol{\theta}_{d_1d_2d_3} = (1\ 2)\boldsymbol{\theta}_{d_1d_3d_2} = \boldsymbol{\Theta}_{d_3d_1d_2} \neq \boldsymbol{\theta}_{d_2d_3d_1}.$$

Any tensor $\boldsymbol{T}$ can be expanded with respect to the basis tensors

$$\boldsymbol{T}_{a_1\ldots a_n} = \sum_{d_1\ldots d_n} T_{d_1\ldots d_n}(\boldsymbol{\theta}_{d_1\ldots d_n})_{a_1\ldots a_n} = T_{a_1\ldots a_n}. \tag{8.19}$$

In the SU(N) transformation O_u, $\boldsymbol{T}_{a_1\ldots a_n}$ is a tensor, but $T_{a_1\ldots a_n}$ is a scalar.

$\boldsymbol{\theta}_{d_1\ldots d_n}$ is the basis tensor of the tensor space $\mathcal{T}$, and $\mathcal{Y}^{[\lambda]}_{\mu}\boldsymbol{\theta}_{d_1\ldots d_n}$ is the basis tensor of the subspace $\mathcal{T}^{[\lambda]}_{\mu}$. Since the dimension of $\mathcal{T}^{[\lambda]}_{\mu}$ is less than that of $\mathcal{T}$, some $\mathcal{Y}^{[\lambda]}_{\mu}\boldsymbol{\theta}_{d_1\ldots d_n}$ may be vanishing or linearly dependent. Our task is to find the complete set of the basis tensors of $\mathcal{T}^{[\lambda]}_{\mu}$, which are linearly independent.

After the application of a Young operator, $\mathcal{Y}\boldsymbol{\theta}_{d_1\ldots d_n}$ is a linear combination of $\boldsymbol{\theta}_{d_1\ldots d_n}$. For example, if the Young tableau $\mathcal{Y}$ is

1	2	4
3	5	

,

$$\begin{aligned}
\mathcal{Y}\boldsymbol{\theta}_{11233} &= [E + (1\ 2) + (1\ 4) + (2\ 4) + (1\ 2\ 4) + (2\ 1\ 4)]\,[E + (3\ 5)] \\
&\quad\cdot\ [E - (1\ 3)]\,[E - (2\ 5)]\,\boldsymbol{\theta}_{11233} \\
&= [E + (3\ 5)]\,[E + (1\ 2) + (1\ 4) + (2\ 4) + (1\ 2\ 4) + (2\ 1\ 4)] \\
&\quad\cdot\ [\boldsymbol{\theta}_{11233} - \boldsymbol{\theta}_{13231} - \boldsymbol{\theta}_{21133} + \boldsymbol{\theta}_{23131}]
\end{aligned}$$

$$\begin{aligned}
&= 2\left[\boldsymbol{\theta}_{11233} + \boldsymbol{\theta}_{31213} + \boldsymbol{\theta}_{13213}\right] + 2\left[\boldsymbol{\theta}_{11332} + \boldsymbol{\theta}_{31312} + \boldsymbol{\theta}_{13312}\right] \\
&\quad - 2\left[\boldsymbol{\theta}_{13231} + \boldsymbol{\theta}_{31231} + \boldsymbol{\theta}_{33211}\right] - 2\left[\boldsymbol{\theta}_{13132} + \boldsymbol{\theta}_{31132} + \boldsymbol{\theta}_{33112}\right] \\
&\quad - \left[\boldsymbol{\theta}_{21133} + \boldsymbol{\theta}_{12133} + \boldsymbol{\theta}_{13123} + \boldsymbol{\theta}_{31123} + \boldsymbol{\theta}_{23113} + \boldsymbol{\theta}_{32113}\right] \\
&\quad - \left[\boldsymbol{\theta}_{21331} + \boldsymbol{\theta}_{12331} + \boldsymbol{\theta}_{13321} + \boldsymbol{\theta}_{31321} + \boldsymbol{\theta}_{23311} + \boldsymbol{\theta}_{32311}\right] \\
&\quad + 4\left[\boldsymbol{\theta}_{23131} + \boldsymbol{\theta}_{32131} + \boldsymbol{\theta}_{33121}\right] .
\end{aligned} \tag{8.20}$$

Usually, the expansion of $\mathcal{Y}\boldsymbol{\theta}_{abcde}$ is quite long. In order to simplify the notations we introduce a graphic method to denote the tensor expansion. $\mathcal{Y}\boldsymbol{\theta}_{a_1\ldots a_n}$ is denoted by a tensor Young tableau where the box filled with j in the Young tableau $\mathcal{Y}$ is now filled with the subscript a_j. For example, the tensor in Eq. (8.20) is denoted by a tensor Young tableau $\begin{array}{|c|c|c|}\hline 1&1&3\\ \hline 2&3\\ \cline{1-2}\end{array}$. Another example is that if the Young tableau $\mathcal{Y}$ is

$$\begin{array}{|c|c|c|}\hline 1&2&6\\ \hline 3&5&7\\ \hline 4\\ \cline{1-1}\end{array}\ ,$$

the tensor $\mathcal{Y}\boldsymbol{\theta}_{a_1\ldots a_7}$ is denoted by a tensor Young tableau

$$\mathcal{Y}^{[\lambda]}_{\mu}\boldsymbol{\theta}_{a_1\ldots a_7} = \begin{array}{|c|c|c|}\hline a_1&a_2&a_6\\ \hline a_3&a_5&a_7\\ \hline a_4\\ \cline{1-1}\end{array}\ .$$

A tensor Young tableau is a linear combination of the basis tensor $\boldsymbol{\theta}_{a_1\ldots a_n}$ which describes a tensor in a given tensor subspace. In different tensor subspaces, the same tensor Young tableau describes different tensors. For example, in two tensor subspaces $\mathcal{Y}^{[2,1]}_1\mathcal{T}$ and $\mathcal{Y}^{[2,1]}_2\mathcal{T}$ for SU(3),

$$\text{Young tableau } \mathcal{Y}^{[2,1]}_1 = \begin{array}{|c|c|}\hline 1&2\\ \hline 3\\ \cline{1-1}\end{array}\ , \qquad \text{Young tableau } \mathcal{Y}^{[2,1]}_2 = \begin{array}{|c|c|}\hline 1&3\\ \hline 2\\ \cline{1-1}\end{array}\ ,$$

the tensor Young tableau $\begin{array}{|c|c|}\hline 1&2\\ \hline 3\\ \cline{1-1}\end{array}$ describes two different tensors,

$$\begin{aligned}
\mathcal{Y}^{[2,1]}_1\boldsymbol{\theta}_{123} &= \begin{array}{|c|c|}\hline 1&2\\ \hline 3\\ \cline{1-1}\end{array} = \boldsymbol{\theta}_{123} - \boldsymbol{\theta}_{321} + \boldsymbol{\theta}_{213} - \boldsymbol{\theta}_{231} \in \mathcal{Y}^{[2,1]}_1\mathcal{T}, \\
\mathcal{Y}^{[2,1]}_2\boldsymbol{\theta}_{132} &= \begin{array}{|c|c|}\hline 1&2\\ \hline 3\\ \cline{1-1}\end{array} = \boldsymbol{\theta}_{132} - \boldsymbol{\theta}_{312} + \boldsymbol{\theta}_{231} - \boldsymbol{\theta}_{213} \in \mathcal{Y}^{[2,1]}_2\mathcal{T}.
\end{aligned}$$

They are related by a permutation (2 3),

$$(2\ 3)\mathcal{Y}^{[2,1]}_1\boldsymbol{\theta}_{123} = \mathcal{Y}^{[2,1]}_2(2\ 3)\boldsymbol{\theta}_{123} = \mathcal{Y}^{[2,1]}_2\boldsymbol{\theta}_{132}\ . \tag{8.21}$$

Please do not get confused between a tensor Young tableau and a Young tableau.

Study the symmetry of the tensor Young tableaux based on the symmetry (6.28) of a Young operator and the Fock condition (6.31)

$$\mathcal{Y}_\mu^{[\lambda]} Q = \delta(Q) \mathcal{Y}_\mu^{[\lambda]}, \qquad \mathcal{Y}_\mu^{[\lambda]} \left\{ E - \sum_\mu (c_\mu \ d_\nu) \right\} = 0. \tag{8.22}$$

Denote by Q_0 a vertical transposition of the Young tableau $\mathcal{Y}_\mu^{[\lambda]}$. Q_0 interchanges two digits in the same column of the Young tableau $\mathcal{Y}_\mu^{[\lambda]}$, say i and j. Right-multiplying Q_0 on $\mathcal{Y}_\mu^{[\lambda]}$, one obtains a minus sign. But, left-multiplying Q_0 on the basis tensor $\boldsymbol{\theta}_{a_1 \ldots a_n}$ interchanges two subscripts a_i and a_j in $\boldsymbol{\theta}_{a_1 \ldots a_n}$. Two digits a_i and a_j are located in the same column of the tensor Young tableau $\mathcal{Y}_\mu^{[\lambda]} Q_0 \boldsymbol{\theta}_{a_1 \ldots a_n}$, namely, two digits in the same column of a tensor Young tableau are antisymmetric. A tensor Young tableau with the repetitive digits in the same column must be 0. The row number of a nonvanishing tensor Young tableau for SU(N) is not larger than N. The Fock condition also gives some relations between the tensor Young tableaux in a tensor subspace. For example,

$$\mathcal{Y}_1^{[2,1]} \boldsymbol{\theta}_{abc} = -\mathcal{Y}_1^{[2,1]} (1\ 3) \boldsymbol{\theta}_{abc}, \qquad \mathcal{Y}_1^{[2,1]} \boldsymbol{\theta}_{abc} = \mathcal{Y}_1^{[2,1]} \left[(2\ 1) + (2\ 3) \right] \boldsymbol{\theta}_{abc},$$

$$\begin{array}{|c|c|} \hline a & b \\ \hline c & \\ \cline{1-1} \end{array} = - \begin{array}{|c|c|} \hline c & b \\ \hline a & \\ \cline{1-1} \end{array}\,, \qquad \begin{array}{|c|c|} \hline a & b \\ \hline c & \\ \cline{1-1} \end{array} = \begin{array}{|c|c|} \hline b & a \\ \hline c & \\ \cline{1-1} \end{array} + \begin{array}{|c|c|} \hline a & c \\ \hline b & \\ \cline{1-1} \end{array}\,. \tag{8.23}$$

Remind that although left-multiplying with a horizontal permutation P of a Young operator $\mathcal{Y}$ does not change the Young operator $\mathcal{Y}$ as well as the tensor Young tableau $\mathcal{Y}\boldsymbol{\theta}_{a_1 \ldots a_n}$, P does not make a symmetry of digits in the tensor Young tableau. For example, the tensor Young tableau $\mathcal{Y}_1^{[2,1]} \boldsymbol{\theta}_{abc}$ is invariant by left-multiplying it with (1 2). The expansion of $\mathcal{Y}_1^{[2,1]} \boldsymbol{\theta}_{abc}$ is symmetric between the first two indices, but not between a and b,

$$\begin{aligned} (1\ 2) \mathcal{Y}_1^{[2,1]} \boldsymbol{\theta}_{abc} &= (1\ 2) \left\{ \boldsymbol{\theta}_{abc} - \boldsymbol{\theta}_{cba} + \boldsymbol{\theta}_{bac} - \boldsymbol{\theta}_{bca} \right\} \\ &= \boldsymbol{\theta}_{bac} - \boldsymbol{\theta}_{bca} + \boldsymbol{\theta}_{abc} - \boldsymbol{\theta}_{cba} = \mathcal{Y}_1^{[2,1]} \boldsymbol{\theta}_{abc} \,. \end{aligned}$$

Now we return to the problem how to find the complete set of the linearly independent basis tensors in the subspace $\mathcal{T}_\mu^{[\lambda]} = \mathcal{Y}_\mu^{[\lambda]} \mathcal{T}$, where $\mathcal{Y}_\mu^{[\lambda]}$ is a standard Young operator. Discuss a useful example first. The general form of a tensor Young tableau in the tensor subspace $\mathcal{T}_1^{[2,1]}$ of the tensor space $\mathcal{T}$ of rank 3 of SU(3) is

$$\begin{array}{|c|c|}\hline a & b \\ \hline c \\ \cline{1-1}\end{array} = \mathcal{Y}_1^{[2,1]}\boldsymbol{\theta}_{abc} = \{E - (1\ 3) + (1\ 2) - (2\ 1)(1\ 3)\}\,\boldsymbol{\theta}_{abc} = \boldsymbol{\theta}_{abc} - \boldsymbol{\theta}_{cba} + \boldsymbol{\theta}_{bac} - \boldsymbol{\theta}_{bca}. \tag{8.24}$$

The tensor Young tableau is vanishing when $a = c$. The tensor Young tableaux with a pair of same digits are

$$\mathcal{Y}_1^{[2,1]}\boldsymbol{\theta}_{112} = \begin{array}{|c|c|}\hline 1 & 1 \\ \hline 2 \\ \cline{1-1}\end{array} = -\begin{array}{|c|c|}\hline 2 & 1 \\ \hline 1 \\ \cline{1-1}\end{array} = 2\boldsymbol{\theta}_{112} - \boldsymbol{\theta}_{211} - \boldsymbol{\theta}_{121},$$

$$\mathcal{Y}_1^{[2,1]}\boldsymbol{\theta}_{113} = \begin{array}{|c|c|}\hline 1 & 1 \\ \hline 3 \\ \cline{1-1}\end{array} = -\begin{array}{|c|c|}\hline 3 & 1 \\ \hline 1 \\ \cline{1-1}\end{array} = 2\boldsymbol{\theta}_{113} - \boldsymbol{\theta}_{311} - \boldsymbol{\theta}_{131},$$

$$\mathcal{Y}_1^{[2,1]}\boldsymbol{\theta}_{122} = \begin{array}{|c|c|}\hline 1 & 2 \\ \hline 2 \\ \cline{1-1}\end{array} = -\begin{array}{|c|c|}\hline 2 & 2 \\ \hline 1 \\ \cline{1-1}\end{array} = \boldsymbol{\theta}_{122} + \boldsymbol{\theta}_{212} - 2\boldsymbol{\theta}_{221},$$

$$\mathcal{Y}_1^{[2,1]}\boldsymbol{\theta}_{133} = \begin{array}{|c|c|}\hline 1 & 3 \\ \hline 3 \\ \cline{1-1}\end{array} = -\begin{array}{|c|c|}\hline 3 & 3 \\ \hline 1 \\ \cline{1-1}\end{array} = \boldsymbol{\theta}_{133} + \boldsymbol{\theta}_{313} - 2\boldsymbol{\theta}_{331},$$

$$\mathcal{Y}_1^{[2,1]}\boldsymbol{\theta}_{233} = \begin{array}{|c|c|}\hline 2 & 3 \\ \hline 3 \\ \cline{1-1}\end{array} = -\begin{array}{|c|c|}\hline 3 & 3 \\ \hline 2 \\ \cline{1-1}\end{array} = \boldsymbol{\theta}_{233} + \boldsymbol{\theta}_{323} - 2\boldsymbol{\theta}_{332},$$

$$\mathcal{Y}_1^{[2,1]}\boldsymbol{\theta}_{223} = \begin{array}{|c|c|}\hline 2 & 2 \\ \hline 3 \\ \cline{1-1}\end{array} = -\begin{array}{|c|c|}\hline 3 & 2 \\ \hline 2 \\ \cline{1-1}\end{array} = 2\boldsymbol{\theta}_{223} - \boldsymbol{\theta}_{322} - \boldsymbol{\theta}_{232},$$

and those with three different digits are

$$\mathcal{Y}_1^{[2,1]}\boldsymbol{\theta}_{123} = \begin{array}{|c|c|}\hline 1 & 2 \\ \hline 3 \\ \cline{1-1}\end{array} = -\begin{array}{|c|c|}\hline 3 & 2 \\ \hline 1 \\ \cline{1-1}\end{array} = \boldsymbol{\theta}_{123} - \boldsymbol{\theta}_{321} + \boldsymbol{\theta}_{213} - \boldsymbol{\theta}_{231},$$

$$\mathcal{Y}_1^{[2,1]}\boldsymbol{\theta}_{132} = \begin{array}{|c|c|}\hline 1 & 3 \\ \hline 2 \\ \cline{1-1}\end{array} = -\begin{array}{|c|c|}\hline 2 & 3 \\ \hline 1 \\ \cline{1-1}\end{array} = \boldsymbol{\theta}_{132} - \boldsymbol{\theta}_{231} + \boldsymbol{\theta}_{312} - \boldsymbol{\theta}_{321},$$

$$\mathcal{Y}_1^{[2,1]}\boldsymbol{\theta}_{213} = \begin{array}{|c|c|}\hline 2 & 1 \\ \hline 3 \\ \cline{1-1}\end{array} = -\begin{array}{|c|c|}\hline 3 & 1 \\ \hline 2 \\ \cline{1-1}\end{array} = \begin{array}{|c|c|}\hline 1 & 2 \\ \hline 3 \\ \cline{1-1}\end{array} - \begin{array}{|c|c|}\hline 1 & 3 \\ \hline 2 \\ \cline{1-1}\end{array}$$
$$= \boldsymbol{\theta}_{213} - \boldsymbol{\theta}_{312} + \boldsymbol{\theta}_{123} - \boldsymbol{\theta}_{132}.$$

A tensor Young tableau is said to be standard if the digit in each column of the tableau increases downward and the digit in each row does not decrease rightward. The dimension of the tensor subspace $\mathcal{T}_1^{[2,1]}$ is 8. The eight linearly independent tensor Young tableaux are all standard.

Theorem 8.2 The standard tensor Young tableaux constitute a complete set of basis tensors in the tensor subspace $\mathcal{T}_\mu^{[\lambda]}$.

Proof For any basis tensor $\boldsymbol{\theta}_{b_1\ldots b_n}$, there is a permutation S to arrange its subscripts in the increasing order:

$$\mathcal{Y}_\mu^{[\lambda]}\boldsymbol{\theta}_{b_1\ldots b_n} = \mathcal{Y}_\mu^{[\lambda]} S \boldsymbol{\theta}_{a_1\ldots a_n}, \qquad a_1 \leq a_2 \leq \ldots \leq a_n.$$

On the other hand, $\mathcal{Y}_\mu^{[\lambda]} S$ belongs to the right-ideal $\mathcal{R}_\mu^{[\lambda]}$, generated by the standard Young operator $\mathcal{Y}_\mu^{[\lambda]}$ in the group algebra of S_n. The basis vectors in $\mathcal{R}_\mu^{[\lambda]}$ are $\mathcal{Y}_\mu^{[\lambda]} R_{\mu\nu}$. Thus, any tensor in the tensor subspace $\mathcal{T}_\mu^{[\lambda]}$ can be expressed as a linear combination of the following tensor Young tableaux:

$$\mathcal{Y}_\mu^{[\lambda]} R_{\mu\nu}\boldsymbol{\theta}_{a_1\ldots a_n}, \qquad a_1 \leq a_2 \leq \ldots \leq a_n. \tag{8.25}$$

Namely, the tensor Young tableaux given in Eq. (8.25) constitute a complete set of basis tensors in the tensor subspace $\mathcal{T}_\mu^{[\lambda]}$. Our task is to show that $\mathcal{Y}_\mu^{[\lambda]} R_{\mu\nu}\boldsymbol{\theta}_{a_1\ldots a_n}$ are standard or vanishing.

$R_{\mu\nu}$ is a permutation transforming the standard Young tableau $\mathcal{Y}_\nu^{[\lambda]}$ to the standard Young tableau $\mathcal{Y}_\mu^{[\lambda]}$ so that

$$\mathcal{Y}_\mu^{[\lambda]} R_{\mu\nu}\boldsymbol{\theta}_{a_1\ldots a_n} = R_{\mu\nu}\left\{\mathcal{Y}_\nu^{[\lambda]}\boldsymbol{\theta}_{a_1\ldots a_n}\right\}. \tag{8.26}$$

The tensor in the curve bracket belongs to the tensor subspace $\mathcal{T}_\nu^{[\lambda]}$, not $\mathcal{T}_\mu^{[\lambda]}$. Equation (8.26) shows that two tensors belonging to different tensor subspaces are related by $R_{\mu\nu}$. It can be checked as follows that those two tensors are described by the same tensor Young tableau. Let

$$R_{\mu\nu} = \begin{pmatrix} \overline{r}_1 & \overline{r}_2 & \ldots & \overline{r}_n \\ 1 & 2 & \ldots & n \end{pmatrix}, \qquad R_{\mu\nu}\boldsymbol{\theta}_{a_1\ldots a_n} = \boldsymbol{\theta}_{b_1\ldots b_n}, \qquad b_j = a_{\overline{r}_j}.$$

Arbitrarily choose a box in the Young pattern $[\lambda]$. The box in the Young tableau $\mathcal{Y}_\mu^{[\lambda]}$ is filled with a digit, say j. The same box in the Young tableau $\mathcal{Y}_\nu^{[\lambda]}$ is filled with $\overline{r}_j$, that in the tensor Young tableau $\mathcal{Y}_\nu^{[\lambda]}\boldsymbol{\theta}_{a_1\ldots a_n}$ is filled with $a_{\overline{r}_j}$, and that in the tensor Young tableau $\mathcal{Y}_\mu^{[\lambda]} R_{\mu\nu}\boldsymbol{\theta}_{a_1\ldots a_n}$ is filled with $b_j = a_{\overline{r}_j}$. Therefore, two tensor Young tableaux are the same.

Since $a_1 \leq a_2 \leq \ldots \leq a_n$ and $\mathcal{Y}_\nu^{[\lambda]}$ is a standard Young tableau, the tensor Young tableau $\mathcal{Y}_\nu^{[\lambda]}\boldsymbol{\theta}_{a_1\ldots a_n}$ satisfies that the digit in each column of $\mathcal{Y}_\nu^{[\lambda]}\boldsymbol{\theta}_{a_1\ldots a_n}$ does not decrease downward and the digit in each row does not decrease rightward. Thus, $\mathcal{Y}_\nu^{[\lambda]}\boldsymbol{\theta}_{a_1\ldots a_n}$ as well as $\mathcal{Y}_\mu^{[\lambda]} R_{\mu\nu}\boldsymbol{\theta}_{a_1\ldots a_n}$ is standard or vanishing depending on whether there are repetitive digits in a column of the tableau. The different standard tensor Young tableaux are linearly independent although $\mathcal{Y}_\mu^{[\lambda]} R_{\mu\nu}\boldsymbol{\theta}_{a_1\ldots a_n}$ with different $R_{\mu\nu}$ may be denoted by the same tensor Young tableau. For example, from Eq. (8.24)

$$\mathcal{Y}_1^{[2,1]} R_{12}\boldsymbol{\theta}_{112} = \begin{array}{|c|c|}\hline 1 & 2 \\ \hline 1 \\ \cline{1-1}\end{array} = 0, \quad \mathcal{Y}_1^{[2,1]} R_{11}\boldsymbol{\theta}_{122} = \mathcal{Y}_1^{[2,1]} R_{12}\boldsymbol{\theta}_{122} = \begin{array}{|c|c|}\hline 1 & 2 \\ \hline 2 \\ \cline{1-1}\end{array},$$

where $R_{11} = E$ and $R_{12} = (2\ 3)$. □

8.1.3 *Chevalley Bases of Generators in SU(N)*

Calculate the Chevalley bases of generators in the self-representation of SU(N). From Eqs. (7.79) and (7.81) one has

$$\begin{aligned}(\boldsymbol{r}_\nu)_j &= \sqrt{2}\left[\boldsymbol{V}_\nu - \boldsymbol{V}_{\nu+1}\right]_j = \sqrt{2}\left[\left(T^{(3)}_{N-j+1}\right)_\nu - \left(T^{(3)}_{N-j+1}\right)_{\nu+1}\right] \\ &= \sqrt{\frac{\nu+1}{\nu}}\delta_{\nu(N-j)} - \sqrt{\frac{\nu-1}{\nu}}\delta_{\nu(N-j+1)}.\end{aligned}$$

In the self-representation of SU(N), the Chevally bases (see Eq. (7.141) where $d_\nu = 1$) are

$$\begin{aligned}H_\nu &= \sum_{j=1}^{\ell} (\boldsymbol{r}_\nu)_j H_j = \sqrt{\frac{2(\nu+1)}{\nu}} T^{(3)}_{\nu+1} - \sqrt{\frac{2(\nu-1)}{\nu}} T^{(3)}_\nu \\ &= T^{(1)}_{\nu\nu} - T^{(1)}_{(\nu+1)(\nu+1)}, \\ E_\nu &= E_{\boldsymbol{\alpha}_{\nu(\nu+1)}} = T^{(1)}_{\nu(\nu+1)} + iT^{(2)}_{\nu(\nu+1)}, \quad F_\nu = T^{(1)}_{\nu(\nu+1)} - iT^{(2)}_{\nu(\nu+1)}.\end{aligned} \tag{8.27}$$

Namely, each of the matrices of H_ν, E_ν, and F_ν contains only a nonvanishing two-dimensional submatrix in the νth and $(\nu+1)$th columns and rows, which is the same as that in the self-representation of SU(2) (see Eq. (7.144)). Their actions on the basis vectors are

$$\begin{aligned}H_\nu\Theta_\nu &= \Theta_\nu, & H_\nu\Theta_{\nu+1} &= -\Theta_{\nu+1}, \\ E_\nu\Theta_{\nu+1} &= \Theta_\nu, & F_\nu\Theta_\nu &= \Theta_{\nu+1}.\end{aligned} \tag{8.28}$$

The remaining actions are vanishing. In the SU(N) transformation, each index in a tensor plays the role of a vector index. The action of a generator on a basis tensor are the sum of the basis tensors, each of which is obtained by the action of the generator on one tensor index according to Eq. (8.28). Thus, the standard tensor Young tableau is the common eigenstate of H_ν with the eigenvalue to be the number of the digit ν filled in the tableau, subtracting the number of the digit $(\nu+1)$. The eigenvalues constitutes the weight $\boldsymbol{m}$ of the standard tensor Young tableau. Two tensor Young tableaux have the same weight if their sets of the filled digits are the same. The action of F_ν on the standard tensor Young tableau is equal to the sum of all possible tensor Young tableaux, each of which is obtained from the original one by replacing one filled digit ν with the digit $(\nu+1)$, and the action of E_ν is the sum of those by replacing $(\nu+1)$ with ν. The obtained Young tableaux may not be standard, but they can be transformed to the sum of the standard tensor Young tableaux by the symmetries (8.22) and (8.23). For example,

$$H_1 \begin{array}{|c|c|}\hline 1 & 1 \\ \hline 3 \\ \cline{1-1}\end{array} = 2 \begin{array}{|c|c|}\hline 1 & 1 \\ \hline 3 \\ \cline{1-1}\end{array}, \qquad H_2 \begin{array}{|c|c|}\hline 1 & 1 \\ \hline 3 \\ \cline{1-1}\end{array} = - \begin{array}{|c|c|}\hline 1 & 1 \\ \hline 3 \\ \cline{1-1}\end{array},$$

$$E_2 \begin{array}{|c|c|}\hline 1 & 1 \\ \hline 3 \\ \cline{1-1}\end{array} = \begin{array}{|c|c|}\hline 1 & 1 \\ \hline 2 \\ \cline{1-1}\end{array}, \qquad E_1 \begin{array}{|c|c|}\hline 1 & 1 \\ \hline 3 \\ \cline{1-1}\end{array} = F_2 \begin{array}{|c|c|}\hline 1 & 1 \\ \hline 3 \\ \cline{1-1}\end{array} = 0,$$

$$F_1 \begin{array}{|c|c|}\hline 1 & 1 \\ \hline 3 \\ \cline{1-1}\end{array} = \begin{array}{|c|c|}\hline 1 & 2 \\ \hline 3 \\ \cline{1-1}\end{array} + \begin{array}{|c|c|}\hline 2 & 1 \\ \hline 3 \\ \cline{1-1}\end{array} = 2 \begin{array}{|c|c|}\hline 1 & 2 \\ \hline 3 \\ \cline{1-1}\end{array} - \begin{array}{|c|c|}\hline 1 & 3 \\ \hline 2 \\ \cline{1-1}\end{array}.$$

8.1.4 *Inequivalent and Irreducible Representations*

Theorem 8.3 The tensor subspace $\mathcal{T}_\mu^{[\lambda]} = \mathcal{Y}_\mu^{[\lambda]}\mathcal{T}$ corresponds to an irreducible representation of SU(N) with the highest weight $\boldsymbol{M}$,

$$\boldsymbol{M} = \sum_{\nu=1}^{N-1} M_\nu \boldsymbol{w}_\nu, \qquad M_\nu = \lambda_\nu - \lambda_{\nu+1}, \tag{8.29}$$

where λ_ν is the box number in the νth row of the Young pattern $[\lambda]$ and $\boldsymbol{w}_\nu$ are the fundamental dominant weights of SU(N). The tensor subspaces are equivalent if and only if their Young patterns are the same. The basis tensors with the same tensor Young tableau in all tensor subspaces $\mathcal{T}_\mu^{[\lambda]}$ with the Young pattern $[\lambda]$ constitute a complete set of basis tensors of the irreducible representation $[\lambda]$ of the permutation group S_n.

Proof The key for proving the representation corresponding to the tensor subspace $\mathcal{T}_\mu^{[\lambda]}$ to be irreducible is whether there is one and only one highest weight state in $\mathcal{T}_\mu^{[\lambda]}$ which satisfies the condition (7.115).

One defines a standard tensor Young tableau to be smaller than another by the concept of "the dictionary order" (see §6.2.2). Compare the filled digits in two standard tensor Young tableaux in $\mathcal{T}_\mu^{[\lambda]}$ from left to right of the first row, and then those of the second row, and so on. For the first different filled digits, a smaller filled digit corresponds to a smaller standard tensor

Young tableau. Obviously, the standard tensor Young tableaux, denoted by ϕ_0, where each box located in the jth row of the tableau is filled with the digit j, is the smallest one in $\mathcal{T}_{\mu}^{[\lambda]}$. ϕ_0 satisfies Eq. (7.115) because the raising operator E_{ν} replaces a digit $(\nu+1)$ filled in the tableau with the digit ν such that the resultant tableau contains two ν in the same column. The highest weight $\boldsymbol{M}$ of ϕ_0 is calculated by Eq. (8.29).

Compare the filled digits in any other standard tensor Young tableau ϕ in $\mathcal{T}_{\mu}^{[\lambda]}$ with ϕ_0, from left to right and row by row. If the first different filled digits occurs at the ith column of the jth row, where the filled digit in ϕ is $k>j$, ϕ is now denoted by $\phi(j,i,k,\alpha)$. The ordinal number α in $\phi(j,i,k,\alpha)$ stands for the case when ϕ with this property is not unique in $\mathcal{T}_{\mu}^{[\lambda]}$. $\phi(j,i,k,\alpha)$ is smaller than $\phi(j',i',k',\alpha')$ if $j>j'$, or $i>i'$ when $j=j'$, or $k<k'$ when $j=j'$ and $i=i'$, or $\alpha<\alpha'$ for the same j, i, and k. $E_{k-1}\phi(j,i,k,\alpha)\neq 0$ because it is a combination of the standard tensor Young tableaux in $\mathcal{T}_{\mu}^{[\lambda]}$ where the smallest one is $\phi(j,i,k-1,\beta)$ with a positive coefficient $n\geq 1$. n is the number of k filled in the jth row of $\phi(j,i,k,\alpha)$. For any linear combination $\boldsymbol{\Phi}$ of the standard tensor Young tableaux with a weight $\boldsymbol{m}$ in $\mathcal{T}_{\mu}^{[\lambda]}$, $E_{k-1}\boldsymbol{\Phi}\neq 0$ if the smallest standard tensor Young tableau in the combination $\boldsymbol{\Phi}$ is $\phi(j,i,k,\alpha)$. Thus, the tensor subspace $\mathcal{T}_{\mu}^{[\lambda]}=\mathcal{Y}_{\mu}^{[\lambda]}\mathcal{T}$ corresponds to an irreducible representation of SU(N) with the highest weight $\boldsymbol{M}$ given in Eq. (8.29). Due to Eqs. (6.67) and (8.26), the remaining conclusions in the Theorem is obvious. □

8.1.5 *Dimensions of Representations of SU(N)*

The dimension $d_{[\lambda]}(\mathrm{SU}(N))$ of the representation $[\lambda]$ of SU(N) is equal to the number of the standard tensor Young tableaux in the tensor subspace $\mathcal{T}_{\mu}^{[\lambda]}$. There is a simpler way, called the hook rule, to calculate the dimensions. Please first review the hook rule for calculating the dimensions of representations of S_n (see §6.2.2).

For a box at the jth column of the ith row in a Young pattern $[\lambda]$, define its content $m_{ij}=j-i$ and its hook number h_{ij} to be the number of the boxes on its right in the ith row of the Young pattern, plus the number of the boxes below it in the jth column, and plus 1. The dimension $d_{[\lambda]}(\mathrm{SU}(N))$ of the representation $[\lambda]$ of SU(N) is expressed by a quotient,

$$d_{[\lambda]}(\mathrm{SU}(N))=\prod_{ij}\frac{N+m_{ij}}{h_{ij}}=\frac{Y_A^{[\lambda]}}{Y_h^{[\lambda]}}. \tag{8.30}$$

$Y_A^{[\lambda]}$ is a tableau obtained from the Young pattern $[\lambda]$ by filling $(N+m_{ij})$ into the box located in its ith row and jth column, and $Y_h^{[\lambda]}$ is a tableau obtained from $[\lambda]$ by filling h_{ij} into that box. The symbol $Y_A^{[\lambda]}$ means the product of the filled digits in it, so does the symbol $Y_h^{[\lambda]}$.

When $[\lambda]=[n]$ is a one-row Young pattern, $\mathcal{T}^{[n]}$ is the set of the totally symmetric tensors, and its dimension is

$$d_{[n]}(\mathrm{SU}(N))=\prod_{j=1}^{n}\frac{N+j-1}{n-j+1}=\frac{(N+n-1)!}{n!(N-1)!}=\begin{pmatrix} n+N-1\\ N-1\end{pmatrix}. \tag{8.31}$$

This formula can be understood that the standard tensor Young tableaux with a one-row Young pattern $[n]$ are characterized by the positions of the $(N-1)$ dividing points between each two neighbored digits.

When $[\lambda]=[\mu,\nu]$ is a two-row Young pattern,

$$\begin{aligned} d_{[\mu,\nu]}(\mathrm{SU}(N)) &= \frac{\begin{array}{|c|c|c|c|c|c|c|}\hline N & N+1 & \cdots & N+\nu-1 & N+\nu & \cdots & N+\mu-1\\ \hline N-1 & N & \cdots & N+\nu-2 & & & \\ \hline\end{array}}{\begin{array}{|c|c|c|c|c|c|c|}\hline \mu+1 & \mu & \cdots & \mu-\nu+2 & \mu-\nu & \cdots & 1\\ \hline \nu & \nu-1 & \cdots & 1 & & & \\ \hline\end{array}} \\ &= \frac{(N+\mu-1)!(N+\nu-2)!(\mu-\nu+1)}{(N-1)!(N-2)!(\mu+1)!\nu!}. \end{aligned} \tag{8.32}$$

$$\begin{aligned} & d_{[n]}(\mathrm{SU}(2))=d_{[n+\nu,\nu]}(\mathrm{SU}(2))=n+1, \\ & d_{[\mu,\nu]}(\mathrm{SU}(3))=(\mu+2)(\nu+1)(\mu-\nu+1)/2. \end{aligned}$$

In fact, the representation D^j of SU(2) given in Chap. 4 is equivalent to $[n]$ with $n=2j$. For SU(3) one has $d_{[1]}(\mathrm{SU}(3))=d_{[1,1]}(\mathrm{SU}(3))=3$, $d_{[2,1]}(\mathrm{SU}(3))=8$, $d_{[3]}(\mathrm{SU}(3))=d_{[3,3]}(\mathrm{SU}(3))=10$, and $d_{[4,2]}(\mathrm{SU}(3))=27$.

For a one-column Young pattern $[1^n]$, $n\le N$, a standard tensor Young tableau is a tableau filled with n digits downward in the increasing order so that its number is the combinatorics of n among N,

$$d_{[1^n]}(\mathrm{SU}(N))=\prod_{j=1}^{n}\frac{N-j+1}{n-j+1}=\frac{N!}{n!(N-n)!}=\begin{pmatrix} N\\ n\end{pmatrix}. \tag{8.33}$$

When $n=N$, there is only one standard tensor Young tableau. Since $\mathcal{Y}^{[1^N]}=\sum_R \delta(R)R$ is an antisymmetrized operator,

$$\left(\mathcal{Y}^{[1^N]}\boldsymbol{\theta}_{12\ldots N}\right)_{a_1\ldots a_N}=\epsilon_{a_1\ldots a_N},\quad \mathcal{Y}^{[1^N]}\boldsymbol{\theta}_{a_1\ldots a_N}=\epsilon_{a_1\ldots a_N}\left(\mathcal{Y}^{[1^N]}\boldsymbol{\theta}_{12\ldots N}\right). \tag{8.34}$$

Define $\boldsymbol{E}=\mathcal{Y}^{[1^N]}\boldsymbol{\theta}_{12\ldots N}$. Due to the Weyl reciprocity,

$$\begin{aligned}
O_u \boldsymbol{E} = \mathcal{Y}^{[1^N]} O_u \boldsymbol{\theta}_{12\dots N} &= \sum_{a_1 \dots a_N} \left(\mathcal{Y}^{[1^N]} \boldsymbol{\theta}_{a_1 \dots a_N}\right) u_{a_1 1} \dots u_{a_N N} \\
&= \left(\mathcal{Y}^{[1^N]} \boldsymbol{\theta}_{12\dots N}\right) \sum_{a_1 \dots a_N} \epsilon_{a_1 \dots a_N} u_{a_1 1} \dots u_{a_N N} \\
&= \left(\mathcal{Y}^{[1^N]} \boldsymbol{\theta}_{12\dots N}\right) \det u = \mathcal{Y}^{[1^N]} \boldsymbol{\theta}_{12\dots N} = \boldsymbol{E}.
\end{aligned}$$

$\boldsymbol{E}$ is an invariant tensor in SU(N), and $[1^N]$ describes the identical representation of SU(N).

The numbers of the standard tensor Young tableaux with the following two Young patterns are evidently equal to each other

$$\begin{aligned}
&[\lambda_1, \lambda_2, \dots, \lambda_{N-1}, \lambda_N] \\
\text{and} \quad &[(\lambda_1 - \lambda_N), (\lambda_2 - \lambda_N), \dots, (\lambda_{N-1} - \lambda_N), 0] ,
\end{aligned} \tag{8.35}$$

because there is only one way to filling the digits in the first λ_N columns of a standard tensor Young tableau with the first Young pattern, namely, the digits in those columns are filled from 1 to N in the increasing order. Two representations with the Young patterns given in Eq. (8.35) will be shown to be equivalent later.

8.1.6 *Subduced Representations with Respect to Subgroups*

SU($N-1$) is a subgroup of SU(N) where the Nth component preserves invariant. In this way one obtains a subgroup chain of SU(N)

$$\text{SU}(N) \supset \text{SU}(N-1) \supset \dots \supset \text{SU}(3) \supset \text{SU}(2) . \tag{8.36}$$

An irreducible representation $[\lambda]$ of SU(N) can be reduced with respect to the subgroups one by one in the subgroup chain. In fact, the basis tensor in the representation $[\lambda]$ of SU(N) is the standard tensor Young tableau where N has to be filled only in the lowest boxes of some columns. Removing the boxes filled with N, one obtains a standard tensor Young tableau of the subduced representations $[\mu]$ of SU($N-1$). For example,

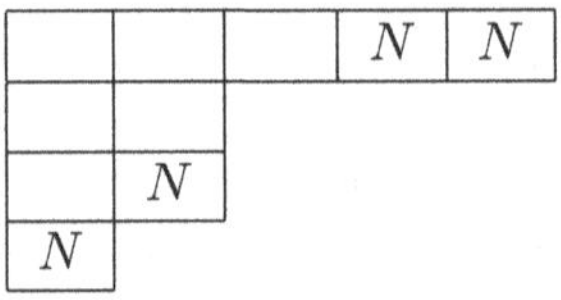

$[\lambda] = [5, 2, 2, 1]$,

$[\mu] = [3, 2, 1]$.

Removing the boxes filled with N from all standard tensor Young tableaux in the representation $[\lambda]$ of SU(N), one obtains all standard tensor Young tableaux in a few representations $[\mu]$ of SU($N-1$). This is the method of reducing the subduced representation $[\lambda]$ of SU(N) with respect to its subgroup S($N-1$):

$$\begin{aligned} &[\lambda] \longrightarrow \bigoplus [\mu], \qquad d_{[\lambda]}(\mathrm{SU}(N)) = \sum_{[\mu]} d_{[\mu]}(\mathrm{SU}(N-1)), \\ &\lambda_N \leq \mu_{N-1} \leq \lambda_{N-1} \leq \mu_{N-2} \leq \ldots \leq \mu_2 \leq \lambda_2 \leq \mu_1 \leq \lambda_1. \end{aligned} \tag{8.37}$$

Remind that in the reduction the multiplicity of each representation $[\mu]$ is not larger than one. By the successive applications of this method one is able to reduce an irreducible representation of SU(N) with respect to the subgroup chain (8.36).

The subduced representations of SU($N+M$) and SU(NM) with respect to the subgroup SU(N)×SU(M) are discussed in §8.4 of [Ma and Gu (2004)]. The Casimir invariants of orders 2 and 3 of SU(N) can be calculated by the method of the subduced representations of SU($N+M$) (see §8.5 in [Ma and Gu (2004)]).

8.2 Orthonormal Irreducible Basis Tensors

Define the inner product in the tensor space $\mathcal{T}$ such that the basis tensors $\boldsymbol{\theta}_{a_1\ldots a_n}$ are orthonormal to each other. The tensor representation of SU(N) with the basis tensors $\boldsymbol{\theta}_{a_1\ldots a_n}$ is the direct product of self-representations, which is unitary. Through the projection of a standard Young operator $\mathcal{Y}_\mu^{[\lambda]}$, the tensor space reduces to its subspace $\mathcal{T}_\mu^{[\lambda]} = \mathcal{Y}_\mu^{[\lambda]}\mathcal{T}$, whose basis tensors are the standard tensor Young tableaux. The standard tensor Young tableaux are the integral combinations of $\boldsymbol{\theta}_{a_1\ldots a_n}$, but they are generally not orthonormal. For example, in the tensor subspace $\mathcal{T}_1^{[2,1]}$ of SU(3), where the general form of the basis tensor is given in Eq. (8.24), $\mathcal{Y}_1^{[2,1]}\boldsymbol{\theta}_{123}$ is not orthogonal to $\mathcal{Y}_1^{[2,1]}\boldsymbol{\theta}_{132}$. $\mathcal{Y}_1^{[2,1]}\boldsymbol{\theta}_{112}$ and $\mathcal{Y}_1^{[2,1]}\boldsymbol{\theta}_{123}$ are normalized to 6 and 4, respectively. Furthermore, the highest weight states both in $\mathcal{T}_1^{[2,1]}$ and in $\mathcal{T}_2^{[2,1]}$ of SU(3) are denoted by the standard tensor Young tableau $\begin{array}{|c|c|}\hline 1 & 1 \\ \hline 2 & \\ \cline{1-1}\end{array}$, but they are not orthogonal,

$$\begin{aligned} \mathcal{Y}_1^{[2,1]}\boldsymbol{\theta}_{112} &= 2\boldsymbol{\theta}_{112} - \boldsymbol{\theta}_{211} - \boldsymbol{\theta}_{121}, \\ \mathcal{Y}_2^{[2,1]}\boldsymbol{\theta}_{121} &= (2\ 3)\ \mathcal{Y}_1^{[2,1]}\boldsymbol{\theta}_{112} = 2\boldsymbol{\theta}_{121} - \boldsymbol{\theta}_{211} - \boldsymbol{\theta}_{112}. \end{aligned} \tag{8.38}$$

Namely, the standard tensor Young tableaux both in one representation of SU(N) and in one representation of S_n are generally not orthonormal.

8.2.1 Orthonormal Basis Tensors in $\mathcal{T}_{\mu}^{[\lambda]}$

Usually, it is by a similarity transformation that a non-unitary representation of a compact Lie group changes to be unitary and the basis states are combined to be orthonormal. As far as the problem of finding the orthonormal basis tensors in an irreducible subspace $\mathcal{T}_{\mu}^{[\lambda]}$ of SU(N) is concerned, one prefers to use the method by applying the lowering operators F_{μ} successively to the highest weight state because the highest weight is single and the highest weight state is orthogonal to any other state. This is nothing but the essence of the method of the block weight diagram. But now, the multiplicity of a weight is easy to count because the standard tensor Young tableaux with the same set of the filled digits have the same weight. The standard tensor Young tableaux with different weights are orthogonal to each other. The modules of the calculated basis states by this method are normalized, instead to 1, to the module of the highest weight state. This method is explained by some examples in SU(3) as follows.

The block weight diagrams of two fundamental representation $(1, 0)$ and $(0, 1)$ are given in Figs. 7.3 (a) and 7.3 (b). The standard tensor Young tableaux with the highest weights in two representations are $\boxed{1}$ and $\begin{array}{|c|}\hline 1\\ \hline 2\\ \hline\end{array}$, respectively. The standard tensor Young tableaux are calculated from the highest weight states by the lowering operators as given in Fig. 8.1 where the block weight diagrams are also listed for comparison. Since two representations $(1, 0)$ and $(0, 1)$ are conjugate to each other, their basis tensors can be related through Eq. (7.153).

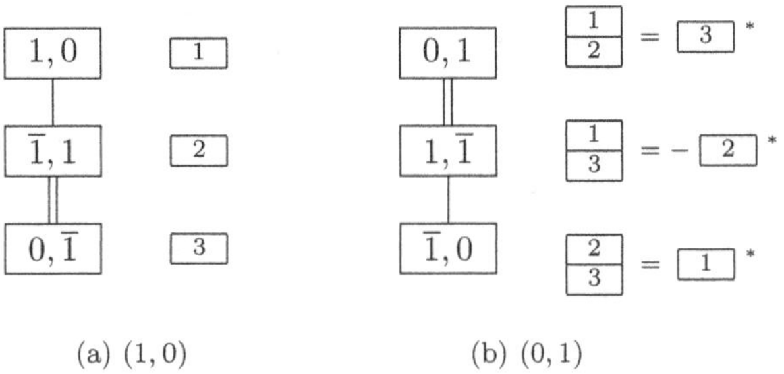

Fig. 8.1 Block weight diagrams and basis tensors of fundamental representations of SU(3).

The Young pattern of the representation of symmetric tensors of rank 3 of SU(3) is $[\lambda] = [3, 0]$. Its highest weight is $\boldsymbol{M} = (3, 0)$. There are three typical standard tensor Young tableaux which are normalized to 36, 12, and 6, respectively,

$$\begin{array}{|c|c|c|}\hline a & a & a \\ \hline\end{array} = 6\left\{\boldsymbol{\theta}_{aaa}\right\},$$

$$\begin{array}{|c|c|c|}\hline a & b & b \\ \hline\end{array} = 2\sqrt{3}\left\{3^{-1/2}\left(\boldsymbol{\theta}_{abb} + \boldsymbol{\theta}_{bab} + \boldsymbol{\theta}_{bba}\right)\right\},$$

$$\begin{array}{|c|c|c|}\hline a & b & c \\ \hline\end{array} = \sqrt{6}\left\{6^{-1/2}\left(\boldsymbol{\theta}_{abc} + \boldsymbol{\theta}_{acb} + \boldsymbol{\theta}_{bac} + \boldsymbol{\theta}_{bca} + \boldsymbol{\theta}_{cab} + \boldsymbol{\theta}_{cba}\right)\right\},$$

where a, b, and c are three different digits. For each set of filled digits there is only one standard tensor Young tableau, so that there is no multiple weight in the representation.

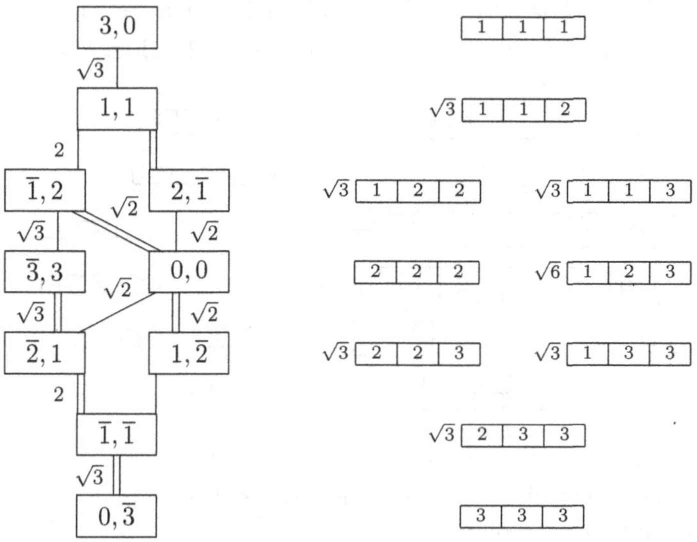

Fig. 8.2 Block weight diagrams and basis tensors of the symmetric representation $[3,0]$ of SU(3).

The highest weight state is $|(3,0),(3,0)\rangle = \begin{array}{|c|c|c|}\hline 1 & 1 & 1 \\ \hline\end{array}$. In the application of the lowering operators F_μ, the obtained tensor Young tableaux should be changed to be standard by the symmetry (8.22). The block weight diagrams and the standard tensor Young tableaux of $[3,0]$ of SU(3) are listed in Fig. 8.2 and some calculations are as follows:

$$\begin{aligned} F_1 \begin{array}{|c|c|c|}\hline 1 & 1 & 1 \\ \hline\end{array} &= \begin{array}{|c|c|c|}\hline 1 & 1 & 2 \\ \hline\end{array} + \begin{array}{|c|c|c|}\hline 1 & 2 & 1 \\ \hline\end{array} + \begin{array}{|c|c|c|}\hline 2 & 1 & 1 \\ \hline\end{array} \\ &= 3 \begin{array}{|c|c|c|}\hline 1 & 1 & 2 \\ \hline\end{array}, \end{aligned}$$

$$F_1 \begin{array}{|c|c|c|}\hline 1 & 1 & 3 \\ \hline\end{array} = \begin{array}{|c|c|c|}\hline 1 & 2 & 3 \\ \hline\end{array} + \begin{array}{|c|c|c|}\hline 2 & 1 & 3 \\ \hline\end{array} = 2 \begin{array}{|c|c|c|}\hline 1 & 2 & 3 \\ \hline\end{array},$$

$$\begin{aligned}|(3,0),(1,1)\rangle &= \sqrt{\frac{1}{3}}\, F_1\, |(3,0),(3,0)\rangle = \sqrt{\frac{1}{3}}\, F_1 \begin{array}{|c|c|c|}\hline 1&1&1\\\hline\end{array} \\ &= \sqrt{3}\begin{array}{|c|c|c|}\hline 1&1&2\\\hline\end{array}\,, \\ |(3,0),(0,0)\rangle &= \sqrt{\frac{1}{2}}\, F_1\, |(3,0),(2,\overline{1})\rangle = \sqrt{\frac{3}{2}}\, F_1 \begin{array}{|c|c|c|}\hline 1&1&3\\\hline\end{array} \\ &= \sqrt{6}\begin{array}{|c|c|c|}\hline 1&2&3\\\hline\end{array}\,.\end{aligned}$$

The conjugate representation of $[3,0]$ is $[3,3]$. Its highest weight state is $\begin{array}{|c|c|c|}\hline 1&1&1\\\hline 2&2&2\\\hline\end{array}$, where $\boldsymbol{M} = (0,3)$. The block weight diagrams and the standard tensor Young tableaux of $[3,3]$ of SU(3) are listed in Fig. 8.3.

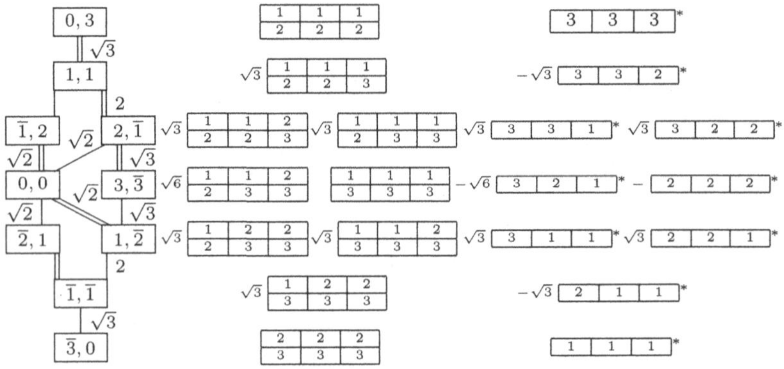

Fig. 8.3 Block weight diagrams and basis tensors of the representation $[3,3]$ of SU(3).

The Young pattern for the mixed symmetric tensors of rank 3 of SU(3) is $[\lambda] = [2,1]$, which is the adjoint representation of SU(3). Its highest weight state is $\begin{array}{|c|c|}\hline 1&1\\\hline 2\\\cline{1-1}\end{array}$, where $\boldsymbol{M} = (1,1)$. The general form of the expansion of the tensor Young tableau in $\mathcal{T}_1^{[2,1]}$ is given in Eq. (8.24). Two typical tensor Young tableaux are

$$\begin{array}{|c|c|}\hline 1&1\\\hline 2\\\cline{1-1}\end{array} = 2\Theta_{112} - \Theta_{211} - \Theta_{121},$$

$$\begin{array}{|c|c|}\hline 1&2\\\hline 3\\\cline{1-1}\end{array} = \Theta_{123} - \Theta_{321} + \Theta_{213} - \Theta_{231}\ .$$

They are normalized to 6 and 4, respectively. The block weight diagrams and the standard tensor Young tableaux of $[2,1]$ of SU(3) are listed in Fig. 8.4. Some calculations related to the multiple weight $(0,0)$ are as follows:

$$
\begin{aligned}
|(1,1),(0,0)_1\rangle &= \sqrt{\frac{1}{2}}\, F_1\, |(1,1),(2,\overline{1})\rangle = \sqrt{\frac{1}{2}}\, F_1 \begin{array}{|c|c|}\hline 1 & 1 \\ \hline 3 \\ \cline{1-1}\end{array} \\
&= \sqrt{\frac{1}{2}} \left\{ \begin{array}{|c|c|}\hline 1 & 2 \\ \hline 3 \\ \cline{1-1}\end{array} + \begin{array}{|c|c|}\hline 2 & 1 \\ \hline 3 \\ \cline{1-1}\end{array} \right\} = \sqrt{2} \begin{array}{|c|c|}\hline 1 & 2 \\ \hline 3 \\ \cline{1-1}\end{array} - \sqrt{\frac{1}{2}} \begin{array}{|c|c|}\hline 1 & 3 \\ \hline 2 \\ \cline{1-1}\end{array}, \\
|(1,1),(0,0)_2\rangle &= \sqrt{\frac{2}{3}} \left\{ F_2\, |(1,1),(\overline{1},2)\rangle - \sqrt{\frac{1}{2}}\, |(1,1),(0,0)_1\rangle \right\} \\
&= \sqrt{\frac{2}{3}} \left\{ F_2 \begin{array}{|c|c|}\hline 1 & 2 \\ \hline 2 \\ \cline{1-1}\end{array} - \begin{array}{|c|c|}\hline 1 & 2 \\ \hline 3 \\ \cline{1-1}\end{array} + \frac{1}{2} \begin{array}{|c|c|}\hline 1 & 3 \\ \hline 2 \\ \cline{1-1}\end{array} \right\} = \sqrt{\frac{3}{2}} \begin{array}{|c|c|}\hline 1 & 3 \\ \hline 2 \\ \cline{1-1}\end{array}.
\end{aligned}
\tag{8.39}
$$

Note that the orthogonal basis tensors are combined by $X_{[2,1]}^{-1} = \begin{pmatrix} 1 & -1/2 \\ 0 & \sqrt{3}/2 \end{pmatrix}$ which is the similarity transformation matrix for the orthogonal bases in the representation $[2,1]$ of the permutation group S_3 (see Eq. (6.85) and Prob. 24 of Chap. 6 in [Ma and Gu (2004)]),

$$
|(1,1),(0,0)_a\rangle = \sqrt{2}\left(X_{[2,1]}^{-1}\right)_{a1} \begin{array}{|c|c|}\hline 1 & 2 \\ \hline 3 \\ \cline{1-1}\end{array} + \sqrt{2}\left(X_{[2,1]}^{-1}\right)_{a1} \begin{array}{|c|c|}\hline 1 & 3 \\ \hline 2 \\ \cline{1-1}\end{array}, \tag{8.40}
$$

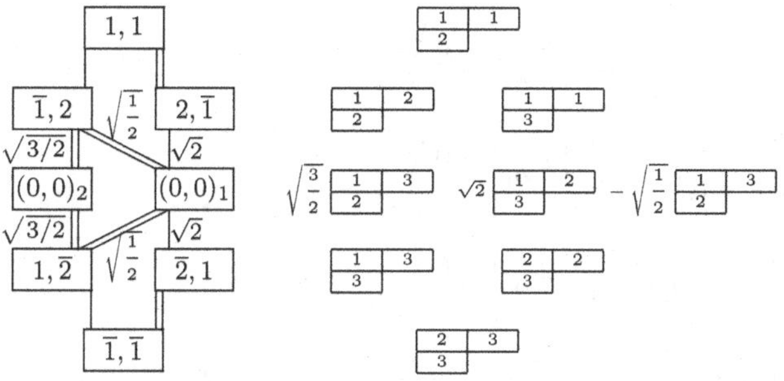

Fig. 8.4 Block weight diagrams and basis tensors of the adjoint representation [2, 1] of SU(3).

The orthonormal basis state with a weight $\boldsymbol{m} = \sum_{\mu=1}^{N-1} \boldsymbol{w}_\mu m_\mu$ in an irreducible representation $[\lambda]$ of $SU(N)$ is denoted by a symbol, usually called the Gelfand bases [Gel'fand et al. (1963)], where $N(N+1)/2$ parameters ω_{ab}, $1 \leq a \leq b \leq N$, are arranged as a regular triangle upside down:

$$
|\omega_{ab}\rangle = \left| \begin{array}{ccccccc}
\omega_{1N} & & \omega_{2N} & \cdots & \omega_{(N-1)N} & & \omega_{NN} \\
& \omega_{1(N-1)} & & \cdots & & \omega_{(N-1)(N-1)} & \\
& & \cdots & \cdots & \cdots & & \\
& & \omega_{12} & & \omega_{22} & & \\
& & & \omega_{11} & & &
\end{array} \right\rangle,
$$

$$\begin{aligned}
&\omega_{aN} = \lambda_a, \qquad \omega_{NN} = \lambda_N = 0, \qquad \omega_{ab} \geq \omega_{a(b-1)} \geq \omega_{(a+1)b}, \\
&m_\mu = -\sum_{d=1}^{\mu+1} \omega_{d(\mu+1)} + 2\sum_{d=1}^{\mu} \omega_{d\mu} - \sum_{d=1}^{\mu-1} \omega_{d(\mu-1)}, \qquad \omega_{a0} = 0.
\end{aligned} \tag{8.41}$$

The representation matrix entries of the Chevalley bases of generators are

$$\begin{aligned}
&H_\mu |\omega_{ab}\rangle = m_\mu |\omega_{ab}\rangle, \qquad E_\mu |\omega_{ab}\rangle = \sum_{\nu=1}^{\mu} A_{\nu\mu}(\omega_{ab})\ |\omega_{ab} + \delta_{a\nu}\delta_{b\mu}\rangle, \\
&A_{\nu\mu}(\omega_{ab}) \\
&\quad = \left\{ -\prod_{t=1}^{\mu-1} \left(\omega_{t(\mu-1)} - \omega_{\nu\mu} - t + \nu - 1\right) \prod_{p=1}^{\mu+1} \left(\omega_{p(\mu+1)} - \omega_{\nu\mu} - p + \nu\right) \right\}^{1/2} \\
&\qquad \times \left\{ \prod_{d\neq\nu, d=1}^{\mu} \left(\omega_{d\mu} - \omega_{\nu\mu} - d + \nu\right)\left(\omega_{d\mu} - \omega_{\nu\mu} - d + \nu - 1\right) \right\}^{-1/2} .
\end{aligned} \tag{8.42}$$

The Gelfand bases for the representation $[2,1]$ of SU(3) are listed as an example,

$$\begin{aligned}
|1\rangle &= \begin{array}{|c|c|}\hline 1 & 1 \\ \hline 2 \\ \cline{1-1}\end{array} = |(1,1)\rangle = \left|\begin{array}{ccccc} 2 & & 1 & & 0 \\ & 2 & & 1 & \\ & & 2 & & \end{array}\right\rangle, \\
|2\rangle &= \begin{array}{|c|c|}\hline 1 & 2 \\ \hline 2 \\ \cline{1-1}\end{array} = |(\overline{1},2)\rangle = \left|\begin{array}{ccccc} 2 & & 1 & & 0 \\ & 2 & & 1 & \\ & & 1 & & \end{array}\right\rangle, \\
|3\rangle &= \begin{array}{|c|c|}\hline 1 & 1 \\ \hline 3 \\ \cline{1-1}\end{array} = |(2,\overline{1})\rangle = \left|\begin{array}{ccccc} 2 & & 1 & & 0 \\ & 2 & & 0 & \\ & & 2 & & \end{array}\right\rangle, \\
|4\rangle &= \sqrt{2}\begin{array}{|c|c|}\hline 1 & 2 \\ \hline 3 \\ \cline{1-1}\end{array} - \sqrt{1/2}\begin{array}{|c|c|}\hline 1 & 3 \\ \hline 2 \\ \cline{1-1}\end{array} = |(0,0)_1\rangle = \left|\begin{array}{ccccc} 2 & & 1 & & 0 \\ & 2 & & 0 & \\ & & 1 & & \end{array}\right\rangle, \\
|5\rangle &= \begin{array}{|c|c|}\hline 2 & 2 \\ \hline 3 \\ \cline{1-1}\end{array} = |(\overline{2},1)\rangle = \left|\begin{array}{ccccc} 2 & & 1 & & 0 \\ & 2 & & 0 & \\ & & 0 & & \end{array}\right\rangle, \\
|6\rangle &= \sqrt{3/2}\begin{array}{|c|c|}\hline 1 & 3 \\ \hline 2 \\ \cline{1-1}\end{array} = |(0,0)_2\rangle = \left|\begin{array}{ccccc} 2 & & 1 & & 0 \\ & 1 & & 1 & \\ & & 1 & & \end{array}\right\rangle, \\
|7\rangle &= \begin{array}{|c|c|}\hline 1 & 3 \\ \hline 3 \\ \cline{1-1}\end{array} = |(1,\overline{2})\rangle = \left|\begin{array}{ccccc} 2 & & 1 & & 0 \\ & 1 & & 0 & \\ & & 1 & & \end{array}\right\rangle, \\
|8\rangle &= \begin{array}{|c|c|}\hline 2 & 3 \\ \hline 3 \\ \cline{1-1}\end{array} = |(\overline{1},\overline{1})\rangle = \left|\begin{array}{ccccc} 2 & & 1 & & 0 \\ & 1 & & 0 & \\ & & 0 & & \end{array}\right\rangle.
\end{aligned} \tag{8.43}$$

8.2.2 *Orthonormal Basis Tensors in* S_n

The highest weight states in $\mathcal{T}_\mu^{[\lambda]}$ of SU(N) with different μ constitute a complete set of the basis tensors for the representation $D^{[\lambda]}(\mathrm{S}_n)$, which is not unitary. Denote by $b_{\mu\mu}^{[\lambda]} = \mathcal{Y}_\mu^{[\lambda]}\boldsymbol{\theta}_{b_1\ldots b_n}$ the highest weight state in $\mathcal{T}_\mu^{[\lambda]}$. Then, due to Eq. (8.26), $b_{\nu\mu}^{[\lambda]} = \mathcal{Y}_\nu^{[\lambda]}R_{\nu\mu}\boldsymbol{\theta}_{b_1\ldots b_n}$ is the highest weight state in $\mathcal{T}_\nu^{[\lambda]}$, and

$$b_{\nu\mu}^{[\lambda]} = R_{\nu\mu}b_{\mu\mu}^{[\lambda]}. \tag{8.44}$$

Letting $X_{[\lambda]}$ be the similarity transformation which changes $D^{[\lambda]}(\mathrm{S}_n)$ to the real orthogonl representation $\overline{D}^{[\lambda]}(\mathrm{S}_n)$ (see Eq. (6.84)), one obtains the orthonormal basis tensors $\phi_{\nu\mu}^{[\lambda]}$ for S_n from Eq. (6.85),

$$\phi_{\nu\mu}^{[\lambda]} = \sum_\rho b_{\rho\mu}\left(X_{[\lambda]}\right)_{\rho\nu} = \sum_\rho \left(X_{[\lambda]}\right)_{\rho\nu} R_{\rho\mu}\mathcal{Y}_\mu^{[\lambda]}\boldsymbol{\theta}_{b_1\ldots b_n}. \tag{8.45}$$

For example, the orthonormal highest weight states for the mixed tensors of SU(3) can be calculated from $X_{[2,1]} = \begin{pmatrix} 1 & 1/\sqrt{3} \\ 0 & 2/\sqrt{3}\end{pmatrix}$ (see Prob. 24 of Chap. 6 in [Ma and Gu (2004)])

$$\begin{aligned}
\phi_{11}^{[2,1]} &= \mathcal{Y}_1^{[2,1]}\boldsymbol{\theta}_{112} = 2\boldsymbol{\theta}_{112} - \boldsymbol{\theta}_{211} - \boldsymbol{\theta}_{121}, \\
\phi_{21}^{[2,1]} &= \sqrt{1/3}\left\{\mathcal{Y}_1^{[2,1]}\boldsymbol{\theta}_{112} + 2\mathcal{Y}_2^{[2,1]}\boldsymbol{\theta}_{121}\right\} = \sqrt{3}\left\{\boldsymbol{\theta}_{121} - \boldsymbol{\theta}_{211}\right\}.
\end{aligned} \tag{8.46}$$

8.3 Direct Product of Tensor Representations

8.3.1 *Outer Product of Tensors*

Let $\boldsymbol{T}^{(1)}_{a_1\ldots a_n}$ and $\boldsymbol{T}^{(2)}_{b_1\ldots b_m}$ be two tensors of rank n and of rank m of SU(N), respectively. Merge them to be one tensor $\boldsymbol{T}^{(1)}_{a_1\ldots a_n}\boldsymbol{T}^{(2)}_{b_1\ldots b_m}$ of rank $(n+m)$, called the outer product of two tensors. Its product space is denoted by $\mathcal{T}$. After the projections of two Young operators acting on two tensors, respectively, $\mathcal{T}$ reduces to a subspace $\mathcal{T}^{[\lambda][\mu]}$:

$$\mathcal{Y}^{[\lambda]}\boldsymbol{T}^{(1)}\mathcal{Y}^{[\mu]}\boldsymbol{T}^{(2)} = \mathcal{Y}^{[\lambda]}\mathcal{Y}^{[\mu]}\boldsymbol{T}^{(1)}\boldsymbol{T}^{(2)} \in \mathcal{Y}^{[\lambda]}\mathcal{Y}^{[\mu]}\mathcal{T} = \mathcal{T}^{[\lambda][\mu]},$$

where the Young patterns $[\lambda]$ and $[\mu]$, whose row numbers are not larger than N, contain n and m boxes, respectively. Here the ordinal index ν for the standard Young operator $\mathcal{Y}_\nu^{[\lambda]}$ is omitted for simplicity. The tensor subspace $\mathcal{T}^{[\lambda][\mu]} \subset \mathcal{T}$ is invariant in the SU(N) transformation and corresponds to the representation $[\lambda] \times [\mu]$ with the dimension $d_{[\lambda]}(\text{SU}(N))d_{[\mu]}(\text{SU}(N))$, where we denote the representation directly by its Young pattern for convenience. Generally, the direct product representation is reducible. It can be reduced as follows. Applying a Young operator $\mathcal{Y}^{[\omega]}$ to $\mathcal{T}$, where $[\omega]$ contains $(n+m)$ boxes and its row number is not larger than N, one has

$$\mathcal{Y}^{[\omega]}\boldsymbol{T}^{(1)}\boldsymbol{T}^{(2)} \in \mathcal{Y}^{[\omega]}\mathcal{T} = \mathcal{T}^{[\omega]} \subset \mathcal{T}.$$

$\mathcal{T}^{[\omega]}$ is invariant in the SU(N) transformation and corresponds to the representation $[\omega]$. If

$$\mathcal{Y}^{[\lambda]}\mathcal{Y}^{[\mu]}t_\alpha\mathcal{Y}^{[\omega]} \neq 0,$$

where t_α is a vector in the group algebra of the permutation group S_{n+m}, there is a subspace corresponding to the representation $[\omega]$ in $\mathcal{T}^{[\lambda][\mu]}$,

$$\mathcal{Y}^{[\lambda]}\mathcal{Y}^{[\mu]}\left\{t_\alpha\mathcal{Y}^{[\omega]}\boldsymbol{T}^{(1)}\boldsymbol{T}^{(2)}\right\} \in \mathcal{Y}^{[\lambda]}\mathcal{Y}^{[\mu]}\mathcal{T} = \mathcal{T}^{[\lambda][\mu]}. \tag{8.47}$$

Remind that the leftmost operator $\mathcal{Y}^{[\lambda]}\mathcal{Y}^{[\mu]}$ determines that the tensor subspace after the projection belongs to $\mathcal{T}^{[\lambda][\mu]}$ because the tensor in the curve bracket belongs to $\mathcal{T}$, and the rightmost operator $\mathcal{Y}^{[\omega]}$ determines the property of the tensor subspace in the SU(N) transformations owing to the Weyl reciprocity.

In comparison with Eq. (6.105) for the outer product of two representations of the permutation group, one can borrow the technique of the Littlewood–Richardson rule to calculate the reduction of the direct product representation of SU(N)

$$[\lambda] \times [\mu] \simeq \bigoplus_{[\omega]} a^{\omega}_{\lambda\mu}[\omega]. \tag{8.48}$$

However, Eq. (8.48) is different from Eq. (6.100) because the representations in Eq. (8.48) are that of the SU(N) group, not that of the permutation group. If a Young pattern $[\omega]$ in Eq. (8.48), which is calculated by the Littlewood–Richardson rule, contains the row number larger than N, $[\omega]$ should be removed from the Clebsch–Gordan series for SU(N). The dimension formula of the reduction (8.48) becomes

$$d_{[\lambda]}(\text{SU}(N))d_{[\mu]}(\text{SU}(N)) = \sum_{[\omega]} a^{\omega}_{\lambda\mu}d_{[\omega]}(\text{SU}(N)). \tag{8.49}$$

For example, the direct product of two adjoint representations $[2,1] \times [2,1]$ of SU(3) and their dimension formula are

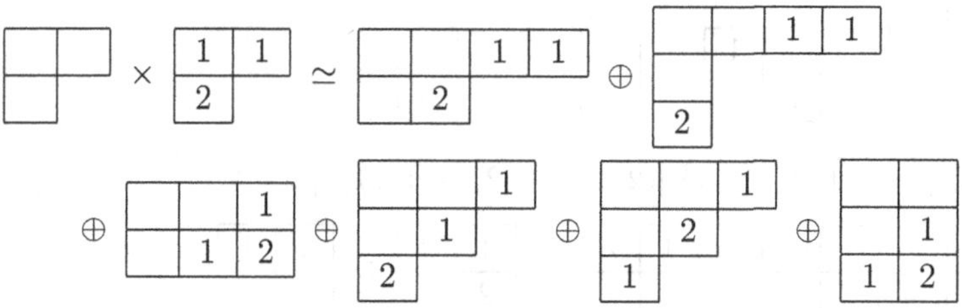

$$8 \times 8 = 27 + 10 + 10^* + 2 \times 8 + 1, \tag{8.50}$$

where 10^* denotes the representation $[3,3]$ which is conjugate with $[3,0]$. Compare the reduction with Example 1 in §6.5.2.

An important example for the reduction is the direct product of the totally antisymmetric tensor representation $[1^N]$ of rank N and an arbitrary representation $[\lambda]$ of SU(N). In the CG series calculated by the Littlewood–Richardson rule, there is only one representation with the row number not larger than N so that two Young patterns are directly adhibited,

$$[1^N] \times [\lambda] \simeq [\lambda'], \qquad \lambda'_j = \lambda_j + 1. \tag{8.51}$$

Since $[1^N]$ is the identical representation, $[\lambda']$ is equivalent to $[\lambda]$, which was mentioned in Eq. (8.35). Therefore, the irreducible representations of SU(N) can be characterized by a Young pattern with the row number less than N, namely by $(N-1)$ parameters where $(N-1)$ is the rank to SU(N).

In order to calculate the Clebsch–Gordan coefficients, one needs to write the expansions of the standard tensor Young tableaux with the highest weights appearing in the Clebsch–Gordan series. In writing an expansion for the highest weight M one first finds out all possible products of two standard tensor Young tableaux in two tensor subspaces where the sum of two weights is M. The coefficient in front of each term can be determined by the condition (7.115) that the expansion is annihilated by each raising operators E_μ. In the following the expansions of the products of standard tensor Young tableaux for the reduction of $[2,1] \times [2,1]$ of SU(3) are listed as examples. The expansions of the products of the basis states are also listed for comparison.

$$\boxed{\begin{array}{cccc} 1 & 1 & 1 & 1 \\ \hline 2 & 2 \end{array}} \sim \boxed{\begin{array}{cc} 1 & 1 \\ \hline 2 \end{array}} \times \boxed{\begin{array}{cc} 1 & 1 \\ \hline 2 \end{array}},$$

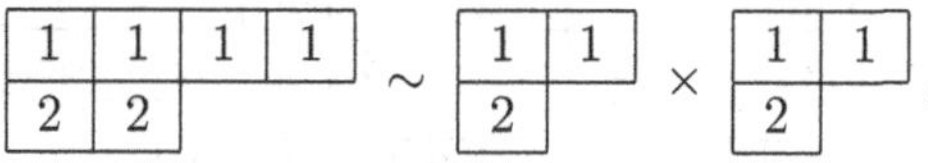

$$\|(2,2),(2,2)\rangle = |(1,1)\rangle|(1,1)\rangle,$$

$$\begin{array}{|c|c|c|c|}\hline 1&1&1&1\\\hline 2\\\cline{1-1} 3\\\cline{1-1}\end{array} \sim \begin{array}{|c|c|}\hline 1&1\\\hline 2\\\cline{1-1}\end{array} \times \begin{array}{|c|c|}\hline 1&1\\\hline 3\\\cline{1-1}\end{array} - \begin{array}{|c|c|}\hline 1&1\\\hline 3\\\cline{1-1}\end{array} \times \begin{array}{|c|c|}\hline 1&1\\\hline 2\\\cline{1-1}\end{array},$$

$$||(3,0),(3,0)\rangle = \sqrt{1/2}\left\{|(1,1)\rangle|(2,\overline{1})\rangle - |(2,\overline{1})\rangle|(1,1)\rangle\right\},$$

$$\begin{array}{|c|c|c|}\hline 1&1&1\\\hline 2&2&2\\\hline\end{array} \sim \begin{array}{|c|c|}\hline 1&1\\\hline 2\\\cline{1-1}\end{array} \times \begin{array}{|c|c|}\hline 1&2\\\hline 2\\\cline{1-1}\end{array} - \begin{array}{|c|c|}\hline 1&2\\\hline 2\\\cline{1-1}\end{array} \times \begin{array}{|c|c|}\hline 1&1\\\hline 2\\\cline{1-1}\end{array},$$

$$||(0,3),(0,3)\rangle = \sqrt{1/2}\left\{|(1,1)\rangle|(\overline{1},2)\rangle - |(\overline{1},2)\rangle|(1,1)\rangle\right\},$$

$$\begin{array}{|c|c|c|}\hline 1&1&1\\\hline 2&2\\\cline{1-2} 3\\\cline{1-1}\end{array}_{\,S} \sim \begin{array}{|c|c|}\hline 1&1\\\hline 2\\\cline{1-1}\end{array} \times \begin{array}{|c|c|}\hline 1&2\\\hline 3\\\cline{1-1}\end{array} + \begin{array}{|c|c|}\hline 1&2\\\hline 3\\\cline{1-1}\end{array} \times \begin{array}{|c|c|}\hline 1&1\\\hline 2\\\cline{1-1}\end{array}$$
$$- \begin{array}{|c|c|}\hline 1&2\\\hline 2\\\cline{1-1}\end{array} \times \begin{array}{|c|c|}\hline 1&1\\\hline 3\\\cline{1-1}\end{array} - \begin{array}{|c|c|}\hline 1&1\\\hline 3\\\cline{1-1}\end{array} \times \begin{array}{|c|c|}\hline 1&2\\\hline 2\\\cline{1-1}\end{array},$$

$$\begin{aligned} ||(1,1),(1,1)\rangle_S = \sqrt{1/20}\,\{&\sqrt{3}\,[\,|(1,1)\rangle|(0,0)_1\rangle + |(0,0)_1\rangle|(1,1)\rangle\,] \\ &+ |(1,1)\rangle|(0,0)_2\rangle + |(0,0)_2\rangle|(1,1)\rangle \\ &- \sqrt{6}\,[\,|(\overline{1},2)\rangle\;|(2,\overline{1})\rangle + |(2,\overline{1})\rangle|(\overline{1},2)\rangle\,]\},\end{aligned}$$

$$\begin{array}{|c|c|c|}\hline 1&1&1\\\hline 2&2\\\cline{1-2} 3\\\cline{1-1}\end{array}_{\,A} \sim \begin{array}{|c|c|}\hline 1&1\\\hline 2\\\cline{1-1}\end{array} \times \begin{array}{|c|c|}\hline 1&2\\\hline 3\\\cline{1-1}\end{array} - \begin{array}{|c|c|}\hline 1&2\\\hline 3\\\cline{1-1}\end{array} \times \begin{array}{|c|c|}\hline 1&1\\\hline 2\\\cline{1-1}\end{array}$$
$$-2\begin{array}{|c|c|}\hline 1&1\\\hline 2\\\cline{1-1}\end{array} \times \begin{array}{|c|c|}\hline 1&3\\\hline 2\\\cline{1-1}\end{array} + 2\begin{array}{|c|c|}\hline 1&3\\\hline 2\\\cline{1-1}\end{array} \times \begin{array}{|c|c|}\hline 1&1\\\hline 2\\\cline{1-1}\end{array}$$
$$- \begin{array}{|c|c|}\hline 1&2\\\hline 2\\\cline{1-1}\end{array} \times \begin{array}{|c|c|}\hline 1&1\\\hline 3\\\cline{1-1}\end{array} + \begin{array}{|c|c|}\hline 1&1\\\hline 3\\\cline{1-1}\end{array} \times \begin{array}{|c|c|}\hline 1&2\\\hline 2\\\cline{1-1}\end{array},$$

$$\begin{aligned} ||(1,1),(1,1)\rangle_A = \sqrt{1/12}\;\{&\,|(1,1)\rangle\;|(0,0)_1\rangle - |(0,0)_1\rangle|(1,1)\rangle \\ &- \sqrt{3}\,[\,|(1,1)\rangle\;|(0,0)_2\rangle - |(0,0)_2\rangle|1,1\rangle\,] \\ &- \sqrt{2}\,[\,|(\overline{1},2)\rangle|(2,\overline{1})\rangle - |(2,\overline{1})\rangle|(\overline{1},2)\rangle\,]\},\end{aligned}$$

$$\begin{array}{|c|c|}\hline 1&1\\\hline 2&2\\\hline 3&3\\\hline\end{array} \sim \begin{array}{|c|c|}\hline 1&1\\\hline 2\\\cline{1-1}\end{array} \times \begin{array}{|c|c|}\hline 2&3\\\hline 3\\\cline{1-1}\end{array} + \begin{array}{|c|c|}\hline 2&3\\\hline 3\\\cline{1-1}\end{array} \times \begin{array}{|c|c|}\hline 1&1\\\hline 2\\\cline{1-1}\end{array} - \begin{array}{|c|c|}\hline 1&1\\\hline 3\\\cline{1-1}\end{array} \times \begin{array}{|c|c|}\hline 2&2\\\hline 3\\\cline{1-1}\end{array}$$
$$- \begin{array}{|c|c|}\hline 2&2\\\hline 3\\\cline{1-1}\end{array} \times \begin{array}{|c|c|}\hline 1&1\\\hline 3\\\cline{1-1}\end{array} - \begin{array}{|c|c|}\hline 1&2\\\hline 2\\\cline{1-1}\end{array} \times \begin{array}{|c|c|}\hline 1&3\\\hline 3\\\cline{1-1}\end{array} - \begin{array}{|c|c|}\hline 1&3\\\hline 3\\\cline{1-1}\end{array} \times \begin{array}{|c|c|}\hline 1&2\\\hline 2\\\cline{1-1}\end{array}$$
$$+2\begin{array}{|c|c|}\hline 1&2\\\hline 3\\\cline{1-1}\end{array} \times \begin{array}{|c|c|}\hline 1&2\\\hline 3\\\cline{1-1}\end{array} + 2\begin{array}{|c|c|}\hline 1&3\\\hline 2\\\cline{1-1}\end{array} \times \begin{array}{|c|c|}\hline 1&3\\\hline 2\\\cline{1-1}\end{array}$$

$$- \begin{array}{|c|c|}\hline 1 & 2 \\ \hline 3 & \multicolumn{1}{c}{} \\ \cline{1-1}\end{array} \times \begin{array}{|c|c|}\hline 1 & 3 \\ \hline 2 & \multicolumn{1}{c}{} \\ \cline{1-1}\end{array} - \begin{array}{|c|c|}\hline 1 & 3 \\ \hline 2 & \multicolumn{1}{c}{} \\ \cline{1-1}\end{array} \times \begin{array}{|c|c|}\hline 1 & 2 \\ \hline 3 & \multicolumn{1}{c}{} \\ \cline{1-1}\end{array} ,$$

$$\begin{aligned} ||(0,0),(0,0)\rangle = \sqrt{1/8}\, \{ & |(1,1)\rangle|(\overline{1},\overline{1})\rangle - |(2,\overline{1})\rangle|(\overline{2},1)\rangle - |(\overline{1},2)\rangle|(1,\overline{2})\rangle \\ & + |(0,0)_1\rangle|(0,0)_1\rangle + |(0,0)_2\rangle|(0,0)_2\rangle - |(1,\overline{2})\rangle|(\overline{1},2)\rangle \\ & - |(\overline{2},1)\rangle|(2,\overline{1})\rangle + |(\overline{1},\overline{1})\rangle|(1,1)\rangle \} . \end{aligned}$$

8.3.2 *Covariant and Contravariant Tensors*

The self-representation of SU(N) is not equivalent to its conjugate representation. The conjugate of a covariant tensor is called a contravariant tensor (see Appendix B). A contravariant tensor $\boldsymbol{T}^{a_1\ldots a_m}$ of rank m of SU(N) and its basis tensors $\boldsymbol{\theta}^{d_1\ldots d_m}$ transform as

$$\begin{aligned} (O_u\boldsymbol{T})^{a_1\ldots a_m} &= \sum_{b_1\ldots b_m} \boldsymbol{T}^{b_1\ldots b_m} \left(u^{-1}\right)_{b_1 a_1} \cdots \left(u^{-1}\right)_{b_m a_m} \\ &= \sum_{b_1\ldots b_m} u^*_{a_1 b_1} \cdots u^*_{a_m b_m} \boldsymbol{T}^{b_1\ldots b_m}, \\ O_u\boldsymbol{\theta}^{d_1\ldots d_m} &= \sum_{b_1\ldots b_m} \left(u^{-1}\right)_{d_1 b_1} \cdots \left(u^{-1}\right)_{d_m b_m} \boldsymbol{\theta}^{b_1\ldots b_m} \\ &= \sum_{b_1\ldots b_m} \boldsymbol{\theta}^{b_1\ldots b_m} u^*_{b_1 d_1} \cdots u^*_{b_m d_m} . \end{aligned} \tag{8.52}$$

The contravariant tensor space is denoted by $\mathcal{T}^*$. The Weyl reciprocity holds for the contravariant tensors so that $\mathcal{T}^*$ can also be decomposed by the projection of the Young operators. The tensor subspace $\mathcal{Y}_\nu^{[\tau]}\mathcal{T}^* = \mathcal{T}_\nu^{[\tau]^*}$ corresponds to the irreducible representation denoted by $[\tau]^*$ which is the conjugate one of the representation $[\tau]$. The basis tensors in $\mathcal{T}_\nu^{[\tau]^*}$ are the standard tensor Young tableaux $\mathcal{Y}_\nu^{[\tau]}\boldsymbol{\theta}^{d_1\ldots d_m}$.

A tensor $\boldsymbol{T}^{b_1\ldots b_m}_{a_1\ldots a_n}$ is called a mixed tensor of rank (n, m) if it contains n subscripts and m superscripts and transforms in SU(N) as

$$\begin{aligned} (O_u\boldsymbol{T})^{b_1\ldots b_m}_{a_1\ldots a_n} &= \sum_{(a')(b')} u_{a_1 a'_1} \cdots u_{a_n a'_n} \boldsymbol{T}^{b'_1\ldots b'_m}_{a'_1\ldots a'_n} \left(u^{-1}\right)_{b'_1 b_1} \cdots \left(u^{-1}\right)_{b'_m b_m} \\ &= \sum_{(a')(b')} u_{a_1 a'_1} \cdots u_{a_n a'_n} u^*_{b_1 b'_1} \cdots u^*_{b_m b'_m} \boldsymbol{T}^{b'_1\ldots b'_m}_{a'_1\ldots a'_n} . \end{aligned} \tag{8.53}$$

The trace tensor of a mixed tensor, called the contraction of a tensor, is the sum of components of the mixed tensor where one covariant and one contravariant indices are taken to be equal and run over from 1 to N. The trace tensor subspace is invariant in SU(N),

$$\begin{aligned} O_u \left(\sum_{c=1}^{N} \boldsymbol{T}^{cb_1\cdots}_{ca_1\cdots} \right) &= \sum_{cdd'} u_{cd} u^*_{cd'} \sum_{(a')(b')} u_{a_1 a'_1} \cdots u^*_{b_1 b'_1} \cdots \boldsymbol{T}^{d'b'_1\cdots}_{da'_1\cdots} \\ &= \sum_{(a')(b')} u_{a_1 a'_1} \cdots u^*_{b_1 b'_1} \cdots \left(\sum_{d} \boldsymbol{T}^{db'_1\cdots}_{da'_1\cdots} \right). \end{aligned} \tag{8.54}$$

The trace tensor of a mixed tensor of rank (n, m) is a tensor of rank $(n-1, m-1)$. There is a special mixed tensor $\boldsymbol{D}^b_a$ of rank $(1,1)$ of SU(N) whose component is the Kronecker δ function

$$\begin{aligned} \boldsymbol{D}^b_a &= \delta^b_a = \begin{cases} 1 & \text{when} \ \ a = b, \\ 0 & \text{when} \ \ a \neq b, \end{cases} \\ (O_u \boldsymbol{D})^b_a &= \sum_{a'b'} u_{aa'} u^*_{bb'} \boldsymbol{D}^{b'}_{a'} = \sum_{a'b'} u_{aa'} u^*_{bb'} \delta^{b'}_{a'} = \delta^b_a = \boldsymbol{D}^b_a. \end{aligned} \tag{8.55}$$

The one-dimensional tensor subspace composed of $\boldsymbol{D}$ is invariant in SU(N) and corresponds to the identical representation of SU(N). A mixed tensor can be decomposed into the sum of a series of traceless tensors with different ranks in terms of the invariant tensor $\boldsymbol{D}^b_a$. For example,

$$\boldsymbol{T}^b_a = \left\{ \boldsymbol{T}^b_a - \boldsymbol{D}^b_a \left(\frac{1}{N} \sum_c \boldsymbol{T}^c_c \right) \right\} + \boldsymbol{D}^b_a \left(\frac{1}{N} \sum_c \boldsymbol{T}^c_c \right). \tag{8.56}$$

The first term is a traceless mixed tensor of rank $(1,1)$ and the trace tensor in the bracket of the second term is a scalar. The decomposition of a mixed tensor into the sum of traceless tensors is straightforward, but tedious. One has to write all possible terms and calculate the coefficients by the traceless conditions. For example,

$$\boldsymbol{\Phi}^d_{ab} = \boldsymbol{T}^d_{ab} + \boldsymbol{D}^d_a \sum_{p=1}^{N} \left\{ c_1 \boldsymbol{T}^p_{bp} + c_2 \boldsymbol{T}^p_{pb} \right\} + \boldsymbol{D}^d_b \sum_{p=1}^{N} \left\{ c_3 \boldsymbol{T}^p_{ap} + c_4 \boldsymbol{T}^p_{pa} \right\}.$$

Solving the traceless conditions $\sum_a \boldsymbol{\Phi}^a_{ab} = 0$ and $\sum_b \boldsymbol{\Phi}^b_{ab} = 0$, one obtain $c_2 = c_3 = -N/(N^2-1)$ and $c_1 = c_4 = 1/(N^2-1)$. The representation of a traceless mixed tensor subspace whose covariant and contravariant parts are projected by two Young operators $\mathcal{Y}^{[\lambda]}_\nu$ and $\mathcal{Y}^{[\mu]}_\rho$, respectively, is denoted by $[\lambda]\backslash[\mu]^*$, which will be shown to be irreducible in the next subsection.

8.3.3 *Traceless Mixed Tensors*

First, we study the condition whether the traceless mixed tensor space denoted by a pair of Young patterns $[\lambda]\backslash[\mu]^*$ is a null space or not. It is a null space if the number of constraints from the traceless conditions is not less than the number of independent tensors. Denote by n and m the row numbers of two Young patterns $[\lambda]$ and $[\mu]$, respectively. For an arbitrary pair of tensor Young tableaux in the space, assume that there are ℓ pairs of repetitive digits in the first columns of the two Young tableaux. Fixing the different digits in the two first columns and the digits in the remaining columns, one changes the pairs of repetitive digits from 1 to N where some tensors are vanishing owing to antisymmetry of indices. The number of independent tensors is

$$\binom{N-(n+m-2\ell)}{\ell}.$$

The traceless condition is written as a sum of tensors to be 0 where the $(\ell-1)$ pairs of digits are fixed and only one pair of digits runs over from 1 to N. The number of the traceless conditions is the number of the possible values of the $(\ell-1)$ pairs of digits, that is,

$$\binom{N-(n+m-2\ell)}{\ell-1}.$$

The condition for the traceless mixed tensor space not to be a null space is that the number of traceless conditions is less than the number of independent tensors, namely, $[N-(n+m-2\ell)]/2 \geq \ell$. The solution is

$$n+m \leq N. \tag{8.57}$$

Second, discuss a totally antisymmetric contravariant tensor subspace $\mathcal{T}^{[1^m]^*}$ of rank m, whose basis tensors are standard tensor Young tableaux $\mathcal{Y}^{[1^m]}\boldsymbol{\theta}^{b_1\ldots b_m}$. Multiplying the basis tensors with a totally antisymmetric tensor of rank N, one obtains

$$\boldsymbol{\Phi}_{a_1\ldots a_{N-m}} = \frac{1}{m!}\sum_{b_1\ldots b_m} \epsilon_{a_1\ldots a_{N-m}b_1\ldots b_m}\mathcal{Y}^{[1^m]}\boldsymbol{\theta}^{b_1\ldots b_m}. \tag{8.58}$$

From the viewpoint of replacement of basis tensors in one tensor subspace, the correspondence of two sets of basis tensors is one-to-one. In fact, only one basis tensor appears in the right-hand side of Eq. (8.58). The number of the basis tensors of two sets are the same

$$\binom{N}{m} = \binom{N}{N-m} .$$

The difference of two sets of basis tensors is only in the arranging order. The representations with respect to two sets of basis tensors are equivalent. From the transformation of the basis tensors in SU(N),

$$\begin{aligned}
O_u \mathbf{\Phi}_{a_1 \ldots a_{N-m}} &= \frac{1}{m!} \sum_{b_1 \ldots b_m} \epsilon_{a_1 \ldots a_{N-m} b_1 \ldots b_m} \left(\mathcal{Y}^{[1^m]} O_u \boldsymbol{\theta}^{b_1 \ldots b_m} \right) \\
&= \frac{1}{m!} \sum_{b_1 \ldots b_m} \sum_{d_1 \ldots d_{N-m}} \sum_{c_1 \ldots c_{N-m}} \epsilon_{d_1 \ldots d_{N-m} b_1 \ldots b_m} \left(u^*_{c_1 d_1} u_{c_1 a_1} \right) \cdots \\
&\quad \times \left(u^*_{c_{N-m} d_{N-m}} u_{c_{N-m} a_{N-m}} \right) \sum_{t_1 \ldots t_m} \mathcal{Y}^{[1^m]} \boldsymbol{\theta}^{t_1 \ldots t_m} u^*_{t_1 b_1} \cdots u^*_{t_m b_m} \\
&= \frac{1}{m!} \sum_{c_1 \ldots c_{N-m}} \sum_{t_1 \ldots t_m} \mathcal{Y}^{[1^m]} \boldsymbol{\theta}^{t_1 \ldots t_m} u_{c_1 a_1} \cdots u_{c_{N-m} a_{N-m}} \\
&\quad \times \left\{ \sum_{d_1 \ldots d_{N-m}} \sum_{b_1 \ldots b_m} u^*_{c_1 d_1} \cdots u^*_{c_{N-m} d_{N-m}} u^*_{t_1 b_1} \cdots u^*_{t_m b_m} \epsilon_{d_1 \ldots d_{N-m} b_1 \ldots b_m} \right\} \\
&= \frac{1}{m!} \sum_{c_1 \ldots c_{N-m}} \sum_{t_1 \ldots t_m} \epsilon_{c_1 \ldots c_{N-m} t_1 \ldots t_m} \mathcal{Y}^{[1^m]} \boldsymbol{\theta}^{t_1 \ldots t_m} u_{c_1 a_1} \cdots u_{c_{N-m} a_{N-m}} \\
&= \sum_{c_1 \ldots c_{N-m}} \mathbf{\Phi}_{c_1 \ldots c_{N-m}} u_{c_1 a_1} \cdots u_{c_{N-m} a_{N-m}},
\end{aligned} \tag{8.59}$$

where the formula for $\epsilon_{a_1 \ldots a_N}$ is used,

$$\begin{aligned}
\sum_{d_1 \ldots d_N} u_{a_1 d_1} \cdots u_{a_N d_N} \epsilon_{d_1 \ldots d_N} &= \epsilon_{a_1 \ldots a_N} \sum_{d_1 \ldots d_N} u_{1 d_1} \cdots u_{N d_N} \epsilon_{d_1 \ldots d_N} \\
&= (\det u) \epsilon_{a_1 \ldots a_N} = \epsilon_{a_1 \ldots a_N} .
\end{aligned} \tag{8.60}$$

Equation (8.59) shows that $\mathbf{\Phi}_{a_1 \ldots a_{N-m}}$ is proportional to the basis tensor $\mathcal{Y}^{[1^{N-m}]} \boldsymbol{\theta}_{a_1 \ldots a_{N-m}}$ in the tensor subspace $\mathcal{T}^{[1^{N-m}]}$. Namely, the representations of two tensor subspaces $\mathcal{T}^{[1^m]^*}$ and $\mathcal{T}^{[1^{N-m}]}$ are equivalent,

$$[1^m]^* \simeq [1^{N-m}]. \tag{8.61}$$

Third, generalize $\mathcal{T}^{[1^m]^*}$ to the traceless tensor subspace $\mathcal{T}^{[1^m]^*}_{[\lambda]}$, corresponding to a pair of Young patterns $[\lambda]\backslash[1^m]^*$ where the row number of $[\lambda]$ is not larger than $N - m$. Denote the basis tensors in $\mathcal{T}^{[1^m]^*}_{[\lambda]}$ by $\mathbf{\Omega}^{b_1 \ldots b_m}_{c \ldots}$ which are traceless between c and each b_j

$$\Omega^{b_1\ldots b_j\ldots b_k\ldots b_m}_{c\ldots} = -\Omega^{b_1\ldots b_k\ldots b_j\ldots b_m}_{c\ldots}, \qquad \sum_{c=1}^{N}\sum_{b_j=1}^{N} \delta^c_{b_j}\Omega^{b_1\ldots b_j\ldots b_m}_{c\ldots} = 0. \quad (8.62)$$

Multiplying the basis tensors with a totally antisymmetric tensor of rank N, one obtains

$$\Phi_{a_1\ldots a_{N-m}c\ldots} = \frac{1}{m!}\sum_{b_1\ldots b_m} \epsilon_{a_1\ldots a_{N-m}b_1\ldots b_m}\Omega^{b_1\ldots b_m}_{c\ldots}. \quad (8.63)$$

$\Phi_{a_1\ldots a_{N-m}c\ldots}$ belongs to a covariant tensor subspace corresponding to a direct product representation, $[1^{N-m}]\times[\lambda]$, which is calculated by the Littlewood–Richardson rule. From the traceless condition (8.62)

$$\begin{aligned}&\sum_{a_1\ldots a_{N-m}c} \epsilon^{a_1\ldots a_{N-m}cd_2\ldots d_m}\Phi_{a_1\ldots a_{N-m}c\ldots}\\ &= \frac{1}{m!}\sum_{a_1\ldots a_{N-m}cb_1\ldots b_m} \epsilon^{a_1\ldots a_{N-m}cd_2\ldots d_m}\epsilon_{a_1\ldots a_{N-m}b_1\ldots b_m}\Omega^{b_1\ldots b_m}_{c\ldots} = 0,\end{aligned}$$

because each term in the sum contains m factors of δ functions including a factor related with c, say $\delta^c_{b_j}$. Thus, in the reduction of $[1^{N-m}]\times[\lambda]$, there is only one nonvanishing term whose Young pattern is obtained by adhibiting $[1^{N-m}]$ and $[\lambda]$ directly,

$$[\lambda]\backslash[1^m]^* \simeq [\lambda'], \qquad \lambda'_k = \lambda_k + 1, \qquad 1 \leq k \leq N-m. \quad (8.64)$$

$[\lambda']$ is the Young pattern obtained by adhibiting $[1^{N-m}]$ and $[\lambda]$ directly.

Similarly, one has $[1^m]\backslash[\lambda]^* \simeq [\lambda']^*$, or equivalently,

$$[\tau]^* \simeq [1^{N-m}]\backslash[\tau']^*, \qquad \tau'_k = \tau_k - 1, \qquad 1 \leq k \leq m, \quad (8.65)$$

where the row number of $[\tau]$ is m, and $[\tau']$ is obtained from $[\tau]$ by removing its first column. The replacement of basis tensors for the equivalent tensor subspaces in Eq. (8.65) is

$$\Phi^{c\ldots}_{a_1\ldots a_{N-m}} = \frac{1}{m!}\sum_{b_1\ldots b_m} \epsilon_{a_1\ldots a_{N-m}b_1\ldots b_m}\mathcal{Y}^{[\tau]}\theta^{b_1\ldots b_m c\ldots}, \quad (8.66)$$

where $\Phi^{c\ldots}_{a_1\ldots a_{N-m}}$ is traceless.

At last, discuss the general case where $\Omega^{b_1\ldots b_m c\ldots}_{d\ldots}$ is the traceless basis tensor in the representation $[\lambda]\backslash[\tau]^*$. The row number of $[\tau]$ is m and that of $[\lambda]$ is not larger than $(N-m)$. The tensor is traceless between each pair of one covariant and one contravariant indices, say d and b_j or d and c. Let

$$\Phi^{c\ldots}_{a_1\ldots a_{N-m}d\ldots} = \frac{1}{m!}\sum_{b_1\ldots b_m} \epsilon_{a_1\ldots a_{N-m}b_1\ldots b_m}\Omega^{b_1\ldots b_m c\ldots}_{d\ldots}. \tag{8.67}$$

The covariant part of $\Phi^{c\ldots}_{a_1\ldots a_{N-m}d\ldots}$ corresponds to the representation $[1^{N-m}] \times [\lambda]$. In its reduction by the Littlewood–Richardson rule, except for the Young pattern by adhibiting $[1^{N-m}]$ and $[\lambda]$ directly, each Young pattern in the Clebsch–Gordan series contains an operation of antisymmetrizing one covariant index, say d, and all new covariant indices a_j. However, $\epsilon^{a_1\ldots a_{N-m}db_2'\ldots b_m'}$ annihilates the antisymmetrized term because the product of two ϵ makes a factor $\delta^d_{b_j}$ which annihilates the traceless tensor $\Omega^{b_1\ldots b_m c\ldots}_{d\ldots}$. Thus, the following two representations are equivalent

$$[\lambda]\backslash[\tau]^* \simeq [\lambda']\backslash[\tau']^*, \qquad \begin{array}{ll} \tau_j' = \tau_j - 1, & 1 \leq j \leq m, \\ \lambda_k' = \lambda_k + 1, & 1 \leq k \leq N-m, \end{array} \tag{8.68}$$

where $[\lambda']$ is the Young pattern obtained by adhibiting $[1^{N-m}]$ and $[\lambda]$ directly, and $[\tau']$ is obtained from $[\tau]$ by removing its first column. Successively applying the replacement (8.68), one is able to transform any traceless mixed tensor subspace denoted by $[\mu]\backslash[\tau]^*$ into a covariant tensor subspace denoted by a Young pattern $[\lambda]$ so that $[\mu]\backslash[\tau]^*$ is irreducible. Furthermore, a contravariant tensor subspace denoted by a Young pattern $[\tau]^*$ is equivalent to a covariant tensor subspace denoted by a Young pattern $[\lambda]$, where

$$[\tau]^* \simeq [\lambda], \qquad \lambda_j = \tau_1 - \tau_{N-j+1}, \qquad 1 \leq j \leq N. \tag{8.69}$$

Adhibiting the Young pattern $[\tau]$ upside down to the Young pattern $[\lambda]$, one obtains an $N \times \tau_1$ rectangle. The relation (8.69) was shown in Figs. 8.1 and 8.3 as examples.

8.3.4 *Adjoint Representation of SU(N)*

As shown in §7.6.1, the highest weight of the adjoint representation of SU(N) is $\boldsymbol{M} = \boldsymbol{w}_1 + \boldsymbol{w}_{N-1}$, corresponding to the Young pattern $[2, 1^{N-2}] \simeq [1]\backslash[1]^*$. In this subsection we are going to discuss the adjoint representation of SU(N) by replacement of tensors.

From the definition (7.8), the adjoint representation of SU(N) satisfies

$$uT_Au^{-1} = \sum_{B=1}^{N^2-1} T_B D^{\text{ad}}_{BA}(u), \tag{8.70}$$

where T_A is the generator in the self-representation of SU(N). T_A is an $N \times N$ traceless Hermitian matrix. A traceless mixed tensor $\boldsymbol{T}^b_a$ of rank

(1,1) has a similar transformation,

$$\left(O_u \boldsymbol{T}\right)_a^b = \sum_{a'b'} u_{aa'} \boldsymbol{T}_{a'}^{b'} \left(u^{-1}\right)_{b'b}. \tag{8.71}$$

$\boldsymbol{T}_a^b$ can be looked as the matrix entry of an $N \times N$ traceless Hermitian matrix at the ath row and the bth column so that it can be expanded with respect to the generators $(T_A)_{ab}$ where $\sqrt{2}\boldsymbol{F}_A$ are the coefficients,

$$\boldsymbol{T}_a^b = \sqrt{2} \sum_{A=1}^{N^2-1} (T_A)_{ab} \boldsymbol{F}_A, \qquad \boldsymbol{F}_A = \sqrt{2} \sum_{ab} (T_A)_{ba} \boldsymbol{T}_a^b. \tag{8.72}$$

From the viewpoint of replacement of tensors, $\boldsymbol{T}_a^b$ and $\boldsymbol{F}_A$ are two tensors in the traceless tensor subspace of rank $(1,1)$ so that they correspond to the same representation $[1]\backslash[1]^*$. Calculate the transformation of $\boldsymbol{F}_A$ in SU(N),

$$\begin{aligned} O_u \boldsymbol{T} &= u \boldsymbol{T} u^{-1} = \sqrt{2} \sum_{A=1}^{N^2-1} u T_A u^{-1} \boldsymbol{F}_A \\ &= \sqrt{2} \sum_{B=1}^{N^2-1} T_B \left(\sum_{A=1}^{N^2-1} D_{BA}^{\text{ad}}(u) \boldsymbol{F}_A \right), \\ O_u \boldsymbol{T} &= \sqrt{2} \sum_{B=1}^{N^2-1} T_B \left(O_u \boldsymbol{F}\right)_B. \end{aligned}$$

Namely,

$$\left(O_u \boldsymbol{F}\right)_B = \sum_{A=1}^{N^2-1} D_{BA}^{\text{ad}}(u) \boldsymbol{F}_A. \tag{8.73}$$

$\boldsymbol{F}_A$ corresponds to the adjoint representation. It shows that the adjoint representation of SU(N) is equivalent to the representation $[1]\backslash[1]^* \simeq [2, 1^{N-2}]$. Both $\boldsymbol{T}_a^b$ and $\boldsymbol{F}_A$ are the tensors transformed according to the adjoint representation of SU(N). Two forms of $\boldsymbol{T}_a^b$ and $\boldsymbol{F}_A$ are commonly used in particle physics.

8.4 SU(3) Symmetry and Wave Functions of Hadrons

As an example of physical applications of group theory, we are going to study the flavor SU(3) symmetry in particle physics in this section. The rank of SU(3) is 2 so that the planar weight diagram is more convenient to demonstrate the basis states in an irreducible representation of SU(3) than the block weight diagram. We will derive the mass relations of hadrons and calculate the wave functions of hadrons in the multiplets of SU(3).

8.4.1 *Quantum Numbers of Quarks*

In the theory of modern particle physics, the "elementary" particles are divided into four classes. The particles participating in the strong interaction are called the hadrons. Those without the strong interaction are called the leptons. The particles mediating the interactions are called the gauge particles. The particles in the fourth class, called the Higgs bosons, are introduced in the theory for providing the static masses of other particles, and are not observed yet in experiments. There are three generations of leptons as well as their anti-particles. They are the charged leptons e, μ, and τ and their neutrinos ν_e, ν_μ, and ν_τ. The gauge particles include the photon mediating the electro-magnetic interaction, the neutral boson Z^0 and charged bosons $W^{\pm}$ mediating the weak interaction, and the gluons mediating the strong interaction. The graviton mediating the gravitational force is being studied in theory and in experiment. Some other particles presented in the supersymmetric theory are still not observed in experiments.

We focus our attention on the hadrons. Among hadrons, the fermions are called the baryons and the bosons are called the mesons. All hadrons are constructed by the more elementary particles called the quarks and antiquarks. In the modern theory there are 18 quarks and 18 antiquarks. Usually, the quarks are described by a visual language, "color" and "flavor" which are not in the common sense on color and flavor. There are three color quantum numbers, say red, yellow, and blue. The quarks with three colors are the bases of the color $\text{SU}(3)_c$ group in the theory of quantum chromodynamics. They participate in the $\text{SU}(3)_c$ gauge interaction mediated by the gluons. The theoretical and experimental researches expect the so-called color confinement that a state with color cannot be observed in the recent experimental energy level. Thus, the quarks in the low energy have to appear in the colorless states, or called the color singlet of $\text{SU}(3)_c$. Namely, three quarks construct the basis states of totally antisymmetric tensors of rank three, and a quark and an antiquark construct the trace state of the mixed tensor of rank $(1,1)$,

$$\sum_{abc} \epsilon^{abc}\, q_a q_b q_c, \qquad \text{and} \qquad \sum_{a=1}^{3} \overline{q}^a q_a. \tag{8.74}$$

Both states correspond to the identical representation of $\text{SU}(3)_c$. The state with three quarks is a baryon state with baryon number 1. The pair of quark and antiquark is a meson state without the baryon number. Therefore, a quark brings the baryon number 1/3 and an antiquark brings the baryon number $-1/3$. Some composite colorless states composed of the states (8.74) are being studied recently. The quark and antiquark states appearing in the following are understood to be the colorless states.

There are six flavor quantum numbers for quarks. They are divided into three generations in weak interaction. Each generation contains two quarks: the up quark u and the down quark d; the charm quark c and the strange quark s; and the top quark t and the bottom quark b. The first quark brings 2/3 electric charge unit and the second $-1/3$ unit. In each generation, their left-hand states constitute a doublet with respect to the weak isospin, and the right-hand states are singlets. The u quark and the d quark are very light, and the s quark is a little bit heavier. They are called the light quarks. The c, b, and t quarks are heavier one by one, and are called the heavy quarks.

The up quark u and the down quark d constitute a doublet in the isospin SU(2) group. Although they have different electric charges, the isospin is conserved approximately in the strong interaction and plays an important role in particle physics. Generalizing the isospin symmetry, one assumes that three light quarks constitute a triplet of the flavor SU(3) group, which is a broken symmetry because the s quark is heavier than u and d quarks. However, the flavor SU(3) symmetry made some historical contributions in discovering new hadrons and predicting their properties. Even recently, the flavor SU(3) symmetry helps the research of hadron physics in some respects. In this section we pay attention to the application of SU(3) to the hadron physics only from the viewpoint of group theory.

Except for color and flavor there are a few inner quantum numbers for quarks such as the baryon number B, the electric charge Q, the isospin T and its third component T_3, the strange number S, and the supercharge Y. They are related by

$$Y = B + S, \qquad Q = T_3 + Y/2. \tag{8.75}$$

The isospin SU(2) is a subgroup of the flavor SU(3) group. T_3 and Y span the Cartan subalgebra of the flavor SU(3) group,

$$T_3 = \frac{1}{2}\begin{pmatrix} 1 & 0 & 0 \\ 0 & -1 & 0 \\ 0 & 0 & 0 \end{pmatrix}, \qquad Y = \frac{2}{\sqrt{3}}T_8 = \frac{1}{3}\begin{pmatrix} 1 & 0 & 0 \\ 0 & 1 & 0 \\ 0 & 0 & -2 \end{pmatrix}. \tag{8.76}$$

The quantum numbers of quarks are listed in Table 8.1. The quantum numbers of antiquarks are changed in sign except for T. In addition, the quarks have some spatial quantum numbers such as the spin and the parity.

Table 8.1 Quantum numbers of light quarks

Quark	B	T	T_3	S	Y	Q
u	1/3	1/2	1/2	0	1/3	2/3
d	1/3	1/2	– 1/2	0	1/3	– 1/3
s	1/3	0	0	– 1	– 2/3	– 1/3

8.4.2 *Planar Weight Diagrams*

Corresponding to the Chevalley bases of SU(3), there are two SU(2) subgroups. One is called the T-spin where the generators are $H_1 = 2T_3$, $E_1 = T_+ = T_1 + iT_2$, and $F_1 = T_- = T_1 - iT_2$. Y is constant in a multiplet of T-spin. The other is called U-spin where the generators are $H_2 = 3Y/2 - T_3$, $E_2 = U_+ = T_6 + iT_7$, and $F_2 = U_- = T_6 - iT_7$. Q is constant in a multiplet of U-spin. Since three quarks are the basis states of the flavor SU(3), one has

$$\begin{array}{llll}
T_+\text{d} = \text{u}, & T_+\text{u} = T_+\text{s} = 0, & T_-\text{u} = \text{d}, & T_-\text{d} = T_-\text{s} = 0, \\
U_+\text{s} = \text{d}, & U_+\text{u} = U_+\text{d} = 0, & U_-\text{d} = \text{s}, & U_-\text{u} = U_-\text{s} = 0. \\
T_+\overline{\text{u}} = -\overline{\text{d}}, & T_+\overline{\text{d}} = T_+\overline{\text{s}} = 0, & T_-\overline{\text{d}} = -\overline{\text{u}}, & T_-\overline{\text{u}} = T_-\overline{\text{s}} = 0, \\
U_+\overline{\text{d}} = -\overline{\text{s}}, & U_+\overline{\text{u}} = U_+\overline{\text{s}} = 0, & U_-\overline{\text{s}} = -\overline{\text{d}}, & U_-\overline{\text{u}} = U_-\overline{\text{d}} = 0.
\end{array} \tag{8.77}$$

In the planar weight diagram for the flavor SU(3), the abscissa axis and the ordinate axis are the eigenvalues of T_3 and T_8, respectively. The simple roots and the fundamental dominant weights are

$$\boldsymbol{r}_1 = \sqrt{2}\boldsymbol{e}_2, \quad \boldsymbol{r}_2 = \sqrt{\frac{3}{2}}\boldsymbol{e}_1 - \frac{\boldsymbol{e}_2}{\sqrt{2}}, \quad \boldsymbol{w}_1 = \frac{\boldsymbol{e}_1}{\sqrt{6}} + \frac{\boldsymbol{e}_2}{\sqrt{2}}, \quad \boldsymbol{w}_2 = \sqrt{\frac{2}{3}}\boldsymbol{e}_1, \tag{8.78}$$

where the unit vectors along the abscissa axis and the ordinate axis are denoted by $\boldsymbol{e}_2$ and $\boldsymbol{e}_1$, respectively, such that $\boldsymbol{r}_\mu$ are positive roots. $T_\pm$ move the basis states along the abscissa axis so that both T_3 and Q change by ± 1, but Y does not change. $U_\pm$ move the basis states along the direction with an angle $2\pi/3$ to the abscissa axis so that T_3 changes by $\mp 1/2$, Y changes by ± 1, but Q does not change.

The hadrons are composed of three quarks or a pair of quark and antiquark, denoted by a suitable standard tensor Young tableaux. In the covariant tensor Young tableaux, 1, 2, and 3 are replaced with u, d, and s, respectively. In the contravariant tensor Young tableaux, $\overline{3}$, $-\overline{2}$, and $\overline{1}$ are replaced with $\overline{\text{s}}$, $-\overline{\text{d}}$, and $\overline{\text{u}}$, respectively. In drawing the planar weight diagram of a given representation $[\lambda]$, one first determines the position of the highest weight state, where the box in the first row is filled with u and that in the second row with d. Then, applying the lowering operators T_- and U_-, one calculates the positions of the other basis states. For a multiple weight, each basis state has definite quantum numbers T, T_3, and Y. All basis states in the representation have the same spin and parity. There are three types of the planar weight diagrams for the flavor SU(3).

(a) For a one-row Young pattern $[\lambda, 0]$, the planar weight diagram is a regular triangle upside down. The length of the edge of the triangle is λ. All weights are single. The conjugate representation of $[\lambda, 0]$ is $[\lambda, \lambda] \simeq [\lambda, 0]^*$ whose planar weight diagram is a regular triangle.

The highest weight state in $[\lambda, 0]$ is described by a standard tensor Young tableau where each box is filled by u so that

$$Y = \lambda/3, \qquad T_3 = \lambda/2. \tag{8.79}$$

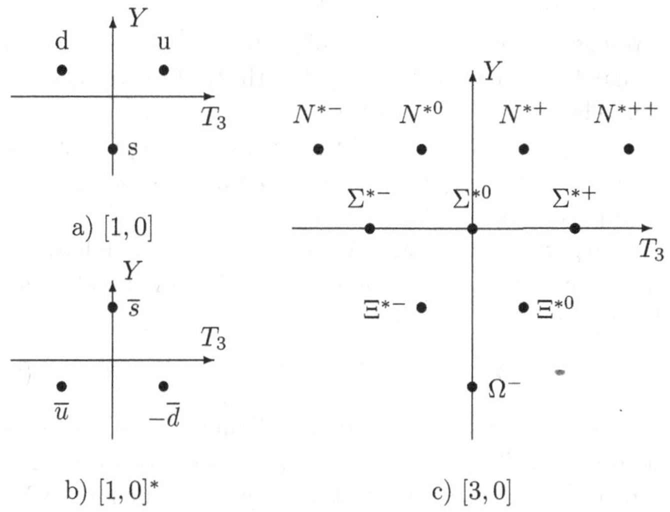

Fig. 8.5 The planar weight diagrams of $[\lambda, 0]$ of SU(3).

From the highest weight state one constructs a T-multiplet by applications of the lowering operator T_-, where u quark is replaced with d quark one by one. In the planar weight diagram, the states in the T-multiplet are located in a horizontal line with $Y = \lambda/3$. Recall that the one-row tensor Young tableau describes a totally symmetric tensor so that the quarks in the tableau can be interchanged symmetrically. From each state in the T-multiplet one constructs the U-multiplets by applications of the lowering operator U_-, where d quark is replaced with s quark one by one. In the planar weight diagram, the states in the U-multiplet are along a line with the angle $2\pi/3$ to the abscissa axis where Q is fixed. All weights in the representation are single. The planar weight diagrams of the representations $[1,0]$, $[1,1] \simeq [1,0]^*$, and $[3,0]$ are listed in Fig. 8.5. $[1,0]$ and $[1,0]^*$ describe the quarks and antiquarks, respectively. $[3,0]$ describe the baryon decuplet, observed in experiment with spin 3/2 and the positive parity. The baryons in the decuplet are denoted by N^*, Σ^*, Ξ^*, and Ω,

$$\begin{array}{ll}
N^{*++} = \begin{array}{|c|c|c|}\hline \mathrm{u} & \mathrm{u} & \mathrm{u} \\ \hline\end{array}\,, & N^{*+} = \sqrt{3}\begin{array}{|c|c|c|}\hline \mathrm{u} & \mathrm{u} & \mathrm{d} \\ \hline\end{array}\,, \\
N^{*0} = \sqrt{3}\begin{array}{|c|c|c|}\hline \mathrm{u} & \mathrm{d} & \mathrm{d} \\ \hline\end{array}\,, & N^{*-} = \begin{array}{|c|c|c|}\hline \mathrm{d} & \mathrm{d} & \mathrm{d} \\ \hline\end{array}\,, \\
\Sigma^{*+} = \sqrt{3}\begin{array}{|c|c|c|}\hline \mathrm{u} & \mathrm{u} & \mathrm{s} \\ \hline\end{array}\,, & \Sigma^{*0} = \sqrt{6}\begin{array}{|c|c|c|}\hline \mathrm{u} & \mathrm{d} & \mathrm{s} \\ \hline\end{array}\,, \\
\Sigma^{*-} = \sqrt{3}\begin{array}{|c|c|c|}\hline \mathrm{d} & \mathrm{d} & \mathrm{s} \\ \hline\end{array}\,, & \Xi^{*0} = \sqrt{3}\begin{array}{|c|c|c|}\hline \mathrm{u} & \mathrm{s} & \mathrm{s} \\ \hline\end{array}\,, \\
\Xi^{*-} = \sqrt{3}\begin{array}{|c|c|c|}\hline \mathrm{d} & \mathrm{s} & \mathrm{s} \\ \hline\end{array}\,, & \Omega^{-} = \begin{array}{|c|c|c|}\hline \mathrm{s} & \mathrm{s} & \mathrm{s} \\ \hline\end{array}\,.
\end{array} \tag{8.80}$$

(b) For a Young pattern $[2\lambda, \lambda] \simeq [\lambda, 0]\backslash[\lambda, 0]^*$, the planar weight diagram is a regular hexagon with the edge length λ. The weights on the edge are single. The multiplicities of the weights increase one by one as their positions go from the edge toward the origin. The representation of $[2\lambda, \lambda]$ is self-conjugate and its planar weight diagram is symmetric in the inversion with respect to the origin.

The highest weight state in $[2\lambda, \lambda]$ is described by a standard tensor Young tableau where each box in the first row is filled by u and each box in the second row is filled by d so that

$$Y = \lambda, \qquad T_3 = \lambda/2. \tag{8.81}$$

From the highest weight state, one constructs a T-multiplet by applications of the lowering operator T_-, where u quark is replaced with d quark one by one. The u quark on the column with two rows cannot be replaced with d quark otherwise two d's are filled in the same column. The states in the T-multiplet are located in a horizontal line with $Y = \lambda$ in the planar weight diagram. From each state in the T-multiplet one constructs the U-multiplets by applications of the lowering operator U_-, where d quark is replaced with s quark one by one. Since the s quarks can be filled in two rows, the multiple weight appears. The basis states in one T-multiplet have the same number of s quarks as each other.

The planar weight diagrams of the representation $[2,1]$ for the baryons and the representation $[1]\backslash[1]^*$ for the mesons are listed in Fig. 8.6. In experiments, the observed baryons in the octet $[2,1]$, called P, N, Σ, Λ, and Ξ, have spin 1/2 and positive parity. The observed mesons in the octet $[1]\backslash[1]^*$ have negative parity. When the spin is 0, they are the scalar mesons, called K, π, η, and $\overline{\text{K}}$. When the spin is 1, they are the vector mesons, called K^*, ρ, ϕ, and $\overline{K}^*$.

The tensor Young tableaux of the $(1/2)^+$ baryon octet and the 0^- meson octet are listed as follows. They can be calculated from the highest weight state by the lowering operators (see Eqs. (8.39) and (8.83)).

$$\begin{array}{ll}
\text{Baryon} & \text{Scalar meson} \\
\text{P} = \begin{array}{l}\text{uu}\\ \text{d}\end{array} & \text{K}^+ = \text{u}\bar{\text{s}} \\
\text{N} = \begin{array}{l}\text{ud}\\ \text{d}\end{array} & \text{K}^0 = \text{d}\bar{\text{s}} \\
\Sigma^+ = \begin{array}{l}\text{uu}\\ \text{s}\end{array} & -\pi^+ = -\,\text{u}\bar{\text{d}} \\
\Sigma^0 = \sqrt{1/2}\left\{2\begin{array}{l}\text{ud}\\ \text{s}\end{array} - \begin{array}{l}\text{us}\\ \text{d}\end{array}\right\} & \pi^0 = \sqrt{1/2}\left\{\text{u}\bar{\text{u}} - \text{d}\bar{\text{d}}\right\} \\
\Sigma^- = \begin{array}{l}\text{dd}\\ \text{s}\end{array} & \pi^- = \text{d}\bar{\text{u}} \\
\Lambda = \sqrt{3/2}\begin{array}{l}\text{us}\\ \text{d}\end{array} & -\eta = -\sqrt{1/6}\left\{\text{u}\bar{\text{u}} + \text{d}\bar{\text{d}} - 2\text{s}\bar{\text{s}}\right\} \\
\Xi^0 = \begin{array}{l}\text{us}\\ \text{s}\end{array} & -\overline{\text{K}^0} = -\,\text{s}\bar{\text{d}} \\
\Xi^- = \begin{array}{l}\text{ds}\\ \text{s}\end{array} & \text{K}^- = \text{s}\bar{\text{u}}
\end{array} \qquad (8.82)$$

$$\begin{aligned}
\pi^0 &= \sqrt{\frac{1}{2}}F_1\left(-\pi^+\right) = \sqrt{\frac{1}{2}}\left\{\boxed{\text{u}}\backslash\boxed{\text{u}}^* - \boxed{\text{d}}\backslash\boxed{\text{d}}^*\right\}, \\
-\eta &= \sqrt{\frac{2}{3}}\left\{F_2\,\text{K}^0 - \sqrt{\frac{1}{2}}\,\pi^0\right\} \\
&= \sqrt{\frac{1}{6}}\left\{2F_2\,\boxed{\text{d}}\backslash\boxed{\text{s}}^* - \boxed{\text{u}}\backslash\boxed{\text{u}}^* + \boxed{\text{d}}\backslash\boxed{\text{d}}^*\right\} \\
&= \sqrt{\frac{1}{6}}\left\{2\,\boxed{\text{s}}\backslash\boxed{\text{s}}^* - \boxed{\text{u}}\backslash\boxed{\text{u}}^* - \boxed{\text{d}}\backslash\boxed{\text{d}}^*\right\}.
\end{aligned} \qquad (8.83)$$

Two sets of the tensor Young tableaux are related by Eqs. (8.23) and (8.58). In fact,

$$
\begin{aligned}
3\begin{array}{|c|c|}\hline u & d \\ \hline s \\ \cline{1-1}\end{array} &= 2\begin{array}{|c|c|}\hline u & d \\ \hline s \\ \cline{1-1}\end{array} + \begin{array}{|c|c|}\hline d & u \\ \hline s \\ \cline{1-1}\end{array} + \begin{array}{|c|c|}\hline u & s \\ \hline d \\ \cline{1-1}\end{array} \\
&= -2\,\boxed{d}\backslash\boxed{d}^{\,*} + \boxed{u}\backslash\boxed{u}^{\,*} + \boxed{s}\backslash\boxed{s}^{\,*}, \\
3\begin{array}{|c|c|}\hline u & s \\ \hline d \\ \cline{1-1}\end{array} &= 2\begin{array}{|c|c|}\hline u & s \\ \hline d \\ \cline{1-1}\end{array} + \begin{array}{|c|c|}\hline s & u \\ \hline d \\ \cline{1-1}\end{array} + \begin{array}{|c|c|}\hline u & d \\ \hline s \\ \cline{1-1}\end{array} \\
&= 2\,\boxed{s}\backslash\boxed{s}^{\,*} - \boxed{u}\backslash\boxed{u}^{\,*} - \boxed{d}\backslash\boxed{d}^{\,*}. \\
\sqrt{2}\begin{array}{|c|c|}\hline u & d \\ \hline s \\ \cline{1-1}\end{array} - \sqrt{\frac{1}{2}}\begin{array}{|c|c|}\hline u & s \\ \hline d \\ \cline{1-1}\end{array} &= \sqrt{\frac{1}{2}}\left\{\boxed{u}\backslash\boxed{u}^{\,*} - \boxed{d}\backslash\boxed{d}^{\,*}\right\}, \\
\sqrt{\frac{3}{2}}\begin{array}{|c|c|}\hline u & s \\ \hline d \\ \cline{1-1}\end{array} &= \sqrt{\frac{1}{6}}\left\{2\,\boxed{s}\backslash\boxed{s}^{\,*} - \boxed{u}\backslash\boxed{u}^{\,*} - \boxed{d}\backslash\boxed{d}^{\,*}\right\}.
\end{aligned}
\tag{8.84}
$$

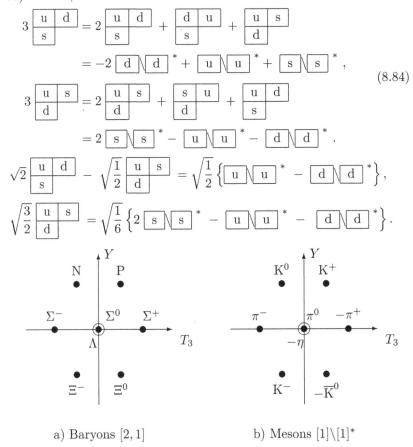

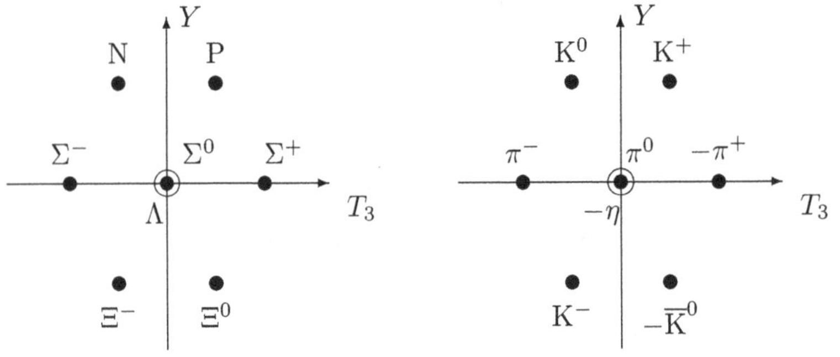

a) Baryons [2, 1] b) Mesons [1]\[1]*

Fig. 8.6 The planar weight diagrams of adjoint representation of SU(3).

(c) For a Young pattern $[\lambda_1, \lambda_2]$, $\lambda_1 > 2\lambda_2 > 0$, the planar weight diagram is a hexagon where the lengths of two unneighbored edges are the same. The length of the top edge is $\lambda_1 - \lambda_2$, and the length of the bottom edge is λ_2. The weights on the edge are single. The multiplicities of the weights increase one by one as their positions go inside until the hexagon becomes a triangle upside down. The weights inside the triangle are $\lambda_2 + 1$. The highest weight state has

$$Y = (\lambda_1 + \lambda_2)/3, \qquad T_3 = (\lambda_1 - \lambda_2)/2. \tag{8.85}$$

The conjugate representation of $[\lambda_1, \lambda_2]$ is $[\lambda_1, \lambda_1 - \lambda_2]$. Their planar weight diagrams are inverted with respect to the origin.

8.4.3 *Mass Formulas*

Present a simple model to study the mass formula of the hadrons in a multiplet of the flavor SU(3), which are made by the quarks and the antiquarks. Assume that the binding energy $-V$ for the hadrons in one multiplet of SU(3) are the same, and the difference of the masses of hadrons comes from the different quarks. The masses of u and d quarks and their antiquarks are m_1 and the masses of s quark and its antiquark are m_2.

The masses in the baryon decuplet are

$$M(N^*) = 3m_1 - V, \qquad M(\Sigma^*) = 2m_1 + m_2 - V,$$
$$M(\Xi^*) = m_1 + 2m_2 - V, \qquad M(\Omega) = 3m_2 - V.$$

Then, one obtains the mass formula

$$M(\Omega) - M(\Xi^*) = M(\Xi^*) - M(\Sigma^*) = M(\Sigma^*) - M(N^*). \tag{8.86}$$

From the experiments, the observed average masses for the isospin multiplets are

$$M_{N^*} = 1232 \text{ MeV}, \qquad M_{\Sigma^*} = 1384.6 \text{ MeV},$$
$$M_{\Xi^*} = 1531.8 \text{ MeV}, \qquad M_{\Omega} = 1672.5 \text{ MeV}.$$

$$M(\Omega) - M(\Xi^*) = 140.7 \text{ MeV},$$
$$M(\Xi^*) - M(\Sigma^*) = 147.2 \text{ MeV},$$
$$M(\Sigma^*) - M(N^*) = 152.6 \text{ MeV}.$$

In the beginning of sixties of the last Century the simple model predicted that a baryon Ω with the spin 3/2, positive parity, supercharge $Y = -2$, and electric charge $Q = -1$ should exist at the mass near 1680 MeV. It was found in 1962 as expected.

This model is too simple to explain the masses of the baryon octet, because both the baryons Σ and Λ are composed of one s quark and two quarks of u and d, but have different masses in experiment. Further analysis shows that the different masses of s quark and the quark of u or d can be demonstrated by a broken mass matrix $\hat{M}$. In addition to the symmetric mass $\hat{M}_0$, $\hat{M} - \hat{M}_0$ has the transformation property like the supercharge Y, called the "33" symmetry broken. Namely, the Hamiltonian contains a mass term $\overline{\psi}\hat{M}\psi$ which is invariant in SU(3). How many parameters appear in the mass term $\overline{\psi}\hat{M}\psi$? Since $\hat{M} - \hat{M}_0$ belongs to the adjoint representation, the parameters come from the reduction of $[2,1] \times [\lambda]$ to $[\lambda]$. When $[\lambda]$ is a one-row Young pattern, $[2,1] \times [\lambda]$ contains one $[\lambda]$ so that there is two mass parameters as shown in Eq. (8.86). When $[\lambda]$ is a two-row Young pattern, $[2,1] \times [\lambda]$ contains two $[\lambda]$ so that a new mass parameter appears.

Gell-Mann, Nishijima, and Okubo expressed the mass operator $\hat{M}$ in the sum of generators and their products,

$$\hat{M} = M_0 + M_1 Y + M_2 \left[Y^2 + cT(T+1)\right],$$

where $T(T+1)$ is the eigenvalue of the operator T^2. Because there are only three mass parameters, the terms of higher order are not needed and a redundant parameter c should be determined. The antisymmetric combination of generators in the polynomial of order 2 is proportional to the linear term of generators (see Eq. (7.5)), and the symmetric combination, as shown in the last formula of Fig. 7.7, contains the representations $[4,2]$, $[2,1]_S$, and $[0,0]$, whose dimensions are 27, 8, and 1, respectively. The parameter c should be determined to exclude the representation $[4,2]$. The mass formula holds for the baryon decuplet such that

$$Y^2 + cT(T+1) = a + bY.$$

For the baryon Ω, $Y = -2$ and $T = 0$, one has $4 = a - 2b$. For the baryon Ξ^*, $Y = -1$ and $T = 1/2$, one has $1 + 3c/4 = a - b$. For the baryon Σ^*, $Y = 0$ and $T = 1$, one has $2c = a$. The solution is $a = -8$, $b = -6$, and $c = -4$. The solution meets the condition from the baryon N^*, where $Y = 1$ and $T = 3/2$. Thus, the Gell-Mann–Nishijima–Okubo mass formula is

$$M(T,Y) = M_0 + M_1 Y + M_2 \left\{Y^2 - 4T(T+1)\right\}. \tag{8.87}$$

For the baryon octet, $M(N) = M_0 + M_1 - 2M_2$, $M(\Sigma) = M_0 - 8M_2$, $M(\Lambda) = M_0$, and $M(\Xi) = M_0 - M_1 - 2M_2$. Then,

$$\frac{M(N) + M(\Xi)}{2} = \frac{M(\Sigma) + 3M(\Lambda)}{4}. \tag{8.88}$$

The prediction fits the experiment data,

$$M(N) = 938.9 \text{ MeV}, \qquad M(\Sigma) = 1193.1 \text{ MeV},$$
$$M(\Lambda) = 1115.7 \text{ MeV}, \qquad M(\Xi) = 1318.1 \text{ MeV}.$$

The left-hand side of Eq. (8.88) is 1128.5 MeV, and the right-hand side is 1135.1 MeV. The formula (8.88) holds approximately for the mass square of the scalar meson octet. The experiment data are

$$m(\pi) = 138.0 \text{ MeV}, \qquad m(K) = 495.7 \text{ MeV}, \qquad m(\eta) = 547.5 \text{ MeV}.$$

The left-hand side of Eq. (8.88) for the mass square is 0.2457 GeV^2, and the right-hand side is 0.2296 GeV^2. The formula is not in good agreement with the vector meson octet because there is a mixture between the meson octet and the meson singlet ω.

8.4.4 *Wave Functions of Mesons*

A meson in low energy is composed of a quark and an antiquark with zero orbit angular momentum. The wave function of a meson is a product of the color, the flavor, and the spinor wave functions. It is not needed to consider the permutation symmetry because the quark and the antiquark are not the identical particles. The mixed tensor of rank $(1,1)$ is decomposed into a traceless tensor and a trace tensor (scalar),

$$\square \times \square^* = \square\backslash\square^* \oplus \mathbf{1}.$$

Due to color confinement, the color wave function of a meson has to be in the colorless state, namely in the singlet of SU$(3)_c$. The flavor wave function of a meson can be in the octet (traceless tensor) or singlet (trace tensor). For the spinor wave functions, the traceless tensor describes the vector mesons and the trace tensor the scalar mesons. The vector meson octet and singlet (ω) with the negative parity and the scalar meson octet and singlet (η') with the negative parity have been observed in experiments.

Denote by (ψ_+, ψ_-) and $(-\overline{\psi_-}, \overline{\psi_+})$ the spinor wave functions for a quark and an antiquark, respectively. The spinor wave functions for the scalar meson and for the vector mesons are

$$\begin{aligned}
&S = 0, \quad && S_3 = 0: \quad && \sqrt{1/2}\left(\overline{\psi_+}\psi_+ + \overline{\psi_-}\psi_-\right), \\
&S = 1, \quad && S_3 = 1: \quad && -\overline{\psi_-}\psi_+, \\
& && S_3 = 0: \quad && \sqrt{1/2}\left(\overline{\psi_+}\psi_+ - \overline{\psi_-}\psi_-\right), \\
& && S_3 = -1: \quad && \overline{\psi_+}\psi_-.
\end{aligned} \tag{8.89}$$

The flavor wave function of the singlet meson is the trace tensor,

$$\sqrt{\frac{1}{3}}\left\{\boxed{\text{u}}\backslash\boxed{\text{u}}^* + \boxed{\text{d}}\backslash\boxed{\text{d}}^* + \boxed{\text{s}}\backslash\boxed{\text{s}}^*\right\}. \tag{8.90}$$

The flavor wave functions of the octet mesons are given in Eq. (8.82). But, in the particle physics, the wave functions are preferred to be expressed in a matrix of three dimensions, where the row index denotes the covariant one and the column index denotes the contravariant one. The basis tensors of the scalar mesons are, for example,

$$\pi^+ = \begin{pmatrix} 0 & 1 & 0 \\ 0 & 0 & 0 \\ 0 & 0 & 0 \end{pmatrix}, \qquad \eta = \sqrt{\frac{1}{6}} \begin{pmatrix} 1 & 0 & 0 \\ 0 & 1 & 0 \\ 0 & 0 & -2 \end{pmatrix}.$$

The traceless tensor is expanded with respect to the basis tensors where the coefficients are written by the names of the mesons,

$$\boldsymbol{M} = \begin{pmatrix} \dfrac{\pi^0}{\sqrt{2}} + \dfrac{\eta}{\sqrt{6}} & \pi^+ & \mathrm{K}^+ \\ \pi^- & -\dfrac{\pi^0}{\sqrt{2}} + \dfrac{\eta}{\sqrt{6}} & \mathrm{K}^0 \\ \mathrm{K}^- & \overline{\mathrm{K}^0} & -\dfrac{2\eta}{\sqrt{6}} \end{pmatrix}, \tag{8.91}$$

$\boldsymbol{M}$ transforms in the flavor SU(3) as follows:

$$\boldsymbol{M} \stackrel{u}{\longrightarrow} u\boldsymbol{M}u^{-1}. \tag{8.92}$$

Through Eq. (8.72), the flavor wave functions can be expressed in those in the real orthogonal representation of eight dimensions,

$$\begin{aligned} &(\boldsymbol{M}_1 + i\boldsymbol{M}_2)/\sqrt{2} = \pi^-, && (\boldsymbol{M}_1 - i\boldsymbol{M}_2)/\sqrt{2} = \pi^+, \\ &(\boldsymbol{M}_4 + i\boldsymbol{M}_5)/\sqrt{2} = \mathrm{K}^-, && (\boldsymbol{M}_4 - i\boldsymbol{M}_5)/\sqrt{2} = \mathrm{K}^+, \\ &(\boldsymbol{M}_6 + i\boldsymbol{M}_7)/\sqrt{2} = \overline{\mathrm{K}^0}, && (\boldsymbol{M}_6 - i\boldsymbol{M}_7)/\sqrt{2} = \mathrm{K}^0, \\ &\boldsymbol{M}_3 = \pi^0, && \boldsymbol{M}_8 = \eta. \end{aligned} \tag{8.93}$$

Similarly, the flavor wave functions of the baryon octet are also expressed in a matrix of three dimensions,

$$\boldsymbol{B} = \begin{pmatrix} \dfrac{\Sigma^0}{\sqrt{2}} - \dfrac{\Lambda}{\sqrt{6}} & -\Sigma^+ & \mathrm{P} \\ \Sigma^- & -\dfrac{\Sigma^0}{\sqrt{2}} - \dfrac{\Lambda}{\sqrt{6}} & \mathrm{N} \\ \Xi^- & -\Xi^0 & \dfrac{2\Lambda}{\sqrt{6}} \end{pmatrix}, \tag{8.94}$$

where the minus sign comes from the definition of the particles as shown in Eq. (8.82).

8.4.5 *Wave Functions of Baryons*

A baryon in low energy is composed of three quarks which are identical particles satisfying the Fermi statistics. Its total wave function has to be antisymmetric in the transposition between the quarks. Assume that the orbital angular momentum of the low energy baryon is vanishing such that its total wave function is a product of the color, the flavor, and the spinor wave functions. Three quarks are described by a tensor of rank 3, which is decomposed by the Young operators,

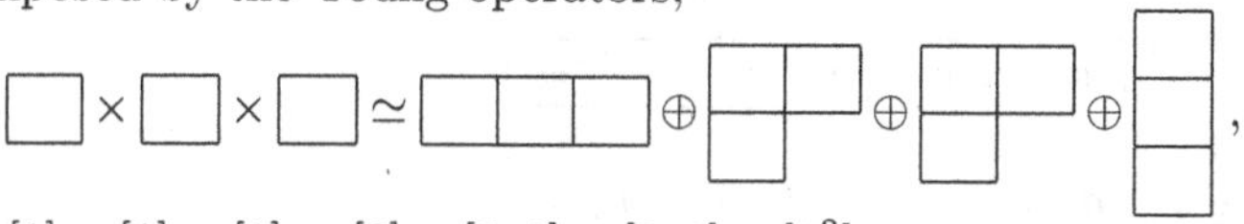

(8.95)

$$[1] \times [1] \times [1] \simeq [3] \oplus [2,1] \oplus [2,1] \oplus [1^3].$$

Those wave functions belong to the representations of the permutation group denoted by the same Young patterns. Due to the color confinement, the color wave function is in the color singlet $[1^3]$ which is totally antisymmetric in the quark transposition. The product of the flavor and the spinor wave functions has to be totally symmetric.

There are three choices for the flavor wave functions. The representation [3] of the flavor SU(3) describes the decuplet which is the totally symmetric states in the permutations. The representation [2, 1] describes the octet which is the mixed symmetric states. The representation $[1^3]$ describes the singlet which is the total antisymmetric states. However, there are only two choices for the spinor wave functions because the representation $[1^3]$ of SU(2) corresponds to the null space. The representation [3] of the spinor SU(2) describes the quadruplet ($S = 3/2$) which is the totally symmetric states in the permutations. The representation [2, 1] of the spinor SU(2) describes the doublet ($S = 1/2$) which is the mixed symmetric states. Since the product of the flavor and the spinor wave functions are totally symmetric, the wave function of flavor decuplet has to multiply that of spinor quadruplet, and the wave function of flavor octet has to multiply that of spinor doublet where a suitable combination is needed such that the multiplied wave functions are combined to be totally symmetric with respect to the permutations. This coincides with the experimental data that the observed low-energy baryons are the baryon decuplet with spin-parity $(3/2)^+$ and the baryon octet with spin-parity $(1/2)^+$.

(a) **The $(3/2)^+$ baryon decuplet.**

The wave function is a product of the flavor and the spinor wave functions. Two examples are given in the following.

$N^{*+}_{1/2}$, $T = 3/2$, $T_3 = 1/2$, $Y = 1$, and $S_3 = 1/2$.

$$\begin{array}{|c|c|c|}\hline u & u & d \\ \hline\end{array} \cdot \begin{array}{|c|c|c|}\hline + & + & - \\ \hline\end{array}$$

$$\begin{aligned} N^{*+}_{1/2} = & \ \frac{1}{3}\left\{u_+u_+d_- + u_+d_+u_- + d_+u_+u_- + u_+u_-d_+ + u_+d_-u_+ \right. \\ & \left. + \ d_+u_-u_+ + u_-u_+d_+ + u_-d_+u_+ + d_-u_+u_+\right\}. \end{aligned}$$

$\Sigma^{*0}_{-1/2}$, $T = 1$, $T_3 = 0$, $Y = 0$, and $S_3 = -1/2$.

$$\begin{array}{|c|c|c|}\hline u & d & s \\ \hline\end{array} \cdot \begin{array}{|c|c|c|}\hline + & - & - \\ \hline\end{array}$$

$$\begin{aligned} \Sigma^{*0}_{-1/2} = & \ \frac{1}{3\sqrt{2}}\left\{u_+d_-s_- + u_+s_-d_- + d_+u_-s_- + d_+s_-u_- \right. \\ & + \ s_+u_-d_- + s_+d_-u_- + u_-d_+s_- + u_-s_+d_- + d_-u_+s_- \\ & + \ d_-s_+u_- + s_-u_+d_- + s_-d_+u_- + u_-d_-s_+ + u_-s_-d_+ \\ & \left. + \ d_-u_-s_+ + d_-s_-u_+ + s_-u_-d_+ + s_-d_-u_+\right\}. \end{aligned}$$

(b) **The $(1/2)^+$ baryon octet.**

Both the flavor and the spinor wave functions are in the mixed symmetry of the permutations. Their product has to be combined as the wave function with total symmetry. The representation matrices of generators of the permutation group S_3 are calculated in Table 6.4. Their direct products and the eigenvectors to the eigenvalue 1 are

$$D^{[2,1]}[(1\ 2)] \times D^{[2,1]}[(1\ 2)] = \begin{pmatrix} 1 & -1 & -1 & 1 \\ 0 & -1 & 0 & 1 \\ 0 & 0 & -1 & 1 \\ 0 & 0 & 0 & 1 \end{pmatrix}, \qquad \begin{pmatrix} 1 \\ 0 \\ 0 \\ 0 \end{pmatrix}, \quad \begin{pmatrix} 0 \\ 1 \\ 1 \\ 2 \end{pmatrix},$$

$$D^{[2,1]}[(1\ 2\ 3)] \times D^{[2,1]}[(1\ 2\ 3)] = \begin{pmatrix} 1 & -1 & -1 & 1 \\ 1 & 0 & -1 & 0 \\ 1 & -1 & 0 & 0 \\ 1 & 0 & 0 & 0 \end{pmatrix}, \qquad \begin{pmatrix} 1 \\ 1 \\ 0 \\ 1 \end{pmatrix}, \quad \begin{pmatrix} 1 \\ 0 \\ 1 \\ 1 \end{pmatrix}.$$

The combination corresponding to the identical representation is the common eigenfunction $\boldsymbol{v}$ to the eigenvalue 1, $\boldsymbol{v}^T = (2, 1, 1, 2)$.

Write the wave function of a proton with $S_3 = -1/2$ as example. A proton has the quantum numbers $T = T_3 = 1/2$ and $Y = 1$. Take the Young tableau $\mathcal{Y} = \begin{array}{|c|c|}\hline 1 & 2 \\ \hline 3 & \multicolumn{1}{|c}{} \\ \cline{1-1}\end{array}$. The basis tensors of the flavor wave functions are

$$\begin{array}{|c|c|}\hline \mathrm{u} & \mathrm{u} \\ \hline \mathrm{d} & \multicolumn{1}{|c}{} \\ \cline{1-1}\end{array} = 2uud - duu - udu, \quad (23)\begin{array}{|c|c|}\hline \mathrm{u} & \mathrm{u} \\ \hline \mathrm{d} & \multicolumn{1}{|c}{} \\ \cline{1-1}\end{array} = 2udu - duu - uud.$$

Similarly, the basis tensors of the spinor wave functions are

$$\begin{array}{|c|c|}\hline + & - \\ \hline - \\ \cline{1-1}\end{array} = (+--)+(-+-)-2(--+),$$

$$(23)\begin{array}{|c|c|}\hline + & - \\ \hline - \\ \cline{1-1}\end{array} = (+--)+(--+)-2(-+-).$$

Thus, the wave function of a proton with $S_3 = -1/2$ is

$$\begin{aligned}
P_{-1/2} = & \begin{array}{|c|c|}\hline u & u \\ \hline d \\ \cline{1-1}\end{array} \left\{ 2\begin{array}{|c|c|}\hline + & - \\ \hline - \\ \cline{1-1}\end{array} + \left[(2\ 3)\begin{array}{|c|c|}\hline + & - \\ \hline - \\ \cline{1-1}\end{array} \right] \right\} \\
& + \left[(2\ 3)\begin{array}{|c|c|}\hline u & u \\ \hline d \\ \cline{1-1}\end{array} \right] \left\{ \begin{array}{|c|c|}\hline + & - \\ \hline - \\ \cline{1-1}\end{array} + 2\left[(2\ 3)\begin{array}{|c|c|}\hline + & - \\ \hline - \\ \cline{1-1}\end{array} \right] \right\} \\
= & \ \{2uud - duu - udu\} \cdot 3\{(+--)-(--+)\} \\
& + \{2udu - duu - uud\} \cdot 3\{(+--)-(-+-)\} \\
= & \ 3\{u_+u_-d_- - 2d_+u_-u_- + u_+d_-u_- - 2u_-u_-d_+ + d_-u_-u_+ \\
& + u_-d_-u_+ + u_-u_+d_- + d_-u_+u_- - 2u_-d_+u_-\}.
\end{aligned} \tag{8.96}$$

The normalization factor should be changed to $\sqrt{1/18}$.

8.5 Exercises

1. Calculate the dimensions of the irreducible representations denoted by the following Young patterns for the SU(3) group and for the SU(6) group, respectively:

$$[3], \quad [2,1], \quad [3,3], \quad [4,2], \quad [5,1].$$

2. Calculate the Clebsch–Gordan series for the following direct product representations, and compare their dimensions by Eq. (8.30) for the SU(3) group and for the SU(6) group, respectively:

(a) $[2,1]\otimes[3,0]$, (b) $[3,0]\otimes[3,0]$, (c) $[3,0]\otimes[3,3]$, (d) $[4,2]\otimes[2,1]$.

3. Try to express each nonzero tensor Young tableau for the irreducible representation $[3,1]$ of SU(3) as the linear combination of the standard tensor Young tableaux.

4. Write the explicit expansion of each standard tensor Young tableau in the tensor subspace $\mathcal{Y}_2^{[3,1]}\mathcal{T}$, where $\mathcal{T}$ is the tensor space of rank 4 for the SU(3) group and the standard Young tableau of the Young operator $\mathcal{Y}_2^{[3,1]}$ is $\begin{array}{|c|c|c|}\hline 1 & 2 & 4 \\ \hline 3 \\ \cline{1-1}\end{array}$.

5. Transform the following traceless mixed tensor representations of the SU(6) group into the covariant tensor representations, respectively, and calculate their dimensions:

$$(1)\ \ [3,2,1]^*, \qquad (2)\ \ [3,2,1]\backslash[3,3]^*, \qquad (3)\ \ [4,3,1]\backslash[3,2]^*.$$

6. Prove the identity:

$$\sum_{A=1}^{N^2-1} \left(T_A\right)_{ac}\left(T_A\right)_{bd} = \frac{1}{2}\delta_a^d\delta_b^c - \frac{1}{2N}\delta_a^c\delta_b^d\ .$$

7. Expand the Gelfand bases in the irreducible representation $[3,0]$ of the SU(3) group with respect to the standard tensor Young tableaux by making use of its block weight diagram given in Fig. 7.3.

8. Expand the Gelfand bases in the irreducible representation $[3,3]$ of the SU(3) group with respect to the standard tensor Young tableaux by making use of its block weight diagram given in Fig. 7.3.

9. Express each Gelfand basis in the irreducible representation $[4,0]$ of the SU(3) group by the standard tensor Young tableau and calculate the nonvanishing matrix entries for the lowering operators F_μ. Draw the block weight diagram and the planar weight diagram for the representation $[4,0]$ of SU(3).

10. Express each Gelfand basis in the irreducible representation $[3,1]$ of the SU(3) group by the standard tensor Young tableau and calculate the nonvanishing matrix entries for the lowering operators F_μ. Draw the block weight diagram and the planar weight diagram for the representation $[3,1]$ of SU(3).

11. Calculate the Clebsch–Gordan series for the direct product representation $[2,1]\times[2,1]$ of the SU(3) group, and expand the highest weight state of each irreducible representation in the Clebsch–Gordan series with respect to the standard tensor Young tableaux.

12. A neutron is composed of one u quark and two d quarks. Construct the wave function of a neutron with spin $S_3 = -1/2$, satisfying the correct permutation symmetry among the identical particles.

Chapter 9

REAL ORTHOGONAL GROUPS

In this chapter we will study the tensor representations and the spinor representations of the SO(N) groups, and then, the irreducible representations of the proper Lorentz group.

9.1 Tensor Representations of SO(N)

The tensor representations are the single-valued representations of SO(N). In this section, the reduction of a tensor space of SO(N) is studied and the orthonormal irreducible basis tensors are calculated.

9.1.1 *Tensors of SO(N)*

Similar to the tensors of SU(N), a tensor of rank n of SO(N) has N^n components and transforms in $R \in$ SO(N),

$$\boldsymbol{T}_{a_1 \ldots a_n} \xrightarrow{R} \left(O_R \boldsymbol{T}\right)_{a_1 \ldots a_n} = \sum_{b_1 \ldots b_n} R_{a_1 b_1} \ldots R_{a_n b_n} \boldsymbol{T}_{b_1 \ldots b_n}. \tag{9.1}$$

A basis tensor $\boldsymbol{\theta}_{d_1 \ldots d_n}$ contains only one nonvanishing component which is equal to 1,

$$\left(\boldsymbol{\theta}_{d_1 \ldots d_n}\right)_{a_1 \ldots a_n} = \delta_{d_1 a_1} \delta_{d_2 a_2} \ldots \delta_{d_n a_n} = \left(\boldsymbol{\theta}_{d_1}\right)_{a_1} \left(\boldsymbol{\theta}_{d_2}\right)_{a_2} \cdots \left(\boldsymbol{\theta}_{d_n}\right)_{a_n}, \tag{9.2}$$

$$O_R \boldsymbol{\theta}_{d_1 \ldots d_n} = \sum_{b_1 \ldots b_n} \boldsymbol{\theta}_{b_1 \ldots b_n} R_{b_1 d_1} \ldots R_{b_n d_n}. \tag{9.3}$$

Any tensor can be expanded with respect to the basis tensors,

$$\boldsymbol{T}_{a_1 \ldots a_n} = \sum_{d_1 \ldots d_n} T_{d_1 \ldots d_n} \left(\boldsymbol{\theta}_{d_1 \ldots d_n}\right)_{a_1 \ldots a_n} = T_{a_1 \ldots a_n}. \tag{9.4}$$

In a permutation of S_n the tensors and the basis tensors are transformed as Eqs. (8.6) and (8.18), respectively. The tensor space is an invariant linear space both in SO(N) and in S_n. The SO(N) transformation is commutable with the permutation (the Weyl reciprocity), so that the tensor space can be reduced by the projection of the Young operators.

The main difference between the tensors of SU(N) and SO(N) is that the transformation matrix $R \in \mathrm{SO}(N) \subset \mathrm{SU}(N)$ is real. As a result, the tensors of SO(N) have the following new characteristics. First, the real part and the imaginary part of a tensor of SO(N) transform separately in Eq. (9.1) so that only the real tensors are needed to be studied. Second, there is no difference between a covariant tensor and a contravariant tensor for the SO(N) transformations. The contraction of a tensor are accomplished between any two indices so that before the projection of a Young operator the tensor space has to be decomposed into a series of traceless tensor subspaces, which are invariant in SO(N). Third, denote by $\mathcal{T}$ the traceless tensor space of rank n. After the projection of a Young operator, $\mathcal{T}_\mu^{[\lambda]} = \mathcal{Y}_\mu^{[\lambda]}\mathcal{T}$ is a traceless tensor subspace with a given permutation symmetry. A similar proof to that in §8.3.3 shows that $\mathcal{T}_\mu^{[\lambda]}$ is a null space if the sum of the numbers of boxes in the first two columns of the Young pattern $[\lambda]$ is larger than N. Fourth, when the row number m of the Young pattern $[\lambda]$ is larger than $N/2$, the basis tensor $\mathcal{Y}_\mu^{[\lambda]}\boldsymbol{\theta}_{d_1\ldots d_m c\ldots}$ can be changed into a dual basis tensor by a totally antisymmetric tensor $\epsilon_{a_1\ldots a_N}$,

$$\begin{aligned} &{}^*\left[\mathcal{Y}^{[\lambda]}\boldsymbol{\theta}\right]_{a_1\ldots a_{N-m}c\ldots} \\ &= \frac{1}{m!}\sum_{a_{N-\tau+1}\ldots a_N} \epsilon_{a_1\ldots a_{N-m}a_{N-m+1}\ldots a_N}\mathcal{Y}^{[\lambda]}\boldsymbol{\theta}_{a_N\ldots a_{N-m+1}c\ldots}. \end{aligned} \tag{9.5}$$

Its inverse transformation is

$$\begin{aligned} &\frac{1}{(N-m)!}\sum_{a_{m+1}\ldots a_N} \epsilon_{b_1\ldots b_m a_{m+1}\ldots a_N}\ {}^*\left[\mathcal{Y}^{[\lambda]}\boldsymbol{\theta}\right]_{a_N\ldots a_{m+1}c\ldots} \\ &= \frac{1}{m!(N-m)!}\sum_{a_1\ldots a_N} \epsilon_{b_1\ldots b_m a_{m+1}\ldots a_N}\epsilon_{a_N\ldots a_{m+1}a_m\ldots a_1}\mathcal{Y}^{[\lambda]}\boldsymbol{\theta}_{a_1\ldots a_m} \\ &= (-1)^{N(N-1)/2}\mathcal{Y}^{[\lambda]}\boldsymbol{\theta}_{b_1\ldots b_m c\ldots}. \end{aligned} \tag{9.6}$$

In fact, the correspondence between two sets of basis tensors are one-to-one and their difference is only in the arranging order. Thus, a traceless tensor subspace $\mathcal{T}_\mu^{[\lambda]}$, where the row number m of the Young pattern $[\lambda]$ is larger than $N/2$, is equivalent to a traceless tensor subspace $\mathcal{T}_\nu^{[\lambda']}$, where the row number of the Young pattern $[\lambda']$ is $N-m < N/2$,

$$[\lambda'] \simeq [\lambda], \qquad \lambda'_j = \begin{cases} \lambda_j, & j \leq (N-m), \\ 0, & j > (N-m), \end{cases} \qquad N/2 < m \leq N. \tag{9.7}$$

Fifth, when N is even and the row number ℓ of $[\lambda]$ is equal to $N/2$, the Young pattern $[\lambda]$ is the same as its dual Young pattern, called the self-dual Young pattern. In order to remove the factor $(-1)^{N(N-1)/2} = (-1)^\ell$ in Eq. (9.6), one introduces a factor $(-i)^\ell$ in the dual relation (9.5),

$$\begin{aligned} {}^*\left[\mathcal{Y}^{[\lambda]}\boldsymbol{\theta}\right]_{a_1\ldots a_\ell c\ldots} &= \frac{(-i)^\ell}{\ell!} \sum_{a_{\ell+1}\ldots a_{2\ell}} \epsilon_{a_1\ldots a_\ell a_{\ell+1}\ldots a_{2\ell}} \mathcal{Y}^{[\lambda]}\boldsymbol{\theta}_{a_{2\ell}\ldots a_{\ell+1}c\ldots}, \\ \mathcal{Y}^{[\lambda]}\boldsymbol{\theta}_{a_1\ldots a_\ell c\ldots} &= \frac{(-i)^\ell}{\ell!} \sum_{a_{\ell+1}\ldots a_{2\ell}} \epsilon_{a_1\ldots a_\ell a_{\ell+1}\ldots a_{2\ell}} {}^*\left[\mathcal{Y}^{[\lambda]}\boldsymbol{\theta}\right]_{a_{2\ell}\ldots a_{\ell+1}c\ldots}. \end{aligned} \tag{9.8}$$

Define

$$\psi^{\pm}_{a_1\ldots a_\ell c\ldots} = \frac{1}{2}\left\{\mathcal{Y}^{[\lambda]}\boldsymbol{\theta}_{a_1\ldots a_\ell c\ldots} \pm {}^*\left[\mathcal{Y}^{[\lambda]}\boldsymbol{\theta}\right]_{a_1\ldots a_\ell c\ldots}\right\}. \tag{9.9}$$

$\psi^+_{a_1\ldots a_\ell b\ldots}$ keeps invariant in the dual transformation and is called the self-dual basis tensor. $\psi^-_{a_1\ldots a_\ell b\ldots}$ changes its sign in the dual transformation and is called the anti-self-dual basis tensor. For example,

$$\psi^{\pm}_{1\ldots\ell} = \frac{1}{2}\left\{\mathcal{Y}^{[1^\ell]}\boldsymbol{\theta}_{1\ldots\ell} \pm (-i)^\ell \mathcal{Y}^{[1^\ell]}\boldsymbol{\theta}_{(2\ell)\ldots(\ell+1)}\right\}. \tag{9.10}$$

When $N = 4$, i.e., $\ell = 2$, one has

$$\begin{aligned} &\mathcal{Y}^{[1,1]}\boldsymbol{\theta}_{ab} = -\mathcal{Y}^{[1,1]}\boldsymbol{\theta}_{ba}, && {}^*\left[\mathcal{Y}^{[1,1]}\boldsymbol{\theta}\right]_{ab} = -\,{}^*\left[\mathcal{Y}^{[1,1]}\boldsymbol{\theta}\right]_{ba}, \\ &{}^*\left[\mathcal{Y}^{[1,1]}\boldsymbol{\theta}\right]_{12} = \mathcal{Y}^{[1,1]}\boldsymbol{\theta}_{34}, && {}^*\left[\mathcal{Y}^{[1,1]}\boldsymbol{\theta}\right]_{13} = \mathcal{Y}^{[1,1]}\boldsymbol{\theta}_{42}, \\ &{}^*\left[\mathcal{Y}^{[1,1]}\boldsymbol{\theta}\right]_{14} = \mathcal{Y}^{[1,1]}\boldsymbol{\theta}_{23}, && {}^*\left[\mathcal{Y}^{[1,1]}\boldsymbol{\theta}\right]_{34} = \mathcal{Y}^{[1,1]}\boldsymbol{\theta}_{12}, \\ &{}^*\left[\mathcal{Y}^{[1,1]}\boldsymbol{\theta}\right]_{24} = \mathcal{Y}^{[1,1]}\boldsymbol{\theta}_{31}, && {}^*\left[\mathcal{Y}^{[1,1]}\boldsymbol{\theta}\right]_{23} = \mathcal{Y}^{[1,1]}\boldsymbol{\theta}_{14}, \end{aligned}$$

$$\begin{aligned} \psi^{\pm}_{12} &= \pm\psi^{\pm}_{34} = \frac{1}{2}\left\{\mathcal{Y}^{[1,1]}\boldsymbol{\theta}_{12} \pm \mathcal{Y}^{[1,1]}\boldsymbol{\theta}_{34}\right\}, \\ \psi^{\pm}_{13} &= \pm\psi^{\pm}_{42} = \frac{1}{2}\left\{\mathcal{Y}^{[1,1]}\boldsymbol{\theta}_{13} \pm \mathcal{Y}^{[1,1]}\boldsymbol{\theta}_{42}\right\}, \\ \psi^{\pm}_{14} &= \pm\psi^{\pm}_{23} = \frac{1}{2}\left\{\mathcal{Y}^{[1,1]}\boldsymbol{\theta}_{14} \pm \mathcal{Y}^{[1,1]}\boldsymbol{\theta}_{23}\right\}. \end{aligned} \tag{9.11}$$

Thus, when the row number of $[\lambda]$ is equal to $N/2$, the representation space $\mathcal{T}^{[\lambda]}_\mu$ is reduced into the self-dual and the anti-self-dual tensor subspaces with the same dimension. Note that the combinations by the Young operators and the dual transformations (9.5) and (9.10) are all real except that the dual transformation (9.10) with $N = 4m + 2$ is complex.

In summary, the traceless tensor subspace $\mathcal{T}_\mu^{[\lambda]}$ corresponds to a representation $[\lambda]$ of $\mathrm{SO}(N)$, where the row number of $[\lambda]$ is less than $N/2$. When the row number ℓ of $[\lambda]$ is equal to $N/2$, $\mathcal{T}_\mu^{[\lambda]}$ is decomposed into the self-dual tensor subspace $\mathcal{T}_\mu^{[(+)\lambda]}$ and anti-self-dual tensor subspace $\mathcal{T}_\mu^{[(-)\lambda]}$ corresponding to the representation $[(\pm)\lambda]$, respectively. Through a similar, but a little bit complicated, proof as that for Theorem 8.3, the representations $[\lambda]$ and $[(\pm)\lambda]$ are irreducible. In fact, there is no further constraint to construct a nontrivial invariant subspace in their representation spaces. The highest weights of the representations $[\lambda]$ and $[(\pm)\lambda]$ are calculated later. All the irreducible representations are real except for $[(\pm)\lambda]$ when $N = 4m+2$. $\mathcal{T}_\mu^{[\lambda]}$ is a null space if the sum of the numbers of boxes in the first two columns of the Young pattern $[\lambda]$ is larger than N. $\mathcal{T}_\mu^{[\lambda]}$ is equivalent to $\mathcal{T}_\nu^{[\lambda']}$ where $[\lambda]$ and $[\lambda']$ are mutually the dual Young patterns (see Eq. (9.7)).

As far as the orthonormal irreducible basis tensors of $\mathrm{SO}(N)$ is concerned, there are two problems. One is how to decompose the standard tensor Young tableaux into a sum of the traceless basis tensors. The decomposition is straightforward, but tedious. The second is how to combine the basis tensors such that they are the common eigenfunctions of H_j and orthonormal to each other. For $\mathrm{SU}(N)$, the standard tensor Young tableaux are the irreducible tensor bases, but not orthonormal. Because the highest weight is simple and the standard tensor Young tableau with the highest weight is orthogonal to any other standard tensor Young tableau in the irreducible representation, the orthonormal basis tensors for $\mathrm{SU}(N)$ can be obtained from the highest weight state by the lowering operators F_μ in terms of the method of the block weight diagram. The merit of the method based on the standard tensor Young tableaux is that the basis tensors are known explicitly and the multiplicity of any weight is equal to the number of the standard tensor Young tableaux with the weight.

For $\mathrm{SO}(N)$, the key in finding the orthonormal irreducible basis tensors is to find the common eigenstates of H_j and the highest weight state in an irreducible representation. For the groups $\mathrm{SO}(2\ell+1)$ and $\mathrm{SO}(2\ell)$, the generators T_{ab} of the self-representation satisfy

$$\begin{aligned} (T_{ab})_{cd} &= -i\left\{\delta_{ac}\delta_{bd} - \delta_{ad}\delta_{bc}\right\}, \\ [T_{ab},\ T_{cd}] &= -i\left\{\delta_{bc}T_{ad} + \delta_{ad}T_{bc} - \delta_{bd}T_{ac} - \delta_{ac}T_{bd}\right\}. \end{aligned} \tag{9.12}$$

The bases H_j in the Cartan subalgebra are

$$H_j = T_{(2j-1)(2j)}, \qquad 1 \le j \le N/2. \tag{9.13}$$

9.1.2 *Irreducible Basis Tensors of* $SO(2\ell+1)$

The Lie algebra of $SO(2\ell+1)$ is B_ℓ. The simple roots of $SO(2\ell+1)$ are

$$\boldsymbol{r}_\mu = \boldsymbol{e}_\mu - \boldsymbol{e}_{\mu+1}, \qquad 1 \leq \mu \leq \ell-1, \qquad \boldsymbol{r}_\ell = \boldsymbol{e}_\ell. \tag{9.14}$$

$\boldsymbol{r}_\mu$ are the longer roots with $d_\mu = 1$ and $\boldsymbol{r}_\ell$ is the shorter root with $d_\ell = 1/2$. From the definition (7.141), the Chevalley bases of $SO(2\ell+1)$ in the self-representation are

$$\begin{aligned}
H_\mu &= T_{(2\mu-1)(2\mu)} - T_{(2\mu+1)(2\mu+2)}, \\
E_\mu &= \tfrac{1}{2}\left\{T_{(2\mu)(2\mu+1)} - iT_{(2\mu-1)(2\mu+1)} - iT_{(2\mu)(2\mu+2)} - T_{(2\mu-1)(2\mu+2)}\right\}, \\
F_\mu &= \tfrac{1}{2}\left\{T_{(2\mu)(2\mu+1)} + iT_{(2\mu-1)(2\mu+1)} + iT_{(2\mu)(2\mu+2)} - T_{(2\mu-1)(2\mu+2)}\right\}, \\
H_\ell &= 2T_{(2\ell-1)(2\ell)}, \\
E_\ell &= T_{(2\ell)(2\ell+1)} - iT_{(2\ell-1)(2\ell+1)}, \\
F_\ell &= T_{(2\ell)(2\ell+1)} + iT_{(2\ell-1)(2\ell+1)}.
\end{aligned} \tag{9.15}$$

$\boldsymbol{\theta}_a$ is not the common eigenvector of H_μ. Generalizing the spherical harmonic basis vectors (4.180) for SO(3), one defines the spherical harmonic basis vectors in the self-representation of $SO(2\ell+1)$

$$\boldsymbol{\phi}_\alpha = \begin{cases} (-1)^{\ell-\alpha+1}\sqrt{1/2}\left(\boldsymbol{\theta}_{2\alpha-1} + i\boldsymbol{\theta}_{2\alpha}\right), & 1 \leq \alpha \leq \ell, \\ \boldsymbol{\theta}_{2\ell+1}, & \alpha = \ell+1, \\ \sqrt{1/2}\left(\boldsymbol{\theta}_{4\ell-2\alpha+3} - i\boldsymbol{\theta}_{4\ell-2\alpha+4}\right), & \ell+2 \leq \alpha \leq 2\ell+1. \end{cases} \tag{9.16}$$

The spherical harmonic basis vectors $\boldsymbol{\phi}_\alpha$ are orthonormal and complete. In the spherical harmonic basis vectors $\boldsymbol{\phi}_\alpha$, the nonvanishing matrix entries of the Chevalley bases are

$$\begin{aligned}
&H_\mu \boldsymbol{\phi}_\mu = \boldsymbol{\phi}_\mu, && H_\mu \boldsymbol{\phi}_{\mu+1} = -\boldsymbol{\phi}_{\mu+1}, \\
&H_\mu \boldsymbol{\phi}_{2\ell-\mu+1} = \boldsymbol{\phi}_{2\ell-\mu+1}, && H_\mu \boldsymbol{\phi}_{2\ell-\mu+2} = -\boldsymbol{\phi}_{2\ell-\mu+2}, \\
&H_\ell \boldsymbol{\phi}_\ell = 2\boldsymbol{\phi}_\ell, && H_\ell \boldsymbol{\phi}_{\ell+2} = -2\boldsymbol{\phi}_{\ell+2}, \\
&E_\mu \boldsymbol{\phi}_{\mu+1} = \boldsymbol{\phi}_\mu, && E_\mu \boldsymbol{\phi}_{2\ell-\mu+2} = \boldsymbol{\phi}_{2\ell-\mu+1}, \\
&E_\ell \boldsymbol{\phi}_{\ell+1} = \sqrt{2}\boldsymbol{\phi}_\ell, && E_\ell \boldsymbol{\phi}_{\ell+2} = \sqrt{2}\boldsymbol{\phi}_{\ell+1}, \\
&F_\mu \boldsymbol{\phi}_\mu = \boldsymbol{\phi}_{\mu+1}, && F_\mu \boldsymbol{\phi}_{2\ell-\mu+1} = \boldsymbol{\phi}_{2\ell-\mu+2}, \\
&F_\ell \boldsymbol{\phi}_\ell = \sqrt{2}\boldsymbol{\phi}_{\ell+1}, && F_\ell \boldsymbol{\phi}_{\ell+1} = \sqrt{2}\boldsymbol{\phi}_{\ell+2},
\end{aligned} \tag{9.17}$$

where $1 \leq \mu \leq \ell-1$. Namely, the diagonal matrices of H_μ and H_ℓ in the spherical harmonic basis vectors $\boldsymbol{\phi}_\alpha$ are

$$\begin{aligned} H_\mu &= \text{diag}\{\underbrace{0,\ldots,0}_{\mu-1},1,-1,\underbrace{0,\ldots,0}_{2\ell-2\mu-1},1,-1,\underbrace{0,\ldots,0}_{\mu-1}\}, \\ H_\ell &= \text{diag}\{\underbrace{0,\ldots,0}_{\ell-1},\ 2,0,-2,\underbrace{0,\ldots,0}_{\ell-1}\}. \end{aligned}$$

The spherical harmonic basis tensor $\phi_{\alpha_1\ldots\alpha_n}$ of rank n for SO$(2\ell+1)$ is the direct product of n spherical harmonic basis vectors $\phi_{\alpha_1}\ldots\phi_{\alpha_n}$. The standard tensor Young tableaux $\mathcal{Y}_\mu^{[\lambda]}\phi_{\alpha_1\ldots\alpha_n}$ are the common eigenstates of H_μ, but generally not orthonormal and traceless. The eigenvalue of H_μ in the standard tensor Young tableaux $\mathcal{Y}_\mu^{[\lambda]}\phi_{\alpha_1\ldots\alpha_n}$ is equal to the number of the digits μ and $(2\ell-\mu+1)$ in the tableau, minus the number of $(\mu+1)$ and $(2\ell-\mu+2)$. The eigenvalue of H_ℓ in the standard tensor Young tableau is equal to the number of ℓ in the tableau, minus the number of $\ell+2$, and then multiplied with 2. The action of F_μ on the standard tensor Young tableau is equal to the sum of all possible tensor Young tableaux, each of which is obtained from the original one by replacing one filled digit μ with the digit $(\mu+1)$, or by replacing one filled digit $(2\ell-\mu+1)$ with the digit $(2\ell-\mu+2)$. The action of F_ℓ on the standard tensor Young tableau is equal to the sum, multiplied with a factor $\sqrt{2}$, of all possible tensor Young tableaux, each of which is obtained from the original one by replacing one filled digit ℓ with the digit $(\ell+1)$ or by replacing one filled digit $(\ell+1)$ with $(\ell+2)$. The actions of E_μ and E_ℓ are opposite. The obtained tensor Young tableaux may be not standard, but they can be transformed to the sum of the standard tensor Young tableaux by the symmetry (8.22).

Two standard tensor Young tableaux with different sets of the filled digits are orthogonal to each other. For an irreducible representation $[\lambda]$ of SO$(2\ell+1)$, where the row number of $[\lambda]$ is not larger than ℓ, the highest weight state corresponds to the standard tensor Young tableau where each box in the αth row is filled with the digit α because every raising operator E_μ annihilates it. The highest weight $\boldsymbol{M}=\sum_\mu \boldsymbol{w}_\mu M_\mu$ is calculated from Eq. (9.17),

$$M_\mu=\lambda_\mu-\lambda_{\mu+1},\qquad 1\le\mu<\ell,\qquad M_\ell=2\lambda_\ell. \tag{9.18}$$

The tensor representation $[\lambda]$ of SO$(2\ell+1)$, where M_ℓ is even is a single-valued representation. It will be known later that the representation with odd M_ℓ is a double-valued representation, called the spinor one.

Although the standard tensor Young tableaux is generally not traceless, the standard tensor Young tableau with the highest weight is traceless because it only contains ϕ_α with $\alpha<\ell+1$ (see Eq. (9.16)). For example, the tensor basis $\boldsymbol{\theta}_1\boldsymbol{\theta}_1$ is not traceless, but $\boldsymbol{\phi}_1\boldsymbol{\phi}_1$ is traceless. Since the highest weight is simple, the highest weight state is orthogonal to any other standard tensor Young tableau in the irreducible representation. Similar to the method of finding the orthonormal basis tensors in SU(N), one is able to find the remaining orthonormal and traceless basis tensors in $[\lambda]$ of SO($2\ell+1$) from the highest weight state by the lowering operators F_μ in terms of the method of the block weight diagram.

As an example, calculate the traceless symmetric tensor of rank 2 in SO(7) whose Lie algebra is B$_3$. Its representation is denoted by the Young pattern $[2,0,0]$, where the highest weight is $\boldsymbol{M} = 2\boldsymbol{w}_1$, or denoted by $(2,0,0)$. The simple roots $\boldsymbol{r}_\mu$ are expressed as

$$\boldsymbol{r}_1 = 2\boldsymbol{w}_1 - \boldsymbol{w}_2, \qquad \boldsymbol{r}_2 = -\boldsymbol{w}_1 + 2\boldsymbol{w}_2 - 2\boldsymbol{w}_3, \qquad \boldsymbol{r}_3 = -\boldsymbol{w}_2 + 2\boldsymbol{w}_3.$$

Two typical standard tensor Young tableaux are

$$\begin{array}{|c|c|}\hline \alpha & \beta \\ \hline\end{array} = \mathcal{Y}^{[2,0,0]}\boldsymbol{\phi}_{\alpha\beta} = \boldsymbol{\phi}_{\alpha\beta} + \boldsymbol{\phi}_{\alpha\beta},$$
$$\begin{array}{|c|c|}\hline \alpha & \alpha \\ \hline\end{array} = \mathcal{Y}^{[2,0,0]}\boldsymbol{\phi}_{\alpha\alpha} = 2\boldsymbol{\phi}_{\alpha\alpha}.$$

The trace tensor is

$$\begin{aligned}\sum_{a=1}^{7} \boldsymbol{\theta}_{aa} &= \frac{1}{2}\left(\boldsymbol{\theta}_1 + i\boldsymbol{\theta}_2\right)\left(\boldsymbol{\theta}_1 - i\boldsymbol{\theta}_2\right) + \frac{1}{2}\left(\boldsymbol{\theta}_1 - i\boldsymbol{\theta}_2\right)\left(\boldsymbol{\theta}_1 + i\boldsymbol{\theta}_2\right) \\ &\quad + \ldots + \boldsymbol{\theta}_7\boldsymbol{\theta}_7 \\ &= -\boldsymbol{\phi}_{17} - \boldsymbol{\phi}_{71} + \boldsymbol{\phi}_{26} + \boldsymbol{\phi}_{62} - \boldsymbol{\phi}_{35} - \boldsymbol{\phi}_{53} + \boldsymbol{\phi}_{44} \\ &= -\begin{array}{|c|c|}\hline 1 & 7 \\ \hline\end{array} + \begin{array}{|c|c|}\hline 2 & 6 \\ \hline\end{array} - \begin{array}{|c|c|}\hline 3 & 5 \\ \hline\end{array} + \frac{1}{2}\begin{array}{|c|c|}\hline 4 & 4 \\ \hline\end{array}.\end{aligned} \tag{9.19}$$

Using the brief symbols, $|\boldsymbol{M},\boldsymbol{m}\rangle = |\boldsymbol{m}\rangle$, one has

$$|\boldsymbol{M}\rangle = |(2,0,0)\rangle = \begin{array}{|c|c|}\hline 1 & 1 \\ \hline\end{array} = 2\boldsymbol{\phi}_{11}.$$

In terms of the method of the block weight diagram, the remaining basis states can be calculated from the highest weight state $|(2,0,0)\rangle$ by the lowering operators F_μ. The calculated results are listed in Fig. 9.1.

There is only one standard tensor Young tableau $\begin{array}{|c|c|}\hline 1 & 2 \\ \hline\end{array}$ with the dominant weight $(0,1,0)$ so that it is a single weight. The representation contains a dominant weight $(0,0,0)$ with the multiplicity 3 because there are four standard tensor Young tableaux with the weight $(0,0,0)$, $\begin{array}{|c|c|}\hline 1 & 7 \\ \hline\end{array}$, $\begin{array}{|c|c|}\hline 2 & 6 \\ \hline\end{array}$, $\begin{array}{|c|c|}\hline 3 & 5 \\ \hline\end{array}$, and $\begin{array}{|c|c|}\hline 4 & 4 \\ \hline\end{array}$, where one combination is the trace tensor.

Through some steps one has

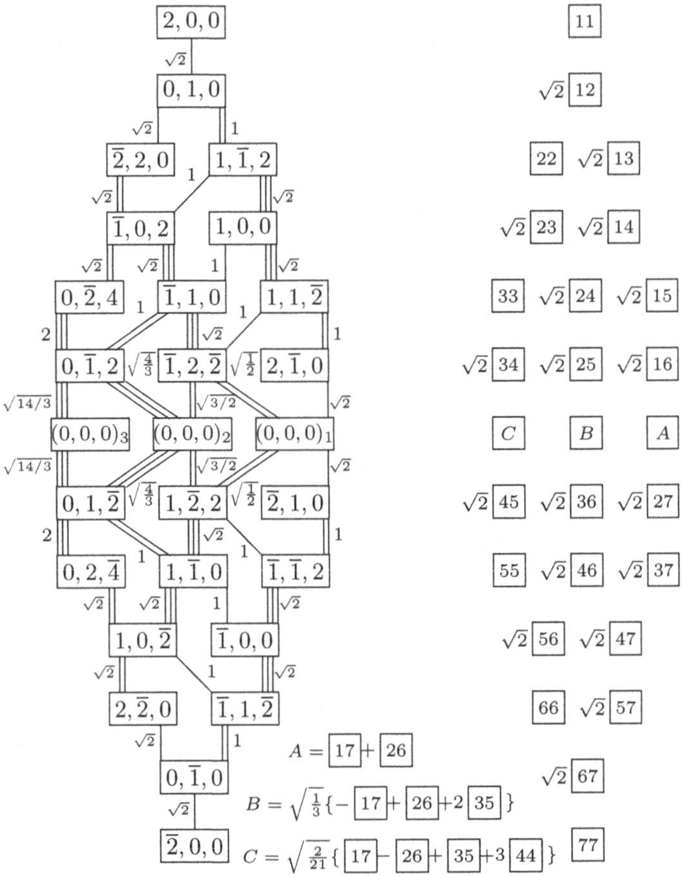

Fig. 9.1 The block weight diagram and the tensor Young tableaux in the representation [2, 0, 0] of SO(7).

$$|(2,\overline{1},0)\rangle = F_2\ |(1,1,\overline{2})\rangle = \sqrt{2}F_2 \ \boxed{1 \mid 5} = \sqrt{2}\ \boxed{1 \mid 6}\ ,$$

$$|(\overline{1},2,\overline{2})\rangle = F_1\ |(1,1,\overline{2})\rangle = \sqrt{2}F_1 \ \boxed{1 \mid 5} = \sqrt{2}\ \boxed{2 \mid 5}\ ,$$

$$|(0,\overline{1},2)\rangle = (1/2)F_3\ |(0,\overline{2},4)\rangle = (1/2)F_3 \ \boxed{3 \mid 3} = \sqrt{2}\ \boxed{3 \mid 4}\ .$$

$|(0,\overline{1},2)\rangle$ belongs to an $\mathcal{A}_3$-quintet. From $|(2,\overline{1},0)\rangle$ and $|(\overline{1},2,\overline{2})\rangle$ one can construct an $\mathcal{A}_1$-triplet and an $\mathcal{A}_2$-triplet, respectively. Since the multiplicity of the dominant weight $(0,0,0)$ is 3. Define that $|(0,0,0)_1\rangle$ belongs to the $\mathcal{A}_1$-triplet, a suitable combination of $|(0,0,0)_1\rangle$ and $|(0,0,0)_2\rangle$ belongs to the $\mathcal{A}_2$-double, and a suitable combination of $|(0,0,0)_1\rangle$, $|(0,0,0)_2\rangle$ and

$|(0,0,0)_3\rangle$ belongs to the $\mathcal{A}_3$-quintet. $|(0,0,0)_2\rangle$ is the $\mathcal{A}_1$-singlet. $|(0,0,0)_3\rangle$ is both the $\mathcal{A}_1$-singlet and the $\mathcal{A}_2$-singlet.

$$\begin{aligned}
&|(0,0,0)_1\rangle = \sqrt{1/2}F_1\ |(2,\overline{1},0)\rangle = F_1 \begin{array}{|c|c|}\hline 1 & 6 \\ \hline \end{array} = \begin{array}{|c|c|}\hline 1 & 7 \\ \hline \end{array} + \begin{array}{|c|c|}\hline 2 & 6 \\ \hline \end{array}, \\
&F_2\ |(\overline{1},2,\overline{2})\rangle = a_1\ |(0,0,0)_1\rangle + a_2\ |(0,0,0)_2\rangle, \\
&F_3\ |(0,\overline{1},2)\rangle = b_1\ |(0,0,0)_1\rangle + b_2\ |(0,0,0)_2\rangle + b_3\ |(0,0,0)_3\rangle, \\
&E_1\ |(0,0,0)_2\rangle = E_1\ |(0,0,0)_3\rangle = E_2\ |(0,0,0)_3\rangle = 0,
\end{aligned}$$

where $a_1^2+a_2^2=2$ and $b_1^2+b_2^2+b_3^2=6$. Applying $E_1F_2=F_2E_1$ to $|(\overline{1},2,\overline{2})\rangle$, one has

$$\begin{aligned}
E_1F_2\ |(\overline{1},2,\overline{2})\rangle &= \sqrt{2}a_1\ |(2,\overline{1},0)\rangle \\
&= F_2E_1\ |(\overline{1},2,\overline{2})\rangle = F_2\ |(1,1,\overline{2})\rangle = \ |(2,\overline{1},0)\rangle.
\end{aligned}$$

The solution is $a_1=\sqrt{1/2}$. Choosing the phase of $|(0,0,0)_2\rangle$ such that a_2 is a positive number, one has $a_2=\sqrt{2-a_1^2}=\sqrt{3/2}$. Applying $E_1F_3=F_3E_1$ and $E_2F_3=F_3E_2$ to $|(0,\overline{1},2)\rangle$, one has

$$\begin{aligned}
&E_1F_3\ |(0,\overline{1},2)\rangle = \sqrt{2}b_1\ |(2,\overline{1},0)\rangle = F_3E_1\ |(0,\overline{1},2)\rangle = 0, \\
&E_2F_3\ |(0,\overline{1},2)\rangle = \left(\sqrt{1/2}b_1 + \sqrt{3/2}b_2\right)\ |(\overline{1},2,\overline{2})\rangle \\
&\quad = F_3E_2\ |(0,\overline{1},2)\rangle = F_3\ |(\overline{1},1,0)\rangle = \sqrt{2}\ |(\overline{1},2,\overline{2})\rangle.
\end{aligned}$$

The solutions are $b_1=0$ and $b_2=\sqrt{4/3}$. Choosing the phase of $|(0,0,0)_3\rangle$ such that b_3 is a positive number, one has $b_3=\sqrt{6-b_2^2}=\sqrt{14/3}$. Thus, the remaining two states with the weight $(0,0,0)$ are

$$\begin{aligned}
|(0,0,0)_2\rangle &= \sqrt{2/3}\left\{F_2\ |(\overline{1},2,\overline{2})\rangle - \sqrt{1/2}\ |(0,0,0)_1\rangle\right\} \\
&= \sqrt{4/3}F_2 \begin{array}{|c|c|}\hline 2 & 5 \\ \hline \end{array} - \sqrt{1/3}\left\{\begin{array}{|c|c|}\hline 1 & 7 \\ \hline \end{array} + \begin{array}{|c|c|}\hline 2 & 6 \\ \hline \end{array}\right\} \\
&= \sqrt{1/3}\left\{-\begin{array}{|c|c|}\hline 1 & 7 \\ \hline \end{array} + \begin{array}{|c|c|}\hline 2 & 6 \\ \hline \end{array} + 2\begin{array}{|c|c|}\hline 3 & 5 \\ \hline \end{array}\right\}, \\
|(0,0,0)_3\rangle &= \sqrt{3/14}\left\{F_3\ |(0,\overline{1},2)\rangle - \sqrt{4/3}\ |(0,0,0)_2\rangle\right\} \\
&= \sqrt{3/7}F_3 \begin{array}{|c|c|}\hline 3 & 4 \\ \hline \end{array} - \sqrt{2/21}\left\{-\begin{array}{|c|c|}\hline 1 & 7 \\ \hline \end{array} + \begin{array}{|c|c|}\hline 2 & 6 \\ \hline \end{array} + 2\begin{array}{|c|c|}\hline 3 & 5 \\ \hline \end{array}\right\} \\
&= \sqrt{2/21}\left\{\begin{array}{|c|c|}\hline 1 & 7 \\ \hline \end{array} - \begin{array}{|c|c|}\hline 2 & 6 \\ \hline \end{array} + \begin{array}{|c|c|}\hline 3 & 5 \\ \hline \end{array} + 3\begin{array}{|c|c|}\hline 4 & 4 \\ \hline \end{array}\right\}.
\end{aligned}$$

In comparison with Eq. (9.19), one knows that three states with the weight $(0,0,0)$ are all traceless.

9.1.3 *Irreducible Basis Tensors of SO(2ℓ)*

The Lie algebra of SO(2ℓ) is D_ℓ. The simple roots of SO(2ℓ) are

$$\boldsymbol{r}_\mu = \boldsymbol{e}_\mu - \boldsymbol{e}_{\mu+1}, \qquad 1 \leq \mu \leq \ell - 1, \qquad \boldsymbol{r}_\ell = \boldsymbol{e}_{\ell-1} + \boldsymbol{e}_\ell. \tag{9.20}$$

The lengths of all simple roots are the same, $d_\mu = 1$. From the definition (7.141), the Chevalley bases of SO(2ℓ) in the self-representation are the same as those of SO(2ℓ + 1) except for $\mu = \ell$,

$$\begin{aligned} H_\ell &= T_{(2\ell-3)(2\ell-2)} + T_{(2\ell-1)(2\ell)}, \\ E_\ell &= \tfrac{1}{2}\left\{T_{(2\ell-2)(2\ell-1)} - iT_{(2\ell-3)(2\ell-1)} + iT_{(2\ell-2)(2\ell)} + T_{(2\ell-3)(2\ell)}\right\}, \\ F_\ell &= \tfrac{1}{2}\left\{T_{(2\ell-2)(2\ell-1)} + iT_{(2\ell-3)(2\ell-1)} - iT_{(2\ell-2)(2\ell)} + T_{(2\ell-3)(2\ell)}\right\}. \end{aligned} \tag{9.21}$$

$\boldsymbol{\theta}_a$ is not the common eigenvector of H_μ. One defines the spherical harmonic basis vectors in the self-representation of SO(2ℓ), which are the generalization of those for SO(4) (see Eq. (9.144)),

$$\boldsymbol{\phi}_\alpha = \begin{cases} (-1)^{\ell-\alpha}\sqrt{1/2}\left(\boldsymbol{\theta}_{2\alpha-1} + i\boldsymbol{\theta}_{2\alpha}\right), & 1 \leq \alpha \leq \ell, \\ \sqrt{1/2}\left(\boldsymbol{\theta}_{4\ell-2\alpha+1} - i\boldsymbol{\theta}_{4\ell-2\alpha+2}\right), & \ell+1 \leq \alpha \leq 2\ell. \end{cases} \tag{9.22}$$

The spherical harmonic basis vectors $\boldsymbol{\phi}_\alpha$ are orthonormal and complete. In the spherical harmonic basis vectors $\boldsymbol{\phi}_\alpha$, the nonvanishing matrix entries of the Chevalley bases are

$$\begin{array}{ll} H_\mu\boldsymbol{\phi}_\mu = \boldsymbol{\phi}_\mu, & H_\mu\boldsymbol{\phi}_{\mu+1} = -\boldsymbol{\phi}_{\mu+1}, \\ H_\mu\boldsymbol{\phi}_{2\ell-\mu} = \boldsymbol{\phi}_{2\ell-\mu}, & H_\mu\boldsymbol{\phi}_{2\ell-\mu+1} = -\boldsymbol{\phi}_{2\ell-\mu+1}, \\ H_\ell\boldsymbol{\phi}_{\ell-1} = \boldsymbol{\phi}_{\ell-1}, & H_\ell\boldsymbol{\phi}_\ell = \boldsymbol{\phi}_\ell, \\ H_\ell\boldsymbol{\phi}_{\ell+1} = -\boldsymbol{\phi}_{\ell+1}, & H_\ell\boldsymbol{\phi}_{\ell+2} = -\boldsymbol{\phi}_{\ell+2}, \\ E_\mu\boldsymbol{\phi}_{\mu+1} = \boldsymbol{\phi}_\mu, & E_\mu\boldsymbol{\phi}_{2\ell-\mu+1} = \boldsymbol{\phi}_{2\ell-\mu}, \\ E_\ell\boldsymbol{\phi}_{\ell+1} = \boldsymbol{\phi}_{\ell-1}, & E_\ell\boldsymbol{\phi}_{\ell+2} = \boldsymbol{\phi}_\ell, \\ F_\mu\boldsymbol{\phi}_\mu = \boldsymbol{\phi}_{\mu+1}, & F_\mu\boldsymbol{\phi}_{2\ell-\mu} = \boldsymbol{\phi}_{2\ell-\mu+1}, \\ F_\ell\boldsymbol{\phi}_{\ell-1} = \boldsymbol{\phi}_{\ell+1}, & F_\ell\boldsymbol{\phi}_\ell = \boldsymbol{\phi}_{\ell+2}, \end{array} \tag{9.23}$$

where $1 \leq \mu \leq \ell - 1$. Namely, the diagonal matrices of H_μ and H_ℓ in the spherical harmonic basis vectors $\boldsymbol{\phi}_\alpha$ are

$$\begin{aligned} H_\mu &= \text{diag}\{\underbrace{0,\ldots,0}_{\mu-1}, 1, -1, \underbrace{0,\ldots,0}_{2\ell-2\mu-2}, 1, -1, \underbrace{0,\ldots,0}_{\mu-1}\}, \\ H_\ell &= \text{diag}\{\underbrace{0,\ldots,0}_{\ell-2}, 1, 1, -1, -1, \underbrace{0,\ldots,0}_{\ell-2}\}. \end{aligned}$$

The spherical harmonic basis tensor $\phi_{\alpha_1 \ldots \alpha_n}$ of rank n for SO(2ℓ) is the direct product of n spherical harmonic basis vectors $\phi_{\alpha_1} \cdots \phi_{\alpha_n}$. The standard tensor Young tableaux $\mathcal{Y}_\mu^{[\lambda]} \phi_{\alpha_1 \ldots \alpha_n}$ are the common eigenstates of H_μ, but generally not orthonormal and traceless. The eigenvalue of H_μ in the standard tensor Young tableaux $\mathcal{Y}_\mu^{[\lambda]} \phi_{\alpha_1 \ldots \alpha_n}$ is equal to the number of the digits μ and $(2\ell - \mu)$ in the tableau, minus the number of $(\mu + 1)$ and $(2\ell - \mu + 1)$. The eigenvalue of H_ℓ in the standard tensor Young tableau is equal to the number of the digits $(\ell - 1)$ and ℓ in the tableau, minus the number of $(\ell + 1)$ and $(\ell + 2)$. The eigenvalues constitutes the weight $\boldsymbol{m}$ of the standard tensor Young tableau. The action of F_μ on the standard tensor Young tableau is equal to the sum of all possible tensor Young tableaux, each of which is obtained from the original one by replacing one filled digit μ with the digit $(\mu + 1)$, or by replacing one filled digit $(2\ell - \mu)$ with the digit $(2\ell - \mu + 1)$. The action of F_ℓ on the standard tensor Young tableau is equal to the sum of all possible tensor Young tableaux, each of which is obtained from the original one by replacing one filled digit $(\ell - 1)$ with the digit $(\ell + 1)$ or by replacing one filled digit ℓ with the digit $(\ell + 2)$. The actions of E_μ and E_ℓ are opposite. The obtained tensor Young tableaux may be not standard, but they can be transformed to the sum of the standard tensor Young tableaux by the symmetry (8.22).

The standard tensor Young tableaux with different weights are orthogonal to each other. For an irreducible representation $[\lambda]$ or $[(+)\lambda]$ of SO(2ℓ), where the row number of $[\lambda]$ is not larger than ℓ, the highest weight state corresponds to the standard tensor Young tableau where each box in the αth row is filled with the digit α because every raising operator E_μ annihilates it. In the standard tensor Young tableau with the highest weight of the representation $[(-)\lambda]$, the box in the αth row is filled with the digit α, but the box in the ℓth row is filled with the digit $(\ell + 1)$. The highest weight $\boldsymbol{M} = \sum_\mu \boldsymbol{w}_\mu M_\mu$ is calculated from Eq. (9.23),

$$\begin{aligned}
&M_\mu = \lambda_\mu - \lambda_{\mu+1}, && 1 \leq \mu < \ell - 1, \\
&M_{\ell-1} = M_\ell = \lambda_{\ell-1}, && \lambda_\ell = 0, \\
&M_{\ell-1} = \lambda_{\ell-1} - \lambda_\ell, \quad M_\ell = \lambda_{\ell-1} + \lambda_\ell, && \text{for} \ \ [(+)\lambda], \\
&M_{\ell-1} = \lambda_{\ell-1} + \lambda_\ell, \quad M_\ell = \lambda_{\ell-1} - \lambda_\ell, && \text{for} \ \ [(-)\lambda].
\end{aligned} \tag{9.24}$$

The tensor representation $[\lambda]$ of SO(2ℓ) where $(M_{\ell-1} + M_\ell)$ is even is a single-valued representation. The representation with odd $(M_{\ell-1} + M_\ell)$, as shown later, is a double-valued representation, called the spinor one.

Although the standard tensor Young tableaux is generally not traceless, the standard tensor Young tableau with the highest weight is traceless because it only contains ϕ_α with $\alpha < \ell + 2$, and furthermore, ℓ and $\ell + 1$ do not appear in the tableau simultaneously (see Eq. (9.22)). Since the highest weight is simple, the highest weight state is orthogonal to any other standard tensor Young tableau in the irreducible representation. One is able to find the remaining orthonormal and traceless basis tensors in the irreducible representation of SO(2ℓ) from the highest weight state by the lowering operators F_μ in terms of the method of the block weight diagrams. But now, the multiplicity of a weight in the representation is easy to obtained by counting the number of the traceless tensor Young tableaux with this weight. The calculation example is left to the readers (see Prob. 5 of Chap. 9 in [Ma and Gu (2004)]) where the orthonormal and traceless symmetric basis tensors of rank 2 for SO(8) are calculated. The calculation is similar to that given in the preceding subsection.

In the following we are going to calculate the standard tensor Young tableaux for the representations $[(\pm)1, 1]$ of SO(4). Compare them with Eq. (9.11). For $[(+)1, 1]$, one has

$$\begin{aligned}
&\begin{array}{|c|}\hline 1\\\hline 2\\\hline\end{array} = \mathcal{Y}^{[1,1]}\phi_{12} = \phi_{12} - \phi_{21} \\
&\quad = (1/2)\left\{-(\theta_1 + i\theta_2)(\theta_3 + i\theta_4) + (\theta_3 + i\theta_4)(\theta_1 + i\theta_2)\right\} \\
&\quad = (1/2)\left\{-\mathcal{Y}^{[1,1]}\theta_{13} - i\mathcal{Y}^{[1,1]}\theta_{23} - i\mathcal{Y}^{[1,1]}\theta_{14} + \mathcal{Y}^{[1,1]}\theta_{24}\right\} \\
&\quad = -\psi^+_{13} - i\psi^+_{14}, \\
&\sqrt{\frac{1}{2}}\, F_2 \begin{array}{|c|}\hline 1\\\hline 2\\\hline\end{array} = \sqrt{\frac{1}{2}}\left(\begin{array}{|c|}\hline 1\\\hline 4\\\hline\end{array} - \begin{array}{|c|}\hline 2\\\hline 3\\\hline\end{array}\right) = \sqrt{\frac{1}{2}}\left\{\phi_{14} - \phi_{41} - \phi_{23} + \phi_{32}\right\} \\
&\quad = \sqrt{1/8}\left\{-(\theta_1 + i\theta_2)(\theta_1 - i\theta_2) + (\theta_1 - i\theta_2)(\theta_1 + i\theta_2)\right. \\
&\qquad \left. - (\theta_3 + i\theta_4)(\theta_3 - i\theta_4) + (\theta_3 - i\theta_4)(\theta_3 + i\theta_4)\right\} \\
&\quad = i\sqrt{1/2}\left\{\mathcal{Y}^{[1,1]}\theta_{12} + \mathcal{Y}^{[1,1]}\theta_{34}\right\} = i\sqrt{2}\psi^+_{12}, \\
&\sqrt{\frac{1}{2}}\, F_2 \left\{\sqrt{\frac{1}{2}}\left(\begin{array}{|c|}\hline 1\\\hline 4\\\hline\end{array} - \begin{array}{|c|}\hline 2\\\hline 3\\\hline\end{array}\right)\right\} = \begin{array}{|c|}\hline 3\\\hline 4\\\hline\end{array} = \phi_{34} - \phi_{43} \\
&\quad = (1/2)\left\{(\theta_3 - i\theta_4)(\theta_1 - i\theta_2) - (\theta_1 - i\theta_2)(\theta_3 - i\theta_4)\right\} \\
&\quad = (1/2)\left\{\mathcal{Y}^{[1,1]}\theta_{31} - i\mathcal{Y}^{[1,1]}\theta_{41} - i\mathcal{Y}^{[1,1]}\theta_{32} - \mathcal{Y}^{[1,1]}\theta_{42}\right\} \\
&\quad = -\psi^+_{13} + i\psi^+_{14}.
\end{aligned} \tag{9.25}$$

For $[(-)\lambda]$, one has

$$
\begin{aligned}
\begin{array}{|c|}\hline 1\\\hline 3\\\hline\end{array} &= \mathcal{Y}^{[1,1]}\phi_{13} = \phi_{13} - \phi_{31} \\
&= (1/2)\left\{-\left(\boldsymbol{\theta}_1 + i\boldsymbol{\theta}_2\right)\left(\boldsymbol{\theta}_3 - i\boldsymbol{\theta}_4\right) + \left(\boldsymbol{\theta}_3 - i\boldsymbol{\theta}_4\right)\left(\boldsymbol{\theta}_1 + i\boldsymbol{\theta}_2\right)\right\} \\
&= (1/2)\left\{-\mathcal{Y}^{[1,1]}\boldsymbol{\theta}_{13} - i\mathcal{Y}^{[1,1]}\boldsymbol{\theta}_{23} + i\mathcal{Y}^{[1,1]}\boldsymbol{\theta}_{14} - \mathcal{Y}^{[1,1]}\boldsymbol{\theta}_{24}\right\} \\
&= -\boldsymbol{\psi}^{-}_{13} + i\boldsymbol{\psi}^{-}_{14}, \\
\sqrt{\frac{1}{2}}\,F_1 \begin{array}{|c|}\hline 1\\\hline 3\\\hline\end{array} &= \sqrt{\frac{1}{2}}\left(\begin{array}{|c|}\hline 1\\\hline 4\\\hline\end{array} + \begin{array}{|c|}\hline 2\\\hline 3\\\hline\end{array}\right) = \sqrt{\frac{1}{2}}\left\{\phi_{14} - \phi_{41} + \phi_{23} - \phi_{32}\right\} \\
&= \sqrt{1/8}\left\{-\left(\boldsymbol{\theta}_1 + i\boldsymbol{\theta}_2\right)\left(\boldsymbol{\theta}_1 - i\boldsymbol{\theta}_2\right) + \left(\boldsymbol{\theta}_1 - i\boldsymbol{\theta}_2\right)\left(\boldsymbol{\theta}_1 + i\boldsymbol{\theta}_2\right)\right. \\
&\qquad \left. + \left(\boldsymbol{\theta}_3 + i\boldsymbol{\theta}_4\right)\left(\boldsymbol{\theta}_3 - i\boldsymbol{\theta}_4\right) - \left(\boldsymbol{\theta}_3 - i\boldsymbol{\theta}_4\right)\left(\boldsymbol{\theta}_3 + i\boldsymbol{\theta}_4\right)\right\} \\
&= i\sqrt{1/2}\left\{\mathcal{Y}^{[1,1]}\boldsymbol{\theta}_{12} - \mathcal{Y}^{[1,1]}\boldsymbol{\theta}_{34}\right\} = i\sqrt{2}\boldsymbol{\psi}^{-}_{12}, \\
\sqrt{\frac{1}{2}}\,F_1\left\{\sqrt{\frac{1}{2}}\left(\begin{array}{|c|}\hline 1\\\hline 4\\\hline\end{array} + \begin{array}{|c|}\hline 2\\\hline 3\\\hline\end{array}\right)\right\} &= \begin{array}{|c|}\hline 2\\\hline 4\\\hline\end{array} = \phi_{24} - \phi_{42} \\
&= (1/2)\left\{\left(\boldsymbol{\theta}_3 + i\boldsymbol{\theta}_4\right)\left(\boldsymbol{\theta}_1 - i\boldsymbol{\theta}_2\right) - \left(\boldsymbol{\theta}_1 - i\boldsymbol{\theta}_2\right)\left(\boldsymbol{\theta}_3 + i\boldsymbol{\theta}_4\right)\right\} \\
&= (1/2)\left\{\mathcal{Y}^{[1,1]}\boldsymbol{\theta}_{31} + i\mathcal{Y}^{[1,1]}\boldsymbol{\theta}_{41} - i\mathcal{Y}^{[1,1]}\boldsymbol{\theta}_{32} + \mathcal{Y}^{[1,1]}\boldsymbol{\theta}_{42}\right\} \\
&= -\boldsymbol{\psi}^{-}_{13} - i\boldsymbol{\psi}^{-}_{14}.
\end{aligned}
\tag{9.26}
$$

9.1.4 *Dimensions of Irreducible Tensor Representations*

The dimension $d_{[\lambda]}(\mathrm{SO}(N))$ of the representation $[\lambda]$ of SO(N) can be calculated by the hook rule [Ma and Dai (1982)]. In this rule the dimension is expressed as a quotient, where the numerator and the denominator are denoted by the symbols $Y_T^{[\lambda]}$ and $Y_h^{[\lambda]}$, respectively:

$$
\begin{aligned}
d_{[(\pm)\lambda]}(\mathrm{SO}(2\ell)) &= \frac{Y_T^{[\lambda]}}{2Y_h^{[\lambda]}}, && \text{where } \lambda_\ell \neq 0, \\
d_{[\lambda]}(\mathrm{SO}(N)) &= \frac{Y_T^{[\lambda]}}{Y_h^{[\lambda]}}, && \text{the remaining cases.}
\end{aligned}
\tag{9.27}
$$

The meaning of two symbols $Y_h^{[\lambda]}$ and $Y_T^{[\lambda]}$ are as follows. The hook path $(i,\ j)$ in the Young pattern $[\lambda]$ is defined to be a path which enters the Young pattern at the rightmost of the ith row, goes leftward in the i row, turns downward at the j column, goes downward in the j column, and leaves from the Young pattern at the bottom of the j column. The inverse hook path $\overline{(i,\ j)}$ is the same path as the hook path $(i,\ j)$ but with the opposite direction. The number of boxes contained in the path $(i,\ j)$, as well as in its inverse, is the hook number h_{ij}. $Y_h^{[\lambda]}$ is a tableau of the Young pattern $[\lambda]$ where the box in the jth column of the ith row is filled with the hook number h_{ij}. Define a series of the tableaux $Y_{T_a}^{[\lambda]}$ recursively by the rule given below. $Y_T^{[\lambda]}$ is a tableau of the Young pattern $[\lambda]$ where each box is filled with the sum of the digits which are respectively filled in the same box of each tableau $Y_{T_a}^{[\lambda]}$ in the series. The symbol $Y_T^{[\lambda]}$ means the product of the filled digits in it, so does the symbol $Y_h^{[\lambda]}$.

The tableaux $Y_{T_a}^{[\lambda]}$ are defined by the following rule:

(a) $Y_{T_0}^{[\lambda]}$ is a tableau of the Young pattern $[\lambda]$ where the box in the jth column of the ith row is filled with the digit $(N + j - i)$.

(b) Let $[\lambda^{(1)}] = [\lambda]$. Beginning with $[\lambda^{(1)}]$, one defines recursively the Young pattern $[\lambda^{(a)}]$ by removing the first row and the first column of the Young pattern $[\lambda^{(a-1)}]$ until $[\lambda^{(a)}]$ contains less than two columns.

(c) If $[\lambda^{(a)}]$ contains more than one column, define $Y_{T_a}^{[\lambda]}$ to be a tableau of the Young pattern $[\lambda]$ where the boxes in the first $(a - 1)$ row and in the first $(a - 1)$ column are filled with 0, and the remaining part of the Young pattern is nothing but $[\lambda^{(a)}]$. Let $[\lambda^{(a)}]$ have r rows. Fill the first r boxes along the hook path $(1,\ 1)$ of the Young pattern $[\lambda^{(a)}]$, beginning with the box on the rightmost, with the digits $(\lambda_1^{(a)} - 1)$, $(\lambda_2^{(a)} - 1)$, $\cdots$, $(\lambda_r^{(a)} - 1)$, box by box, and fill the first $(\lambda_i^{(a)} - 1)$ boxes in each inverse hook path $\overline{(i,\ 1)}$ of the Young pattern $[\lambda^{(a)}]$, $1 \le i \le r$, with -1. The remaining boxes are filled with 0. If a few -1 are filled in the same box, the digits are summed. The sum of all filled digits in the pattern $Y_{T_a}^{[\lambda]}$ with $a > 0$ is 0.

The calculation method (9.27) is explained through some examples.

Ex. 1 The dimension of the representation $[3, 3, 3]$ of SO(7).

$$Y_T^{[3,3,3]} = \begin{array}{|c|c|c|}\hline 7 & 8 & 9 \\ \hline 6 & 7 & 8 \\ \hline 5 & 6 & 7 \\ \hline \end{array} + \begin{array}{|c|c|c|}\hline 2 & 2 & 2 \\ \hline -2 & & \\ \hline -3 & -1 & \\ \hline \end{array} + \begin{array}{|c|c|c|}\hline & & \\ \hline & 1 & 1 \\ \hline & -2 & \\ \hline \end{array} = \begin{array}{|c|c|c|}\hline 9 & 10 & 11 \\ \hline 4 & 8 & 9 \\ \hline 2 & 3 & 7 \\ \hline \end{array},$$

$$d_{[3,3,3]}(\mathrm{SO}(7)) = \frac{\begin{array}{|c|c|c|}\hline 9 & 10 & 11 \\ \hline 4 & 8 & 9 \\ \hline 2 & 3 & 7 \\ \hline\end{array}}{\begin{array}{|c|c|c|}\hline 5 & 4 & 3 \\ \hline 4 & 3 & 2 \\ \hline 3 & 2 & 1 \\ \hline\end{array}} = 11 \times 9 \times 7 \times 2 = 1386.$$

Ex. 2 The representation of one-row Young pattern $[n]$ of $\mathrm{SO}(N)$.

$$\begin{aligned}
d_{[n]}(\mathrm{SO}(N)) &= \frac{\begin{array}{|c|c|c|c|}\hline N & N+1 & \cdots & N+n-1 \\ \hline\end{array} + \begin{array}{|c|c|c|c|}\hline -1 & \cdots & -1 & n-1 \\ \hline\end{array}}{\begin{array}{|c|c|c|c|}\hline n & n-1 & \cdots & 1 \\ \hline\end{array}} \\
&= \begin{array}{|c|c|c|c|c|}\hline N-1 & N & \cdots & N+n-3 & N+2n-2 \\ \hline\end{array} \;/n! \\
&= \frac{(N+n-3)!(N+2n-2)}{(N-2)!n!} \\
&= \frac{N+2n-2}{N-2}\binom{N+n-3}{n} .
\end{aligned} \tag{9.28}$$

$$\begin{aligned}
d_{[n]}(\mathrm{SO}(3)) &= 2n+1 , \\
d_{[n]}(\mathrm{SO}(4)) &= (n+1)^2 , \\
d_{[n]}(\mathrm{SO}(5)) &= (n+1)(n+2)(2n+3)/6 .
\end{aligned}$$

Ex. 3 The representation of two-row Young pattern $[n,m]$ of $\mathrm{SO}(N)$.

$$\begin{aligned}
Y_T^{[n,m]} &= \begin{array}{|c|c|c|c|c|c|c|c|}\hline N & N+1 & \cdots & N+m-2 & N+m-1 & \cdots & N+n-2 & N+n-1 \\ \hline N-1 & N & \cdots & N+m-3 & N+m-2 & & & \\ \hline\end{array} \\
&\quad + \begin{array}{|c|c|c|c|c|c|c|c|c|}\hline -1 & -1 & \cdots & -1 & -1 & \cdots & -1 & m-1 & n-1 \\ \hline -2 & -1 & \cdots & -1 & & & & & \\ \hline\end{array} \\
&\quad + \begin{array}{|c|c|c|c|c|c|c|}\hline & & & & & & \\ \hline & -1 & \cdots & -1 & m-2 & & \\ \hline\end{array} \\
&= \begin{array}{|c|c|c|c|c|c|c|c|}\hline N-1 & \cdots & N+m-3 & N+m-2 & \cdots & N+n-4 & N+n+m-3 & N+2n-2 \\ \hline N-3 & \cdots & N+m-5 & N+2m-4 & & & & \\ \hline\end{array},
\end{aligned}$$

$$Y_h^{[n,m]} = \begin{array}{|c|c|c|c|c|c|}\hline n+1 & \cdots & n-m+2 & n-m & \cdots & 1 \\ \hline m & \cdots & 1 & & & \\ \hline\end{array},$$

$$\begin{aligned}
d_{[n,m]}(\mathrm{SO}(N)) = {} & \frac{(n-m+1)(N+n-4)!(N+m-5)!}{(n+1)!m!(N-2)!(N-4)!} \\
& \times\ (N+n+m-3)(N+2n-2)(N+2m-4).
\end{aligned} \tag{9.29}$$

$$\begin{aligned}
d_{[n,m]}(SO(4)) &= (n-m+1)(n+m+1), \\
d_{[n,m]}(SO(5)) &= (n-m+1)(n+m+2)(2n+3)(2m+1)/6.
\end{aligned}$$

The factor 2 in the denominator of Eq. (9.27) for SO(4) was considered.

For the representation with one-column Young pattern $[1^n]$, there is no traceless condition so that when $n < N/2$,

$$d_{[1^n]}(\mathrm{SO}(N)) = d_{[1^n]}(\mathrm{SU}(N)) = \binom{N}{n} = \frac{N!}{n!(N-n)!}. \tag{9.30}$$

9.1.5 *Adjoint Representation of* SO(N)

As shown in §7.6.2, the highest weight of the adjoint representation of SO(N) is $\boldsymbol{M} = \boldsymbol{w}_2$, corresponding to the Young pattern $[1,1,0,\ldots,0]$. In this subsection we are going to discuss the adjoint representation of SO(N) by replacement of tensors.

The $N(N-1)/2$ generators T_{ab} in the self-representation of SO(N) construct the complete bases of N-dimensional antisymmetric matrices. Denote T_{ab} by T_A for convenience, $1 \leq A \leq N(N-1)/2$. Tr $(T_A T_B) = 2\delta_{AB}$. For $R \in$SO(N), one has from Eq. (7.8)

$$RT_AR^{-1} = \sum_{B+1}^{N(N-1)/2} T_B D^{\text{ad}}_{BA}(R) \; . \tag{9.31}$$

The antisymmetric tensor $\boldsymbol{T}_{ab}$ of rank 2 of SO(N) satisfies a similar relation in the SO(N) transformation R

$$(O_R\boldsymbol{T})_{ab} = \sum_{cd} R_{ac}\boldsymbol{T}_{cd}\left(R^{-1}\right)_{db} = \left(R\boldsymbol{T}R^{-1}\right)_{ab}.$$

$\boldsymbol{T}_{ab}$ can be looked like an antisymmetric matrix and expanded with respect to $(T_A)_{ab}$

$$\boldsymbol{T}_{ab} = \sum_{A=1}^{N(N-1)/2} (T_A)_{ab}\,\boldsymbol{F}_A, \qquad \boldsymbol{F}_A = \frac{1}{2}\sum_{ab} (T_A)_{ba}\,\boldsymbol{T}_{ab}, \tag{9.32}$$

where the coefficient $\boldsymbol{F}_A$ is a tensor which transforms in the SO(N) transformation R as

$$\begin{aligned}
(O_R\boldsymbol{T})_{ab} &= \left(R\boldsymbol{T}R^{-1}\right)_{ab} = \sum_A \left(RT_AR^{-1}\right)_{ab}\boldsymbol{F}_A \\
&= \sum_B (T_B)_{ab}\left\{\sum_A D^{\text{ad}}_{BA}(R)\boldsymbol{F}_A\right\}, \\
(O_R\boldsymbol{T})_{ab} &= \sum_B (T_B)_{ab} O_R\boldsymbol{F}_B.
\end{aligned}$$

Thus, $\boldsymbol{F}_A$ transforms according to the adjoint representation of SO(N)

$$(O_R\boldsymbol{F})_B = \sum_A D^{\text{ad}}_{BA}(R)\boldsymbol{F}_A. \tag{9.33}$$

The adjoint representation of SO(N) is equivalent to the antisymmetric tensor representation $[1,1]$ of rank 2. $[1,1] \simeq [1]$ for SO(3). The adjoint representation of SO(N) where $N = 3$ or $N > 4$ is irreducible so that SO(N), except for $N = 2$ and 4, is a simple Lie group. The SO(2) group is Abelian. The adjoint representation of SO(4) is reducible, and the direct product of two SU(2) is homomorphic onto SO(4) through a two-to-one correspondence (see §9.5).

9.1.6 *Tensor Representations of O(N)*

The group O(N) is a mixed Lie group, whose group space falls into two disjoint regions corresponding to det $R = 1$ and det $R = -1$. Its invariant subgroup SO(N) has a connected group space corresponding to det $R = 1$. The set of elements related to the other connected piece where det $R = -1$ is the coset of SO(N). The property of O(N) can be characterized completely by SO(N) and a representative element in the coset.

For an odd N, the representative element in the coset is usually chosen to be $\sigma = -\mathbf{1}$. σ is self-inverse and commutable with every element in O(N) so that the representation matrix $D(\sigma)$ in an irreducible representation of O(N) is a constant matrix

$$D(\sigma) = c\mathbf{1}, \qquad D(\sigma)^2 = \mathbf{1}, \qquad c = \pm 1. \tag{9.34}$$

Denote by R the element in SO(N) and by $R' = \sigma R$ the element in the coset. From each irreducible representation $D^{[\lambda]}(SO(\text{N}))$ one obtains two induced irreducible representations $D^{[\lambda]\pm}(O(\text{N}))$,

$$D^{[\lambda]\pm}(R) = D^{[\lambda]}(R), \qquad D^{[\lambda]\pm}(\sigma R) = \pm D^{[\lambda]}(R). \tag{9.35}$$

Two representations $D^{[\lambda]\pm}(O(\text{N}))$ are inequivalent because the characters of σ in two representations are different.

For an even $N = 2\ell$, $\sigma = -\mathbf{1}$ belongs to SO(N), and the representative element in the coset is usually chosen to be τ. τ is a diagonal matrix where the diagonal entries are 1 except for $\tau_{NN} = -1$. $\tau^2 = \mathbf{1}$, but τ is not commutable with some elements in O(N). Any tensor Young tableau $\mathcal{Y}^{[\lambda]}_\mu \boldsymbol{\theta}_{a_1 \ldots a_n}$ is an eigentensor of τ with the eigenvalue 1 or -1 depending on whether there are even or odd number of filled digits N in the tableau. In the spherical harmonic basis tensors, τ interchanges the filled digits ℓ and $(\ell+1)$ in the tensor Young tableau $\mathcal{Y}_{\mu}^{[\lambda]}\phi_{\alpha_1\ldots\alpha_n}$. Thus, the representation matrix $D^{[\lambda]}(\tau)$ is known.

Denote by R the element in SO(2ℓ) and by $R'=\tau R$ the element in the coset. From each irreducible representation $D^{[\lambda]}(\mathrm{SO}(2\ell))$ where the row number of $[\lambda]$ is less than ℓ, one obtains two induced irreducible representations $D^{[\lambda]\pm}(\mathrm{O}(2\ell))$,

$$D^{[\lambda]\pm}(R)=D^{[\lambda]}(R), \qquad D^{[\lambda]\pm}(\tau R)=\pm D^{[\lambda]}(\tau)D^{[\lambda]}(R). \tag{9.36}$$

Two representations $D^{[\lambda]\pm}(\mathrm{O}(N))$ are inequivalent because the characters of τ in two representations are different.

When the row number of $[\lambda]$ is $\ell=N/2$, there are two inequivalent irreducible representations $D^{[(\pm)\lambda]}$ of SO(2ℓ) whose basis tensors are given in Eq. (9.9). Two terms in Eq. (9.9) contain different numbers of the subscripts N such that τ changes the tensor Young tableau in $[(\pm)\lambda]$ to that in $[(\mp)\lambda]$, namely, the representation spaces of both $D^{[(\pm)\lambda]}(\mathrm{SO}(2\ell))$ are not invariant in O(2ℓ). Only their direct sum is an invariant space corresponding to an irreducible representation $D^{[\lambda]}$ of O(2ℓ),

$$D^{[\lambda]}(R)=D^{[(+)\lambda]}(R)\oplus D^{[(-)\lambda]}(R), \qquad D^{[\lambda]}(\tau R)=D^{[\lambda]}(\tau)D^{[\lambda]}(R), \tag{9.37}$$

where the representation matrix $D^{[\lambda]}(\tau)$ can be calculated by interchanging the filled digits ℓ and $(\ell+1)$ in the tensor Young tableau $\mathcal{Y}_{\mu}^{[\lambda]}\phi_{\alpha_1\ldots\alpha_n}$ (see Eqs. (9.25) and (9.26)). Two representations with different signs of $D^{[\lambda]}(\tau)$ are equivalent because they can be related by a similarity transformation $X=\begin{pmatrix}\mathbf{1} & 0\\ 0 & -\mathbf{1}\end{pmatrix}$.

9.2 Γ Matrix Groups

Dirac introduced four γ matrices, which are the generalization of the Pauli matrices. Similar to the Pauli matrices, four γ matrices also satisfy the anticommutative relations. In terms of the tool of γ matrices, Dirac established the equation of motion for the relativistic particle with spin 1/2, called the Dirac equation. In the language of group theory, Dirac found the spinor representation of the Lorentz group. The tool of γ matrices is generalized for finding the spinor representations of SO(N) in this section. The set of products of the γ matrices forms the matrix group Γ. Its group algebra is called the Clifford algebra in mathematics.

9.2.1 *Property of* Γ *Matrix Groups*

Define N matrices γ_a satisfying the anticommutative relations

$$\{ \gamma_a \ , \ \gamma_b \} = \gamma_a \gamma_b + \gamma_b \gamma_a = 2\delta_{ab}\mathbf{1}, \qquad 1 \leq a, \ b \leq N, \tag{9.38}$$

namely, $\gamma_a^2 = \mathbf{1}$ and $\gamma_a \gamma_b = -\gamma_b \gamma_a$ when $a \neq b$. The inverse of a product of γ_a matrices is the same product but in the opposite order. The set of all products of the γ_a matrices, in the multiplication rule of matrices, satisfies four axioms of a group and forms a group, denoted by Γ_N. In a product of γ_a matrices two γ_b with the same subscript can be moved together and eliminated by Eq. (9.38) so that Γ_N is a finite matrix group.

Choose a faithful irreducible unitary representation of Γ_N to be its self-representation. As shown in Eq. (9.49), the representation does exist. Thus, from Eq. (9.38) γ_a is unitary and Hermitian,

$$\gamma_a^\dagger = \gamma_a^{-1} = \gamma_a. \tag{9.39}$$

The eigenvalue of γ_a is 1 or -1. Let

$$\gamma_\chi^{(N)} = \gamma_1 \gamma_2 \cdots \gamma_N, \qquad \left(\gamma_\chi^{(N)}\right)^2 = (-1)^{N(N-1)/2} \ \mathbf{1}. \tag{9.40}$$

When N is odd, $\gamma_\chi^{(N)}$ is commutable with every γ_a matrix so that it is a constant matrix owing to the Schur theorem,

$$\gamma_\chi^{(N)} = \begin{cases} \pm\mathbf{1} & \text{when} \ \ N = 4m + 1, \\ \pm i\mathbf{1} & \text{when} \ \ N = 4m - 1. \end{cases} \tag{9.41}$$

Namely, $\gamma_\chi^{(4m+1)}$ is equal to either 1 or -1, which is not a new element in Γ_{4m+1}. Two groups with different $\gamma_\chi^{(4m+1)}$ are isomorphic through a one-to-one correspondence, say

$$\gamma_a \longleftrightarrow \gamma_a', \qquad 1 \leq a \leq 4m, \qquad \gamma_{4m+1} \longleftrightarrow -\gamma_{4m+1}'. \tag{9.42}$$

Furthermore, for a given $\gamma_\chi^{(4m+1)}$, $\gamma_{4m+1}^{(4m+1)}$ can be expressed as a product of other γ_a matrices. Thus, all elements both in Γ_{4m} and in Γ_{4m+1} can be expressed as the products of matrices γ_a, $1 \leq a \leq 4m$ so that they are isomorphic. On the other hand, since $\gamma_\chi^{(4m-1)}$ is equal to either $i\mathbf{1}$ or $-i\mathbf{1}$, Γ_{4m-1} is isomorphic onto a group composed of Γ_{4m-2} and $i\Gamma_{4m-2}$,

$$\Gamma_{4m+1} \approx \Gamma_{4m}, \qquad \Gamma_{4m-1} \approx \{\Gamma_{4m-2}, \ i\Gamma_{4m-2}\}. \tag{9.43}$$

9.2.2 The Case $N = 2\ell$

First, calculate the order $g^{(2\ell)}$ of $\Gamma_{2\ell}$. Obviously, if $R \in \Gamma_{2\ell}$, $-R$ belongs too. Choosing one element in each pair of elements $\pm R$, one obtains a set $\Gamma'_{2\ell}$ containing $g^{(2\ell)}/2$ elements. Denote by S_n a product of n different γ_a. Since the number of different S_n contained in the set $\Gamma'_{2\ell}$ is equal to the combinatorics of n among 2ℓ, then

$$g^{(2\ell)} = 2\sum_{n=0}^{2\ell} \binom{2\ell}{n} = 2(1+1)^{2\ell} = 2^{2\ell+1}. \tag{9.44}$$

Second, for any element $S_n \in \Gamma_{2\ell}$ except for $\pm\mathbf{1}$, one is able to find a matrix γ_a which is anticommutable with S_n. In fact, if n is even and γ_a appears in the product S_n, one has $\gamma_a S_n = -S_n\gamma_a$. If n is odd, there must exist at least one γ_a which does not appear in the product S_n so that $\gamma_a S_n = -S_n \gamma_a$. Thus, the trace of S_n is 0,

$$\text{Tr } S_n = \text{Tr}\left(\gamma_a^2 S_n\right) = -\text{Tr}\left(\gamma_a S_n \gamma_a\right) = -\text{Tr } S_n = 0.$$

Namely, the character of the element S in the self-representation of $\Gamma_{2\ell}$ is

$$\chi(S) = \begin{cases} \pm d^{(2\ell)} & \text{when } S = \pm\mathbf{1}, \\ 0 & \text{when } S \neq \pm\mathbf{1}, \end{cases} \tag{9.45}$$

where $d^{(2\ell)}$ is the dimension of γ_a. Since the self-representation of $\Gamma_{2\ell}$ is irreducible,

$$2\left(d^{(2\ell)}\right)^2 = \sum_{S\in\Gamma_{2\ell}} |\chi(S)|^2 = g^{(2\ell)} = 2^{2\ell+1},$$
$$d^{(2\ell)} = 2^\ell. \tag{9.46}$$

Due to Eqs. (9.39) and (9.45), $\det \gamma_a = 1$ when $\ell > 1$.

Third, since $\gamma_\chi^{(2\ell)}$ is anticommutable with every γ_a, one is able to define $\gamma_f^{(2\ell)}$ by multiplying $\gamma_\chi^{(2\ell)}$ with a factor such that $\gamma_f^{(2\ell)}$ satisfies Eq. (9.38),

$$\gamma_f^{(2\ell)} = (-i)^\ell \gamma_\chi^{(2\ell)} = (-i)^\ell \gamma_1\gamma_2\ldots\gamma_{2\ell}, \qquad \left(\gamma_f^{(2\ell)}\right)^2 = \mathbf{1}. \tag{9.47}$$

In fact, $\gamma_f^{(2\ell)}$ may be defined to be the matrix $\gamma_{2\ell+1}$ in $\Gamma_{2\ell+1}$.

Fourth, the matrices in the set $\Gamma'_{2\ell}$ are linearly independent. Otherwise, there is a linear relation $\sum_S C(S)S = 0$, where $S \in \Gamma'_{2\ell}$. Multiplying it with $R^{-1}/d^{(2\ell)}$ and taking the trace, one obtains that any coefficient $C(R) = 0$. Thus, the set $\Gamma'_{2\ell}$ contains $2^{2\ell}$ linearly independent matrices of dimension

$d^{(2\ell)} = 2^\ell$ so that they constitute a complete set of basis matrices. Any matrix M of dimension $d^{(2\ell)}$ can be expanded with respect to $S \in \Gamma'_{2\ell}$,

$$M = \sum_{S\in\Gamma'_{2\ell}} C(S)S, \qquad C(S) = \frac{1}{d^{(2\ell)}}\text{Tr}\left(S^{-1}M\right). \tag{9.48}$$

Fifth, due to Eq. (9.38), $\pm S$ construct a class. $\mathbf{1}$ and $-\mathbf{1}$ construct two classes, respectively. The $\Gamma_{2\ell}$ group contains $2^{2\ell}+1$ classes. It is a one-dimensional representation that arbitrarily chosen n matrices γ_a correspond to 1 and the remaining matrices γ_b correspond to -1. The number of the one-dimensional inequivalent representations is

$$\sum_{n=0}^{2\ell} \begin{pmatrix} 2\ell \\ n \end{pmatrix} = 2^{2\ell}.$$

The remaining irreducible representation of $\Gamma_{2\ell}$ has to be $d^{(2\ell)}$-dimensional, which is faithful. The γ_a matrices in the representation is called the irreducible γ_a matrices. The irreducible γ_a matrices may be chosen as follows [Georgi (1982)]. Expressed γ_a as a direct product of ℓ two-dimensional matrices, which are the Pauli matrices σ_a and the unit matrix $\mathbf{1}$:

$$\begin{aligned}
\gamma_{2n-1} &= \underbrace{\mathbf{1}\times\ldots\times\mathbf{1}}_{n-1}\times\sigma_1\times\underbrace{\sigma_3\times\ldots\times\sigma_3}_{\ell-n}, \\
\gamma_{2n} &= \underbrace{\mathbf{1}\times\ldots\times\mathbf{1}}_{n-1}\times\sigma_2\times\underbrace{\sigma_3\times\ldots\times\sigma_3}_{\ell-n}, \\
\gamma_f^{(2\ell)} &= \underbrace{\sigma_3\times\ldots\times\sigma_3}_{\ell}.
\end{aligned} \tag{9.49}$$

Since $\gamma_f^{(2\ell)}$ is diagonal, the forms in Eq. (9.49) are called the reduced spinor representation. Remind that the eigenvalues ± 1 are arranged mixed in the diagonal line of $\gamma_f^{(2\ell)}$.

At last, there is an equivalent theorem for the γ_a matrices.

Theorem 9.1 (The equivalent theorem) Two sets of $d^{(2\ell)}$-dimensional matrices γ_a and $\overline{\gamma}_a$, both of which satisfy the anticommutative relation (9.38) where $N = 2\ell$, are equivalent

$$\overline{\gamma}_a = X^{-1}\gamma_a X, \qquad 1 \le a \le 2\ell. \tag{9.50}$$

The similarity transformation matrix X is determined up to a constant factor. If the determinant of X is restricted to be 1, there are $d^{(2\ell)}$ choices for the factor: $\exp\left(-i2n\pi/d^{(2\ell)}\right)$, $0 \le n < d^{(2\ell)}$.

Proof The irreducible representations of $\Gamma_{2\ell}$ constructed by two sets of matrices γ_a are equivalent because the characters of any element $S \in \Gamma_{2\ell}$ are equal to each other (see Eq. (9.45)).

Assume that there are two similarity transformation matrices X and Y,

$$\overline{\gamma}_a = X^{-1}\gamma_a X, \qquad \overline{\gamma}_a = Y^{-1}\gamma_a Y.$$

Thus, YX^{-1} can commute with every γ_a so that $Y = cX$. □

Being an important application of the equivalent theorem, the charge conjugation matrix $C^{(2\ell)}$ used in particle physics is defined based on the theorem. From irreducible unitary matrices γ_a satisfying the anticommutative relation (9.38), define $\overline{\gamma}_a = -(\gamma_a)^T$, where T denotes the transpose of the matrix. $\overline{\gamma}_a$ also satisfy Eq. (9.38) so that $\overline{\gamma}_a$ are equivalent to γ_a,

$$\left(C^{(2\ell)}\right)^{-1}\gamma_a C^{(2\ell)} = -(\gamma_a)^T, \qquad \left(C^{(2\ell)}\right)^\dagger C^{(2\ell)} = \mathbf{1}, \qquad \det C^{(2\ell)} = 1. \tag{9.51}$$

$\gamma_f^{(2\ell)}$ satisfies

$$\left(C^{(2\ell)}\right)^{-1}\gamma_f^{(2\ell)} C^{(2\ell)} = (-i)^\ell (\gamma_1)^T (\gamma_2)^T \ldots (\gamma_{2\ell})^T = (-1)^\ell \left(\gamma_f^{(2\ell)}\right)^T. \tag{9.52}$$

Taking the transpose of Eq. (9.51), one has

$$\begin{aligned}\gamma_a &= -\left(C^{(2\ell)}\right)^T \gamma_a^T \left[\left(C^{(2\ell)}\right)^{-1}\right]^T \\ &= \left[\left(C^{(2\ell)}\right)^T \left(C^{(2\ell)}\right)^{-1}\right] \gamma_a \left[\left(C^{(2\ell)}\right)^T \left(C^{(2\ell)}\right)^{-1}\right]^{-1}.\end{aligned}$$

Thus, $\left(C^{(2\ell)}\right)^T \left(C^{(2\ell)}\right)^{-1} = \lambda^{(2\ell)}\mathbf{1}$, $\left(C^{(2\ell)}\right)^T = \lambda^{(2\ell)} C^{(2\ell)}$, and

$$C^{(2\ell)} = \lambda^{(2\ell)}\left(C^{(2\ell)}\right)^T = \left(\lambda^{(2\ell)}\right)^2 C^{(2\ell)}, \qquad \lambda^{(2\ell)} = \pm 1.$$

The constant $\lambda^{(2\ell)}$ can be determined as follows. Remind that $\Gamma'_{2\ell}$ is a complete set of $d^{(2\ell)}$-dimensional matrices and is composed of S_n, $0 \le n \le 2\ell$, where S_n is a product of n different γ_a matrices. Since

$$\left(S_n C^{(2\ell)}\right)^T = \lambda^{(2\ell)} C^{(2\ell)} (S_n)^T = \lambda^{(2\ell)} (-1)^{n(n+1)/2} \left(S_n C^{(2\ell)}\right), \tag{9.53}$$

$S_n C^{(2\ell)}$ is either symmetric or antisymmetric. The number of S_n as well as $S_n C^{(2\ell)}$ is the combinatorics of n among 2ℓ. Because the number of the symmetric matrices of dimension $d^{(2\ell)}$ is larger than that of the antisymmetric ones, $S_\ell C^{(2\ell)}$ has to be symmetric,

$$\left(C^{(2\ell)}\right)^T = \lambda^{(2\ell)} C^{(2\ell)}, \qquad \lambda^{(2\ell)} = (-1)^{\ell(\ell+1)/2} . \tag{9.54}$$

The charge conjugation matrix $C^{(2\ell)}$ satisfies Eqs. (9.51) and (9.54). In the reduced spinor representation (9.49), one has

$$C^{(4m)} = \underbrace{(\sigma_1 \times \sigma_2) \times (\sigma_1 \times \sigma_2) \times \ldots \times (\sigma_1 \times \sigma_2)}_{m}, \quad C^{(4m+2)} = \sigma_2 \times C^{(4m)}. \tag{9.55}$$

In the particle physics, the strong space–time reflection matrix $B^{(2\ell)}$ is also used,

$$\begin{aligned} &B^{(2\ell)} = \gamma_f^{(2\ell)} C^{(2\ell)} = (-1)^\ell \lambda^{(2\ell)} \left(B^{(2\ell)}\right)^T, \qquad \det B^{(2\ell)} = 1 \\ &\left(B^{(2\ell)}\right)^{-1} \gamma_a B^{(2\ell)} = (\gamma_a)^T, \qquad \left(B^{(2\ell)}\right)^\dagger B^{(2\ell)} = \mathbf{1}. \end{aligned} \tag{9.56}$$

9.2.3 *The Case* $N = 2\ell + 1$

Since $\gamma_f^{(2\ell)}$ and (2ℓ) matrices γ_a in $\Gamma_{2\ell}$, $1 \leq a \leq 2\ell$, satisfy the antisymmetric relations (9.38), they can be defined to be the $(2\ell+1)$ matrices γ_a in $\Gamma_{2\ell+1}$. In this definition, $\gamma_\chi^{(2\ell+1)}$ in $\Gamma_{2\ell+1}$ has been chosen,

$$\gamma_{2\ell+1} = \gamma_f^{(2\ell)}, \qquad \gamma_\chi^{(2\ell+1)} = \gamma_1 \ldots \gamma_{2\ell+1} = (i)^\ell \mathbf{1}. \tag{9.57}$$

Obviously, the dimension $d^{(2\ell+1)}$ of the matrices in $\Gamma_{2\ell+1}$ is the same as $d^{(2\ell)}$ in $\Gamma_{2\ell}$,

$$d^{(2\ell+1)} = d^{(2\ell)} = 2^\ell. \tag{9.58}$$

When N is odd, the equivalent theorem has to be modified because the multiplication rule of elements in $\Gamma_{2\ell+1}$ includes Eq. (9.41). A similarity transformation cannot change the sign of $\gamma_\chi^{(2\ell+1)}$. Namely, the equivalent condition for two sets of γ_a and $\overline{\gamma}_a$ has to include a new condition $\gamma_\chi = \overline{\gamma}_\chi$, in addition to those given in Theorem 9.1.

Letting $\overline{\gamma}_a = -(\gamma_a)^T$, one has

$$\begin{aligned} \overline{\gamma}_\chi^{(2\ell+1)} &= \overline{\gamma}_1 \ldots \overline{\gamma}_{2\ell+1} = -\left\{\gamma_{2\ell+1} \ldots \gamma_1\right\}^T \\ &= (-1)^{\ell+1} \left\{\gamma_\chi^{(2\ell+1)}\right\}^T = (-1)^{\ell+1} \gamma_\chi^{(2\ell+1)}, \end{aligned} \tag{9.59}$$

namely, $C^{(4m-1)}$ satisfying Eq. (9.51) exists, but $C^{(4m+1)}$ does not. In the same reason, $B^{(4m+1)}$ satisfying Eq. (9.56) exists, but $B^{(4m-1)}$ does not. In fact, due to Eq. (9.52), $\gamma_{2\ell+1} = \gamma_f^{(2\ell)}$ satisfies Eq. (9.51) when $N = 4m - 1$, but does not when $N = 4m + 1$. Thus,

$$\begin{aligned} C^{(4m-1)} &= C^{(4m-2)}, & \left(C^{(4m-1)}\right)^T &= (-1)^m C^{(4m-1)}, \\ B^{(4m+1)} &= B^{(4m)} = \gamma_f^{(4m)} C^{(4m)}, & \left(B^{(4m+1)}\right)^T &= (-1)^m B^{(4m+1)}. \end{aligned} \tag{9.60}$$

Note that $C^{(4m-1)}$, $B^{(4m+1)}$, $C^{(4m-2)}$ and $C^{(4m)}$ have the same symmetry in transpose.

9.3 Spinor Representations of SO(N)

9.3.1 *Covering Groups of* SO*(N)*

From a set of N irreducible unitary matrices γ_a satisfying the anticommutative relation (9.38), define

$$\overline{\gamma}_a = \sum_{b=1}^{N} R_{ab}\gamma_b, \qquad R \in \mathrm{SO}(N). \tag{9.61}$$

Since R is a real orthogonal matrix, $\overline{\gamma}_a$ satisfy

$$\overline{\gamma}_a\overline{\gamma}_b + \overline{\gamma}_b\overline{\gamma}_a = \sum_{cd} R_{ac}R_{bd}\left\{\gamma_c\gamma_d + \gamma_d\gamma_c\right\} = 2\sum_c R_{ac}R_{bc}\mathbf{1} = 2\delta_{ab}\mathbf{1}.$$

Due to Eq. (9.38) and $\sum_a R_{1a}R_{2a} = 0$,

$$\begin{aligned} \sum_{a_1a_2} R_{1a_1}R_{2a_2}\gamma_{a_1}\gamma_{a_2} &= \frac{1}{2}\sum_{a_1\neq a_2} R_{1a_1}R_{2a_2}\left(\gamma_{a_1}\gamma_{a_2} - \gamma_{a_2}\gamma_{a_1}\right), \\ \overline{\gamma}_1\overline{\gamma}_2\cdots\overline{\gamma}_N &= \sum_{a_1\ldots a_N} R_{1a_1}\cdots R_{Na_N}\gamma_{a_1}\gamma_{a_2}\cdots\gamma_{a_N} \\ &= \sum_{a_1\ldots a_N} R_{1a_1}\cdots R_{Na_N}\epsilon_{a_1\ldots a_N}\gamma_1\gamma_2\cdots\gamma_N \\ &= (\det\ R)\,\gamma_1\gamma_2\cdots\gamma_N = \gamma_1\gamma_2\cdots\gamma_N. \end{aligned}$$

From the equivalent theorem, γ_a and $\overline{\gamma}_a$ can be related through a unitary similarity transformation $D(R)$ with determinant 1,

$$D(R)^{-1}\gamma_a D(R) = \sum_{d=1}^{N} R_{ad}\gamma_d, \qquad \det\ D(R) = 1, \tag{9.62}$$

where $D(R)$ is determined up to a constant

$$\exp\left(-i2n\pi/d^{(N)}\right), \qquad 0 \leq n < d^{(N)}. \tag{9.63}$$

The set of $D(R)$ defined in Eq. (9.62), in the multiplication rule of matrices, satisfies four axioms of a group and forms a Lie group G_N'. There is a

$d^{(N)}$-to-one correspondence between the elements in G'_N and the elements in SO(N), and the correspondence is invariant in the multiplication of elements. Therefore, G'_N is homomorphic onto SO(N). Since the group space of SO(N) is doubly-connected, its covering group is homomorphic onto it by a two-to-one correspondence. Thus, the group space of G'_N must fall into several disjoint pieces, where the piece containing the identical element E forms an invariant subgroup G_N of G'_N. G_N is a connected Lie group and is the covering group of SO(N). Since the group space of G_N is connected, based on the property of the infinitesimal elements, a discontinuous condition will be found to pick up G_N from G'_N.

Let R be an infinitesimal element. Expand R and $D(R)$ with respect to the infinitesimal parameters ω_{ab},

$$R_{cd} = \delta_{cd} - i \sum_{a<b} \omega_{ab} \left(T_{ab}\right)_{cd} = \delta_{cd} - \omega_{cd},$$
$$D(R) = \mathbf{1} - i \sum_{a<b} \omega_{ab} S_{ab},$$

where T_{ab} are the generators in the self-representation of SO(N) given in Eq. (7.85) and S_{ab} are the generators in G_N. From Eq. (9.62) one has

$$[\gamma_c,\ S_{ab}] = \sum_d \left(T_{ab}\right)_{cd} \gamma_d = -i \left\{\delta_{ac}\gamma_b - \delta_{bc}\gamma_a\right\}. \tag{9.64}$$

The solution is

$$S_{ab} = \frac{1}{4i} \left(\gamma_a\gamma_b - \gamma_b\gamma_a\right). \tag{9.65}$$

S_{ab} is Hermitian because $D(R)$ is unitary.

For simplifying the notations, define

$$C = \begin{cases} B^{(N)} & \text{when } N = (4\ell + 1), \\ C^{(N)} & \text{when } N \neq (4\ell + 1). \end{cases} \tag{9.66}$$

Thus, $C^{-1} S_{ab} C = -\left(S_{ab}\right)^T = -S^*_{ab}$,

$$C^{-1} D(R) C = \left\{D(R^{-1})\right\}^T = D(R)^*. \tag{9.67}$$

The discontinuous condition (9.67) restricts the factor in $D(R)$ (see Eq. (9.63)) such that there is a two-to-one correspondence between $\pm D(R)$ in G_N and R in SO(N) through the relations (9.62) and (9.67). Namely, G_N is the covering group of SO(N),

$$\text{SO}(N) \sim G_N. \tag{9.68}$$

When $N = 3$, $G_3 \approx \mathrm{SU}(2)$. G_N is called the fundamental spinor representation, or briefly, the spinor representation, denoted by $D^{[s]}(\mathrm{SO}(N))$. S_{ab} is called the spinor angular momentum operators. The irreducible tensor representation $[\lambda]$ is a single-valued representation of $\mathrm{SO}(N)$, but a non-faithful one of G_N. The faithful representation of G_N is a double-valued one of $\mathrm{SO}(N)$.

Since the products S_n span a complete set of the $d^{(N)}$-dimensional matrices, it can be decided by checking the commutative relations of S_n with the generators S_{ab} whether there is a nonconstant matrix commutable with all S_{ab}. The result is that only $\gamma_\chi^{(N)}$ is commutable with all S_{ab}. $\gamma_\chi^{(2\ell+1)}$ is a constant matrix so that the fundamental spinor representation $D^{[s]}(\mathrm{SO}(2\ell+1))$ is irreducible. Due to Eqs. (9.67) and (9.60), $D^{[s]}(\mathrm{SO}(N))$ is self-conjugate when $N = 2\ell + 1$ and is real when $N = 8k \pm 1$.

$\gamma_\chi^{(2\ell)}$ is not a constant matrix so that the fundamental spinor representation $D^{[s]}(\mathrm{SO}(2\ell))$ is reducible. Through a similarity transformation X, $\gamma_f^{(2\ell)}$ is changed to $\sigma_3 \times \mathbf{1}$ and $D^{[s]}(\mathrm{SO}(2\ell))$ is reduced into the direct sum of two irreducible representations,

$$X^{-1} D^{[s]}(R) X = \begin{pmatrix} D^{[+s]}(R) & 0 \\ 0 & D^{[-s]}(R) \end{pmatrix}. \tag{9.69}$$

Two representations $D^{[\pm s]}(\mathrm{SO}(2\ell))$ can be proved to be inequivalent by reduction to absurdity. In fact, if $Z^{-1} D^{[-s]}(R) Z = D^{[+s]}(R)$ and $Y = \mathbf{1} \oplus Z$, then all generators $(XY)^{-1} S_{ab} XY$ are commutable with $\sigma_1 \times \mathbf{1}$, but their product is not commutable with it,

$$2^\ell (XY)^{-1} \left(S_{12} S_{34} \ldots S_{(2\ell-1)(2\ell)} \right) XY = Y^{-1} \left[X^{-1} \gamma_f^{(2\ell)} X \right] Y = \sigma_3 \times \mathbf{1}.$$

It leads to contradiction.

Introduce two projective operators $P_\pm$,

$$\begin{aligned} & P_\pm = \frac{1}{2}\left(\mathbf{1} \pm \gamma_f^{(2\ell)}\right), && P_\pm D^{[s]}(R) = D^{[s]}(R) P_\pm, \\ & X^{-1} P_+ X = \begin{pmatrix} \mathbf{1} & 0 \\ 0 & 0 \end{pmatrix}, && X^{-1} P_+ D^{[s]}(R) X = \begin{pmatrix} D^{[+s]}(R) & 0 \\ 0 & 0 \end{pmatrix}, \\ & X^{-1} P_- X = \begin{pmatrix} 0 & 0 \\ 0 & \mathbf{1} \end{pmatrix}, && X^{-1} P_- D^{[s]}(R) X = \begin{pmatrix} 0 & 0 \\ 0 & D^{[-s]}(R) \end{pmatrix}. \end{aligned} \tag{9.70}$$

From Eq. (9.52) one has

$$C^{-1} D^{[s]}(R) P_\pm C = \begin{cases} D^{[s]}(R)^* P_\pm & \text{when } N = 4m, \\ D^{[s]}(R)^* P_\mp & \text{when } N = 4m+2. \end{cases} \tag{9.71}$$

Two inequivalent representations $D^{[\pm s]}(R)$ are conjugate to each other when $N = 4m + 2$, are self-conjugate when $N = 4m$, and are real when $N = 8k$ owing to Eq. (9.54). The dimension of the irreducible spinor representations of SO(N) is

$$d_{[s]}[\mathrm{SO}(2\ell+1)] = 2^{\ell}, \qquad d_{[\pm s]}[\mathrm{SO}(2\ell)] = 2^{\ell-1}. \tag{9.72}$$

9.3.2 *Fundamental Spinors of* SO(N)

In a SO(N) transformation R, $\mathbf{\Psi}$ is called the fundamental spinor of SO(N) if it transforms by the fundamental spinor representation $D^{[s]}(R)$:

$$\left(O_R\mathbf{\Psi}\right)_{\mu} = \sum_{\nu} D^{[s]}_{\mu\nu}(R)\mathbf{\Psi}_{\nu}, \qquad O_R\mathbf{\Psi} = D^{[s]}(R)\mathbf{\Psi}, \tag{9.73}$$

where $\mathbf{\Psi}$ is a column matrix with $d_{[s]}$ components.

The Chevalley bases $H_\mu(S)$ $E_\mu(S)$ and $F_\mu(S)$ with respect to the spinor angular momentum can be obtained from Eqs. (9.15) and (9.21) by replacing T_{ab} with S_{ab}. In the chosen forms of γ_a given in Eq. (9.49), the Chevalley bases for the SO($2\ell+1$) group (B$_\ell$ Lie algebra) are

$$\begin{aligned}
H_\mu(S) &= \underbrace{\mathbf{1}\times\ldots\times\mathbf{1}}_{\mu-1}\times\frac{1}{2}\left\{\sigma_3\times\mathbf{1}-\mathbf{1}\times\sigma_3\right\}\times\underbrace{\mathbf{1}\times\ldots\times\mathbf{1}}_{\ell-\mu-1},\\
H_\ell(S) &= \underbrace{\mathbf{1}\times\ldots\times\mathbf{1}}_{\ell-1}\times\sigma_3,\\
E_\mu(S) &= \underbrace{\mathbf{1}\times\ldots\times\mathbf{1}}_{\mu-1}\times\left\{\sigma_+\times\sigma_-\right\}\times\underbrace{\mathbf{1}\times\ldots\times\mathbf{1}}_{\ell-\mu-1} = F_\mu(S)^T,\\
E_\ell(S) &= \underbrace{\sigma_3\times\cdots\times\sigma_3}_{\ell-1}\times\sigma_+ = F_\ell(S)^T,
\end{aligned} \tag{9.74}$$

where $1 \le \mu < \ell$. The Chevalley bases for the SO(2ℓ) group (D$_\ell$ Lie algebra) are the same as those for SO($2\ell+1$) except for $\mu = \ell$,

$$\begin{aligned}
H_\ell(S) &= \underbrace{\mathbf{1}\times\ldots\times\mathbf{1}}_{\ell-2}\times\frac{1}{2}\left\{\sigma_3\times\mathbf{1}+\mathbf{1}\times\sigma_3\right\},\\
E_\ell(S) &= -\underbrace{\mathbf{1}\times\ldots\times\mathbf{1}}_{\ell-2}\times\left\{\sigma_+\times\sigma_+\right\} = F_\ell(S)^T.
\end{aligned} \tag{9.75}$$

The basis spinor $\chi[\boldsymbol{m}]$ of SO(N) is also expressed as a direct product of ℓ two-dimensional basis spinors $\chi(\alpha)$,

$$\chi[\boldsymbol{m}] = \chi_{(\alpha_1,\alpha_2,\ldots,\alpha_\ell)} = \chi(\alpha_1)\chi(\alpha_2)\ldots\chi(\alpha_\ell), \tag{9.76}$$

$$\chi(+) = \begin{pmatrix} 1 \\ 0 \end{pmatrix}, \qquad \chi(-) = \begin{pmatrix} 0 \\ 1 \end{pmatrix}. \tag{9.77}$$

When N is even, the fundamental spinor space is decomposed into two subspaces by the project operators $P_\pm$, $\mathbf{\Psi}_\pm = P_\pm \mathbf{\Psi}$, corresponding to irreducible spinor representations $D^{[\pm s]}$. The basis spinor in the representation space of $D^{[+s]}$ contains even number of factors $\chi(-)$, and that of $D^{[-s]}$ contains odd number of $\chi(-)$. The highest weight states $\chi[\boldsymbol{M}]$ and their highest weights $\boldsymbol{M}$ are

$$\begin{array}{lll}
\underbrace{\chi(+)\ldots\chi(+)}_{\ell-1}\chi(+), & \boldsymbol{M} = (\underbrace{0,\ldots,0}_{\ell-1},1), & [s] \text{ of } \mathrm{SO}(2\ell+1), \\
\underbrace{\chi(+)\ldots\chi(+)}_{\ell-1}\chi(+), & \boldsymbol{M} = (\underbrace{0,\ldots,0}_{\ell-2},0,1), & [+s] \text{ of } \mathrm{SO}(2\ell), \\
\underbrace{\chi(+)\ldots\chi(+)}_{\ell-1}\chi(-), & \boldsymbol{M} = (\underbrace{0,\ldots,0}_{\ell-2},1,0), & [-s] \text{ of } \mathrm{SO}(2\ell).
\end{array} \tag{9.78}$$

The remaining basis states can be calculated by the applications of lowering operators $F_\mu(S)$.

9.3.3 *Direct Products of Spinor Representations*

The spinor representation is unitary so that

$$O_R \mathbf{\Psi}^\dagger = \mathbf{\Psi}^\dagger D^{[s]}(R)^{-1}. \tag{9.79}$$

$$\begin{aligned}
\mathbf{\Psi}^\dagger \mathbf{\Psi} &= \sum_\mu \mathbf{\Psi}_\mu^* \mathbf{\Psi}_\mu = \sum_{\mu\nu} \mathbf{\Psi}_\mu^* \delta_{\mu\nu} \mathbf{\Psi}_\nu, \\
O_R \left(\mathbf{\Psi}^\dagger \mathbf{\Psi}\right) &= \mathbf{\Psi}^\dagger D^{[s]}(R)^{-1} D^{[s]}(R) \mathbf{\Psi} = \mathbf{\Psi}^\dagger \mathbf{\Psi}.
\end{aligned} \tag{9.80}$$

$\mathbf{\Psi}^* \mathbf{\Psi}$ is invariant in the SO(N) transformations. It is a scalar of SO(N). In the language of group theory, the products of $\mathbf{\Psi}_\mu^*$ and $\mathbf{\Psi}_\nu$ span an invariant space, corresponding to the direct product representation $D^{[s]*} \times D^{[s]}$ of SO(N). In the reduction of $D^{[s]*} \times D^{[s]}$ there is an identical representation where the Clebsch–Gordan coefficients are $\delta_{\mu\nu}$. Generally,

$$\begin{aligned}
O_R \left(\mathbf{\Psi}^\dagger \gamma_{a_1} \ldots \gamma_{a_n} \mathbf{\Psi}\right) &= \mathbf{\Psi}^\dagger D^{[s]}(R)^{-1} \gamma_{a_1} \ldots \gamma_{a_n} D^{[s]}(R) \mathbf{\Psi} \\
&= \sum_{b_1 \ldots b_n} R_{a_1 b_1} \ldots R_{a_n b_n} \mathbf{\Psi}^\dagger \gamma_{b_1} \ldots \gamma_{b_n} \mathbf{\Psi}.
\end{aligned} \tag{9.81}$$

$\mathbf{\Psi}^\dagger \gamma_{a_1} \ldots \gamma_{a_n} \mathbf{\Psi}$ is an antisymmetric tensor of rank n of SO(N) corresponding to the Young pattern $[1^n]$. n is obviously not larger than N, otherwise the repetitive γ_a can be moved together and eliminated.

When $N = 2\ell + 1$, $\gamma_f^{(2\ell+1)}$ is a constant matrix so that the product of $(N-n)$ matrices γ_a can be changed to a product of n matrices γ_a. Thus, the rank n of the tensor (9.81) is less than $N/2$, and the Clebsch–Gordan series is

$$[s]^* \times [s] \simeq [s] \times [s] \simeq \ [0] \oplus [1] \oplus [1^2] \oplus \ldots \oplus [1^\ell], \quad \text{for SO}(2\ell+1). \quad (9.82)$$

The matrix entries of the product of γ_a are the Clebsch–Gordan coefficients. The highest weight in the product space is $\boldsymbol{M} = (0, \ldots, 0, 2)$, corresponding to the representation $[1^\ell]$.

When $N = 2\ell$, due to the property of the projective operators $P_\pm$,

$$\begin{aligned} &P_+ P_- = P_- P_+ = 0, \qquad P_\pm P_\pm = P_\pm, \qquad \gamma_f^{(2\ell)} P_\pm = \pm P_\pm, \\ &P_\mp \gamma_{a_1} \cdots \gamma_{a_{2m}} P_\pm = 0, \qquad P_\pm \gamma_{a_1} \cdots \gamma_{a_{2m+1}} P_\pm = 0, \end{aligned} \quad (9.83)$$

the product of $(N-n)$ matrices γ_a can still be changed to a product of n matrices γ_a. If $n = \ell$, one has $\gamma_1\gamma_2 \cdots \gamma_\ell = (-i)^\ell \gamma_{2\ell}\gamma_{2\ell-1} \cdots \gamma_{\ell+1}\gamma_f^{(2\ell)}$

$$\gamma_1\gamma_2 \cdots \gamma_\ell P_\pm = \frac{1}{2} \left\{\gamma_1\gamma_2 \cdots \gamma_\ell \pm (-i)^\ell \gamma_{2\ell}\gamma_{2\ell-1} \cdots \gamma_{\ell+1}\right\} P_\pm. \quad (9.84)$$

If $N = 4m$,

$$\begin{aligned} &[\pm s]^* \times [\pm s] \simeq [\pm s] \times [\pm s] \simeq [0] \oplus [1^2] \oplus [1^4] \oplus \ldots \oplus [(\pm)1^{2m}], \\ &[\mp s]^* \times [\pm s] \simeq [\mp s] \times [\pm s] \simeq [1] \oplus [1^3] \oplus [1^5] \oplus \ldots \oplus [1^{2m-1}]. \end{aligned} \quad (9.85)$$

If $N = 4m + 2$,

$$\begin{aligned} &[\pm s]^* \times [\pm s] \simeq [\mp s] \times [\pm s] \simeq [0] \oplus [1^2] \oplus [1^4] \oplus \ldots \oplus [1^{2m}] \\ &[\mp s]^* \times [\pm s] \simeq [\pm s] \times [\pm s] \simeq [1] \oplus [1^3] \oplus [1^5] \oplus \ldots \oplus [(\pm)1^{2m+1}]. \end{aligned} \quad (9.86)$$

The self-dual and anti-self-dual representations occur in the reduction of the direct product $[\pm s] \times [\pm s]$, but not in the reduction of $[+s] \times [-s]$. The highest weights are $\boldsymbol{M} = (0, \ldots, 0, 0, 2)$ in the product space $[+s] \times [+s]$, $\boldsymbol{M} = (0, \ldots, 0, 2, 0)$ in the product space $[-s] \times [-s]$, and $\boldsymbol{M} = (0, \ldots, 0, 1, 1)$ in the product space $[+s] \times [-s]$.

9.3.4 *Spinor Representations of Higher Ranks*

In the SO(3) group, $D^{1/2}$ is the fundamental spinor representation, and the spinor representations D^j of higher ranks can be obtained from the reduction of the direct product of the fundamental spinor representation and a tensor representation,

$$D^{1/2} \times D^{\ell} \simeq D^{\ell+1/2} \oplus D^{\ell-1/2}. \tag{9.87}$$

The spinor representations of higher ranks of SO(N) can be obtained in the same way.

A spinor $\mathbf{\Psi}_{a_1 \ldots a_n}$ with the tensor indices is called a spin–tensor if it transforms in $R \in$ SO(N) as follows

$$(O_R \mathbf{\Psi})_{a_1 \ldots a_n} = \sum_{b_1 \ldots b_n} R_{a_1 b_1} \ldots R_{a_n b_n} D^{[s]}(R) \mathbf{\Psi}_{b_1 \ldots b_n}. \tag{9.88}$$

The tensor part of the spin–tensor can be decomposed like a tensor. Namely, the tensor is decomposed into a direct sum of the traceless tensors with different ranks and each traceless tensor subspace can be reduced by the projection of the Young operators. Thus, the reduced subspace of the traceless tensor part of the spin–tensor is denoted by a Young pattern $[\lambda]$ or $[(\pm)\lambda]$ where the row number of $[\lambda]$ is not larger than $N/2$. However, this subspace of the spin–tensor corresponds to the direct product of the fundamental spinor representation $[s]$ and the irreducible tensor representation $[\lambda]$ or $[(\pm)\lambda]$, and it is still reducible. This is the generalization of Eq. (9.87). It is required to find a new restriction to pick up the irreducible subspace like the subspace of $D^{\ell+1/2}$ in Eq. (9.87) for SO(3). The restriction comes from the so-called trace of the second kind of the spin–tensor which is invariant in the SO(N) transformations:

$$\begin{aligned}
\mathbf{\Phi}_{a_1 \ldots a_{i-1} a_{i+1} \ldots a_n} &= \sum_{b=1}^{N} \gamma_b \mathbf{\Psi}_{a_1 \ldots a_{i-1} b a_{i+1} \ldots a_n}, \\
(O_R \mathbf{\Phi})_{a_1 \ldots a_{i-1} a_{i+1} \ldots a_n} &= \sum_{b_1 \ldots b_n b'} R_{a_1 b_1} \ldots R_{a_n b_n} \left[\sum_b \gamma_b R_{bb'} \right] \\
&\quad \cdot D^{[s]}(R) \mathbf{\Psi}_{b_1 \ldots b_{i-1} b' b_{i+1} \ldots b_n} \\
&= \sum_{b_1 \ldots b_n} R_{a_1 b_1} \ldots R_{a_n b_n} D^{[s]}(R) \left[\sum_{b'} \gamma_{b'} \mathbf{\Psi}_{b_1 \ldots b_{i-1} b' b_{i+1} \ldots b_n} \right] \\
&= \sum_{b_1 \ldots b_n} R_{a_1 b_1} \ldots R_{a_n b_n} D^{[s]}(R) \mathbf{\Phi}_{b_1 \ldots b_{i-1} b_{i+1} \ldots b_n}.
\end{aligned}$$

The irreducible subspace of SO(N) contained in the spin–tensor space, in addition to the projection of a Young operator, satisfies the usual traceless conditions of tensors and the traceless conditions of the second kind:

$$\sum_b \psi_{a \cdots b \cdots b \cdots c} = 0, \qquad \sum_b \gamma_b \psi_{a \cdots b \cdots c} = 0. \tag{9.89}$$

The highest weight $\boldsymbol{M}$ of the irreducible representation is the highest weight in the direct product space. The irreducible representation is denoted by $[s, \lambda]$ for $\text{SO}(2\ell+1)$

$$\begin{aligned} &[s] \times [\lambda] \simeq [s, \lambda] \oplus \ldots \\ &\boldsymbol{M} = [(\lambda_1 - \lambda_2), \ldots, (\lambda_{\ell-1} - \lambda_\ell), (2\lambda_\ell + 1)], \end{aligned} \tag{9.90}$$

and $[\pm s, \lambda]$ for $\text{SO}(2\ell)$,

$$\begin{aligned} &[+s] \times [\lambda] \quad \text{or} \quad [+s] \times [(+)\lambda] \simeq [+s, \lambda] \oplus \ldots \\ &\boldsymbol{M} = [(\lambda_1 - \lambda_2), \ldots, (\lambda_{\ell-1} - \lambda_\ell), (\lambda_{\ell-1} + \lambda_\ell + 1)], \\ &[-s] \times [\lambda] \quad \text{or} \quad [-s] \times [(-)\lambda] \simeq [-s, \lambda] \oplus \ldots \\ &\boldsymbol{M} = [(\lambda_1 - \lambda_2), \ldots, (\lambda_{\ell-1} + \lambda_\ell + 1), (\lambda_{\ell-1} - \lambda_\ell)], \\ &[+s] \times [(-)\lambda] \simeq [-s, \lambda_1, \lambda_2, \ldots, \lambda_{\ell-1}, (\lambda_\ell - 1)] \oplus \ldots \\ &\boldsymbol{M} = [(\lambda_1 - \lambda_2), \ldots, (\lambda_{\ell-1} + \lambda_\ell), (\lambda_{\ell-1} - \lambda_\ell + 1)], \\ &[-s] \times [(+)\lambda] \simeq [+s, \lambda_1, \lambda_2, \ldots, \lambda_{\ell-1}, (\lambda_\ell - 1)] \oplus \ldots \\ &\boldsymbol{M} = [(\lambda_1 - \lambda_2), \ldots, (\lambda_{\ell-1} - \lambda_\ell + 1), (\lambda_{\ell-1} + \lambda_\ell)]. \end{aligned} \tag{9.91}$$

These irreducible representations $[s, \lambda]$ of $\text{SO}(2\ell+1)$ and $[\pm s, \lambda]$ of $\text{SO}(2\ell)$ are called the spinor representations of higher ranks. Remind that the row number of the Young pattern $[\lambda]$ in the spinor representation of higher rank is not larger than ℓ, otherwise the space is null. For example, in the space of the spinor representation $[s, 1^n]$ of $\text{SO}(2\ell+1)$, the number of spin–tensors before taking the traceless conditions is $d^{(2\ell+1)}$ times the combinatorics of n among $(2\ell+1)$. There is no traceless condition of the first kind. The number of traceless conditions of the second kind is $d^{(2\ell+1)}$ times the combinatorics of $(n-1)$ among $(2\ell+1)$. Thus, the space is null when $n > \ell$ because the number of the traceless conditions is not less than the number of tensors.

The remaining representations in the Clebsch–Gordan series (9.90) and (9.91) are calculated by the method of dominant weight diagram. For example, when $[\lambda]$ is a one-row Young diagram, one has

$$\begin{aligned} &SO(2\ell+1): \quad [s] \times [\lambda, 0, \ldots, 0] \simeq [s, \lambda, 0, \ldots, 0] \oplus [s, \lambda - 1, 0, \ldots, 0], \\ &SO(2\ell): \quad [\pm s] \times [\lambda, 0, \ldots, 0] \simeq [\pm s, \lambda, 0, \ldots, 0] \oplus [\mp s, \lambda - 1, 0, \ldots, 0]. \end{aligned} \tag{9.92}$$

$[\mp s, \lambda - 1, 0, \ldots, 0]$ appears in the second reduction because the factor γ_b in Eq. (9.89) is anticommutable with γ_f in $P_\pm$.

9.3.5 *Dimensions of the Spinor Representations*

The dimension of a spinor representation $[s, \lambda]$ of SO$(2\ell + 1)$ or $[\pm s, \lambda]$ of SO(2ℓ) can be calculated by the hook rule [Dai (1983)]. In this rule the dimension is expressed as a quotient multiplied with the dimension of the fundamental spinor representation, where the numerator and the denominator of the quotient are denoted by the symbols $Y_S^{[\lambda]}$ and $Y_h^{[\lambda]}$, respectively:

$$d_{[s,\lambda]}(\mathrm{SO}(2\ell+1)) = 2^{\ell}\frac{Y_S^{[\lambda]}}{Y_h^{[\lambda]}}, \qquad d_{[\pm s,\lambda]}(\mathrm{SO}(2\ell)) = 2^{\ell-1}\frac{Y_S^{[\lambda]}}{Y_h^{[\lambda]}}. \tag{9.93}$$

The concepts of a hook path $(i,\ j)$ and an inverse hook path $\overline{(i,\ j)}$ are discussed in §9.1.4. The number of boxes contained in the hook path $(i,\ j)$ is the hook number h_{ij} of the box in the jth column of the ith row. $Y_h^{[\lambda]}$ is a tableau of the Young pattern $[\lambda]$ where the box in the jth column of the ith row is filled with the hook number h_{ij}. Define a series of the tableaux $Y_{S_a}^{[\lambda]}$ recursively by the rule given below. $Y_S^{[\lambda]}$ is a tableau of the Young pattern $[\lambda]$ where each box is filled with the sum of the digits which are respectively filled in the same box of each tableau $Y_{S_a}^{[\lambda]}$ in the series. The symbol $Y_S^{[\lambda]}$ means the product of the filled digits in it, so does the symbol $Y_h^{[\lambda]}$.

The tableaux $Y_{S_a}^{[\lambda]}$ are defined by the following rule:

(a) $Y_{S_0}^{[\lambda]}$ is a tableau of the Young pattern $[\lambda]$ where the box in the jth column of the ith row is filled with the digit $(N-1+j-i)$.

(b) Let $[\lambda^{(1)}] = [\lambda]$. Beginning with $[\lambda^{(1)}]$, we define recursively the Young pattern $[\lambda^{(a)}]$ by removing the first row and the first column of the Young pattern $[\lambda^{(a-1)}]$ until $[\lambda^{(a)}]$ contains less than two rows.

(c) If $[\lambda^{(a)}]$ contains more than one row, define $Y_{S_a}^{[\lambda]}$ to be a tableau of the Young pattern $[\lambda]$ where the boxes in the first $(a-1)$ row and in the first $(a-1)$ column are filled with 0, and the remaining part of the Young pattern is nothing but $[\lambda^{(a)}]$. Let $[\lambda^{(a)}]$ have r rows. Fill the first $(r-1)$ boxes along the hook path $(1,\ 1)$ of the Young pattern $[\lambda^{(a)}]$, beginning with the box on the rightmost, with the digits $\lambda_2^{(a)}$, $\lambda_3^{(a)}$, $\cdots$, $\lambda_r^{(a)}$, box by box, and fill the first $\lambda_i^{(a)}$ boxes in each inverse hook path $\overline{(i,\ 1)}$ of the Young pattern $[\lambda^{(a)}]$, $2 \leq i \leq r$, with -1. The remaining boxes are filled with 0. If a few -1 are filled in the same box, the digits are summed. The sum of all filled digits in the pattern $Y_{S_a}^{[\lambda]}$ with $a > 0$ is 0.

Ex. 1 Dimension of the representation $[+s, 3, 3, 3]$ of SO(8).

$$Y_S^{[3,3,3]} = \begin{array}{|c|c|c|}\hline 7 & 8 & 9 \\ \hline 6 & 7 & 8 \\ \hline 5 & 6 & 7 \\ \hline \end{array} + \begin{array}{|c|c|c|}\hline & 3 & 3 \\ \hline -1 & -1 & \\ \hline -2 & -1 & -1 \\ \hline \end{array} + \begin{array}{|c|c|c|}\hline & & \\ \hline & & 2 \\ \hline & -1 & -1 \\ \hline \end{array} = \begin{array}{|c|c|c|}\hline 7 & 11 & 12 \\ \hline 5 & 6 & 10 \\ \hline 3 & 4 & 5 \\ \hline \end{array},$$

$$Y_h^{[3,3,3]} = \begin{array}{|c|c|c|}\hline 5 & 4 & 3 \\ \hline 4 & 3 & 2 \\ \hline 3 & 2 & 1 \\ \hline \end{array},$$

$$d_{[+s(3,3,3)]}(\mathrm{SO}(8)) = \ 2^3 \times 11 \times 7 \times 5^2 = 15400.$$

Ex. 2 Dimension of the representations $[\pm s, n]$ of SO(2ℓ) and $[s, n]$ of SO($2\ell + 1$).

$$\begin{aligned}
Y_S^{[n]} &= \begin{array}{|c|c|c|c|}\hline N-1 & N & \cdots & N+n-2 \\ \hline \end{array} = \frac{(N+n-2)!}{(N-2)!}, \\
Y_h^{[n]} &= \begin{array}{|c|c|c|c|}\hline n & n-1 & \cdots & 1 \\ \hline \end{array} = n!, \\
d_{[\pm s,n]}(\mathrm{SO}(2\ell)) &= 2^{\ell-1}\frac{(2\ell+n-2)!}{n!(2\ell-2)!} = 2^{\ell-1}\begin{pmatrix} 2\ell+n-2 \\ n \end{pmatrix}, \\
d_{[s,n]}(\mathrm{SO}(2\ell+1)) &= 2^{\ell}\frac{(2\ell+n-1)!}{n!(2\ell-1)!} = 2^{\ell}\begin{pmatrix} 2\ell+n-1 \\ n \end{pmatrix}.
\end{aligned} \tag{9.94}$$

Ex. 3 Dimension of the representations $[\pm s, 1^n]$ of SO(2ℓ) and $[s, n]$ of SO($2\ell + 1$).

$$Y_S^{[1^n]} = \begin{array}{|c|}\hline N-1 \\ \hline N-2 \\ \hline \vdots \\ \hline N-n+1 \\ \hline N-n \\ \hline \end{array} + \begin{array}{|c|}\hline 1 \\ \hline 1 \\ \hline \vdots \\ \hline 1 \\ \hline -n+1 \\ \hline \end{array} = \begin{array}{|c|}\hline N \\ \hline N-1 \\ \hline \vdots \\ \hline N-n+2 \\ \hline N-2n+1 \\ \hline \end{array}, \tag{9.95}$$

$$d_{[\pm s,1^n]}(SO(2\ell)) = 2^{\ell-1}\ \frac{(2\ell)!(2\ell-2n+1)}{n!(2\ell-n+1)!},$$

$$d_{[\pm s,1^n]}(SO(2\ell+1)) = 2^{\ell}\ \frac{(2\ell+1)!(2\ell-2n+2)}{n!(2\ell-n+2)!}.$$

9.4 Rotational Symmetry in N-Dimensional Space

9.4.1 *Orbital Angular Momentum Operators*

The relations between the rectangular coordinates x_a and the spherical coordinates r and θ_b in an N-dimensional space are

$$\begin{aligned} x_1 &= r\cos\theta_1 \sin\theta_2 \ldots \sin\theta_{N-1}, \\ x_2 &= r\sin\theta_1 \sin\theta_2 \ldots \sin\theta_{N-1}, \\ x_b &= r\cos\theta_{b-1} \sin\theta_b \ldots \sin\theta_{N-1}, \qquad 3 \leq b \leq N-1, \\ x_N &= r\cos\theta_{N-1}, \\ \sum_{a=1}^{N} & x_a^2 = r^2. \end{aligned} \tag{9.96}$$

The unit vector along $\boldsymbol{x}$ is usually denoted by $\hat{\boldsymbol{x}} = \boldsymbol{x}/r$. The volume element of the configuration space is

$$\begin{aligned} &\prod_{a=1}^{N} dx_a = r^{N-1} dr d\Omega, \qquad d\Omega = \prod_{a=1}^{N-1} (\sin\theta_a)^{a-1} d\theta_a, \\ &0 \leq r < \infty, \quad -\pi \leq \theta_1 \leq \pi, \quad 0 \leq \theta_b \leq \pi, \quad 2 \leq b \leq N-1. \end{aligned} \tag{9.97}$$

The orbital angular momentum operators L_{ab} are the generators of the transformation operators P_R for the scalar function, $R \in$SO(N),

$$L_{ab} = -L_{ba} = -ix_a \frac{\partial}{\partial x_b} + ix_b \frac{\partial}{\partial x_a}, \qquad L^2 = \sum_{a<b=2}^{N} L_{ab}^2, \tag{9.98}$$

where the natural units are used, $\hbar = c = 1$. From the second Lie theorem, L_{ab} satisfy the same commutation relations as Eq. (9.12) of the generators T_{ab} in the self-representation of SO(N).

9.4.2 *Spherical Harmonic Functions*

The Chevalley bases $H_\mu(L)$, $E_\mu(L)$, and $F_\mu(L)$ can be obtained from Eqs. (9.15) and (9.21) by replacing T_{ab} with L_{ab}. Because

$$L_{ab} x_d = \sum_{c=1}^{N} x_c \left(T_{ab}\right)_{cd}, \qquad O_R \boldsymbol{\theta}_d = \sum_{c=1}^{N} \boldsymbol{\theta}_c R_{cd}, \tag{9.99}$$

the common eigenfunctions X_α of $H_\mu(L)$ can be obtained from ϕ_α given in Eq. (9.16) for SO$(2\ell+1)$ and in Eq. (9.22) for SO(2ℓ) by replacing the basis vector $\boldsymbol{\theta}_a$ with the rectangular coordinate x_a.

The spherical harmonic function $Y_{\boldsymbol{m}}^{[\lambda]}(\hat{\boldsymbol{x}})$ is the eigenfunction of the orbital angular momentum operators $H_\mu(L)$ for a single particle,

$$H_\mu(L)Y_{\boldsymbol{m}}^{[\lambda]}(\hat{\boldsymbol{x}}) = m_\mu Y_{\boldsymbol{m}}^{[\lambda]}(\hat{\boldsymbol{x}}), \qquad L^2 Y_{\boldsymbol{m}}^{[\lambda]}(\hat{\boldsymbol{x}}) = C_2([\lambda])Y_{\boldsymbol{m}}^{[\lambda]}(\hat{\boldsymbol{x}}), \quad (9.100)$$

where $C_2([\lambda])$ is the Casimir invariant of order 2. For an irreducible tensor representation of SO(N), $C_2([\lambda])$ or $C_2([(\pm)\lambda])$ can be calculated directly from Eq. (7.137)

$$C_2([\lambda]) = \sum_{\mu=1}^{\ell} \lambda_\mu(\lambda_\mu + N - 2\mu), \qquad N = 2\ell + 1 \ \text{ or } \ 2\ell. \quad (9.101)$$

Since there is only one coordinate vector $\boldsymbol{x}$, the representation $[\lambda]$ for the spherical harmonic function $Y_{\boldsymbol{m}}^{[\lambda]}(\hat{\boldsymbol{x}})$ has to be the totally symmetric representation denoted by the one-row Young pattern $[\lambda] = [\lambda, 0, \ldots, 0]$. $C_2([\lambda, 0, \ldots, 0]) = \lambda(\lambda + N - 2)$ for SO(N). When $\lambda = 1$, the highest weight state $Y_{(1,0\ldots,0)}^{[1,0\ldots,0]}(\hat{\boldsymbol{x}})$ is proportional to X_1,

$$X_1 = \frac{(-1)^t}{\sqrt{2}}\left(x_1 + ix_2\right), \qquad t = \begin{cases} \ell, & N = 2\ell + 1, \\ \ell - 1, & N = 2\ell. \end{cases} \quad (9.102)$$

Generally, the highest weight state $Y_{\boldsymbol{M}}^{[\lambda]}(\hat{\boldsymbol{x}})$ with $[\lambda] = [\lambda, 0, \ldots, 0]$ and $\boldsymbol{M} = (\lambda, 0, \ldots, 0)$ is (see Prob. 6 of Chap. 9 in [Ma and Gu (2004)])

$$\begin{aligned} Y_{\boldsymbol{M}}^{[\lambda]}(\hat{\boldsymbol{x}}) &= C_{N,\lambda}\left(\frac{X_1}{r}\right)^\lambda = C_{N,\lambda}\left\{\frac{(-1)^t\left(x_1 + ix_2\right)}{r\sqrt{2}}\right\}^\lambda, \\ C_{N,\lambda}^2 &= \begin{cases} \dfrac{(2\lambda + 2\ell - 1)!}{2^{\lambda+2\ell}\pi^\ell \lambda!(\lambda + \ell - 1)!}, & N = 2\ell + 1, \\ \dfrac{2^{\lambda-1}(\lambda + \ell - 1)!}{\pi^\ell \lambda!}, & N = 2\ell. \end{cases} \end{aligned} \quad (9.103)$$

The remaining spherical harmonic function $Y_{\boldsymbol{m}}^{[\lambda]}(\hat{\boldsymbol{x}})$ with the weight $\boldsymbol{m}$ can be calculated by the lowering operators $F_\mu(L)$. Remind that $r^\lambda Y_{\boldsymbol{m}}^{[\lambda]}(\hat{\boldsymbol{x}})$ is called the harmonic polynomial which is a homogeneous polynomial of degree λ with respect to the rectangular coordinates x_a and satisfies the Laplace equation

$$\nabla_{\boldsymbol{x}}^2 \mathcal{Y}_{\boldsymbol{m}}^{[\lambda]}(\boldsymbol{x}) = 0, \qquad \mathcal{Y}_{\boldsymbol{m}}^{[\lambda]}(\boldsymbol{x}) = r^\lambda Y_{\boldsymbol{m}}^{[\lambda]}(\hat{\boldsymbol{x}}), \qquad \nabla_{\boldsymbol{x}}^2 = \sum_{a=1}^{N} \frac{\partial^2}{\partial x_a^2}. \quad (9.104)$$

9.4.3 Schrödinger Equation for a Two-body System

An isolated n-body quantum system is invariant in the translation of space–time and the spatial rotation. After separating the motion of the center-of-mass (see §4.9), there is only one Jacobi coordinate vectors for a two-body system in N-dimensions, denoted by $\boldsymbol{R}_1 \equiv \boldsymbol{x}$ for simplicity. Remind that a factor of the square root of mass has been included in $\boldsymbol{R}_1$ (see Eq. (4.217)). The eigenfunction of angular momentum of the system has to be proportional to the spherical harmonic function $Y_{\boldsymbol{m}}^{[\lambda]}(\hat{\boldsymbol{x}})$, where $[\lambda] = [\lambda, 0, \ldots, 0]$ is a one-row Young pattern

$$\psi_{\boldsymbol{m}}^{[\lambda]}(\boldsymbol{x}) = \phi^{[\lambda]}(r) Y_{\boldsymbol{m}}^{[\lambda]}(\hat{\boldsymbol{x}}) = \left\{ r^{-\lambda} \phi^{[\lambda]}(r) \right\} \mathbf{Y}_{\boldsymbol{m}}^{[\lambda]}(\boldsymbol{x}), \tag{9.105}$$

where $\phi^{[\lambda]}(r)$ is the radial function. Since the system is spherically symmetric, one can discuss the wave function with the highest weight. The calculation is simplified by making use of the harmonic polynomials,

$$\begin{aligned}
&\nabla_{\boldsymbol{x}}^2 \left[\phi^{[\lambda]}(r) Y_{\boldsymbol{M}}^{[\lambda]}(\hat{\boldsymbol{x}}) \right] = \nabla_{\boldsymbol{x}}^2 \left[r^{-\lambda} \phi^{[\lambda]}(r) \mathbf{Y}_{\boldsymbol{M}}^{[\lambda]}(\boldsymbol{x}) \right] \\
&= \mathbf{Y}_{\boldsymbol{M}}^{[\lambda]}(\boldsymbol{x}) \nabla_{\boldsymbol{x}}^2 \left\{ r^{-\lambda} \phi^{[\lambda]}(r) \right\} + 2 \nabla_{\boldsymbol{x}} \left\{ r^{-\lambda} \phi^{[\lambda]}(r) \right\} \cdot \nabla_{\boldsymbol{x}} \mathbf{Y}_{\boldsymbol{M}}^{[\lambda]}(\boldsymbol{x}) \\
&= \mathbf{Y}_{\boldsymbol{M}}^{[\lambda]}(\boldsymbol{x}) \left\{ r^{1-N} \frac{d}{dr} r^{N-1} \frac{d}{dr} \left[r^{-\lambda} \phi^{[\lambda]}(r) \right] \right\} \\
&\quad + 2 \frac{d}{dr} \left\{ r^{-\lambda} \phi^{[\lambda]}(r) \right\} \left\{ \frac{\boldsymbol{x}}{r} \cdot \nabla_{\boldsymbol{x}} \mathbf{Y}_{\boldsymbol{M}}^{[\lambda]}(\boldsymbol{x}) \right\} \\
&= Y_{\boldsymbol{M}}^{[\lambda]}(\hat{\boldsymbol{x}}) \left\{ \frac{d^2}{dr^2} \phi^{[\lambda]}(r) + \frac{N - 2\lambda - 1}{r} \frac{d}{dr} \phi^{[\lambda]}(r) - \frac{\lambda(N - \lambda - 2)}{r^2} \phi^{[\lambda]}(r) \right\} \\
&\quad + 2 \left\{ \frac{-\lambda}{r} \phi^{[\lambda]}(r) + \frac{d}{dr} \phi^{[\lambda]}(r) \right\} \frac{\lambda}{r} Y_{\boldsymbol{M}}^{[\lambda]}(\hat{\boldsymbol{x}}) \\
&= Y_{\boldsymbol{M}}^{[\lambda]}(\hat{\boldsymbol{x}}) \left\{ \frac{d^2}{dr^2} \phi^{[\lambda]}(r) + \frac{N-1}{r} \frac{d}{dr} \phi^{[\lambda]}(r) - \frac{\lambda(N + \lambda - 2)}{r^2} \phi^{[\lambda]}(r) \right\}.
\end{aligned}$$

Substituting $\psi_{\boldsymbol{M}}^{[\lambda]}(\boldsymbol{x})$ into the Schrödinger equation in the coordinate system of the center-of-mass

$$-\frac{1}{2} \nabla_{\boldsymbol{x}}^2 \psi_{\boldsymbol{M}}^{[\lambda]}(\boldsymbol{x}) + V(r) \psi_{\boldsymbol{M}}^{[\lambda]}(\boldsymbol{x}) = E \psi_{\boldsymbol{M}}^{[\lambda]}(\boldsymbol{x}), \tag{9.106}$$

one obtains the radial equation

$$\begin{aligned}
&-\frac{1}{2} \left\{ \frac{d^2}{dr^2} \phi^{[\lambda]}(r) + \frac{N-1}{r} \frac{d}{dr} \phi^{[\lambda]}(r) - \frac{\lambda(N + \lambda - 2)}{r^2} \phi^{[\lambda]}(r) \right\} \\
&= \{ E - V(r) \} \phi^{[\lambda]}(r).
\end{aligned} \tag{9.107}$$

9.4.4 Schrödinger Equation for a Three-body System

After separating the motion of the center-of-mass (see §4.9), there are two Jacobi coordinate vectors in a three-body system. $\boldsymbol{R}_1 \equiv \boldsymbol{x}$ describes the mass-weighted separation from the first particle to the center-of-mass of the last two particles, and $\boldsymbol{R}_2 \equiv \boldsymbol{y}$ describes the mass-weighted separation of last two particles. There are $3N$ degrees of freedom for a three-body system, where N degrees of freedom describe the motion of center-of-mass, $N(N-1)/2-(N-2)(N-3)/2=2N-3$ degrees of freedom describe the global rotation of the system, and the remaining three degrees of freedom describe the internal motion. The internal variables are denoted by $\xi_1 = \boldsymbol{x}\cdot\boldsymbol{x}$, $\xi_2=\boldsymbol{y}\cdot\boldsymbol{y}$, and $\xi_3=\boldsymbol{x}\cdot\boldsymbol{y}$, which are invariant in the global rotation.

Since there are two coordinate vectors, the angular momentum states of the system are described by a two-row Young pattern $[\lambda_1,\lambda_2] = [\lambda_1,\lambda_2,0,\ldots,0]$, $\lambda_1\geq\lambda_2\geq 0$. The eigenfunction of the angular momentum is expressed as the combination of the products of $\boldsymbol{Y}^{[\omega]}_{\boldsymbol{m}}(\boldsymbol{x})$ and $\boldsymbol{Y}^{[\tau]}_{\boldsymbol{m}'}(\boldsymbol{y})$, where $[\omega]$ and $[\tau]$ are both one-row Young patterns and $[\lambda_1,\lambda_2]$ has to be contained in the reduction of $[\omega]\times[\tau]$. From the calculation given in the preceding subsection, it is convenient to express the eigenfunction of the angular momentum by the harmonic polynomial $\boldsymbol{Y}^{[\lambda]}_{\boldsymbol{m}}$ instead of the spherical harmonic function $Y^{[\lambda]}_{\boldsymbol{m}}$.

The Clebsch–Gordan (CG) series of $[\omega]\times[\tau]$ of SO(N) consists of two parts. One is calculated by the Littlewood–Richardson rule, just like the reduction of $[\omega]\times[\tau]$ of SU(N). The other comes from the trace operation between two indices belonging to two representations, respectively. Without loss of generality, one assumes $\omega\geq\tau$,

$$\begin{aligned}
&([\omega]\times[\tau])_{LR} \simeq [\omega+\tau,0]\oplus[\omega+\tau-1,1]\oplus\ldots\oplus[\omega,\tau] \\
&\qquad\simeq \bigoplus_{s=0}^{\tau}\,[\omega+\tau-s,s], \\
&[\omega]\times[\tau]\simeq([\omega]\times[\tau])_{LR}\oplus([\omega-1]\times[\tau-1])_{LR}\oplus\cdots \\
&\quad\oplus([\omega-\tau]\times[0])_{LR}\simeq\bigoplus_{s=0}^{\tau}\bigoplus_{t=0}^{\tau-s}\,[\omega+\tau-s-2t,s].
\end{aligned} \tag{9.108}$$

For SO(3), $s=0$ or 1 and $[\lambda,1]\simeq[\lambda,0]$. For SO(4), the representation with $s\neq 0$ is reduced into the direct sum of a self-dual representation and an anti-self-dual representation.

As far as the independent basis eigenfunctions of the angular momentum is concerned, one has to exclude the functions in the form $(\boldsymbol{x}\cdot\boldsymbol{y})F(\boldsymbol{x},\boldsymbol{y})$

because the factor $(\boldsymbol{x}\cdot\boldsymbol{y})$ is an internal variable and can be incorporated into the radial functions. The function, which belongs to a representation with $t \neq 0$ in the reduction (9.108), is nothing but that in the form $(\boldsymbol{x}\cdot\boldsymbol{y})F(\boldsymbol{x},\boldsymbol{y})$. Thus, the independent eigenfunction with the angular momentum $[\lambda_1, \lambda_2]$ is expressed as the combination of $\boldsymbol{Y}^{[\omega]}_{\boldsymbol{m}}(\boldsymbol{x})\boldsymbol{Y}^{[\tau]}_{\boldsymbol{m}'}(\boldsymbol{y})$, where $([\omega]\times[\tau])_{LR}$ contains the representation $[\lambda_1, \lambda_2]$, namely, $\lambda_1 + \lambda_2 = \omega + \tau$, $\lambda_2 \leq \omega$ and $\lambda_2 \leq \tau$. Denoting ω by q, one has $\tau = \lambda_1 + \lambda_2 - q$ and $\lambda_2 \leq q \leq \lambda_1$. The Young pattern $[\lambda_1, \lambda_2]$ appears in the reduction,

$$([q]\times[\lambda_1+\lambda_2-q])_{LR} \simeq \bigoplus_{s=0}^{\min\{q,(\lambda_1+\lambda_2-q)\}} [\lambda_1+\lambda_2-s, s]. \tag{9.109}$$

The number of the independent basis eigenfunctions of the angular momentum $[\lambda_1, \lambda_2]$ is equal to $(\lambda_1 - \lambda_2 + 1)$. Each independent basis eigenfunction is related to a tableau of $[\lambda_1, \lambda_2]$ where q boxes in the left of the first row are filled with x and the remaining boxes are filled with y. For example, for $[\lambda_1, \lambda_2] = [8, 3]$ and $q = 5$, the tableau is

x	x	x	x	x	y	y	y
y	y	y					

(9.110)

Fortunately, due to the spherical symmetry of the system, only the wave function with the highest weight $\boldsymbol{M} = (\lambda_1 - \lambda_2, \lambda_2, 0, \ldots, 0)$ is needed in the derivation of the radial equation and the normalization does not matter. The wave function with the highest weight is denoted by $Q^{[\lambda_1,\lambda_2]}_q(\boldsymbol{x},\boldsymbol{y})$ satisfying

$$\begin{aligned}
&H_1(L)Q^{[\lambda_1,\lambda_2]}_q(\boldsymbol{x},\boldsymbol{y}) = (\lambda_1-\lambda_2)Q^{[\lambda_1,\lambda_2]}_q(\boldsymbol{x},\boldsymbol{y}),\\
&H_2(L)Q^{[\lambda_1,\lambda_2]}_q(\boldsymbol{x},\boldsymbol{y}) = \lambda_2 Q^{[\lambda_1,\lambda_2]}_q(\boldsymbol{x},\boldsymbol{y}),\\
&H_\nu(L)Q^{[\lambda_1,\lambda_2]}_q(\boldsymbol{x},\boldsymbol{y}) = E_\mu(L)Q^{[\lambda_1,\lambda_2]}_q(\boldsymbol{x},\boldsymbol{y}) = 0,\\
&3 \leq \nu \leq \ell, \qquad 1 \leq \mu \leq \ell.
\end{aligned} \tag{9.111}$$

The partners of $Q^{[\lambda_1,\lambda_2]}_q(\boldsymbol{x},\boldsymbol{y})$ can be calculated, if necessary, by the lowering operators $F_\mu(L)$ where $L_{ab} = L_{ab}(\boldsymbol{x}) + L_{ab}(\boldsymbol{y})$. Being the highest weight state, $Q^{[\lambda_1,\lambda_2]}_q(\boldsymbol{x},\boldsymbol{y})$ is proportional to a product where each block in the first row of the tableau (9.110) corresponds to a factor $\boldsymbol{Y}^{[1,0,\ldots,0]}_{(1,0,\ldots,0)}$ and each block in the second row corresponds to a factor $\boldsymbol{Y}^{[1,0,\ldots,0]}_{(\overline{1},1,0,\ldots,0)}$. Two boxes in the same column have to correspond to an antisymmetric combination, namely,

$$Q_q^{[\lambda_1,\lambda_2]}(\boldsymbol{x},\boldsymbol{y}) \propto \left[\boldsymbol{Y}^{[1,0,\ldots,0]}_{(1,0,\ldots,0)}(\boldsymbol{x})\right]^{q-\lambda_2} \left[\boldsymbol{Y}^{[1,0,\ldots,0]}_{(1,0,\ldots,0)}(\boldsymbol{y})\right]^{\lambda_1-q}$$
$$\cdot \left\{\boldsymbol{Y}^{[1,0,\ldots,0]}_{(1,0,\ldots,0)}(\boldsymbol{x})\boldsymbol{Y}^{[1,0,\ldots,0]}_{(\overline{1},1,0,\ldots,0)}(\boldsymbol{y}) - \boldsymbol{Y}^{[1,0,\ldots,0]}_{(\overline{1},1,0,\ldots,0)}(\boldsymbol{x})\boldsymbol{Y}^{[1,0,\ldots,0]}_{(1,0,\ldots,0)}(\boldsymbol{y})\right\}^{\lambda_2},$$
$$\boldsymbol{Y}^{[1,0,\ldots,0]}_{(1,0,\ldots,0)}(\boldsymbol{x}) \propto X_1 = x_1 + ix_2, \qquad \boldsymbol{Y}^{[1,0,\ldots,0]}_{(\overline{1},1,0,\ldots,0)}(\boldsymbol{x}) \propto X_2 = x_3 + ix_4,$$
$$\boldsymbol{Y}^{[1,0,\ldots,0]}_{(1,0,\ldots,0)}(\boldsymbol{y}) \propto Y_1 = y_1 + iy_2, \qquad \boldsymbol{Y}^{[1,0,\ldots,0]}_{(\overline{1},1,0,\ldots,0)}(\boldsymbol{y}) \propto Y_2 = y_3 + iy_4.$$

Introducing a factor $[(q-\lambda_2)!(\lambda_1-q)!]^{-1}$ only for simplification in the derivation of the radial equation, one obtains

$$Q_q^{[\lambda_1,\lambda_2]}(\boldsymbol{x},\boldsymbol{y}) = \frac{X_1^{q-\lambda_2}Y_1^{\lambda_1-q}}{(q-\lambda_2)!(\lambda_1-q)!}\left(X_1Y_2 - X_2Y_1\right)^{\lambda_2}. \tag{9.112}$$

$Q_q^{[\lambda_1,\lambda_2]}(\boldsymbol{x},\boldsymbol{y})$ is a homogeneous polynomial of degree q and degree $(\lambda_1+\lambda_2-q)$ with respect to the components of $\boldsymbol{x}$ and $\boldsymbol{y}$, respectively, and satisfies Eq. (9.111) and the Laplace equations

$$\nabla_{\boldsymbol{x}}^2 Q_q^{[\lambda_1,\lambda_2]}(\boldsymbol{x},\boldsymbol{y}) = \nabla_{\boldsymbol{y}}^2 Q_q^{[\lambda_1,\lambda_2]}(\boldsymbol{x},\boldsymbol{y}) = \nabla_{\boldsymbol{x}}\cdot\nabla_{\boldsymbol{y}} Q_q^{[\lambda_1,\lambda_2]}(\boldsymbol{x},\boldsymbol{y}) = 0. \tag{9.113}$$

For SO(3), λ_2 in Eq. (9.112) has to be equal to 0 or 1, and $X_2 = x_3$, $Y_2 = y_3$. For SO(4), the representation $[\lambda_1,\lambda_2]$ is reduced into the direct sum of a self-dual representation $[(+)\lambda_1,\lambda_2]$ and an anti-self-dual representation and $[(-)\lambda_1,\lambda_2]$. $Q_q^{[(+)\lambda_1,\lambda_2]}(\boldsymbol{x},\boldsymbol{y})$ has the same form as Eq. (9.112), but $Q_q^{[(-)\lambda_1,\lambda_2]}(\boldsymbol{x},\boldsymbol{y})$ is obtained from Eq. (9.112) by replacing X_2 with $X_3 = x_3 - ix_4$ and Y_2 with $Y_3 = y_3 - iy_4$.

The Schrödinger equation for a three-body system in the coordinate system of the center-of-mass is

$$-\frac{1}{2}\left\{\nabla_{\boldsymbol{x}}^2 + \nabla_{\boldsymbol{y}}^2\right\}\psi_{\boldsymbol{M}}^{[\lambda_1,\lambda_2]}(\boldsymbol{x},\boldsymbol{y}) = \left[E - V(\xi_1,\xi_2,\xi_3)\right]\psi_{\boldsymbol{M}}^{[\lambda_1,\lambda_2]}(\boldsymbol{x},\boldsymbol{y}). \tag{9.114}$$

Let

$$\psi_{\boldsymbol{M}}^{[\lambda_1,\lambda_2]}(\boldsymbol{x},\boldsymbol{y}) = \sum_{q=\lambda_2}^{\lambda_1} \phi_q^{[\lambda_1,\lambda_2]}(\xi_1,\xi_2,\xi_3)Q_q^{[\lambda_1,\lambda_2]}(\boldsymbol{x},\boldsymbol{y}). \tag{9.115}$$

The action of the Laplace operator on $\psi_{\boldsymbol{M}}^{[\lambda_1,\lambda_2]}(\boldsymbol{x},\boldsymbol{y})$ is divided into three parts. The first part is its action on the radial functions $\phi_q^{[\lambda_1,\lambda_2]}$, which can be calculated by replacement of variables (see Eq. (9.117)). The second part is its action on the basis eigenfunctions $Q_q^{[\lambda_1,\lambda_2]}(\boldsymbol{x},\boldsymbol{y})$ which is vanishing.

The third part is its mixed action

$$2\left\{\left(\frac{\partial}{\partial \xi_1}\phi_q^{[\lambda_1,\lambda_2]}\right)2\boldsymbol{x}+\left(\frac{\partial}{\partial \xi_3}\phi_q^{[\lambda_1,\lambda_2]}\right)\boldsymbol{y}\right\}\cdot\nabla_{\boldsymbol{x}}Q_q^{[\lambda_1,\lambda_2]}(\boldsymbol{x},\boldsymbol{y})$$
$$+2\left\{\left(\frac{\partial}{\partial \xi_2}\phi_q^{[\lambda_1,\lambda_2]}\right)2\boldsymbol{y}+\left(\frac{\partial}{\partial \xi_3}\phi_q^{[\lambda_1,\lambda_2]}\right)\boldsymbol{x}\right\}\cdot\nabla_{\boldsymbol{y}}Q_q^{[\lambda_1,\lambda_2]}(\boldsymbol{x},\boldsymbol{y}).$$

From Eq. (9.112) one has

$$\begin{aligned}
\boldsymbol{x}\cdot\nabla_{\boldsymbol{x}}Q_q^{[\lambda_1,\lambda_2]}&=qQ_q^{[\lambda_1,\lambda_2]},\\
\boldsymbol{y}\cdot\nabla_{\boldsymbol{y}}Q_q^{[\lambda_1,\lambda_2]}&=(\lambda_1+\lambda_2-q)Q_q^{[\lambda_1,\lambda_2]},\\
\boldsymbol{y}\cdot\nabla_{\boldsymbol{x}}Q_q^{[\lambda_1,\lambda_2]}&=(\lambda_1-q+1)Q_{q-1}^{[\lambda_1,\lambda_2]},\\
\boldsymbol{x}\cdot\nabla_{\boldsymbol{y}}Q_q^{[\lambda_1,\lambda_2]}&=(q-\lambda_2+1)Q_{q+1}^{[\lambda_1,\lambda_2]}.
\end{aligned}\tag{9.116}$$

Thus, the general radial equation for the radial function $\phi_q^{[\lambda_1,\lambda_2]}$ is (see [Gu et al. (2001c)])

$$\begin{aligned}
&\left\{\nabla_{\boldsymbol{x}}^2+\nabla_{\boldsymbol{y}}^2\right\}\phi_q^{[\lambda_1,\lambda_2]}+4q\frac{\partial}{\partial\xi_1}\phi_q^{[\lambda_1,\lambda_2]}+4(\lambda_1+\lambda_2-q)\frac{\partial}{\partial\xi_2}\phi_q^{[\lambda_1,\lambda_2]}\\
&\quad+2(\lambda_1-q)\frac{\partial}{\partial\xi_3}\phi_{q+1}^{[\lambda_1,\lambda_2]}+2(q-\lambda_2)\frac{\partial}{\partial\xi_3}\phi_{q-1}^{[\lambda_1,\lambda_2]}\\
&\quad=-2\left(E-V\right)\phi_q^{[\lambda_1,\lambda_2]},\\
&\left\{\nabla_{\boldsymbol{x}}^2+\nabla_{\boldsymbol{y}}^2\right\}\phi_q^{[\lambda_1,\lambda_2]}=\left\{4\xi_1\frac{\partial^2}{\partial\xi_1^2}+4\xi_2\frac{\partial^2}{\partial\xi_2^2}+2N\left(\frac{\partial}{\partial\xi_1}+\frac{\partial}{\partial\xi_2}\right)\right.\\
&\quad\left.+\left(\xi_1+\xi_2\right)\frac{\partial^2}{\partial\xi_3^2}+4\xi_3\left(\frac{\partial^2}{\partial\xi_1\partial\xi_3}+\frac{\partial^2}{\partial\xi_2\partial\xi_3}\right)\right\}\phi_q^{[\lambda_1,\lambda_2]}.
\end{aligned}\tag{9.117}$$

This method can be generalized to a quantum multiple-body system (see [Gu et al. (2003b)]).

9.4.5 *Dirac Equation in $(N+1)$-dimensional Space–time*

The transformation between two inertial systems in four-dimensional space–time is the Lorentz transformation. There are two common used sets of coordinates and metric tensors for the four-dimensional space–time. One is (x_0, x_1, x_2, x_3) with the Minkowski metric tensor η =diag$(1,-1,-1,-1)$ where $x_0=ct$ (see [Bjorken and Drell (1964)]). The other is (x_1, x_2, x_3, x_4) with the Euclidian metric tensor $\delta_{\mu\nu}$ where $x_4=ict$ (see [Schiff (1968); Marshak et al. (1969)]). The Lorentz transformation matrices for two sets are related by a similarity transformation,

$$\mathcal{A}^T\eta\mathcal{A} = \eta, \qquad A^T A = \mathbf{1}, \qquad X^{-1}\mathcal{A}X = A,$$
$$X = \begin{pmatrix} 0 & 0 & 0 & -i \\ 1 & 0 & 0 & 0 \\ 0 & 1 & 0 & 0 \\ 0 & 0 & 1 & 0 \end{pmatrix}, \qquad X^{-1} = \begin{pmatrix} 0 & 1 & 0 & 0 \\ 0 & 0 & 1 & 0 \\ 0 & 0 & 0 & 1 \\ i & 0 & 0 & 0 \end{pmatrix}. \tag{9.118}$$

In this textbook we adopt the second set and generalize it to the $(N+1)$-dimensional space–time. The formula for the first set can be found in [Gu et al. (2002)].

The Dirac equation in $(D+1)$-dimensional space–time can be expressed as (see [Schiff (1968)])

$$\sum_{\mu=1}^{N+1} \gamma_\mu \left(\frac{\partial}{\partial x_\mu} - ieA_\mu\right) \mathbf{\Psi}(\boldsymbol{x}, t) + M\mathbf{\Psi}(\boldsymbol{x}, t) = 0, \tag{9.119}$$

where M is the mass of the particle, and $(N+1)$ matrices γ_μ satisfy the anticommutative relation:

$$\gamma_\mu\gamma_\nu + \gamma_\nu\gamma_\mu = 2\delta_{\mu\nu}\mathbf{1}. \tag{9.120}$$

For simplicity, the natural units $\hbar = c = 1$ are employed in this subsection. Discuss the special case where only the time component of A_{N+1} is nonvanishing and spherically symmetric:

$$eA_{N+1} = iV(r), \qquad A_a = 0 \qquad \text{when } \ 1 \leq a \leq N. \tag{9.121}$$

The Hamiltonian $H(\boldsymbol{x})$ of the system is expressed as

$$\begin{aligned} &i\frac{\partial}{\partial t}\mathbf{\Psi}(\boldsymbol{x}, t) = H(\boldsymbol{x})\mathbf{\Psi}(\boldsymbol{x}, t), \\ &H(\boldsymbol{x}) = \sum_{a=1}^{N} \gamma_{N+1}\gamma_a \frac{\partial}{\partial x_a} + V(r) + \gamma_{N+1}M. \end{aligned} \tag{9.122}$$

The orbital angular momentum operator L_{ab} is given in Eq. (9.98). The spinor operator S_{ab} is given in Eq. (9.65). The total angular momentum operator is $J_{ab} = L_{ab} + S_{ab}$. There are three Casimir operators of order 2 for the total, orbital, and spinor wave functions, respectively,

$$J^2 = \sum_{a<b=2}^{N} J_{ab}^2, \qquad L^2 = \sum_{a<b=2}^{N} L_{ab}^2, \qquad S^2 = \sum_{a<b=2}^{N} S_{ab}^2. \tag{9.123}$$

Due to the spherical symmetry, J_{ab} is commutable with the Hamiltonian $H(\boldsymbol{x})$. There is another conservative operator κ which is commutable with both J_{ab} and $H(\boldsymbol{x})$ (see [Schiff (1968)])

$$\begin{aligned}\kappa &= -i\gamma_{N+1}\sum_{a<b=2}^{N}\gamma_a\gamma_b L_{ab} + \frac{1}{2}\gamma_{N+1}(N-1)\\ &= \gamma_{N+1}\left\{J^2 - L^2 - S^2 + (N-1)/2\right\}.\end{aligned} \tag{9.124}$$

The set of mutually commutable operators consists of $H(\boldsymbol{x})$, J^2, κ, S^2, and the Chevalley bases $H_\mu(J)$. Their common eigenfunctions are calculated from the products of the spherical harmonic functions $Y^{[\lambda]}_{\boldsymbol{m}}(\hat{\boldsymbol{x}})$ and the fundamental basis spinor $\chi(\boldsymbol{m}_s)$ by the Clebsch–Gordan coefficients, where $[\lambda] = [\lambda, 0, \ldots, 0]$. Due to the spherical symmetry, the eigenfunction with the highest weight is only interesting for deriving the radial equation. Introduce $(2\ell+1)$ unitary matrices β_a, which are 2^ℓ-dimensional and satisfy

$$\beta_a\beta_b + \beta_b\beta_a = 2\delta_{ab}\mathbf{1}. \tag{9.125}$$

The concrete forms of β_a are the same as those γ_a given in Eq. (9.49).

(a) *The case of* $N = 2\ell + 1$
Let

$$\gamma_{N+1} = \sigma_3 \times \mathbf{1}, \qquad \gamma_a = \sigma_2 \times \beta_a, \qquad 1 \le a \le 2\ell + 1. \tag{9.126}$$

The spinor operator S_{ab} and the κ operator become the block matrices

$$\begin{aligned}&S_{ab} = \mathbf{1} \times \overline{S}_{ab}, \qquad &&\overline{S}_{ab} = -i\left(\beta_a\beta_b - \beta_b\beta_a\right)/4,\\ &\kappa = \sigma_3 \times \overline{\kappa}, &&\overline{\kappa} = -i\sum_{a<b=2}^{N}\beta_a\beta_b L_{ab} + \frac{N-1}{2}.\end{aligned} \tag{9.127}$$

They are commutable with γ_{N+1}. The relation between S_{ab} and $\overline{S}_{ab}$ is similar to that between the spinor operators for the Dirac spinors and for the Pauli spinors. At the level of the Pauli spinors, the fundamental spinor $\chi[\boldsymbol{m}]$ belongs to the fundamental spinor representation $[s]$ with the highest weight $(0, \ldots, 0, 1)$. Due to Eq. (9.92), there are two sets of the wave functions belonging to the representation $[j] \equiv [s, \lambda, 0, \ldots, 0]$ of $SO(2\ell+1)$ whose highest weight is $\boldsymbol{M} = (\lambda, 0, \ldots, 0, 1)$. Calculating from Eq. (7.137), one has $C_2([j]) = \lambda(\lambda+2\ell) + \ell(2\ell+1)/4$, $C_2([\lambda]) = \lambda(\lambda+2\ell-1)$, $C_2([s]) =$

$\ell(2\ell+1)/4$, and

$$\begin{aligned} &C_2([j]) - C_2([\lambda]) - C_2([s]) + \ell = \lambda + \ell = |K|, \\ &C_2([j]) - C_2([\lambda+1]) - C_2([s]) + \ell = -\lambda - \ell = -|K|. \end{aligned} \tag{9.128}$$

Thus, two eigenfunctions of the total angular momentum with the highest weight $\boldsymbol{M}$ of $[j]$ and the different eigenvalues of $\overline{\kappa}$ are calculated by the Clebsch–Gordan coefficients (see Prob. 14 of Chap. 9 in [Ma and Gu (2004)], also see Eq. (4.189))

$$\begin{aligned} \phi_{|K|,[j]}(\hat{\boldsymbol{x}}) &= Y^{[\lambda]}_{(\lambda,0,\ldots,0)}(\hat{\boldsymbol{x}})\chi[(0,\ldots,0,1)] \\ &= C_{(2\ell+1),\lambda}\left\{\frac{(-1)^{\ell}(x_1+ix_2)}{r\sqrt{2}}\right\}^{\lambda}\chi[(0,\ldots,0,1)], \\ \phi_{-|K|,[j]}(\hat{\boldsymbol{x}}) &= \sum_{\boldsymbol{m}} Y^{[\lambda+1]}_{\boldsymbol{m}}(\hat{\boldsymbol{x}})\chi[\boldsymbol{M}-\boldsymbol{m}]C^{[s],[\lambda+1]}_{\boldsymbol{M}-\boldsymbol{m},\boldsymbol{m},[j],\boldsymbol{M}} \\ &= \frac{C_{(2\ell+1),(\lambda+1)}(-1)^{\ell\lambda}\sqrt{\lambda+1}}{(r)^{\lambda+1}2^{\lambda/2}\sqrt{2\ell+2\lambda+1}}(x_1+ix_2)^{\lambda} \\ &\quad\times\left\{x_{2\ell+1}\chi[(0,\ldots,0,1)]+(x_{2\ell-1}+ix_{2\ell})\chi[(0,\ldots,0,1,\overline{1})]\right. \\ &\quad+(x_{2\ell-3}+ix_{2\ell-2})\chi[(0,\ldots,0,1,\overline{1},1)]+\ldots \\ &\quad\left.+(x_3+ix_4)\chi[(1,\overline{1},0,\ldots,0,1)]+(x_1+ix_2)\chi[(\overline{1},0,\ldots,0,1)]\right\}, \end{aligned}$$

$$\overline{\kappa}\phi_{K,[j]}(\hat{\boldsymbol{x}}) = K\phi_{K,[j]}(\hat{\boldsymbol{x}}). \tag{9.129}$$

Both $\phi_{\pm|K|,[j]}(\hat{\boldsymbol{x}})$ are annihilated by every raising operator $E_{\mu}(J)$. Remind that two coefficients in $\phi_{\pm|K|,[j]}(\hat{\boldsymbol{x}})$ are the same except for a factor r,

$$C_{N,(\lambda+1)}\sqrt{\frac{\lambda+1}{N+2\lambda}} = C_{N,\lambda}. \tag{9.130}$$

Introducing two operators

$$\boldsymbol{\beta}\cdot\hat{\boldsymbol{x}} = r^{-1}\sum_{a=1}^{N}\beta_a x_a, \qquad \boldsymbol{\beta}\cdot\bigtriangledown = \sum_{a=1}^{N}\beta_a\frac{\partial}{\partial x_a}, \tag{9.131}$$

one has

$$(\boldsymbol{\beta}\cdot\hat{\boldsymbol{x}})^2 = \mathbf{1}, \qquad (\boldsymbol{\beta}\cdot\hat{\boldsymbol{x}})\,\phi_{K,[j]}(\hat{\boldsymbol{x}}) = \phi_{-K,[j]}(\hat{\boldsymbol{x}}),$$

$$\begin{aligned}
&(\boldsymbol{\beta}\cdot\bigtriangledown)\, r^{-(N-1)/2} f(r)\phi_{K,[j]}(\hat{\boldsymbol{x}}) \\
&= (\boldsymbol{\beta}\cdot\hat{\boldsymbol{x}})^2 (\boldsymbol{\beta}\cdot\bigtriangledown)\, r^{-(N-1)/2} f(r)\phi_{K,[j]}(\hat{\boldsymbol{x}}) \\
&= (\boldsymbol{\beta}\cdot\hat{\boldsymbol{x}}) \left[\frac{\partial}{\partial r} + \frac{i}{r}\sum_{a<b} \beta_a\beta_b L_{ab}\right] r^{-(N-1)/2} f(r)\phi_{K,[j]}(\hat{\boldsymbol{x}}) \\
&= r^{-(N-1)/2}\left[\frac{df(r)}{dr} - \frac{Kf(r)}{r}\right]\phi_{-K,[j]}(\hat{\boldsymbol{x}}).
\end{aligned} \tag{9.132}$$

At the level of the Dirac spinors, the wave function $\boldsymbol{\Psi}_{K,[j]}(\boldsymbol{x})$ with the highest weight in the irreducible representation $[j]$ and with the different eigenvalues K of κ can be expressed as

$$\begin{aligned}
&\boldsymbol{\Psi}_{K,[j]}(\boldsymbol{x},t) = r^{-\ell}e^{-iEt}\begin{pmatrix} F_K(r)\phi_{K,[j]}(\hat{\boldsymbol{x}}) \\ iG_K(r)\phi_{-K,[j]}(\hat{\boldsymbol{x}})\end{pmatrix}, \\
&\kappa\Psi_{K,(j)}(\boldsymbol{x}) = K\Psi_{K,(j)}(\boldsymbol{x}), \qquad K = \pm(\lambda+\ell).
\end{aligned} \tag{9.133}$$

Its partners can be calculated from it by the lowering operators $F_\mu(J)$. The factor i in the radial function $iG_K(r)$ can be removed by replacing σ_2 in Eq. (9.126) with σ_1.

Substituting $\boldsymbol{\Psi}_{K,[j]}(\boldsymbol{x},t)$ into the Dirac equation (9.122) one obtains the radial equation

$$\begin{aligned}
&\frac{dG_K(r)}{dr} + \frac{K}{r}G_K(r) = [E - V(r) - M]F_K(r), \\
&-\frac{dF_K(r)}{dr} + \frac{K}{r}F_K(r) = [E - V(r) + M]G_K(r).
\end{aligned} \tag{9.134}$$

Equation (9.134) holds for SO(3) although the algebra of SO(3) is A_1.

(b) *The case of $N = 2\ell$ with $\ell > 2$*

The spinor representation of SO(2ℓ) is reduced into two fundamental spinor representations $[\pm s]$ with the highest weights $(0,\ldots,0,1)$ and $(0,\ldots,0,1,0)$, respectively. Let

$$\gamma_{N+1} = \beta_{2\ell+1}, \qquad \gamma_a = \beta_a, \qquad 1 \le a \le 2\ell. \tag{9.135}$$

γ_{N+1} is a diagonal matrix where half of the diagonal elements are equal to $+1$ and the remaining to -1. Because the spinor operator S_{ab} and the operator κ are commutable with γ_{N+1}, each of them becomes a direct sum of two matrices, acting on the fundamental spinors $\chi_\pm(\boldsymbol{m})$, respectively,

$$\gamma_{N+1}\chi_+(\boldsymbol{m}) = \chi_+(\boldsymbol{m}), \qquad \gamma_{N+1}\chi_-(\boldsymbol{m}) = -\chi_-(\boldsymbol{m}). \tag{9.136}$$

For a given total angular momentum there are two kinds of representations of SO(2ℓ) at the Pauli spinor level, $[j_\pm] = [\pm s, \lambda, 0, \ldots, 0]$. Because of Eq. (9.92), two sets of the wave functions for each of representations $[j_\pm]$ are calculated to have different eigenvalues of $\overline{\kappa}$. From Eq. (7.137) one calculates that $C_2([\pm s]) = \ell(2\ell-1)/4$, $C_2([\lambda]) = \lambda(\lambda+2\ell-2)$, $C_2([j_\pm]) = \lambda(\lambda+2\ell-1)+\ell(2\ell-1)/4$, and

$$\begin{aligned}
&C_2([j_\pm]) - C_2([\lambda]) - C_2([\pm s]) + \ell - 1/2 = \lambda + \ell - 1/2 = |K|, \\
&C_2([j_\pm]) - C_2([\lambda+1]) - C_2([\pm s]) + \ell - 1/2 = -\lambda - \ell + 1/2 = -|K|.
\end{aligned} \tag{9.137}$$

Two wave functions with the highest weight $\boldsymbol{M}_+ = (\lambda, 0 \ldots, 0, 1)$ of $[j_+]$ are

$$\begin{aligned}
\phi_{|K|,[j_+]}(\hat{\boldsymbol{x}}) &= Y^{[\lambda]}_{(\lambda,0,\ldots,0)}(\hat{\boldsymbol{x}})\chi_+[(0,\ldots,0,1)] \\
&= C_{(2\ell),\lambda}\left\{\frac{(-1)^{\ell-1}(x_1+ix_2)}{r\sqrt{2}}\right\}^\lambda \chi_+[(0,\ldots,0,1)], \\
\phi_{-|K|,[j_+]}(\hat{\boldsymbol{x}}) &= \sum_{\boldsymbol{m}} Y^{[\lambda+1]}_{\boldsymbol{m}}(\hat{\boldsymbol{x}})\chi_-[\boldsymbol{M}_+ - \boldsymbol{m}]C^{[-s],[\lambda+1]}_{\boldsymbol{M}_+-\boldsymbol{m},\boldsymbol{m},[j_+],\boldsymbol{M}_+} \\
&= C_{(2\ell),(\lambda+1)}\frac{(-1)^{(\ell-1)\lambda}\sqrt{\lambda+1}}{(r\sqrt{2})^{\lambda+1}\sqrt{\ell+\lambda}}(x_1+ix_2)^\lambda \\
&\quad \times \Big\{(x_{2\ell-1}+ix_{2\ell})\chi_-[(0,\ldots,0,1,0)] \\
&\quad + (x_{2\ell-3}+ix_{2\ell-2})\chi_-[(0,\ldots,0,1,\overline{1},0)] \\
&\quad + (x_{2\ell-5}+ix_{2\ell-4})\chi_-[(0,\ldots,0,1,\overline{1},0,1)] + \ldots \\
&\quad + (x_3+ix_4)\chi_-[(1,\overline{1},0,\ldots,0,1)] \\
&\quad + (x_1+ix_2)\chi_-[(\overline{1},0,\ldots,0,1)]\Big\}.
\end{aligned} \tag{9.138}$$

Two wave functions with the highest weight $\boldsymbol{M}_- = (\lambda, 0 \ldots, 0, 1, 0)$ of $[j_-]$ are

$$\begin{aligned}
\phi_{|K|,[j_-]}(\hat{\boldsymbol{x}}) &= Y^{[\lambda]}_{(\lambda,0,\ldots,0)}(\hat{\boldsymbol{x}})\chi_-[(0,\ldots,0,1,0)] \\
&= C_{(2\ell),\lambda}\left\{\frac{(-1)^{\ell-1}(x_1+ix_2)}{r\sqrt{2}}\right\}^\lambda \chi_-[(0,\ldots,0,1,0)], \\
\phi_{-|K|,[j_-]}(\hat{\boldsymbol{x}}) &= \sum_{\boldsymbol{m}} Y^{[\lambda+1]}_{\boldsymbol{m}}(\hat{\boldsymbol{x}})\chi_+[\boldsymbol{M}_- - \boldsymbol{m}]C^{[+s],[\lambda+1]}_{\boldsymbol{M}_--\boldsymbol{m},\boldsymbol{m},[j_-],\boldsymbol{M}_-} \\
&= \frac{C_{(2\ell),(\lambda+1)}(-1)^{(\ell-1)\lambda}\sqrt{\lambda+1}}{(r\sqrt{2})^{\lambda+1}\sqrt{\ell+\lambda}}(x_1+ix_2)^\lambda \\
&\quad \times \Big\{(x_{2\ell-1}-ix_{2\ell})\chi_+[(0,\ldots,0,1)]
\end{aligned}$$

$$
\begin{aligned}
&+ (x_{2\ell-3} + ix_{2\ell-2})\chi_+[(0,\ldots,0,1,0,\overline{1})] \\
&+ (x_{2\ell-5} + ix_{2\ell-4})\chi_+[(0,\ldots,0,1,\overline{1},1,0)] + \ldots \\
&+ (x_3 + ix_4)\chi_+[(1,\overline{1},0,\ldots,0,1,0)] \\
&+ (x_1 + ix_2)\chi_+[(\overline{1},0,\ldots,0,1,0)]\Big\}.
\end{aligned} \tag{9.139}
$$

The Clebsch–Gordan coefficients in Eqs. (9.138) and (9.139) are calculated by the condition that the highest weight state is annihilated by every raising operators $E_\mu(J)$. The coefficients can be found in Prob. 14 of Chap. 9 of [Ma and Gu (2004)]. Due to Eq. (9.130), Eq. (9.132) still holds for the case with $N = 2\ell$. At the level of the Dirac spinors, the wave functions $\boldsymbol{\Psi}_{K,[j]}(\boldsymbol{x})$ with the highest weights in the irreducible representation $[j_\pm]$ and with the different eigenvalues K of κ can be expressed as

$$
\begin{aligned}
&\boldsymbol{\Psi}_{|K|,[j_+]}(\hat{\boldsymbol{x}}) = r^{-\ell+1/2}e^{-iEt}\left\{F_{|K|}(r)\phi_{|K|,[j_+]}(\hat{\boldsymbol{x}}) + G_{|K|}(r)\phi_{-|K|,[j_+]}(\hat{\boldsymbol{x}})\right\}, \\
&\boldsymbol{\Psi}_{-|K|,[j_-]}(\hat{\boldsymbol{x}}) \\
&\quad = r^{-\ell+1/2}e^{-iEt}\left\{F_{-|K|}(r)\phi_{-|K|,([j_-]}(\hat{\boldsymbol{x}}) + G_{-|K|}(r)\phi_{|K|,[j_-]}(\hat{\boldsymbol{x}})\right\}, \\
&\kappa\boldsymbol{\Psi}_{\pm|K|,[j_\pm]}(\boldsymbol{x}) = \pm|K|\Psi_{\pm|K|,[j_\pm]}(\boldsymbol{x}), \qquad |K| = \lambda + \ell - 1/2.
\end{aligned} \tag{9.140}
$$

Their partners can be calculated from them by the lowering operators $F_\mu(J)$.

Substituting $\boldsymbol{\Psi}_{K[j_\pm]}(\boldsymbol{x})$ into the Dirac equation (9.122) one obtains the radial equations, which are the same as Eq. (9.134). When $D = 4$, SO(4) $\sim$ SU(2)$\times$SU(2), and the representations $[j_\pm]$ belong to two different SU(2) groups, respectively. When $D = 2$, the SO(2) group is an Abelian group, and the radial equation (9.134) holds for SO(2) but $K = \pm 1/2$, $\pm 3/2$, ... [Dong et al. (2003); Dong et al. (1998c)]. The radial equations (9.134) are the ordinary differential equations for the variable r, and can be solved easily. Their solutions for the Coulomb potential are given in [Gu et al. (2002)]. The Levinson theorem for the Dirac equation in $(N+1)$-dimensional space–time is given in [Gu et al. (2003e)].

9.5 The SO(4) Group and the Lorentz Group

The Lorentz group is an important symmetric group in physics. The Lorentz group is a noncompact Lie group, whose representations can be calculated from those of the compact Lie group SO(4). The study of the Lorentz group provides a typical example in mathematics for studying the noncompact Lie groups.

9.5.1 *Irreducible Representations of SO(4)*

The Lie algebra of SO(4) is $D_2 \approx A_1 \oplus A_1$. In this subsection the SO(4) group is evidently decomposed into a direct product of two SU(2) groups. Based on this decomposition, the parameters of SO(4) are chosen and the irreducible representations of SO(4) are calculated analytically.

Six generators of SO(4) in its self-representation are given in Eq. (7.85). Through the suitable combinations one has

$$T_1^{(\pm)} = \frac{1}{2}\left(T_{23} \pm T_{14}\right) = \frac{1}{2}\begin{pmatrix} 0 & 0 & 0 & \mp i \\ 0 & 0 & -i & 0 \\ 0 & i & 0 & 0 \\ \pm i & 0 & 0 & 0 \end{pmatrix} \tag{9.141}$$

and those by the cycle (1 2 3). $T_a^{(\pm)}$ can be expressed as

$$\begin{aligned} T_1^{(+)} &= \frac{1}{2}\sigma_2 \times \sigma_1, \qquad & T_2^{(+)} &= \frac{-1}{2}\sigma_2 \times \sigma_3, \qquad & T_3^{(+)} &= \frac{1}{2}\mathbf{1}_2 \times \sigma_2, \\ T_1^{(-)} &= \frac{-1}{2}\sigma_1 \times \sigma_2, \qquad & T_2^{(-)} &= \frac{-1}{2}\sigma_2 \times \mathbf{1}_2, \qquad & T_3^{(-)} &= \frac{1}{2}\sigma_3 \times \sigma_2. \end{aligned} \tag{9.142}$$

The generators are divided into two sets, each of which satisfies the commutative relations of generators in SU(2),

$$\left[T_a^{(\pm)},\, T_b^{(\pm)}\right] = i\sum_{c=1}^{3} \epsilon_{abc} T_c^{(\pm)}, \qquad \left[T_a^{(+)},\, T_b^{(-)}\right] = 0. \tag{9.143}$$

This property becomes clearer if one makes a similarity transformation N,

$$\begin{gathered} N^{-1}T_a^{(+)}N = (\sigma_a/2) \times \mathbf{1}_2, \qquad N^{-1}T_a^{(-)}N = \mathbf{1}_2 \times (\sigma_a/2), \\ N = \frac{1}{\sqrt{2}}\begin{pmatrix} -1 & 0 & 0 & 1 \\ -i & 0 & 0 & -i \\ 0 & 1 & 1 & 0 \\ 0 & i & -i & 0 \end{pmatrix}. \end{gathered} \tag{9.144}$$

Note that Eq. (9.22) is a generalization of Eq. (9.144). An arbitrary element R in SO(4) becomes

$$\begin{aligned} R = & \exp\left(-i \sum_{a<b}^{4} \omega_{ab}T_{ab}\right) \\ = & \exp\left\{-i \sum_{a=1}^{3} \left(\omega_a^{(+)}T_a^{(+)} + \omega_a^{(-)}T_a^{(-)}\right)\right\} \\ = & \exp\left\{-i\omega^{(+)}\hat{\boldsymbol{n}}^{(+)} \cdot \boldsymbol{T}^{(+)}\right\} \exp\left\{-i\omega^{(-)}\hat{\boldsymbol{n}}^{(-)} \cdot \boldsymbol{T}^{(-)}\right\} \\ = & N\left\{u(\hat{\boldsymbol{n}}^{(+)}, \omega^{(+)}) \times u(\hat{\boldsymbol{n}}^{(-)}, \omega^{(-)})\right\} N^{-1}, \end{aligned} \tag{9.145}$$

where

$$\begin{aligned} \omega_1^{(\pm)} &= \omega_{23} \pm \omega_{14} = \omega^{(\pm)}n_1^{(\pm)}, \\ \omega_2^{(\pm)} &= \omega_{31} \pm \omega_{24} = \omega^{(\pm)}n_2^{(\pm)}, \qquad \omega^{(\pm)} = \left\{\sum_{a=1}^{3} \left(\omega_a^{(\pm)}\right)^2\right\}^{1/2}. \\ \omega_3^{(\pm)} &= \omega_{12} \pm \omega_{34} = \omega^{(\pm)}n_3^{(\pm)}, \end{aligned} \tag{9.146}$$

Thus, R is expressed evidently as the direct product of two unimodular unitary matrices of dimension 2. R keeps invariant if two unitary matrices change their signs simultaneously. Namely, Eq. (9.145) gives a one-to-two correspondence between the element R in SO(4) and the element $u \times u'$ in SU(2) $\times$ SU(2)$'$, and the correspondence preserves invariant in the multiplication of elements, so that

$$\mathrm{SO}(4) \sim \mathrm{SU}(2) \otimes \mathrm{SU}(2)'. \tag{9.147}$$

The parameters of the group SO(4) are chosen to be those of SU(2) $\times$ SU(2)$'$ where the group space of SU(2)$'$ is shortened by half such that there is a one-to-one correspondence between the set of parameters and the group element at least in the region where the measure is not vanishing. Namely, the group space of SO(4) is

$$\begin{aligned} &0 \le \omega^{(+)} \le 2\pi, \qquad &0 \le \omega^{(-)} \le \pi, \\ &0 \le \theta^{(\pm)} \le \pi, \qquad &-\pi \le \varphi^{(\pm)} \le \pi, \end{aligned} \tag{9.148}$$

where $\theta^{(\pm)}$ and $\varphi^{(\pm)}$ are the polar angle and the azimuthal angle of $\hat{\boldsymbol{n}}^{(\pm)}$, respectively. Due to the shortening, two end points of a diameter correspond to one element so that the group space of SO(4) is doubly-connected

$$\begin{aligned} &R(\hat{\boldsymbol{n}}^{(+)}, \omega^{(+)}; \hat{\boldsymbol{n}}^{(-)}, \omega^{(-)}) = R(-\hat{\boldsymbol{n}}^{(+)}, (2\pi - \omega^{(+)}); -\hat{\boldsymbol{n}}^{(-)}, (2\pi - \omega^{(-)})), \\ &R(\hat{\boldsymbol{n}}^{(+)}, \omega^{(+)}; \hat{\boldsymbol{n}}^{(-)}, \pi) = R(-\hat{\boldsymbol{n}}^{(+)}, (2\pi - \omega^{(+)}); -\hat{\boldsymbol{n}}^{(-)}, \pi). \end{aligned} \tag{9.149}$$

The covering group of SO(4) is SU(2) × SU(2)′. The weight function for the group integral of SO(4) is

$$\begin{aligned} dR = & \left(8\pi^4\right)^{-1} \sin^2\left(\omega^{(+)}/2\right)\sin^2\left(\omega^{(-)}/2\right)\sin\theta^{(+)}\sin\theta^{(-)} \\ & \times\ d\omega^{(+)}d\omega^{(-)}d\theta^{(+)}d\theta^{(-)}d\varphi^{(+)}d\varphi^{(-)}. \end{aligned} \tag{9.150}$$

An irreducible representation of SO(4) is a direct product of representations of two SU(2),

$$D^{jk}\left(\hat{\boldsymbol{n}}^{(+)},\omega^{(+)};\hat{\boldsymbol{n}}^{(-)},\omega^{(-)}\right) = D^{j}\left(\hat{\boldsymbol{n}}^{(+)},\omega^{(+)}\right)\times D^{k}\left(\hat{\boldsymbol{n}}^{(-)},\omega^{(-)}\right). \tag{9.151}$$

D^{jk} is $(2j+1)(2k+1)$-dimensional. Its row (column) index is denoted by two letters $(\mu\nu)$ and its generator I^{jk}_{ab} is expressed in terms of I^{j}_{a} of SU(2),

$$\begin{aligned} & I_a^{jk(+)} = I_a^j \times \mathbf{1}_{2k+1}, \qquad I_a^{jk(-)} = \mathbf{1}_{2j+1}\times I_a^k, \\ & I_{ab}^{jk} = \sum_{c=1}^{3} \epsilon_{abc}\left(I_c^{jk(+)} + I_c^{jk(-)}\right) = \sum_{c=1}^{3} \epsilon_{abc}\left(I_c^j \times \mathbf{1}_{2k+1} + \mathbf{1}_{2j+1}\times I_c^k\right), \\ & I_{a4}^{jk} = I_a^{jk(+)} - I_a^{jk(-)} = I_a^j \times \mathbf{1}_{2k+1} - \mathbf{1}_{2j+1}\times I_a^k. \end{aligned} \tag{9.152}$$

The reduction of the direct product of two D^{jk} can be calculated by the CG series for SU(2) such as

$$\begin{aligned} & \left(C^{j_1j_2}\right)^{-1}\left(D^{j_10}(R)\times D^{j_20}(R)\right)\left(C^{j_1j_2}\right) = \bigoplus_{J=|j_1-j_2|}^{j_1+j_2} D^{J0}(R), \\ & \left(C^{k_1k_2}\right)^{-1}\left(D^{0k_1}(R)\times D^{0k_2}(R)\right)\left(C^{k_1k_2}\right) = \bigoplus_{K=|k_1-k_2|}^{k_1+k_2} D^{0K}(R), \\ & D^{j_1k_1}(R)\times D^{j_2k_2}(R) \simeq \bigoplus_{J=|j_1-j_2|}^{j_1+j_2}\ \bigoplus_{K=|k_1-k_2|}^{k_1+k_2} D^{JK}(R). \end{aligned} \tag{9.153}$$

In fact, $I_a^{jk(\pm)}$ are related directly with the Chevalley bases,

$$\begin{aligned} & H_1 = 2I_3^{jk(-)}, \qquad H_2 = 2I_3^{jk(+)}, \\ & E_1 = I_1^{jk(-)} + iI_2^{jk(-)}, \qquad E_2 = I_1^{jk(+)} + iI_2^{jk(+)}. \end{aligned} \tag{9.154}$$

The highest weight in D^{jk} is $\boldsymbol{M} = (2k, 2j)$, and its Young pattern is

$$\begin{aligned}
&D^{jj} \simeq [2j, 0], && j = k, \\
&D^{jk} \simeq [(+)(j+k), (j-k)], && j-k > 0 \text{ is integer}, \\
&D^{jk} \simeq [(-)(j+k), (k-j)], && k-j > 0 \text{ is integer}, \\
&D^{jk} \simeq [+s, (j+k-1/2), (j-k-1/2)], && j-k-1/2 > 0 \text{ is integer}, \\
&D^{jk} \simeq [-s, (j+k-1/2), (k-j-1/2)], && k-j-1/2 > 0 \text{ is integer}.
\end{aligned} \tag{9.155}$$

The identical representation of SO(4) is $D^{0\,0}$, and its self-representation is equivalent to $D^{\frac{1}{2}\,\frac{1}{2}}$. The fundamental spinor representation of SO(4) is $D^{\frac{1}{2}\,0} \oplus D^{0\,\frac{1}{2}}$ and its generators (see Eq. (9.49)) are

$$\begin{aligned}
S_{23} &= (\sigma_2 \times \sigma_2)/2, & S_{14} &= -(\sigma_1 \times \sigma_1)/2, \\
S_{31} &= -(\sigma_1 \times \sigma_2)/2, & S_{24} &= -(\sigma_2 \times \sigma_1)/2, \\
S_{12} &= (\sigma_3 \times \mathbf{1})/2, & S_{34} &= (\mathbf{1} \times \sigma_3)/2.
\end{aligned} \tag{9.156}$$

Through the combination (9.141), $S_a^{\pm}$ are the generators of $D^{\frac{1}{2}\,0}$ and $D^{0\,\frac{1}{2}}$, respectively. In fact, through a similarity transformation X, one has

$$\begin{gathered}
X = \begin{pmatrix} 1 & 0 & 0 & 0 \\ 0 & 0 & 1 & 0 \\ 0 & 0 & 0 & 1 \\ 0 & -1 & 0 & 0 \end{pmatrix}, \\
X^{-1} S_a^{+} X = \begin{pmatrix} \sigma_a/2 & 0 \\ 0 & 0 \end{pmatrix}, \qquad X^{-1} S_a^{-} X = \begin{pmatrix} 0 & 0 \\ 0 & \sigma_a/2 \end{pmatrix}.
\end{gathered} \tag{9.157}$$

9.5.2 *Single-valued Representations of* O(4)

The group space of SO(4) falls into two pieces depending on whether the determinant of the element R is 1 or -1. The piece with det $R = 1$ constructs an invariant subgroup SO(4). As discussed in §9.1.6, the representative element in the coset is usually chosen to be τ which is a diagonal matrix where the diagonal entries are 1 except for $\tau_{44} = -1$. The transformation rule of generators in the action of τ can be calculated from that in the self-representation, namely from the multiplication rule of the elements of O(4),

$$\begin{aligned}
&\tau T_{ab} \tau^{-1} = T_{ab}\,, && \tau T_{a4} \tau^{-1} = -T_{a4}, \\
&\tau T_a^{(\pm)} \tau^{-1} = T_a^{(\mp)}, && \tau \left(T^{(\pm)}\right)^2 \tau^{-1} = \left(T^{(\mp)}\right)^2,
\end{aligned} \tag{9.158}$$

where

$$\left(T^{(\pm)}\right)^2 = \sum_{a=1}^{3} \left(T_a^{(\pm)}\right)^2. \tag{9.159}$$

The tensor representation D^{jk} of SO(4), where $j+k$ is an integer, is single-valued. Denote by P_R the transformation operators of O(4) for scalar functions and by L_{ab} its generators. Similarly, $L_a^{\pm}$ and $\left(L^{(\pm)}\right)^2$ can be obtained by Eqs. (9.141) and (9.159). If the basis function $\Psi_{\mu\nu}^{jk}$ belongs to the representation D^{jk} of SO(4),

$$\begin{aligned} P_R \Psi_{\mu\nu}^{jk} &= \sum_{\mu'\nu'} \Psi_{\mu'\nu'}^{jk} D_{\mu'\nu',\mu\nu}^{jk}(R) \\ &= \sum_{\mu'\nu'} \Psi_{\mu'\nu'}^{jk} D_{\mu'\mu}^{j}(\hat{\boldsymbol{n}}^{(+)}, \omega^{(+)}) D_{\nu'\nu}^{k}(\hat{\boldsymbol{n}}^{(-)}, \omega^{(-)}), \end{aligned} \tag{9.160}$$

$\Psi_{\mu\nu}^{jk}$ is the common eigenfunction of $L_3^{(\pm)}$ and $\left(L^{(\pm)}\right)^2$,

$$\begin{aligned} &L_3^{(+)} \Psi_{\mu\nu}^{jk} = \mu \Psi_{\mu\nu}^{jk}, &\qquad &L_3^{(-)} \Psi_{\mu\nu}^{jk} = \nu \Psi_{\mu\nu}^{jk}, \\ &\left(L^{(+)}\right)^2 \Psi_{\mu\nu}^{jk} = j(j+1) \Psi_{\mu\nu}^{jk}, &\qquad &\left(L^{(-)}\right)^2 \Psi_{\mu\nu}^{jk} = k(k+1) \Psi_{\mu\nu}^{jk}. \end{aligned} \tag{9.161}$$

In the transformation of τ, $P_\tau \Psi_{\mu\nu}^{jk}$ is still the common eigenfunction of $L_3^{(\pm)}$ and $\left(L^{(\pm)}\right)^2$, but the eigenvalues are interchanged between μ and ν and between $j(j+1)$ and $k(k+1)$,

$$\begin{aligned} &P_\tau \Psi_{\mu\nu}^{jk} = \Phi_{\nu\mu}^{kj}, &\qquad &P_\tau \Phi_{\nu\mu}^{kj} = \Psi_{\mu\nu}^{jk}, \\ &L_3^{(+)} \Phi_{\nu\mu}^{kj} = \nu \Phi_{\nu\mu}^{kj}, &\qquad &L_3^{(-)} \Phi_{\nu\mu}^{kj} = \mu \Phi_{\nu\mu}^{kj}, \\ &\left(L^{(+)}\right)^2 \Phi_{\nu\mu}^{kj} = k(k+1) \Phi_{\nu\mu}^{kj}, &\qquad &\left(L^{(-)}\right)^2 \Phi_{\nu\mu}^{kj} = j(j+1) \Phi_{\nu\mu}^{kj}, \end{aligned}$$

where $P_\tau^2 = P_E = \mathbf{1}$ is used. $\Phi_{\nu\mu}^{kj}$ belongs to the representation D^{kj} of SO(4).

When $j \neq k$, neither of the two representation spaces of D^{jk} and D^{kj} of SO(4) is invariant in O(4), only their direct sum is invariant. Namely, from two irreducible representations D^{jk} and D^{kj} of SO(4) one induces an irreducible representation Δ^{jk} of O(4). In addition to the indices μ and ν, a new index $\alpha = \pm$ has to be added for the row (column) indices of Δ^{jk} to distinguish the two representation subspaces,

$$\begin{aligned} &\Delta_{\mu\nu+,\mu'\nu'+}^{jk}(R) = D_{\mu\nu,\mu'\nu'}^{jk}(R), &\qquad &\Delta_{\nu\mu-,\nu'\mu'-}^{jk}(R) = D_{\nu\mu,\nu'\mu'}^{kj}(R), \\ &\Delta_{\mu\nu+,\nu'\mu'-}^{jk}(R) = \Delta_{\nu\mu-,\mu'\nu'+}^{jk}(R) = 0, &\qquad &\Delta_{\mu\nu\alpha,\nu'\mu'\beta}^{jk}(\tau) = \delta_{(-\alpha)\beta}\delta_{\mu\mu'}\delta_{\nu\nu'}, \end{aligned} \tag{9.162}$$

where $R \in \mathrm{SO}(4)$. The representation by changing the sign of $\Delta^{jk}(\tau)$ leads to an equivalent one.

When $j = k$, define

$$\psi_{\mu\nu}^{jj\pm} \sim \Psi_{\mu\nu}^{jj} \pm \Phi_{\mu\nu}^{jj}, \qquad P_\tau \psi_{\mu\nu}^{jj\pm} = \pm \psi_{\nu\mu}^{jj\pm}.$$

Thus, $\psi_{\mu\nu}^{jj\pm}$ span two different spaces both of which are invariant in O(4). Namely, from one irreducible representation D^{jj} of SO(4) one induces two inequivalent irreducible representations $\Delta^{jj\pm}$ of O(4),

$$\Delta^{jj\pm}(R) = D^{jj}(R), \quad R \in \mathrm{SO}(4), \qquad \Delta_{\mu\nu,\nu'\mu'}^{jj\pm}(\tau) = \pm\delta_{\mu\mu'}\delta_{\nu\nu'}. \quad (9.163)$$

9.5.3 *The Lorentz Group*

The Lorentz transformation A is a transformation between two inertial systems in a four-dimensional space–time (see Eq. (9.118)),

$$A^T A = AA^T = \mathbf{1}. \quad (9.164)$$

The matrix entries of A satisfy the condition

$$\begin{array}{l} A_{ab} \text{ and } A_{44} \text{ are real,} \\ A_{a4} \text{ and } A_{4a} \text{ are imaginary,} \end{array} \qquad a \text{ and } b = 1,\ 2,\ 3. \quad (9.165)$$

This condition preserves invariant in the product of two Lorentz transformations. The set of all such orthogonal matrices A, in the multiplication rule of matrices, constitutes the homogeneous Lorentz group, denoted by O(3,1) or L_h. The orthogonal condition (9.164) gives

$$\det A = \pm 1, \qquad A_{44}^2 = 1 + \sum_{a=1}^{3} |A_{a4}|^2 \geq 1. \quad (9.166)$$

These two discontinuous constraints divide the group space of L_h into four disjointed pieces. The elements in the piece to which the identical element belongs constitute an invariant subgroup L_p of L_h, called the proper Lorentz group. The element in L_p satisfies

$$\det A = 1, \qquad A_{44} \geq 1. \quad (9.167)$$

Since there is no upper limit of A_{44}, the group space of L_p is an open region in the Euclidean space, and L_p is a noncompact Lie group. The representative elements in L_p and its three cosets are usually chosen to be the identical element E, the space inversion σ, the time inversion τ, and the space–time inversion ρ:

$$\begin{aligned}
&E = \text{diag}\,(1,\ 1,\ 1,\ 1) \in \mathrm{L}_+^{\uparrow} = \mathrm{L}_p, && \det A = 1, && A_{44} \geq 1, \\
&\sigma = \text{diag}\,(-1,\ -1,\ -1,\ 1) \in \mathrm{L}_-^{\uparrow}, && \det A = -1, && A_{44} \geq 1, \\
&\tau = \text{diag}\,(1,\ 1,\ 1,\ -1) \in \mathrm{L}_-^{\downarrow}, && \det A = -1, && A_{44} \leq -1, \\
&\rho = \text{diag}\,(-1,\ -1,\ -1,\ -1) \in \mathrm{L}_+^{\downarrow}, && \det A = 1, && A_{44} \leq -1.
\end{aligned} \tag{9.168}$$

Four elements constitute the inversion group V_4 of order 4.

9.5.4 *Irreducible Representations of* L_p

Discuss the generators in the self-representation of L_p. Let A be an infinitesimal element of L_p:

$$\begin{aligned}
&A = \mathbf{1} - i\alpha X, && A^T = \mathbf{1} - i\alpha X^T, \\
&\mathbf{1} = A^T A = \mathbf{1} - i\alpha \left(X + X^T\right), && X^T = -X, \\
&1 = \det A = 1 - i\alpha \mathrm{Tr} X, && \mathrm{Tr} X = 0.
\end{aligned}$$

Thus, X is a traceless antisymmetric matrix. Expand X with respect to the generators T_{ab} in the self-representation of SO(4):

$$\begin{aligned}
A = &\ \mathbf{1} - i \sum_{a<b=2}^{3} \omega_{ab} T_{ab} - i \sum_{a=1}^{3} \omega_{a4} T_{a4} \\
= &\ \mathbf{1} - i \sum_{a=1}^{3} \left(\Omega_a T_a^{(+)} + \Omega_a^* T_a^{(-)} \right),
\end{aligned} \tag{9.169}$$

where ω_{ab} is real, ω_{a4} is imaginary, and

$$\Omega_a = \frac{1}{2} \sum_{b,c=1}^{3} \epsilon_{abc} \omega_{bc} + \omega_{a4}. \tag{9.170}$$

Except for the imaginary parameters, the generators in the self-representations of SO(4) and L_p are completely the same, so are the generators in the corresponding irreducible representations of two groups. The finite-dimensional inequivalent irreducible representations of L_p are also denoted by $D^{jk}(L_p)$, whose generators are the same as those for SO(4) (see Eq. (9.152))

$$I_{ab}^{jk} = \sum_{c=1}^{3} \epsilon_{abc} \left\{ I_c^{jk(+)} + I_c^{jk(-)} \right\}, \qquad I_{a4}^{jk} = I_a^{jk(+)} - I_a^{jk(-)},$$
$$I_a^{jk(+)} = I_a^j \times \mathbf{1}_{2k+1} , \qquad I_a^{jk(-)} = \mathbf{1}_{2j+1} \times I_a^k, \tag{9.171}$$

where I_a^j and I_a^k are the generators in the representations of SU(2). Two representations $D^{jk}(\mathrm{SO}(4))$ and $D^{jk}(\mathrm{L}_p)$ have the same dimensions and are simultaneously single-valued or double-valued. The reductions of their direct product representations are also the same. However, the global properties of the two groups are very different because the parameters ω_{a4} in L_p are imaginary. The finite-dimensional irreducible representation D^{jk} of L_p, except for the identical representation, is not unitary. There are infinite-dimensional unitary representations of L_p [Adams et al. (1987)].

The transformation generated by T_{ab} is obviously a pure rotation, belonging to the subgroup SO(3). But now, its transformation matrix is four-dimensional,

$$R(\boldsymbol{e}_3, \varphi) = \exp\{-i\varphi T_{12}\} = \begin{pmatrix} \cos\varphi & -\sin\varphi & 0 & 0 \\ \sin\varphi & \cos\varphi & 0 & 0 \\ 0 & 0 & 1 & 0 \\ 0 & 0 & 0 & 1 \end{pmatrix}. \tag{9.172}$$

The transformation generated by T_{34} with the parameter $\omega_{34} = i\omega$ is a Lorentz boost along the direction of z-axis with the relative velocity v

$$\begin{aligned} A(\boldsymbol{e}_3, i\omega) = \exp\{-i(i\omega)T_{34}\} &= \begin{pmatrix} 1 & 0 & 0 & 0 \\ 0 & 1 & 0 & 0 \\ 0 & 0 & \cosh\omega & -i\sinh\omega \\ 0 & 0 & i\sinh\omega & \cosh\omega \end{pmatrix}, \\ v = \tanh\omega, \qquad \cosh\omega &= \left(1 - v^2\right)^{-1/2}, \\ \sinh\omega &= v\left(1 - v^2\right)^{-1/2}, \end{aligned} \tag{9.173}$$

where the natural units $\hbar = c = 1$ are employed. In comparison with the Lorentz boost in the physical textbook, the transformation matrix becomes its inverse because the viewpoint of transformation of the system is used here, instead of that of the coordinate frame in the usual physical textbook.

Similar to the Euler angles in SO(3), a new set of parameters of L_p is expected such that each element A of L_p can be expanded as a product of some pure rotations and a Lorentz boost along the z-direction. From a given A matrix in L_p, calculate ω from $A_{44} = \cosh\omega$. Extracting a factor $-i\sinh\omega$ from A_{a4}, one obtains a unit vector $\hat{\boldsymbol{n}}(\theta, \varphi)$ in the three-dimensional space, whose rectangular coordinates are $(i/\sinh\omega)(A_{14}, A_{24}, A_{34})$:

$$\begin{array}{ll} A_{44} = \cosh\omega, & iA_{14}/\sinh\omega = \sin\theta\cos\varphi, \\ iA_{24}/\sinh\omega = \sin\theta\sin\varphi, & iA_{34}/\sinh\omega = \cos\theta. \end{array} \tag{9.174}$$

Thus, the parameters θ and φ are determined. Since

$$R(\mathbf{e}_3, \varphi)R(\mathbf{e}_2, \theta)A(\mathbf{e}_3, i\omega)\begin{pmatrix} 0 \\ 0 \\ 0 \\ 1 \end{pmatrix} = \begin{pmatrix} A_{14} \\ A_{24} \\ A_{34} \\ A_{44} \end{pmatrix},$$

$\{R(\mathbf{e}_3, \varphi)R(\mathbf{e}_2, \theta)A(\mathbf{e}_3, i\omega)\}^{-1} A$ is a pure rotation $R(\alpha, \beta, \gamma)$, where the Euler angles α, β, and γ can be calculated. Thus, an arbitrary Lorentz transformation A in L_p is expressed as a product

$$\begin{aligned} A = A(\varphi, \theta, \omega, \alpha, \beta, \gamma) &= R(\varphi, \theta, 0)A(\mathbf{e}_3, i\omega)R(\alpha, \beta, \gamma), \\ R(\alpha, \beta, \gamma) &= R(\mathbf{e}_3, \alpha)R(\mathbf{e}_2, \beta)R(\mathbf{e}_3, \gamma), \end{aligned} \tag{9.175}$$

where

$$\begin{array}{lll} 0 \leq \omega < \infty, & 0 \leq \theta \leq \pi, & 0 \leq \beta \leq \pi, \\ -\pi \leq \varphi \leq \pi, & -\pi \leq \alpha \leq \pi, & -\pi \leq \gamma \leq \pi. \end{array} \tag{9.176}$$

The geometrical meaning of the expression (9.175) is evident. Two rotations in the twosides of $A(\mathbf{e}_3, i\omega)$ transform two frames before and after the Lorentz transformation A such that two z-axes are changed to the direction of the relative motion, and the remaining axes are changed to be parallel to each other, respectively. After two rotations, the Lorentz transformation A is simplified into $A(\mathbf{e}_3, i\omega)$.

The representation matrix of A in $D^{jk}(\mathrm{L}_p)$ is easy to be written. Due to Eq. (9.170), Ω_a is real for a pure rotation and imaginary for a Lorentz boost,

$$\begin{aligned} D^{jk}(\alpha, \beta, \gamma) &= D^j(\alpha, \beta, \gamma) \times D^k(\alpha, \beta, \gamma), \\ D^{jk}(\mathbf{e}_3, i\omega) &= \exp(\omega I_3^j) \times \exp(-\omega I_3^k), \end{aligned} \tag{9.177}$$

where I_3^j and I_3^k are diagonal. Thus,

$$\begin{aligned} D^{jk}_{\mu\nu,\mu'\nu'}(\varphi, \theta, \omega, \alpha, \beta, \gamma) &= \sum_{\rho\tau} e^{-i(\mu+\nu)\varphi} d^j_{\mu\rho}(\theta) d^k_{\nu\tau}(\theta) e^{(\rho-\tau)\omega} \\ &\times e^{-i(\rho+\tau)\alpha} d^j_{\rho\mu'}(\beta) d^k_{\tau\nu'}(\beta) e^{-i(\mu'+\nu')\gamma}. \end{aligned} \tag{9.178}$$

9.5.5 *The Covering Group of* L_p

For a matrix Lie group G, there exists an exponential mapping of G if its every element can be written in an exponential function of matrix where the exponent belongs to its real Lie algebra. Equation (9.145) shows that there exists an exponential mapping of SO(4). Based on Eq. (9.145) the covering group of SO(4) is found. In mathematics, there exists an exponential mapping of a Lie group G if every element in G belongs to a one-parameter Lie subgroup. The exponential mapping exists for a compact Lie group, but is not necessary to exist for a noncompact Lie group. For example, the set of two-dimensional unimodular complex matrices, in the multiplication rule of matrices, constitutes a group $\mathrm{SL}(2,C)$. $\mathrm{SL}(2,C)$ is noncompact and its exponential mapping does not exist, because the following elements cannot be written in an exponential form in the condition that the exponent belongs to the real Lie algebra of $\mathrm{SL}(2,C)$,

$$\begin{pmatrix} -1 & -2 \\ 0 & -1 \end{pmatrix} = -e^{\sigma_1+i\sigma_2}, \qquad \begin{pmatrix} -1-i & -1 \\ -1 & -1+i \end{pmatrix} = -e^{\sigma_1+i\sigma_3}.$$

In this subsection we are going to find the covering group of L_p first, and then, to show that every element in L_p can be written in an exponential function of matrix where the exponent belongs to its real Lie algebra, namely, there exists an exponential mapping of L_p although L_p is noncompact and some elements in L_p cannot be diagonalized.

The self-representation of SO(4) can be changed to $D^{\frac{1}{2}\frac{1}{2}}$ by the similarity transformation N (see Eq. (9.145)), so can that of L_p because the generators in the two representations are the same.

$$\begin{aligned} A = &\ A(\varphi,\theta,\omega,\alpha,\beta,\gamma) \\ = &\ N\left\{u(\varphi,\theta,0)\times u(\varphi,\theta,0)\right\}\left\{\exp\left(\omega\sigma_3/2\right)\times\exp\left(-\omega\sigma_3/2\right)\right\} \\ &\cdot\ \left\{u(\alpha,\beta,\gamma)\times u(\alpha,\beta,\gamma)\right\}N^{-1} \\ = &\ N\left\{M\times\left(\sigma_2M^*\sigma_2\right)\right\}N^{-1}, \\ M = &\ u(\varphi,\theta,0)\exp\left(\omega\sigma_3/2\right)u(\alpha,\beta,\gamma)\in \mathrm{SL}(2,C), \end{aligned} \tag{9.179}$$

where $D^{1/2}(\alpha,\beta,\gamma)=u(\alpha,\beta,\gamma)$ is the element of SU(2). The determinant of M is 1 so that M belongs to $\mathrm{SL}(2,C)$. Equation (9.179) gives a map from each element A in L_p to at least one element M in $\mathrm{SL}(2,C)$. If one element A in L_p maps two elements M and M' in $\mathrm{SL}(2,C)$ through Eq. (9.179), then,

$$\mathbf{1} = N^{-1}\left(AA^{-1}\right)N = MM'^{-1}\times\sigma_2\left(MM'^{-1}\right)^*\sigma_2.$$

Namely, MM'^{-1} is a constant matrix $c\mathbf{1}$. Since det $M =$ det $M' = 1$, one has $c^2 = 1$. There is a one-to-two correspondence between an element A in L_p and two elements $\pm M$ in SL$(2, C)$. The correspondence preserves invariant in multiplication of elements. Since the orders of L_p and SL$(2, C)$ both are 6, the one-to-two correspondence shows that SL$(2, C)$ is homomorphic onto L_p (see Prob. 18 of Chap. 9 in [Ma and Gu (2004)]). SL$(2, C)$ is the covering group of L_p.

We are going to show that every M in SL$(2, C)$, if neglecting the possible minus sign, can be written in an exponential function of matrix where the exponent is traceless. The product of two eigenvalues of M is equal to 1. M can be diagonalized if its two eigenvalues are different,

$$Y^{-1}MY = \begin{pmatrix} e^{-i\tau} & 0 \\ 0 & e^{i\tau} \end{pmatrix} = \exp\left\{-i\begin{pmatrix} \tau & 0 \\ 0 & -\tau \end{pmatrix}\right\}, \quad Y \in \text{SL}(2, C). \quad (9.180)$$

$M = \exp\left(-i\tau Y\sigma_3 Y^{-1}\right)$ and τ is a complex number. Except for $\mathbf{1}$, M can be changed to a ladder matrix if its two eigenvalues both are 1,

$$Z^{-1}MZ = \begin{pmatrix} 1 & -2 \\ 0 & 1 \end{pmatrix} = \exp\left\{\begin{pmatrix} 0 & -2 \\ 0 & 0 \end{pmatrix}\right\}, \qquad Z \in \text{SL}(2, C). \qquad (9.181)$$

$M = \exp\left\{-Z\left(\sigma_1 + i\sigma_2\right) Z^{-1}\right\}$. If two eigenvalues of M are both -1, one can diagonalize $-M$ instead of M, because $\pm M$ correspond to the same A in L_p. Therefore, removing the possible minus sign, $M = \exp(-iB)$ where B is a traceless matrix and can be expanded with respect to the Pauli matrices,

$$M = \exp\left(-i\boldsymbol{\Omega} \cdot \boldsymbol{\sigma}/2\right). \qquad (9.182)$$

Substituting Eq. (9.182) into (9.179), the arbitrary element A in L_p is expressed as

$$\begin{aligned} A = &\ N\left\{\exp\left(-i\boldsymbol{\Omega} \cdot \boldsymbol{\sigma}/2\right) \times \exp\left(-i\boldsymbol{\Omega}^* \cdot \boldsymbol{\sigma}/2\right)\right\} N^{-1} \\ = &\ \exp\left(-i\boldsymbol{\Omega} \cdot \boldsymbol{T}^{(+)}\right) \exp\left(-i\boldsymbol{\Omega}^* \cdot \boldsymbol{T}^{(-)}\right) \\ = &\ \exp\left\{-i\sum_{a=1}^{3} \left(\Omega_a T_a^{(+)} + \Omega_a^* T_a^{(-)}\right)\right\} \\ = &\ \exp\left\{-i\sum_{a<b} \omega_{ab} T_{ab} - i\sum_{a=1}^{3} \omega_{a4} T_{a4}\right\}, \end{aligned} \qquad (9.183)$$

where ω_{ab} is real and ω_{a4} is pure imaginary,

$$\omega_{ab} = \frac{1}{2} \sum_{c=1}^{3} \epsilon_{abc} \left(\Omega_c + \Omega_c^* \right), \qquad \omega_{a4} = \frac{1}{2} \left(\Omega_a - \Omega_a^* \right). \tag{9.184}$$

Thus, every element in L_p is written in an exponential function of matrix where the exponent belongs to the real Lie algebra of L_p. When the parameters are infinitesimal, Eq. (9.183) returns to Eq. (9.169). The representation matrix of A in D^{jk} is

$$\begin{aligned} D^{jk}(A) &= \exp\left\{ -i \sum_{a<b} \omega_{ab} I_{ab}^{jk} - i \sum_{a=1}^{3} \omega_{a4} I_{a4}^{jk} \right\} \\ &= \exp\left\{ -i \sum_{a=1}^{3} \varOmega_a I_a^j \right\} \times \exp\left\{ -i \sum_{a=1}^{3} \varOmega_a^* I_a^k \right\}. \end{aligned} \tag{9.185}$$

In principle, two sets of parameters can be transformed to each other, although the transformation is quite complicated. In fact, from the parameters (φ, θ, ω, α, β, γ) one calculates M by Eq. (9.179) and determines the parameters ω_{ab} and ω_{a4} by Eqs. (9.182) and (9.184). Conversely, from the parameters ω_{ab} and ω_{a4} one calculates M by Eq. (9.182) and A by Eq. (9.179), and then, determines the parameters $(\varphi, \theta, \omega, \alpha, \beta, \gamma)$ by Eqs. (9.174) and (9.175).

9.5.6 *Classes of* L_p

The classes of L_p can be discussed from the classes of $\text{SL}(2, C)$ owing to $\text{L}_p \sim \text{SL}(2, C)$. When two eigenvalues of $M \in \text{SL}(2, C)$ are different, M is conjugate to an element given in Eq. (9.180). Since $\Omega_3 = 2\tau$, one has $\omega_{12} = \varphi = \tau + \tau^*$ and $\omega_{34} = i\omega = \tau - \tau^*$. $\pm M$ belong to two classes in $\text{SL}(2, C)$ but correspond to one element in L_p belonging to a class of L_p characterized by the parameters φ and ω, or by the representative element $A(\varphi, 0, \omega, 0, 0, 0)$:

$$A = \begin{pmatrix} \cos\varphi & -\sin\varphi & 0 & 0 \\ \sin\varphi & \cos\varphi & 0 & 0 \\ 0 & 0 & \cosh\omega & -i\sinh\omega \\ 0 & 0 & i\sinh\omega & \cosh\omega \end{pmatrix} \qquad \begin{aligned} &-\pi \leq \varphi \leq \pi, \\ &0 \leq \omega < \infty. \end{aligned} \tag{9.186}$$

$M = \pm \mathbf{1}$ constitutes two classes in $\text{SL}(2, C)$ but correspond to one class in L_p. If two eigenvalues of M both are 1, except for $\mathbf{1}$, M is conjugate to an element given in Eq. (9.181). The set of $-M$ constitute another class in $\text{SL}(2, C)$. Two classes of $\text{SL}(2, C)$ correspond to one class of L_p. Since

$$u(-\pi,\pi/4,0)\begin{pmatrix}\sqrt{2}+1 & 0\\ 0 & \sqrt{2}-1\end{pmatrix}u(\pi,3\pi/4,0)=\begin{pmatrix}1 & -2\\ 0 & 1\end{pmatrix}$$
$$=\exp\left\{\begin{pmatrix}0 & -2\\ 0 & 0\end{pmatrix}\right\}=\exp\left\{-\sigma_1-i\sigma_2\right\}=\exp\left\{-i\sum_{a=1}^{3}\Omega_a\sigma_a/2\right\},$$

where $\Omega_1=-2i$, $\Omega_2=2$, and $\Omega_3=0$. The representative element in the class of L_p is $A(-\pi,\pi/4,\omega,\pi,3\pi/4,0)$ with $\cosh\omega=3$,

$$A=e^{-i2T_{31}-2T_{14}}=\begin{pmatrix}1 & 0 & 2 & 2i\\ 0 & 1 & 0 & 0\\ -2 & 0 & -1 & -2i\\ -2i & 0 & -2i & 3\end{pmatrix}. \tag{9.187}$$

A is a proper Lorentz transformation matrix with all four eigenvalues to be 1 (see Prob. 18 of Chap. 9 in [Ma and Gu (2004)]).

9.5.7 *Irreducible Representations of* L_h

The irreducible representations of L_h can be obtained from those of L_p and the properties of τ and ρ of two cosets $\mathrm{L}_-^{\downarrow}$ and $\mathrm{L}_+^{\downarrow}$. Another representative element $\sigma=\tau\rho$ in the coset $\mathrm{L}_-^{\uparrow}$ is calculable. Since ρ is commutable with every element in L_h and $\rho^2=E$, it takes a constant matrix in an irreducible representation of L_h. The property of τ has been given in Eq. (9.158).

From the single-valued representation $D^{jk}(\mathrm{L}_p)$, its induced representation with respect to L_h can be calculated like those of O(4). Introduce four irreducible representations of the inversion group V_4 of order 4,

$$\begin{aligned}&V^{(1)}(\tau)=V^{(2)}(\tau)=V^{(1)}(\rho)=V^{(4)}(\rho)=1,\\ &V^{(3)}(\tau)=V^{(4)}(\tau)=V^{(2)}(\rho)=V^{(3)}(\rho)=-1.\end{aligned} \tag{9.188}$$

When $j=k$, there are four induced representations $\Delta^{jj\lambda}(\mathrm{L}_h)$,

$$\begin{aligned}&\Delta^{jj\lambda}(A)=D^{jj}(A), \qquad A\in\mathrm{L}_p,\\ &\Delta^{jj\lambda}_{\mu\nu,\nu'\mu'}(\tau)=V^{(\lambda)}(\tau)\delta_{\mu\mu'}\delta_{\nu\nu'},\\ &\Delta^{jj\lambda}(\rho)=V^{(\lambda)}(\rho)\,\mathbf{1}, \qquad 1\le\lambda\le 4.\end{aligned} \tag{9.189}$$

When $j\neq k$ but $j+k$ is an integer, only the direct sum of two representation spaces $D^{jk}(\mathrm{L}_p)$ and $D^{kj}(\mathrm{L}_p)$ is invariant for the group L_h, and there are two induced representations $\Delta^{jk\pm}(\mathrm{L}_h)$,

$$\begin{aligned}&\Delta^{jk\pm}(A) = D^{jk}(A) \oplus D^{kj}(A), \qquad A \in \mathrm{L}_p, \qquad \Delta^{jk\pm}(\rho) = \pm\mathbf{1},\\&\Delta^{jk\pm}_{\mu\nu\alpha,\nu'\mu'\beta}(\tau) = \delta_{(-\alpha)\beta}\delta_{\mu\mu'}\delta_{\nu\nu'}, \qquad \alpha,\ \beta = \pm 1,\end{aligned} \tag{9.190}$$

where two representation spaces of $D^{jk}(\mathrm{L}_p)$ and $D^{kj}(\mathrm{L}_p)$ are distinguished by the subscripts α and β. Since the diagonal entries of $\Delta^{jk\pm}(\tau)$ are all 0, the representation by changing the sign of $\Delta^{jk\pm}(\tau)$ leads to an equivalent one.

For the double-valued representation $D^{jk}(\mathrm{L}_p)$, we only discuss the Dirac spinor representation $D(\mathrm{L}_h)$, which is the induced representation of $D^{\frac{1}{2}\frac{1}{2}}(\mathrm{L}_p)$. Introduce

$$\overline{\gamma}_\mu = \sum_{\nu=1}^{4} A_{\mu\nu}\gamma_\nu, \qquad 1 \leq \mu \leq 4.$$

Since $A \in \mathrm{L}_h$ is an orthogonal matrix, $\overline{\gamma}_\mu$ also satisfy the anticommutative relations (9.38). From Theorem 9.1 two sets of γ_μ and $\overline{\gamma}_\mu$ are equivalent and they are related by a unimodular similarity transformation $D(A)$,

$$D(A)^{-1}\gamma_\mu D(A) = \sum_{\nu=1}^{4} A_{\mu\nu}\gamma_\nu, \qquad \det D(A) = 1. \tag{9.191}$$

The set of $D(A)$ forms a multiple-valued representation of L_h where the generators are

$$I_{\mu\nu} = \frac{-i}{4}\left(\gamma_\mu\gamma_\nu - \gamma_\nu\gamma_\mu\right). \tag{9.192}$$

Introduce the charge conjugate matrix C

$$C^{-1}\gamma_\mu C = -\gamma_\mu^T, \quad C^\dagger C = \mathbf{1}, \quad C^T = -C\,, \quad \det C = 1. \tag{9.193}$$

Due to $C^{-1}I_{\mu\nu}C = -I_{\mu\nu}^T$, a new constraint is added to restrict the representation to be double-valued,

$$C^{-1}D(A)C = \left\{D(A^{-1})\right\}^T, \qquad A \in \mathrm{L}_p. \tag{9.194}$$

Remind that the right-hand side of Eq. (9.194) is not equal to $D(A)^*$. In physics, Eq. (9.194) becomes Eq. (9.195) for the elements in L_h

$$C^{-1}D(A)C = \frac{A_{44}}{|A_{44}|}\left\{D(A^{-1})\right\}^T, \qquad A \in L_h. \tag{9.195}$$

The representation matrices of some representative elements are

$$D(\sigma) = \pm i\gamma_4, \qquad D(\tau) = \pm \gamma_4\gamma_5, \qquad D(\rho) = \pm i\gamma_5 \, . \tag{9.196}$$

$D(A)$ satisfies Eqs. (9.191) and (9.195). There is a two-to-one correspondence between $\pm D(A)$ and A in L_h. The set of $D(A)$ is called the Dirac spinor representation which is the covering group of L_h. For L_p, the Dirac spinor representation is reduced to two irreducible representations of dimension 2, both of which are isomorphic onto the covering group SL$(2, C)$ of L_p. The Dirac spinor representation is not unitary and satisfies (see Prob. 17 of Chap. 9 in [Ma and Gu (2004)])

$$\gamma_4 D(A)^\dagger \gamma_4 = \frac{A_{44}}{|A_{44}|} \, D(A)^{-1}. \tag{9.197}$$

9.6 Exercises

1. Calculate the dimensions of the irreducible representations of the SO(8) group denoted by the following Young patterns: (a) $[4, 2]$, (b) $[3, 2]$, (c) $[4, 4]$, (d) $[(+)3, 2, 1, 1]$, (e) $[(+)3, 3, 1, 1]$.
2. Calculate the Clebsch–Gordan series for the subduced representations of the following irreducible representations of the SU(N) group with respect to the subgroup SO(N), and then check the results by their dimensions for $N = 7$:

$$\begin{array}{llllll}
[2] \, , & [3] \, , & [2,1] \, , & [4] \, , & [3,1] \, , & [2,2] \\
{[2,1^2]} \, , & [5] \, , & [4,1] \, , & [3,2] \, , & [3,1,1] \, , & [2,2,1] \, , \\
{[2,1^3]} \, , & [6] \, , & [5,1] \, , & [4,2] \, , & [4,1,1] \, , & [3,3] \, , \\
{[3,2,1]} \, , & [3,1^3] \, , & [2^3] \, , & [2^2,1^2] \, , & [2,1^4] \, . &
\end{array}$$

3. Calculate the Clebsch–Gordan series in the reductions of the following direct products of the irreducible tensor representations, and check the results by their dimensions for the SO(7) group:

$$(1) \ \ [2] \otimes [2], \qquad (2) \ \ [2] \otimes [1,1], \qquad (3) \ \ [3] \otimes [2,1].$$

4. Calculate the orthonormal bases in the irreducible representation space $[2, 2]$ of SO(5) by the method of the block weight diagram, and then express the orthonormal bases by the standard tensor Young tableaux in the traceless tensor space of rank 4 for SO(5).

5. Calculate the orthonormal bases in the irreducible representation space $[2,0,0,0]$ of SO(8) by the method of the block weight diagram, and then, express the orthonormal bases by the standard tensor Young tableaux in the traceless symmetric tensor space of rank 2 for SO(8).

6. Calculate the spherical harmonic functions $Y_{\boldsymbol{m}}^{[\lambda]}$ in an N-dimensional space.

7. Calculate the dimension of the irreducible spinor representation of the SO(7) group denoted by the following Young patterns:

$$\begin{array}{lll} (1)\ \ [s,4,2], & (2)\ \ [s,3,2], & (3)\ \ [s,4,4], \\ (4)\ \ [s,3,1,1], & (5)\ \ [s,3,2,2]. & \end{array}$$

8. Calculate the Clebsch–Gordan series for the direct product of the tensor representation $[\lambda]$ and the fundamental spinor representation $[s]$ of $SO(2\ell+1)$ or $[\pm s]$ of $SO(2\ell)$, where $[\lambda]$ is a one-row Young pattern or a one-column Young pattern.

9. Calculate the basis states in the fundamental spinor representation $[s]$ of SO(7).

10. Calculate the basis states in the fundamental spinor representations $[\pm s]$ of SO(8).

11. Expand the eigenfunction of the total angular momentum with the highest weight in terms of the product of the spherical harmonic function $Y_{\boldsymbol{m}}^{[\lambda]}(\hat{\boldsymbol{x}})$ and the spinor basis $\chi(\boldsymbol{m})$.

12. Discuss the classes in the SO(4) group and calculate their characters in the irreducible representation D^{jk}.

13. Calculate six parameters of the following proper Lorentz transformation A, and write its representation matrix in the irreducible representation $D^{jk}(A)$ of the proper Lorentz group L_p:

$$A(\varphi,\theta,\omega,\alpha,\beta,\gamma) = \begin{pmatrix} 1 & 0 & 0 & 0 \\ 0 & \sqrt{3}/2 & (\cosh\omega)/2 & -i(\sinh\omega)/2 \\ 0 & -1/2 & \sqrt{3}(\cosh\omega)/2 & -i\sqrt{3}(\sinh\omega)/2 \\ 0 & 0 & i\sinh\omega & \cosh\omega \end{pmatrix}.$$

14. Prove that the Dirac spinor representation satisfies

$$\gamma_4 D(A)^\dagger \gamma_4 = \frac{A_{44}}{|A_{44}|}\, D(A)^{-1}.$$

Chapter 10

THE SYMPLECTIC GROUPS

In §7.4.3 we have introduced the fundamental property of the USp(2ℓ) group and the Sp($2\ell, R$) group. In this chapter we will study the irreducible representations of USp(2ℓ) and their applications to physics.

10.1 Irreducible Representations of USp(2ℓ)

10.1.1 *Decomposition of the Tensor Space of* USp(2ℓ)

The element u in USp(2ℓ) is a transformation matrix in a (2ℓ)-dimensional complex space,

$$x_a \xrightarrow{u} x_a' = \sum_b u_{ab} x_b, \tag{10.1}$$

where the index a is taken to be j or $\overline{j}$, $1 \leq j \leq \ell$, in the following order,

$$a = 1,\ \overline{1},\ 2,\ \overline{2},\ \ldots,\ \ell,\ \overline{\ell}. \tag{10.2}$$

The u matrix satisfies

$$u^T J u = J, \qquad u^\dagger = u^{-1}, \tag{10.3}$$

where

$$J_{ab} = \begin{cases} 1 & \text{when } a = j, \quad b = \overline{j}, \\ -1 & \text{when } a = \overline{j}, \quad b = j, \\ 0 & \text{the remaining cases}, \end{cases} \tag{10.4}$$

$$J = \mathbf{1}_\ell \times (i\sigma_2) = -J^{-1} = -J^T, \qquad \det\ J = 1.$$

Thus, the self-representation of USp(2ℓ) is self-conjugate

$$u^* = J^{-1} u J. \tag{10.5}$$

The tensor $\boldsymbol{T}_{a_1\ldots a_n}$ of rank n and the basis tensors $\boldsymbol{\theta}_{a_1\ldots a_n}$ with respect to USp(2ℓ) are defined as

$$\begin{aligned}
\boldsymbol{T}_{a_1\ldots a_n} &\xrightarrow{u} (O_u\boldsymbol{T})_{a_1\ldots a_n} = \sum_{b_1\ldots b_n} u_{a_1b_1}\ldots u_{a_nb_n}\boldsymbol{T}_{b_1\ldots b_n}, \\
\boldsymbol{\theta}_{a_1\ldots a_n} &\xrightarrow{u} O_u\boldsymbol{\theta}_{a_1\ldots a_n} = \sum_{b_1\ldots b_n} \boldsymbol{\theta}_{b_1\ldots b_n}u_{b_1a_1}\ldots u_{b_na_n}.
\end{aligned} \tag{10.6}$$

Due to Eq. (10.5) the contravariant tensor is equivalent to the covariant one. In comparison with the tensors of SO(N), there also exists two invariant tensors of USp(2ℓ): J_{ab} and $\epsilon_{a_1\ldots a_{2\ell}}$. In fact, denote by $\boldsymbol{J}$ an antisymmetric tensor of rank 2 whose component is equal to J_{ab},

$$(O_u\boldsymbol{J})_{ab} = \sum_{cd} u_{ac}u_{bd}\boldsymbol{J}_{cd} = \left(u\boldsymbol{J}u^T\right)_{ab} = \boldsymbol{J}_{ab}. \tag{10.7}$$

$\boldsymbol{J}_{ab}$ is invariant in USp(2ℓ). Through a contraction of two antisymmetric indices j and $\overline{j}$, the trace tensor of a tensor of rank n for USp(2ℓ)

$$\sum_{a_ra_s} J_{a_ra_s}\boldsymbol{T}_{a_1\ldots a_{r-1}a_ra_{r+1}\ldots a_{s-1}a_sa_{s+1}\ldots},$$

is a tensor of rank $(n-2)$,

$$\begin{aligned}
&\sum_{a_ra_s} (O_uJ_{a_ra_s}\boldsymbol{T})_{a_1\ldots a_{r-1}a_ra_{r+1}\ldots a_{s-1}a_sa_{s+1}\ldots a_n} \\
&\quad = \sum_{a_ra_sb_1\ldots b_n} J_{a_ra_s}u_{a_1b_1}\ldots u_{a_nb_n}\boldsymbol{T}_{b_1\ldots b_n} \\
&\quad = \sum_{b_1\ldots b_n} u_{a_1b_1}\ldots u_{a_{r-1}b_{r-1}}u_{a_{r+1}b_{r+1}}\ldots u_{a_{s-1}b_{s-1}}u_{a_{s+1}b_{s+1}}\ldots u_{a_nb_n} \\
&\quad\quad \times \left(J_{b_rb_s}\boldsymbol{T}_{b_1\ldots b_{r-1}b_rb_{r+1}\ldots b_{s-1}b_sb_{s+1}\ldots}\right).
\end{aligned} \tag{10.8}$$

The subspace of the trace tensors is invariant in USp(2ℓ). A tensor space can be decomposed into the direct sum of a series of subspaces of traceless tensors

$$\sum_{ab} J_{ab}\boldsymbol{T}_{\ldots a\ldots b\ldots} = 0. \tag{10.9}$$

For example, similar to Eq. (8.56), a tensor of rank 2 is decomposed into a sum of a traceless tensor of rank 2 and a scalar

$$\boldsymbol{T}_{ab} = \left\{\boldsymbol{T}_{ab} - \boldsymbol{J}_{ab}\left(\frac{1}{2\ell}\sum_{cd} J_{cd}\boldsymbol{T}_{cd}\right)\right\} + \boldsymbol{J}_{ab}\left(\frac{1}{2\ell}\sum_{cd} J_{cd}\boldsymbol{T}_{cd}\right). \tag{10.10}$$

The Weyl reciprocity holds for the tensors of USp(2ℓ) so that the tensor space can be reduced by the projection of the Young operators. Denote by $\mathcal{T}$ the space of traceless tensors of rank n of USp(2ℓ). Projecting by a Young operator $\mathcal{Y}_\mu^{[\lambda]}$, one obtains a traceless tensor subspace $\mathcal{Y}_\mu^{[\lambda]}\mathcal{T} = \mathcal{T}_\mu^{[\lambda]}$, where the basis tensors are taken to be the tensor Young tableaux.

We first study the condition whether the traceless tensor space $\mathcal{T}_\mu^{[\lambda]}$ is a null space or not. Without loss of generality, assume that the row number of $[\lambda]$ is r, and there are m pairs of digits j and $\overline{j}$ in the first column of an arbitrary tensor Young tableau in $\mathcal{T}_\mu^{[\lambda]}$. Fixing the unpaired digits in the first column and the digits in the remaining columns, one changes the pairs of digits from 1 to ℓ, where some tensor Young tableaux may be vanishing owing to antisymmetry of indices. The number of linearly independent tensor Young tableaux is the combinatorics of m among $[\ell-(r-2m)]$. The traceless condition is written as Eq. (10.9), where the $(m-1)$ pairs of digits are fixed and only one pair of digits runs over from 1 to ℓ. The number of traceless conditions is the number of possible values of the $(m-1)$ pairs of digits, that is the combinatorics of $(m-1)$ among $[\ell-(r-2m)]$. The traceless tensor space $\mathcal{T}_\mu^{[\lambda]}$ is a null space if the number of traceless conditions is not less than the number of independent tensors, namely, $[\ell-(r-2m)]/2 < m$, and then, $r > \ell$. Thus, the traceless tensor space $\mathcal{T}_\mu^{[\lambda]}$ is a null space if the row number r of the Young pattern $[\lambda]$ is larger than ℓ.

The other invariant tensor of USp(2ℓ) is the totally antisymmetric tensor $\epsilon_{a_1\cdots a_{2\ell}}$ of rank 2ℓ. This tensor violates the traceless condition (10.9). It is irrelevant to the reduction of the tensor space for USp(2ℓ). The traceless tensor subspace $\mathcal{Y}_\mu^{[\lambda]}\mathcal{T}$ is minimal because there is no further constraint to construct a nontrivial invariant subspace in $\mathcal{Y}_\mu^{[\lambda]}\mathcal{T}$. Thus, the irreducible representation of USp(2ℓ) is self-conjugate and denoted by the Young pattern $[\lambda]$ with the row number not larger than ℓ.

10.1.2 *Orthonormal Irreducible Basis Tensors*

The generators in the self-representation of USp(2ℓ) are given in Eq. (7.106)

$$T_{jk}^{(2)} \times \mathbf{1}_2, \qquad T_{jk}^{(1)} \times \sigma_d, \qquad T_{jj}^{(1)} \times \sigma_d/\sqrt{2},$$

where $1 \leq d \leq 3$, $1 \leq j < k \leq \ell$, and $T_{jk}^{(r)}$ are the generators in the self-representation of SU(ℓ). The generators satisfy the orthonormal condition $\text{Tr}(T_A T_B) = \delta_{AB}$. The Chevalley bases of USp(2ℓ) can be calculated by Eq. (7.141),

$$\begin{aligned}
H_\mu &= \left\{T^{(1)}_{\mu\mu} - T^{(1)}_{(\mu+1)(\mu+1)}\right\} \times \sigma_3, & H_\ell &= T^{(1)}_{\ell\ell} \times \sigma_3, \\
E_\mu &= T^{(1)}_{\mu(\mu+1)} \times \sigma_3 + iT^{(2)}_{\mu(\mu+1)} \times \mathbf{1}_2\ , & E_\ell &= T^{(1)}_{\ell\ell} \times (\sigma_1 + i\sigma_2)/2, \\
F_\mu &= T^{(1)}_{\mu(\mu+1)} \times \sigma_3 - iT^{(2)}_{\mu(\mu+1)} \times \mathbf{1}_2\ , & F_\ell &= T^{(1)}_{\ell\ell} \times (\sigma_1 - i\sigma_2)/2,
\end{aligned} \tag{10.11}$$

where $1 \leq \mu < \ell$. The nonvanishing action of the generators on the basis vectors $\boldsymbol{\theta}_a$ are

$$\begin{aligned}
&H_\mu \boldsymbol{\theta}_\mu = \boldsymbol{\theta}_\mu, && H_\mu \boldsymbol{\theta}_{\mu+1} = -\boldsymbol{\theta}_{\mu+1}, && H_\mu \boldsymbol{\theta}_{\overline{\mu+1}} = \boldsymbol{\theta}_{\overline{\mu+1}}, \\
&H_\mu \boldsymbol{\theta}_{\overline{\mu}} = -\boldsymbol{\theta}_{\overline{\mu}}, && H_\ell \boldsymbol{\theta}_\ell = \boldsymbol{\theta}_\ell, && H_\ell \boldsymbol{\theta}_{\overline{\ell}} = -\boldsymbol{\theta}_{\overline{\ell}}, \\
&E_\mu \boldsymbol{\theta}_{\mu+1} = \boldsymbol{\theta}_\mu, && E_\mu \boldsymbol{\theta}_{\overline{\mu}} = -\boldsymbol{\theta}_{\overline{\mu+1}}, && E_\ell \boldsymbol{\theta}_{\overline{\ell}} = \boldsymbol{\theta}_\ell, \\
&F_\mu \boldsymbol{\theta}_\mu = \boldsymbol{\theta}_{\mu+1}, && F_\mu \boldsymbol{\theta}_{\overline{\mu+1}} = -\boldsymbol{\theta}_{\overline{\mu}}, && F_\ell \boldsymbol{\theta}_\ell = \boldsymbol{\theta}_{\overline{\ell}}.
\end{aligned} \tag{10.12}$$

Namely, $\boldsymbol{\theta}_a$ is the common eigenvector of H_μ and H_ℓ, but its arranged order is not convenient for the calculation of the raising and lowering operators. The minus signs in the actions of E_μ and F_μ are also not convenient. Define

$$\phi_\mu = \boldsymbol{\theta}_\mu, \qquad \phi_{\ell+\mu} = (-1)^{\mu+1} \boldsymbol{\theta}_{\overline{\ell-\mu+1}}, \qquad 1 \leq \mu \leq \ell, \tag{10.13}$$

namely the order of ϕ_α, $1 \leq \alpha \leq 2\ell$, is

$$1,\ 2,\ \ldots,\ \ell,\ \overline{\ell},\ \overline{\ell-1},\ \ldots,\ \overline{2},\ \overline{1}.$$

Thus, Eq. (10.12) becomes

$$\begin{aligned}
&H_\mu \phi_\mu = \phi_\mu, && H_\mu \phi_{\mu+1} = -\phi_{\mu+1}, \\
&H_\mu \phi_{2\ell-\mu} = \phi_{2\ell-\mu}, && H_\mu \phi_{2\ell-\mu+1} = -\phi_{2\ell-\mu+1}, \\
&H_\ell \phi_\ell = \phi_\ell, && H_\ell \phi_{\ell+1} = -\phi_{\ell+1}, \\
&E_\mu \phi_{\mu+1} = \phi_\mu, && E_\mu \phi_{2\ell-\mu+1} = \phi_{2\ell-\mu}, \\
&E_\ell \phi_{\ell+1} = \phi_\ell, && F_\mu \phi_\mu = \phi_{\mu+1}, \\
&F_\mu \phi_{2\ell-\mu} = \phi_{2\ell-\mu+1}, && F_\ell \phi_\ell = \phi_{\ell+1},
\end{aligned} \tag{10.14}$$

where $1 \leq \mu < \ell$.

The basis tensor $\phi_{\alpha_1 \ldots \alpha_n}$ is the direct product of the basis vectors ϕ_α. The action of a generator on the basis tensor is equal to the sum of its action on each basis vector in the product. The standard tensor Young tableaux $\mathcal{Y}^{[\lambda]}_\mu \phi_{\alpha_1 \ldots \alpha_n}$ are the common eigenstates of H_μ, but generally not orthonormal and traceless. The eigenvalue of H_μ in the standard tensor Young tableau $\mathcal{Y}^{[\lambda]}_\mu \phi_{\alpha_1 \ldots \alpha_n}$ is the number of the digits μ and $(2\ell - \mu)$ filled in the tableau, minus the number of the digits $(\mu + 1)$ and $(2\ell - \mu + 1)$.

The eigenvalue of H_ℓ in the standard tensor Young tableau is equal to the number of the digit ℓ in the tableau, minus the number of the digit $\ell+1$. The eigenvalues constitutes the weight $\boldsymbol{m}$ of the standard tensor Young tableau. Two standard tensor Young tableaux are orthogonal if their weights are different. The action of F_μ on the standard tensor Young tableau is equal to the sum of all possible tensor Young tableaux, each of which is obtained from the original one by replacing one filled digit μ with the digit $(\mu+1)$, or by replacing one filled digit $(2\ell-\mu)$ with the digit $(2\ell-\mu+1)$. The action of F_ℓ on the standard tensor Young tableau is equal to the sum of all possible tensor Young tableaux, each of which is obtained from the original one by replacing one filled digit ℓ with the digit $(\ell+1)$. The actions of E_μ and E_ℓ are opposite. The obtained tensor Young tableaux may be not standard, but they can be transformed to the sum of the standard tensor Young tableaux by the symmetry (8.22).

The row number of $[\lambda]$ in the traceless tensor subspace $\mathcal{T}_\mu^{[\lambda]}$ is not larger than ℓ, $[\lambda]=[\lambda_1,\lambda_2,\ldots,\lambda_\ell]$. Through a similar proof as that for Theorem 8.3, there is one and only one traceless standard tensor Young tableau in $\mathcal{T}_\mu^{[\lambda]}$ which is annihilated by every raising operators E_μ [see Eq. (7.115)]. This traceless standard tensor Young tableau, where each box in its αth row is filled with the digit α, corresponds to the highest weight in $\mathcal{T}_\mu^{[\lambda]}$, and the highest weight $\boldsymbol{M}=\sum_\mu \boldsymbol{w}_\mu M_\mu$ is calculated from Eq. (10.14),

$$M_\mu=\lambda_\mu-\lambda_{\mu+1}, \qquad M_\ell=\lambda_\ell, \qquad 1\leq\mu<\ell. \tag{10.15}$$

It means that the irreducible representation of USp(2ℓ) is denoted by the Young pattern $[\lambda]$ with the row number not larger than ℓ. The remaining basis tensors in $\mathcal{T}_\mu^{[\lambda]}$ can be calculated from the standard tensor Young tableau with the highest weight by the lowering operators F_μ in the method of block weight diagram, where the multiplicity of a weight can be obtained by counting the number of the traceless standard tensor Young tableaux with the weight in $\mathcal{T}_\mu^{[\lambda]}$. The calculated standard tensor Young tableaux are traceless and orthonormal. Obviously, they are normalized to what the highest weight state is normalized to.

The irreducible representations of Sp(2ℓ, R) can be obtained from those of USp(2ℓ) by replacing some parameters to be pure imaginary (see the discussion below Eq. (7.106)) but preserving the generators invariant.

The basis tensors in the representations $[1,1,0]$ and $[1,1,1]$ are calculated in Probs. 5 and 6 of Chap. 10 of [Ma and Gu (2004)]. The tensor in $[1,1,0]$ is antisymmetric and can be decomposed into a traceless tensor and a trace tensor (scalar), as shown in Eq. (10.10). The representation

$[1,1,0]$ contains a single dominant weight $(0,1,0)$ and a double dominant weight $(0,0,0)$. The traceless standard tensor Young tableaux with the weight $(0,0,0)$ are

$$\sqrt{\frac{1}{2}}\left(\begin{array}{|c|}\hline 1 \\ \hline 6 \\ \hline\end{array} + \begin{array}{|c|}\hline 2 \\ \hline 5 \\ \hline\end{array}\right), \qquad \sqrt{\frac{1}{6}}\left(-\begin{array}{|c|}\hline 1 \\ \hline 6 \\ \hline\end{array} + \begin{array}{|c|}\hline 2 \\ \hline 5 \\ \hline\end{array} + 2\begin{array}{|c|}\hline 3 \\ \hline 4 \\ \hline\end{array}\right),$$

which are orthogonal to the trace tensor

$$\begin{array}{|c|}\hline 1 \\ \hline 6 \\ \hline\end{array} - \begin{array}{|c|}\hline 2 \\ \hline 5 \\ \hline\end{array} + \begin{array}{|c|}\hline 3 \\ \hline 4 \\ \hline\end{array}.$$

Please notice the different definitions in the basis vectors ϕ_α between Eq. (10.13) of this textbook and Eq. (10.26) in [Ma and Gu (2004)]. Namely, ϕ_5 changes a sign for USp(6). The representation $[1,1,1]$ contains two single dominant weights $(0,0,1)$ and $(1,0,0)$.

In the following the basis tensors in the adjoint representation $[2,0,0]$ of USp(6) are calculated. The tensors in the representation are symmetric so that all basis tensors are traceless. The simple roots $\boldsymbol{r}_\mu$ of USp(6) are expressed with respect to the fundamental dominant weights $\boldsymbol{w}_\nu$,

$$\mathbf{r}_1 = 2\mathbf{w}_1 - \mathbf{w}_2, \quad \mathbf{r}_2 = -\mathbf{w}_1 + 2\mathbf{w}_2 - \mathbf{w}_3, \quad \mathbf{r}_3 = -2\mathbf{w}_2 + 2\mathbf{w}_3. \tag{10.16}$$

There are two typical tensor Young tableaux

$$\begin{array}{|c|c|}\hline \alpha & \alpha \\ \hline\end{array} = 2\phi_{\alpha\alpha}, \qquad \begin{array}{|c|c|}\hline \alpha & \beta \\ \hline\end{array} = \phi_{\alpha\alpha} + \phi_{\beta\alpha}.$$

They are normalized to 4 and 2, respectively. The highest weight in $[2,0,0]$ is $(2,0,0)$. The block weight diagram and the basis tensors of $[2,0,0]$ of USp(6) are listed in Fig. 10.1. Please compare Fig. 10.1 with Fig. 9.1. Since $[2,0,0]$ is the adjoint representation of USp(2ℓ), all positive roots can be written from Fig. 10.1. In addition to the three simple roots given in Eq. (10.16), the remaining positive roots are

$$\begin{aligned}
&\boldsymbol{\alpha}_1 = \boldsymbol{w}_1 + \boldsymbol{w}_2 - \boldsymbol{w}_3 = \boldsymbol{r}_1 + \boldsymbol{r}_2, && \boldsymbol{\alpha}_2 = -\boldsymbol{w}_1 + \boldsymbol{w}_3 = \boldsymbol{r}_2 + \boldsymbol{r}_3, \\
&\boldsymbol{\alpha}_3 = \boldsymbol{w}_1 - \boldsymbol{w}_2 + \boldsymbol{w}_3 = \boldsymbol{r}_1 + \boldsymbol{r}_2 + \boldsymbol{r}_3, && \boldsymbol{\alpha}_4 = -2\boldsymbol{w}_1 + 2\boldsymbol{w}_2 = 2\boldsymbol{r}_2 + \boldsymbol{r}_3, \\
&\boldsymbol{\alpha}_5 = \boldsymbol{w}_2 = \boldsymbol{r}_1 + 2\boldsymbol{r}_2 + \boldsymbol{r}_3, && \boldsymbol{\alpha}_6 = 2\boldsymbol{w}_1 = 2\boldsymbol{r}_1 + 2\boldsymbol{r}_2 + \boldsymbol{r}_3.
\end{aligned} \tag{10.17}$$

The calculations related with the multiple weight $(0,0,0)$ are given as follows.

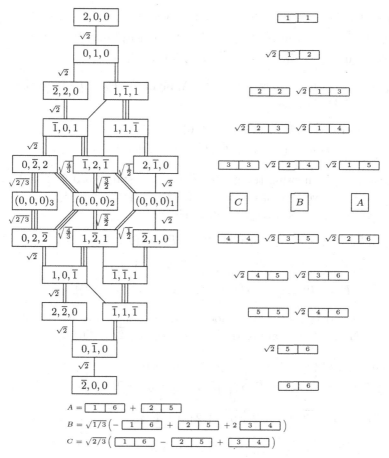

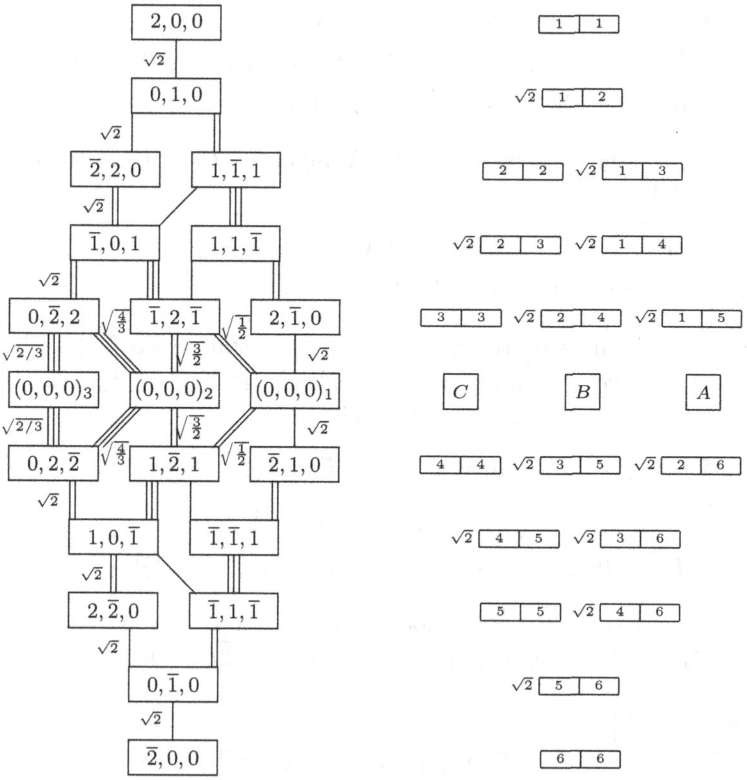

$$A = \begin{array}{|c|c|}\hline 1 & 6 \\ \hline\end{array} + \begin{array}{|c|c|}\hline 2 & 5 \\ \hline\end{array}$$

$$B = \sqrt{1/3}\left(-\begin{array}{|c|c|}\hline 1 & 6 \\ \hline\end{array} + \begin{array}{|c|c|}\hline 2 & 5 \\ \hline\end{array} + 2\begin{array}{|c|c|}\hline 3 & 4 \\ \hline\end{array}\right)$$

$$C = \sqrt{2/3}\left(\begin{array}{|c|c|}\hline 1 & 6 \\ \hline\end{array} - \begin{array}{|c|c|}\hline 2 & 5 \\ \hline\end{array} + \begin{array}{|c|c|}\hline 3 & 4 \\ \hline\end{array}\right)$$

Fig. 10.1 The block weight diagram and the basis tensors in $[2,0,0]$ of USp(6).

$$\begin{aligned}
&|(2,\bar{1},0)\rangle = F_2|(1,1,\bar{1})\rangle = \sqrt{2}\begin{array}{|c|c|}\hline 1 & 5 \\ \hline\end{array},\\
&|(\bar{1},2,\bar{1})\rangle = F_1|(1,1,\bar{1})\rangle = F_3|(\bar{1},0,1)\rangle = \sqrt{2}\begin{array}{|c|c|}\hline 2 & 4 \\ \hline\end{array},\\
&|(0,\bar{2},2)\rangle = \sqrt{1/2}F_2|(\bar{1},0,1)\rangle = \begin{array}{|c|c|}\hline 3 & 3 \\ \hline\end{array}.
\end{aligned}$$

From three states, an $\mathcal{A}_1$-triplet, an $\mathcal{A}_2$-triplet, and an $\mathcal{A}_3$-triplet are constructed, respectively. There are three standard tensor Young tableaux with the weight $(0,0,0)$ so that the weight is triple. Assume

$$|(0,0,0)_1\rangle = \sqrt{1/2}F_1\ |(2,\bar{1},0)\rangle = F_1\begin{array}{|c|c|}\hline 1 & 5 \\ \hline\end{array} = \begin{array}{|c|c|}\hline 1 & 6 \\ \hline\end{array} + \begin{array}{|c|c|}\hline 2 & 5 \\ \hline\end{array},$$

$$\begin{aligned}
&F_2\ |(\overline{1},2,\overline{1})\rangle = a_1\ |(0,0,0)_1\rangle + a_2\ |(0,0,0)_2\rangle, \\
&F_3\ |(0,\overline{2},2)\rangle = b_1\ |(0,0,0)_1\rangle + b_2\ |(0,0,0)_2\rangle + b_3\ |(0,0,0)_3\rangle, \\
&E_1\ |(0,0,0)_2\rangle = E_1\ |(0,0,0)_3\rangle = E_2\ |(0,0,0)_3\rangle = 0,
\end{aligned}$$

where $a_1^2 + a_2^2 = 2$ and $b_1^2 + b_2^2 + b_3^2 = 2$. Applying $E_1F_2 = F_2E_1$ to the basis state $|(\overline{1},2,\overline{1})\rangle$, one has

$$\begin{aligned}
E_1F_2\ |(\overline{1},2,\overline{1})\rangle &= \sqrt{2}a_1\ |(2,\overline{1},0)\rangle \\
&= F_2E_1\ |(\overline{1},2,\overline{1})\rangle = F_2\ |(1,1,\overline{1})\rangle = \ |(2,\overline{1},0)\rangle.
\end{aligned}$$

Thus, $a_1 = \sqrt{1/2}$. Choosing the phase of the basis state $|(0,0,0)_2\rangle$ such that a_2 is real positive, one has $a_2 = \sqrt{3/2}$. Applying $E_1F_3 = F_3E_1$ and $E_2F_3 = F_3E_2$ to the basis state $|(0,\overline{2},2)\rangle$, one has

$$\begin{aligned}
&E_1F_3\ |(0,\overline{2},2)\rangle = \sqrt{2}\ b_1\ |(2,\overline{1},0)\rangle = F_3E_1\ |(0,\overline{2},2)\rangle = 0, \\
&E_2F_3\ |(0,\overline{2},2)\rangle = \left(\sqrt{1/2}\ b_1 + \sqrt{3/2}\ b_2\right)\ |(\overline{1},2,\overline{1})\rangle \\
&\qquad = F_3E_2\ |(0,\overline{2},2)\rangle = \sqrt{2}\ F_3\ |(\overline{1},0,1)\rangle = \sqrt{2}\ |(\overline{1},2,\overline{1})\rangle.
\end{aligned}$$

Thus, $b_1 = 0$ and $b_2 = 2/\sqrt{3}$. Choosing the phase of the basis state $|(0,0,0)_3\rangle$ such that b_3 is real positive, one has $b_3 = \sqrt{2/3}$. Then,

$$\begin{aligned}
|(0,0,0)_2\rangle &= \sqrt{2/3}\left\{F_2\ |(\overline{1},2,\overline{1})\rangle - \sqrt{1/2}\ |(0,0,0)_1\rangle\right\} \\
&= \sqrt{4/3}\ F_2\ \boxed{2\,|\,4} - \sqrt{1/3}\left\{\boxed{1\,|\,6} + \boxed{2\,|\,5}\right\} \\
&= \sqrt{1/3}\left\{-\boxed{1\,|\,6} + \boxed{2\,|\,5} + 2\boxed{3\,|\,4}\right\}, \\
|(0,0,0)_3\rangle &= \sqrt{3/2}\left\{F_3\ |(0,\overline{2},2)\rangle - \sqrt{4/3}\ |(0,0,0)_2\rangle\right\} \\
&= \sqrt{3/2}\ F_3\ \boxed{3\,|\,3} - \sqrt{2/3}\left\{-\boxed{1\,|\,6} + \boxed{2\,|\,5} + 2\boxed{3\,|\,4}\right\} \\
&= \sqrt{2/3}\left\{\boxed{1\,|\,6} - \boxed{2\,|\,5} + \boxed{3\,|\,4}\right\}.
\end{aligned}$$

10.1.3 *Dimensions of Irreducible Representations*

The dimension of an irreducible representation $[\lambda]$ of USp(2ℓ) can be calculated by the hook rule. In this rule, the dimension is expressed as a quotient, where the numerator and the denominator are denoted by the symbols $Y_P^{[\lambda]}$ and $Y_h^{[\lambda]}$, respectively:

$$d_{[\lambda]}(\mathrm{Sp}(2\ell)) = \frac{Y_P^{[\lambda]}}{Y_h^{[\lambda]}}. \tag{10.18}$$

We still use the concept of the hook path $(i,\ j)$ in the Young pattern $[\lambda]$, which enters the Young pattern at the rightmost of the ith row, goes leftward in the i row, turns downward at the j column, goes downward in the j column, and leaves from the Young pattern at the bottom of the j column. The inverse hook path $\overline{(i,\ j)}$ is the same path as the hook path $(i,\ j)$ except for the opposite direction. The number of boxes contained in the hook path $(i,\ j)$ is the hook number h_{ij} of the box in the jth column of the ith row. $Y_h^{[\lambda]}$ is a tableau of the Young pattern $[\lambda]$ where the box in the jth column of the ith row is filled with the hook number h_{ij}. Define a series of the tableaux $Y_{P_a}^{[\lambda]}$ recursively by the rule given below. $Y_P^{[\lambda]}$ is a tableau of the Young pattern $[\lambda]$ where each box is filled with the sum of the digits which are respectively filled in the same box of each tableau $Y_{P_a}^{[\lambda]}$ in the series. The symbol $Y_P^{[\lambda]}$ means the product of the filled digits in it, so does the symbol $Y_h^{[\lambda]}$.

The tableaux $Y_{P_a}^{[\lambda]}$ are defined by the following rule:

(a) $Y_{P_0}^{[\lambda]}$ is a tableau of the Young pattern $[\lambda]$ where the box in the jth column of the ith row is filled with the digit $(2\ell + j - i)$.

(b) Let $[\lambda^{(1)}] = [\lambda]$. Beginning with $[\lambda^{(1)}]$, we define recursively the Young pattern $[\lambda^{(a)}]$ by removing the first row and the first column of the Young pattern $[\lambda^{(a-1)}]$ until $[\lambda^{(a)}]$ contains less than two rows.

(c) If $[\lambda^{(a)}]$ contains more than one row, define $Y_{P_a}^{[\lambda]}$ to be a tableau of the Young pattern $[\lambda]$ where the boxes in the first $(a-1)$ rows and in the first $(a-1)$ columns are filled with 0, and the remaining part of the Young pattern is nothing but $[\lambda^{(a)}]$. Let $[\lambda^{(a)}]$ have r rows. Fill the first $(r-1)$ boxes along the hook path $(1,\ 1)$ of the Young pattern $[\lambda^{(a)}]$, beginning with the box on the rightmost, with the digits $\lambda_2^{(a)}, \lambda_3^{(a)}, \cdots, \lambda_r^{(a)}$, box by box, and fill the first $\lambda_i^{(a)}$ boxes in each inverse hook path $\overline{(i,\ 1)}$ of the Young pattern $[\lambda^{(a)}]$, $2 \leq i \leq r$, with -1. The remaining boxes are filled with 0. If a few -1 are filled in the same box, the digits are summed. The sum of all filled digits in the pattern $Y_{S_a}^{[\lambda]}$ with $a > 0$ is 0.

The calculation method (10.18) is explained through some examples.

Ex. 1 The dimension of the representation $[3,3,3]$ of USp(6).

$$Y_P^{[3,3,3]} = \begin{array}{|c|c|c|}\hline 6 & 7 & 8 \\ \hline 5 & 6 & 7 \\ \hline 4 & 5 & 6 \\ \hline \end{array} + \begin{array}{|c|c|c|}\hline & 3 & 3 \\ \hline -1 & -1 & \\ \hline -2 & -1 & -1 \\ \hline \end{array} + \begin{array}{|c|c|c|}\hline & & \\ \hline & & 2 \\ \hline & -1 & -1 \\ \hline \end{array} = \begin{array}{|c|c|c|}\hline 6 & 10 & 11 \\ \hline 4 & 5 & 9 \\ \hline 2 & 3 & 4 \\ \hline \end{array},$$

$$d_{[3,3,3]}[\mathrm{USp}(6)] = \frac{\begin{array}{|c|c|c|}\hline 6 & 10 & 11 \\ \hline 4 & 5 & 9 \\ \hline 2 & 3 & 4 \\ \hline \end{array}}{\begin{array}{|c|c|c|}\hline 5 & 4 & 3 \\ \hline 4 & 3 & 2 \\ \hline 3 & 2 & 1 \\ \hline \end{array}} = 11 \times 5 \times 3 \times 2 = 330.$$

Ex. 2 The representation of one-row Young pattern $[n]$ of USp(2ℓ). The tensors in the representation are symmetric, so that all standard tensor Young tableaux are traceless.

$$d_{[n]}[\mathrm{USp}(2\ell)] = d_{[n]}[\mathrm{SU}(2\ell)] = \binom{n+2\ell-1}{n}. \tag{10.19}$$

Ex. 3 The representation of one-column Young pattern $[1^n]$ of USp(2ℓ).

$$Y_P^{[1^n]} = \begin{array}{|c|}\hline 2\ell \\ \hline 2\ell-1 \\ \hline \vdots \\ \hline 2\ell-n+2 \\ \hline 2\ell-n+1 \\ \hline \end{array} + \begin{array}{|c|}\hline 1 \\ \hline 1 \\ \hline \vdots \\ \hline 1 \\ \hline -n+1 \\ \hline \end{array} = \begin{array}{|c|}\hline 2\ell+1 \\ \hline 2\ell \\ \hline \vdots \\ \hline 2\ell-n+3 \\ \hline 2\ell-2n+2 \\ \hline \end{array},$$

$$d_{[1^n]}[\mathrm{USp}(2\ell)] = \frac{(2\ell+1)!(2\ell-2n+2)}{n!(2\ell-n+2)!}. \tag{10.20}$$

Ex. 4 The representation of two-row Young pattern $[n,m]$ of USp(2ℓ).

$$\begin{aligned} Y_P^{[n,m]} &= \begin{array}{|c|c|c|c|c|c|}\hline 2\ell & \cdots & 2\ell+m-1 & \cdots & 2\ell+n-2 & 2\ell+n-1 \\ \hline 2\ell-1 & \cdots & 2\ell+m-2 \\ \cline{1-3} \end{array} \\ &+ \begin{array}{|c|c|c|c|c|c|}\hline 0 & \cdots & 0 & \cdots & 0 & m \\ \hline -1 & \cdots & -1 \\ \cline{1-3} \end{array} \\ &= \begin{array}{|c|c|c|c|c|c|}\hline 2\ell & \cdots & 2\ell+m-1 & \cdots & 2\ell+n-2 & 2\ell+n+m-1 \\ \hline 2\ell-2 & \cdots & 2\ell+m-3 \\ \cline{1-3} \end{array}, \end{aligned}$$

$$Y_h^{[n,m]} = \begin{array}{|c|c|c|c|c|c|}\hline n+1 & \cdots & n-m+2 & n-m & \cdots & 1 \\ \hline m & \cdots & 1 \\ \cline{1-3} \end{array},$$

$$d_{[n,m]}[\mathrm{USp}(2\ell)] = \frac{(n-m+1)(2\ell+n+m-1)(2\ell+n-2)!(2\ell+m-3)!}{(n+1)!m!(2\ell-1)!(2\ell-3)!}. \tag{10.21}$$

For the groups USp(4) and USp(6), one has

$$\begin{aligned} d_{[n,m]}[\mathrm{USp}(4)] &= (n-m+1)(n+m+3)(n+2)(m+1)/6, \\ d_{[n,m]}[\mathrm{USp}(6)] &= (n-m+1)(n+m+5)(n+4)(n+3)(n+2) \\ &\quad \times (m+3)(m+2)(m+1)/720. \end{aligned}$$

10.2 Physical Application

The Hamiltonian equation of a classical system with ℓ degrees of freedom is

$$\frac{dq_j}{dt} = \frac{\partial H}{\partial p_j}, \qquad \frac{dp_j}{dt} = -\frac{\partial H}{\partial q_j}. \tag{10.22}$$

Arranging the coordinates q_j and the momentums p_j in the order,

$$x_a = (q_1,\ p_1,\ q_2,\ p_2,\ \cdots,\ q_\ell,\ p_\ell), \tag{10.23}$$

one obtains the coordinates x_a in the (2ℓ)-dimensional phase space. The Hamiltonian equation can be expressed in a unified form

$$\frac{dx_a}{dt} = \sum_b J_{ab} \frac{\partial H}{\partial x_b}, \qquad \frac{dx}{dt} = J \frac{\partial H}{\partial x}, \tag{10.24}$$

called the symplectic form of the Hamiltonian equation. If dx_a satisfy the Hamiltonian equation (10.24), after the symplectic transformation

$$dz_a = \sum_b R_{ab} dx_b, \qquad R_{ab} = \frac{\partial z_a}{\partial x_b}, \qquad R^T J R = J, \tag{10.25}$$

dz_a still satisfy the Hamiltonian equation (10.24),

$$\begin{aligned} \frac{dz_a}{dt} &= \sum_b R_{ab} \frac{dx_b}{dt} = \sum_{bc} R_{ab} J_{bc} \sum_r \frac{\partial z_r}{\partial x_c} \frac{\partial H}{\partial z_r} \\ &= \sum_r \left(R J R^T\right)_{ar} \frac{\partial H}{\partial z_r} = \sum_r J_{ar} \frac{\partial H}{\partial z_r}. \end{aligned}$$

The method of Runge–Kutta is commonly used in the numerical calculations by computer. This method does not reflect the characteristic of the equation of motion so that the calculation error will be accumulated. If the calculation is repeated in a tremendous number, the accumulated error will make a big deviation of the calculation data from the real orbit, for example, the calculation in cyclotron reaction and in satellites. If each step in the numerical calculation reflects the characteristic of the Hamiltonian equation, say each step satisfies the symplectic transformation,

$$z_a = z_a^{(0)} + \tau \sum_b J_{ab} \frac{\partial H(x)}{\partial z_b}, \qquad x_a = \left(z_a + z_a^{(0)} \right) /2, \tag{10.26}$$

where τ is the length of the step, the accumulated error will decrease greatly. The group leaded by Professor Feng Kang studied deeply this problem. Please see his paper [Feng (1991)] in detail.

10.3 Exercises

1. Prove that the determinant of R in Sp$(2\ell, R)$ and the determinant of u in USp(2ℓ) are both $+1$.

2. Count the number of independent real parameters of R in Sp$(2\ell, R)$ and u in USp(2ℓ) directly from their definitions (7.95) and (10.3).

3. Express the simple roots of USp(2ℓ) by the vectors $\mathbf{V}_a$ given in Eq. (7.79) for the SU$(\ell+1)$ group, and then, write their Cartan–Weyl bases of generators in the self-representation of USp(2ℓ).

4. Calculate the dimensions of the irreducible representations of the USp(6) group denoted by the following Young patterns:

   (1) $[4,2]$, (2) $[3,2]$, (3) $[4,4]$, (4) $[3,3,2]$, (5) $[4,4,3]$.

5. Calculate the orthonormal bases in the irreducible representation $[1,1,0]$ of the USp(6) group by the method of the block weight diagram, and then, express the orthonormal bases by the standard tensor Young tableaux in the traceless tensor space of rank 2 for USp(6).

6. Calculate the orthonormal bases in the irreducible representation $[1,1,1]$ of the USp(6) group by the method of the block weight diagram, and then, express the orthonormal bases by the standard tensor Young tableaux in the traceless tensor space of rank 3 for USp(6).

7. Calculate the Clebsch–Gordan series for the reduction of the direct product representation $[1,1,0] \times [1,1,0]$ of the USp(6) group and the highest weight states for the representations contained in the series by the method of the standard tensor Young tableau.

Appendix A

Identities on Combinatorics

There are two binomial identities for a complex z and a positive integer a

$$\begin{aligned}
(1+z)^a &= \sum_{n=0}^{a} z^n \binom{a}{n}, \qquad \binom{a}{n} = \frac{a!}{n!(a-n)!}, \\
(1+z)^{-a} &= \sum_{n=0}^{\infty} (-1)^n z^n \binom{a+n-1}{n}.
\end{aligned} \tag{A.1}$$

Hereafter, the summation index runs over the region where the denominator is finite. Since

$$\begin{aligned}
(1+z)^{a+b} &= \sum_m z^m \binom{a+b}{m} = (1+z)^a (1+z)^b \\
&= \sum_n z^n \binom{a}{n} \sum_\ell z^\ell \binom{b}{\ell} = \sum_m z^m \sum_n \binom{a}{n} \binom{b}{m-n},
\end{aligned}$$

one has

$$\begin{aligned}
&\sum_p \binom{u}{p} \binom{v}{r-p} = \binom{u+v}{r}, \\
&\sum_p \{p!(v-r+p)!(u-p)!(r-p)!\}^{-1} = \frac{(u+v)!}{u!v!r!(u+v-r)!}.
\end{aligned} \tag{A.2}$$

Since

$$\begin{aligned}
(1+z)^{a-b} &= \sum_m (-1)^m z^m \binom{b-a+m-1}{m} \\
&= (1+z)^a (1+z)^{-b} = \sum_n z^n \binom{a}{n} \sum_\ell (-1)^\ell z^\ell \binom{b+\ell-1}{\ell}
\end{aligned}$$

$$= \sum_m (-1)^m z^m \sum_n (-1)^n \binom{a}{n} \binom{b+m-n-1}{m-n},$$

one has

$$\begin{aligned} &\sum_p (-1)^p \binom{u}{p} \binom{v-p}{r-p} = \binom{v-u}{r}, \\ &\sum_p \frac{(-1)^p (v-p)!}{p!(u-p)!(r-p)!} = \frac{(v-u)!(v-r)!}{u!r!(v-u-r)!}. \end{aligned} \tag{A.3}$$

Since

$$\begin{aligned} (1+z)^{-a-b} &= \sum_m (-1)^m z^m \binom{a+b+m-1}{m} \\ &= (1+z)^{-a}(1+z)^{-b} \\ &= \sum_n (-1)^n z^n \binom{a+n-1}{n} \sum_\ell (-1)^\ell z^\ell \binom{b+\ell-1}{\ell} \\ &= \sum_m (-1)^m z^m \sum_n \binom{a+n-1}{n} \binom{b+m-n-1}{m-n}, \end{aligned}$$

one has

$$\begin{aligned} &\sum_p \binom{u+p-1}{p} \binom{v+r-p-1}{r-p} = \binom{u+v+r-1}{r}, \\ &\sum_p \frac{(u+p-1)!(v+r-p-1)!}{p!(r-p)!} = \frac{(u+v+r-1)!(u-1)!(v-1)!}{r!(u+v-1)!}. \end{aligned} \tag{A.4}$$

Appendix B

Covariant and Contravariant Tensors

Let G be a group whose elements R are $N \times N$ matrices. R can be looked like a coordinate transformation in an N-dimensional space,

$$x_a \xrightarrow{R} x'_a = \sum_d R_{ad} x_d. \tag{B.1}$$

A covariant tensor field $\boldsymbol{T}(x)_{a_1 \ldots a_n}$ of rank n with respect to the group G contains n subscripts and N^n components, which transform in $R \in G$ as

$$[O_R \boldsymbol{T}(x)]_{a_1 \ldots a_n} = \sum_{d_1 \ldots d_n} R_{a_1 d_1} \ldots R_{a_n d_n} \boldsymbol{T}(R^{-1} x)_{d_1 \ldots d_n}. \tag{B.2}$$

A covariant tensor field becomes a covariant tensor if its components are independent of coordinates x_a. A covariant tensor field can be expanded with respect to the covariant basis tensors $\boldsymbol{\theta}_{b_1 b_2 \ldots b_n}$

$$(\boldsymbol{\theta}_{b_1 \ldots b_n})_{a_1 \ldots a_n} = \delta_{a_1 b_1} \ldots \delta_{a_n b_n}, \tag{B.3}$$

$$\begin{aligned} \boldsymbol{T}(x) &= \sum_{b_1 \ldots b_n} T(x)_{b_1 \ldots b_n} \boldsymbol{\theta}_{b_1 \ldots b_n}, \\ \boldsymbol{T}(x)_{a_1 \ldots a_n} &= \sum_{b_1 \ldots b_n} T(x)_{b_1 \ldots b_n} (\boldsymbol{\theta}_{b_1 \ldots b_n})_{a_1 \ldots a_n} = T(x)_{a_1 \ldots a_n}. \end{aligned} \tag{B.4}$$

The coefficient $T(x)_{a_1 \ldots a_n}$ is equal to the tensor component $\boldsymbol{T}(x)_{a_1 \ldots a_n}$ in values, but transforms in R like a scalar. The tensor transformation is carried out by the basis tensor $\boldsymbol{\theta}_{b_1 \ldots b_n}$,

$$\begin{aligned} [O_R T(x)]_{a_1 \ldots a_n} &= [P_R T(x)]_{a_1 \ldots a_n} = T(R^{-1} x)_{a_1 \ldots a_n}, \\ O_R \boldsymbol{\theta}_{b_1 \ldots b_n} &= Q_R \boldsymbol{\theta}_{b_1 \ldots b_n} = \sum_{d_1 \ldots d_n} \boldsymbol{\theta}_{d_1 \ldots d_n} R_{d_1 b_1} \ldots R_{d_n b_n}, \end{aligned} \tag{B.5}$$

because

$$
\begin{aligned}
[O_R \boldsymbol{\theta}_{b_1\ldots b_n}]_{a_1\ldots a_n} &= \sum_{d_1\ldots d_n} R_{a_1d_1}\cdots R_{a_nd_n}\,[\boldsymbol{\theta}_{b_1\ldots b_n}]_{d_1\ldots d_n} \\
&= R_{a_1b_1}\cdots R_{a_nb_n} = \sum_{d_1\ldots d_n} [\boldsymbol{\theta}_{d_1\ldots d_n}]_{a_1\ldots a_n} R_{d_1b_1}\cdots R_{d_nb_n}.
\end{aligned}
$$

A contravariant tensor field $\boldsymbol{T}(x)^{a_1\cdots a_n}$ of rank n with respect to the group G contains n subscripts and N^n components, which transform in $R \in G$ as

$$
[O_R \boldsymbol{T}(x)]^{a_1\ldots a_n} = \sum_{d_1\ldots d_n} \boldsymbol{T}(R^{-1}x)^{d_1\ldots d_n} \left(R^{-1}\right)_{d_1a_1}\cdots\left(R^{-1}\right)_{d_na_n}. \tag{B.6}
$$

A contravariant tensor field becomes a contravariant tensor if its components are independent of coordinates x_a. A contravariant tensor field can be expanded with respect to the contravariant basis tensors $\boldsymbol{\theta}^{b_1\cdots b_n}$,

$$
\left(\boldsymbol{\theta}^{b_1\ldots b_n}\right)^{a_1\ldots a_n} = \delta_{a_1b_1}\cdots\delta_{a_nb_n}, \tag{B.7}
$$

$$
\begin{aligned}
\boldsymbol{T}(x) &= \sum_{b_1\ldots b_n} T(x)^{b_1\ldots b_n}\boldsymbol{\theta}^{b_1\ldots b_n}, \\
\boldsymbol{T}(x)^{a_1\ldots a_n} &= \sum_{b_1\ldots b_n} T(x)^{b_1\ldots b_n}\left(\boldsymbol{\theta}^{b_1\ldots b_n}\right)^{a_1\ldots a_n} = T(x)^{a_1\ldots a_n}.
\end{aligned} \tag{B.8}
$$

The coefficient $T(x)^{a_1\cdots a_n}$ is equal to the tensor component $\boldsymbol{T}(x)^{a_1\cdots a_n}$ in values, but transforms in R like a scalar. The tensor transformation is carried out by the basis tensor $\boldsymbol{\theta}^{b_1\cdots b_n}$,

$$
\begin{aligned}
&[O_R T(x)]^{a_1\ldots a_n} = [P_R T(x)]^{a_1\ldots a_n} = T(R^{-1}x)^{a_1\ldots a_n}, \\
&O_R\boldsymbol{\theta}_{b_1\ldots b_n} = Q_R\boldsymbol{\theta}_{b_1\ldots b_n} = \sum_{d_1\ldots d_n}\left(R^{-1}\right)_{b_1d_1}\cdots\left(R^{-1}\right)_{b_nd_n}\boldsymbol{\theta}_{d_1\ldots d_n}.
\end{aligned} \tag{B.9}
$$

A mixed tensor field $\boldsymbol{T}(x)^{b_1\cdots b_m}_{a_1\cdots a_n}$ of rank (n, m) with respect to the group G contains n subscripts, m superscripts, and N^{n+m} components, which transform in $R \in G$ as

$$
\begin{aligned}
[O_R\boldsymbol{T}(x)]^{b_1\ldots b_m}_{a_1\ldots a_n} = &\sum_{c_1\ldots c_n d_1\ldots d_m} R_{a_1c_1}\cdots R_{a_nc_n}\boldsymbol{T} \\
&\times (R^{-1}x)^{d_1\ldots d_m}_{c_1\ldots c_n}\left(R^{-1}\right)_{d_1a_1}\cdots\left(R^{-1}\right)_{d_na_n}.
\end{aligned} \tag{B.10}
$$

Appendix C

The Space Groups

230 space groups are listed with both Schroenflies notations (Sch) and the international notations for space groups (INSG). The star in the ordinal number denotes that the space group is symmorphic. The subscripts in the symbol for INSG are moved to a bracket for convenience. For example, the symbol $F \pm 2_{\frac{1}{4}\frac{1}{4}0} 2'_{0\frac{1}{4}\frac{1}{4}}$ for the space group D_{2h}^{24} is replaced with $F \pm 2(\frac{1}{4}\frac{1}{4}0)2'(0\frac{1}{4}\frac{1}{4})$.

Table C.1 Triclinic crystal system

Ordinal	Sch.	INSG	Ordinal	Sch.	INSG
*1	C_1^1	$P1$	*2	C_i^1	$P\overline{1}$

Table C.2 Monoclinic crystal system

Ordinal	Sch.	INSG	Ordinal	Sch.	INSG
*3	C_2^1	$P2$	*10	C_{2h}^1	$P \pm 2$
4	C_2^2	$P2(0\frac{1}{2}0)$	11	C_{2h}^2	$P \pm 2(0\frac{1}{2}0)$
*5	C_2^3	$A2$	*12	C_{2h}^3	$A \pm 2$
*6	C_s^1	$P\overline{2}$	13	C_{2h}^4	$P \pm 2(\frac{1}{2}00)$
7	C_s^2	$P\overline{2}(\frac{1}{2}00)$	14	C_{2h}^5	$P \pm 2(\frac{1}{2}\frac{1}{2}0)$
*8	C_s^3	$A\overline{2}$	15	C_{2h}^6	$A \pm 2(\frac{1}{2}00)$
9	C_s^4	$A\overline{2}(\frac{1}{2}00)$			

Table C.3 Orthorhombic crystal system

Ordinal	Sch.	INSG	Ordinal	Sch.	INSG
*16	D_2^1	$P22'$	*25	C_{2v}^1	$P2\overline{2}'$
17	D_2^2	$P22'(0\frac{1}{2}0)$	26	C_{2v}^2	$P2(00\frac{1}{2})\overline{2}'$
18	D_2^3	$P22'(\frac{1}{2}\frac{1}{2}0)$	27	C_{2v}^3	$P2\overline{2}'(00\frac{1}{2})$
19	D_2^4	$P2(00\frac{1}{2})2'(\frac{1}{2}\frac{1}{2}0)$	28	C_{2v}^4	$P2\overline{2}'(\frac{1}{2}00)$
20	D_2^5	$A22'(\frac{1}{2}00)$	29	C_{2v}^5	$P2(00\frac{1}{2})\overline{2}'(\frac{1}{2}0\frac{1}{2})$
*21	D_2^6	$A22'$	30	C_{2v}^6	$P2\overline{2}'(0\frac{1}{2}\frac{1}{2})$
*22	D_2^7	$F22'$	31	C_{2v}^7	$P2(00\frac{1}{2})\overline{2}'(\frac{1}{2}00)$
*23	D_2^8	$I22'$	32	C_{2v}^8	$P2\overline{2}'(\frac{1}{2}\frac{1}{2}0)$
24	D_2^9	$I2(00\frac{1}{2})2'(\frac{1}{2}\frac{1}{2}0)$	33	C_{2v}^9	$P2(00\frac{1}{2})\overline{2}'(\frac{1}{2}\frac{1}{2}\frac{1}{2})$

Table C.3 Orthorhombic crystal system (continued)

Ordinal	Sch.	INSG	Ordinal	Sch.	INSG
34	C_{2v}^{10}	$P2\overline{2}'(\frac{1}{2}\frac{1}{2}\frac{1}{2})$	55	D_{2h}^{9}	$P\pm 22'(\frac{1}{2}\frac{1}{2}0)$
*35	C_{2v}^{11}	$C2\overline{2}'$	56	D_{2h}^{10}	$P\pm 2(\frac{1}{2}\frac{1}{2}0)2'(\frac{1}{2}0\frac{1}{2})$
36	C_{2v}^{12}	$C2(00\frac{1}{2})\overline{2}'$	57	D_{2h}^{11}	$P\pm 2(00\frac{1}{2})2'(0\frac{1}{2}0)$
37	C_{2v}^{13}	$C2\overline{2}'(00\frac{1}{2})$	58	D_{2h}^{12}	$P\pm 22'(\frac{1}{2}\frac{1}{2}\frac{1}{2})$
*38	C_{2v}^{14}	$A2\overline{2}'$	59	D_{2h}^{13}	$P\pm 2(00\frac{1}{2})2'(0\frac{1}{2}\frac{1}{2})$
39	C_{2v}^{15}	$A2\overline{2}'(0\frac{1}{2}0)$	60	D_{2h}^{14}	$P\pm 2(\frac{1}{2}\frac{1}{2}0)2'(\frac{1}{2}\frac{1}{2}0)$
40	C_{2v}^{16}	$A2\overline{2}'(\frac{1}{2}00)$	61	D_{2h}^{15}	$P\pm 2(\frac{1}{2}0\frac{1}{2})2'(\frac{1}{2}\frac{1}{2}0)$
41	C_{2v}^{17}	$A2\overline{2}'(\frac{1}{2}\frac{1}{2}0)$	62	D_{2h}^{16}	$P\pm 2(00\frac{1}{2})2'(\frac{1}{2}\frac{1}{2}\frac{1}{2})$
*42	C_{2v}^{18}	$F2\overline{2}'$	63	D_{2h}^{17}	$A\pm 22'(\frac{1}{2}00)$
43	C_{2v}^{19}	$F2\overline{2}'(\frac{1}{4}\frac{1}{4}\frac{1}{4})$	64	D_{2h}^{18}	$A\pm 22'(\frac{1}{2}\frac{1}{2}0)$
*44	C_{2v}^{20}	$I2\overline{2}'$	*65	D_{2h}^{19}	$A\pm 22'$
45	C_{2v}^{21}	$I2\overline{2}'(\frac{1}{2}\frac{1}{2}0)$	66	D_{2h}^{20}	$A\pm 2(\frac{1}{2}00)2'$
46	C_{2v}^{22}	$I2\overline{2}'(\frac{1}{2}00)$	67	D_{2h}^{21}	$A\pm 22'(0\frac{1}{2}0)$
*47	D_{2h}^{1}	$P\pm 22'$	68	D_{2h}^{22}	$A\pm 2(\frac{1}{2}00)2'(0\frac{1}{2}0)$
48	D_{2h}^{2}	$P\pm 2(\frac{1}{2}\frac{1}{2}0)2'(0\frac{1}{2}\frac{1}{2})$	*69	D_{2h}^{23}	$F\pm 22'$
49	D_{2h}^{3}	$P\pm 22'(00\frac{1}{2})$	70	D_{2h}^{24}	$F\pm 2(\frac{1}{4}\frac{1}{4}0)2'(0\frac{1}{4}\frac{1}{4})$
50	D_{2h}^{4}	$P\pm 2(\frac{1}{2}\frac{1}{2}0)2'(0\frac{1}{2}0)$	*71	D_{2h}^{25}	$I\pm 22'$
51	D_{2h}^{5}	$P\pm 22'(\frac{1}{2}00)$	72	D_{2h}^{26}	$I\pm 22'(\frac{1}{2}\frac{1}{2}0)$
52	D_{2h}^{6}	$P\pm 2(\frac{1}{2}\frac{1}{2}0)2'(\frac{1}{2}\frac{1}{2}\frac{1}{2})$	73	D_{2h}^{27}	$I\pm 2(\frac{1}{2}0\frac{1}{2})2'(\frac{1}{2}\frac{1}{2}0)$
53	D_{2h}^{7}	$P\pm 22'(\frac{1}{2}0\frac{1}{2})$	74	D_{2h}^{28}	$I\pm 22'(\frac{1}{2}00)$
54	D_{2h}^{8}	$P\pm 2(\frac{1}{2}0\frac{1}{2})2'(00\frac{1}{2})$			

Table C.4 Tetragonal crystal system

Ordinal	Sch.	INSG	Ordinal	Sch.	INSG
*75	C_{4}^{1}	$P4$	96	D_{4}^{8}	$P4(00\frac{3}{4})2'(\frac{1}{2}\frac{1}{2}0)$
76	C_{4}^{2}	$P4(00\frac{1}{4})$	*97	D_{4}^{9}	$I42'$
77	C_{4}^{3}	$P4(00\frac{1}{2})$	98	D_{4}^{10}	$I4(00\frac{1}{4})2'$
78	C_{4}^{4}	$P4(00\frac{3}{4})$	*99	C_{4v}^{1}	$P4\overline{2}'$
*79	C_{4}^{5}	$I4$	100	C_{4v}^{2}	$P4\overline{2}'(\frac{1}{2}\frac{1}{2}0)$
80	C_{4}^{6}	$I4(00\frac{1}{4})$	101	C_{4v}^{3}	$P4(00\frac{1}{2})\overline{2}'(00\frac{1}{2})$
*81	S_{4}^{1}	$P\overline{4}$	102	C_{4v}^{4}	$P4(00\frac{1}{2})\overline{2}'(\frac{1}{2}\frac{1}{2}\frac{1}{2})$
*82	S_{4}^{2}	$I\overline{4}$	103	C_{4v}^{5}	$P4\overline{2}'(00\frac{1}{2})$
*83	C_{4h}^{1}	$P\pm 4$	104	C_{4v}^{6}	$P4\overline{2}'(\frac{1}{2}\frac{1}{2}\frac{1}{2})$
84	C_{4h}^{2}	$P\pm 4(00\frac{1}{2})$	105	C_{4v}^{7}	$P4(00\frac{1}{2})\overline{2}'$
85	C_{4h}^{3}	$P\pm 4(\frac{1}{2}00)$	106	C_{4v}^{8}	$P4(00\frac{1}{2})\overline{2}'(\frac{1}{2}\frac{1}{2}0)$
86	C_{4h}^{4}	$P\pm 4(0\frac{1}{2}\frac{1}{2})$	*107	C_{4v}^{9}	$I4\overline{2}'$
*87	C_{4h}^{5}	$I\pm 4$	108	C_{4v}^{10}	$I4\overline{2}'(00\frac{1}{2})$
88	C_{4h}^{6}	$I\pm 4(\frac{1}{4}\frac{1}{4}\frac{1}{4})$	109	C_{4v}^{11}	$I4(00\frac{1}{4})\overline{2}'(\frac{1}{2}00)$
*89	D_{4}^{1}	$P42'$	110	C_{4v}^{12}	$I4(00\frac{1}{4})\overline{2}'(\frac{1}{2}0\frac{1}{2})$
90	D_{4}^{2}	$P42'(\frac{1}{2}\frac{1}{2}0)$	*111	D_{2d}^{1}	$P\overline{4}2'$
91	D_{4}^{3}	$P4(00\frac{1}{4})2'$	112	D_{2d}^{2}	$P\overline{4}2'(00\frac{1}{2})$
92	D_{4}^{4}	$P4(00\frac{1}{4})2'(\frac{1}{2}\frac{1}{2}0)$	113	D_{2d}^{3}	$P\overline{4}2'(\frac{1}{2}\frac{1}{2}0)$
93	D_{4}^{5}	$P4(00\frac{1}{2})2'$	114	D_{2d}^{4}	$P\overline{4}2'(\frac{1}{2}\frac{1}{2}\frac{1}{2})$
94	D_{4}^{6}	$P4(00\frac{1}{2})2'(\frac{1}{2}\frac{1}{2}0)$	*115	D_{2d}^{5}	$P\overline{4}2''$
95	D_{4}^{7}	$P4(00\frac{3}{4})2'$	116	D_{2d}^{6}	$P\overline{4}2''(00\frac{1}{2})$

Table C.4 Tetragonal crystal system (continued)

Ordinal	Sch.	INSG	Ordinal	Sch.	INSG
117	D_{2d}^{7}	$P\bar{4}2''(\frac{1}{2}\frac{1}{2}0)$	130	D_{4h}^{8}	$P\pm4(\frac{1}{2}00)2'(\frac{1}{2}0\frac{1}{2})$
118	D_{2d}^{8}	$P\bar{4}2''(\frac{1}{2}\frac{1}{2}\frac{1}{2})$	131	D_{4h}^{9}	$P\pm4(00\frac{1}{2})2'$
*119	D_{2d}^{9}	$I\bar{4}2''$	132	D_{4h}^{10}	$P\pm4(00\frac{1}{2})2'(00\frac{1}{2})$
120	D_{2d}^{10}	$I\bar{4}2''(00\frac{1}{2})$	133	D_{4h}^{11}	$P\pm4(0\frac{1}{2}\frac{1}{2})2'(0\frac{1}{2}0)$
*121	D_{2d}^{11}	$I\bar{4}2'$	134	D_{4h}^{12}	$P\pm4(0\frac{1}{2}\frac{1}{2})2'(0\frac{1}{2}\frac{1}{2})$
122	D_{2d}^{12}	$I\bar{4}2'(0\frac{1}{2}\frac{1}{4})$	135	D_{4h}^{13}	$P\pm4(00\frac{1}{2})2'(\frac{1}{2}\frac{1}{2}0)$
*123	D_{4h}^{1}	$P\pm42'$	136	D_{4h}^{14}	$P\pm4(00\frac{1}{2})2'(\frac{1}{2}\frac{1}{2}\frac{1}{2})$
124	D_{4h}^{2}	$P\pm42'(00\frac{1}{2})$	137	D_{4h}^{15}	$P\pm4(0\frac{1}{2}\frac{1}{2})2'(\frac{1}{2}00)$
125	D_{4h}^{3}	$P\pm4(\frac{1}{2}00)2'(0\frac{1}{2}0)$	138	D_{4h}^{16}	$P\pm4(0\frac{1}{2}\frac{1}{2})2'(\frac{1}{2}0\frac{1}{2})$
126	D_{4h}^{4}	$P\pm4(\frac{1}{2}00)2'(0\frac{1}{2}\frac{1}{2})$	*139	D_{4h}^{17}	$I\pm42'$
127	D_{4h}^{5}	$P\pm42'(\frac{1}{2}\frac{1}{2}0)$	140	D_{4h}^{18}	$I\pm42'(00\frac{1}{2})$
128	D_{4h}^{6}	$P\pm42'(\frac{1}{2}\frac{1}{2}\frac{1}{2})$	141	D_{4h}^{19}	$I\pm4(\frac{1}{4}\frac{1}{4}\frac{1}{4})2'(\frac{1}{2}00)$
129	D_{4h}^{7}	$P\pm4(\frac{1}{2}00)2'(\frac{1}{2}00)$	142	D_{4h}^{20}	$I\pm4(\frac{1}{4}\frac{1}{4}\frac{1}{4})2'(\frac{1}{2}0\frac{1}{2})$

Table C.5 Trigonal crystal system

Ordinal	Sch.	INSG	Ordinal	Sch.	INSG
*143	C_{3}^{1}	$P3$	*156	C_{3v}^{1}	$P3\bar{2}'$
144	C_{3}^{2}	$P3(00\frac{1}{3})$	*157	C_{3v}^{2}	$P3\bar{2}''$
145	C_{3}^{3}	$P3(00\frac{2}{3})$	158	C_{3v}^{3}	$P3\bar{2}'(00\frac{1}{2})$
*146	C_{3}^{4}	$R3$	159	C_{3v}^{4}	$P3\bar{2}''(00\frac{1}{2})$
*147	C_{3i}^{1}	$P\bar{3}$	*160	C_{3v}^{5}	$R3\bar{2}'$
*148	C_{3i}^{2}	$R\bar{3}$	161	C_{3v}^{6}	$R3\bar{2}'(\frac{1}{2}\frac{1}{2}\frac{1}{2})$
*149	D_{3}^{1}	$P32''$	*162	D_{3d}^{1}	$P\bar{3}2''$
*150	D_{3}^{2}	$P32'$	163	D_{3d}^{2}	$P\bar{3}2''(00\frac{1}{2})$
151	D_{3}^{3}	$P3(00\frac{1}{3})2''$	*164	D_{3d}^{3}	$P\bar{3}2'$
152	D_{3}^{4}	$P3(00\frac{1}{3})2'$	165	D_{3d}^{4}	$P\bar{3}2'(00\frac{1}{2})$
153	D_{3}^{5}	$P3(00\frac{2}{3})2''$	*166	D_{3d}^{5}	$R\bar{3}2'$
154	D_{3}^{6}	$P3(00\frac{2}{3})2'$	167	D_{3d}^{6}	$R\bar{3}2'(\frac{1}{2}\frac{1}{2}\frac{1}{2})$
*155	D_{3}^{7}	$R32'$			

Table C.6 Hexagonal crystal system

Ordinal	Sch.	INSG	Ordinal	Sch.	INSG
*168	C_{6}^{1}	$P6$	182	D_{6}^{6}	$P6(00\frac{1}{2})2'$
169	C_{6}^{2}	$P6(00\frac{1}{6})$	*183	C_{6v}^{1}	$P6\bar{2}'$
170	C_{6}^{3}	$P6(00\frac{5}{6})$	184	C_{6v}^{2}	$P6\bar{2}'(00\frac{1}{2})$
171	C_{6}^{4}	$P6(00\frac{1}{3})$	185	C_{6v}^{3}	$P6(00\frac{1}{2})\bar{2}'(00\frac{1}{2})$
172	C_{6}^{5}	$P6(00\frac{2}{3})$	186	C_{6v}^{4}	$P6(00\frac{1}{2})\bar{2}'$
173	C_{6}^{6}	$P6(00\frac{1}{2})$	*187	D_{3h}^{1}	$P\bar{6}2''$
*174	C_{3h}^{1}	$P\bar{6}$	188	D_{3h}^{2}	$P\bar{6}2''(00\frac{1}{2})$
*175	C_{6h}^{1}	$P\pm6$	*189	D_{3h}^{3}	$P\bar{6}2'$
176	C_{6h}^{2}	$P\pm6(00\frac{1}{2})$	190	D_{3h}^{4}	$P\bar{6}2'(00\frac{1}{2})$
*177	D_{6}^{1}	$P62'$	*191	D_{6h}^{1}	$P\pm62'$
178	D_{6}^{2}	$P6(00\frac{1}{6})2'$	192	D_{6h}^{2}	$P\pm62'(00\frac{1}{2})$
179	D_{6}^{3}	$P6(00\frac{5}{6})2'$	193	D_{6h}^{3}	$P\pm6(00\frac{1}{2})2'(00\frac{1}{2})$
180	D_{6}^{4}	$P6(00\frac{1}{3})2'$	194	D_{6h}^{4}	$P\pm6(00\frac{1}{2})2'$
181	D_{6}^{5}	$P6(00\frac{2}{3})2'$			

Table C.7 Cubic crystal system

Ordinal	Sch.	INSG	Ordinal	Sch.	INSG
∗195	T^1	$P3'22'$	213	O^7	$P3'4(\frac{1}{4}\frac{3}{4}\frac{1}{4})2''(\frac{3}{4}\frac{1}{4}\frac{1}{4})$
∗196	T^2	$F3'22'$	214	O^8	$I3'4(\frac{1}{4}\frac{3}{4}\frac{1}{4})2''(\frac{3}{4}\frac{1}{4}\frac{1}{4})$
∗197	T^3	$I3'22'$	∗215	T_d^1	$P3'\bar{4}\bar{2}''$
198	T^4	$P3'2(\frac{1}{2}0\frac{1}{2})2'(\frac{1}{2}\frac{1}{2}0)$	∗216	T_d^2	$F3'\bar{4}\bar{2}''$
199	T^5	$I3'2(\frac{1}{2}0\frac{1}{2})2'(\frac{1}{2}\frac{1}{2}0)$	∗217	T_d^3	$I3'\bar{4}\bar{2}''$
∗200	T_h^1	$P\bar{3}'22'$	218	T_d^4	$P3'\bar{4}(\frac{1}{2}\frac{1}{2}\frac{1}{2})\bar{2}''(\frac{1}{2}\frac{1}{2}\frac{1}{2})$
201	T_h^2	$P\bar{3}'2(\frac{1}{2}\frac{1}{2}0)2'(0\frac{1}{2}\frac{1}{2})$	219	T_d^5	$F3'\bar{4}(00\frac{1}{2})\bar{2}''(00\frac{1}{2})$
∗202	T_h^3	$F\bar{3}'22'$	220	T_d^6	$I3'\bar{4}(\frac{1}{4}\frac{1}{4}\frac{1}{4})\bar{2}''(\frac{1}{4}\frac{1}{4}\frac{3}{4})$
203	T_h^4	$F\bar{3}'2(\frac{1}{4}\frac{1}{4}0)2'(0\frac{1}{4}\frac{1}{4})$	∗221	O_h^1	$P\bar{3}'42''$
∗204	T_h^5	$I\bar{3}'22'$	222	O_h^2	$P\bar{3}'4(\frac{1}{2}00)2''(00\frac{1}{2})$
205	T_h^6	$P\bar{3}'2(\frac{1}{2}0\frac{1}{2})2'(\frac{1}{2}\frac{1}{2}0)$	223	O_h^3	$P\bar{3}'4(\frac{1}{2}\frac{1}{2}\frac{1}{2})2''(\frac{1}{2}\frac{1}{2}\frac{1}{2})$
206	T_h^7	$I\bar{3}'2(0\frac{1}{2}0)2'(00\frac{1}{2})$	224	O_h^4	$P\bar{3}'4(0\frac{1}{2}\frac{1}{2})2''(\frac{1}{2}\frac{1}{2}0)$
∗207	O^1	$P3'42''$	∗225	O_h^5	$F\bar{3}'42''$
208	O^2	$P3'4(\frac{1}{2}\frac{1}{2}\frac{1}{2})2''(\frac{1}{2}\frac{1}{2}\frac{1}{2})$	226	O_h^6	$F\bar{3}'4(00\frac{1}{2})2''(00\frac{1}{2})$
∗209	O^3	$F3'42''$	227	O_h^7	$F\bar{3}'4(0\frac{1}{4}\frac{1}{4})2''(\frac{1}{4}\frac{1}{4}0)$
210	O^4	$F3'4(\frac{1}{4}\frac{1}{4}\frac{1}{4})2''(\frac{1}{4}\frac{1}{4}\frac{1}{4})$	228	O_h^8	$F\bar{3}'4(\frac{1}{2}\frac{1}{4}\frac{1}{4})2''(\frac{1}{4}\frac{1}{4}\frac{1}{2})$
∗211	O^5	$I3'42''$	∗229	O_h^9	$I\bar{3}'42''$
212	O^6	$P3'4(\frac{3}{4}\frac{1}{4}\frac{3}{4})2''(\frac{1}{4}\frac{3}{4}\frac{3}{4})$	230	O_h^{10}	$I\bar{3}'4(\frac{1}{4}\frac{3}{4}\frac{1}{4})2''(\frac{3}{4}\frac{1}{4}\frac{1}{4})$

Bibliography

Adams, B. G., Cizek, J. and Paldus, J. (1987). Lie algebraic methods and their applications to simple quantum systems, *Advances in Quantum Chemistry*, Vol. **19**, Academic Press, New York.

Andrews, G. E. (1976). The Theory of Partitions, Encyclopedia of Mathematics and Its Applications, Vol. **2**, Ed. Gian–Carlo Rota, Addison–Wesley.

Bayman, B. F. (1960). Some Lectures on Groups and their Applications to Spectroscopy, Nordita.

Berenson, R. and Birman, J. L. (1975). Clebsch–Gordan coefficients for crystal space group, *J. Math. Phys.* **16**, 227.

Biedenharn, L. C., Giovannini, A. and Louck, J. D. (1970). Canonical definition of Wigner coefficients in $U(n)$, *J. Math. Phys.* **11**, 2368.

Biedenharn, L. C. and Louck, J. D. (1981). Angular Momentum in Quantum Physics, Theory and Application, *Encyclopedia of Mathematics and its Application*, Vol. **8**, Ed. G. C. Rota, Addison–Wesley, Massachusetts.

Bjorken, J. D. and Drell, S. D. (1964). Relativistic Quantum Mechanics, McGraw–Hill Book Co., New York.

Boerner, H. (1963). Representations of Groups, North–Holland, Amsterdam.

Bourbaki, N. (1989). Elements of Mathematics, Lie Groups and Lie Algebras, Springer–Verlag, New York.

Bradley, C. J. and Cracknell, A. P. (1972). The Mathematical Theory of Symmetry in Solids, Clarendon Press, Oxford.

Bremner, M. R., Moody, R. V. and Patera, J. (1985). Tables of Dominant Weight Multiplicities for Representations of Simple Lie Algebras, Pure and Applied Mathematics, A Series of Monographs and Textbooks 90, Marcel Dekker, New York.

Burns, G. and Glazer, A. M. (1978). Space Groups for Solid State Scientists, Academic Press, New York.

Chen, J. Q., Wang, P. N., Lu Z. M. and Wu, X. B. (1987). Tables of the Clebsch–Gordan, Racah and Subduction Coefficients of $SU(n)$ Groups, World Scientific, Singapore.

Chen, J. Q., Ping, J. L. and Wang, F. (2002). Group Representation Theory for Physicists, 2nd edition, World Scientific, Singapore.

Chen, J. Q. and Ping, J. L. (1997). Algebraic expressions for irreducible bases of icosahedral group, *J. Math. Phys.* **38**, 387.
Cotton, F. A. (1971). Chemical Applications of Group Theory, Wiley, New York.
de Swart, J. J. (1963). The octet model and its Clebsch–Gordan coefficients, *Rev. Mod. Phys.* **35**, 916.
Dai A. Y. (1983). A graphic rule for dimensions of irreducible spinor representations of SO(N), J. Lanzhou Univ. (Natural Sciences) **19**, No. 2, 33 (in Chinese).
Deng Y. F. and Yang, C. N. (1992). Eigenvalues and eigenfunctions of the Hückel Hamiltonian for Carbon–60, *Phys. Lett. A* **170**, 116.
Dirac, P. A. M. (1958). The Principle of Quantum Mechanics, Clarendon Press, Oxford.
Dong, S. H., Hou, X. W. and Ma, Z. Q. (1998a). Irreducible bases and correlations of spin states for double point groups, *Inter. J. Theor. Phys.* **37**, 841.
Dong, S. H., Xie, M. and Ma, Z. Q. (1998b). Irreducible bases in icosahedral group space, *Inter. J. Theor. Phys.* **37**, 2135.
Dong, S. H., Hou, X. W. and Ma, Z. Q. (1998c). Relativistic Levinson theorem in two dimensions, *Phys. Rev. A* **58**, 2160.
Dong, S. H., Hou, X. W. and Ma, Z. Q. (2001). Correlations of spin states for icosahedral double group, *Inter. J. Theor. Phys.* **40**, 569.
Dong, S. H. and Ma, Z. Q. (2003). Exact solutions to the Dirac equation with a Coulomb potential in $2 + 1$ dimensions, *Phys. Lett. A* **312**, 78.
Duan, B., Gu, X. Y. and Ma, Z. Q. (2001). Precise calculation for energy levels of a helium atom in P states, *Phys. Lett. A* **283**, 229.
Duan, B., Gu, X. Y. and Ma, Z. Q. (2002). Numerical calculation of energies of some excited states in a helium atom, *Eur. Phys. J. D* **19**, 9.
Dynkin, E. B. (1947). The structure of semisimple algebras, *Usp. Mat. Nauk. (N. S.)*, **2**, 59. Transl. in *Am. Math. Soc. Transl. (I)*, **9**, 308, 1962.
Eckart, C. (1934). The kinetic energy of polyatomic molecules, *Phys. Rev.* **46**, 383.
Edmonds, A. R. (1957). Augular Momentum in Quantum Mechanics, Princeton University Press, Princeton.
Elliott J. P. and Dawber, P. G. (1979). Symmetry in Physics, McMillan Press, London.
Feng Kang (1991). The Hamiltonian way for computing Hamiltonian dynamics, *Applied and Industrial Mathematics*, Ed. R. Spigler, Kluwer Academic Publishers, p. 17.
Fronsdal, C. (1963). Group theory and applications to particle physics, 1962, *Brandies Lectures*, Vol. **1**, 427. Ed. K. W. Ford, Benjamin, New York.
Gao, S. S. (1992). Group Theory and its Applications in Particle Physics, Higher Education Press, Beijing (in Chinese).
Gel'fand, I. M., Minlos, R. A. and Shapiro, Z. Ya. (1963). Representations of the Rotation and Lorentz Groups and Their Applications, Transl. from Russian by G. Cummins and T. Boddington, Pergamon Press, New York.
Gel'fand, I. M. and Zetlin, M. L. (1950). Matrix Elements for the Unitary Groups, *Dokl. Akad. Nauk* **71**, 825.

Gell-Mann, M. and Ne'eman, Y. (1964). The Eightfold Way, Benjamin, New York.

Georgi, H. (1982). Lie Algebras in Particle Physics, Benjamin, New York.

Gilmore, R. (1974). Lie Groups, Lie Algebras and Some of Their Applications, Wiley, New York.

Gu, X. Y., Duan, B. and Ma, Z. Q. (2001a). Conservation of angular momentum and separation of global rotation in a quantum N-body system, *Phys. Lett. A* **281**, 168.

Gu, X. Y., Duan, B. and Ma, Z. Q. (2001b). Independent eigenstates of angular momentum in a quantum N-body system, *Phys. Rev. A* **64**, 042108(1-14).

Gu, X. Y., Duan, B. and Ma, Z. Q. (2001c). Quantum three-body system in D dimensions, *J. Math. Phys.* **43**, 2895.

Gu, X. Y., Ma, Z. Q. and Dong, S. H. (2002). Exact solutions to the Dirac equation for a Coulomb potential in $D+1$ dimensions, *Inter. J. Mod. Phys. E* **11**, 335.

Gu, X. Y., Ma, Z. Q. and Duan, B. (2003a). Interdimensional degeneracies for a quantum three-body system in D dimensions, *Phys. Lett. A* **307**, 55.

Gu, X. Y., Ma, Z. Q. and Sun, J. Q. (2003b). Quantum four-body system in D dimensions, *J. Math. Phys.* **44**, 3763.

Gu, X. Y., Ma, Z. Q. and Sun, J. Q. (2003c). Interdimensional degeneracies in a quantum isolated four-body system, *Phys. Lett. A* **314**, 156.

Gu, X. Y., Ma, Z. Q. and Sun, J. Q. (2003d). Interdimensional degeneracies in a quantum N-body system, *Europhys. Lett.* **64**, 586.

Gu, X. Y., Ma, Z. Q. and Dong, S. H. (2003e). The Levinson theorem for the Dirac equation in $D+1$ dimensions, *Phys. Rev. A* **67**, 062715(1-12).

Hamermesh, M. (1962). Group Theory and its Application to Physical Problems, Addison–Wesley, Massachusetts.

Han, Q. Z. and Sun, H. Z. (1987). Group Theory, Peking University Press, Beijing (in chinese).

Heine, V. (1960). Group Theory in Quantum Mechanics, Pergamon Press, London.

Hirschfelder, J. O. and Wigner, E. P. (1935). Separation of rotational coordinates from the Schrödinger equation for N particles, *Proc. Natl. Acad. Sci. U.S.A.* **21**, 113.

Hou, Bo-Yu, Hou, Bo-Yuan and Ma Z. Q. (1990a). Clebsch–Gordan coefficients, Racah coefficients and braiding fusion of quantum $s\ell(2)$ enveloping algebra I, *Commun. Theor. Phys.* **13**, 181.

Hou, Bo-Yu, Hou, Bo-Yuan and Ma Z. Q. (1990b). Clebsch–Gordan coefficients, Racah coefficients and braiding fusion of quantum $s\ell(2)$ enveloping algebra II, *Commun. Theor. Phys.* **13**, 341.

Hou, Bo-Yuan and Hou, Bo-Yu (1997). Differential Geometry for Physicists, World Scientific, Singapore.

Hou, X. W., Xie, M., Dong, S. H. and Ma, Z. Q. (1998). Overtone spectra and intensities of tetrahedral molecules in boson-realization models, *Ann. Phys. (N.Y.)* **263**, 340.

Hsiang, W. T. and Hsiang, W. Y. (1998). On the reduction of the Schrödinger's equation of three-body problem to a system of linear algebraic equations,

preprint.
Hsiang, W. Y. (1998). On the kinematic geometry of many body system, preprint.
Itzykson, C. and Nauenberg, M. (1966). Unitary groups: Representations and decompositions, *Rev. Mod. Phys.* **38**, 95.
Joshi, A. W. (1977). Elements of Group Theory for Physicists, Wiley.
Kazdan, D. and Lusztig, G. (1979). Representations of Coxeter groups and Hecke algebras, *Invent. Math.* **53**, 165.
Koster, G. F. (1957). Space Groups and Their Representations in Solid State Physics, Eds. F. Seitz and D. Turnbull, Academic Press, New York, **5**, 174.
Kovalev, O. V. (1961). Irreducible Representations of Space Groups, translated from Russian by A. M. Gross, Gordon & Breach.
Lipkin, H. J. (1965). Lie Groups for Pedestrians, North–Holland, Amsterdam.
Littlewood, D. E. (1958). The Theory of Group Characters, Oxford University Press, Oxford.
Liu, F., Ping, J. L. and Chen, J. Q. (1990). Application of the eigenfunction method to the icosahedral group, *J. Math. Phys.* **31**, 1065.
Ma, Z. Q. (1993). Yang–Baxter Equation and Quantum Enveloping Algebras, World Scientific, Singapore.
Ma, Z. Q. and Dai, A. Y. (1982). A graphic rule for dimensions of irreducible tensor representations of SO(N), J. Lanzhou Univ. (Natural Sciences) **18**, No. 2, 97 (in Chinese).
Ma, Z. Q. and Gu, X. Y. (2004). Problems & Solutions in Group Theory for Physicists, World Scientific, Singapore.
Ma, Z. Q., Hou, X. W. and Xie, M. (1996). Boson-realization model for the vibrational spectra of tetrahedral molecules, *Phys. Rev. A* **53**, 2173.
Marshak, R. E., Riazuddin, and Ryan, C. P. (1969). Theory of Weak Interactions in Particle Physcs, John Wiley & Sons, Inc., New York.
Miller, Jr. W., (1972). Symmetry Groups and Their Applications, Academic Press, New York.
Racah, G. (1951). Group Theory and Spectroscopy, Lecture Notes in Princeton.
Ren S. Y. (2006). Electronic States in Crystals of Finite Size, Quantum confinement of Bloch waves, Springer, New York (2006).
Roman, P. (1964). Theory of Elementary Particles, North-Holland, Amsterdam.
Rose, M. E. (1957). Elementary Theory of Angular Momentum, Wiley, New York.
Salam, A. (1963). The Formalism of Lie Groups, in *Theoretical Physics*, Director: A. Salam, International Atomic Energy Agency, Vienna, 173.
Schiff, L. I. (1968). Quantum Mechanics, Third Edition, McGraw-Hill, New York.
Serre, J. P. (1965). Lie Algebras and Lie Groups, Benjamin, New York.
Tinkham, M. (1964). Group Theory and Quantum Mechanics, McGraw-Hill, New York.
Tong, D. M., Zhu, C. J. and Ma, Z. Q. (1992). Irreducible representations of braid groups, *J. Math. Phys.* **33**, 2660.
Tung, W. K. (1985). Group Theory in Physics, World Scientific, Singapore.
Weyl, H. (1931). The Theory of Groups and Quantum Mechanics, translated from German by H. P. Robertson, Dover Publications.
Weyl, H. (1946). The Classical Groups, Princeton University Press, Princeton.

Wigner, E. P. (1959). Group Theory and its Applications to the Quantum Mechanics of Atomic Spectra, Academic Press, New York.

Wybourne, B. G. (1974). Classical Groups for Physicists, Wiley, New York.

Yamanouchi, T. (1937). On the construction of unitary irreducible representation of the symmetric group, *Proc. Phys. Math. Soc. Jpn* **19**, 436.

Zou, P. C. and Huang, Y. C. (1995). Proof of "irreducible postulation" and its applications, *High Ener. Phys. Nucl. Phys.* **19**, 375.

Index